Surface Processes and Landforms

Second Edition

DON J. EASTERBROOK

 Prentice Hall

Prentice Hall, Upper Saddle River, New Jersey 07458

Library of Congress Cataloging-in-Publication Data
Easterbrook, Don J.
    Surface processes / Don J. Easterbrook. -- 2nd ed.
        p.   cm.
    Includes bibliographical references and index.
    ISBN 0-13-860958-6
    1. Geomorphology.   I. Title.
    GB401.5.E26   1999
    551.41--dc21   98-45003
                                                    CIP

Executive Editor: Robert A. McConnin
Executive Managing Editor: Kathleen Schiaparelli
Assistant Managing Editor: Lisa Kinne
Manufacturing Manager: Trudy Pisciotti
Art Director: Jayne Conte
Cover Design: Bruce Kenselaar
Cover Photo: IR image of the Malaspina glacier. Photo by U.S. Department of Agriculture.
Unless otherwise indicated all photos are courtesy of Don J. Easterbrook

Printed in the United States of America
10 9 8 7 6 5

ISBN 0-13-860958-6

Brief Contents

Contents

CHAPTER SIXTEEN

Shorelines   428

CHAPTER SEVENTEEN

Eolian Processes and Landforms   471

CHAPTER EIGHTEEN

Dating Geomorphic Features   494

Preface

In completing one discovery we never fail to get an imperfect knowledge of others of which we could have no idea before, so that we cannot solve one doubt without creating several new ones.

Joseph Priestly, 1786

Geomorphology is the study of surface processes and landforms. Although it is commonly considered a subdiscipline of geology, it is really an interdisciplinary science, based upon the application of physics, chemistry, and mathematics to natural systems at the earth's surface, overlapping with climatology, hydrology, soil science, geography, meteorology, and others. Geomorphology is not just the study of surface processes, but also includes the evolution of landforms and interpretation of their origin. Time often plays an important role in the development of landforms.

During the past decade, geomorphology has been revolutionized by many rapidly-evolving tools--computers, sophisticated electronic measuring devices, lasers, mass spectrometers, new methods of dating landforms and deposits, and others. These have led to better methods of measuring and understanding surface processes and more precise recognition of time as a factor in relict landforms.

Some universities teach geomorphology courses at the senior level so that students have relatively broad backgrounds in geology and related sciences, whereas others teach a more basic geomorphology course at the sophomore or junior level, followed by more advanced courses in fluvial, glacial, and other topics. In recent years, the increasing relevance of geomorphology to environmental concerns has placed additional emphasis on applied aspects of geomorphology. With all this in mind, I have attempted to make this book as widely useful as possible by including both basic material for students without a great deal of background in geology, and more advanced material for those students with more extensive backgrounds. For students desiring even more in-depth material, I would hope that they might use this text as a tool, making use of the many references to extend their pursuit of knowledge on topics of interest.

So much has been written on the multitude of subjects that together make up the field of geomorphology that the list of references perused in writing this text totaled more than ten thousand. Although reassuring to realize that such a wealth of information is available, the dilemma arises—which references and information are the most significant? In the same manner as no perfect blend of coffee exists, so too no perfect blend of references exists because of the infinite variety of opinions among geomorphologists about what is important and what isn't. Therefore, the references included in this text should be looked upon as a sample of the rich literature that exists, and I hope that they will serve to encourage students to use them as a starting point for further in-depth reading. To help students become more aware of the evolutionary nature of many concepts, a number of classic papers are included in the reference lists. Many contain material now considerably refined, but are useful reminders of important initial concepts. References in this, the second edition, have been updated and many new ones added.

ACKNOWLEDGMENTS

Although responsibility for the accuracy of presentation of concepts in this text rests solely with the author, I would like to thank the following esteemed colleagues who have generously given of their time and expertise in reviewing the text and in providing many valuable comments, new information, and photos: I want to especially thank Ken Hamblin, Brigham Young University, for sharing his remarkable collection of photographs of geomorphic features and satellite photos. Many of the new photos in this edition were taken from his collection, which he so generously made available. Many of the satellite images in the text were brought to my attention by Ken, who spent much time sorting through the huge NASA, NOAA, Earth Satellite Corp., and Earth Observation Satellite Co. satellite photo collections and U. S. Geological Survey, Department of Agriculture, and Canadian Dept. of Energy air photos.

For this, the second edition, Hugh French, University of Ottawa, was especially helpful in reviewing the new chapter on periglacial features and in providing new material and photographs. Bernard Hallet, University of Washington, reviewed the periglacial chapter and contributed many useful comments. Dori Kovanen, University of British Columbia, and Jaakko Putkonen, University of Washington, supplied new reference material for the new periglacial chapter.

Robert P. Sharp, California Institute of Technology; Peter Birkeland, University of Colorado; Allan Gillespie, University of Washington; Ed Evenson, Lehigh University; Martin L. Stout, California State University, Los Angeles; Scott Burns, Portland State University; Marie Morisawa, SUNY Binghamton; and Julie Brigham-Grette, University of Massachusetts, Amherst, reviewed the first edition and contributed many helpful comments.

Others who provided useful reviews include: Richard Bonnett, Marshall University, John Rockaway, University of Missouri-Rolla, Leland Dexter, Northern Arizona University, James Walters, University of Northern Iowa, Lallie Fay Scott, Northeastern Oklahoma State University, Jon Harbor, Purdue University, Kenneth Johnson, Skidmore College, Darrel Schmitz, Mississippi State University, Ross Sutherland, University of Hawaii-Manoa, Bruce Pape, Central Michigan University, Lawrence Dingman, University of New Hampshire, James baer, Brigham Young University, Robert D. Hall, Indiana University, Purdue University, John D. Vitek, Oklahoma State University; John C. Kraft, University of Delaware; Steven Esling, Southern Illinois University; Wakefield Dort, University of Kansas; Indianapolis; Charles G. Higgins, University of California, Davis; Ernst H. Kastning, Radford University; Katherine H. Price, DePauw University; Richard V. Birdseye, University of Southwestern Louisiana; Phillip R. Kernmerly, Austin Peay State University; Michael E. Ritter, University of Wisconsin, Stevens Point; Patricia Humberston, Youngstown State University; B. Ray Knox, Southeast Missouri State University; William J. Wayne, University of Nebraska; Richard L. Bowen, University of Southern Mississippi; and Dale Monsebroten, Eastern Kentucky University.

Many people assisted in seeking and acquiring references, editing and proofreading text, and keeping track of the huge mass of material that went into the manuscript. For the second edition, they include Dori Kovanen, Tracy Minick, and Kathy Thumlert, along with Ellen Easterbrook who contributed several drawings and helped with the acquisition of permission to use photos and diagrams from various sources. Bob McConnin at Prentice Hall guided the project to completion and Ray Robinson was responsible for production of the text. Cynthia Carlstad, Kim Farley, Russell Nance, Harriet Beale, Vicki Critchlow, Melanie Carpenter, Kim Brown, Sherri Rodgers, and Stacy Weber helped with the first edition.

I would also like to acknowledge the outstanding air photographs of geomorphic features taken by David Rahm. Prior to his death in a plane crash in 1975, we spent many hours together in the air, seeking out new landforms to photograph and admiring nature's fine touch in sculpting the earth's surface.

DJE

Introduction

Henry Mts., Utah where G. K. Gilbert made important geomorphology studies in 1877.

GEOMORPHOLOGY is the study of the origin and evolution of topographic features by physical and chemical processes operating at or near the earth's surface. The name is derived from the Greek terms *geo*, meaning earth, *morph*, meaning form, and *logos*, meaning discourse.

The study of surface processes and landforms relies heavily on geologic principles. Yet, like other sciences, geomorphology also depends on the application of basic principles of physics, chemistry, biology, and mathematics to natural systems. As an interdisciplinary science, Geomorphology overlaps with other disciplines, such as hydrology, glaciology, geography, statistics, climatology, soil science, computer modeling, aerodynamics, and others.

HISTORIC CONTRIBUTIONS TO GEOMORPHOLOGY

The evolution of ideas that set the stage for twentieth-century geomorphology dates back to the fifth century B.C. when the gradual, almost haphazard development of geomorphic thought consisted largely of occasional observations about the relationship between modern processes and relict topographic features. Herodotus, (485?–425 B.C.) realized that Egyptian rocks containing marine fossils had been formed beneath the sea, and he recognized that the growth of the Nile delta must have taken thousands of years. Aristotle (384–322 B.C.) believed that land could be raised from beneath the sea and that areas of dry land could become submerged. He also described erosion by streams and the deposition of alluvium in deltas. Lucretius, (99–55 B.C.) recognized the decay of rocks by weathering. Strabo (54 B.C.–A.D. 25) found examples of the elevation and subsidence of land, described the transporting action of water, and explained the principles of delta growth. Seneca (A.D. 3–65) observed the erosion of valleys by streams, and Avicenna (Arabic name Ibn-Sina; A.D. 980–1037) saw that mountains could be created by crustal upheaval and sculpted by long-term erosion. The implication that landscapes are carved by erosion from the products of upheaval was a very important geomorphic concept.

Curiously, the Greek philosophers never pursued their observations far enough to establish a science, nor did the pragmatic Romans, despite their opportunities for wide travel. During the Dark Ages after the fall of the Roman Empire, the slight stimulus given to geomorphology by the Greeks and Romans was almost entirely lost. Feudal Europe provided little atmosphere for scientific thinking. The only "voice in the wilderness" through all those centuries came from Persia in the Arabic writings of Avicenna.

The Renaissance saw occasional flashes of insight. For example, daVinci (1452–1519) found marine fossils in rocks, recognized evidence for changes between land and sea, and attributed landscape sculpture to stream erosion. Agricola (German name George Bauer; 1494–1555) explained mountains that are produced by weathering and mass movement. Steno (1638–87) saw that flowing water is the chief agent of erosion. However, until the nineteenth century, further advancements in geology were hampered by a strict, biblical interpretation of creation according to Genesis, which allowed little opportunity for the slow development of landforms. Many scientists and philosophers believed that landforms were created specially and instantaneously by catastrophic processes, and thoughts to the contrary were considered heresy.

Despite the widespread biblical notion of instantaneous creation of the earth and its surface features, a number of writers did continue to express keen perceptions on stream erosion of the landscape by evolutionary processes over long periods of time. For example, Targioni-Tozzetti (1712–84) wrote in 1752 that streams erode their valleys differentially according to rock type and lose their erosive ability as they approach sea level. Leclerc (1707–88) suggested that over long periods of time, stream erosion could reduce the land to sea level. Guettard (1715–86) discussed stream erosion, stream transport, and the buildup of floodplains. He recognized the importance of wave action in the destruction of chalk cliffs. Desmarest (1725–1805) provided evidence that landscapes, such as the stream-worn valleys of central France, evolve through successive erosional stages. In 1787, De Saussure (1740–99) wrote about the erosive and depositional action of alpine glaciers.

Despite such recognition of the effects of fluvial erosion, the issue became further obscured by the teachings and influence of Werner (1749–1817), who postulated a global ocean from which sedimentary and volcanic rocks were precipitated. The followers of this belief, known as "Neptunists," claimed that the surface features of the earth were formed by submarine deposition and erosion and by the rush of ocean water as it receded (although the Neptunists could not explain what happened to such a huge volume of seawater). In 1802, Lamarck (1744–1829) published his work *"Hydrogeologie"* in which he asserted that mountains had been carved by running water. However, Cuvier (1769–1832) further clouded the issue by proposing great catastrophes, the last of which was the biblical flood, that resulted in unconformities in rocks and carved out the earth's landscape as the ocean waters drained. Cuvier's beliefs were quite popular, and followers became known as "catastrophists."

Opposition to the Neptunist and catastrophist views grew during the latter part of the eighteenth century, largely as a result of continuing observations by Scottish geologist Hutton (1726–97). In his work *"Theory of the Earth, with Proofs and Illustrations"* published in 1788, Hutton contended that granitic and volcanic rocks had *not* been precipitated from a global ocean, as believed by the Neptunists, but rather formed by subsurface heat and fusion of rocks later pushed up to the surface. From his extensive fieldwork, Hutton concluded that landforms are shaped by the slow, continuous action of running water that erodes the land, and that the sediment deposited by rivers in the sea provides the material from which sedimentary rocks are produced. Uplift of such rocks creates new land which, in turn, is again eroded by streams that carry new sediment to the sea. Hutton saw "no vestige of a beginning, no prospect of an end" to this continuing process. In his work, Hutton took direct opposition to the previously held ideas of catastrophic creation of the features of the earth's surface, thereby paving the way for the ultimate rejection of the theory that the earth's landscape was formed by the mythical biblical flood. Hutton argued that the present is the key to the past and that by observing present processes, one could extrapolate into the past to infer the origin of landforms made by ancient processes. Hutton's concept of **uniformitarianism** —that the natural physical and chemical processes which can be observed today have operated in the past and will continue to do so in the future—forms the basis for the interpretation of geomorphic features, as it does for all scientific principles.

Unfortunately, as a result of Hutton's difficult style of writing, his work was not widely read at first. However, upon his death, his friend Playfair (1748–1819), professor of mathematics at Edinburgh University, saw the validity of Hutton's concepts, and in a masterful exposition of Hutton's ideas, he rewrote Hutton's work in a much more lucid style as *"Illustrations of the Huttonian Theory of the Earth,"* published in 1802. Playfair also made many original contributions of his own. His work on rivers was especially significant in that he recognized that streams cut their own valleys, rather than inheriting them from some ancient cataclysmic event. He established beyond reasonable doubt the concept that streams systematically carve their own drainage basins, a concept that later became known as Playfair's law. He also perceived that the gradient of streams is adjusted toward equilibrium with velocity, discharge of water, and the amount of sediment carried, and that the constituent parts of a river system are mutually adjusted. The age of reason came to geology with Playfair's clear exposition of Hutton's ideas. The scientific world finally recognized the antiquity of the earth, and the way was cleared for progress in geomorphology. The stigma of religious heresy was diminished by the popular textbooks of Lyell (1797–1875), a widely traveled geologist who published many editions of a highly regarded book, *"The Principles of Geology,"* from 1833 to 1875. Lyell was a strong promoter of uniformitarianism and a vehement opponent of catastrophism, especially "Noah's flood."

Meanwhile, other geologists and naturalists continued their observations. Surrell (1813–87), through his work on torrential streams (1841), gained early insight into the relation of erosion and deposition to the longitudinal profile of a stream. Jukes (1811–69) showed conformation of drainage patterns to geologic structures and introduced the concept of subsequent valleys. Reade (1832–1909) proved by measurement the importance of solution as a factor in erosion. Sternberg (1825–85) related processes of stream transport and the longitudinal profiles of streams to the character of the bedload.

Largely as a result of Lyell's views of marine erosion, the idea of widespread marine dissection became popular. Ramsay (1814–91) emphasized wave action as a means of planing off large erosional surfaces, and Lesley (1819–1903) explained differential erosion of the folded Appalachian Mountains. Von Richthofen (1833–1905) wrote that marine abrasion is the primary agent that prepares the erosion surfaces upon which transgressive deposits are eventually laid down. In 1868, Geikie (1835–1924) wrote *Denudation Now in Progress,* in which calculations were made to show the superiority of subaerial erosion, acting upon a large surface, over marine erosion acting only on the shoreline.

Europe became the scene of increased geomorphic inquiry with the discovery of evidence for a former ice age. This discovery stimulated interest not only in glaciation, but in other processes as well, so that near the end of the nineteenth century, enough source material was available for the organization of the science of physiography (as geomorphology was then called). In 1832, Bernhardi (1802–87) published the first evidence of ancient continental glaciation in a large area now free of ice (northern Germany). Although former expansion of alpine glaciers was inferred earlier by residents of glaciated mountainous regions, and later documented by Venetz (1788–1859) in Switzerland and Esmark in Norway, Bernhardi deserves credit for the first theory of a large-scale, continental ice age. In 1836, Agassiz (1807–73), professor of natural history at Neuchatel and an expert on fossil fish, and de Charpentier (1786–1855) traveled together to study glacial features in the vicinity of Mont Blanc. Agassiz had been skeptical of the theory of former large-scale expansion of glaciers in the Alps, but de Charpentier convinced him that glaciers had, in fact, been much enhanced in former times. Earlier, Bernhardi had published evidence for glaciation of northern Europe by continental glaciers from Scandinavia, and de Charpentier had published similar evidence for the Alps. Thus, Agassiz was not the original author of the concept of former ice ages, but his 1840 publication *Etudes sur les Glaciers* did much to popularize the concept. In his travels and lectures in Britain and North America, Agassiz spread the idea and convinced many geologists. In America, T. C. Chamberlin (1843–1928) and Leverett (1859–1943) documented the extension of a large continental glacier from Canada into the United States.

Best known for his concept of evolution published in *On the Origin of Species,* in 1859, not many people are aware that Charles Darwin (1809–82) was also an unusually capable geologist. Despite having had little formal education in geology, his geological observations during the voyage of the *Beagle* (1831–35) resulted in major publications on the structure and distribution of coral reefs (1842), observations on volcanic islands (1844), and the geology of parts of South America (1846).

At the University of Edinburgh, he attended lectures of Robert Jameson, a champion of Werner's Neptunist theory, but found them very disappointing. However, he did spend most of August 1831 looking at the geology of Wales with

Adam Sedgwick, who later defined the Cambrian System based on these rocks. As he later noted in his autobiography:

On this tour I had a striking instance of how easy it is to overlook phenomena, however conspicuous, before they have been observed by anyone. We spent many hours . . . examining all the rocks with extreme care . . . but neither of us saw a trace of the wonderful glacial phenomena all around us.

During the voyage of the *Beagle*, Darwin spent many days investigating both geological and biological phenomena as the *Beagle* circumnavigated the globe. Shortly after a very strong earthquake in Chile on February 20, 1835, Darwin noted that shorelines in the area had been uplifted several feet, planting the idea that the Earth's crust was not nearly so stable as presumed. He developed his hypothesis on the evolution of coral reefs to form atolls before he actually saw the reefs in the Pacific that confirmed his theory.

No other work of mine was begun in so deductive a spirit as this; for the whole theory was thought out on the west coast of S. America before I had seen a true coral reef. I had therefore only to verify and extend my views by a careful examination of living reefs. But it should be observed that I had during the two previous years been incessantly attending to the effects on the shores of S. America of the intermittent elevation of the land, together with the denudation and deposition of sediment. This necessarily led me to reflect much on the effects of subsidence, and it was easy to replace in imagination the continued deposition of sediment by the upward growth of coral. To do this was to form my theory of the formation of barrier reefs and atolls.

Although Darwin's theory of coral reefs is his most famous geological work, he also made other geologic contributions of great importance. He observed surface rupture and displacement during earthquakes, fossilization of extinct organisms now found in high mountains, alteration of rocks in contact with hot lava, cleavage and foliation in metamorphic rocks, crustal uplift of mountains, evidence for differing climates in the past based on fossils and glacial deposits, and past fluctuations in sea level.

During the last half of the nineteenth century, geologists played an important role in the opening of the American West. They were among the earliest explorers in many parts of the West and produced a series of geologic surveys that described spectacular geomorphic features and surface processes that were neither random nor static. Among the most exceptional of these pioneering geologists were J. W. Powell, C. E. Dutton, and G. K. Gilbert.

Powell (1834–1902) was the first to float down the Colorado River in the Grand Canyon, committing his party to the gorge not knowing whether or not it contained waterfalls. During his journey down the turbulent Colorado in the Grand Canyon, Powell was struck both by the ability of the river to cut its canyon through thousands of feet of bedrock and by the amount of sediment that the river carried. He realized that the sediment in the muddy waters came not only from the Colorado, but also from all of the tributaries in its drainage basin. He reasoned that long, continuous transfer of such vast

quantities of sediment would eventually reduce the entire drainage basin of the Colorado to a land of low relief. Powell deduced that because streams cannot flow uphill, they cannot erode below a lower limit. Thus, he conceived the idea of **base level**, the level below which a land surface cannot be eroded. The ultimate base level is the sea. Powell concluded that as streams erode their channels ever nearer to base level, their ability to erode further is reduced concomitantly with reduction of their gradients, and that therefore the completion of degradation would require vast amounts of time and stability of both the land and the level of the sea. Powell followed two unconformities for miles in the rocks along the cliffs of the Grand Canyon. Both unconformities truncated Precambrian rocks along a surface of very low relief that must have represented long periods of erosion, nearly to base level. Many years later, Sharp (1940) presented evidence that these erosional surfaces were eroded by streams above base level, rather than by advancing ocean waves.

Dutton (1841–1912) and Gilbert (1843–1918) followed in Powell's footsteps and applied the concept of base level to the superbly exhibited erosional processes at work on the Colorado Plateau. They recognized the manner in which streams systematically adjust to rock resistance and geologic structure, and they saw that streams erode at rates commensurate with elevation above local base levels, such as falls, rapids, lakes, and stream junctions with main trunk streams.

Gilbert laid the foundation for many basic geomorphic principles with his classic monograph on stream sculpture of the Henry Mountains of Utah. This critical analysis of erosional processes established the concepts of graded streams, lateral planation, stream competence, the origin of pediments, the evolution of drainage patterns, the retreat of divides, the relationship between erosion rates and rock structure, and the effect of geologic structure on drainage. He later published a paper on Pleistocene shoreline features and isostatic rebound of pluvial Lake Bonneville, the ancestor of Great Salt Lake; a paper on geomorphic evidence for block faulting in the Great Basin of Nevada and eastern California; and a study of hydraulic mining debris in California that quantified relationships between stream load and discharge, velocity, and gradient.

The concept that landforms evolve sequentially with time, passing through a series of characteristic stages, was anticipated long ago by Desmarest and recognized in the work of Powell, Dutton, and Gilbert. Davis (1850–1934) welded these ideas into a comprehensive concept known as the **cycle of erosion**, which he described in his work *Geographical Essays* published in 1909. In its simplest form, the cycle of erosion consisted of the incessant dissection of an initial surface high above base level, with the stream-carved landscapes evolving through a sequence of stages until the land is reduced to a low plain (a peneplain) near base level. Davis named the stages in the cycle **youth, maturity,** and **old age** in analogy with the life history of an individual. At the end of this succession of stages, he visualized a **peneplain,** a vast, gently undulating, erosional surface of low relief, sloping gently to base level. This ideal case required stability of both sea level and the crust of the earth, but Davis realized the many departures from the ideal and explained them fully. However, some geologists, apparently not familiar with his many publications, objected that he did not take these variations into consideration. The contention of the German geomorphologist, Penck (1888–1923)—that slopes are controlled by uplift rates—opened up opposing views. In addition, many geomorphologists attempted to apply the concept of cyclic erosion to flat surfaces in mountainous areas, creating much vigorous debate of Davis's concepts in the years that followed.

During the first half of the twentieth century, the early quantitative work of Gilbert was followed by many years of rational, deductive research in geomorphology, emphasized by Davis, Bryan, Johnson, Blackwelder, Cotton, vonEngeln, Fenneman, Lobek, Mackin, and many others. However, the need for quantification became increasingly apparent and quantitative approaches to geomorphic phenomena began to flourish during the period 1930–65 as a result of the work of Bagnold, Sharp, Rubey, Hjulstrom, Bascomb, Horton, Strahler, Leopold, Wolman, and others. Today geomorphology continues to become increasingly quantitative and analytical, and in recent years, it has turned away from singly deductive approaches toward more quantitative analyses, especially with the advent of computer data processing and modeling.

REFERENCES

Agassiz, L., 1840, On glaciers, and the evidence of their having once existed in Scotland, Ireland, and England: Proceedings of the Geological Society of London, v. 3, p. 327–332.

Bagnold, R. A., 1941, The physics of blown sand and desert dunes: Methuen and Co., London, 265 p.

Bascom, W. H., 1959, Ocean waves: Scientific American, v. 201, p. 74–84.

Bascom, W. H., 1964, Waves and beaches: Doubleday and Co., Garden City, N.J., 267 p.

Beckinsale, R. P., and Chorley, R. J., 1968, History of geomorphology: in Fairbridge, R. W., ed., The encyclopedia of geomorphology, Reinhold, N.Y., p. 410–416.

Blackwelder, E., 1915, Post-Cretaceous history of the mountains in central Wyoming: Journal of Geology, v. 23, p. 97–117; 193–217; 307–340.

Blackwelder, E., 1928, Mudflow as a geological agent in semiarid mountains: Geological Society of America Bulletin, v. 39, p. 465–484.

Blackwelder, E., 1931, Pleistocene glaciation in the Sierra Nevada and Basin Ranges: Geological Society of America Bulletin, v. 42, p. 865–922.

Bretz, J H., 1913, Glaciation of the Puget Sound region: Washington Division of Mines and Geology Bulletin, v. 8, 244 p.

Bretz, J H., 1923, The channeled scablands of the Columbia Plateau: Journal of Geology, v. 31, p. 617–649.

Bryan, K., 1922, Erosion and sedimentation in the Papago country, Arizona with a sketch of the geology: U.S. Geological Survey Bulletin 730, p. 19–90.

Bryan, K., 1940, The retreat of slopes: Association of American Geographers Annals, v. 30, p. 254–268.

Chamberlain, T. C., 1895, The classification of American glacial deposits: Journal of Geology, v. 3, p. 270–277.

Chorley, R. J., 1963, Diastrophic background to twentieth-century geomorphological thought: Geological Society of America Bulletin, v. 74, p. 953–970.

Chorley, R. J., 1965, A re-evaluation of the geomorphic system of W. M. Davis: in Chorley, R. J., and Haggett, P., eds., Frontiers in geographical teaching: Methuen and Co., London, p. 21–38.

Chorley, R. J., 1974, Walther Penck: in Dictionary of Scientific Biography: Charles Scribner's Sons, N.Y., v. 10, p. 506–509.

Chorley, R. J., and Beckinsale, R. P., 1980, G. K. Gilbert's geomorphology: Geological Society of America Special Paper 183, p.129–142.

Chorley, R. J., Beckinsale, R. P., and Dunn, A. J., 1973, The history of the study of landforms, v. 2, The life and work of William Morris Davis: Methuen and Co., London, 874 p.

Chorley, R. J., Dunn, A. J., and Beckinsale, R. P., 1964, The history of the study of landforms: v. 1, Geomorphology before Davis: Methuen and Co., London, 678 p.

Cotton, C. A., 1940, Classification and correlation of river terraces: Journal of Geomorphology, v. 3, p. 27–37.

Dana, J. D., 1874, Corals and coral islands: Dodd Mead, N.Y., 406 p.

Dana, J. D., 1885, The origin of coral reefs and islands: American Journal of Science, v. 30, p. 89–105; p. 169–191.

Darwin, C., 1842, The structure and distribution of coral reefs: Smith Elder, London, 344 p.

Darwin, C., 1844, Geological observations on the volcanic islands visited during the voyage of the H.M.S. Beagle: Smith Elder, London.

Darwin, C., 1846, Geological observations on South America: Smith Elder, London.

Darwin, C., 1962, The voyage of the Beagle: Doubleday, N.Y., 524 p.

Davis, W. M., 1889, The rivers and valleys of Pennsylvania: National Geographic Magazine, v. 1, p. 183–253 .

Davis, W. M., 1899, The geographical cycle: Geographical Journal, v. 14, p. 481–504.

Davis, W. M., 1909, Geographical essays: Ginn and Company, Boston, (reprinted 1954 by Dover Publications, Inc., N.Y.), 777 p.

Davis, W. M., 1932, Piedmont benchlands and the primarrumpfe: Geological Society of America Bulletin, v. 43, p. 399–440.

Davis, W. M., 1938, Sheetfloods and streamfloods: Geological Society of America Bulletin, v. 49, p. 1337–1416.

DeBeer, G., 1964, Geological results of the voyage of the Beagle: in Charles Darwin: A scientific biography, Doubleday, New York, p. 56–77.

Dutton, C. E., 1880, Geology of the High Plateaus of Utah: U.S. Geographical and Geological Survey, Rocky Mt. Region, Washington.

Dutton, C. E., 1882, Tertiary history of the Grand Canyon district, with atlas: U.S. Geological Survey Monograph 2, 264 p.

Fenneman, N. M., 1936, Cyclic and non-cyclic aspects of erosion: Science, v. 83, p. 87–94.

Flemal, R. C., 1971, The attack on the Davisian system of geomorphology: a synthesis: Journal of Geological Education, v. 19, p. 3–13.

Forbes, J. D., 1843, Travels through the Alps of Savoy: Oliver and Boyd, Edinburgh, U.K.

Geikie, A., 1905, The founders of geology: St. Martin's Press, Inc., New York.

Gilbert, G. K., 1877, Report on the geology of the Henry Mts., Utah: U.S. Geographical and Geological Survey, Rocky Mt. Region, U.S. Department of the Interior, chapter 5, Land sculpture, p. 93–144.

Gilbert, G. K., 1890, Lake Bonneville: U.S. Geological Survey Monograph 1, 438 p.

Gilbert, G. K., 1909, The convexity of hilltops: Journal of Geology, v. 17, p. 344–350.

Gilbert, G. K., 1914, The transportation of debris by running water: U.S. Geological Survey Professional Paper 86.

Herbert, S., 1985, Darwin, the young geologist: in Kohn, D. ed., The Darwinian heritage, Princeton University Press, Princeton, N. J., p. 483–518.

Hinds, N. E. A., 1943, Geomorphology, the evolution of landscape: Prentice-Hall, Inc., N.Y., 894 p.

Hjulstrom, F., 1935, Studies on the morphological activity of rivers as illustrated by the river Fryis: University of Upsala Geological Institute Bulletin, v. 25, p. 221–527.

Holmes, C. D., 1941, Till fabric: Geological Society of America Bulletin, v. 52, p. 1299–1354.

Holmes, C. D., 1955, Geomorphic development in humid and arid regions—a synthesis: American Journal of Science, v. 253, p. 377–390.

Holmes, C. D., 1964, Equilibrium in humid-climate physiographic processes: American Journal of Science, v. 262, p. 436–445.

Horton, R. E., 1945, Erosional development of streams and their drainage basins: hydrophysical approach to quantitative morphology: Geological Society of America Bulletin, v. 56, p. 275–370.

Johnson, D. W., 1919, Shore processes and shoreline development: John Wiley and Sons, Inc., N.Y.

Johnson, D. W., 1925, The New England-Acadian shoreline: John Wiley and Sons, Inc., N.Y.

Johnson, D. W., 1932a, Streams and their significance: Journal of Geology, v. 40, p. 481–497.

Johnson, D. W., 1932b, Rock fans of arid regions: American Journal of Science, v. 23, p. 389–420.

Johnson, D. W., 1941, The function of meltwater in cirque formation: Journal of Geomorphology, v. 4, p. 252–262.

Leopold, L. B., and Wolman, M. G., 1957, River channel patterns: braided, meandering, and straight: U.S. Geological Survey Professional Paper 282-B.

Leverett, F., 1899, The Illinois glacial lobe: U.S. Geological Survey Monograph 38, 817 p.

Lobeck, A. K., 1939, Geomorphology: McGraw-Hill, N.Y.

Lyell, Charles, 1872, Principles of geology: Appleton and Co., N.Y.

Mackin, J. H., 1936, The capture of the Greybull River: American Journal of Science, v. 31, p. 373–385.

Mackin, J. H., 1941, Study of drainage changes near Wind Gap; a study in map interpretation: Journal of Geomorphology, v. 4, p. 24–53.

Mackin, J. H., 1948, Concept of the graded stream: Geological Society of America Bulletin, v. 59, p. 463–512.

Mackin, J. H., 1963, Rational and empirical methods of investigation in geology: in Albritton, C. C., Jr., ed., The Fabric of geology: Reading, Addison-Wesley Publishing Company, p. 135–163.

Matthes, F. E., 1930, Geologic History of the Yosemite Valley: U.S. Geological Survey Professional Paper 160.

Matthes, F. E., 1900, Glacial sculpture of the Bighorn Mts., Wyoming: U.S. Geological Survey 21st Annual Report, Part 2, p. 167–190.

McGee, W. J., 1897, Sheetflood erosion: Geological Society of America Bulletin, v. 8, p. 87–112.

Milankovich, M., 1941, Kanon der Erdbestrahlung und seine Andwendung aug das eiszeitenproblem: Belgrade, Yugoslavia, Royal Serbian Academy Special Publication 133, 633 p.

Nye, J. F., 1952, The mechanics of glacier flow: Journal of Geology, v. 2, p. 82–93.

Nye, J. F., 1953, The flow law of ice from measurements in glacier tunnels, laboratory experiments and the Jungfraufirn borehole experiment: Proceedings of the Royal Society of London, Series A, p. 477–489.

Penck, W., 1924, Die morphologische Analyse: J. Englehorn's Nachfolgelr, Stuttgart.

Playfair, John, 1802, Illustrations of the Huttonian theory of the earth: William Creech, Edinburgh, U.K.

Powell, J. W., 1875, Exploration of the Colorado River of the West and its tributaries: Smithsonian Institution, Washington D.C., 291 p.

Powell, J. W., 1877. Report on geographical and geological survey of the Rocky Mt. region: Washington, D.C., U.S. Government Printing Office.

Powell, J. W., 1878, Report on the lands of the arid region of the U.S. west with more detailed account of the lands of Utah: Washington, D.C., U.S. Government Printing Office.

Rubey, W. W., 1938, The force required to move particles on a stream bed: U.S. Geological Survey Professional Paper 189-E, p. 121–141.

Russell, I. C., 1885, Geological history of Lake Lahontan, a Quaternary lake of northwestern Nevada: U.S. Geological Survey Monograph 11, 288 p.

Russell, I. C., 1889, Quaternary history of Mono Valley, California: U.S. Geological Survey 8th Annual Report, p. 261–394.

Schumm, S. A., 1956, Evolution of drainage systems and slopes in badlands at Perth Amboy, New Jersey: Geological Society of America Bulletin, v. 67, p. 597–646.

Sharp, R. P., 1940, Ep-Archean and Ep-Algonkian erosion surfaces, Grand Canyon, Arizona: Geological Society of America Bulletin, v. 51, p. 1235–1270.

Sharp, R. P., 1940, Geomorphology of the Ruby-East Humboldt Range, Nevada: Geological Society of America Bulletin, v. 51, p. 337–372.

Sharp, R. P., 1948, The constitution of valley glaciers: Journal of Glaciology, v. 1, p. 182–189.

Sharp, R. P., 1949, Studies of superglacial debris on valley glaciers: American Journal of Science, v. 247, p. 289–315.

Sharp, R. P., 1954, Glacial flow: a review: Geological Society of America Bulletin, v. 65, p. 821–838.

Sharp, R. P., 1960, Pleistocene glaciation in the Trinity Alps of northern California: American Journal of Science, v. 258, p. 305–340.

Sharp, R. P., 1964, Wind-driven sand in Coachella Valley, California: Geological Society of America Bulletin, v. 75, p. 785–804.

Stone, I., 1980, The origin: A biographical novel of Charles Darwin: Garden City, Doubleday and Co., N.Y., 765 p.

Strahler, A. N., 1945, Hypotheses of stream development in the folded Appalachians of Pennsylvania: Geological Society of America Bulletin, v. 56, p. 45–88.

Strahler, A. N., 1950, Equilibrium theory of erosional slopes approached by frequency distribution analysis: American Journal of Science, v. 248, p. 673–696.

Strahler, A. N., 1952, Dynamic basis of geomorphology: Geological Society of America Bulletin, v. 63, p. 923–938.

Stegner, W., 1954, Beyond the 100th meridian: Houghton Mifflin, Boston, MA, 438 p.

Tarr, R. S., and Butler, R. S., 1909, The Yakutat Bay region, Alaska: U.S. Geological Survey Professional Paper 64, 178 p.

Tarr, R. S., and Martin, L., 1914, Alaskan glacier studies: Washington, D.C., National Geographic Society, 498 p.

Terzaghi, K., 1950, Mechanisms of landslides: in Paige, S., ed., Application of geology to engineering practice: Geological Society of America, Berkey Volume, p. 83–123.

Willis, B., 1898, Drift phenomena of Puget Sound: Geological Society of America Bulletin, v. 9, 111–162.

CHAPTER TWO

Basic Concepts

Streams, natural equilibrium systems.

FUNDAMENTAL PRINCIPLES

The study of geomorphology is based on a number of fundamental principles, some of which are shared with other scientific disciplines and some of which are unique to geomorphology. Because the Earth's topographic features are a mixture of landforms being formed at the present time and others that have been shaped in the past by processes no longer active, geomorphology embraces the investigation of both the mechanics of modern processes and the historic influence of geologic time. The former includes an understanding of the physics and chemistry of surface processes that generate landforms, and the latter adds the element of time to landforms that evolve over periods of time too long to study in the context of modern processes. The origin of landforms can be related to a particular geologic process, or set of processes, and the landforms thus developed evolve with time through a sequence of forms having distinct characteristics at successive stages (Davis, 1909).

Geomorphic Process

Each **geomorphic process** bestows distinctive features on the landscape and develops characteristic assemblages of landforms from which the origin of the forms can be identified. In the same way that fingerprints can be used to identify a person, the uniqueness of landforms can be used to "fingerprint" the process that created them.

The surface processes responsible for most of the earth's topographic features are weathering, mass wasting, running water, ground water, glaciers, waves, wind, tectonism, and volcanism. In this book, we will first consider the mechanics, and in some cases the chemistry, of each of these processes, followed by analysis of the landforms that they create and the evolution of such forms with time.

Evolution of Landforms

Landforms evolve with time through a continuous sequence of forms having typical features at successive stages of development, largely as a result of continuous changes in processes and rates as time goes on. For example, a **nivation basin,** formed under a permanent snowfield below the snowline, enlarges until enough snow and ice accumulate to form a glacier, which then develops a steep upper headwall and creates a scoured basin. The changes are progressive, and the forms evolve continuously.

Geologic Structure

Some landforms are a result of tectonic disturbance of the earth's crust. Such landforms are produced either by directly offsetting the land surface or by secondary erosion of rocks of differing resistance. Most of the earth's major features, such as mountain ranges and the configuration of the continents and ocean basins, are the result of large-scale motion of crustal plates, discussed in Chapter 8.

Direct offsetting of the land surface is known as *neotectonism,* referring to very recent crustal activity. For example, an **escarpment** created by fault movement is a landform caused directly by tectonic movement.

Not quite so obvious is the etching out of valleys and ridges by erosion of rocks of differing resistance whose spatial configuration is determined by the geologic structure. For example, the rocks of the numerous ridges and valleys of the Appalachian Mountains in the eastern United States were deformed at the end of the Paleozoic era, some 250 million years ago, when they were folded into their present structures. More recent erosion has etched out the weaker beds into valleys and the more resistant ones into ridges whose position and orientation are determined by the earlier crustal deformation.

Geomorphic Systems

The understanding of surface processes and geomorphic systems involves application of a number of concepts common to all scientific disciplines. These concepts include uniformitarianism, equilibrium, and positive feedback.

Uniformitarianism

In order to understand the origin of relict landforms created by ancient processes, one must understand modern processes and use this knowledge to extrapolate into the past. Implicit in this approach is the assumption that basic physical and chemical processes that apply to the present apply equally to the past and the future, a concept championed by Hutton and Lyell as **uniformitarianism.** This concept is often stated as, "The present is the key to the past," meaning that the clue to recognition of the origin of relict landforms lies in studying present-day processes. Some care must be exercised in the application of uniformitarianism because rates of surface processes may have varied in the past and because processes may have existed in the past but may not be operative at present. For example, imagine the difficulty in interpreting relict glacial landforms if glaciers were not present today. However, in most cases, landforms may be understood by extrapolation of present processes backward in time.

Equilibrium Systems

An **equilibrium system** is one in which a delicate balance exists between opposing forces such that any change in the variables that control the system produces a change in one or more of the other variables to generate a new balance. In 1884, Le Chatelier expressed the concept that a change in any of the variables governing the equilibrium of a chemical system will cause a compensating change among other variables that will reestablish stability in the system. This general principle, now known as **equilibrium** (also known as *homeostasis, self-regulation,* or *negative feedback*) was anticipated seven years before Le Chatelier by American geomorphologist G. K. Gilbert. He applied the concept of negative feedback to explain streams that transport as much load as they are capable of carrying without either eroding or depositing any of it, a condition he referred to as **graded.** If a **graded stream** has a steeper slope than that required to develop the velocity it needs for transporting the load, the increased velocity accompanying the steeper slope will lead to entrainment of more channel material, and erosion of the bed will lower the slope until the velocity is adjusted to just transport the load. A change in slope from steep to more gentle will have the opposite effect—increasing the slope by deposition of material in the channel. In this way, a graded stream produces a smooth, longitudinal **profile of equilibrium.** Every reach of a graded stream is part of an equilibrium system, and any interruption of the variables controlling the system will introduce waves of erosion and deposition that move through the system until a new equilibrium is produced. In geomorphic equilibrium systems, *negative feedback* processes counteract external changes to the system and stabilize the system by making adjustments within it. For example, if coarse debris, beyond the capacity of the stream to transport, is introduced into a stream, deposition occurs for a time, thereby increasing the slope of the channel and increasing the velocity of flow until it is sufficient to transport the newly added coarse material, at which time a new steady state is attained.

Gilbert applied the concept of equilibrium not only to streams but also to hillslopes and pediments in the western United States (Gilbert, 1877, 1890, 1909). He identified equilibrium in stream channels, equilibrium of hillslopes (which he referred to as **graded slopes**), and equilibrium control of pediments. Each of these systems responds to changes in controlling variables by evoking a change in the system, which reestablishes a new balance.

W. M. Davis clearly recognized the importance of Gilbert's work, especially his perception of equilibrium in streams and slopes, and he incorporated it into his concept of the cycle of erosion (Davis, 1909). Mackin also placed great emphasis on the **profile of equilibrium** in his classic 1948 paper on the graded stream:

> *Its diagnostic characteristic is that any change in any of the controlling factors will cause a displacement of the equilibrium in a direction that will tend to absorb the effect of the change.* (Mackin, 1948)

Time is a critical element in geomorphologic equilibrium because most natural geomorphic systems are dynamic and equilibrium must be considered in the framework of long or short periods of geologic time. In his definition of a graded stream, Mackin wrestled with the problem of changes in a fluvial system over short periods of time, and he framed the equilibrium in streams as perpetuated "over a period of years" in order to deal with changes in streams that occur hourly, daily, weekly, and so on. Most of these short-term changes balance out in the long run and do not really produce a lasting geologic effect.

Thoughtful perusal of the writings of Gilbert, Dutton, Powell, Davis, and Mackin clearly shows that all of them were aware of the concept of changing equilibrium in different time frameworks. The labeling of various short-term and long-term time structures of equilibrium by Schumm (1977) has helped to clarify some of the confusion about geomorphic systems that formerly existed. These states of equilibrium are illustrated graphically in Figure 2–1 and are discussed in the following paragraphs.

Steady-state equilibrium occurs under conditions that change very little with time. An example is an erosional surface of low relief in the latter stages of the cycle of erosion. Figures 2–1A and 2–1B illustrate two variations of this type of equilibrium. In Figures 2–1A and 2–1B, conditions are constant, but in Figure 2–1B, small variations in form oscillate about a constant average condition, and short-term changes offset one another, so that the overall trend does not deviate significantly.

Declining equilibrium occurs when the rate of change declines with time to successively lower rates (Figure 2–1C). A related type of equilibrium, known as **dynamic equilibrium,** (Figure 2–1D) consists of small variations about a changing average condition. Short-term oscillations do not entirely offset one another, so that the average trend changes progressively. This type of equilibrium is characteristic of graded streams and graded slopes in the cycle of erosion where the rate of erosion of streams slowly changes as topographic relief of an area diminishes with time.

Positive Feedback Systems

Positive feedback occurs when a change in the input variable causes magnification of change in the direction of the initial adjustment. For example, the bend of a stream meander causes the water to enhance erosion on the outside of the bend, which causes the bend to enlarge, which in turn causes accelerated erosion on the outside of the bend, until finally two meanders intersect and destroy one another. Positive feedback systems thus carry the source of their own destruction and eventually accelerate to self-elimination.

A. Steady-state equilibrium (constant conditions)

B. Steady-state equilibrium (oscillating conditions)

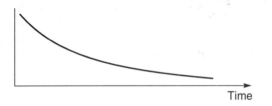

C. Declining equilibrium

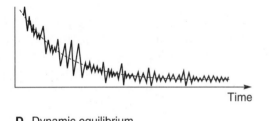

D. Dynamic equilibrium

E. Metastable equilibrium (episodic erosion)

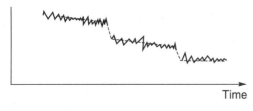

F. Dynamic metastable equilibrium
(episodic erosion)

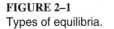

FIGURE 2–1
Types of equilibria.

RATIONAL AND EMPIRICAL APPROACHES TO GEOMORPHOLOGY

The **rational approach** to geomorphology concentrates on discerning cause-and-effect relationships in understanding surface phenomena. It may, or may not, be based on quantitative data. Even though the rational approach may depend heavily on quantitative and experimental data, the objective lies not in the numbers themselves but in what cause-and-effect relationship they demonstrate. Data are logically evaluated at each step, and the outcome depends largely on critical thinking and reasoning.

The **empirical approach** in geomorphology is a somewhat different approach to the understanding of geomorphic phenomena, emphasizing measurement and numerical data. The empirical approach may be complexly interwoven with the rational approach, and a distinction between the two is not always easy to discern. For example, consider that the size of particles being transported by a stream obviously decreases downstream. Because particle size can easily be measured, the decrease in particle size downstream can readily be quantified, and these data could then serve as the basis for formulating a mathematical equation showing the rate of average decrease in particle size per unit length of stream for that river. If the approach is purely empirical, the investigation might end with the numbers and the equation that show how much particle size decreases downstream. This empirical equation, however, does not establish the cause of the decrease in particle size downstream; it only describes the rate of the decrease. Of greater interest, is the question, *Why* does particle size decrease downstream? Emphasis then shifts from the assembling of numerical data to the search for cause-and-effect relationships.

Although the rational and empirical approaches are different, neither is better nor worse than the other. They are merely aimed at different goals and often may complement one another. For example, the plotting of data points on a graph may show relationships that would not otherwise be discernible, and such a graph may lead to a rational explanation of cause and effect between variables.

The rational approach to analysis encompasses critical evaluation of various lines of evidence, each bit of evidence is scrutinized for accuracy and relevance to the problem at hand (Mackin, 1963). Irrelevant evidence is rejected, and such data points do not make their way onto a graph. Attention is focused on evidence most critical to demonstrating a conclusion. Quantitative data may play an important role in the analysis, but each number is analyzed qualitatively for relevance to the problem before being used. In contrast, the empirical method stresses a numerical approach that includes *all* available numerical data, regardless of cause-and-effect relationships. Plotting a large number of points on a graph may lead to a numerical equation. Mackin (1963) makes the comparison of the empirical method as a "shotgun" approach, because little in the way of evaluation of individual points need be made before they are "shot" onto a graph. In contrast, he compares the rational approach to a "rifle" that uses "bullets" (data points) that are carefully evaluated for their relevance before plotting. By plotting *all* data related to an analysis, the empirical approach has the advantage of showing relationships that might not otherwise be obvious. However, Mackin (1963) cautions that use of solely the empirical method may lead to erroneous conclusions because cause-and-effect relationships are not necessarily taken into account.

Multiple Working Hypotheses

An especially powerful approach to geomorphic investigations of surface processes and of the origin and evolution of landforms is the *method of multiple working hypotheses*. This method begins with the collection of observations and formulation of alternate hypotheses that might logically explain the observations, not unlike the detective in a mystery who determines all the possible suspects who might have been responsible for a crime. The first step usually points out the need for further facts that might reasonably be expected to result from the reality of each hypothesis; this step also directs the hunt for additional information. The second step tests each hypothesis by attempting to verify the deduced conclusions by further focusing on critical facts; then the hypotheses are modified to produce the most probable one. Sometimes the observed facts do not allow positive verification of any single hypothesis, but in such cases a definite conclusion may be drawn if all other possible hypotheses can be discounted.

The method of multiple working hypotheses may be broken down into several steps:

1. *Observation of facts.* The first step is to collect as many relevant observations and facts as possible. Among the possible landform data are shape, size, orientation, composition, and association with other features.

2. *Development of multiple working hypotheses.* Several alternate interpretations may be possible to explain the observed data. Once all possibilities have been identified, attention can be focused on critical information needed to reject hypotheses that do not fit the observed facts.

3. *Testing hypotheses.* After rejecting hypotheses adverse to the observed facts, the next step is to focus on the search for new information that might support or reject any of the remaining hypotheses. Each hypothesis is then critically tested against all the known facts, which focuses on needs for certain critical evidence and sharpens concentration

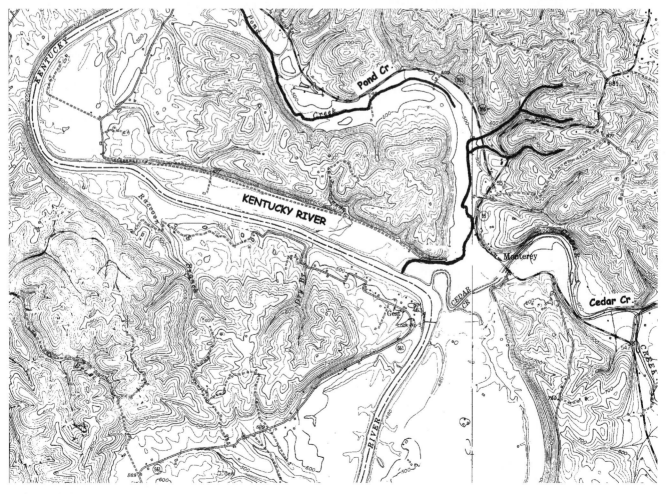

FIGURE 2–2A
Present drainage of Cedar Creek and the Kentucky River, Kentucky. (Modified from Lockport quadrangle, U.S. Geological Survey)

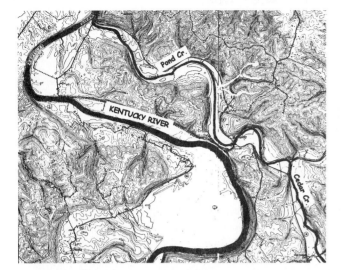

FIGURE 2–2B
Former courses of Cedar Creek and the Kentucky River.

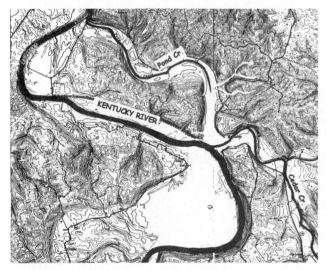

FIGURE 2–2C
Intercision of meanders of Cedar Creek and the Kentucky River.

upon the central issues. Ideally, only one hypothesis will be consistent with all the available information, but sometimes more than one hypothesis satisfies all of the facts, and a unique solution is not possible with available data.

The stream changes shown in Figure 2–2 illustrate the method of multiple working hypotheses applied to problem solving in geomorphology. The valley north of Cedar Creek presents several anomalies. It is partially occupied by northward-flowing Pond Creek, but the southern portion contains a small, unnamed creek flowing in the opposite direction, separated from Pond Creek by an in-valley divide. The problem is, What is the origin of this valley?

The first step in the analysis is to identify all of the possible alternatives. Pond Creek obviously did not erode the valley, because it occupies only a portion of the valley. The same can be said for the small, southward-flowing creek in the valley, so both of these streams can be eliminated as possible alternatives. They are flowing in the valley only as a result of inheritance after the valley was abandoned by some previous occupant. The only other streams that could have possibly eroded the valley are the Kentucky River and Cedar Creek. Thus, if either one of these can be eliminated, the origin could be attributed to the other. Testing each hypothesis against the available evidence leads to the solution. Although the valley is wide enough to accommodate the Kentucky River, if the Kentucky were ever in the valley we would have to explain how it abandoned that channel in favor of its present course. It could not have been captured by headward erosion of a tributary because, like most streams, tributaries of the Kentucky have steeper gradients than their trunk streams, and thus intersection by a tributary stream as a result of headward erosion would leave the tributary high above the Kentucky. With no viable method of getting the Kentucky out of the valley and into its present course, we can eliminate the Kentucky as the maker of the valley. That leaves only Cedar Creek as an alternative. To verify this hypothesis, we focus our attention on the mechanism of such a channel change. If lower Cedar Creek had ever flowed in the valley, how did the creek abandon it? At the time Cedar Creek flowed northward (Figure 2–2B), a meander of Cedar could have intersected a meander of the Kentucky River at Monterey diverting Cedar Creek into the Kentucky and abandoning the valley to the north (Figure 2–2C). Subsequent collection of local drainage then could have led to the origin of Pond Creek and the small, southward-flowing creek. Thus, not only is Cedar Creek the only "surviving" hypothesis, but a mechanism for the drainage diversion is also viable.

REFERENCES

Chamberlin, T. C., 1897, The method of multiple working hypotheses: Journal of Geology, v. 5, p. 837–848.

Davis, W. M., 1909, Geographical Essays: Ginn and Company, Boston, MA, (reprinted 1954 by Dover Publications, N.Y.), 777 p.

Gilbert, G. K., 1877, Report on the geology of the Henry Mts., Utah: U.S. Geographical and Geological Survey, Rocky Mt. Region.

Gilbert, G. K., 1886, The inculcation of scientific method by example, with an illustration drawn from the Quaternary geology of Utah: American Journal of Science, v. 31, p. 284–299.

Gilbert, G. K., 1890, Lake Bonneville: U.S. Geological Survey Monograph 1, 438 p.

Gilbert, G. K., 1909, The convexity of hilltops: Journal of Geology, v. 17, p. 344–350.

Johnson, D. W., 1933, Role of analysis in scientific investigation: Geological Society of America Bulletin, v. 44, p. 461–494.

Mackin, J. H., 1941, Study of drainage changes near Wind Gap; a study in map interpretation: Journal of Geomorphology, v. 4, p. 24–53.

Mackin, J. H., 1948, Concept of the graded stream: Geological Society of America Bulletin. v. 59, p. 463–512.

Mackin, J. H., 1963, Rational and empirical methods of investigation in geology: *in* Albritton, C. C., Jr., ed., The fabric of geology: Addison-Wesley Publishing Company, Reading, MA, p. 135–163.

Powell, J. W., 1877, Report on geographical and geological survey of the Rocky Mt. region: Washington, D.C., U.S. Government Printing Office.

Schumm, S. D., 1977, The fluvial system: John Wiley and Sons, N.Y., 338 p.

Weathering

Exfoliation of granite on Half Dome, Yosemite National Park, California. (Photo by F. E. Matthes, U.S. Geological Survey)

INTRODUCTION

WEATHERING is the disintegration and decomposition of rocks and minerals at the earth's surface as a result of physical and chemical action. Weathering comprises all the processes that destroy bedrock and convert it into fragments, ions in solution, or colloids. These changes take place largely *in situ*, and movements of weathered materials are local and confined largely to the outcrop. Even the unconsolidated rock residue may continue to break down in place until an end product, essentially in equilibrium with the environment, is formed. On the other hand, if the weathered rock is eroded away as rapidly as it is formed, a stable end product may never appear. Weathering prepares the way for erosion by weakening the rock and making it more susceptible to mass movements and removal by the other agents of erosion.

The processes operating at or near the ground surface involve the three states of matter: liquid water and its solutes, mineral and organic solids, and atmospheric gases. These substances may interpenetrate, react, or mingle to such an extent that their fundamental properties are changed. In addition, not all the substances are inanimate, for this is a zone of abundant plant and animal life—visible and microscopic organisms that make their demands on the environment and exert an influence that is exceedingly varied and almost always complex. In other words, this zone reflects the interactions between and among the lithosphere, the atmosphere, the hydrosphere, and the biosphere.

An important factor during weathering is that rocks, created largely at depth, are inherently less stable at the surface and thus are vulnerable to the attack of the atmosphere, hydrosphere, and biosphere. The geomorphic fate of any mineral aggregate depends upon its physical and chemical properties, such as strength, permeability, structure, and chemical reactivity. These are determined by the composition, size, shape, intergranular structure, and intragranular structure of the minerals (Barshad, 1964; Bohn et al., 1979; Deju, 1971; Garrels and Christ, 1965; Lindsay, 1979; Marshall, 1977). Such multiple-mineral frameworks, like a common chain, have weak links, which are critical in determining total vulnerability to the various agents of change.

The rate of weathering is not constant everywhere but varies according to differences in intensity of processes at any given point. The kind of weathering that predominates on the surface also varies from place to place. Because weathering refers only to the breaking up of rock in place, erosion and transportation are not considered weathering processes. An especially important factor in mechanical weathering is the substantial increase in surface area as rocks disintegrate, which facilitates chemical reactions and chemical weathering.

Generally, the topography in arid climates is characterized by angularity and sharp breaks in slope, steep cliffs with accumulations of loose debris at the base, and extensive exposure of bedrock. In contrast, topography in humid climates is characterized by smooth, rolling hills with gently rounded slopes covered with a deep accumulation of weathered material and little exposure of bedrock. These differences are brought about largely because weathering processes vary in humid, arid, polar, and alpine climates. This is not to say that weathering processes do not occur in each environment but rather that one process may dominate in a particular environment. For example, the cation chemistry of mountain streams indicates that chemical decomposition and pedogenic clay can occur even in high alpine regions (Reynolds and Johnson, 1972; Caine, 1979; Birkeland et al., 1987).

MECHANICAL WEATHERING

Mechanical weathering refers to disintegration or breaking up of rock by physical processes without changes in chemical or mineral composition. Such weathering is brought about by stresses originating within rocks and by stresses applied externally. Internal stresses accomplish disruptive expansion when the overburden is removed from rocks that have been under high confining pressure at depth. The release of pressure by removal of overburden is known as **unloading.** Internal stresses also cause disruptive volume changes with intense heating and cooling. Stresses may also develop from the wedging action of ice, salt crystals, and plants; the plucking action of colloids, and the earth-moving activities of organisms.

Unloading

Rocks exposed at the earth's surface often originated at great depths in environments of great pressure. Estimates of the depths of erosion in mountainous areas range up to 30 km (Table 3–1) (Bateman and Eaton, 1967; Whitney and McLelland, 1973; Stevens, 1974; Clark and Jager, 1969). Zeitler et al. (1982) calculated that the rocks of Nanga Parbat in the Himalayas of Pakistan were buried to a depth of about 5 km as recently as 400,000 years ago and have been uplifted and eroded at a rate of about one cm/year since then.

High confining pressure from the weight of overlying rock compresses rocks elastically to a smaller volume without permanent deformation because the confinement also increases the strength of the rocks. Pressures produced by the magnitude of burial suggested in Table 3–1 may reach up to 8 kilobars, capable of causing about 0.8-percent expansion upon release of pressure (Birch, 1966). However, because the rocks

TABLE 3–1
Depth of erosion in alpine areas.

Mountains	Depth of Erosion
Sierra Nevada	8–16 km
Appalachian Mountains	8 km
Adirondack Mountains	24 km
New Zealand Alps	16–24 km
European Alps	30 km

contract in the cooler temperatures at shallower depths, only a portion of expansion from pressure release shows up in the upper few kilometers (Skinner, 1966; Haxby and Turcotte, 1976; Bruner, 1984).

When rock is uplifted and exposed at the surface by erosion, the high pressure is much reduced, and expansion may take place rapidly enough to produce fracturing. An unwelcome result of extremely rapid unloading is rock-burst, a process familiar to miners and quarrymen. Occasionally, a freshly quarried block or part of a mined face will fly apart with explosive violence in a spectacular demonstration of the expansion caused by unloading. However, in natural situations, the overburden is removed more slowly by erosion. Yet, for many rocks in which the strain cannot be absorbed by movement along existing fractures, expansion is rapid enough to cause new joints to appear. As the rock expands, new fractures form joints parallel to the earth's surface in the direction of least pressure.

Such jointing is most common in massive granite and thick sandstone and is usually best developed near the surface, becoming more widely spaced with depth. Measurements of the spacing of joints in New England granite suggest that the distance between fractures increases at an exponential rate with depth below the surface (Jahns, 1943) (Figure 3–1).

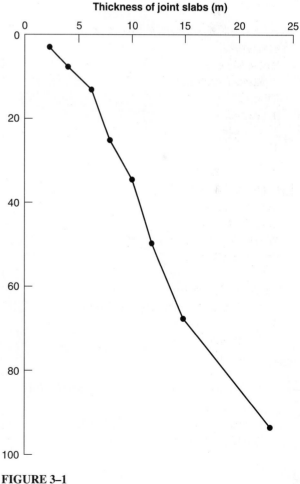

FIGURE 3–1
Changes in the spacing of joints with increasing depth.

Joints of this type are called *sheeting joints* or *exfoliation fractures* (Twidale, 1973) in contrast to joints of tectonic origin. They break the rock into sheetlike slabs or broad, overlapping lenses (Figure 3–2). The spalling of fresh rock surfaces into sheets that are gradually removed is a weathering process called **exfoliation.** In massive rocks, the exfoliation fractures are independent of, and may cut across, internal structures such as foliation, lithologic contacts, inclusions, fabric (Jahns, 1943), or bedding (Bradley, 1963). Similar fracturing along bedding planes in properly oriented sedimentary rocks may be caused by unloading (Dapples, 1959). Joints parallel to canyon walls in massive Navajo sandstone were found to be limited to the outer 10 m of rock encountered in tunnels at Glen Canyon dam, Arizona (Bradley, 1963). Exfoliation along vertical walls in massive sandstone or granite may produce broad arches (Figure 3–3).

The close relationship between sheet jointing and surface topography is shown in glaciated mountains where sheet jointing is concave upward, parallel to cirque walls and floors, in contrast to convex-upward sheet jointing on nearby convex summits (Ollier, 1969).

Exfoliation domes, such as Half Dome in Yosemite National Park (Chapter-opening photo), are thought to have been developed by concentric exfoliation of granite (Matthes, 1930), resulting from physical expansion upon unloading. Sheeting joints in Yosemite are a few centimeters apart near the surface (Figure 3–2), increasing to a few meters at depth and dying out within about 30 m (100 ft) below the surface (Matthes, 1930).

Freezing and Thawing

When water freezes, its volume is increased by about 9 percent. The expansion of the ice being frozen in a confined space thus exerts great pressure against the sides of the material enclosing the ice. Ice is a threat to plumbing and automobile radiators in cold regions, and it was used to winter-quarry rock in the days when rock was split by pouring water down strategically drilled holes and waiting for it to freeze and pry apart the rock. The success of this method is guaranteed by plugging the holes (Taber, 1943). Among the terms used to describe this process are *frost wedging, frost action, ice wedging,* and *freeze-and-thaw* (Brace et al., 1972; Russell, 1943; Thorn, 1979; Whalley et al., 1982; White, 1976; Winkler, 1968).

When distilled water in a perfectly sealed chamber is cooled to a temperature of −22°C, it freezes and expands in

FIGURE 3–2
Exfoliation of granite in the Sierra Nevada, California, resulting from expansion of the surface upon unloading of overlying rock. (Photo by W. K. Hamblin)

FIGURE 3–3
Exfoliation arch in Jurasic sandstone, southern Utah. (Photo by W. K. Hamblin)

volume by 9.05 percent. This increase in volume can create pressures of 2045 atmospheres, almost 30,000 lb/in^2 (2100 kg/cm^2), which is about the crushing strength of granite. Above this critical temperature, the water is kept liquid by the pressure-melting effect of its own expansion, and by the heat of fusion of water (79.7 cal/g) that must be absorbed by the surrounding rock before additional freezing can occur. High pressures are seldom attained in nature because perfectly closed systems are rare and vulnerable to destruction before the maximum pressure is reached. Grawe (1936), in a review of misconceptions about frost weathering, pointed out that the wedging apart of the walls of a natural rock chamber (such as a joint) is a matter of overcoming the tensile strength of the rock, which is considerably lower than the crushing strength. Tensile strengths rarely exceed 3500 lb/in^2 (250 kg/cm^2). Moreover, air (a compressible gas) is inevitably included with the water and helps to take up some of the strain.

Some closed-system effect is probably present in the wedging operations described above. Supercooled water (about −5°C or −6°C) commonly forms in cracks (Grawe, 1936). The freezing of a plug of ice at the surface of the water may close the system enough that expansion with continued cooling develops pressures in excess of the tensile strength of the rock, and thus the crack is widened and deepened.

Growth pressures of ice crystals are another aspect of the process. Taber (1943) believed that the wedging effect due to crystal growth can be more powerful than the mere expansion of freezing water against a plug of ice. He suggested that this is particularly true of large crystals grown slowly by a prolonged freeze. Ice crystals forming on the walls of a crack tend to grow outward, normal to the walls, until the crack is filled (Shumskii, 1964; Taber, 1943). At this stage, continued growth may develop pressure in excess of the tensile strength of the rock.

Water in very small cracks may remain liquid, even at very low temperatures, as a result of the pressure-melting effect and the capillary adhesion of water to rock, and it may retain its hydraulic properties at temperatures of −60°C to −125°C (Mellor, 1970). In such instances, water under high pressure may be forced into microfractures in rocks as capillary films which pry apart the cracks ahead of the freezing front (Tharp, 1987; Walder and Hallet, 1985). This process, known as *hydrofracturing,* has been shown to extend to depths of 12 to 15 m in areas where freezing rarely exceeds one meter.

Water in permeable materials may be segregated in various ways as it freezes. For example, efflorescence of ice at the surface of moist soil produces a crust of columnar or platy crystals that are nourished by soil moisture rising from below

(Taber, 1943; Shumskii, 1964). The growth of these crystals as needle ice may lift soil particles and stones several inches above the surface. Similar efflorescence at the surface of permeable rock may cause flaking. Ice tends to segregate beneath the larger particles scattered through a stony soil. The lower parts of the stones freeze sooner than the surrounding soil matrix because the stones are better conductors of heat and provide bridges for the penetration of a wave of freezing. The ice masses grow at the expense of water in the adjacent soil, and by their growth they thrust the rock particles upward.

Soil water freezes most easily in large openings (Everett, 1961), whereas water in small pores is kept liquid by the effect of pressure. Thus, ice-crystal nuclei in large openings may be nourished by water migrating from the surrounding soil where pores are smaller. The growth of the crystalline masses creates lens-like segregations that cause heaving of the overlying soil (Taber, 1930, 1943). Similar crystal growth in permeable rocks may develop a plexus of cracks and cause disintegration.

Optimum conditions for the wedging effect of freezing require a supply of water, many alternations of freezing and thawing, and yet enough sustained freezing at temperatures well enough below 0°C that masses of ice will grow, and freezing will penetrate into the ground. This wedging effect is most spectacular at high latitudes and high altitudes (particularly above the timberline) where these conditions are met.

In areas of high latitude or high altitude, alternate freezing and thawing of water at or near the Earth's surface may occur almost every day under certain conditions. During the daytime, water seeps into cracks and crevices in the rock, where it may freeze at night. Expansion of the ice then causes wedging of the rock. Repeated freezing and thawing, with resulting shattering of rock (Figure 3–4), is especially common in high mountains above the timberline, where meltwater produced during the day percolates into cracks and refreezes at night. A layer of angular rub-

ble can be produced by the wedging effect of freezing water. Fields of boulders in rocky parts of the Arctic and on broad, alpine summits form **block fields** or **felsenmeer** (German for "rock sea") as striking examples of widespread freezing and thawing.

Growth of Crystals

The mechanical effects of crystallization of minerals upon evaporation of water within the soil or at and near the surface of an outcrop are somewhat similar to those discussed above for the formation of ice crystals in soil and rock. Figure 3–5 shows the shattering of fence posts by the expansion of salt crystals that have grown from salt spray seeping into the wood and crystallizing as the water evaporates. Precipitation of sulfates, chlorides, and carbonates containing mobile ions of potassium, sodium, and magnesium may cause cracking and flaking off of small pieces of rock.

Seawater, concentrated mostly on the surface of coastal outcrops as spray, may cause flaking and disruption of mineral grains as the salts crystallize with evaporation. In arid and semiarid regions, saline waters may become saturated and even supersaturated within the small passages of permeable rocks. Evaporation of the water near the surface causes crystallization of magnesium sulfate, anhydrite, sodium-calcium sulfate, calcium carbonate, and sodium carbonate.

The growth of crystals may cause considerable disaggregation of the surface of rock, particularly among permeable sed-

FIGURE 3–5
Fence post shattered by expansion of salt crystals.
Saltwater spray from Great Salt Lake, Utah seeps into the wood and when the moisture evaporates, salt crystals grow and break apart the wood. (Photo by W. K. Hamblin)

FIGURE 3–4
Boulder shattered by freezing and thawing of ice in crevices over a 12 year period, Breidamerkurjokull, Iceland.

imentary rocks such as sandstone (Emery, 1960; Smith, 1977). Moreover, hydration of some salts produces a disruptive volume change as water of crystallization is taken into the structure (for example, anhydrite to gypsum, $CaSO_4 + 2H_2O \rightarrow CaSO_4 2H_2O$) (Evans, 1970; Schwartz, 1989). The additional water may be brought to the rock in rain showers or as dew or water vapor.

The effects of hydration pressures are especially significant when hydration is followed by temperature change and hydration again over short time periods (Winkler and Wilhelm, 1970). The hydration/dehydration cycle of $Na_2SO_4 10H_2O$ (mirabilite) to Na_2SO_4 (thenardite) is capable of disaggregating granite saturated with a solution of sodium sulfate for 17 hr at room temperature and then dried for 7 hr at 105°C (Kessler, et al., 1940; Evans, 1970).

The effects of hydrating crystals are often noticeable on historic stone monuments (Johnson et al., 1972; Winkler and Wilhelm, 1970; Winkler, 1975). For example, rock obelisks with engraved hieroglyphics persisted for more than 3000 years in the arid climate of Egypt where the average rainfall is about one inch per year (2.5 cm/yr). One of a pair of obelisks was moved to Central Park in New York City in 1879 where, despite efforts at preservation, the humid climate has made the hieroglyphics illegible. Meanwhile, the obelisk that remained in Egypt was unaltered (Winkler, 1965). However, weathering does occur even in the arid climate of Egypt. The Great Pyramid of Cheops near Cairo was constructed of four different rock types: resistant granite and dense limestone, and less-resistant silty limestone and limestone made of tiny fossils. The granite and dense limestone remain virtually unweathered, but the less-resistant limestones have been deeply weathered where the facing of resistant rocks have been removed and the stepped sides of the pyramids are covered with weathering debris (Figure 3–6a). Some of the deeply weathered older pyramids are now almost completely covered with their own weathered debris (Figure 3–6b). Comparison of the obelisks in New York and Egypt show the effects of climate on weathering rates, and comparison of the various rock types making up the pyramids shows the effects of lithology on weathering rates.

Thermal Expansion and Contraction

Extreme temperature changes, such as those accompanying forest and brush fires, are sufficient to cause spalling and general fragmentation of rock. A common example may occur when campers who have used crystalline rocks to line a campfire are rudely interrupted by the explosion of outer shells of the rock suddenly spalling off.

The process was verified experimentally by Blackwelder (1927), who caused the spalling of a variety of igneous rocks by rapidly heating them to temperatures between 300°C and 375°C. However, the temperatures reached in Blackwelder's experiments were well above the hottest temperatures to which rocks have been raised by solar heating (for example, 70°C to 80°C in deserts). The range of temperature in the ex-

FIGURE 3–6
Weathering of Egyptian pyramids. (A) Early stages.
(B) Accumulation of weathered material on steps.
(C) Pyramid buried in its own weathered debris.
(Photos by W. K. Hamblin)

periments (nearly 300°C) also exceeds diurnal ranges, and the rapidity of the experimental heating and chilling was far greater than can be expected from solar heating. The necessity to change temperature rapidly through a large range during the experiments made for a stringent test, and the results suggested that diurnal heating and cooling do not cause enough expansion and contraction to spall slabs from outcrops or break crystalline rocks into granules.

Rocks do not conduct heat well, so a thin outer layer of rock exposed to the intense heat of a fire becomes much hotter than the rock beneath and expands more rapidly, causing it to spall off in thin sheets or small fragments. Because the most common minerals have different coefficients of thermal expansion, and because some minerals expand differently along different crystallographic axes, strong thermal gradients produce internal strain between crystals, which leads to cracking and disintegration. Forest and brush fires may explain the occurrence of undecomposed rock spalls in parts of the western United States, noted by Blackwelder (1927) and others. Birkeland (1984) pointed out that angular stones on the rock glaciers of Mount Sopris, Colorado, have thick, red, oxidation weathering rinds, whereas stones in a nearby forest that burned several decades earlier were rounded and lacked the thick oxidation rinds of the unburned area.

Similar spalling of rocks by intense heat occurs on mountain peaks struck by lightning, but the affected areas are smaller. The effects of fire cracking and fire spalling are best seen in areas where intense forest or brush fires have occurred and other weathering processes are not as important in breaking down the rocks.

The effects of less intense thermal expansion and contraction by solar insolation have been debated for many years, but at present, the significance of diurnal or seasonal temperature changes remains equivocal (Blackwelder, 1933; Peel, 1974). Tarr (1915) showed that the stresses of daily expansion-contraction in granites can be absorbed readily by mineral grains because their elastic strengths are well above the stresses produced by solar heating. However, Griggs (1936a, b) wondered about the effects of a large number of repeated stresses that might gradually weaken the rock and cause a condition similar to metal fatigue. To test this possibility, he first experimented with specimens of granite, syenite, felsite, and sandstone by subjecting them to rapid and severe temperature changes. He found these samples unaffected by 3.8 "years" of simulated weathering. Next, he repeated the experiment with a polished specimen of coarse-grained granite. Griggs believed that the large grain size would provide the best opportunity to develop intergranular stresses and bring about granulation. The specimen was carefully photographed before the experiment, and all existing fractures were noted. The granite block was heated in a furnace for 5 min from 32°C to 176°C; then it was cooled for 10 min by an electric fan, which brought it rapidly back to the original temperature. In this experiment, the rock was subjected to 89,400 cycles, equivalent to 244 years of unnaturally severe diurnal change.

Experiments with metal fatigue show that with an increase in stress, the number of cycles necessary to produce failure decreases exponentially. Hence, the rapid heating-cooling, estimated by Griggs to develop stresses 20 times greater than normal, signified far more than 244 experimental "years." Re-

markably, the specimen showed absolutely no signs of weakening. Close inspection of enlargements of photos taken at the end of the experiment showed that no new fractures had appeared, and all the original cracks remained exactly as they were at the beginning.

In a sequel to this experiment, Griggs heated the same specimen as before, but he cooled it with a spray of water instead of with the fan. After 2 1/2 simulated years, the surface lost its polish, new cracks appeared, existing cracks became wider and deeper, and flaking began. These results suggested that thermal changes play a minor role and that at best they may be of slight aid in disintegrating rocks already weakened by chemical weathering.

The degree to which disintegration depends on the presence of water and the degree to which it is caused by solar thermal expansion-contraction remain debatable. Ollier (1963, 1969) found shattered, angular fragments of rocks in the deserts of Australia and suggested that thermal expansion-contraction was responsible because no other explanation could be found. Others (for example, Roth, 1965; Gray, 1965; Rice, 1976) continue to contend that solar thermal expansion and contraction is a viable weathering process and that laboratory experiments may not hold the final answer because of difficulties in simulating long periods of geologic time and the total number of temperature fluctuations.

Forest and brush fires are known to cause intense spalling of boulders, especially granite. After fires, the bases of boulders are littered with spalled fragments (Evenson et al., 1990). Granites are particularly vulnerable to fire spalling because the coefficient of thermal expansion of quartz is about three times that of feldspar, resulting in generation of intense stress between grains during fires.

Wetting and Drying

The process of wetting and drying is responsible for the slaking (disintegration) of shale. Even though the clay minerals of shale are relatively resistant to chemical weathering (being products of weathering in the first place), some of them swell when wet and shrink when dry to the extent that the shale cracks apart into innumerable tiny chips and spalls (Mugridge and Young, 1983). Such litter is familiar on or near outcrops of shale and helps to explain the relative weakness of shale under processes of erosion. Shale is not only soft to begin with, and therefore easily abraded, but also may be so weakened by disintegration that the erosive agent needs only to sweep away the chips. Clayey siltstone and sandstone may also be disintegrated by this process.

Much of the effectiveness of wetting and drying is due to volume changes produced by the swelling properties of clay minerals. Clay minerals are hydrated layered silicates of aluminum, iron, and magnesium with atomic structures consisting of sandwiches of silica tetrahedra sheets and alumina octahedra sheets. The ratio of silica to alumina sheets determines the ability of a clay to take water into its atomic structure. For example, kaolinite is composed of one silica sheet and one alumina sheet held together by ionic bonds, giving it a 1:1 ratio of silica to alumina sheets. The bonding between the sheets is strong enough to keep the atomic lattice fixed and prevent expansion of the clay when wetted. Other clays, such as the smectite group, consist of one alumina sheet sandwiched between two silica sheets, giving it a 2:1 ratio of silica to alumina sheets. The sheets are tightly bonded, as in the 1:1 sheets, but isomorphous substitution of Fe^{++} and Mg^{++} for Al^{+++} is common in the alumina octahedron sheets. This gives a net negative charge that is partly balanced by hydrated ions in the space between adjacent sheets, leading to easy exchange of cations and water. Such clay minerals thus expand or contract as water layers are added or removed. Montmorillonite, the most common of the 2:1 clays of the smectite group, is noted for its swelling characteristics. The amount of swelling is dependent largely on the nature of the cations present—the expansion of Na-montmorillonite may be as much as 1000 percent of the original volume, whereas swelling in Ca-montmorillonite is much less, ranging from about 50 to 100 percent.

Another phenomenon related to wetting and drying is the increase in air pressure in pore spaces. Upon wetting, the air pressure in pores in dry clay may increase enough to cause disaggregation. The rate of wetting is important for this process to be effective; if clay is wetted slowly, air can be expelled slowly and pore air pressure does not become high.

Many pedestal rocks in the southwestern United States originate by the slaking of shale to the point where an overlying resistant caprock becomes undermined and isolated by erosion until it resembles the head of a mushroom (Figure 3–7). Bryan (1925) observed the essentials of pedestal formation at the base of the Vermilion and Echo Cliffs near the Marble Canyon section of the Colorado River where pedestals are formed from the Moenkopi Shale and capped by the Shinarump Conglomerate. During storms, rainwater either falls to the ground or soaks the pedestal. The occasional soakings of the pedestal in this way account for streaks of desert varnish on the underside of the caprock. In the case of the

pedestals described by Bryan, soaking of the top of the pedestal by water transmitted through the permeable caprock is part of the process, and local disintegration by the growth of salt crystals may be a minor contributing process. The same type of slaking that accounts for the formation of shale pedestals may also produce niches along outcrops of shale in the faces of cliffs, thus undermining the overlying formations and aiding in cliff retreat.

Colloidal Plucking

During **colloidal plucking,** small fragments are pulled off from rock surfaces by soil colloids in contact with them. The drying colloids contract and exert a strong tensile stress across the surface to which they adhere. For example, a thin film of drying gelatin can pluck flakes of glass from the surface of a container. Reiche (1950) suggested that soil colloids may behave similarly, although the importance of the process remains to be determined.

Gravitational Impact

The role of gravity in weathering is an indirect one because movement of material is generally considered erosion or transportation. However, on steep mountain cliffs, blocks loosened by freezing and thawing or other processes may tumble downslope and break off additional blocks upon impact. The loosening of blocks by the force of impact of falling material contributes to mechanical weathering, but later movement is considered erosion.

Organic Activity

The importance of plants and animals in rock weathering is difficult to measure directly. Thus, although geomorphologists recognize the importance of organic processes in chemical weathering, they do not always agree on the effectiveness of organic activity in physical weathering.

Given sufficient moisture and suitable growing conditions, plants can grow almost anywhere at the earth's surface. As roots grow and extend along fractures or bedding planes, they pry the rock apart. Because the force exerted on rock from root growth is difficult to measure, some geomorphologists have discounted the significance of root growth, but cracked and heaved concrete sidewalks, seen so commonly around the bases of trees, as well as rocks pried apart by roots, attest to the force of root growth. Large roots can wedge boulder-size pieces apart. Overturning and uprooting of trees by windstorms further contribute to the breaking up of the rock in which the trees were rooted. Growth pressures at the tips of plant rootlets may be sufficient to loosen mineral grains and pry them apart. The depth of penetration of roots below the ground surface varies considerably, many extending for 3 to 7 m (10 to 23 ft).

Even the tiny roots of plants as small as lichens may discover cracks within and between grains. Rock fragments can

FIGURE 3–7
Pedestal block formed by differential weathering of shale beneath a resistant sandstone caprock, near Lee's Ferry, Arizona. (Photo by W. K. Hamblin)

be pulled loose from surfaces by lichens, which contract during dry periods, and if lichen is knocked off a rock surface, it often takes some of the rock with it.

Physical weathering by animals is largely a matter of mixing unconsolidated materials. Worms, ants, rodents, and other burrowing animals are notable for their earthmoving activities. The mixing of weathered material by burrowing organisms brings unweathered fragments into contact with chemical weathering agents, water and air. Grazing animals reduce the amount of vegetation, and large animals compact the soil by repeated treading of particular areas, both of which affect infiltration and soil moisture.

CHEMICAL WEATHERING

Chemical weathering is the decomposition of rocks by surface processes that change the chemical composition of the original material. It involves chemical reactions between the elements in the constituent minerals of a rock and the atmosphere and/or groundwater. Chemical weathering differs from mechanical weathering, in that the chemically weathered product has a composition different from that of the original material, whereas mechanically weathered rock is physically broken up into fragments with no chemical change.

Although the two processes are distinctly different, they often operate together with differing significance in different environments. The mechanical disintegration of rock greatly increases surface area, thus preparing the material for much enhanced chemical reactivity. For example, the surface area of a cube 16 cm on a side successively divided only four times increases from 1536 cm² to 24,576 cm² (240 to 3800 in²) (Figure 3–8).

Most rocks exposed at the earth's surface originated in chemical environments quite different from those of the weathering environment. Thus, rocks at the surface are commonly out of equilibrium with conditions there and react chemically to form more stable compounds.

Role of Water

Water, the most abundant substance at the earth's surface, plays a highly important role in weathering because it acts as a medium for exchange of elements between rocks and the atmosphere, and, in addition, may take part directly in chemical reactions. Water also removes the weathering products to expose fresh rock for continued weathering. The rate and intensity of chemical weathering is thus greatly influenced by the amount of precipitation in an area. In general, chemical weathering is most active in wet climates.

Water is liquid at most surface temperatures and is virtually the only significant inorganic fluid in nature. No place on Earth is devoid of precipitation; some rain falls in even the most arid climates. Just as remarkable as the abundance and

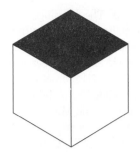

A. A cube of rock 16 cm on a side exposes a surface area of 1536 cm².

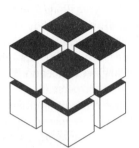

B. Three additional joints, dividing the block into eight cubes, would increase the surface area to 3072 cm².

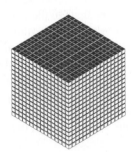

C. If joints 1 cm apart cut the rock, the surface area exposed to weathering would be increased to 24,576 cm².

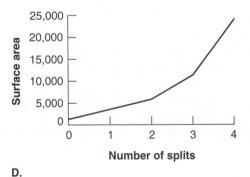

D.

FIGURE 3–8
Increase in surface area with decrease in grain size. (A) A cube of rock 16 cm on a side exposes a surface area of 1536 cm². (B) Cutting of the cube by three joints increases the surface area to 3072 cm². (C) Cutting of each cube 1 cm apart increases the surface area to 24,576 cm².

mobility of water are its physical and chemical properties, which govern its behavior. These properties are all vitally important to most geomorphic (and biologic) processes as they operate on the earth. Interestingly, most of the properties of water can be explained in terms of the peculiar structure of the water molecule (H_2O), which is dipolar and capable of linking with other water molecules or with both positively and negatively charged ions.

The H_2O molecule consists of hydrogen atoms covalently bonded (sharing electrons) with an oxygen atom in such a way that two of the electrons of oxygen are available at one side of the molecule for coordination with two units of positive charge. The two hydrogen atoms with their electrons buried in the oxygen structure each expose a portion of their nucleus (a proton) to their surroundings.

The water molecule is a tetrahedron with oxygen at the center and a pair of hydrogen atoms at two of the corners. Thus, the tetrahedral structure is dipolar because of the region of strong positive charges opposite a region of strong negative charges. This structure allows linkage with other water molecules at the four points of available charge by means of a hydrogen bond. Each of the four hydrogen bonds of a water molecule involves sharing a hydrogen atom between two oxygen atoms. Through such linkages, water molecules become even more strongly dipolar than they are individually.

The important properties of water may be explained in terms of its dipolarity and the hydrogen bond. Water is strongly dielectric because of its dipolarity; that is, small dipolar aggregates of molecules become oriented in an electric field, neutralizing the field. This accounts for water's ability to insulate dissolved ions from one another, in this way minimizing their tendency to build crystalline structures, and partly explains water's powerful solvent capacity for many substances. Water is also a very poor electrolyte because of its almost negligible dissociation. Only 10^{-7} mols of H ion and 10^{-7} mols of OH ion are released by dissociation of a liter of chemically pure water at 20°C.

Another important aspect of the dissolving power of water is its ability to pull incompletely bonded ions from the surfaces of minerals, again because of the affinity of dipolar water for all ions—positive or negative—and because of its ability to create a hydrogen bond of moderate strength. Ions are pulled from the crystal structure by clusters of water molecules (Figure 3–9), which then surround the dissolved ions as insulating envelopes of so-called bound water. This limits the solvent capacity of water because part of the H_2O is tied to the dissolved ions and is no longer available for continued solution.

Water has a high boiling point and freezes readily in comparison to other compounds of similar molecular weight, again because of the ready formation of hydrogen bonding. It boils with difficulty because this linkage must first be broken. To complete the transformation of one gram of liquid water already at the boiling point to vapor, 539.4 calories of heat energy are needed. This is the second highest heat of vaporization of any substance. The heat of fusion is also very high (second only to that of ammonia), requiring 79.7 cal/g to melt ice at 0°C. Hydrogen bonding also explains water's re-

markable heat capacity or specific heat. In fact, a calorie, the fundamental unit of heat energy, is taken as the amount of heat necessary to raise the temperature of one gram of water 1°C. All other substances, except ammonia, have specific heats below that of water—that is, some fraction of a calorie. Thus, as water cools (or even freezes), heat must be liberated, and as water is warmed (or boiled), vast quantities of heat energy must be supplied. For this reason, water is a tremendous reservoir of heat energy and has a profound moderating effect on climate. The spectacular ranges in temperature from night to day in low-latitude deserts are possible only because of the very low humidity of the air.

Water is different from almost all other liquids in its property of expanding upon freezing. Like other liquids, fresh water contracts with cooling due to a decrease in thermal agitation that allows a closer grouping of the molecules. At 4°C, water freezes into the hydrogen-bound mineral (ice) which, with a specific gravity of 0.9, can float on water. This unusual property of expansion with freezing also explains the reverse process of pressure melting. Pressure, applied at temperatures at or slightly below freezing, induces a collapse of the open structure of ice into the more compact liquid state.

Hydrogen bonding gives water a strong internal cohesion, which is shown by its remarkable surface tension—the greatest of all commonly occurring liquids. A striking manifestation of water's powerful surface tension is the almost spherical form of raindrops, which literally bounce upon impact, splattering into smaller spheres. This almost elastic behavior of raindrops gives them some of the characteristics of a bullet. Thus, raindrops exert strong pressures upon impact and bring about disruption, excavation, and compaction of loose particles at the surface.

Another property related to surface tension is the adhesiveness, or wetting ability, of water. Again, this property occurs because the hydrogen-bounded dipoles are attracted by the electrostatic charge available at all mineral surfaces. The phenomenon of capillarity—the ability of thin columns of water to ascend tiny tubes, pores, or other openings—is a good illustration of water's internal coherence and its adhesiveness.

Hydrogen bonding also helps to establish water's viscosity, imparting to it an enhanced ability to entrain available mineral particles, which can then be used most effectively as corrasive tools. Other properties of water are its thermal conductivity—the greatest of all liquids except mercury—and its transparency in thin sheets. A little careful reflection will reveal that these and the above-mentioned properties are of the utmost importance to geomorphology.

Solution

Solution, the disruption of a mineral in water into constituent ions or into molecules, is a fundamental weathering process. In fact, virtually all chemical weathering involves some solution during which the original material is broken down

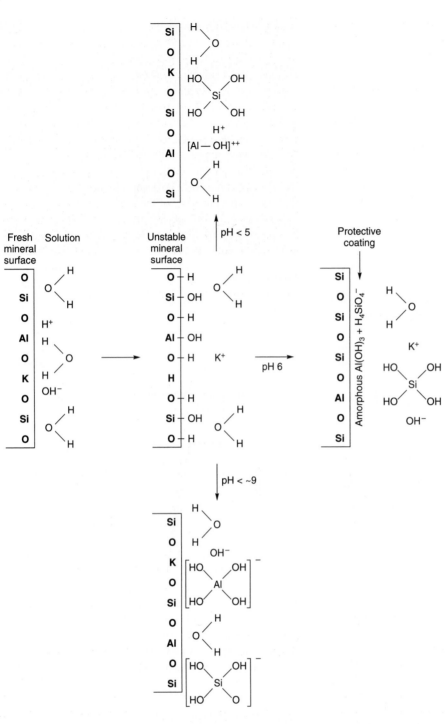

FIGURE 3–9
Reaction of H₂O with a K-feldspar surface under different pH conditions. (From Birkeland, 1984)

chemically. Solution imparts extreme mobility to ions formerly held in mineral structures and makes the ions available for reactions in other chemical weathering processes.

Simple solution depends on the dielectric properties of strongly dipolar clusters of water molecules, which, upon entering the mineral structure, break down the attractive forces between ions by insulating them from one another. The attraction of dipolar water for incompletely bonded ions at the surface of the solute permits the entry of water into the structure. Ions are captured by the formation of hydrogen bonds with the water dipoles. The dissolved ions, enveloped by clus-

ters of water molecules, are prevented from reassociation in the solution by dissipation of their charge over the molecular envelope and by random motion in the liquid.

The increase in the solubility of crystalline materials with rising temperature results from the increased molecular motion imparted by warming. Warming also relaxes the packing of the crystal structures and reduces the attraction between ions.

Circulation hastens solution by removing dissolved ions from the vicinity of the mineral surface and replacing them with fresh water. In chemical weathering, natural flushing has a similar and equally important effect.

Even the least soluble minerals can be made to dissolve in time, as long as the solution is kept dilute by circulation. However, solution is not accomplished by pure water. Natural water always contains enough dissolved ions to alter its solvent capacity radically, and the original solvent, rainwater, is typically acid with dissolved carbon dioxide. Carbonated waters forming in the atmosphere and at the surface are far more powerful solvents than water alone.

In some solution reactions, all of the released ions may remain in solution (congruent dissolution), whereas in others, some of the released ions recombine to precipitate new compounds (incongruent dissolution). Some examples of congruent dissolution of soluble minerals follow.

$$Halite$$
$$NaCl \rightarrow Na^+ + Cl^-$$
$$Gypsum$$
$$CaSO_4\,2H_2O \rightarrow Ca^+ + SO_4^- + 2H_2O$$
$$Quartz$$
$$SiO_2 + 2H_2O \rightarrow H_4SiO_4$$

The dissolving of calcite is another important congruent dissolution reaction:

Calcite	+	Water	+	Carbon Dioxide	→	Bicarbonate
$CaCO_3$	+	H_2O	+	CO_2	→	$Ca(HCO_3)_2$
$CaCO_3$	+	H_2CO_3	→	Ca^{++}	+	$2(H_2CO_3)^-$

This reaction is significant because atmospheric CO_2 gas dissolves easily in rainwater to form a weak acid, carbonic acid, which plays an important role in chemical weathering, especially in karst regions (see Chapter 7).

Incongruent dissolution, in which a new crystalline compound is formed, includes breakdown of various aluminosilicates, such as feldspars, to form clay minerals (see the sections on hydrolysis and carbonation, which follow in this chapter). Minerals composing the common rocks of the earth's crust are soluble to varying degrees in water. The mineral halite is readily dissolved in water, and calcite ($CaCO_3$) is soluble in water that contains CO_2. Thus, limestone and marble are particularly susceptible to solution (Figure 3–10), and humid regions underlain by these rocks are strongly affected by groundwater solution. Silicate minerals, the most common minerals at the earth's surface, vary in their solubility. Silica is soluble under most conditions at or near the surface, varying somewhat depending on the pH of fluids (Figure 3–11) and depending on whether the silica is amorphous or in quartz (Morey et al., 1962, 1964). Amorphous silica is more soluble than quartz.

FIGURE 3–10
Etching of limestone by solution, Guilin, China. (Photo by W. K. Hamblin)

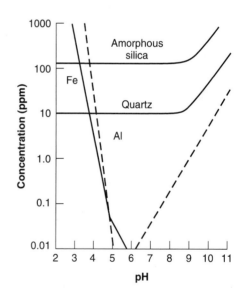

FIGURE 3–11
Relationship of pH and solubility of silica, iron, and aluminum. (From Birkeland, 1984)

Oxidation-Reduction

Oxidation occurs when an ion in a mineral structure loses an electron to an oxygen ion. In the weathering of common iron-bearing, rock-forming minerals, the typical oxidation reaction is the conversion of ferrous iron (Fe^{++}) to ferric iron (Fe^{+++}) by combination with oxygen in the presence of water (Rowell, 1981). Ferric oxides are generally very insoluble and thus precipitate as solid compounds.

The effect of leaving iron objects outside in wet conditions for any length of time is well known. The reddish brown coating of rust that appears on the surface of iron is a good example of the process of oxidation. Even dew has sufficient dissolved oxygen to oxidize metallic iron and to change ferrous iron to ferric iron. The rate of reaction is quite rapid and commonly leaves yellow-brown, brown, or red-brown coatings on natural rock surfaces.

During oxidation, ferrous silicates, such as the common rock-forming minerals pyroxene, amphibole, olivine, and biotite, form (among other compounds) hematite and hydrous iron oxides (limonite, goethite). As a result, rocks containing these minerals often turn brown, reddish brown, or yellow-brown during weathering. For example, the Fe^{++} released by the weathering of fayalite, the iron-rich end member of the olivine family, oxidizes to form Fe_2O_3 (hematite), a nearly insoluble ferric oxide.

Fayalite Olivine	Water	Oxygen	Hematite	Dissolved silicic acid
$2Fe_2SiO_4$ +	$4H_2O$ +	O_2 $\rightarrow$	$2Fe_2O_3$ +	$2H_4SiO_4$

Oxidation takes place more rapidly in the presence of water because most water has available oxygen in solution. Under reducing conditions, which may be oxygen deficient, ferrous iron may remain soluble. Water may enter into the composition of the new minerals, as when hydrous oxides are formed.

Iron	Water	Oxygen $\rightarrow$	Hydrous iron oxide
$4Fe^{++}$ +	$3H_2O$ +	O_2 $\rightarrow$	$2Fe_2O_3 \, 3H_2O$

Oxidation is often the earliest form of weathering on newly exposed rocks and forms weathering rinds that increase in thickness with age (Colman, 1981, 1982b; Colman and Pierce, 1981). Such rinds form more rapidly in humid than in arid climates as a result of the difference in availability of water to the oxidizing surfaces.

Removal of iron or other elements from ferromagnesian silicate minerals (olivine, biotite, hornblende, and pyroxene), ferrous carbonates (siderite and ankerite), ferrous sulfides (pyrite), or other minerals by oxidation, disrupts the structures of the minerals and makes them more vulnerable to chemical weathering by other processes. Oxidation-reduction is not limited to iron but also occurs in other metals.

Reduction, the opposite of oxidation, changes elements such as iron from the ferric to the ferrous state. However, oxygen is generally readily available in the aerated zone above the water table and at the earth's surface. Thus, oxidation is more common in the weathering zone than reduction. Reducing environments are generally restricted to regions below the water table (Loughnan, 1969). Oxidation-reduction is sensitive to pH, and concentrations of organic material, as in swamps, may cause local reducing conditions.

Iron oxides are commonly responsible for the red-brown colors produced by intense weathering and may harden in iron-rich soils to make laterite and ferricrete (sometimes known as *ironstone*), which may be preserved in the geologic record for many millions of years.

Ionic Exchange

Ionic exchange reactions involve the substitution of ions (usually cations) in minerals by ions in solution without rearrangement of the mineral structure. The most commonly exchangeable cations are H^+, K^+, Na^+, Ca^{++}, Mg^{++}, Fe^{+++}, Si^{++++}, and Al^{+++}. The ability of one cation to replace another depends on the ionic radius and ionic charge of the cations. Cations with similar ionic radii and ionic charge may substitute readily for one another; thus, Na and K are interchangeable, as are Ca and Mg and Si and Al. The relative ease with which some common cations may substitute for another is shown in the following list (Feth et al., 1964):

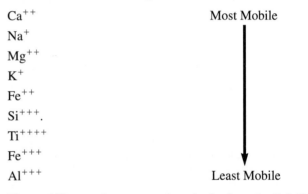

Ca^{++} Most Mobile
Na^+
Mg^{++}
K^+
Fe^{++}
Si^{+++}.
Ti^{++++}
Fe^{+++}
Al^{+++} Least Mobile

The mobility can be expressed as the **ionic potential** *(IP)*, which is the ratio of the ionic charge *(Z)* to the ionic radius *(r)*.

$$IP = \frac{Z}{r}$$

The most mobile ions have ionic potentials of less than three and can remain in solution as ions, whereas ions having ionic potentials greater than three precipitate out as hydroxides. Other factors, such as pH (potential hydrogen), Eh (oxidation potential), leaching, and fixing of ions, may also affect ion mobility.

Ion	Ionic Potential (Zr)
K^+	0.75
Na^+	1.0
Ca^{++}	2.0
Fe^{++}	2.7
Mg^{++}	3.0
Fe^{+++}	4.7
Al^{+++}	5.9

Exchangeable cations are loosely held by adsorption on the surfaces of colloids. The Na^+, Ca^{++}, and K^+ within feldspar frameworks are too tightly bound to be displaced without disruption of the structure, but the same ions and others, such as Mg^{++} and Fe^{++}, are relatively loosely held within, between, or upon the layered structures of clay minerals. In these relatively accessible positions, they are easily replaced. The cation population of clay minerals (like zeolites used in softening water) reflects the concentration of cations in the surrounding waters. Illitic clays, $K(Al,Fe,Mg)_2$ $(Si-Al)_4 O_{10}$ $(OH)_2$ (Na, Ca, Mg), and especially montmorillonite, $(Al, Mg)_2$ $(Si-Al)_4O_{10}$ $(OH)_2$ (Na, Ca, M), are excellent hosts for replaceable cations. Kaolinite, $Al_2 Si_2O_5 (OH)_4$ provides no structural exchange sites, but colloidal particles may adsorb limited amounts of cations on their surfaces.

Each clay mineral has a different cation exchange capacity, the capability for adsorbing cations, expressed as milliequivalents per 100 grams of clay. Clays with high cation exchange capacities react readily with available cations, a factor that is important in fixing certain ions (such as metals) in the soil where they lessen the potential for health hazards. The pH and composition of water in contact with clays are important in determining which cations will be exchanged. H^+ and Ca^{++} are adsorbed by clays more readily than K^+, Na^+, or Mg^{++} under humid conditions, whereas Ca^{++} and Mg^{++} are commonly exchanged under arid, well-drained conditions.

Hydrolysis

Hydrolysis is the chemical addition of H^+ and OH^- ions in water into the internal structure of a mineral to produce a new mineral. The small, highly charged hydrogen ion is the aggressor. It enters the structures of rock-forming minerals and displaces other cations, disrupting the structure in the process. Consequently, hydrolysis is favored by the supply of hydrogen ions and by the removal of the products of replacement. The following conditions favor continued hydrolysis:

1. Repeated leaching by fresh water
2. Introduction of H^+ ions, which will:
   a. Combine with OH^- ions, thereby removing them as water
   b. Displace cations from their structures
3. Precipitation of ions as relatively insoluble compounds
4. Removal of ions in organic complexes
5. Absorption and assimilation of the products by living plants and animals
6. Absorption of the products by colloidal substances

A fundamental source of H^+ ions is water. The pH of the water determines how many H^+ and OH^- ions are available for hydrolysis. At neutrality, water has a pH of 7 under standard conditions (at 20°C). Because pH is the negative logarithm of hydrogen ion concentration, 10^{-7} mols of H^+ and 10^{-7} mols of OH^- occur in each liter of water. A pH less than 7 indicates acidity due to an increase in hydrogen ion con-

centration. For example, a pH of 5 means that 10^{-5} mols of H^+ occur per liter of solution, or 100 times as many potential hydrogen ions as at neutrality. This is balanced by an equal decrease in OH^- concentration, that is, 10^{-9} mols of OH^-. A pH of 10 means that with only 10^{-10} mols of H^+, the solution has 1000 times less available hydrogen than at neutrality and 1000 times more OH^-. Alkaline solutions, characterized by a deficiency in H^+ and a complementary surplus of OH^-, are solutions with a pH greater than 7.

Hydrolysis of common rock-forming minerals in neutral water is shown by their **abrasion pH** (Stevens and Carron, 1948; Grant, 1969).

$$\text{abrasion pH} = \frac{f(Na + K + Ca + Mg)}{\text{clay minerals}}$$

When a mineral is abraded in water, metallic cations combine with OH^-, and anionic complexes associate with $H^{+\prime}$. Because most minerals release strongly ionizing cations into the solution, pH is typically alkaline. However, minerals with cations that tend to form poorly soluble hydroxides hydrolize to a slightly acid solution. The important effect is that the mineral structure is disrupted by hydrogen ions, and as long as the water supply is renewed, the products can be removed and the supply of hydrogen ions can be replenished. Hydrolysis by nearly neutral water is effective with continued leaching, but it can be enormously more effective with an additional source of hydrogen ions. Weathering processes that supply additional hydrogen ions to the water are special agents of hydrolysis.

The hydrolysis of potash feldspar is particularly important in nature because feldspar is so abundant.

K-feldspar	Water	Kaolinite	Silicic acid
$2KAlSi_3O_8 + 2H^+ + 9H_2O \rightarrow H_4Al_2Si_2O_9 + 4H_4SiO_4 + 2K^+$			

The reaction will continue as long as H^+ ions and cations in the mineral are present and the water does not reach saturation with the existing cations. These conditions are usually met where ground water is percolating downward through the weathering zone. Prolonged continuation of the process eventually removes all of the available cations from the mineral, leaving behind only the poorly mobile constituent (aluminum), which remains as an aluminum oxide, Al_3O_4 (bauxite).

Hydrolysis by Plants

Hydrogen ions released in the metabolism of living plants are concentrated at their roots where they are exchanged for nutrient metallic cations. The atmosphere of hydrogen ions about a rootlet is available to attack the fresh mineral directly or to be adsorbed by colloidal clays and organic colloids. These moist, sticky colloids may bridge the distance between the root tip and the mineral fragment and may allow the transfer of H^+ from root tip to colloid and eventually to the mineral with a return flow of adsorbed metallic cations from the mineral to the plant. Colloidal clays carrying abundant adsorbed H^+ are called acid clays and are known to be powerful agents of hydrolysis (Keller and Fredrickson, 1952).

Humic acid, an incompletely understood complex of organic acids derived from decayed vegetation (humus), is another source of H^+ ions. The various humic acids are weak and poorly ionizing, yet they appear important in lowering pH and in removing cationic products of weathering as so-called humates. Humus is also a source of chelators, organic ring compounds capable of powerfully bonding available metallic cations by holding the cations within their ring structures. In fact, chelators may be important in transporting cations into the roots of living plants. Although the chelators are not sources of hydrogen ions, they do provide a means of removing cations, and thus, they facilitate hydrolysis. The capture of metallic cations by chelators is a weathering process called **chelation** and is discussed later in this chapter.

Carbonation/Carbonates

In the process of chemical weathering called **carbonation,** minerals that contain calcium, sodium, potassium, or magnesium are changed to carbonates by the action of carbonic acid. Pure water has few ions free to react with other ions. However, the atmosphere contains 0.03 percent CO_2, which dissolves readily in water to form weak carbonic acid capable of dissolving many compounds more readily than pure water. Carbon dioxide is also added to ground water by biogenic activity in the soil.

Carbon Dioxide	Water	Carbonic Acid	Bicarbonate Ion
CO_2	$H_2O \rightarrow$	H_2CO_3	$\rightarrow H^+ + HCO_3^-$

Carbon dioxide is more soluble in colder water and under higher pressures. Thus, changes in pressure and/or temperature lead to changes in dissolved CO_2 content and acidity of the water (Figure 3–12). The concentration of CO_2 increases until $CaCO_3$ is precipitated in the soil, leading to the development of carbonate-rich horizons (see K-horizons, caliche, and calcrete in the section on soils later in this chapter). The equilibrium of the carbonate-bicarbonate system is reversible, depending on conditions.

$$Ca^{++} + 2HCO_3^- \rightleftarrows CaCO_3 + H_2CO_3$$

An increase in CO_2 content in soil air, a decrease in pH, or a decrease in ion concentration (by dilution) will drive the reaction to the left; carbonate will dissolve to Ca^{++} and HCO_{3-} in ground water. A decrease in CO_2 pressure, an increase in pH, or an increase in ion concentration will drive the reaction to the right, and calcium carbonate will precipitate.

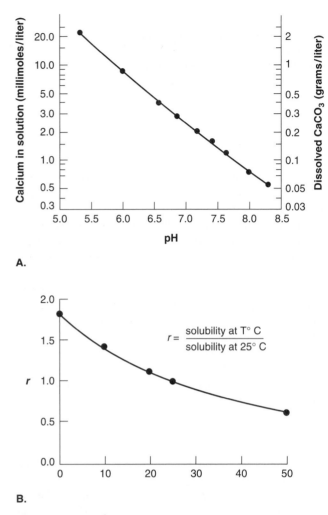

FIGURE 3–12
Effect of pH and temperature on solubility of calcium carbonate. (From Birkeland, 1984, after Arkley, 1963)

Hydrolysis and Carbonation

We have already seen that carbon dioxide is readily soluble in rainwater to form a weak solution of carbonic acid. Conditions that favor the reaction are greater concentrations of CO_2 and relatively cool temperatures. Partial pressures of CO_2 in soil air may be 10 to 100 times more concentrated than in atmospheric air (Buckman and Brady, 1969), largely as a result of CO_2 production by the activity and decay of plants (including such lower forms as fungi and bacteria). The solubility of CO_2, like that of other gases, is inversely proportional to temperature. Climate, therefore, helps control the concentration of H_2CO_3 by regulating both temperature and vegetation.

Hydrogen ions are provided by the dissociation of H_2CO_3 as follows:

$$2H_2CO_3 \rightarrow 3H^+ + HCO^- + CO^{--}$$

The additional H^+ supplied by this reaction suppresses the dissociation of H_2O and therefore limits OH^-. Cations, released by the breakdown of mineral structures attacked by H^+, are removed as carbonates and bicarbonates. Thus, the intensified hydrolysis in the presence of carbonic acid is facilitated by the complementary process of carbonation.

Hydration

Hydration, as the name suggests, means "combination with water." The hydration most familiar to geologists is exemplified by the change from anhydrite ($CaSO_4$) to gypsum ($CaSO_4 \cdot 2H_2O$). A complex ion, Ca_2H_2O, is formed by the coordination of two molecules of H_2O about a Ca^{++} ion. This complex cation combines with SO_4 anionic radicals to build a structure with monoclinic symmetry. The growth of Ca^{++} to $Ca_2H_2O^{++}$ destroys the orthorhombic symmetry of anhydrite and causes wholesale structural expansion.

Such expansion is an important mechanical effect of chemical weathering, and the granular disintegration of coarse-grained igneous and metamorphic rocks is strongly affected by it. The importance of hydration to chemical weathering of rocks is limited by the abundance in rocks of minerals susceptible to hydration.

Similar complexing with water molecules is more important at the surfaces of colloids, particularly clays, where cations are available for bonding. These bonds can be broken only with relatively strong heating. Differential thermal analysis of clays and other hydrated minerals is based upon the thermal energy required to dehydrate bound water from minerals.

Another form of hydration, easily accomplished and easily reversed, is the adsorption of H_2O molecules onto the surfaces of colloids by weak, residual charges. Variable amounts of planar water adsorbed between phyllosilicate layers of clay minerals account for the ready swelling of some clay minerals when they are wetted. Montmorillonite is a good example. The ephemeral nature of this form of hydration is shown by partial dehydration at low temperatures found during laboratory differential thermal analysis.

Combination of silicate wreckage or metallic cations with OH^- during hydrolysis and during oxidation is usually termed *hydration.* Chemists might object to this term, but the custom of calling OH compounds *hydrated* minerals is well established in mineralogy.

Hydration often occurs along with carbonation and other chemical processes during decomposition. A very important example of hydration and carbonation operating together is the weathering of feldspar, the most abundant mineral in the earth's crust, to clay. The chemical decomposition proceeds according to the following reaction:

The clay (kaolinite) thus formed in this reaction is a hydrous aluminum silicate. Potassium carbonate, a result of carbonation, and silica are carried away in solution. The phase diagram for the system potassium-alumina-silicawater is shown in Figure 3–13.

Chelation

Chelation is the equilibrium reaction between a metal ion and a complexing agent, characterized by the formation of more than one bond between the metal and a molecule of the complexing agent, and resulting in the formation of a ring structure incorporating the metal ion (Lehman, 1963). Although the exact mechanism by which chelating agents attack a mineral surface is not well known, the importance of chelation in weathering is considerable, perhaps as important as hydrolysis in some instances (Birkeland, 1984; Schatz et al., 1954; Schatz, 1963). Metal ions, such as Fe and Al, which are normally immobile during weathering, become involved

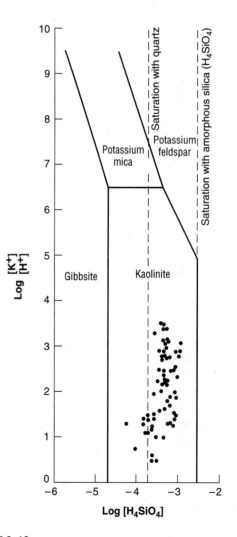

FIGURE 3–13
Phase diagram for the system potassium-alumina-silicawater at 25°C and 1 atm. (From Feth et al., 1964)

		Carbon		Potassium	
Orthoclase	+ Water	+ dioxide $\rightarrow$	Kaolinite	+ carbonate	+ Silica
$2KAlSi_3O_8$	$+\ 2H_2O$	$+\ CO_2$	$\rightarrow Al_2Si_2O_5(OH)_4$	$+\ K_2CO_3$	$+\ 4SiO_2$

in the reactions under conditions that normally would not result in their mobilization.

Most chelating agents are organic substances produced by biological processes in the soil and by lichens growing on rocks. Schalscha et al. (1967) demonstrated the ability of chelating agents to extract metals from ground-up minerals in laboratory studies. When the ground-up minerals were allowed to react with solutions containing chelating agents, cations were released into the solution at rates greater than expected by the effects of hydrogen ions alone. Their studies showed little correlation between pH and the amount of iron released. Hydrogen ions, released from the organic molecule during chelating reactions, can then participate in hydrolysis reactions.

Lichens play an important role in weathering. They secrete chelating agents that affect the rocks on which they grow. Jackson and Keller (1970) found more extensive weathering of lichen-covered basalt than of lichen-free basalt in Hawaii (Figure 3–14). Lichen-free rock showed virtually no weathering rind (compared to a moderate one for the lichen-covered rock) and only slight enrichment in iron, silicon, and calcium with depletion of aluminum and virtually constant titanium. In sharp contrast, iron in the lichen-covered rock was enriched six times, and aluminum, silicon, titanium, and calcium were sharply depleted. The conclusion reached in this study was that lichens produce chelating agents that greatly enhance weathering.

Other organic substances may also produce chelating agents, such as alteration of humus into fulvic acid (Wright and Schnitzer, 1963; Tan, 1980), but these processes are not well understood.

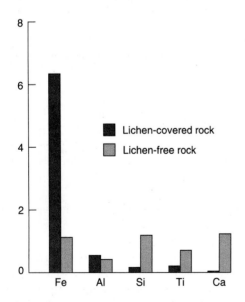

FIGURE 3–14
Concentration ratio of weathered crust to fresh rock for lichen-covered rock and fresh rock. (Plotted from data in Jackson and Keller, 1970)

CONTROLS OF RATE AND CHARACTER OF WEATHERING

Weathering processes respond to the physical and chemical environment. They are controlled by the availability of water and its solutes, by conditions of temperature and pressure, and by the circulation of water. Temperature is the primary control of freezing and thawing, and temperatures above freezing determine solubilities, the rates of chemical reactions, and the reactivity of substances (with the exception of the solubility of gases, which is directly proportional to temperature). The solubility of CO_2 is directly proportional to atmospheric pressure and, like that of other gases, inversely proportional to temperature. The amount and circulation of water determine the concentration of ions in solution and the rate at which they will be brought to a mineral undergoing weathering. The capacity of water to remove the products of weathering also depends upon its amount and circulation. Moreover, all of these controls regulate the biologic world, which in turn has its own special physical and chemical roles in weathering.

Parent Material

The physical and chemical properties of bedrock undergoing weathering have a pronounced effect on the rate of weathering. Differential weathering of diverse rocks in the same environment is a convincing indication of the influence of parent material on weathering (Plate 1B). Differences in resistance of the parent materials are controlled by the resistance of the constituent minerals to chemical and physical weathering. The extent to which water can enter and remain within a mass of rock or soil is important in determining resistance to both physical and chemical weathering. For example, quartz is more resistant to both chemical and mechanical weathering than most other minerals. Hence, rocks such as sandstone and quartzite, which consist largely of quartz, weather more slowly than many other rock types. For many rocks and minerals, however, resistance is a matter of how the material resists the local environment because substances that are highly stable in one environment may be unstable in another. Rocks such as limestone and marble, which consist largely of calcite, are notably soluble in humid climates where carbonated waters are available, but they are among the most resistant rocks of arid regions.

Physical features of rocks, such as jointing, bedding, and porosity, also affect rates of weathering because these factors allow percolation of water through the rock. A summary of the weathering characteristics of some common rock-forming minerals is given in Table 3–2.

Mineral Stability

In general, easily soluble minerals, such as halite, gypsum, calcite, aragonite, and dolomite, are least resistant to chemi-

TABLE 3–2
Susceptibility of silicate minerals to weathering.

Olivine (forsterite), anorthite	Most Susceptible
Pyroxine (augite)	
Amphibole (hornblende)	
Biotite	
Albite	
Orthoclase	
Muscovite	
Quartz	Least Susceptible

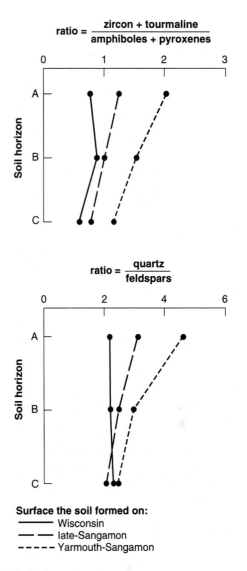

Surface the soil formed on:
—— Wisconsin
— — late-Sangamon
- - - - - Yarmouth-Sangamon

FIGURE 3–15
Ratios of resistant to less-resistant minerals from soils of different age. Note that less-resistant minerals decrease with age. (From Ruhe, 1956)

cal weathering. Minerals that are the products of chemical weathering, particularly where the weathering has been extreme, are among the most resistant (for example, illite, kaolinite, montmorillonite, limonite, hematite, and gibbsite). Certain accessory minerals, such as rutile, ilmenite, and corundum, are very resistant to weathering because of their inertness and their extremely low solubilities, even in the finely divided state (Figure 3–15). These minerals may become concentrated in the clay fraction of soils (Jackson et al., 1948; Reiche, 1950).

Silicate minerals account for approximately 84 percent of the atoms in rock-forming minerals. The relative resistance of silicate minerals to weathering is fundamentally a matter of the strength of bonding. Compared to other elements in common rock-forming minerals, silicon is a small, highly charged cation with high ionic potential. Its positive charge of four is concentrated over the (theoretically) spherical and relatively limited surface of the silicon. Oxygen, the only important anion in silicate minerals, has a moderate ionic potential. A negative charge of two is distributed over a relatively large ionic surface. Among rock-forming silicate minerals, silicon is able to bind the oxygen anions more powerfully than other cations and forms, in coordination with four oxygen anions, a silica tetrahedron (Figure 3–16) of great integrity. The small silicon cations (ionic radius of 0.42 angstrom) is surrounded by four large oxygen anions (ionic radius of 1.40 angstrom) in tetrahedral coordination. The bonding of silicon and oxygen in the silica tetrahedron is partly ionic and partly covalent, but the structure depends on the attraction of oppositely charged ions and their ability to pack as efficiently as thermal agitation permits. Thus, positive ions surround themselves with negative ions, and vice versa. As a result, ions of similar charge are effectively insulated from one another, and a maximum electrostatic (ionic) bonding pervades the structure. Ionic bonding is important because chemical weathering is largely a matter of ionization in water and because the products of decomposition are assembled from available ions. The silica tetrahedron has a negative charge of four, capable of attaining electrical balance with available cations.

Silicon-oxygen bonding increases where oxygen atoms are shared between adjacent tetrahedra, so that each shared oxygen atom is the corner of a pair of tetrahedra. Tetrahedra linked through sharing of common oxygen atoms may then form single chains, double chains, sheets, or three-dimensional frameworks (Figure 3–16). The weakest bonds in such silicate structures are at cationic bridges between unshared tetrahedral oxygens and within tetrahedra where aluminum has substituted for silicon. Thus, anorthite ($CaAl_2SiO_8$) is more easily weathered than albite ($NaAlSi_3O_8$) because of the greater substitution of aluminum for silicon in anorthite. Chain structures of amphiboles and pyroxene break down, not only at the cationic bridges where the chains are bound together, but also at places where tetrahedral aluminum has been substituted for silicon along the chains. The important conclusion drawn from this is that the breakdown of silicate minerals is not between silicon-oxygen bonds but rather at places where the bonding is weaker. The wreckage of silicate decomposition may still include silica tetrahedra, out of which may reassemble clay minerals, hydroxides of iron and

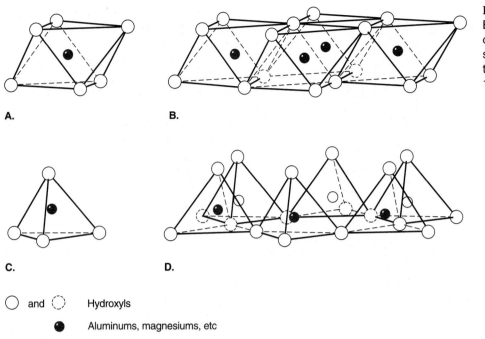

FIGURE 3–16
Basic silicate structures: (A) octahedral unit, (B) octahedral sheet, (C) tetrahedral unit, (D) tetrahedral sheet. (From Grim, 1968)

○ and ⊙ .... Hydroxyls

● .... Aluminums, magnesiums, etc

aluminum, carbonates and bicarbonates of calcium and magnesium, free quartz, and other minerals.

The susceptibility of silicate minerals to weathering has been studied in the field and in the laboratory, resulting in recognition of a more or less orderly progression of stability in weathering (Table 3–2). Goldich (1938) first suggested that the stability of silicate minerals to weathering was related to the order in which the minerals crystallize from an igneous melt, that is, the inverse order of Bowen's reaction series.

Keller (1954) calculated the energies of the cation/oxygen bonds and found the following:

Bond	Energy	Bond Strength
Si–O	<3000 kg cal	Strongest
Al–O	>2000 kg cal	
Base ions–O	>1000 kg cal	Weakest

The energies of total bonds show a fairly good correlation with mineral stability during weathering. Curtis (1976) showed a correlation of total energy released by mineral decomposition and mineral stability.

Olivine, which crystallizes at the highest temperature among the silicate minerals, is composed of individual SiO_4 tetrahedra held together by ionic bonding of iron and magnesium. It weathers readily because the individual SiO_4 tetrahedra are held together by relatively weak cation-oxygen bonds, which are broken during weathering. The single-chain silicates (pyroxenes), double-chain silicates (amphiboles), sheet silicates (biotite), and three-dimensional silicates (feldspar and quartz) are successively more resistant to weathering because SiO_4 tetrahedra are bound together by stronger Si-O and Al-O bonds.

Among the plagioclase feldspars, Na-plagioclase (albite) is more stable during weathering than Ca-plagioclase (anorthite) because isomorphous substitution of Al^{+++} for Si^{++++} in the SiO_4 tetrahedra as the plagioclase becomes more calcic replaces Si-O bonds with weaker Al-O bonds.

Grain size also plays an important role in weathering of minerals. Although several smaller crystals have a greater surface area than a single large crystal (and therefore should react more readily), coarse-grained crystalline rocks often show more strongly developed weathering characteristics than fine-grained rocks (Figure 3–17) (Birkeland, 1973, 1984; Burke and Birkeland, 1979; Colman and Pierce, 1981). However, sometimes just the reverse is true. For example, Figure 3–18

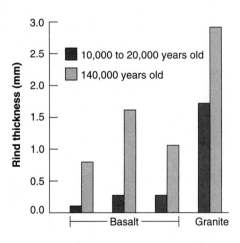

FIGURE 3–17
Thickness of weathering rinds on fine-grained basalt and coarse-grained granite. Note that rinds are much greater for both young and old granite. (From Birkeland, 1984)

FIGURE 3–18
Coarse feldspar crystals etched into relief by weathering of the matrix around them, Sierra Nevada, California.

shows coarse phenocrysts of feldspar etched into relief by more rapid weathering of finer grains.

Because weathering attacks rocks at the points of weakest resistance, the presence of a single weak mineral in a rock often leads to more rapid decomposition of the whole rock. For example, biotite weathers rather easily because of expansion due to the oxidation of iron, and thus biotite-rich rocks commonly break up expeditiously during weathering. Rocks with high amounts of glass also weather quickly.

Parent material has a significant influence on weathering in environments where the weathering residues are so young, so resistant to further weathering, or so continuously removed that they are little more than a weathered version of the parent material. The importance of parent material diminishes as the weathering residue continues to change to a product (usually a soil) significantly different from the parent material. Extended time reduces the significance of parent material in weathering.

Climate

Climate establishes the temperature and moisture conditions at and near the earth's surface. For this reason, climate, of all the controlling factors of weathering, is the most significant long-term control of weathering processes. Climate deter-

mines the relative importance of individual weathering processes. For example, processes such as the wedging effect of freezing water may be absent in many climates too warm or too dry for much freezing of water (or too cold to allow melting), while other processes, such as fire spalling, may be suppressed by the overwhelming importance of chemical weathering. Similar rocks may yield contrasting products after deep weathering in different climates. In fact, the long-term effects of climatic control are so overriding that similar soils may develop from a variety of parent materials under a given climate, and the worldwide pattern of soil types reflects climatic zonation with great fidelity.

The amount and distribution of precipitation are especially important because water promotes chemical weathering. In moist tropical regions where rainfall is high and fairly evenly distributed throughout the year, water is in contact with weathering material for long periods of time, and chemical weathering produces intense leaching and deep weathering (Figures 3–19, 3–20, and 3–21).

Arid regions receive little rainfall, and the rain often falls in a few intense showers during a year. The high intensity of the precipitation results in low infiltration into the soil, thus limiting the time during which water remains in contact with rock material and reducing the effectiveness of chemical weathering processes.

Infiltration of water into the weathering profile depends on the rate at which precipitation strikes the surface and on the permeability of the ground. A certain amount of time is necessary for the downward percolation of water into the ground, so if large amounts of water are contributed suddenly, infiltration will not be able to dispose of the water as effectively as it could if the same amount of water accumulated slowly. Precipitation greater than infiltration capacity produces runoff.

As water is added to the ground, infiltration capacity diminishes at a rate dependent upon how rapidly the pores, structural cracks, and other openings become clogged with water and sediment. Thus, during a period of high precipitation, runoff increases as infiltration capacity decreases.

The importance of climate to weathering processes is shown by the clear relationships between soils and climate with varying latitude (Figure 3–22). In tropical regions, where rainfall is abundant and temperatures are high, weathering is intense, and deep, red soils with high amounts of iron and aluminum oxides are developed. In mid-latitude deserts, where temperatures are high but precipitation is low, rates of chemical weathering are low, and thin, carbonate-enriched soils are formed. In humid higher-latitude areas, where temperatures are low and moisture is moderate, soils of intermediate thickness and weathering intensity form. Near the polar regions, where both temperature and moisture are low, decomposition is slow.

Vegetation

Vegetation is dependent largely on climate and thus is not a completely independent factor. The amount and kind of vegetation affect the ratio of the amount of water that soaks into

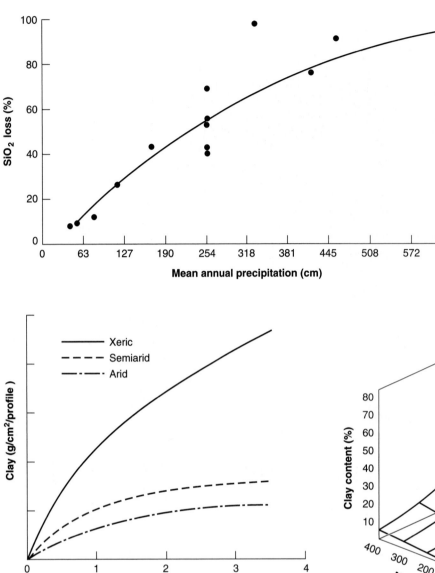

FIGURE 3–19
Increasing silica loss in basalt
with higher precipitation. (From
Hay and Jones, 1972)

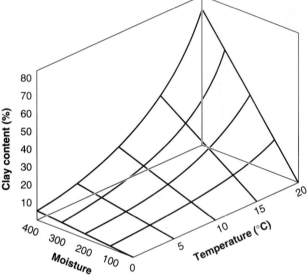

FIGURE 3–20
Differences in amount of clay developed under three
different climates. (From McFadden, 1982)

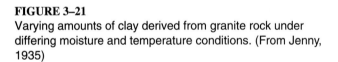

FIGURE 3–21
Varying amounts of clay derived from granite rock under
differing moisture and temperature conditions. (From Jenny,
1935)

the ground to the amount that runs off on the surface. Heavily vegetated areas inhibit runoff, and water soaks into the ground, where it promotes chemical weathering and soil development.

Vegetation operates in several ways to reduce runoff. Plants consume a certain amount of the water that falls upon the surface, but the water must first get into the soil before it can be taken up through the roots. Raindrops are intercepted on leaves, stems, blades of grass, and twigs, and are either absorbed into the vegetal tissue or held until they slowly evaporate or trickle away. In addition, the vegetation and organic debris at ground level baffle the flow of water in its attempt to run off.

In areas of bare slopes (as in deserts), sparse vegetation offers little impediment to runoff, and only a small percent-

age of precipitation soaks into the ground. In these areas, water is not readily available, and chemical weathering is not nearly as effective as in more humid climates.

Vegetation is also an important soil-forming factor. Retention of water by vegetation promotes chemical weathering in the soil, and root action promotes mechanical breaking up of bedrock. Organic acids chemically attack the rock, and organic material from decaying plants is added to the soil.

Organic material accumulates at the surface as forest or grassland litter and as roots underground. The decay of such material produces organic acids (a source of hydrogen ions) and organic colloids capable of ion exchange. Some chelating

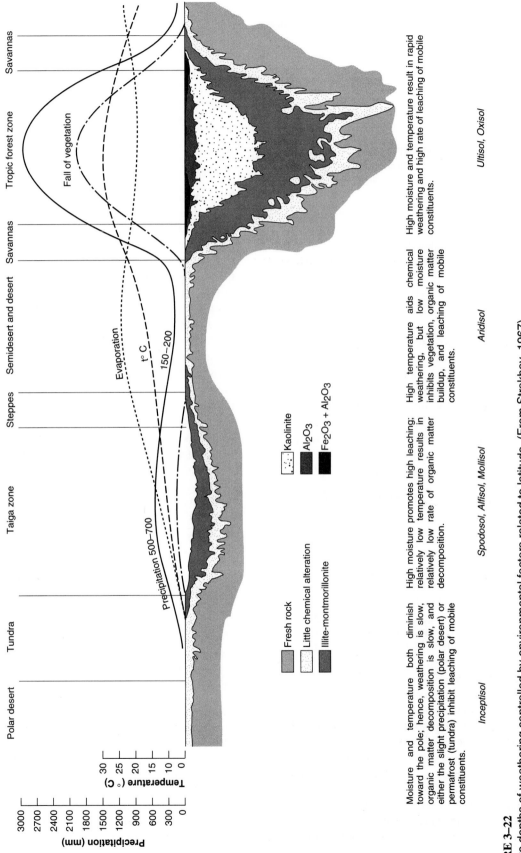

FIGURE 3–22
Relative depths of weathering controlled by environmental factors related to latitude. (From Strakhov, 1967)

agents may be products of decay, whereas others are secreted by growing plants. Nutrient cations are extracted by plants in exchange for hydrogen ions at root tips. The capacity of certain crops to deplete the soil of nutrient ions is well known to agriculture and is the basis of the fertilizer industry.

Topography

The most important elements of topography that affect weathering are altitude and slope. In high mountains above the timberline, freezing-thawing is an important weathering agent. Although mechanical weathering is prominent, chemical weathering also occurs in alpine environments, and significant soil formation occurs (Reynolds and Johnson, 1972; Caine, 1979; Burns and Tonkin, 1982; Birkeland et al., 1987). However, at lower altitudes, vegetative cover is heavier, freeze-thaw activity is diminished, and chemical weathering is enhanced. Topography affects surface and subsurface drainage by providing slopes of varying length, position, and steepness for the transmission of water and by controlling the position and slope of the water table (Muhs, 1982). This, in turn, affects the amount of water available for weathering processes.

Flat or gently sloping areas not only retain their soils easily but also weather relatively deeply because of the more effective infiltration of water. Infiltration is inversely related to slope—the steeper the slope, the more effectively the surface water can flow away instead of infiltrating into the ground. The amount of runoff is directly related to slope, up to an angle of about 45°. Slopes steeper than this intercept less rainfall per square inch and hence provide less runoff per unit area. For example, a vertical cliff furnishes very little runoff. Finally, each patch of ground, with its characteristic slope, must handle the influx of water and sediment from above and must react to their movement downhill.

Where slopes are steep and precipitation runs off quickly, water is retained only briefly for chemical weathering or for plant growth, and vegetation is thus relatively sparse. On more gentle slopes, where more moisture is retained in the weathering mantle, weathered products accumulate on the surface, and vegetation is heavier. Soils develop on gentle to moderate slopes, whereas steep slopes or cliffs in the same areas may expose only bare rock.

Aspect, the direction in which a slope faces, influences the microclimate next to the ground, especially where strong differences in insolation, exposure to wind, and exposure to wind-driven precipitation are present. Strong differences in moisture and temperature may result, and these may influence the vegetation, which affects weathering and soil moisture. In the Northern Hemisphere, south-facing slopes receive more direct insolation than north-facing slopes, which receive insolation at a lower oblique angle. In regions of winter snow, north-facing slopes retain snow cover longer into the spring and summer and are likely to have higher soil-moisture content longer into the summer. In temperate to semiarid areas, south-facing slopes have higher light intensities, higher air

and soil temperatures, higher evaporation rates, lower soil moisture, and less vegetation than north-facing slopes (Lotspeich and Smith, 1953). These conditions, in turn, lead to differences in organic carbon in the soil, pH, organic acids, and various other factors important to weathering processes (Birkeland, 1984). The result of these differences may be responsible for asymmetry of slopes and differing soil thicknesses (Figure 3–23).

Time

The length of time that weathering has been taking place in any region has a direct bearing on the amount of weathering that has occurred (Figure 3–24). If all the surface of the earth had been exposed to weathering for the same length of time, time would be a constant factor. However, stripping away of the weathered mantle by glaciations during the Pleistocene, tectonic uplift of areas previously below sea level, and various other changes of the earth's surface result in differences in the length of time that weathering has affected a given area.

Differences in the weathering of bedrock in areas covered by large continental glaciers during the last glacial stage and areas that lay beyond the glaciers are particularly striking. In many parts of the world, rock polished and grooved by glaciers remains relatively untouched by weathering, whereas uninterrupted weathering in regions beyond the ice has produced deep soils.

Given certain parent materials, and a characteristic climate, topography, and vegetation, the visible effects of weathering will increase with time until an equilibrium is established in which all the products of weathering are removed at the same rate as they are formed. Inasmuch as the slope of a surface usually controls the rate at which weathering products are eroded away, two extreme examples could be:

1. The face of a vertical cliff upon which only a thin layer of weathered grains exists

2. Thick, well-developed residual soil on a very gently sloping surface

Weathering Rates

Rates of weathering are functions of the five controlling factors: parent material, climate, vegetation, topography, and time. These factors combine in so many ways that a large range in weathering rates exists and makes generalizing about rates of weathering difficult. Even when parent materials, climate, topography, and organisms are fairly well stabilized, doubt may remain that the rate of weathering is constant. As fresh rock becomes insulated by its weathered portions, weathering probably decelerates, and equilibrium is approached asymptotically. Recognition of equilibrium states is difficult, and without this knowledge, estimates of rates of weathering are difficult to make.

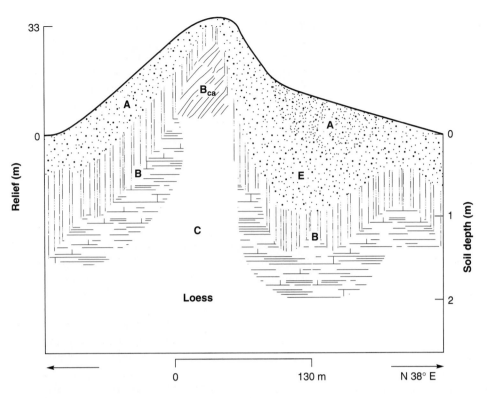

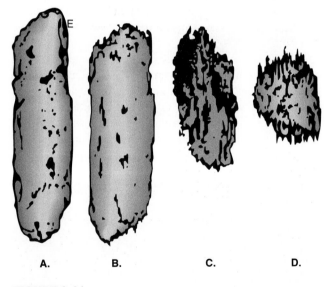

FIGURE 3–24
Comparison of slightly etched hypersthene grains (A and B) of late Wisconsin age (10,000 to 20,000 years old) with highly etched grains (C and D) of Illinoian age (<150,000 years old). (From Birkeland, 1984)

Various attempts have been made to measure weathering rates of rocks of known age. Short-term weathering rates can be estimated from:

1. Experimental laboratory studies
2. Studies of weathering of dated tombstones, monuments, and building stones

3. Studies of dissolved elements in streams

Longer-term weathering rates can be approximated from studies of dated geologic samples (for example, stones on Pleistocene moraines, terraces, and fans dated by various radiometric methods).

Laboratory experiments, which accelerate weathering processes by use of ground-up powders and elevated temperatures, can be carried out over known time periods (Schaller and Vlisidis, 1959; Correns, 1963; Wollast, 1967; Nickel, 1973; Lagache, 1965, 1976; Parham, 1969; Petrovic et al., 1976; Holdren and Berner, 1979). These experiments yield quantitative results but leave open questions such as the following:

1. Can the short time span of the experiments be extrapolated to longer-term geologic processes?
2. Can natural weathering conditions be replicated in the laboratory (Petrovic et al., 1976; Holdren and Berner, 1979; Berner, 1978)?

However, an important conclusion from experimental studies is that the rates of many chemical processes decrease exponentially; that is, the rates decrease with time.

Observations of weathering of dated tombstones, monuments, and building stones yield estimates of weathering rates over historic time spans (Cann, 1974; Goodchild, 1890; Matthias, 1967; Rahn, 1971; Winkler, 1966, 1975). Studies of the solution weathering of limestone and marble tombstones (Figures 3–25 and 3–26) suggest that carbonate solution takes place at a constant rate over historic time, depending on the

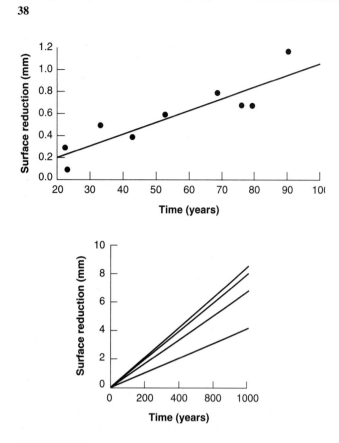

FIGURE 3–25
Weathering of limestone tombstones with time. (From Colman, 1981, after Cann, 1974, and Goodchild, 1980)

solubility of the carbonate and climatic conditions (Goodchild, 1890; Cann, 1974; Meierding, 1981; Winkler, 1966, 1975). However, Matthias (1967) found that the rate of surface reduction of arkosic sandstone tombstones decreased with time until about 1900 when the rate accelerated (Figure 3–27), presumably as a result of changes in atmospheric conditions during the Industrial Revolution.

A complicating factor in establishing rates of weathering is that different parts of the same rock may weather at different rates. Figure 3–27A and 3–27B show accelerated weathering near the base of vertical stones as a result of differences in moisture retention on the faces.

Studies of weathering rates of geologic samples over time periods of many thousands of years have the advantage of relating to natural weathering conditions, but time control may be difficult. A number of attempts at relating weathering phenomena to dated material have been made (see reviews in Birkeland, 1973, 1984; Burke and Birkeland, 1979; Ollier, 1969; Winkler, 1975; Colman, 1981; Colman and Pierce, 1981).

Colman (1981) and Colman and Pierce (1981) studied weathering rind thickness on basalt and andesite pebbles ranging in age from about 10,000 to several hundred thousand years. They found that weathering rates decreased with time in the western United States (Figure 3–29). They attributed the decreasing rates to formation of weathering

FIGURE 3–26
Weathering of tombstones in Philadelphia: (A) engraved in 1804, (B) engraving totally obliterated. (Photo by W. K. Hamblin)

A.

B.

FIGURE 3–27
Accelerated weathering at the base of stones. (A) Enhanced weathering at the base of a tombstone in Ohio, (B) "mushroom rock," Death Valley, California (Photo by W. K. Hamblin)

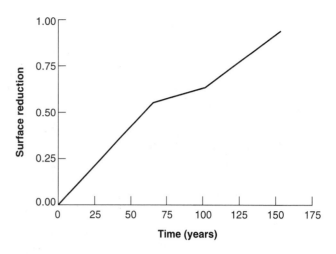

FIGURE 3–28
Weathering of arkosic tombstone with time. (From Colman, 1981, after Matthias, 1967)

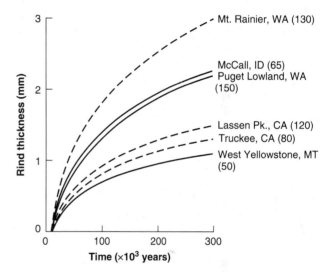

FIGURE 3–29
Variation in thickness of weathering rinds on volcanic pebbles with time. Numbers in parentheses are mean annual precipitation (cm). (After Colman and Pierce, 1981)

residues that restricted access of water and inhibited ionic diffusion. The rate of formation of weathering rinds followed an equation of the form:

$$r = k \ln(t)$$
$$\text{or } r = kt^{1/2}$$

where r = rind thickness
k = a constant
t = time

In other words, the rate of formation of the rind thickness (residual layer) is inversely proportional to its thickness (Colman, 1981). An equilibrium thickness of the weathering residual may be attained when the decrease in the rate of rind formation equal the rate of its destruction by other processes.

Other studies of dated volcanic rocks and ash (Chinn, 1981; Hay, 1960; Hay and Jones, 1972; Jackson and Keller, 1970; Ruxton, 1968) also suggest that the rate of ion loss (expressed as SiO_2, CaO, MgO, Na_2O, and K_2O) decreases with time.

Interdependence of Factors of Weathering

Weathering factors are not independent of one another. Therefore, attempts to measure the effects of one while holding the others constant are complicated by interactions between and among the factors. Rates of weathering are variable according to the combinations of parent materials, climate, topography, and vegetation.

Parent materials might at first seem independent of the other variables, but their effect on topography is the basis of sculpture by differential erosion. Moreover, fresh rocks and minerals do not persist long at the surface. Yet, the continued breakdown of weathered rock and intermixed organic material consists of the weathering of parent materials that have already been conditioned by time, climate, topography, and vegetation.

Microclimate depends to a degree upon topography, vegetation, and parent materials. Topography also depends on parent materials, climatically controlled erosional processes, and vegetationally controlled erosional resistance. Vegetation depends on climate, parent materials, and topography.

EFFECTS OF WEATHERING

Weathering processes, working alone or in various combinations, produce a number of interesting geomorphic features. Chemical weathering by combinations of hydrolysis, hydration, and oxidation causes clay formation, granular disintegration, weathering pits, spheroidal weathering, tor formation, deep chemical weathering and stripping, cavernous weathering, and soil formation. Each of these effects is discussed in the following sections.

Clay Minerals

Clay minerals are hydrated silicates of aluminum, iron, and magnesium arranged in sheets of silica and alumina octahedra. They are known as phyllosilicates or layer silicates because of the way in which silica tetrahedra and alumina octahedra are linked together in sheets.

Silica tetrahedral sheets consist of silica tetrahedrons (Figure 3–16) linked together by the sharing of oxygen ions making up the base of the tetrahedra. Alumina octahedral sheets consist of six O or OH ions around a central Al, Fe, or Mg ion (Figure 3–16).

The silica tetrahedral sheets and alumina octahedral sheets may be arranged in various "sandwiches" to form diverse clay minerals. Three structural groups of the two basic sheets may form the following:

1. One sheet of silica tetrahedra attached to one sheet of alumina octahedra (Figure 3–30) to form a 1:1 layer phyllosilicate

2. One sheet of alumina octahedra sandwiched between two sheets of silica tetrahedra (Figure 3–30) to form a 2:1 layer phyllosilicate

3. One sheet of alumina octahedra between adjacent 2:1 layers to form a 2:1:1 layer phyllosilicate

The clay mineral groups most important to weathering are the 1:1 and 2:1 layer phyllosilicates. Two of the common 1:1 layer phyllosilicates are kaolinite and halloysite, which are structurally similar except that halloysite has a layer of water between successive layers. Both consist of one silica tetrahedral sheet and one alumina octahedral sheet held together by ionic bonds that are strong enough to inhibit invasion of cations or water between crystal units (Figure 3–31) and prevent expansion of the structure in the presence of water. Thus, kaolinite has low cation exchange capacity and does not swell when wetted.

Clays having the 2:1 layer phyllosilicate structure include smectites and micas. Isomorphous substitution is common in the smectite group and allows for considerable variation in chemical composition, generally by substitution of Fe^{++} and Mg^{++} for Al^{+++} in the alumina octahedral sheet. Because the alumina octahedral sheet is relatively far away from the cations that bond the layers together, the bonds are not very strong, and exchange of water and cations can occur easily. Thus, smectites expand readily upon wetting (they are often referred to as *swelling* clays) and have a high cation exchange capacity. The most important clay mineral of the smectite group is montmorillonite.

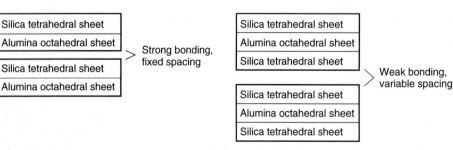

A.   **B.**

FIGURE 3–30
Basic structure of clay minerals: (A) structure of 1:1 phyllosilicates such as kaolinite, (B) structure 2:1 phyllosilicates such as the smectite group (including montmorillonite) and the mica group. (From Birkeland, 1984)

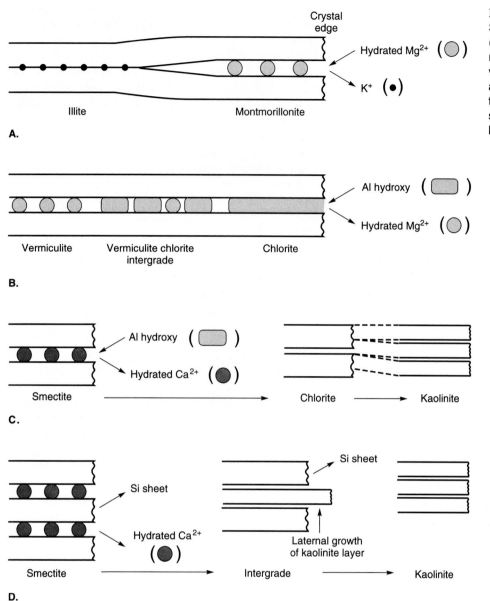

FIGURE 3–31
Some clay alteration processes: (A) alteration of illite to montmorillonite, (B) alteration of vermiculite to chlorite, (C) alteration of smectite to chlorite to kaolinite, (D) alteration of smectite to kaolinite. (From Birkeland, 1984)

The most significant members of the mica group of 2:1 layer phyllosilicates are illite and vermiculite. The structure of illite is similar to that of montmorillonite except that Al^{+++} substitutes for Si^{++++} in the silica tetrahedral sheet, producing a negative charge. The crystal units are bonded together by K^+, which is relatively close to the negative charge, and this results in stronger bonding than in montmorillonite. Thus, illites have cation exchange capacities and swell characteristics intermediate between those of kaolinite and montmorillonite. Vermiculite is similar to illite in that the important isomorphous substitutions occur in the silica tetrahedral sheet, but the interlayer bonding (by Ca^{++} or Mg^{++} rather than K^+) is weaker and limited swelling may occur.

Chlorite, a 2:1:1 layer phyllosilicate that is common in soils, consists of alternating 2:1 layers and alumina octahedral sheets. Bonding between layers is strong and expansion does not occur. Mixed-layer clays, which result from interlaying of 2:1 and 2:1:1 layers, are quite common but have such variation in composition and structure that they are difficult to identify.

Because clay minerals differ in their stability, and because natural chemical environments often change with time, initially formed clay minerals may alter from one clay species to another. Jackson et al. (1948) suggested that clay minerals form in a progressive order dependent upon their relative resistance to weathering, and they assigned weathering stability numbers for various clay minerals. Starting with a mineral of the mica group, the sequence would proceed with increasing weathering through vermiculite to montmorillonite, ending with kaolinite. However, in addition to changes with time, climatic changes influence the formation of clays, and some clay minerals, such as kaolinite, may form directly from weathering of primary minerals without going through the sequence proposed by Jackson et al. (1948). For example, Keller (1978, 1982) found kaolinite forming directly from the breakdown of feldspar. Given sufficient time and appropriate weathering conditions, granite will decompose completely to clay (Figure 3–32).

Granular Disintegration

Granular disintegration of coarse-grained crystalline rocks produces a litter of slightly weathered constituent grains known as **grus**. The grus is literally pried from the rock by the loosening of oxidized ferromagnesian minerals by hydration and by the swelling of clays derived from the weathering of aluminum silicates. It is a familiar sight around outcrops and boulders in deserts where the soil may consist of little more than a layer of grus. Roth (1965) noted a correlation between water content and weathering on a large, isolated boulder undergoing granular disintegration in the Mohave Desert, California. The water content was found to increase inward from the surface and to be greatest on the north side of the boulder, the shadiest and most deeply weathered face.

The granular disintegration of clastic sedimentary rocks, particularly sandstone, occurs through weathering and solution of the cement. It may be aided by the efflorescence of salts, the growth of ice crystals, and the expansion of original sedimentary clay or clay weathered from constituent aluminum silicates.

Weathering Pits and Etched Knobs

Weathering pits—saucer-like depressions etched into rock by granular disintegration and the spalling of thin rock flakes—are common in granitic terrains (Figure 3–33). Smith (1941) described a sequence of typical weathering pits, located on a granite dome in South Carolina, which range from faint con-

FIGURE 3–32
Granite cobble completely weathered to clay easily carved with a knife, Corsica, France.

A.

B.
FIGURE 3–33
Weathering pits in granite. (A) North Dome, Yosemite National Park (photo by U.S. Geological Survey), (B) water retained in weathering pit, Sierra Nevada, California.

cavities developed by flaking to huge basins 40 ft (12 m) in diameter and 3 ft (1 m) deep. Microscopic study of the minerals revealed coatings of kaolinite on the feldspars; and filling of mineral sutures with kaolinite, chlorite, and iron oxides. The rock is rendered slightly porous by decomposition of biotite and epidote. Fractures developed by spalling are coated with iron oxide stain, and water lines in the larger pits are recorded by accumulations of iron oxides and organic matter.

In pits such as these, chemical weathering during times of wetting (Figure 3–33b) induces spalling and granular disintegration. If the pit fills only partially with water (due to the large size of the pit or the presence of an outlet below the rim), the walls will be only partially etched, leaving the rim overhanging. Weathered material may be flushed out and removed in solution during wet periods and blown away during dry periods. Ollier and Tuddenham (1962) suggested that black, clayey residues in weathering pits on Ayers Rock, Australia, might perform colloid plucking of mineral fragments as they dry into hard, curled flakes. Weathering pits, etched crystals, residual knobs, and other related phenomena (Figure 3–34) have been recognized in the Sierra Nevada and

FIGURE 3–35
Spheroidal weathering in granite, Stevens Pass, Washington.

FIGURE 3–34
Residual knobs in granite. (A) Mafic inclusion standing above weathered granite, (B) Fine-grained aplite dike etched out of enclosiing coarse-grained granite.

the Rocky Mountains by Burke and Birkeland (1979) and are common in many other alpine areas.

Spheroidal Weathering

Spheroidal weathering is a type of chemical weathering that produces rounded, boulder-like forms. When feldspars are converted to clay by chemical weathering, an increase in volume takes place, and the interlocking texture of the minerals may be broken up. Percolation of water along intersecting joint planes helps cause decomposition of feldspars from the joint planes inward. Corners of joint-bounded blocks are progressively rounded off and are split off in rounded concentric shells when exposed at the surface, resulting in production of spherically weathered forms (Figure 3–35).

Spheroidal weathering converts initially angular blocks of rock into oval forms characterized by weathered shells arranged concentrically, like the skins of an onion, about a relatively fresh core. The shells are produced by expansion of the exterior of the rock mass to the point where it springs loose from the interior along a smooth, curved fracture.

Blackwelder (1925) observed that the shells of spheroidal boulders are rotten from decomposition, in sharp contrast to the relatively fresh interiors. He also noted the sharp contrast between the paucity of spheroidal boulders in the driest deserts and their abundance in damp environments, and he stressed the common occurrence of spheroidal weathering underground—well beyond the reach of thermal changes. Samples taken along the radii of selected cores of spheroidally-weathered igneous rock displayed chemical alteration progressing from the surface inward (Chapman and Greenfield, 1949). Microscopic examination of thin sections of shells of weathered basalt showed distortion of the vesicles, largely by the inward jamming of crystals involved in a general swelling of the shells. Alteration products included hydrated oxides, micaceous minerals, epidote, and clays. Similar analyses of weathered spalls on spheroidal boulders of subgraywacke (Simpson, 1964) showed an expansion of the matrix due to the formation of vermiculite.

Farmin (1937) emphasized the importance of unloading in the exfoliation of spheroidal, pebble-like inclusions contained beyond the reach of chemical alteration in breccia dikes. He suggested that other examples of spheroidal weathering might also be due to exfoliation by unloading of joint blocks that may have been weakened by chemical weathering. On the other hand, spheroidal weathering is common in rocks that have never been deeply buried, such as near-surface basalts, and, in view of the obvious chemical alteration of the shells, little doubt exists that most spheroidal weathering is primarily the result of oxidation, hydrolysis, and hydration, which cause expansion and separation of the shells.

Tors

Tors have long aroused the curiosity of geomorphologists. Although the word tor has been applied to various features, it is presently used for bold, isolated outcrops that rise abruptly from their surroundings as steep-sided mounds or towers of rock several meters to tens of meters high. Tors are typical of uplands, particularly well-drained summits and ridges, and are rarely found in valleys. They are residual features, most commonly developed in massive rocks by the removal of the surrounding rock, which is usually of the same composition, implying that many, if not most, tors have been exhumed from a mantle of deeply weathered rock.

The origin of British tors is explained by Linton (1955) as follows:

1. Massive rock such as granite, characterized by zones of relatively close jointing and zones of widely spaced jointing, undergoes chemical weathering along the joints.

2. Percolating ground water converts the smaller joint blocks to spheroids and then to grus in the zone of aeration (above the water table). The tor is outlined underground as the surrounding rock rots almost completely, while the core-stones of the more massive tor become somewhat rounded.

3. The tor is exhumed by erosion of the surrounding rotten rock. Such tors may stand upon a rock platform interpreted as protected from deep weathering by the groundwater zone of saturation. This basal platform reflects the water table during the period of underground weathering.

The supporting evidence includes:

1. The development of tors on summits and spurs where the distance to the water table should have been greatest

2. The discovery of clearly outlined potential tors in quarries excavated through decomposed rock

Cavernous Weathering

Cavernous weathering of rock faces is one of the most peculiar of rock-weathering phenomena. It is responsible for the development of alcoves and closely spaced holes similar to those in Swiss cheese or the cells in a honeycomb (Figure 3–36). The holes are known as *tafoni* (from Corsica), and their close assemblage is called *stone lace* or *stone lattice.* This particular type of cavernous weathering has been called *alveolar weathering* (after the alveoli of lung tissue) or *honeycomb weathering.*

Cavernous weathering is typical of rocks susceptible to spalling and granular disintegration, but its occurrence is not ubiquitous. Even in places where cavernous weathering is conspicuous, most available rock faces are smooth or interrupted very little by cavities.

Each hole is enlarged by the spalling of thin flakes from its walls and by granular disintegration (Bradley et al., 1978; Bryan, 1922; Blackwelder, 1928; Calkin and Calilleaux, 1962; Dragovitch, 1967; Martini, 1978; Mustoe, 1982; Ollier, 1965b; Segerstrorn and Henriquez, 1964). Material weathered from the walls and roof of a hole collects on the floor as a telltale residue. Most of the residue is probably blown away periodically, but other processes, such as flushing by rainwater and excavation by rodents or lizards, may be locally

FIGURE 3–36
Honeycomb weathering, Lummi Island, Washington.

important in its removal. In many cases, cavernous weathering involves the breaching of a mineral coating (case hardening), typically composed of iron and manganese oxides. In some instances, the holes are located along and at the junctions of fractures, but in many cases, no structural control of their distribution is apparent.

The precise origin of cavernous weathering remains somewhat obscure and may vary from place to place. Most of the spalling and granular disintegration is probably due to hydrolysis and hydration (Segerstrom and Henriquez, 1964). Mustoe (1982) performed microscopic and chemical analyses on well-developed honeycomb weathering in arkosic sandstone exposed just above tide level and showed that mineral grains were disaggregated rather than decomposed and that thin walls separating adjacent cavities were due to protective organic coatings of microscopic algae. Although attributing the honeycomb weathering in his study to crystallization of salt during evaporation of wave splash from seawater, Mustoe also recognized that this phenomenon does little to explain occurrences where no seawater is present.

Large niches and alcoves weathered out of cliffs also occur. Ollier and Tuddenham (1962) describe the basal sapping of Ayers Rock, an inselberg of arkosic conglomerate in central Australia, where cavernous weathering has broken through a thick outer crust and developed caves that penetrate as much as 30 m into the rock.

Deep Chemical Weathering and Stripping

Deep chemical weathering, to depths of hundreds of meters, apparently is related to conditions of regional denudation during succeeding periods of stripping. For example, the slopes of the harbor basin at Hong Kong consist of largely unweathered masses of granite that protrude sporadically above a surface of altered grus and core-stones behind an irregular weathering front that extends to depths of 100 m (Ruxton and Berry, 1957). Landscape sculpture is a matter of exhumation of core-stones and bedrock by rainwash and mass movements of the deeply weathered mantle. In such a situation, exposures of relatively unweathered bedrock may act as local base levels, which prevent the deep stripping of weathered material immediately upslope.

Ollier (1960, 1965b) applied the concept of deep chemical weathering followed by stripping to the **inselbergs** (residual bedrock knobs) of Uganda and to granite landscapes in Australia. He suggested that granite landscapes with widely separated, scattered outcrops may be exhumed, whereas those with abundant, closely spaced outcrops probably owe most of their sculpture to surface processes. Thomas (1965) applied the same interpretation to inselbergs of Nigeria, and he developed an evolutionary sequence in which the weathering front and the land surface migrate downward as inselbergs and tors emerge. Thomas suggested that exfoliation can occur underground by physical expansion of solid rock masses to-

ward the surrounding rotten rock. In this way, he explained the exhumation of domed inselbergs that are typical exfoliation domes. Once exposed, the inselbergs are destroyed by marginal collapse of joint blocks and exfoliation sheets, leaving tors perched for a time at their summits and eventually a talus-bordered residual, called a *castle kopje.*

The concept of deep weathering followed by exhumation goes far toward explaining the occurrence of many isolated towers and hills that rise abruptly from their gentler surroundings. The British tors and the African inselbergs may be explained in this way. Previously, the inselbergs had been puzzling to geomorphologists, who sought to explain them as steep-sided residuals created strictly by fluvial erosion in the late stages of erosion. On the other hand, rock towers in exposed positions may look like tors but may not be the products of differential weathering underground. They may be the survivors of thoroughly frost-shattered masses of rock. Some buttes and mesas in arid lands may be closely analogous to inselbergs, but their origin as residuals after advanced fluvial dissection is betrayed by the way they continue to weather and by the pattern of debris and stream channels about their base.

Soils

Perhaps the most important effect of weathering is the development of soils. Plants depend on soil for growth, and humans depend directly or indirectly on plants for food. The nature of soils changes with the factors that control the rate and character of weathering. In addition, the composition of a soil generally changes with depth in a systematic manner. The geologic definition of **soil** is "a weathering residue that has become differentiated with depth into horizons."

Climate appears to be a most important factor in the development of soils. Russian scientists advanced the ideas that similar soils are developed in similar climates more or less independently of the nature of parent material and that soils developed on the same parent material differ if the climate varies from place to place. For example, soil on a steep slope is usually different from soil on flat ground in the same area. Soil formed under a forest cover is different from soil formed under a cover of grass.

Time also affects soil formation. Soils formed very recently show differences from soils developed over a long period of time.

Soil profiles may be differentiated vertically with depth into horizons lying above the parent material. The horizons are developed by the following:

1. Accumulation of organic material at or near the surface

2. Leaching of the parent material to the point where large amounts of one or more constituent minerals and their weathering products have been removed

3. Accumulation of organic material (humus) in the upper parts of the profile

4. Accumulation of weathering products at depth

An upper zone of leaching with the addition of humus and a lower zone of mineral and colloid accumulation occur in a typical soil. These are called *zones of eluviation* and illu*viation,* respectively, corresponding to the A and B horizons. Partially weathered parent material is known as the C *horizon.* All of these horizons need not be well developed (or even present) in a soil.

O Horizon

Accumulation of organic material on the ground beneath vegetation forms a layer known as the **O horizon,** characterized by decomposing plant material with little mineral content. The horizon may be further subdivided based on the degree of decomposition of the plant material.

A Horizon

When vegetation becomes established on fresh parent rock, accumulation of organic matter begins, and the parent rock begins to decompose. Organic material, added by decaying leaves and other plant matter, gives the **A horizon** a dark color. Humified organic matter and organic acids attack the parent rock, and percolation of water downward through the soil removes certain elements from the A horizon and carries them downward into the B horizon. Hence, the A horizon is sometimes known as the **zone of leaching.** Leaching carries dissolved chemicals from the surface downward, leaving the leached A horizon relatively loose and friable. Its open texture is also developed by the prying action and decay of roots and by the accumulation of humus at the surface. The ratio of various elements to one another in the weathered horizon and in the parent material can be used to define a leaching factor (Jenny, 1941) as follows:

$$\text{Leaching factor} = \frac{(K_2O + Na_2O)/SiO_2 \text{ (weathered horizon)}}{(K_2O + Na_2O)/SiO_2 \text{ (parent material)}}$$

In some soil profiles, a light-colored horizon, characterized by leaching of iron and aluminum and containing less organic material than the overlying O and/or A horizons, is recognized as an E horizon.

B Horizon

The **B horizon,** commonly referred to as the **zone of accumulation,** lies below the O, A, or E horizon and consists of material accumulated by downward percolation from the overlying layer. The material in the B horizon has undergone more weathering than in the underlying C horizon, and the nature of the parent material is difficult to recognize.

A recognizable **Bt horizon** develops when and where clay and hydrated oxides of iron and aluminum begin to coat other particles and fill the spaces between them, making the B horizon relatively dense and impermeable. A clayey B horizon swells and becomes sticky when wet and cracks upon drying to develop a crude columnar structure. It often has a red-brown color as a result of accumulation of iron oxides. The clay content is markedly higher than in the adjacent horizons (Figure 3–37), largely as a result of weathering in overlying horizons and translocation downward, and from weathering of minerals to clay in the B horizon. The translocated clay often occurs as films that coat grain surfaces or fill voids. However, the clay films can be destroyed by shrinking and swelling of the clay or by precipitation of calcium carbonate, so their absence is not as significant as their presence.

In arid or semiarid regions, evaporation of soil water may precipitate calcium carbonate (caliche) as coatings, nodules, or crusts in the B horizon. The accumulation of $CaCO_3$ is especially noticeable under arid conditions and is referred to as a **Bk horizon.** If carbonate coats all of the grains and makes up more than 50 percent of the horizon, it is referred to as a **K horizon** (Gile et al., 1981; Gile et al., 1965, 1966). The terms **caliche** and **calcrete** are also commonly applied to such carbonate-rich horizons (Bretz and Horberg, 1949; Brown, 1956; Gardner, 1972; Machette, 1987; Reeves, 1970). If the CO_2 partial pressure is high (as from plant activity), HCO_3^- forms and is carried downward in the profile along with Ca^{++} released from weathering of calcium-bearing minerals. $CaCO_3$ precipitates as a result of decrease in CO_2 partial pressure below the root zone and as Ca^{++} and HCO_3^- are concentrated during loss of water. Several stages in the accumulation of carbonate can be recognized, beginning with carbonate coating of pebbles and continuing until as much as 90 percent of the horizon may consist of carbonate (Figure 3–38).

C Horizon

The **C horizon** consists of slightly decayed rock which, although it is weathered parent material, has not undergone leaching or accumulation to the degree of the overlying horizons. It is characterized by angular blocks or chips, grus, or spheroidal core-stones lying on unweathered parent material (the **R horizon**). Oxidation of the parent material is less than that found in the B horizon, but it may be visible as a **Cox horizon.** The composition of the original material is still identifiable, although changes have occurred.

Soil Maturity

Immature soils reflect the composition of the material from which they were derived. For example, an immature soil derived from granite differs from a soil developed on limestone. A soil may eventually reach maturity or a time-independent, steady state of development as soil formation decelerates. Development of the C horizon is impeded with depth by the insulating effect of the overlying weathered material and by the inhibition of weathering processes. At depth, temperature fluctuations are reduced; the wedging effect of freezing

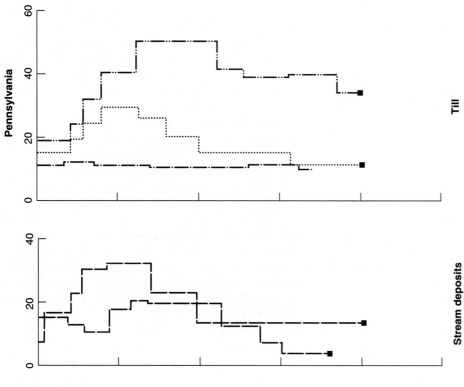

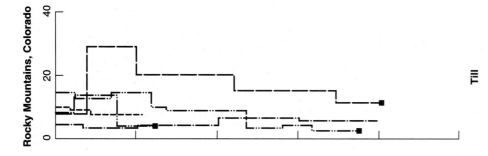

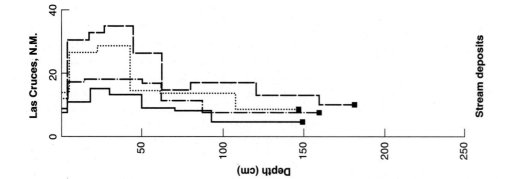

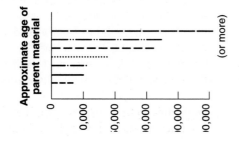

FIGURE 3–37

Variation in amount of clay with depth for soils of different ages. (From Birkeland, 1984)

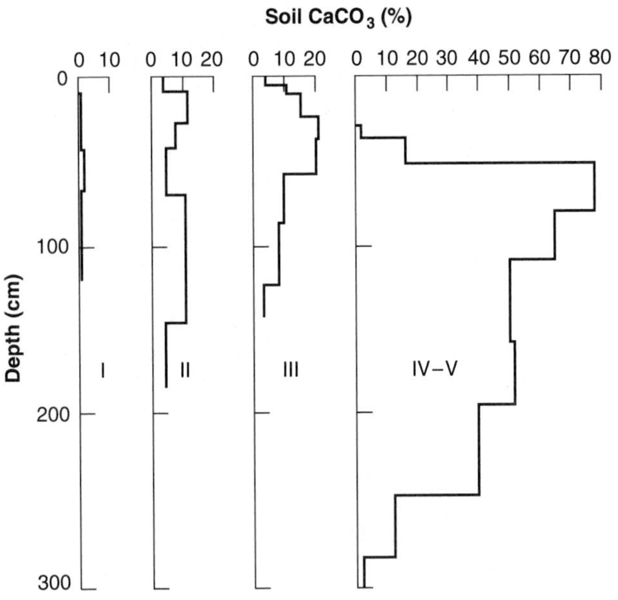

Soil CaCO₃ (%)

FIGURE 3–38
Calcium carbonate distribution with depth for soils of different ages. Roman numerals refer to stage of carbonate buildup in the soil. (From Birkeland, 1984)

FIGURE 3–39
Soil developed on loess, Columbia Plateau, Washington.

water may be ineffective; organic activity becomes negligible; descending waters may lose their original dissolving power; many mineral grains become coated with clays and colloids; and surface area available to chemical weathering is less abundant.

Before equilibrium is reached, a developing soil may give some indication of its age in one or more of its properties. On this basis, comparisons of immature soils on glacial deposits have been made. Care is taken to make such comparisons among similar parent materials in similar environments. Such properties as depth of oxidation, depth of leaching of soluble carbonates, thickness of clay accumulation, degree of weathering of stones in the B horizon, and thickness of humus are either obvious or recognizable with simple tests. All of these properties are proportional to age, and their measurement thus establishes the relative ages of soils.

A mature soil is one in which a series of horizons having well-defined characteristics has been produced by weathering processes (Figure 3–39). In such a soil, a succession of distinctive horizons forms from the surface downward to the parent material. At maturity, the A horizon encroaches on the B horizon, while the B horizon encroaches on the C horizon, which in turn grows downward into the unweathered parent material. Erosion at the surface prevents the A horizon from thickening indefinitely and regulates the ultimate thickness of a soil. Soil thickness is inversely proportional to the intensity of erosion which, in most cases, is itself proportional to slope. Deep soils

are formed in inactive erosional environments, whereas only shallow soils or no soil at all may represent the equilibrium conditions of active environments such as steep slopes.

Mature soils reflect the dominating influences of climate and vegetation. Differences in parent material are subordinated as weathering proceeds, and similar residues are produced from various outcrops. In fact, a large-scale map of soils (for example, a world map or a map of the United States) shows, at best, only a fair correspondence of soils with geology, but it exhibits a remarkably close correlation of soils with climatic and vegetational patterns.

Although all soils are of interest to geomorphologists, mature and immature soils are particularly useful in evaluating gradational processes. Azonal soils may be formed from the other soils by accelerated erosion (for example, sheetwash or vigorous mass movements). Azonal soils are typical of incipient weathering where materials have been freshly exposed by (1) the withdrawal of ice or water, (2) faulting, (3) erosion, or (4) active aggradation. They are characteristic of such landscape features as landslide scars, active alluvial fans, playas, fresh glacial tills, beaches, newly deglaciated surfaces, badland slopes, fault scarps, and so on.

Polygenetic soils are soils that record changes in one or more of the soil-forming factors. Most important are changes in climate-vegetation and changes in topography-drainage. Climatic change is indicated where a soil, characteristic of an earlier climate, has had superimposed upon it some of the soil characteristics of a later climate without obliteration of the earlier-formed characteristics. Changes from humid to more arid climates are more likely to be recorded than the reverse type of change because of the ready solubility of such indicators of aridity as $CaCO_3$ and gypsum. An interpretation of climatic change in soils was made by Bryan and Albritton (1943), who found evidence in soils of west Texas for the alternate accumulation and solution of caliche. They inferred fluctuations from aridity to humidity to aridity. These authors also cite Russian examples of the conversion of lateritic soils (warm, seasonally wet climates) to Chernozems (semiarid climate).

Changes in topography affect soils mainly through their influence on drainage. Dissection of an upland surface commonly improves the drainage of soil water and brings about the flushing of the existing soil. In this way, deeply weathered, poorly drained tills rich in residual clay may be converted to better-drained, siltier residues as the upper part of the clay horizon is flushed away. Changes in surface drainage induced by local changes in topography may also upset the balance between weathering and erosion and cause (1) truncation of profiles by erosional stripping or (2) thickening of profiles by slow sedimentation.

Composite soils are soils weathered into two or more parent materials. They are common in the glaciated parts of the central United States where a soil profile may penetrate till through blankets of overlying loess or glacial outwash. Simonson (1954) warned that some deeply weathered, clayey soils might actually be a composite of loess and till, neither of which by itself has weathered to the full extent.

Paleosols

Paleosols are ancient soils that have been removed from the zone of soil formation, usually by burial by younger sediments (Figure 3–40). The preservation of buried soils depends upon the original degree of horizon differentiation and protection from damage by the processes responsible for their burial. Some potentially damaging environmental changes accompanying burial are waterlogging, repeated freezing and thawing, mass movements of the soil, fluvial erosion, glacial erosion, baking by lava, and mixing by burrowing animals. Certain parts of the soil horizons are vulnerable to destruction, whereas others are persistent and might remain as evidence of former soil development. Accumulation of organic matter in the A horizon is relatively nonpersistent; the pattern of gains and losses in clay minerals and sesquioxides is fairly persevering; cementation and the development of a characteristic structure are persistent; and color changes depend on the endurance of the horizons in which the changes have occurred (Simonson, 1954). Thus, the interpretation of paleosols demands some care.

Paleosols delineate unconformities and represent the survival of a surface long enough to have been weathered to a distinctive soil type before eventual burial. For this reason alone, they are important to an analysis of the se-

FIGURE 3–40
Buried paleosols in loess near the Seine River, France. The dark layers represent ancient soils later buried by influxes of wind-blown silt.

quence of landscape evolution. In addition, many paleosols are definitive enough to be important clues to the environmental conditions (particularly the climate) at the time of their formation. The least that can be said of a paleosol is that a surface and a climate capable of soil formation existed at that time and that favorable conditions lasted long enough for a soil to form. At best, the paleosol may give a rather definite picture of climate and accompanying vegetation. Mature paleosols are most desirable for climatic interpretations.

Paleosols have been vital to the recognition, classification, and mapping of deposits left by continental glaciation in the north central United States. In many places, these paleosols have aided interpretation of interglacial climates as, for example, in the upper Mississippi Valley where buried soils are a bright red color and contain some carbonate at the base of the solum relative to the known, low-clay-content, noncalcic surface soils. These paleosols suggest an approach to red-yellow podzolic soils, in contrast to the gray-brown podzolic soils of the surface, which are lacking in carbonate. From this evidence, Simonson (1954) suggested a Yarmouth interglacial climate slightly warmer and drier than the present climate of the upper Mississippi Valley.

Soil Classification

Soil classification systems have been developed to group soils of similar properties and to provide a systematic means of mapping soils. Early attempts at soil classification were strongly influenced by Soviet soil scientists and were defined in part by genetic criteria. Soils are grouped primarily on the

TABLE 3–3
Soil orders.

Entisols	Soils without distinct horizons and with only faint imprint of pedologic processes (*ent* = recent)
Inceptisols	Soils with weakly developed horizons (*incept* = inception)
Vertisols	Soils with >35 percent swelling clay content, subject to extensive cracking upon drying (*vert* = turn, invert)
Aridisols	Dry, desert soils with pedogenic horizons, low organic content, carbonate horizons (*aridus* = dry, arid)
Mollisols	Soils with dark, organic-rich A horizon, high base saturation, usually moist (*molli* = soft, mollify)
Alfisols	Soils with organic content similar to that of mollisols, >35 percent base saturation, usually moist, clayey B horizon, common in humid forests (*alfi* = from pedalfer)
Ultisols	Soils with <35 percent base saturation, usually moist, clayey B horizon (*ulti* = ultimate)
Spodosols	Soils with accumulations of amorphous materials in subsurface (*spodo* = wood ash)
Oxisols	Deep soils with a thick B horizon of kaolin and hydrated Fe and Al oxides, common in tropics (*oxi* = oxide)
Histosols	Organic soils, peat (*histo* = tissue, histology)

TABLE 3–4
Suborder names.

acr	Most strongly weathered (*acr* = highest)
alb	Light-colored soils with clay and iron removed (*albus* = white)
alt	Cool, high altitude or latitude (*alt* = high)
aqu	Soils that are wet for long periods (*aqua* = water)
arg	Soils with clay accumulation (*argilla* = clay)
bor	Northern, cool (*boreas* = northern, cool)
ferr	Contains iron (*ferrum* = iron)
fibr	Least decomposed organic matter (*fibra* = fiber)
fluv	Alluvial (*fluvius* = river)
hum	Presence of organic material (*humus* = earth)
ochr	Little organic material (*ochros* = pale)
orth	Common, typical (*orthos* = true)
psamm	Sandy (*psammos* = sand)
rend	Calcareous parent material (*rend* = Rendzina)
sapr	Most decomposed organic material (*sapros* = rotten)
trop	Continually warm climate (*tropos* = tropical)
ud	Humid (*udus* = humid)
umbr	Dark, much organic material (*umbra* = shade)
ust	Dry climate with summer rain (*ustus* = burnt)
xer	Dry with winter rains (*xeros* = dry)

basis of characteristics of their profiles into basic units known as *soil series*.

Many classifications have been proposed, but none have received worldwide recognition and acceptance. The system widely used in North America for many years included various genetic factors (climate and vegetation), but that system was abandoned by the U.S. Department of Agriculture (Soil Survey Staff, 1975) and was replaced with a totally new, descriptive system emphasizing observable physical and chemical properties of soil profiles. The classification scheme, known variously as the SCS *system* or the *7th Approximation, is* covered in a 754-page document that makes extensive use of Greek and Latin terms. Ten soil orders are recognized and are subdivided into 47 suborders and into numerous great groups, subgroups, families, and series. The re-

sulting possible combinations of terms is colossal and overwhelming to someone just beginning to learn the system. However, because the system is based on Greek and Latin, many of the terms can be decoded from knowing the root words of a complex name.

The ten soil orders, listed in Table 3–3, are distinguished on the basis of characteristics of horizons in the soil profile. They are given names that describe some of the soil properties, for example, *aridisol* from the Latin aridius, meaning "dry" or "arid," and sol, meaning "soil." Most of the ten orders can be identified in the field, but to classify a soil in detail usually requires laboratory data.

A full discussion of such a complex system would require many pages and thus lies beyond the scope of this book. However, Table 3–4 shows some of the basic elements of the system. Those interested in greater detail may refer to the original publication by the Soil Survey Staff (1975) or to a discussion of it in Birkeland (1984).

REFERENCES

Anand, R. R., Gilkes, R. J., Armitage, T. M., and Hillyer, J. W., 1985, Feldspar weathering in lateritic saprolite: Clays and Clay Mineralogy, v. 33, p. 31–43.

April, R., Newton, R., and Coles, L. T., 1986, Chemical weathering in two Adirondack watersheds: Past and present-day rates: Geological Society of America Bulletin, v. 97, p. 1232–1238.

Arkley, R., 1963, Calculation of carbonate and water movement in soil from climatic data: Soil Science, v. 96, p. 239–248.

Atkinson, H., and Wright, J., 1957, Chelation and the vertical movement of soil constituents: Soil Science, v. 84, p. 1–11.

Bachman, G. O., and Machette, M. N., 1977, Calcic soils and calcretes in the southwestern United States: U. S. Geological Survey Open-File Report 77-794, 163 p.

Barshad, I., 1964, Chemistry of soil development: *in* F.E. Bear, ed., Chemistry of the soil: Reinhold Publishing, N.Y., p. 1–70.

Barton, D. C., 1916, The disintegration of granite in Egypt: Journal of Geology, v. 24, p.382–393

Bateman, P.C., and Eaton, J. P., 1967, Sierra Nevada batholith: Science, v. 158, p. 1407–1417.

Berner, R. A., 1978, Rate control of mineral dissolution under earth surface conditions: American Journal of Science, v. 278, no. 9, p. 1235–1252.

Berner, R.A., and Holdren, G.R.,Jr., 1977, Mechanism of feldspar weathering: Some observational evidence: Geology, v. 5, p. 369–372.

Berner, R.A., and Holdren, G.R., Jr., 1979, Mechanisms of feldspar weathering—II. Observations of feldspars from soils: Geochimica et Cosmochimica Acta, v. 43, p. 1173–1186

Berry, M., 1987, Morphological and chemical characteristics of soil catenas on Pinedale and Bull Lake moraine slopes in the Salmon River Mountains of Idaho: Quaternary Research, v. 28, p. 210–225.

Birkeland, P. W., 1969, Quaternary paleoclimatic implications of soil clay mineral distribution in a Sierra Nevada-Great Basin transect: Journal of Geology, v. 77, p. 289–302.

Birkeland, P. W., 1973, Use of relative age-dating methods in a stratigraphic study of rock glacier deposits, Mt. Sopris, Colorado: Arctic and Alpine Research, v. 5, p. 401–416.

Birkeland, P. W., 1978, Soil development as an indication of relative age of Quaternary deposits, Baffin Island, N. W. T., Canada: Arctic and Alpine Research, v. 10, p. 733–747.

Birkeland, P. W., 1984 Soils and geomorphology: Oxford University Press, N.Y., 372 p.

Birkeland, P. W., and Burke, R., 1988, Soil catena chronosequences on eastern Sierra Nevada moraines, California, USA: Arctic and Alpine Research, v. 20., p. 473–484.

Birkeland, P. W., and Janda, R. J., 1971, Clay mineralogy of soils developed from Quaternary deposits of the eastern Sierra Nevada: Geological Society of America Bulletin, v. 82, p. 2495–2514.

Birkeland, P. W., Burke, R. M., and Shroba, R. R., 1987, Holocene alpine soils in gneissic cirque deposits, Colorado Front Range: U.S. Geological Survey Bulletin 1590-E, 21 p.

Birkeland, P. W., Berry, M., and Swanson, D., 1991, Use of soil catena field data for estimating relative ages of moraines: Geology, v. 19, p. 281–283.

Birkeland, P. W., Burke, R. M., and Walker, A. L., 1980, Soils and subsurface rock-weathering features of Sherwin and pre-Sherwin glacial deposits, eastern Sierra Nevada, California: Geological Society of America Bulletin, v. 91, p. 238–244.

Blackwelder, E., 1925, Exfoliation as a phase of rock weathering: Journal of Geology, v. 33, p. 793–806.

Blackwelder, E., 1927, Fire as an agent in rock weathering: Journal of Geology, v. 35, p. 135–140.

Blackwelder, E., 1933, The insolation hypothesis of rock weathering: American Journal of Science, v. 226, p. 97–113.

Blair, R. W., 1986, Development of natural sandstone arches in southeastern Utah: in Gardiner, V., ed., International Geomorphology, Part II, John Wiley and Sons, London, p. 597–604.

Bohn, H. I., Mc.Neal, B. L., and O'Conner, G. A., 1979, Soil chemistry: John Wiley and Sons, N.Y., 329p.

Brace, W. F., Silver, E., Hadley, H., and Goetze, C., 1972, Cracks and pores: A closer look: Science, v. 178, p. 162–164.

Bradley, W. C., 1963, Large-scale exfoliation in massive sandstones of the Colorado Plateau: Geological Society of America Bulletin, v. 75, p. 519–528,

Bradley, W. C., 1980, Role of salts in development of granitic tafoni, south Australia: A reply: Journal of Geology, v. 88, p. 121–122.

Bradley, W. C., Hutton, J. T., and Twidale, C. R., 1978, Role of salts in development of granitic tafoni, south Australia: Journal of Geology, v. 86, p. 647–654.

Brantley, S., Crane, S., Crerar, D., Hellmann, R., and Stallard, R., 1986, Dissolution at dislocation etch pits in quartz: Geochemica et Cosmochemica Acta, v. 50, p. 2349–2361.

Bretz, J. H., and Horberg, L., 1949, Caliche in southeastern New Mexico: Journal of Geology, v. 57, p. 491–511.

Brown, C. H., 1956, The origin of caliche on the northeastern Llano Estacado, Texas: Journal of Geology, v. 64, p. 1–15.

Bruner, W. M., 1984, Crack growth during unroofing of crustal rocks: Effects on thermo-elastic behavior and near-surface stresses: Journal of Geophysical Research, v. 89, p. 4167–4184.

Bryan, W. M., 1922, The hot-water supply of the Hot Springs, Arkansas: Journal of Geology, v. 30, no. 6, p. 425–449.

Bryan, K., 1925, Pedestal rocks: Engineering and Mineral Journal-Press, v. 119, no. 4, p. 172–173.

Bryan, K. and Albritton, C. C., 1943, Soil phenomena as evidence of climatic changes: American Journal of Science, v. 241, no. 8, p. 469–490.

Buckman, H. O., and Brady, N. C., 1969, The nature and properties of soils: Macmillan, Toronto, 653 p.

Buol, S. W., Hole, F. D., and McCracken, R. J., 1973, Soil genesis and classification: Iowa State University Press, Ames, Iowa, 360p.

Burke, R. M., and Birkeland, P. W., 1979, Reevaluation of multiparameter relative dating techniques and their application to the glacial sequence along the eastern Sierra Nevada: Quaternary Research, v. 11, p. 21–51.

Burns, S. F., and Tonkin, P. J., 1982, Soil-geomorphic models and the spatial distribution and development of alpine soils: in C. E. Thorn, ed., Space and time in geomorphology: Allen & Unwin, London, p. 25–43.

Caine, N., 1979, Rock weathering rates at the soil surface in an alpine environment: Catena, v. 6, p. 131–144.

Calkin, P., and Cailleaux, A., 1962, A quantitative study of cavernous weathering (tafonis) and its application to glacial chronology in Victoria Valley, Antarctica: Zeitschrift für Geomorphology, v. 6, p. 317–324.

Cann, J. H., 1974, A field investigation into rock weathering and soil forming processes: Journal of Geological Education, v. 22, no. 5, p. 226–230.

Carroll, D., 1959, Ion exchange in clays and other minerals: Geological Society of America Bulletin, v. 70, p. 749–780.

Carroll, D., 1970, Rock weathering: Plenum Press, N.Y., 203p.

Carson, M. A., and Kirby, M. J., 1972, Hillslope form and process: Cambridge University Press, Oxford, England, 475 p.

Chapman, R. W., and Greenfield, M. A., 1949, Spheroidal weathering of igneous rocks: American Journal of Science, v. 247, p. 407–429.

Chinn, T. H. J., 1981, Use of rock weathering-rind thickness for Holocene absolute age-dating in New Zealand: Arctic and Alpine Research, v. 13, p. 33–45.

Ciolkosz, E. J., Carter, B. J., Hoover, M. T., Cronce, R. C., Waltman, W. J., and Dobos., R. R., 1990, Genesis of soils and landscapes in the Ridge and Valley province of central Pennsylvania: Geomorphology, v. 3, p. 245–261.

Clark, S. P., and Jager, E., 1969, Denudation rate in the Alps from geochronology and heat flow: American Journal of Science, v. 267, p. 1143–1160.

Colman, S. M., 1981, Rock-weathering rates as functions of time: Quaternary Research, v. 15, p. 250–264.

Colman, S. M., 1982a, Chemical weathering of basalts and andesite: Evidence from weathering rinds: U. S. Geological Survey Professional Paper 1246, 51p.

Colman, S. M., 1982b Clay mineralogy of weathering rinds and possible implications concerning the sources of clay minerals in soils: Geology, v. 10, p. 370–375.

Colman, S. M., and Pierce, K. L., 1981, Weathering rinds on andesitic and basaltic stones as a Quaternary age indicator, western U. S.: U. S. Geological Survey Professional Paper 1210.

Correns, C. W., 1963, Experiments on the decomposition of silicates and discussion of chemical weathering: in Clays and Clay Minerals, v. 10—National Conference of Clays and Clay Minerals, 10th, 1961, Macmillan, N.Y., p. 443–459.

Cronan, C., 1985, Chemical weathering and solution chemistry in acid forest soils: Differential influence of soil type, biotic processes, and H+ deposition: in Drever, J., ed., The Chemistry of Weathering, Reidel Publishing, Netherlands p. 175–195.

Curtis, C. D., 1976, Chemistry of rock weathering: Fundamental reactions and controls: in E. Derbyshire, ed., Geomorphology and climate: John Wiley and Sons, N.Y., p. 25–57.

Dale, T. N., 1923, Commercial granites of New England: U. S. Geological Survey Bulletin 738, 488p.

Dapples, E. C., 1959, The behavior of silica in diagenesis: in H. A. Ireland, ed., Social Economic Paleontologists and Mineralists Special Publication, no. 7, p. 36–54.

Deju, R. A., 1971, A model of chemical weathering of silicate minerals: Geological Society of America Bulletin, v. 82, p. 1055–1062.

Dixon, J. C., 1986, Solute movement on hillslopes in the alpine environment of the Colorado Front Range: in Abrahams, A. D., Hillslope Processes, Allen and Unwin, Boston, MA, p. 139–159.

Douglas, G. R., McGreevey, J. P., and Whalley, W. B., 1983, Rock weathering by frost shattering processes: in Proceedings of the 4th International Permafrost Conference: National Academy of Science, p. 244–248.

Dragovich, D., 1967, Flaking, a weathering process operating on cavernous rock surfaces: Geological Society of America Bulletin, v. 78, p. 801–804.

Drever J. I., and Smith, C.L., 1978, Cyclic wetting and drying of the soil zone as an influence on the chemistry of groundwater in arid terrains: American Journal of Science, v. 278, p. 1448–1454.

Eggleton, R. A., Foudoulis, C., and Varkevisser, D., 1987, Weathering of basalt: Changes in rock chemistry and mineralogy: Clays and Clay Minerals, v. 35, p. 161–169.

Eggler, D. H., Larson, E. E., and Bradley, W. C., 1969, Granites, grusses, and the Sherman erosion surface, southern Laramie Range, Wyoming: American Journal of Science, v. 267, p. 510–522.

Emery, K. O., 1960, Weathering of the Great Pyramid: Journal of Sedimentary Petrology, v. 301, p. 140–143.

Evans, I. S., 1970, Salt crystallization and rock weathering: A review: Rev. Geomorph. Dynamique, v. 19, p. 153–177.

Evenson, E. B., 1990, Gillespie, A. R., Stephens, G. C., 1990, Extensive boulder spalling resulting from a range fire at the Pinedale type locality, Freemont Lake, Wyoming: Geological Society of America, Abstracts with Program, p. 110.

Everett, D.H., 1961, The thermodynamics of frost damage to porous solids: Trans. Faraday Society, v. 57, p. 1541–1551.

Farmin, R., 1937, Hypogene exfoliation in rock masses: Journal of Geology, v. 45, no. 6, p. 625–635.

Feth, J., Robertson, C., and Polzer, W., 1964, Sources of mineral constituents in water from granitic rocks, Sierra Nevada, California, and Nevada: U.S. Geological Survey Water Supply Paper 1535-I.

Folk, R.L., and Patton, E.B., 1982, Buttressed expansion of granite and development of grus in central Texas: Zeitschrift für Geomorphologie, v. 26, p. 17–32.

Friedman, I., 1968, Hydration rind dates rhyolite flows: Science, v. 159, p. 878–8789.

Gardner, L. R., 1972, Origin of the Mormon Mesa caliche: Geological Society of America Bulletin, v. 83, p. 143–156.

Garrels, R. M., and Christ, C. L., 1965, Solutions, minerals, and equilibria: Harper and Row, N.Y., 450 p.

Gerrard, A. J., 1981, Soils and landforms: Allen and Unwin, London, 219p.

Gibbs, R. J., 1967, The geochemistry of the Amazon River system, part 1: Geological Society of America Bulletin, v. 78, p. 1203–1232.

Gile, L. H., 1975, Holocene soils and soil-geomorphic relations in an arid region of southern New Mexico: Quaternary Research, v. 5, p. 321–360.

Gile, L. H., Hawley, J. W., and Grossman, R. B., 1981, Soils and geomorphology in the Basin and Range area of southern New Mexico; guidebook to the Desert Project: New Mexico Bureau of Mines and Mineral Resources Memoir 39.

Gile, L. H., Peterson, F. F., and Grossman, R. B., 1965, The K horizon: A master soil horizon of carbonate accumulation: Soil Science, v. 99, p. 74–82.

Gile, L. H., Peterson, F. F., and Grossman, R. B., 1966, Morphological and genetic sequences of carbonate accumulation in desert soils: Soil Science, v. 101, p. 347–360.

Goldich, S., 1938, A study of rock weathering: Journal of Geology, v. 46, p. 17–58.

Goodchild, J. G., 1890, Notes on some observed rates of weathering of limestones: Geological Magazine, v. 27, p. 463–466.

Goudie, A. S., 1973, Duricrusts in tropical and subtropical landscapes: Oxford University Press, London, 174p.

Goudie, A. S., 1989, Weathering processes: in Thomas, D. S. G., ed., Arid Zone Geomorphology, John Wiley and Sons, N.Y., p. 1–12.

Grant, W. H., 1969, Abrasion pH, an index of weathering: Clays and Clay Minerals, v. 17, p. 151–155.

Grawe, O. R., 1936, Ice as an agent of rock weathering; a discussion: Journal of Geology, v. 44, no. 2, p. 173.182.

Gray, W. M., 1965, Surface spalling by thermal stresses in rocks: in Rock mechanics symposium: Toronto, Proceedings, Canada Department of Mines and Technology Surveys.

Griggs, D., 1936a, The factor of fatigue in rock exfoliation: Journal of Geology, v. 44, p. 783–796.

Griggs, D., 1936b, Deformation of rocks under high confining pressures: Journal of Geology, v. 44, p. 541–577.

Grim, R. E., 1968, Clay mineralogy: McGraw-Hill, N.Y., 596p.

Hall, R., and Horn, L., 1993, Rates of hornblende etching in soils in glacial deposits of the northern Rocky Mts. (Wyoming-Montana, U.S.A.): Influence of climate and characteristics of the parent material: Chemical Geology, v. 105, p. 17–29.

Hall, R., and Michaud, D., 1988, The use of hornblende etching, clast weathering, and

soils to date alpine glacial and periglacial deposits: A study from southwestern Montana: Geological Society of America Bulletin, v. 100, p. 458–467.

Harden, J. W., 1982, A quantitative index of soil development from field descriptions: Examples from a chronosequence in central California: Geoderma, v. 28, p. 1–28.

Harden, J. W., 1987 Soils developed in granitic alluvium near Merced, California: U. S. Geological Survey Bulletin 1590-A, 65 p.

Harden, J. W., and Taylor E. M., 1983, A quantitative comparison of soil development in four climatic regimes: Quaternary Research, v. 20, p. 342–359.

Harden, J. W., Taylor, E. M., Hill, C., Mark, R., McFadden, L., Reheis, M., Sauer, J., and Wells, S., 1991, Rate of soil development from four chronosequences in the southwestern Great Basin: Quaternary Research, v. 35, p. 383–399.

Haxby, W. F., and Turcotte, D. L., 1976, Stresses induced by the addition or removal of overburden and associated thermal effects: Geology, v. 4, p. 181–184.

Hay, R. L., 1960, Rate of clay formation and mineral alteration in a 4000-year-old volcanic ash soil on St. Vincent, B. W. I.: American Journal of Science, v. 258, p. 354–368.

Hay, R. L., and Jones, B. F., 1972, Weathering of basaltic tephra on the island of Hawaii: Geological Society of America Bulletin, v. 83, p. 317–332.

Helgeson, J., Murphy, W., and Aagaard, P., 1984, Thermodynamic and kinetic constraints on reaction rates among numerals and aqueous solution: II Rate constants, effective surface area, and the hydrolysis of feldspar: Geochemica et Cosmochemica Acta, v. 48, p. 2405–2432.

Hess, P. C., 1966, Phase equilibria of some minerals in the K_2O-Na_2O-Al_2O_3- SiO_2 -H_2O system at 25C and 1 atmosphere: American Journal of Science, v. 264, p. 289–309.

Holdren, G. R. Jr., and Berner, R. A., 1979, Mechanism of feldspar weathering—Part. 1, Experimental studies: Geochimica et. Cosmochimica Acta. v. 43, p. 1161–1172.

Holdren, G. R., and Speyer, P. M., 1985, Reaction rate surface area relationships during the early stages of weathering—I. Initial observation: Geochimica et. Cosmochimica Acta. v. 49, p. 674–681.

Holdren, G. R., and Speyer, P. M., 1986, Stoichiometry of alkali feldspar dissolution at room temperature and various pH values: in Colman, S. M., and Dethier, D. P., eds., Rates of chemical weathering of rocks and minerals: Academic Press, Orlando, FL, p. 61–81.

Isherwood, C., and Street, A., 1976, Biotite-induced grussification of the Boulder Creek Granodiorite, Boulder County, Colorado. Geological Society of America Bulletin, v. 87, p. 366–370.

Jackson, M. L. and Pennington, R. P., 1948, Segregation of clay minerals of polycomponent soil clays: Soil Science of America Proceedings, v. 12, p. 452–457.

Jackson, M. L., and Sherman, G. D., 1953, Chemical weathering of minerals in soils: Advances in Agronomy, v. 5, p. 219–318.

Jackson, M. L., Bourbeau, G. A., Pennington, R. P., Tyler, S. A., Willis, A. L., 1948, Fundamental generalizations—Pt. 1 of Weathering sequence of clay-size minerals in soils and sediments: Journal of Physical and Colloid Chemistry, v. 52, p. 1237–1260.

Jackson, M. L., Hseung, Y., Corey, R., Evans, E., and Heuval, R., 1952, Weathering sequence of clay size minerals in soils and sediments: Soil Science Society of American Proceedings, v. 16, p. 3–6.

Jackson, T. A., and Keller, W. D., 1970, A comparative study of the role of lichens and "inorganic" processes in the chemical weathering of recent Hawaiian lavas: American Journal of Science, v. 269, p. 446–466.

Jahns, R. H., 1943, Sheet structure in granite: Its origin and use as a measure of glacial erosion in New England: Journal of Geology, v. 51, p. 71–98.

Jenny, H., 1935, The clay content of the soil as related to climatic factors, particularly temperature: Soil Science, v. 40, p. 111–128.

Jenny, H., 1941, Factors of Soil Formation: McGraw-Hill, N.Y., 281 p.

Jenny, H., 1980, The soil resource-origin and behavior: Springer-Verlag, N.Y., 377 p.

Johnson, D., Keller, E., and Rockwell, T., 1990, Dynamic pedogenesis: New Views on some key soil concepts, and a model for interpreting Quaternary soils: Quaternary Research, v. 33, p. 306–319.

Johnson, N. M., Reynolds, R. C., and Likens, G. E., 1972, Atmospheric sulfur: Its effect on the chemical weathering of New England: Science, v. 177, p. 514–516.

Jones, D. E., and Holtz, W. G., 1973, Expansive soils—the hidden disaster: American Society of Civil Engineers, v. 43, p. 49–51.

Keller, W. D., 1954, Bonding energies of some silicate minerals: American Mineralogist, v. 39, p. 783–793.

Keller, W. D., 1964, Processes of origin and alteration of clay minerals: in C.I. Rich and G.W. Kunze, eds., Soil clay mineralogy, Univ. of North Carolina Press, Chapel Hill, NC, p. 3–76.

Keller, W. D., 1978, Kaolinization of feldspar as displayed in scanning electron micrographs: Geology, v. 6, p. 184–188.

Keller, W., 1982, Kaolin—A most diverse rock in genesis, texture, physical properties and uses: Geological Society of America Bulletin, v. 93, p. 27–36.

Keller, W. D., and Fredrickson, A. F., 1952, Role of plants and colloidal acids in the mechanism of weathering: American Journal of Science, v. 250, p. 594–608.

Kessler, D. W., Insley, H., and Sligh, W. H., 1940, Physical mineralogical and durability studies on the buildings and monuments of the U.S.: U.S. National Bureau of Standards, Journal of Research, v. 25, p. 161–206.

Knuepfer, P., and McFadden L., eds., 1990, Soils and landscape evolution: Proceedings of the 21st Binghampton Symposium in Geomorphology, 378 p.

Lagache, M., 1965, Contribution to the study of feldspar alteration in water, between 100–200 degrees C, under varying CO_2 pressures, and its application to the synthesis of clay minerals: Society of Fr. Mineralogy and Crystallography, Bulletin, v. 88, p. 223–253.

Lagache, M., 1976, New data on the kinetics of the dissolution of alkali feldspars at 200 degrees C in CO_2 charged water: Geochimica et Cosmochimica Acta, v. 40, p. 157–161.

Lasaga, A., 1984, Chemical kinetics of water-rock interactions: Journal of Geophysical Research, v. 89, p. 4006–4025.

Lehman, D. S., 1963, Some principles of chelation chemistry: Soil Science Society of America Proceedings, v. 27, p. 167–170.

Lindsay, W. L., 1979, Chemical equilibria in soils: John Wiley and Sons, N.Y., 449 p.

Linton, D. L., 1955, The problem of tors: Geographical Journal, v. 121, p. 470–486.

Locke, W. W., 1979, Etching of hornblende grains in arctic soils: An indicator of relative age and paleoclimate: Quaternary Research, v. 11, p. 197–212.

Lotspeich, F. B. and Smith, H. W., 1953, The Palouse catena- Pt. 1 of Soils of the Palouse loess (Wash.): Soil Science, v. 76, no. 6, p. 467–480.

Loughnan, F., 1969, Chemical weathering of the silicate minerals: American Elsevier, N.Y., 154 p.

Loughnan, F., and Bayliss P., 1961, The mineralogy of the bauxite deposits near Weipa, Queensland: American Mineralogist, v. 46, p. 209–217.

Lyon, T. L., Buckman, H. O., and Grady, N. C., 1952, The nature and properties of soils: Macmillan Publishing, N.Y., 591 p.

Machette, M. N., 1987, Calcic soils of the southwestern United States: in Weide, D. L., and Faber, M. L., Soils and Quaternary Geology of the southwestern U. S., Geological Society of America Special Paper 203, p. 1–21.

Manley, E. P., and Evans, L. J., 1986, Dissolutions of feldspars by low-molecular-weight aliphatic and aromatic acids: Soil Science, v. 141, p. 106–112.

Marion, G. W., Schlesinger, W. H., and Fonteyn, P. J., 1985, CALDEP: A regional model for soil $CaCo_3$ (caliche) deposition in southwestern deserts: Soil Science, v., 139, p. 468–481.

Marshall, C. E., 1977, The physical chemistry and mineralogy of soils—v. II: Soils in place: John Wiley and Sons, N.Y., 313p.

Martini, I.P., 1978, Tafoni weathering, with examples from Tuscany, Italy: Zeitschrift für Geomorphologie, v. 22, p. 44–67.

Matthes, F. E., 1930, Geologic history of the Yosemite Valley: U. S. Geological Survey Professional Paper 160, 137 p.

Matthias, G. F., 1967, Weathering rates of Portland arkose tombstones: Journal of Geological Education, v. 15, no. 4, p. 140–144.

Mayer, L., McFadden, L. D., and Harden, J. W., 1988, Distribution of calcium carbonate in desert soils: A model: Geology, v. 16, p. 303–306.

McFadden, L. D., 1982, The impacts of temporal and spatial climatic changes on alluvial soils genesis in southern California: [PhD Thesis]: Tucson, University of Arizona.

McFadden, L. D., 1988, Climatic influences on rates and processes of soil development in Quaternary deposits of southern California: in Reinhardt, J., and Sigleo, W., eds., Paleosols and weathering through geologic time, Geological Society of America Special Paper 216, p. 153–177.

McFadden, L. D., and Hendricks, D. M., 1985, Changes in the content and composition of pedogenic iron oxhydroxides in a chronosequence of soils in southern California: Quaternary Research, v. 23, p. 189–204.

McFadden, L. D., and Weldon, R. J., 1987, Rates and processes of soil development on Quaternary terraces in Cajon Pass, California: Geological Society of America Bulletin, v. 98, p. 280–293.

McLennan, S. M., 1993, Weathering and global denudation: Journal of Geology, v. 101, p. 295–303.

Meierding, T. C., 1981, Marble tombstone weathering rates: A transect of the United States: Physical Geography, v. 2, p. 1–18.

Mellor, M., 1970, Phase composition of pore water in cold rocks: U. S. Corps of Army Engineers cold regions research and engineering laboratory research report 292, 61 p.

Mielenz, R., and King, M., 1955, Physical-chemical properties and engineering performance of clays: California Division of Mines Bulletin, v. 169, p. 196–254.

Morey, G., Fournier, R., and Rowe, J., 1962, The solubility of quartz in water in the temperature interval from 25 degrees C to 300 degrees C: Geochimica et Cosmochimica Acta, v. 26, p. 1029–1043.

Morey G., Fournier, R., and Rowe, J., 1964, The solubility of amorphous silica at 25 degrees C: Journal of Geophysical Research, v. 69, p. 1995–2002.

Morton, A. C., 1984, Stability of detrital heavy minerals in Tertiary sandstones from the North Sea Basin: Clay Minerals, v. 19, p. 287–308.

Mugridge, S.J., and Young, H.R., 1983, Disintegration of shale by cyclic wetting and drying and frost action: Canadian Journal of Earth Sciences, v. 20, p. 568–576.

Muhs, D. R., 1982, The influence of topography on the spatial variability of soils in Mediterranean climates: in C. E. Thorn, ed., Space and time in geomorphology: Allen & Unwin, London, p. 269–284.

Mustoe, G. E., 1982, The origin of honeycomb weathering: Geological Society of America Bulletin, v. 93, p. 108–115.

Nesbitt, H. W., and Young, G. M., 1984, Prediction of some weathering trends of plutonic and volcanic rocks based on thermodynamic and kinetic considerations: Geochimica et Cosmochimica Acta, v. 48, p. 1523–1534.

Nesbitt, H. W., and Young, G. M., 1989, Formation and diagenesis of weathering profiles: Journal of Geology, v. 97, p. 129–147.

Nickel, E., 1973, Experimental dissolution of light and heavy minerals in comparison with weathering and interstratal solution: Contributions to Sedimentology, v. 1, p. 1–68.

Ollier, C. D., 1960, The inselbergs of Uganda: Zeitschrift für Geomorphologie, Neve Folge Band 4, Heft 1, p. 43–52.

Ollier, C. D., 1962, Slope development at Coober Pedy, South Australia: Geological Society of Australia, Journal v. 9, no. 1, p. 93–105.

Ollier, C. D., 1963, Insolation weathering: Examples from central Australia: American Journal of Science, v. 261, p. 376–381.

Ollier, C. D., 1965a, Dirt cracking, a type of insolation weathering: Australian Journal of Science, v. 27, no. 8, p. 236–237.

Ollier, C. D., 1965b, Some features of granite weathering in Australia: Zeitschrift für Geomorphologie, v. 9, p. 285–304.

Ollier, C. D., 1969, Weathering: Oliver and Boyd, Edinburgh, U.K., 309p.

Ollier, C. D., 1973, Catenas in different climates: in Derbyshire, E., ed., Geomorphology and climate: John Wiley and Sons, N.Y., p. 137–169.

Ollier, C. D., and Ash, J. E., 1983, Fire and rock breakdown: Zeitschrift für Geomorphologie, v. 27, p. 363–374.

Ollier, C. D., and Tuddenham, W. G., 1962, Inselbergs of central Australia: Zeitschrift für Geomorphologie v. 5, p. 257–276.

Owens, L. B., and Watson, J. P., 1979, Rates of weathering and soil formation on granite in Rhodesia: Soil Science Society of America Proceedings, p. 43, p. 160–1966.

Parham, W.E., 1969, Formation of halloysite from feldspar: Low temperature, artificial weathering versus natural weathering: Clays and Clay Minerals, v. 17, p. 13–22.

Parker, A., 1970, An index of weathering for silicate rocks: Geological Magazine, v. 107, p. 501–504.

Pavich, M., 1986, Processes and rates of sapprolite production and erosion on a foliated granite rock of the Virginia Piedmont: in Colman, S., and Dethies, D., eds., Rates of chemical weathering of rocks and minerals, Orlando, Florida, Academic Press, p. 551–590.

Pavich, M., 1989, Regolith residence time and the concept of surface age of the Piedmont "peneplain:" Geomorphology, v. 2, p. 181–196.

Pavich, M., Brown, L., Valette-Silver, J., Klein, J., and Middleton, R., 1985, [10]Be analysis of a Quaternary weathering profile in the Virginia Piedmont: Geology, v. 13, p. 39–41.

Pavich, M., Leo, G., Obermeier, S., and Estabrook, J., 1989, Investigations of the characteristics, origin, and residence time of the upland residual mantle of the Piedmont of Fairfax, County, VA, U. S. Geological Survey Professional Paper 1352.

Peel, R. F., 1974, Insolation weathering: Some measurements of diurnal temperature changes in exposed rocks in the Tibesti region, central Sahara: Zeitschrift für Geomorphologie Supplementband 21, p. 19–28.

Petrovic, R., Berner, R. A., and Goldhaber, M. B., 1976, Rate control in dissolution of alkali feldspars—Pt. 1, Study of residue feldspar grains by X-ray photoelectron spectroscopy: Geochimica et Cosmochimica Acta, v. 40, p. 537–548.

Plummer, N., Wigley, T., and Parkhurst, D., 1978, The kinetics sof calcite dissolution in CO_2 water systems at 5° to 60° and 0.0 to 1 atm. CO_2: American Journal of Science, v. 278, p. 179–216.

Poldervaart, A., ed., 1955, Chemistry of the Earth's crust: in Crust of Earth, Geological Society of America Special Paper, v. 62, p. 119–144.

Raeside, J. D., 1959, Stability of index minerals in soils with particular reference to quartz, zircon, and garnet: Journal of Sedimentary Petrology, v. 29, p. 493–502.

Rahn, P. H., 1971, The weathering of tombstones and its relationship to the topography of New England: Journal of Geological Education, v. 19, p. 112–118.

Reeves, C. C., Jr., 1970, Origin, classification, and geologic history of caliche on the southern High Plains, Texas and eastern New Mexico: Journal of Geology, v. 78, p. 352–362.

Reheis, M. C., 1987, Climatic supplication of alternating clay and carbonate formation in semiarid soils of south-central Montana: Quaternary Research v. 27, p. 270–282.

Reiche, P., 1943, Graphical representation of chemical weathering: Journal of Sedimentary Petrology, v. 13, p. 58–68.

Reiche, P., 1950, Survey of weathering processes and products: second edition: New Mexico Univ. Press, Albuquerque, N.M., 95p.

Reid, J.M., MacLeod, D.A., and Cresser, M., 1981, The assessment of chemical weathering rates within an upland catchment in North-East Scotland: Earth Surf. Proceedings and Landforms, v. 6, p.447–57.

Reinhardt, J., and Sigleo, W., eds., 1988, Paleosols and weathering through geologic time: Geological Society of America Special Paper, 216 p.

Retallack, G. J., 1988, Field recognition of paleosols: in Reinhardt, J., and Sigleo, W., eds., Paleosols and weathering through geologic time, Geological Society of America Special Paper, v. 216, p. 1–20.

Retallack, G. J., 1990, Soils of the past: An introduction to paleopedology: Unwin Hyman, Boston, 520 p.

Reynolds, R. C., and Hower, J., 1970, The nature of interlaying in mixed-layer illite-montmorillonites: Clays and Clay Minerals, v. 18, p. 25–36.

Reynolds, R. C., Jr., and Johnson, N. M., 1972, Chemical weathering in the temperate glacial environment of the northern Cascade Mountains: Geochimica et Cosmochimica Acta, v. 36, p. 537–554.

Rice, A., 1976, Insolation warmed over: Geology, p. 61–62.

Roth, E. S., 1965, Temperature and water content as factors in desert weathering: Journal of Geology, v. 73, p. 454–468.

Roth, E. S., 1965, Temperature and water content as factors in desert weathering: Journal of Geology, v. 73, no. 3, p. 454–468.

Rowell, D. L., 1981, Oxidation and reduction: in D.J. Greenland and M.H.B. Hayes, eds., The chemistry of soil processes: John Wiley and Sons, N.Y.

Ruhe, R. V., 1956, Geomorphic surfaces and the nature of soils: Soil Science, v. 82, p. 441–455.

Ruhe, R. V., 1965, Quaternary paleopedology: in Wright, H., and Frey, D., eds., The Quaternary of the United States, Princeton University Press, p. 735–764.

Ruhe, R. V., 1969, Quaternary landscapes in Iowa: Iowa State University Press, Ames, Iowa, 255 p.

Russell, R., 1943, Freeze-thaw frequencies in the United States: American Geophysical Union Transactions, v. 24, p. 125–133.

Ruxton, B. P., 1968, Measures of the degree of chemical weathering of rocks: Journal of Geology, v. 76, no. 5, p. 518–527.

Ruxton, B. P., and Berry, L., 1957, Weathering of granite and associated erosional features in Hong Kong: Geological Society of America Bulletin, v. 68, p. 1263–1292.

Ruxton, B. P., and Berry, L., 1961, weathering profiles and geomorphic position on granite in two tropical regions: Reviews Geomorph. Dynamique, v. 12, p. 16–31.

Sawhney, B. L., 1977, Interstratification in layer silicates: in J. B. Dixon and S. B. Weed, eds., Minerals in soil environments, Soil Science Society of America, p. 405–434.

Schalascha, E. B., Appelt, H., and Schatz, A., 1967, Chelation as a weathering mechanism —I. Effect of complexing agents on the solubilization of iron from minerals and granodiorite: Geochimica et Cosmochimica Acta, v. 31, p. 587–596.

Schaller, W. T. and Vlisidis, A. C., 1959, Spontaneous oxidation of a sample of powdered siderite: American Mineralogist, v. 44, nos. 3–4, p. 433–435.

Schatz A., Cheronis N., Schatz, V., and Trelawney, G., 1954, Chelation (sequestration) as a biological weathering factor in pedogenesis: Pennsylvania Academy of Science Proceedings, v. 28, p. 44–57.

Schatz, A., 1963, Chelation in nutrition, soil microörganisms and soil chelation. The pedogenic action of lichens and lichen acids: Journal Agriculture and Food Chemistry, v. 11, p. 112–118.

Schlesinger, W. H., The formation of caliche in soils of the Mohjave Desert, California: Geochimica et Cosmochimica Acta, v. 49, p. 57–66.

Schwartz, S. E., 1989, Acid deposition: Unraveling a regional phenomenon: Science, v. 243, p. 753–763.

Schwertmann, U., and Taylor, R. M., 1977, Iron oxides: in J.B. Dixon and S.B. Weed, eds., Minerals in soil environments, Soil Science Society of America, Madison, Wisconsin.

Seed, H., Woodward, R., and Lundgren, R., 1964, Clay mineralogical aspects of the Atterberg limits: American Society of Civil Engineers Proceedings, v. 90, no. SM4, p. 107–131.

Segerstrom, K. and Henriquez, H., 1964, Cavities, or "tefoni", in rock faces of the Atacama desert, Chile: U.S. Geological Survey, Prof. Paper 501-C, p. 121–125.

Selby, M.J., 1982, Hillslope materials and processes: Oxford University Press, N.Y.

Senstius, M., 1958, Climax forms of chemical rock-weathering: American Scientist, v. 46, p. 355–367.

Shoji, S., Yamada, I., and Kurashima, K., 1981, Mobilities and related factors of chemical elements in the topsoils of andosols in Tohuku, Japan: 2. Chemical fractions and factors influencing the mobilities of major chemical elements: Soil Science, v. 132, p. 330–346.

Shumskii, P. A., 1964, Principles of structural glaciology; the petrography of fresh-water ice as a method of glaciological investigation: Dover Publications, N.Y., 497 p.

Siegel, D. I., and Pfannkuch, H. O., 1984, Silicate mineral dissolution at pH4 and near standard temperature and pressure: Geochimica et Cosmochimica Acta, v. 48, p. 197–201.

Siever, R., and Woodward, N., 1979, Dissolution kinetics and the weathering of mafic minerals: Geochimica et Cosmochimica Acta, v. 43, p. 717–724.

Simmons, G., and Richter, D., 1976, Microcracks in rocks: in Strens, R.G.J., ed., The Physics and chemistry of minerals and rocks, John Wiley and Sons, London, p. 105–137.

Simmons, G., Todd, T., and Baldridge, W. S., 1975, Toward a quantitative relationship between elastic properties and cracks in low porosity rocks: American Journal of Science, v. 275, p. 318–345.

Simonson, R. W. and Hutton, C. E., 1954, Distribution curves for loess (Iowa-Mo): American Journal of Science, v. 252, p. 99–105.

Simonson, R. W., 1954, Identification and interpretation of buried soils (Mississippi Valley): American Journal of Science, v. 252, p. 705–732.

Simpson, D. 1964, Exfoliation in the upper Pocahontas Sandstone, Mercer County, West Virginia: American Journal of Science, v. 262, p. 545–551.

Skinner, B. J., 1966, Thermal expansion: in Clark, S. P., Jr., ed., Handbook of physical constants: Geological Society of America Memoir 97, p. 75–96.

Smith, B. J., 1977, Rock temperature measurements from the northwest Sahara and their implications for rock weathering: Catena, v. 4, p. 41–63.

Smith, L. L., 1941, Weather pits in granite of the southern piedmont: Journal of Geomorphology, v. 4, p. 117–127.

Soil Survey Staff, 1975, Soil taxonomy: A basic system of soil classification for making and interpreting soil surveys: U. S. Department of Agriculture Handbook 436, 754 p.

Sperling, C. H., B., and Cooke, R. U., 1985, Laboratory simulation of rock weathering by salt crystallization and hydration processes in hot, arid environments: Earth Surface Processes and Landforms, v. 10, p. 541–555.

Stevens, G. R., 1974, Rugged landscape: The geology of central New Zealand: A.H. and A. W. Reed, Wellington, New Zealand, 286 p.

Stevens, R., and Carron, M., 1948, Simple field test for distinguishing minerals by abrasion pH: American Mineralogist, v. 33, p.31–49.

Strakhov, N. M., 1967, Principles of lithogenesis (translated by J. P. Fitzsimmons): Oliver and Boyd, Ltd., Edinburgh, U.K., 245 p.

Tabor, S., 1930, The mechanics of frost heaving: Journal of Geology, v. 38, p. 303–317.

Tabor, S., 1943, Perennially frozen ground in Alaska: Its origin and history: Geological Society of America Bulletin, v. 54, p. 1433–1548.

Tan, K. H., 1980, The release of silicon, aluminum, and potassium during decomposition of soil minerals by humic acid: Soil Science, v. 129, p. 5–11.

Tarr, W. A., 1915, A study of some heating tests, and of the light they throw on the cause of the disaggregation of granite: Economic Geology, v. 10, p. 348–367.

Tedrow, J. C. F., 1977, Soils of the polar landscapes: Rutgers University Press, New Brunswick, N.J., 638p.

Tharp, T. M., 1987, Conditions for crack propagation by frost wedging Geological Society of America Bulletin, v. 99, p. 94–102.

Thomas, M. F., 1965, Some aspects of the geomorphology of tors and domes in Nigeria: Zeitschrift für Geomorphologie, v. 9, p. 63–81.

Thomas, M. F., 1974, Tropical geomorphology: A study of weathering and landform development in warm climates: Macmillan Publishing, London, 332 p.

Thorn, C. E., 1979, Bedrock freeze-thaw weathering regime in an alpine environment, Colorado Front Range: Earth Surface Proceedings, v. 4, p. 211–228.

Todd, T. W., 1968, Paleoclimatology and the relative stability of feldspar minerals under atmospheric conditions: Journal of Sedimentary Petrology, v. 38, p. 832–844.

Trudgill, S. T., 1976, Rock weathering and climate: quantitative and experimental aspects: in Derbyshire, E., ed., Geomorphology and climate, John Wiley and Sons, N.Y., p. 59–99.

Twidale, C. R., 1968, Weathering: in Fairbridge, R. W., ed., Encyclopedia of Geomorphology, Reinholdt Book Co., N.Y., p. 1228–1232.

Twidale, C. R., 1973, On the origin of sheet jointing: Rock Mechanics, v. 5, p. 163–187.

Twidale, C. R., 1982, Granite landforms: Elsevier, Amsterdam, Netherlands, 372 p.

Ugolini, F. C., 1986, Processes and rates of weathering in cold and polar desert environments: in Colman, S. M., and Dethier, D. P., eds., Rates of chemical weathering of rocks and minerals, Academic Press, Orlando, Fla., p. 193–235.

Velbel, M. A., 1985, Geochemical mass balances and weathering rates in forested watersheds of the southern Blue Ridge: American Journal of Science, v. 285, p. 9094–930.

Violante, P., and Wilsen, M. J., 1983, Mineralogy of some Italian andesols with specific reference to the origin of the clay fraction: Geoderma, v. 29, p. 157–174.

Walder, J., and Hallet, B., 1985, A theoretical model of the fracture of rock during freezing: Geological Society of America Bulletin, v. 96, p. 336–346.

Weinert, H., 1965, Climate factors affecting the weathering of igneous rocks: Agriculture and Meterology, v. 2, p. 27–42.

Whalley, W.B., Doutlas, G.R., and McGreevey, J. P., 1982, Crack propagation and associated weathering in igneous rocks: Zeitschrift für Geomorphologie, v. 26, p. 33–54.

White, S. E., 1976, Is frost action really only hydration shattering? A review: Arctic and Alpine Research, v. 8, p. 1–6.

Whitney, P. R., and McClelland, J. M., 1973, Origin of coronas in metagabros of the Adirondack Mts., New York: Contributions to Mineralogy and Petrology, v. 39, p. 81–98.

Winkler, E. M., 1965, Weathering rates as exemplified by Cleopatra's Needle in New York City: Journal of Geological Education, v. 13, p. 50–52.

Winkler, E. M., 1966, Corrosion rates of carbonate rocks for construction: Engineering Geology, v. 3, nos. 1–2, p. 52–58.

Winkler, E. M., 1966, Important agents of weathering for building and monumental stone: Engineering Geology, v. 1, no. 5, p. 381–400.

Winkler, E. M., 1968, Frost damage to stone and concrete: Geological considerations: Engineering Geology, v. 2, p. 315–323.

Winkler, E. M., 1975, Stone; properties, durability in man's environment: Applied Mineralogy, Springer-Verlag, N.Y., 230 p.

Winkler, E. M., 1980, Role of salts in development of granitic tafoni, South Australia: Discussion: Journal of Geology, v. 88, p. 119–120.

Winkler, E. M., and Wilhelm, E. J., 1970, Salt bursts by hydration pressures in architectural stone in urban atmosphere: Geological Society of America Bulletin, v. 81, p. 567–572.

Wollast, R., 1967, Kinetics of the alteration of K-feldspar in buffered solutions at low temperature: Geochimica et Cosmochimica Acta, v. 31, p. 635–648.

Wright J., and Schnitzer, M., 1963, Metallo-organic interactions associated with podsolisation: Soil Science Society of American Proceedings, v. 27, p. 171–176.

Young, A. R. M., 1987, Salt as an agent in the development of cavernous weathering: Geology, v. 15, p. 962–966.

Zeitler, P.K., Johnson, N.M., Naeser, C.W., and Tahirkheli, R.A.K., 1982, Fission-track evidence for Quaternary upift of the Nanga Parbat region, Pakistan: Nature, v. 298, p.255–257.

Mass Wasting

Earthflow, Saddle Mts., Washington. (Photo by D. A. Rahm)

INTRODUCTION

The force of gravity inevitably, continuously, sometimes catastrophically, wears down mountains and valley sides. Gravity drags everything constantly downward, causing downslope movement of material that varies from slow, subtle, continuous creeping to rapid, devastating landslides. All downslope movements of soil and rock under the direct influence of gravity are collectively known as mass wasting. Many varieties of mass wasting occur, but all involve the breaking away of rock or unconsolidated material from its previous position on a slope. Later in the chapter, we will examine each mass wasting process, but first let us consider the general causative factors of downslope movements.

FACTORS AFFFECTING DOWNSLOPE MOVEMENTS

Mass wasting is the downslope movement of rock debris in response to gravity. Downslope movement occurs when gravitational driving force exceeds the frictional resistance of the material resting on the slope. The magnitude of the driving force in mass wasting processes depends upon two things: the mass of the material involved in movement and the slope angle. The driving force, which tends to make the material move downslope, is opposed by friction between the material at its base or by the friction between grains within a material. When the driving force exceed the frictional resistance, a portion of the material breaks off from the main mass and begins to move. Whether or not movement takes place is determined by the steepness of the slope, by friction, and by the physical properties of the material, that is, whether it behaves as an elastic solid, a plastic solid, or a fluid.

Driving Forces

Effects of Slope

Although other factors in mass wasting processes may play significant roles in triggering downslope movement, the ultimate driving force is gravity. The force of gravity acts vertically downward toward the Earth's center, but it may be resolved into two components, one parallel to the sloping surface and the other perpendicular to it (Figure 4–1). As the slope angle increases, the gravitational component of force parallel to the surface increases, varying from zero on a horizontal surface to a maximum on a vertical slope.

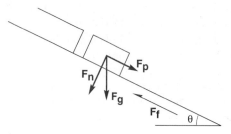

A.

FIGURE 4–1
Gravitational driving forces and resisting frictional forces on a mass resting on an inclined plane.

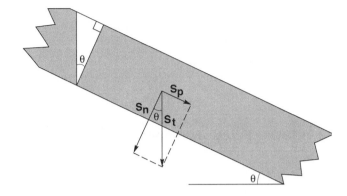

FIGURE 4–3
Gravitational stress components acting on a mass resting on an inclined plane.

FIGURE 4–2
Gravitational forces acting on a block of rock on a slope.

In Figures 4–1 and 4–2, the gravitational force vector F_g may be resolved into components F_p, the component of gravity acting parallel to the slope, and F_n, the component normal (90°) to the slope. The magnitude of the component parallel to the slope (Floodplain) is related to the total gravitational force F_g as follows:

$$F_p = F_g \sin \theta$$

where F_g = the weight in pounds or kilograms
          θ = slope angle

As the slope angle increases, $\sin \theta$ increases, so F_p will vary from zero on a horizontal surface to a maximum (F_g) on a vertical slope.

The normal component F_n, perpendicular to the slope, is

$$F_n = F_g \cos \theta$$

The importance of F_n is that as it increases, the frictional force between an object and the surface it rests upon also increases. The value of $\cos \theta$ decreases as the angle θ increases, so F_n is at a maximum on a horizontal surface and decreases progressively to zero on a vertical slope.

The gravitation forces acting with a mass of material resting on a slope may be treated in a similar fashion as a block resting on an inclined plane. The force acting on a plane at some distance x below the surface acts on a surface area, so a more useful parameter is **stress** (force per unit area), usually expressed as lb/in^2 (psi), lb/ft^2 (psf), dynes/cm^2, or newtons/m^2. A convenient way of expressing stress on a mass of material on an inclined slope is to multiply its density d (lb/ft^3 or kg/m^3) times the vertical distance x from the surface to the desired point of calculation.

In Figure 4–3, the stress component parallel to the inclined plane (S_p) is

$$(d)(x) = S_t \text{ (total stress)}$$
$$S_p = St_t \sin \gamma = (d)(x) \sin \theta$$

The stress component normal to the inclined plane S_n is

$$S_n = S_t \cos \theta = (d)(x) \cos \theta$$

The downslope component S_p of the total stress S_t is the shear stress, and the perpendicular component S_n is the **normal stress.**

The component of gravity perpendicular to the slope also varies with slope angle, only just the opposite of the previous case. As the slope angle increases, the component of the gravitational force perpendicular to the slope decreases, varying from zero on a vertical slope to a maximum on a horizontal surface. The importance of the perpendicular slope component is that as it increases, the frictional force between an object and the surface beneath it also increases.

Several kinds of stresses act on material resting on hillslopes. Tensile stresses tend to pull a rock or soil mass apart and change the shape of the material. The volume in a soil may also change if fractures and pores are opened. Shear stresses deform a body by sliding one part over the other, thereby changing the shape of the body but not its volume. Compressive stresses reduce the void space of a rock mass and lower its volume. Any deformation or any change in shape or volume caused by application of stress is known as **strain.**

RESISTING FORCES ON SLOPES

Physical Characteristics of Material

Resisting forces act in opposition to the gravitational forces just described and thus tend to oppose deformation or motion. The resistance of rock material to applied stress is governed largely by the physical properties of the material. If a body returns to its original size and shape after it has been stressed, it is **elastic;** if it retains the size and shape imparted by stress, it is **plastic;** if it behaves as a fluid, it is **viscous.** Many natural materials are not purely elastic, plastic, or viscous but instead exhibit various combinations of properties.

Elasticity

Elastic behavior in material follows several basic rules:

1. The same amount of stress always produces the same strain.
2. Sustaining a given stress produces constant strain.
3. Removing a stress results in total recovery of strain.

In his 1679 experiments, English physicist Robert Hooke found that strain is proportional to the stress applied, up to a point called the *elastic limit* of the material. Beyond this limit, deformation is no longer recoverable and the material becomes plastic. Plastic yielding in most rocks is very limited, and the rocks fail under stress by fracturing just beyond the elastic limit. However, wet, unconsolidated material can behave as a plastic.

The relationship between stress and strain may be expressed by Hooke's law:

$$\frac{\text{stress}}{\text{strain}} = \text{a constant}$$

Such a constant remains the same for a given material under constant conditions and is known as a *modulus*. Four elastic moduli are known:

$$\text{Young's modulus} = \frac{\text{longitudinal stress}}{\text{change in length/original length}}$$

$$\begin{array}{c}\text{Rigidity modulus} \\ \text{(or shear modulus)}\end{array} = \frac{\text{shear stress}}{\text{angular deformation}}$$

$$\begin{array}{c}\text{Bulk modulus} \\ \text{(or} \\ \text{compressibility} \\ \text{modulus)}\end{array} = \frac{\text{stress}}{\text{change in volume}}$$

$$\text{Poisson's ratio} = \frac{\text{change in width/original width}}{\text{change in length/original length}}$$

The geologic significance of these moduli is that rocks under stress may behave differently according to the type of stress applied and the physical nature of the material.

Plasticity

Plasticity occurs when stress is applied to a material until the bonds holding it together are broken. Once the threshold yield stress is exceeded, deformation occurs at a uniform rate under constant stress. Such deformation is most common in unconsolidated materials (especially clay) or in very weak rocks such as gypsum or salt.

Slow, continuous plastic deformation is known as **creep**. It is most common in rock materials that undergo changes in water content, temperature, or loads from overlying material. Creep is an important process on slopes and may be significant in clay-rich sediment and in plastic rocks such as salt or evaporite deposits.

Viscosity

Viscosity is the internal frictional resistance of a fluid to flowage. Liquids that flow at rates proportional to applied stresses are known as *Newtonian fluids.* The most geologically important, naturally occurring viscous masses, such as mudflows and debris flows consisting of solid grains suspended in liquid, are non-Newtonian because their viscosity changes as applied stress changes. Mudflows and debris flows often begin as solid debris slides that become saturated downslope and undergo viscous flow, changing progressively from plastic to viscous behavior.

Changes in Physical Characteristics

Water content plays an important role in the physical properties of unconsolidated sediments, especially clay, which changes its response to stress as water content changes. In a dry condition, clay becomes brick-like and behaves as an elastic solid that will fail by brittle fracture. As water content increases past the plastic limit, the clay will deform plastically; and as the water content increases past the liquid limit, the shear strength of the clay goes to zero, and the clay becomes a viscous fluid.

Strength of Material

The maximum resistance of material to stress is known as its **strength,** defined as the amount of stress required to cause failure (rupture) of the material. The most common types of resisting strengths are shear strength, tensile strength, and compressive strength.

Strengths of materials vary considerably depending on the physical properties of the material, and the resisting strength of a material may vary for different kinds of stress. Noncohesive, unconsolidated material behaves differently than cohesive, unconsolidated material does, and both differ from solid rock.

Shear Strength   The factors important to **shear strength** (resistance to movement) and shear failure were recognized by French physicist Charles Coulomb in 1773. He found that failure occurs along a surface if the shear stress acting in that plane is large enough to overcome the cohesive strength of the material and its resistance to movement. The cohesive strength of a material is its strength when the stress normal to the shear plane is zero. Shear strength is equal to the stress normal to the shear plane multiplied by the coefficient of internal friction of the material. Coulomb's law expresses the relationship between failure, internal friction, and cohesion as follows:

$$\tau = c + S_n \tan \phi$$

where
- τ = shearing stress
- c = cohesion
- S_n = normal stress
- ϕ = *angle of internal friction* or shearing resistance

In 1882, Mohr modified Coulomb's concept and proposed that when a rock is subjected to compressive stress, shear fracturing occurs parallel to two planes for which shearing stress is at a maximum and normal pressure is at a minimum. Shearing failure takes place when internal friction is overcome at a point where the optimum ratio occurs between shearing components, internal friction, and cohesion. The shearing angle is constant for a specific material at a given temperature and pressure and is determined by the ratio of compressional to tensile strength.

Friction   The primary force that resists the effect of gravity in moving material downslope is *friction*. Friction is produced when one body slides past another and is defined as the mechanical resistance to the relative motion of adjacent masses of material. Two types of friction, sliding and internal, are important in mass wasting processes:

Sliding friction between contiguous masses separated by a well-defined plane where friction varies with roughness of the surface, area of contact points, moisture at the contact, and any upward-pushing fluid pressure. Friction produced by two bodies sliding past one another depends primarily upon the roughness of the sliding surfaces, the area of the surfaces in contact, and the gravitational force pressing the bodies together. Smooth surfaces can slide past one another more easily than rough surfaces, and the greater the area in contact, the more frictional resistance created. Slope also plays a role here because as the slope angle increases, the component of gravitational force perpendicular to the slope decreases, and this results in less friction between adjacent masses.

Several different kinds of sliding friction have been recognized, including *static* friction, the resistance to initial motion, and *dynamic* friction, the resistance to sliding once motion has begun. Dynamic friction is usually slightly less than static friction.

Friction between two sliding masses depends primarily upon the roughness of the sliding surfaces and the normal stress pressing the bodies together. Smooth surfaces can slide past one another more easily than rough surfaces, and frictional resistance to sliding increases with the pressure of two bodies against one another.

The force of friction F_f between two bodies is

$$F_f = F_n \times C_f$$

where F_n = force normal to the surface
   C_f = coefficient of sliding friction (also known as the **angle of friction,** tan ϕ, which depends largely on the roughness of the surface.

Figure 4–4 illustrates the relationship between these factors. If force F_p parallel to a potential sliding surface is applied to a block whose weight is F_n, motion is resisted by the force of friction F_f. The resultant force on the block is F_s, which forms an angle ϕ between the resultant force and the normal force. When F_p increases so that the block is just on the verge of

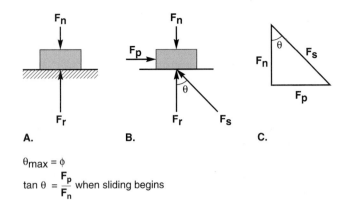

$\theta_{max} = \phi$

$\tan \theta = \dfrac{F_p}{F_n}$ when sliding begins

FIGURE 4–4
Relationship of gravitational and frictional forces.

motion, ϕ will be at its maximum, and the angle of friction (coefficient of friction) is tan $\phi = F_p/F_n$.

Internal friction is the friction between individual grains within a mass of material. This factor relates primarily to unconsolidated material where friction between grains serves to retard movement of grains relative to one another. Thus, when the gravitational driving force acts on a material, internal friction between grains impedes movement within the material.

Unconsolidated material slides until it reaches a stable slope angle called the **angle of repose**. The tangent of the angle of repose approximates the **angle of internal friction,** but it is usually slightly greater (Van Burkalow, 1945). The angle of internal friction of a material varies with the size, shape, and surface roughness of the grains, the density of packing, the moisture content, and, if the material is saturated, the pore pressures of fluid between grains.

Effect of water   Water content is important in determining internal friction. Children playing in a sandbox quickly realize that, whereas dry sand promptly slides to low angles, slightly wet sand stands in nearly vertical walls. Moist sand has surprisingly more cohesion than dry sand, as long as the pore spaces are not filled with water. The **cohesion** of the sand—shear strength not related to interparticle friction—results from the surface tension of films of water, which bind the grains together. Cohesion explains why sand castles can be built of dampened sand, whereas this is impossible if the sand is dry. However, if all the pores between grains become filled with water, surface tension is destroyed, and the sand may collapse with little or no cohesion. At best, it can be piled into gently sloping, subaqueous mounds and will support only limited weight placed upon it; but if the pore water is drained, the sand gains internal strength as the grain-to-grain pressures are augmented by the downward seepage pressures of the water column.

The reverse procedure—flushing water up through the sand—may rob the deposit of all its strength. In this case, the upward pressure of pore water reacts against the downward pressure of the grains and pushes them apart. A mass on the surface of the sand would disappear as the sand becomes "quick." Quicksand, in fact, is associated with springs in which waters ascending through the saturated deposit have just this effect. Cohesion, dependent upon grain-to-grain pressures, can be reduced by an increase in pore-water pressure. Saturation of permeable channel materials in cutbanks during times of flooding can increase pore-water pressures if drawdown of the water after the flood exceeds the draining of the bank. The pore water is stranded in the bank materials and acts as a hydrostatic column in which pore-water pressure increases systematically downward. Seepage pressures also act laterally toward the face of the cut. The consequent reduction of cohesion, particularly in the lower parts of the bank, explains many channel-side failures following floods.

Water content also plays an important role in the physical properties of clay. When dry, clay is like a brick, but as water content increases, it will deform plastically and can be molded and shaped. As the water content increases even more, the strength of the clay goes to zero, and the clay acts like a fluid.

The most significant role of water on internal friction is the effect of fluid pore pressure in reducing the gravitational component perpendicular to a slope. The gravitational component acting at points of contact between grains, produces internal friction within a material. In dry material, the component of gravitational force acting perpendicular to grain surfaces is supported entirely at the points of contact between grains, and pores are filled with air at atmospheric pressure, so that water pressure in the pores is zero. In a saturated sand, part of the gravitational force perpendicular to a surface is supported by the pore water (Figure 4–5), thus reducing the force exerted on the grains. Imagine a car with a flat tire. If you pump up the tire, the air pressure lifts the car up. If you filled the tire with water under pressure, the same thing would happen. Now imagine a mass of rock resting on a thin film of water under pressure. The water pressure carries part of the load and lessens the perpendicular gravitational force on the underlying surface. The transfer of load from grains to pore water decreases the normal force at points of grain-to-grain contact. Because pore-water pressures decrease effective perpendicular gravitational force, saturated soil on a slope has less internal frictional strength than a dry mass. Consequent-

ly, slope failures are commonly associated with heavy rains that saturate the soil.

Water not only reduces grain-to-grain friction but can also seep along bedding or joint planes, where it pushes upward on the overlying rock and decreases frictional resistance between slabs of rock and the surfaces they rest upon. In addition to its role in decreasing internal friction, water in saturated material adds weight that may trigger a landslide on a slope that was stable when dry.

Another effect of water is its absorption in expandable clays, such as montmorillonite. Such clays are known as *quick clays* because they absorb water readily, swell, and lose their internal strength.

Effective Normal Stress

The component of gravitational stress normal to a shear surface, or points of contact of grains within material, increases the resistance to shear and helps hold the material together. In dry material, **normal stress**—the component of stress acting perpendicular to grain surfaces—is supported entirely at the points of contact between grains, and pores are filled with air at atmospheric pressure, so that pore-water pressure is zero. In a fully saturated sand, part of the normal stress is transferred from the grains to the pore water, and the pore-water pressure becomes positive. The transfer of load from grains to pore water decreases the normal force at points of grain-to-grain contact. The effective normal stress S_e then becomes

$$S_e = S_n - S_p$$

where S_n = normal stress
  S_p = pore-water pressure

The Coulomb equation then becomes

$$\tau = c + S_e \tan \theta$$

Because positive pore-water pressures decrease effective normal stress, saturated soil on a slope has less internal frictional strength than a dry mass. Consequently, slope failures are commonly associated with heavy rains that saturate the soil.

Cohesion

Cohesion binds grains together by electrostatic forces between clay particles or by chemical cementation. As the cohesion of a material increases, its total shear strength increases according to the Coulomb equation. Moist silt and clay often stand in vertical bluffs without failure because of their cohesion, whereas non-cohesive dry sand does not.

Figure 4–6 shows graphical plots of measurements of the shear stress required to cause failure in material under various normal stresses. As the effective normal stress increases, the shear strength also increases, defining a straight line, the Coulomb failure line, which passes through the origin of both axes. The angle between the line and the abscissa is the angle

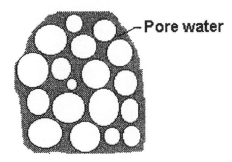

FIGURE 4–5
Effect of pore water pressure on internal friction. Pore water pressure pushes the grains apart and reduces the friction between them.

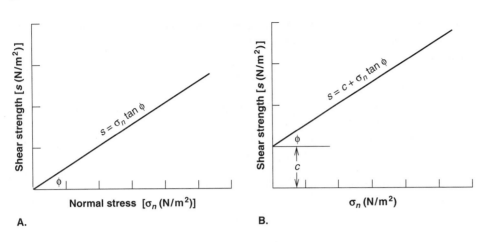

A.

B.

FIGURE 4–6
Effect of cohesion on the relationship between shear strength and normal stress. The graphs illustrate shear strength vs. effective stress for (A) cohesionless material and (B) cohesive material.

of internal friction ϕ. As shown in Figure 4–6(A), noncohesive material has zero shear strength when effective normal stress is zero. The value for cohesion (c) is the residual shear strength c' when effective normal stress is zero; it may be determined from the position of a line drawn parallel to the abscissa at the intersection of the failure line and the ordinate, as shown in Figure 4–6(B).

Some materials may exhibit a higher initial peak strength but a lowered residual strength after the application of stress. The magnitude of stress required to cause failure in such cases may not be constant for that material because earlier strain may have rearranged particles and may have affected internal shear strength.

The shear strength of cohesive material, such as clay, is inversely proportional to its plasticity index and liquid limit. Clay changes from nonplastic to plastic to viscous as water content increases. High water content can cause both cohesive and noncohesive material to behave as a viscous fluid. In 1911, Atterberg devised several tests to distinguish transitions from solid to plastic to liquid states, and these have since become known as *Atterberg limits,* composed of the following:

1. The shrinkage limit
2. The plastic limit
3. The liquid limit expressed as water content present when a material changes its response to stress

Between zero percent water content and the plastic limit, the material behaves as a solid; between the plastic limit and the liquid limit, the material behaves as a plastic. The range of water content between the plastic limit and the liquid limit is the plasticity index *PI*:

$$PI = LL - PL$$

where   *LL* = liquid limit
          PL = plastic limit

Above the liquid limit the material behaves as a liquid. Figure 4–7 shows the relationships of these limits. The type of clay minerals present in a material has a significant effect on Atterberg limits and plays an important role in determining plasticity. Nonexpansive clays, such as illite and kaolinite,

Stages of consistency

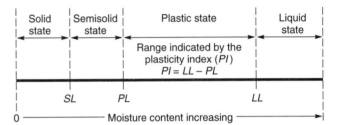

FIGURE 4–7
Atterberg limits and the states of soil consistency defined by them.

have lower plasticity indices than expansive clays like smectite, which has a wide range of plasticity and liquid limits as a result of differences in exchangeable cations. Calcium-rich montmorillonite has lower plasticity indices than sodium-rich smectite.

Plasticity increases with an increase in clay-sized particles. The ratio of the plasticity index to the percent clay may be used to define the activity of a clay. Shear strength decreases as the activity increases.

$$\text{activity} = \frac{PI}{\text{percent clay}}$$

The internal strength of clayey soils is directly related to their water content. Some sensitive clays have water contents above their liquid limits as a result of an open internal structure capable of retaining water in excess of the liquid limit. Such a structure is unstable but may persist as a solid with some shear strength. However, sudden disruption of the internal structure by earthquakes or other shocks causes the excess water to be released, and the solid material instantly becomes fluid and can cause serious slope failures.

Measurement of Shear Strength

Laboratory tests may be performed on material to determine its shear strength, but natural conditions are often difficult to replicate. Some tests commonly carried out in engineering laboratories include direct shear tests, triaxial compression tests, and uniaxial (unconfined) compression tests.

Direct shear testing involves application of shear stress to a sample until failure is obtained for various values of stress normal to the shear plane. Such data provide coordinates for construction of a shear strength envelope and measurement of the internal angle of friction. This test does not provide for measurement of pore pressure or confining pressure, both of which may affect the behavior of material. Only small samples can be tested in the laboratory, and measurements may not be accurate indicators of the true strength of a large mass in a natural setting.

Triaxial tests are performed in a cell that allows control of compressive and confining stresses. Compressive strengths for a range of confining pressures are measured and used to construct a Mohr diagram, which allows determination of the internal angle of friction and shear strength.

Several attempts have been made to estimate rock strength in the field. Most such schemes consider such variables as the following:

1. Strength

2. Incidence of discontinuities (such as joints)

3. Spacing, orientation, width, continuity, and roughness of discontinuities

4. Water content

5. Weathering

An example of one scheme (Selby, 1980) is shown in Figure 4–8. A rock mass may be categorized as very strong, strong, moderate, weak, or very weak based on numerical ratings r for several of the parameters just mentioned. The sum of weighted values then gives a total rating, which allows placing the mass in one of the strength categories.

Effect of Vegetation

Roots of trees and shrubs furnish an interlocking network that strengthens unconsolidated materials. Root mats commonly overhang undercut slopes because the mats are more resistant to erosion than the material in which they are growing. This effect is especially apparent where vegetation has been removed and plant roots have rotted out, leaving the soil much more vulnerable to slope failure. As we shall see later in this chapter, one of the ways to stabilize unstable slopes is to plant vegetation.

Parameter	1 Very Strong	2 Strong	3 Moderate	4 Weak	5 Very Weak
Intact rock strength	r: 20	r: 18	r: 14	r: 10	r: 5
Weathering	Unweathered	Slightly weathered	Moderately weathered	Highly weathered	completely weathered
	r: 10	r: 9	r: 7	r: 5	r: 3
Spacing of joints	>3 m	3–1 m	1–0.3 m	300–50 mm	<50 mm
	r: 30	r: 28	r: 21	r: 15	r: 8
Joint orientations	Very factorable. Steep dips into slope, cross joints interlock	Favorable. Moderate dips into slope	Fair. Horizontal dips or nearly vertical (hard rocks only)	Unfavorable. Moderate dips out of slope	Very unfavorable. Steep dips out of slope
	r: 20	r: 18	r: 14	r: 9	r: 5
Width of joints	<0.1 mm	0.1–1 mm	1–5 mm	5–20 mm	>20 mm
	r: 7	r: 6	r: 5	r: 4	r: 2
Continuity of joints	None continuous	Few continuous	Continuous, no infill	Continuous, thin infill	Continuous, thick infill
	r: 7	r: 6	r: 5	r: 4	r: 1
Outflow of groundwater	None	Trace	Slight <251 l/min/10 m²	Moderate 25–125 l/min/10 m²	Great >125 l/min/10 m²
	r: 6	r: 5	r: 4	r: 3	r: 1
Total rating (R)	100–91	90–71	70–51	50–26	<26

FIGURE 4–8
Strength of rock masses (After Selby, 1980).

Although vegetation generally increases slope stability, plants do add weight to slope and thus increase the driving forces. However, this effect is usually not as significant as the stabilizing effect of roots.

SLOPE STABILITY

Causes of Slope Failure

Slope failures provide effective and widespread mechanisms by which hillslopes are modified. Movement may take place by creep, failing, sliding, or flowing, or by some combination of these processes. Slope failures take place because the sum of the gravitational stress (plus any added stress or residual stress) exceeds the resisting strength of the slope material. The stability of a slope may be defined by a factor of safety F.

$$F = \frac{F_f}{F_g}$$

where F_f = sum of resisting frictional forces
             F_g = driving gravitational force

If the factor of safety F exceeds a value of one, the slope is stable; if it is less than one, the slope is unstable; and if it equals one, the slope is on the brink of instability.

A number of factors may upset the balance of gravitational and resisting forces. Terzaghi (1950) viewed this loss of equilibrium as due to external causes, internal causes, or both. External causes produce an increase in the shearing stress without any change in the shearing resistance of the slope materials. Internal causes involve a decrease in the shearing resistance of material without any change in shearing stress. External and internal causes may not always be independent of one another, but in most cases, one is clearly dominant as the cause of movement.

External causes of slope instability include all those mechanisms responsible for overcoming internal shear strength of a mass, thereby causing it to fail. An increase in the weight of slope material means that shearing stress is increased, leading to a decrease in the stability of a slope, which may ultimately give rise to a slide. This increase in weight can be brought about by natural or artificial (human) activity. For example, removal of support from the toe of a slope, by either erosion or excavation, is a frequent cause of slides, as is overloading of the top of a slope. Earthquakes or other shocks and vibrations constitute another external cause of failure. Earthquake vibrations in granular materials cause particle accelerations that not only increase the external stresses on slope material but can also cause compaction with reduction of pore spaces, which can increase pore-water pressures.

Internal causes of slope instability include processes acting within a mass, which reduce its shear strength below the external forces, resulting in slope failure. The effects of ground water often constitute the most important factor in slope failures. Saturation of a mass increases its weight, increases pore pressures, produces volume changes in clay, gen-

erates desiccation cracks by wetting and drying, and decomposes minerals. For example, slopes submerged during floods or high lake levels are comparatively stable because the water pressure acting on the surface of the slope helps to support it. If, however, the water level falls rapidly (as in drawdown of a lake or dropping of water levels from a flood), the stabilizing influence of the water against the bank disappears. If the slopes consist of cohesive soils, the water table will not be lowered at the same rate as the body of water, and the slope will be temporarily overloaded with excess pore water, which may lead to slope failure. Thus, rapid drawdown may be a critical factor in slope stability as a result of the increase in pore-water pressures, which reduces the effective shear strength. The following list summarizes the processes that increase or decrease shear stress:

Processes Increasing Shear Stress

Loading of slopes
    Saturation with water
    Addition of fill material or buildings
Earthquakes
Removal of lateral support
    Erosion from rivers, waves, and weathering
    Excavation associated with human activity
Removal of support from beneath
    Collapse of limestone caves and mines

Processes Decreasing Shear Strength

Increased pore pressure
    Saturation with water
    Rapid drawdown of water level
Fissuring
Solution of cementing material

The mechanisms of the various mass movements and the behavior of the participating materials offer a logical basis for classification of slope failures. Many types of mass movements are gradational from one to another—earth materials may behave as fluids, elastic solids, or plastic solids, depending on environmental conditions.

The difference between flow and slip is an important distinction. *Flow* is continuous, internal deformation without the development of a slip surface, whereas *slip* is any movement that breaks away along one or more definite surfaces of shearing (Sharpe, 1938). The two types of movement are transitional into one another, but, in most natural examples, one of the types often appears dominant. Small-scale shearing, such as might take place between soil particles, is regarded as flow. Thus, flow, in this context, does not always involve viscous movement —material moving en masse as a plastic solid will also flow. Mass movements of elastic solids occur through breakaway of material on a slope, after which it may slide, roll, bounce, or fall.

Some solids that are brittle at the surface may flow plastically under pressure. Ice is an excellent example of such a substance; crevasses at the surface of a glacier show the fracture of a brittle solid carried irregularly along by plastic underflow. Certain poorly consolidated materials on slopes behave similarly.

The strengths of surface materials include a broad spectrum, ranging from strong, well-indurated rocks on one hand, to weak muddy slurries or loose, dry sand on the other. The strengths of materials behaving as elastic or plastic solids can be measured in terms of the stress necessary to cause rupture or flow. However, strengths differ according to the kind of stress applied to a given material. **Compressive strength** is a measure of resistance to compression, **tensile strength** measures resistance to tension, and **shear strength** establishes the yield point under a shearing stress. Tensile strengths are lowest of all, and compressive strengths are greatest for most materials. Thus, the tendency of materials to pull away or shear away from a slope depends on their tensile strengths and shear strengths.

Fluids, by definition, have no shear strength at all and flow under their own weight. Therefore, even small amounts of fluid material will flow downslope. On the other hand, droplets may be held by surface tension, and thin films of water may actually impart considerable strength to an aggregate. As the amount of interstitial water diminishes, the remaining water molecules exert electrostatic attraction and operate as a powerful "interstitial glue." This explains why limp, fluid mud sets as it dries and becomes very hard with thorough drying.

Because the fluidization of materials at and near the surface depends upon saturation with water, another important property of material is its infiltration capacity, the limiting rate at which water can enter the ground. It can be measured, like rainfall, in units of depth per unit of time (for example, in/hr). Any downward flow of water into openings must be accompanied by the displacement of subsurface water or air. As Horton (1933) pointed out, the infiltration capacity of surficial material is controlled by the following:

1. Soil texture
2. Soil structure
3. Vegetal cover
4. Biologic structures in the soil, including plant roots and root perforations, earthworm, insect, and rodent perforations, humus, and vegetal debris
5. Moisture content of the soil
6. Condition of the soil surface (for example, whether newly cultivated, baked, or sun-cracked)
7. Temperature (temperatures below freezing effectively seal pore openings with ice)

TYPES OF MASS WASTING

The mechanisms of movement and the behavior of participating material provide a logical basis for classification of mass movements. Thus, the first part of the name of a type of mass movement (Figure 4–9) describes the kind of material, and the second part describes the mechanism of movement.

Slow, Continuous Movements

Creep

Creep is "the slow downslope movement of superficial soil or rock debris, usually imperceptible except to observations of long duration" (Sharpe, 1938). It includes soil creep, rock creep, and talus creep, and its special forms include the migration of ice-charged breccia as rock glaciers and the flowage of wet, plastic muds as **solifluction** under the influence of freezing and thawing of saturated ground.

Although creep may appear to be continuous, it is actually the sum of numerous minute, discrete movements of slope material (**colluvium**) under the influence of gravity. Many of these small movements are facilitated by heaving of the ground—expansion and contraction caused by freezing and thawing, wetting and drying, or other volumetric changes. Expansion displaces soil particles perpendicular to the ground surface, but during subsequent contraction, the particles settle vertically downward under gravity. Although return of particles upon contraction of the surface is theoretically vertically downward, the direction on natural slopes is probably somewhere between vertical and normal to the slope. Repeated cycles of expansion and contraction add a lateral component to movement of individual particles resting on inclined surfaces, which results in net downslope movement (Figure 4–10).

Soil creep is reflected in the tilting of surface objects such as fence posts, tombstones, retaining walls, and other human-made structures (Figure 4–11). The curvature of trees that attempt to straighten up at their growing tips as their trunks are tilted downslope (Figure 4–12) has long been cited as evidence of creep. This view was challenged by Phipps (1974) on the basis of random tilting of certain types of trees, but straight-growing conifers with their bases consistently bent downslope appear to be valid indications of creep.

The recognition of "float"—boulders and other rock fragments that have crept downhill from their source outcrops (Figure 4–2)—is a sign of creep that aids in the discovery of the upslope limits of a rock type. Weak strata (including strata weakened by weathering) are frequently bent downslope at the base of the creeping mantle of material (Figure 4–11).

The downhill tilting of objects, such as telephone poles and fence posts, and the downhill bending of strata suggest that soil creep is most rapid and most effective at the surface and dies out with depth in the ground. Most of the processes that contribute to soil creep occur at the surface. Culling (1965) likens the movements of individual particles in the soil to a restricted Brownian movement upon which is superimposed a pervasive downhill component of gravitational stress that determines a net downhill transportation. Rapid creep occurs on steep, convex segments of a slope where gravitational stress is large and the surface material is under tension.

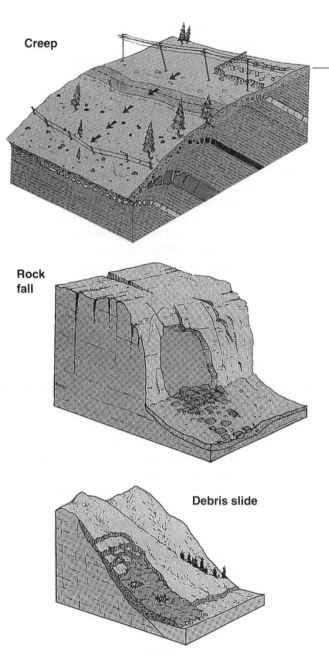

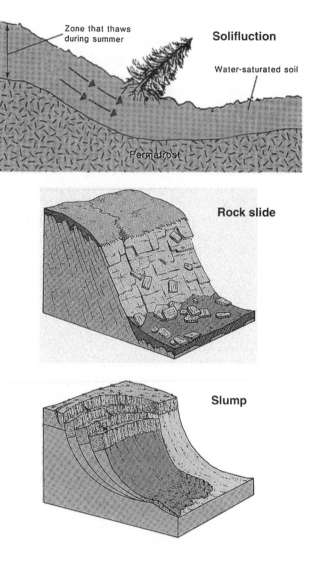

FIGURE 4–9
Types of mass movement. (Modified from Hamblin and Christianson, 1995)

Creep is slowed by a decrease in slope, especially at concavities where the surface material comes under compression.

Mechanisms of soil creep include differential expansion-contraction, displacement of particles by organisms, downhill release of particles by weathering, and heaving by freezing of the ground. For most soil-covered slopes, enough precipitation occurs to occasionally liquefy parts of the soil and cause flowage or small-scale solifluction. Volume changes within the surface material may cause small displacements of particles with recovery in the downhill direction. Also, according to Sharpe (1938), expansion is differential, being most successful in the downhill direction and least successful uphill against gravity. Contraction, on the other hand, is hampered in the uphill direction and facilitated in the downhill direction. Thus, the changes in volume with the wetting and drying of clayey soils may cause a slow downhill creep.

Sharpe (1938) cites the experiments of Mosely in 1858 and Davison in 1888, who measured the temperature-controlled creep of solids down inclines in response to diurnal expansion-contraction. On a short-term basis, the direction of creep movement may be random (Fleming and Johnson, 1975; Finlayson, 1981), but the net long-term direction is downslope.

Displacement of particles by organisms includes prying of the soil by the roots of trees and other large plants that may sway in the wind and exert the necessary leverage (Kirkby, 1967). The growth and the decay of roots dislodge grains and create perforations, which are filled by collapse. Burrowing by insects, rodents, and worms generally results in a downhill displacement of material brought to the surface and gradual collapse of the burrows. Earthworms abound in moist, vegetated soils where they may feed on organic matter. They may number from 10,000 or more per acre to nearly a million per

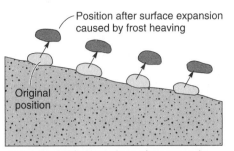

Position after surface expansion
caused by frost heaving

Original
position

As the ground freezes, the surface expands
upward at right angles to the original position.

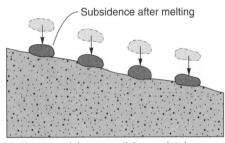

Subsidence after melting

As the ground thaws, particles are let down
vertically.

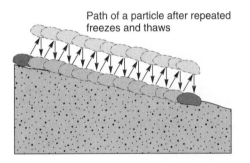

Path of a particle after repeated
freezes and thaws

FIGURE 4–10
Net downslope movement of surficial material by repeated
cycles of expansion and contraction caused by freezing and
thawing.

acre where organic matter is plentiful. A population of
150,000 earthworms in an acre of soil may raise 10 or 15 tons
of casts to the surface annually (Holmes, 1955). The perforations left from these activities gradually collapse downhill,
and most of the material brought to the surface accumulates
downslope from the excavations. Rodents and ants, also active in moist, well-vegetated soils, may be the important earth-
movers in sparsely vegetated surface material where
earthworms are lacking. Large animals cause a significant
amount of creep through the dislodgment of loose particles
and the downhill displacement of unconsolidated or plastic
materials under their feet. Many grassy hillsides are completely covered with terracettes, ample testimony to the effectiveness and persistence of grazing animals. The soil on
some slopes of the Columbia Plateau has been displaced several meters (typically 0.5 to 3 m) in the downslope direction
since the 1870s when cattle were introduced into the area
(Rahm, 1962).

The growth of ice lenses below the surface causes a sur-
faceward swelling or heave. Thawing brings about downhill
collapse. Particles thrust away from the surface by the growth
of supporting ice needles come to rest slightly downhill with
every thaw. The ice needles grow normal to the slope, but the
particles fall back along a plumb line. Their downward migration is augmented if, after the ice needles melt, the particles
roll or slide beyond their point of first contact with the ground.

In addition to the mechanisms described thus far, other
displacements of particles cause minor reshufflings similar
to those during the creep of sand in a tilted box that is repeatedly tapped (Culling, 1965). Earthquake shocks or other
vibrations have such an effect on the surface material. Less
abrupt, internal strains, whether local or pervasive, may produce similar effects on a more leisurely schedule. Plastic deformation of rock or surface material under its own weight
may cause gradual sagging of competent units and squeezing of incompetent units toward the free face of the slope
where support is lacking.

As yet, too few data are available to give a complete picture
of rates of soil creep, although the process is largely imperceptible and rates no doubt vary by several orders of magnitude (for example, from fractions of millimeters to meters per
year), depending on slope angle, susceptibility of the materials, intensity of the processes, and water content (Schumm,
1967; Kirkby, 1967). Detailed measurements indicate that
creep rates decrease with depth and effectively terminate below
a depth of about 20 cm. They range from 0.1 to 15 mm/yr on
vegetated soil, but in areas of high freeze/thaw frequency, rates
may be as high as 500 mm/yr (Caine, 1981; Gardner, 1979;
Selby, 1982; Young, 1960; Kirkby, 1967).

Rock creep is demonstrated by the presence of isolated joint
blocks some distance downslope from their parent outcrops.
Mechanisms for the creep of these and other boulders are suggested by the availability of processes and the behavior of the
underlying materials. For example, Balk (1939) suggested that
frost wedging and prying by vegetation were mechanisms for
the detachment and further creep of exfoliation slabs across
bedrock in the Adirondack Mountains of New York. He pointed out how loss of material through continued weathering of the
block could shift its center of gravity enough to cause toppling
or rotation. Differential expansion-contraction may also contribute to the process. A heavy block may take advantage of
the plasticity of weak, deformable materials, such as underlying clay, or it may submit passively to the general creep of the
regolith in which it is partially immersed and gain a free ride.

Solifluction

Solifluction is the downslope flowage of saturated soil. It was
described as early as 1906 and recognized as of great importance in cold climates where the soil is periodically saturated
by thawing of ice in the ground. When frozen ground thaws
from the surface downward, the meltwater can't permeate
downward through the still-frozen soil. As a result, the soil becomes saturated until it flows like a viscous fluid, commonly

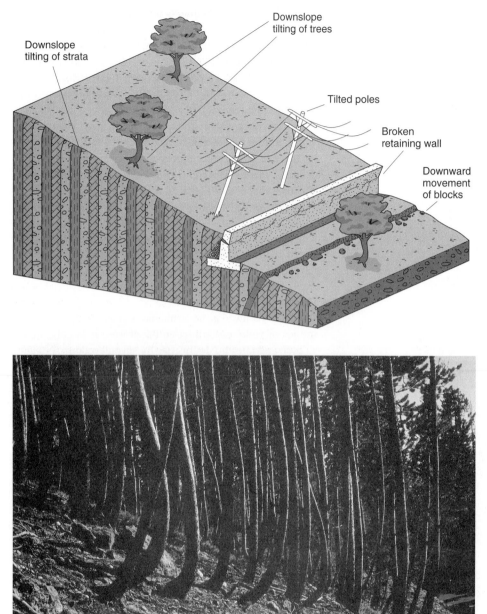

FIGURE 4–11
Downslope movement of surface features by creep.

FIGURE 4–12
Bases of trees bent downslope by creep, Yellowstone National Park. (Photo by W. K. Hamblin)

generating lobes of saturated material having the consistency of wet concrete. Solifluction is not limited to latitudes where the ground freezes, but it is best developed there because of the abundance of meltwater, impermeability of frozen ground beneath a thawing layer, lack of vegetation, and general absence of other processes that might destroy the evidence of solifluction. A more detailed discussion of solifluction may be found in the chapter on periglacial processes and landforms.

Rapid, Discontinuous Movements

Slope failures that consist of rapid downward sliding, falling, or flowing of a mass of rock, weathered material, or a mixture of the two are commonly referred to as **landslides**

(Varnes, 1958), although the term is also used for a more specific type of failure. They differ from creep in that movement of a discrete mass takes place rapidly along a well-defined surface of failure, whereas creep, solifluction, and related processes move slowly, and more or less continuously, on slopes over a broad area, with transitional boundaries between moving and stationary material (Terzaghi, 1950).

Rockslides

Rockslides occur where blocks of rock break away and slide down a planar surface. The two most common discontinuities in rock that serve as planes of failure are *jointing* and *bedding* (Figures 4–13, 4–14, and 4–15; Plate 1D). Joints are pla-

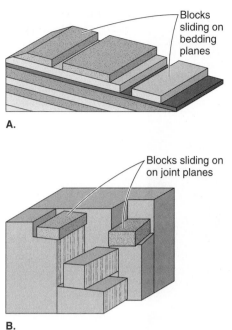

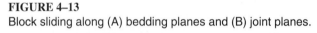

FIGURE 4–13

Block sliding along (A) bedding planes and (B) joint planes.

FIGURE 4–14

Sliding of joint-bounded blocks of andesite, Snoqualmie Pass, Washington. Note the large block in the center of the photo that has pulled away from the rock above and is held on the slope only by friction.

nar cracks without any relative displacement of blocks on opposite sides, typically forming in sets with different orientations. Thus, intersecting joint planes commonly isolate joint-bounded blocks held in place only by friction between the block and the underlying rock (Figure 4–14).

Sliding in solid rock is determined largely by the occurrence, orientation, and nature of discontinuities in the rock (Terzaghi, 1962). The physical characteristics of the opposing joint surfaces play an important role in determining the overall frictional resistance of the rock mass. If the two sides of a joint surface are pressed tightly together and if the faces are rough, the friction along the joint surface is high. However, if the joint is not tightly held together and/or if the surfaces are relatively smooth, the frictional resistance is low. The shearing strength of rock with random jointing can be obtained from the Coulomb equation. The angle of shearing resistance depends on the nature of interlocking between blocks on either side of the surface of sliding. Terzaghi concluded that the critical slope angle for slopes developed on hard, massive rocks with a random joint pattern is about 70° if the joint surfaces are not affected by pore pressures.

Some joints are tightly closed, but others may be open. Not only do open joints allow water to enter and increase pore-water pressure, but they also are subject to weathering and filling with material of low strength. If the filling is thick, the rock surfaces on opposite sides of the joint are not in contact with one another, and the resistance to sliding of the rock mass is reduced to the strength of the joint-filling material.

The physical characteristics of the opposing joint surfaces play an important role in determining the overall strength of the rock mass. If the joint surface is tight and if the faces are rough, the effective angle of friction along the joint surface and the shear strength of the rock mass are high. However, if the joint is open and filled with low strength material, the overall strength is low (Barton, 1978). The following list illustrates the relationship between the physical characteristics of a joint surface and the shear strength of the rock mass:

Nature of Joint Surface

Irregular, rough, stepped	Greatest shear strength
Smooth, stepped	
Slickensided, stepped	
Rough, irregular, undulating	
Smooth, undulating	
Slickensided, undulating	
Rough, irregular, planar	
Smooth, planar	
Slickensided, planar	Least shear strength

The shear strength r along a joint surface may be expressed by the following equation (Barton, 1976):

$$\tau = \theta_n \tan \left[JRC \log \left(\frac{JCS}{\tau_n} \right) + \theta_b \right]$$

FIGURE 4–15
Sliding of rocks along an inclined bedding plane, Punta del Inca, Argentina. The mass of jumbled blocks in the lower right corner slid along the bedding plane in the center of the photo.

where τ_n = effective normal stress
   JRC = joint roughness coefficient
   JCS = joint wall compressive strength
   θ_b = basic friction angle

The joint roughness coefficient ranges from 0 to 20 (from the smoothest to the roughest surface), and the joint wall compressive strength is equal to the unconfined compressive strength of the rock if the joint is unweathered. If the joints are weathered, the strength may be reduced by as much as 75 percent.

Because joints offer little resistance to tension, they reduce the effective shear strength of a rock mass. Where joints dip toward a freestanding face, they impart a high degree of instability to the rock mass. The critical slope angle at which sliding may be initiated is determined by the joint orientation relative to the slope angle.

As with various other types of slope failure, block sliding in rock is often associated with times of heavy rainfall, suggesting that water plays an important role in destabilizing the slope. The long-held idea that water acted as a lubricant along the plane of failure was challenged by Terzaghi (1950), who noted that in humid climates enough water is present most of the time to cause lubrication, but often sliding occurs only after heavy rainfall. The most important effect of water is that it increases pore pressures. This increase reduces the effective normal stress on potential sliding surfaces according to the Coulomb equation, and this in turn reduces shear strength S_s.

$$S_s = c + (S_n - S_p) \tan \theta$$

where S_s = shear strength
   c = cohesion
   S_n = normal stress
   S_p = fluid pore pressure
   θ = the angle of internal friction

Heavy rains cause pore pressures to rise in soil and along potential sliding surfaces, and the resulting decrease in normal stress lowers shear strength to the point where failure can occur.

An example of massive sliding along joint planes is the Turtle Mountain slide (Figure 4–16), which destroyed much of the town of Frank, Alberta, in 1903. Joints cutting across limestone beds acted as sliding surfaces. Reduction of support along the base of the mountain by coal mining may also have contributed to the slide. When gravitational shearing stresses exceeded the shearing resistance along the joint surfaces, approximately 30 million cubic meters (40 million cubic yards) of rock slid down Turtle Mountain into the valley below.

Bedding planes also serve as potential sliding surfaces in stratified rocks, especially when bedding is undercut and terminates in the air. An example of failure along a bedding plane is the 1925 Gros Ventre slide in Wyoming (Figure 4–17). The thick, massive Tensleep sandstone overlies the relatively weak Amsden shale, both dipping about 20° toward the valley floor on the south side of the valley. The sandstone bed was cut through and left unsupported by deep erosion of the Gros Ventre River, so that the only strength of the rock mass was friction at the contact between the sandstone and underlying shale. After a period of heavy rainfall during the winter of 1925, wetting of the shale diminished its shearing resistance, and a huge mass of sandstone broke away from the slope, sliding along the shale bed and carrying 38 million

FIGURE 4–16
(A) Turtle Mt. slide, Frank, Alberta. (Her majesty the Queen in right of Canada, reproduced from the collection of the National Air Photo Library with the permission of Natural Resources Canada.) (B) View from the valley floor. (Photo by L. F. Hintze)

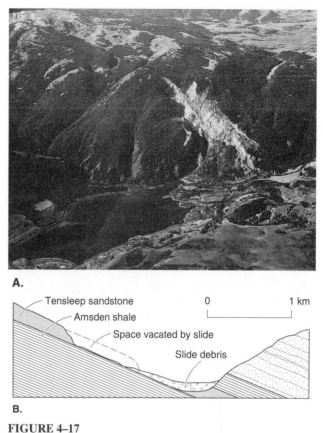

FIGURE 4–17
(A) 1925 Gros Ventre landslide, Wyoming. (Photo by Austin Post). (B) Geologic cross section (Modified from Alden, 1928; Keefer and Love, 1956)

m³ (50 million yd³) of rock into the valley below. The slide created a dam 68 m (225 ft) high, which impounded a lake 6 to 8 km (4 to 5 mi) long, and momentum carried slide debris 120 m (400 ft) up the opposite side of the valley (Hayden, 1956).

Another example of a rock slide along a bedding plane was the disastrous 1963 Vaiont slide in northern Italy. The Vaiont Dam, the highest thin-arch, concrete dam in the world at over 65 m (875 ft), was constructed in an area of low slope stability. Sedimentary beds dip steeply into the valley below (Figure 4–18) and were cut through by deep erosion of the river. This erosion removed lateral support of the rocks.

Filling of the reservoir aggravated an already-precarious situation. Hydraulic pressure of ground water rose in the rocks of the reservoir walls, buoying up the rocks and reducing frictional resistance. Swelling of clays in some layers further decreased their strength.

In 1960, a 700,000 m³ block of rock slid from the slopes of the south valley wall into the reservoir, and significant rock creep, probably induced by the weight of rock on underlying weak clay beds, was detected over a large area. Monitoring stations set up to measure creep rates showed that in 1960–61 the creep rate occasionally reached 25 to 30 cm (10 to 12 in.) per week, and the total volume of rock affected by the creep was estimated at about 200 million m³. The creep rates later declined to about one cm per week. Heavy rains fell in the Vaiont Valley in late summer and early fall of 1963, raising the reservoir level by more than 20 meters and saturating the rocks. This added more weight to the already unstable slopes and further increased pore-water pressures. The creep rate shot up to 20 to 40 cm (11 to 22 in.)/day, signaling the possibility of imminent slope failure.

On October 8, 1963, engineers discovered that all the creep monitoring stations were moving together and that a far larger area of the hillside was moving than had been previously thought. The gates of two reservoir outlet tunnels were opened in an attempt to reduce the water level in the reservoir, but the water level continued to rise. The huge creeping mass of

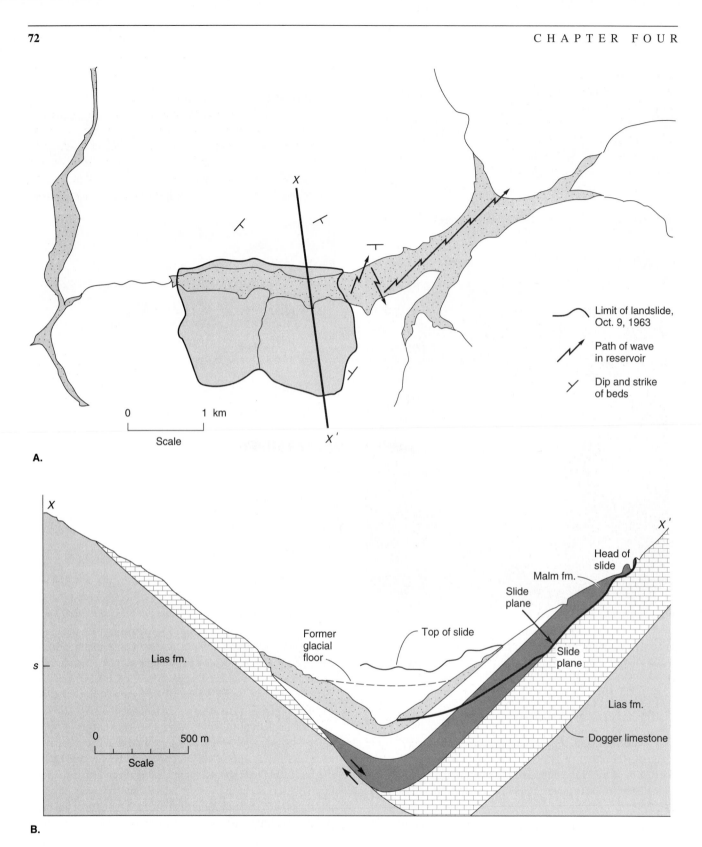

FIGURE 4–18
Geology of the 1963 Vaiont landslide, northern Italy: (A) map, (B) geologic cross section, X–X. (Modified from Kiersch, 1964)

rock began to reach the reservoir, and, on the night of October 9, disaster struck. A mass of rock 2 km (1.2 mi) long, 1.6 km (1 mi) wide, and 150 m (~500 ft) thick, involving over 240 million m³ of rock, broke loose from the south side of the valley and crashed into the reservoir. The slide splashed water 260 m (~850 ft) up the valley wall on the opposite side of the valley, and 100-meter-high waves (330 ft) crashed over the Vaiont dam into the valley below. In less than a minute, the slide filled the reservoir with rock debris 270 meters (890 ft) deep, up to 180 meters (~600 ft) above the water level for 2 km (1.2 mi) upvalley. As the huge wave of water rushed down the valley below, the wall of water was still more than 70 m (230 ft) high 1.5 km downstream. At the junction with the Piave River, the wave of water flowed *upstream* into the Piave valley for more than 2 km. Within about five: minutes, entire towns were wiped out (Figure 4–19) and three thousand people were killed.

Despite the tremendous force exerted on the Vaiont dam, it did not fail. All of the destruction downvalley came from the water displaced by the slide. Sadly, the Vaiont tragedy could have been avoided by consideration of the evidence for slope instability prior to construction of the dam (Kiersch, 1964).

Debris Slides

Debris slides occur when masses of unconsolidated material break loose and slide over the underlying surface (Plate 1C). Both debris slides and rockslides move on predefined slide planes —the principal difference between them is in the nature of the sliding material. Debris slides are especially common where thin, unconsolidated sediment mantling sloping bedrock surfaces becomes saturated and separates from the underlying rock surface'

Rockfalls

Rockfalls (Figure 4–9) occur when rock breaks loose from steep slopes and falls through the air to the ground below.

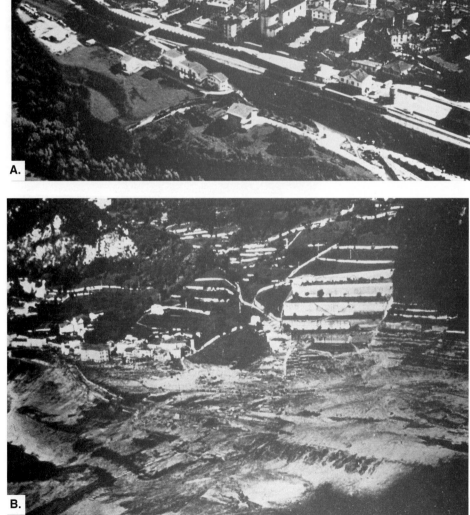

FIGURE 4–19
Destruction of an Italian village by a giant flood wave produced by the Vaiont landslide into a reservoir. (A) is the village before the wave; (B) is the same area after the wave. (Photos from the U.S. Geological Survey collection)

Movement of material is largely by free fall through the air but may also include some bouncing or rolling. Rockfalls occur when masses of rock break loose from bedrock and fall through the air to the slope below. Movement is mostly by free fall through the air, but it may also include some bouncing or rolling. Rockfalls are most common on near-vertical slopes where the bedrock is well jointed, providing many joint-bounded blocks that may break loose. The process is accelerated by artificial undercutting of rock, such as in the steep rock faces of excavations.

A spectacular example of a rockfall occurred July 10, 1996 in Yosemite National Park when a huge slab of granite about 150 m (500 ft) wide, ~10 m (~30 ft) long, and 6 m (20 ft) thick plunged about 500 m (~1500 ft) off the valley wall to the valley floor below. The rock mass, which weighed some 162,000 tons, accelerated to about 160 mph before crashing onto the valley floor where it generated an air blast that leveled 2000 trees.

Toppling is a particular type of rockfall that includes forward rotation of a block around a fixed hinge line, usually at the base of the block; it may involve large masses of rock. Toppling occurs when the center of gravity of a block lies in a vertical plane beyond its base (Figure 4–20) (Hoek, 1971; Caine, 1982).

Continued rockfall produces **talus**—an accumulation of loose fragments in cones or aprons at the base of steep slopes —reflecting the traffic of fragments from above (Figure 4–21). Talus may sometimes form a ridge of rock debris separated from the base of the source slope by a depression. Such ridges, known var-

A.

B.

FIGURE 4–21
(A) Talus at the base of rocky cliffs, Banff National Park, Canada. (B) Cross section of talus banked against rock cliff, Vantage, Washington.

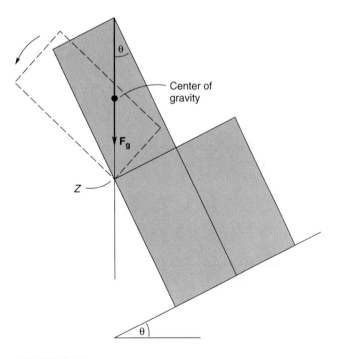

FIGURE 4–20
Toppling of rock mass. The block will rotate and fall when the plane of the center of gravity (F_g) lies outside of the base of the block at Z.

iously as **protalus ramparts** (Bryan, 1934), **winter talus ridges**, or **nivation ridges** (Behre, 1933), develop where seasonal snowbanks at the base of the slope shunt the rock debris beyond the base of the slope. They may blend into the main talus slope as a benchlike extension wherever the original depression behind the rampart later becomes filled with rock debris.

Talus accumulates by processes of rockfall, but upon accumulation it is subject to talus creep. As long as the talus s abundantly fed from above, it may be expected to maintain as steep a slope as the fragments can sustain— the angle of repose. However, the talus continues to gradually creep downslope under the influence of gravity.

The angle of repose of talus involves a balance between frictional resistance and gravitational stresses. As one might expect, the angle of repose is steeper in proportion to the compaction of particles, their angularity, surface roughness, and degree of sorting. Well-rounded, well-sorted, and loosely packed particles with smooth surfaces have unstable shapes, few points of contact, and relatively low coefficients of friction. The angle of repose is inversely proportional to the size of particles if the deposit is well sorted. Poorly sorted deposits gain support from the tendency of smaller particles to lodge between large ones, and factors such as sorting and particle shape influence stability more strongly than size alone. This may help to explain some observations (for example, Behre, 1933) that steep talus is typically coarse. Talus with contours that are convex outward has a gentler angle of repose than does talus with contours that are concave outward (as within the crater of a cinder cone). The reason is that in the convex-outward talus the fragments fan out away from their source, whereas in the concave-outward talus they lodge together and provide mutual support.

Slump

Slump is the downward and outward sliding of a mass of material along a curved, usually concave-upward, shearing plane, with some backward rotation of the land surface at the top of the failure (Figure 4–22). The arc is a locus along which shearing resistance is exceeded by the driving moment of the slump block as it rotates on the plane of failure.

A slump block, composed of sediment or rock, slips along a concave-upward surface, with rotation about an axis parallel to the slope. As a result, the original surface of the block becomes less steep, and the top of the slumped block is rotated backward (Figures 4–22 and 4–23) because of the geometry of movement along the concave-upward surface. The main slump block often breaks up into a series of secondary slumps to form a staircase whose treads have been rotated backward (Figures 4–22, 4–23, and 4–24) where undrained depressions, swamps, and small ponds may collect. Earthflows or mudflows commonly form at the base of slumps.

The surface of failure is concave upward, but it seldom corresponds to a circular arc of uniform curvature. If the shearing strength of the material is smaller in the horizontal direction than in the vertical, the arc may flatten downslope; or if the shearing strength is greater in the horizontal direction than in the vertical, the arc may increase downslope. Slumping of clay often generates a spoon-shaped slip surface. Slumps may be produced in relatively homogeneous material either as a result of increased shearing stress caused by undercutting of slopes or as a result of reduced shearing

resistance brought about by loss of cohesion, often because of water saturation. Rotational slides such as slumps develop from tensional stresses in the upper part of a slope. Concentric tension cracks are common at the head of a slump prior to its failure, and a concave scarp usually is left at the head of a slump. Because this scarp is very steep and unsupported, multiple failures may progress headward along a common basal failure surface. Skempton and Hutchinson (1969) described "bottle neck slides," which are retrogressive failures peculiar to quick clays. They start as a rotational failure of stiff clay along the banks of an incised stream and are followed by additional failures of weaker clay, which moves through the original opening in the bank. If rotational sliding in clay is prevented by an underlying resistant layer, translational sliding along the contact may develop instead of slumping (Skempton and Hutchinson, 1969).

Causes of slumping include factors that increase the driving moment and/or reduce the moment of shearing resistance. Earthquake shocks, loading of slopes, undercutting, deep weathering, thorough wetting, swelling of clays, freezing and thawing, and excess pore-water pressures are common causes of slumping. Thorough wetting is a particularly important cause, which explains why slumps frequently accompany heavy rainfall and earthflows. Cement material can be dissolved, clays are made plastic, weight is added to the slope, and pore-water pressures increase. A familiar type of slump occurs on cutbanks of stream channels as a result of rapid drawdown after a flood. During the flood, the banks are saturated with pore water and become unsupported when the flood- waters recede. The pore-water pressure at the base of the cut is increased proportionally to the height of the water column above the interconnected pores of the bank materials, and in many cases, this increase is enough to cause failure. The backward rotation of slump blocks encourages further slumping by providing traps for water on the tread of each block.

Debris Avalanches

Debris avalanches are unsorted, incoherent mixtures of earth or rock with water or ice (or both) that move rapidly down steep slopes after a mass of snow or saturated ground has broken loose. They include the familiar snow avalanches, which are rapidly moving mixtures of debris and ice. Their scars resemble long, narrow funnels, slightly flared at the top and ending in a hummocky accumulation of debris at the base. Inasmuch as many debris avalanches are triggered by the saturation of heavily vegetated slopes (which prevent the slightly more leisurely process of mudflow), their tracks through forests are very conspicuous, and their deposits are a chaotic mixture of broken timber, smaller vegetation, and other debris.

In alpine country, snow avalanches may be significant movers of debris. Of course, many snow avalanches are composed largely of snow, but in many others the debris content is merely disguised by billows of snow that bloom from the flowing mass as earth and rock are swept along and carried

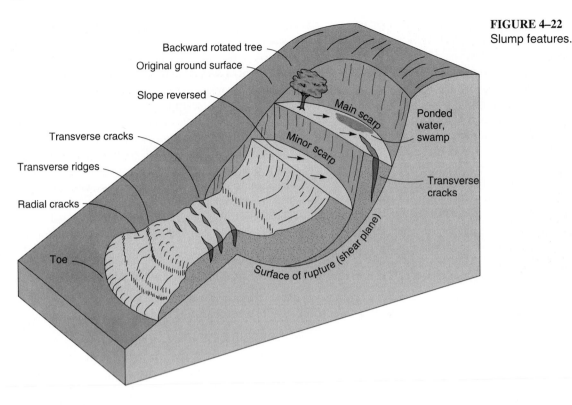

FIGURE 4–22
Slump features.

effectively within. When the avalanche deposit melts away, the debris is lowered haphazardly to the ground, and no distinctive topographic form remains.

Large, rapid avalanches may displace enough air to create winds of hurricane strength, which may blow down timber adjacent to their tracks. They may also gain mobility by flowing across cushions of entrapped air. The erosive power of avalanches is shown by troughlike avalanche chutes on rock walls (the couloirs of the mountaineer). These chutes may appear smooth and round bottomed from a distance, but they are usually pockmarked and ragged in detail from the bombardment of falling rock (Blackwelder, 1942). Avalanche chutes descending into forests are common in high mountain areas where avalanches begin above the timberline.

A debris avalanche occurred January 10, 1962, on the slopes of 22,205-ft (~6700-m) Nevado Huascaran, the highest mountain in Peru. The avalanche, triggered by thaw, started as large masses of a glacier caved away from the rim of Huascaran's north summit. The impact of this initial mass jarred loose three million tons of ice from a glacier below, and a mixture of ice, water, mud, and rock pitched forth at speeds greater than 100 km/hr (60 mi/hr). Accompanied by powerful winds, the avalanche ricocheted from canyon walls as it fell nearly four vertical km (2.5 mi) to the valley of the Rio Santa 15 km (9 mi) away. The debris covered this track in seven minutes, and, even at the last, where it overwhelmed

the large town of Ranrahirca, the front of the avalanche moved at an estimated 100 km/hr (60 mi/hr). Afterward, the Rio Santa flowed against a dam of debris, and Ranrahirca, along with several other towns, lay beneath as much as 18 m (60 ft) of icy mud. Approximately 3500 Peruvians lost their lives in the avalanche (McDowell and Fletcher, 1962). Another huge avalanche that began high on the slopes at Nevado Huascardn occurred in 1970, moving down a valley at high velocities and over a distance of 14.5 km (9 mi). It destroyed virtually everything in its path, including the village of Yungay with its 20,000 inhabitants (Eriksen et al., 1970; Browning, 1973; Schuster and Krizak, 1978).

Some debris avalanches are very rapid, fluid movements of dry particles. Fluidization depends upon the interaction of the particles and entrapment and compression of air (Varnes, 1958; Shreve, 1968; Kent, 1966). Snow avalanches have been seen gliding on cushions of air, and certain very large rock fragment flows may have behaved similarly. The Blackhawk slide (Figure 4–25), a large, rocky flow in southeastern California was emplaced virtually dry (Shreve, 1966, 1968), yet it moved a substantial distance and spread out as a surprisingly thin lobe. Shreve proposed that the Blackhawk slide was launched over a ledge and rode on a layer of compressed air trapped beneath the slide debris. The trapped, compressed air acted as a nearly frictionless base that allowed the slide material to spread out as an astonishingly thin lobe.

FIGURE 4–23
Slumping along the banks of the Missouri River, Sioux City, Iowa. (Photo by D. A. Rahm)

Large, similar, rocky avalanches occur elsewhere in the world. The corrugations on their surfaces, and their steep fronts suggest that although these movements may have started as rockfalls or rockslides, they continued as debris avalanches with little frictional resistance and at unusually high speeds. Geological evidence suggests that these large avalanches may have ridden on a cushion of virtually frictionless, compressed air. Accounts of powerful air blasts associated with slides at Frank, Alberta, and Madison Canyon, Montana, give an idea of the forceful displacement of air (Kent, 1966).

Sturzstroms

When a large rock mass moves down a steep slope, it may lose its integrity as the mass disintegrates and thus may travel as broken debris, rolling, sliding, and/or flowing with great velocity. Such slope failures are known as **sturzstroms**. These huge volumes of rock debris may move as dry or wet aggregates. The Huascaran sturzstrom contained a good deal of water and behaved as a fast-moving mudflow in its lower route.

Hsu (1975) proposed that sturzstrom movement is primarily a flow phenomenon, reviving a conclusion made earlier. However, the mechanism differs from normal viscous flow in that particles are disseminated in a dust cloud and the kinetic energy of the flow is conveyed from fragment to fragment as the particles bang together. The body moves until all of the original energy is consumed by friction between the particles. In this type of flow, the internal friction is very low because fragments are dispersed in the dust cloud, giving the mass extraordinary mobility.

In 1863, fused rock was discovered at the site of a 2.1-km³ (~1.3-mi³) landslide at Kofels, Austria, where a large

area of a mountainside slid into the Otz Valley, crashing against the opposite wall of the valley with so much force that it shattered and pulverized the rock. At first, the fused rocks were believed to be volcanic or due to meteorite impact, but Preuss (1974, 1986) proposed that it was the result of fusion at the base of a giant landslide. The fused rock, known as *frictionite* (Erismann et al., 1979), was the first indication of the energy involved in such giant slope failures (Heuberger et al., 1984; Heuberger, 1989; Masch et al., 1985; Erismann, 1979; Milton, 1964). Shatter cones and shocked quartz in the slide material suggest considerable impact (Surenian, 1988).

Earthquake-generated Slope Failures

Earthquakes are capable of initiating many types of slope failure. However, some earthquakes set off deep-seated landslides in rock. A good example is the Hebgen slide (Figure 4–26) (Witkind, 1964; Hadley, 1964). On August 17, 1959, a strong earthquake hit the Madison Valley west of Yellowstone National Park. Campers at a campground in the valley were struck by a great blast of air. The mother of a family saw her husband "grasp a tree for support, then saw him lifted off his feet by the air blast and strung him out like a flag before he let go." Before losing consciousness, she saw one of her children blown past her and a car tumbling over and over (Witkind, 1964). The air blast was produced by a huge landslide, 30 m³ (40 yds³), that slid into the Madison Canyon where it dammed the Madison River (Figure 4–26A) and buried 26 people in a campground. The cause of the landslide was breaking of a dolomite buttress (Figure 4–26B) that had supported the weak schist and gneiss making up the hillslope above.

Other examples of earthquake-generated landslides occur in the Nooksack valley of northwest Washington where a remarkable clustering of shallow-focus earthquake epicenters are associated with an extraordinarily high incidence of very large, deep-seated, bedrock landslides (Engebretson et al., 1995, 1996; Kovonen and Easterbrook, 1996). The largest landslide originated from failure of an entire mountainside and filled the valley below for 12 km (7.2 mi) to depths of 95 m (315 ft.) with an estimated volume of 2.83 ×10⁸ m³ (3.7 ×10⁸ yds³). Four other nearby landslides each cover about 10 km² (4 mi²) with six others less than one km²·(0.4 mi²).

Flows

When the internal shearing resistance of a material approaches zero, the material is capable of flow as a viscous fluid. The viscosity of a fluid is the ratio of applied shear stress to the rate of shear. Flow is usually defined as continuous, permanent deformation of a material under applied stress; it does not occur until the yield strength of the material has been ex-

FIGURE 4–24
Multiple slumps from mesas near Albuquerque, New Mexico.

ceeded. True liquids, such as water, have zero yield strength and will flow under any applied shear stress.

The most common cause of loss of internal shearing resistance is saturation of clayey sediments, but dry flows are also possible under certain circumstances. Many flows begin as some type of non-viscous slope failure; they pick up water during their movement and later become viscous flows. Because of the great variations in consistency, a number of types of flows are recognized, all of which involve areas of overlap. Few viscous slope failures are uniquely distinct throughout their entire mass.

Most classifications of flowing sediment-water mixtures are based either upon the type and rate of movement, for flows observed in motion, or upon morphology and sediment type, for flows not observed in motion. Sharpe (1938) distinguished debris avalanches, mudflows, earthflows, solifluction, creep, and streamflow on the basis of velocity and sediment concentration. A later, widely used, classification by Varnes (1978)

added block streams, dry/wet sand or silt flows, and rapid earthflow. Numerous other schemes have since been proposed, mostly variations of the basic elements used by Sharpe and Varnes. Among other elements suggested are mud flood, water flood, debris torrents, hyperconcentrated flow, and debris flow (VanDine, 1985). Pierson and Costa (1987) proposed a classification, based on flow rheology, that recognizes normal streamflow, hyperconcentrated streamflow, slurry flow, and granular flow. They suggested two types of slurry flow (inertial and viscous) as replacements for debris flow; fluidized granular flow for sturtzstrom; inertial granular flow for grain flow; and viscous granular flow for earthflow (Figure 4–27).

Earthflows

Earthflows are the downslope, viscous flow of saturated, fine-grained materials, moving at speeds ranging from bare-

FIGURE 4–25
Blackhawk slide, California. (Photo by Wallace, U.S. Geological Survey)

ly perceptible to rapid. Typical rapid earthflows move at speeds of 0. 17 to 17 km/hr (0. 1 to 10 mi/hr). Although earthflows may resemble mudflows, they are generally slower moving and are often covered with nonfluid material carried along by flow from within. They differ from solifluction flows in that they are more rapid. Materials susceptible to earthflow include clay, fine sand and silt, and fine-grained, pyroclastic material. The velocity of earthflows is governed largely by water content: The higher the water content, the higher the velocity. Some earthflows continue to move slowly for years.

Earthflows generally begin when pore pressures in a fine-grained mass increase until enough of the weight of the material is supported by pore water to significantly decrease the internal shearing strength of the material. Flowage often generates a bulging lobe (Chapter opening photo and Figure 4–28), which advances with a slow, rolling motion (Skempton and Hutchinson, 1969).

As lobes of an earthflow spread out, drainage of the mass increases, and the margins dry out, thus lowering the velocity (Scheffel, 1920). The slowing of the flow causes it to thicken, especially in its terminal area. These types of thick, lobate earthflows are typical and usually extend for only short distances downslope from their origin. The thick, slow-moving, bulbous variety of earthflows that do not spread laterally very much are not as spectacular as more rapid types of flows, but they are far more common. They develop a sag at their heads and are usually associated with headward slumping at the source.

Earthflows occur much more frequently during periods of heavy precipitation, which saturates the ground and adds water to the slope materials (Sharpe and Dosch, 1942). Fissures that develop during movement of clayey material facilitate infiltration of water into earthflows. Water increases pore-water pressure and reduces the internal shearing strength of the material.

Debris Flows

Debris flows (the slurry flows of Pierson and Costa, 1987) include rapid flowage of saturated masses of clay-to-boulder-size material mixed with water or ice or both water and ice. Debris flows are distinguished from mudflows primarily on the basis of viscosity of flow. They commonly follow heavy rainfall or sudden thawing of frozen ground.

As the water concentration of a mass increases, its yield strength decreases rapidly. The rate of decrease in yield strength is more rapid in fine-grained material than in coarse mixtures where internal friction between grains and grain interlocking play a role (Rodine, 1974).

Debris flows have the consistency of wet concrete and sufficient yield strength to display plastic flow behavior and to form steep, lobate fronts (Figure 4–29) and marginal levees (Figure 4–30) (Carter, 1975). Pore water is often 25 to 45 percent, so that hydrostatic pore-water pressure in the saturated material of the debris flow partially carries the weight of the solid phase (Carson, 1976) and allows flowage of the mass when the yield strength is exceeded. Flow velocities vary considerably, ranging from about 1 to 15 m/s (3 to 50 ft/s). Velocities generally depend on slope angle, sediment/pore-water ratio, and grain-size distribution. If the shear strength of the debris flow is high enough, large boulders can be suspended in debris flows or rolled along by the flow. As the pore fluid drains out of the flowing mass, grain-to-grain contact increases internal friction, and the flow eventually stops.

When the silt/clay content of a debris flow is relatively high and/or the shear rate and water content are relatively low, the flow behavior resembles that of a plastic (Bagnold, 1954; Thomas, 1963; Johnson and Rahn, 1970). When the water content and shear rates are high, viscous forces are replaced by inertial forces, and momentum is transferred by particle collisions. Flows by this mechanism have been termed *inertial slurry flows* by Pierson and Costa (1987). With increasing water content, debris flows grade into hyperconcentrated flow in which particles are carried by turbulent water flow (Qian et al., 1980).

Mudflows

Flowage of heterogeneous mixtures of rock debris saturated with water produces **mudflows** (Figure 4–31), which often begin as landslides or slumps along the upper parts of a hillslope. When the material becomes thoroughly mixed with

A.

FIGURE 4–26
(A) The Madison slide, triggered by the 1959 West Yellowstone earthquake (Photo by W. K. Hamblin), (B) geologic cross section through the slide (From Hadley, 1964)

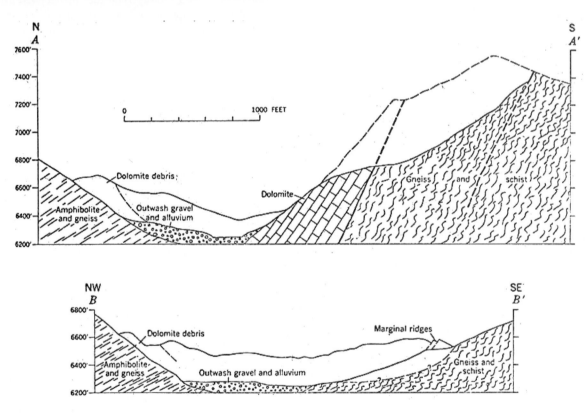

B.

water, it flows as a viscous mass. The rate of the movement depends in part on the angle of slope and in part on the amount of water mixed with the mud. All gradations from stiff mud to muddy water exist. Wet debris flows range from rapid flows of water—saturated debris to streams so heavily clogged with debris that they are turbid throughout and viscous enough to dam themselves locally.

A mudflow may start without the addition of water from any external source. If the bonds between clay particles and adsorbed films of water are broken, cohesion is destroyed, and the pore water reacts against reconsolidation of the separated particles and keeps the mass fluid. Certain clays with this behavior are extra sensitive or "quick." They may be solid after the formation of a cohesive bond between particles, but

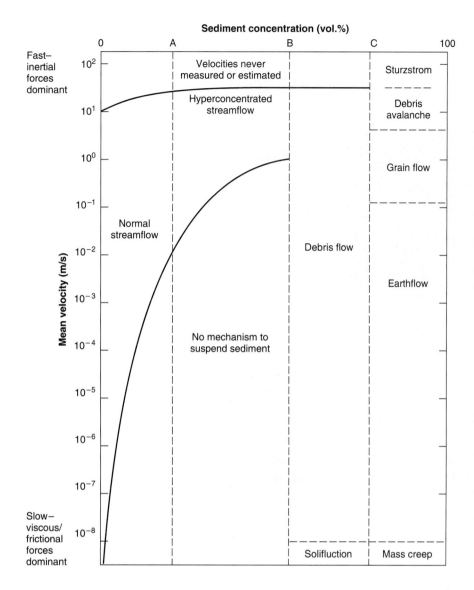

Sediment concentration (vol.%)

Fast–inertial forces dominant

Velocities never measured or estimated

Hyperconcentrated streamflow

Normal streamflow

No mechanism to suspend sediment

Debris flow

Sturzstrom

Debris avalanche

Grain flow

Earthflow

Solifluction

Mass creep

Slow–viscous/frictional forces dominant

Mean velocity (m/s)

FIGURE 4–27
Types of flow (From Pierson and Costa, 1987)

Fluid type	Newtonian	Non-Newtonian		
Interstitial fluid	Water	Water + fines		Water + air + fines
Flow category	Streamflow		Slurry flow	Granular flow
Flow behavior	Liquid	Plastic		

they become fluid with dramatic suddenness if the bonds are broken by remolding or by vibrations.

Some clays are thixotropic colloids; they may be solid (gels) until agitated, whereupon they suddenly liquefy (become sots). Spontaneous liquefaction may also occur in fine sand and silt below the water table if vibrations upset the packing of grains and allow the pore water to react against granular reconsolidation (Terzaghi, 1950). Terzaghi recounts the experience of H. L. Munro who had the misfortune to be standing on the edge of a clay cliff about 15 m (16 yd) high when it collapsed into the gully below. He recounts:

"After reaching the bottom I was thrown about in such a manner that at one time I found myself facing upstream toward what had been the top of the gully . . . . The appearance of the stream was that of a huge, rapidly tumbling, and moving mass of moist clayey earth . . . . At no time was it smooth looking, evenly flowing, or very liquid. Although I rode in and on the mass for some time my clothes afterwards did not show any serious signs of

moisture or mudstains .... As I was carried further down the gully away from the immediate effect of the rapid succession of collapsing slices near its head ... it became possible to make short scrambling dashes across its surface toward the solid ground without sinking much over the ankles." (Terzaghi, 1950)

Similar flows have occurred in clays of the St. Lawrence Valley, coastal Maine, and Scandinavia (Sharpe, 1938; Varnes, 1958). Rapid earthflows of this type have drained large volumes of material (for example, thousands to millions of cubic meters) from slopes and the outer margins of clay terraces. The basins that remain are characterized by arcuate scarps at their heads, where chunks of clay have slumped inward, and by irregular, hummocky floors and bottleneck outlets.

Because of their density and viscosity, the finer mixtures of water and sediment exert buoyant forces and tractive forces more powerful than those of analogous streamflow. Their competence is shown by large quantities of boulders, which become important components of the debris wherever they are available. Mudflows are characterized by masses of wet, slippery mud, in contrast to the coarser sandy or gravelly matrix of other debris flows. They gain particular competence to

transport enormous boulders. According to Blackwelder (1928), house-size boulders have glided for kilometers across alluvial fans before the transporting mudflow finally congealed. Optimum conditions for the initiation of mudflows include the following:

1. A supply of unconsolidated debris capable of saturation (such as alluvium or soil)
2. Infrequent runoff of high intensity, as with a cloudburst or a period of rapid snowmelt
3. Lack of vegetational constraint
4. Favorable topography, such as steep slopes and, perhaps, a convergence of slopes in the source area

These conditions are met on a large scale in high alpine areas, on the ashy slopes of volcanoes, and in arid landscapes.

The track of a typical mudflow may explain the processes by which it was made. An upper zone of erosion is marked by a scarred area at the head where saturated masses of debris moved away from the slope. Downslope, the track becomes more linear and consists of a channel between marginal embankments or levees. Particularly where coarse

FIGURE 4–29
Debris flows, east of Hebgen Lake, Montana. (Photo by W. K. Hamblin)

debris is prevalent, the mudflow levees may be remarkably steep—even steeper than the angle of repose for the loose fragments—because of the mortaring effect of the mud. Large mudflow levees may be more than 3 m (10 ft) high, and bouldery levees from 1 to 3 m high are common. A depositional zone through the lower portions of the track is characterized by spreading of the debris in lobes, which may end in steep, bouldery fronts or which may taper down to the ground as thin layers of mud. Sharp (1942) observed the growth of levees along mudflows in Alaska where boulders, concentrated at the snout of the flow, were bulldozed aside. Levees may also result from the trapping and plastering of material along the margins of the channel, where friction is greatest and velocities are least. Side pressures may also develop within the heavy traffic of boulders as the swiftest ones, carried along by the central current, shoulder past their slower neighbors and knock them even farther aside. Intermittent surges of debris account for the continued growth of some levees. The size of mudflows varies from a few inches to several miles.

Sharp and Nobles (1953) reported eyewitness accounts, velocity measurements, and observations from moving pictures of a representative mudflow that descended 1500 m (5000 ft) over a distance of 25 km (15 mi) and moved 0.9 million m^3 (1.2 million yd^3) of material. Wavelike surges at intervals of a few seconds to tens of minutes were typical during the peak of activity. They were attributed to the following:

1. Periodic sloughing of debris in the source area
2. Temporary choking of the channel
3. Caving of undercut banks
4. Friction between the moving debris and the channel

During the most active period, surges moved at velocities near 3 m/s (10 ft/s), and a maximum velocity of 4.4 m/s (14.5 ft/s) was measured. The flow material behaved like freshly mixed concrete and splashed about in the more turbulent parts of the stream, spattering nearby objects. Its water content was estimated to be from 15 to 20 percent, and a dry sample of the debris was made to flow again by mixing it to 16 percent water content and releasing it on a 7° slope. Viscosities were estimated to be from 2×10^3 to 6×10^3 poises, or about 2 to 6×10^5 times the viscosity of water. The deposits from this mudflow contained abundant vesicles (representing entrapped air bubbles) and a voluminous sampling of the local organic debris. Features such as these may aid in the differentiation of mudflow deposits from glacial till, which they often resemble.

Lahars

Lahars are volcanic mudflows, commonly spawned by volcanic activity. They are especially dangerous because they can travel long distances at high velocity with little warning. An example of the devastation that can be created by a lahar is the destruction of the city of Armero, Columbia, which was destroyed by a lahar from the volcano Nevado del Ruiz 50 km (31 mi) away (Figure 4–32). A 130-ft (40-m) wall of mud roared down the Lagunilla. River Valley on November 13, 1985, killing 23,000 people.

Ash began falling on the town at 5:30 p.m., and an hour later, a hard rain fell mingled with the ash, accompanied by a strong smell of sulfur, (McDowell, 1986). However, the local radio station and public address system at the town church reassured people to stay calm. Several chilling eyewitness accounts depict the tragedy that followed. Shortly after 11 p.m., a mudflow buried the village. According to 16-year-old Slaye Molina,

"People screamed, 'The world is ending!' We ran upstairs to our terrace, but we saw another house collapse. So we rushed outside, though my grandmother had just had an operation and could not run. A friend took my hand and dragged me faster toward a hill.

FIGURE 4–30
Debris flow, Mendosa Valley, Argentina.

FIGURE 4–31
Mudflow, near Pahsimeroi River, Idaho. (Photo by W. K. Hamblin)

FIGURE 4–31
Mudflow, near Pahsimeroi River, Idaho. (Photo by W. K. Hamblin)

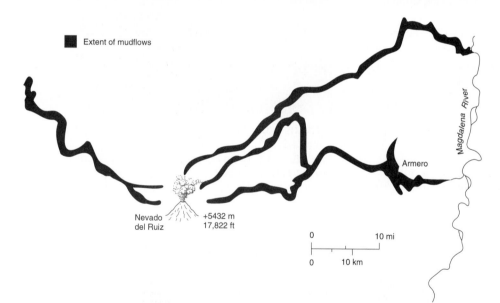

■ Extent of mudflows

Magdalena River

Armero

Nevado del Ruiz   +5432 m
                  17,822 ft

0 ———————— 10 mi

0 ———————— 10 km

FIGURE 4–32
Extent of lahars from Nevado del Ruiz, Columbia, that killed 20,000 people in Armero November 13, 1985.

I looked back and saw my grandmother and aunt and uncle embracing each other. I do not want to be selfish, but only thought about saving myself. I ran. The mud would catch up, and we would run faster. When we reached the hill, we saw Armero disappear in 15 minutes." (McDowell, 1986)

Eften Torre Vergara, a young industrial worker, remembered being caught by the flow:

"It came in circles like batter in a blender. It turned me inside the mud, then brought me out. I rode on top of the thing at very high speed, holding on to a car. And then it left me on the side of a hill."

By the time two flows merged and gushed from the canyon 3 km (2 mi) from Armero, the lahar had become a watery mass 40 m (130 ft) deep and traveling 40 km/hr (24 mi/hr). Ironically, an earlier lahar had been noted in the records of the Mariquita church, dated 1845: "Stupendous and Deplorable Happening, By the Holy Will of the Highest." The clergy described earthquake-induced mudflow on the Lagunilla "so that . . . their successors would not forget . . . (McDowell, 1986).

Armero was clearly in the path of possible mudflows from Ruiz. In November, a warning was issued by the Colombia Instituto of Geologic Mining Research, predicting a 67-percent chance that eruptions would cause mud-flows and noting that Armero could be evacuated in two hours without danger. Volcanic tremors began November 10 and continued for three days. On November 12, people noticed a strong sulfur smell. At 3:00 p.m., November 13, a technical emergency committee meeting in a town, 70 km (43.5 mi) from Armero, recommended the evacuation of Armero. Instead, Radio Armero urged calm and was still playing cheerful music when power failed and mud engulfed the station (McDowell, 1986).

The disastrous lahars at Nevado del Ruiz, although not commonplace, are not rare in the vicinity of active volcanoes. Among notable examples are the lahars associated with the May 1980 eruption of Mount St. Helens in Washington, where mudflows traveled more than 120 km (72 mi) from the volcano (Figures 4–33 and 4–34). Lahars. flowed down almost all of the valleys heading on the volcano, destroying 27 bridges in the main northern tributary, damaging or destroying many houses and other structures, and burying about half of state highway 504. About 34 million cubic meters (45 million cubic yards) of mud were deposited in the Columbia River, 45 km (28 mi) from the volcano, blocking shipping in the main channel.

When lahars have run their course, they leave behind deposits of poorly sorted mixtures of mud, sand, and coarse clasts that enclose many logs and much other debris caught up in the flow (Figure 4–35). Thus, lahars leave a depositional record of part of the geologic history of a volcano that can be radiocarbon dated.

Debris Torrents

Debris torrents have been described by Swanston (1970) as "a special type of debris flow occurring in main drainage channels . . . caused by short debris avalanches in steep walled tributary gullies . . ." They have also been referred to as "channel scour events" (Dyrness, 1967). However, neither of these statements fully describes most debris torrents.

Debris torrents are intermediate between debris avalanches and floods. They commonly consist of huge masses of logs, sediment, and water hurtling down steep channels at high velocity with very destructive effects (Figure 4–36). An eyewitness account of a debris torrent recounts an explosion of logs and debris that were tossed 25 ft (8 m) in the air as the slurry of debris clogged up at a bridge:

> "Then it came down and two seconds later this mass of debris was rolling across the road. Looked like 50 bulldozers just pushing a big bunch of spaghetti (logs), just going end over end." (Art Larson, court testimony)

Observations of drainage basins made soon after debris torrents occurred have revealed several characteristics of torrents. The debris often consists almost entirely of logs, mostly logging slash left in the channels from previous timber-har-

FIGURE 4–33
Lahar extending down the Toutle Valley from Mt. St. Helens, May 18, 1980. Some lahars continued downvalley for 120 km (72 mi).

vesting operations. Stream channels are scoured to bedrock with trimlines averaging 5 to 10 m (15 to 30 ft) above the floor of the channels. Debris torrents have been reported to occur in surges, characterized by a buildup of water, followed by surges of logs and water. The point of origin of a debris torrent is often easily identifiable because of the sharp contrast between the scoured channel below and the debris-choked channel above.

Recurrence intervals have been calculated from precipitation records for storms during which debris torrents occurred. These records indicate that unusually large amounts of precipitation are not the critical causative factor in triggering debris torrents. The increasing frequency of debris torrents since timber harvesting and the direct relationship of the failures to logging-related activities (especially roads) confirm this conclusion.

Recurrence intervals for storms associated with debris torrents in the Cascade Range foothills of Washington has ranged from 2 to 15 years. Hourly intensity recurrence intervals has ranged from 10 to 15 years.

FIGURE 4–34
Surface of lahar from Mt. St. Helens in the Toutle Valley, May 18, 1980.

The negative effect of logging on slope stability has been well documented in the literature. Following logging operations, large amounts of logging slash are often left on the slope. This debris eventually finds its way into stream channels by soil creep and sliding. Once in the channel, it awaits rapid movement downstream by floodwaters or debris torrents. The slash does not cause debris torrents, but once a debris torrent is generated, the slash is readily available for incorporation in the debris torrent. In studies of 12 drainages in western Washington, logging debris was responsible for almost all of the damage associated with the torrents (Figure 4–36) (Easterbrook, 1983, 1993; Syverson et al., 1985). Logging slash that accumulates in channels acts as temporary storage areas until the debris is overwhelmed by the high momentum of debris torrents.

Removal of forest canopy causes less water to be intercepted by trees, and thus less water is lost to evapotranspiration. Consequently, a greater volume of water runs off and infiltrates the soil. The resulting increased infiltration leads to saturation of soils, an increase in pore-water pressure, reduced internal friction, and lower strength of slope materials.

Tree roots have a stabilizing effect on slope by binding soil together, making it more resistant to erosion. Where soils are thin, tree roots often penetrate through the soil and into cracks in the bedrock, thereby anchoring the soil to bedrock and making it more resistant to erosion. After an area is clear-cut, small roots take about two years to decay, and large roots take about ten years. The root systems of the brush that naturally revegetates a logged area do not come close to making up for the lost strength of the tree root system. About 12 years after the clear-cut, some root strength has returned (Ziemer and Swanston, 1977) as a result of alder growth, but after about 12 years, the root strength decreases again as the area is revegetated by conifers.

The point of origin of debris torrents is often sliding failure at logging roads. Reasons for the road failures fall into two categories:

1. failure of the road fill because of saturation of the fill material which increases pore-water pressures, and

2. diversion of runoff along the road (analogous to a gutter on a roof) and then draining of this water onto the slope or into the channel.

Failure of the road-fill wedges is well shown where portions of roads have collapsed into stream gullies, initiating debris torrents downstream. That the road-fill failure caused the debris torrents is clear: The stream channels are scoured below the road-fill avalanches, but are not scoured above failures. That debris torrents seem to occur more frequently now than in the past because of human activity is demonstrated by comparing air photos taken when logging was just beginning in the basin with air photos taken later. Early air photos show no debris avalanche scars or evidence of former debris torrents in the main channel. In the following years, the basin was logged extensively, and many miles of road were built. The first signs of problems appear in 1978 photos that show at least nine debris avalanche scars originating at logging roads. Debris torrents scoured the drainage in 1979, 1980, and 1983, and 1983 photos show at least 31 debris avalanche scars originating at logging road failures (Figure 4–37).

The increasing frequency of debris torrents is attributable to poor land management. The presence of roads has been identified as the most important causative factor in initiation of debris torrents. Logging, particularly if the slope is not replanted with conifers, is also detrimental to slope stability. The destabilizing effects of road building and logging may persist for 50 years or more after slopes are logged.

EVOLUTION OF HILLSLOPES

The curved, sloping surfaces that make up most landscapes are shaped by the rate at which weathering produces rock debris and the rate at which the weathered material is

FIGURE 4–35
Poorly sorted debris and logs in a lahar from Mt. Baker in the Nooksack Valley, Washington.

moved downslope by mass wasting processes. Since the early work of Gilbert (1909), geomorphologists have recognized that as surface processes gradually sculpt hillslopes, the form of the slope profile at any time depends largely on the rates of weathering and mass wasting. However, whether or not slopes evolve with time due to systematic changes in rates of weathering and erosion has long been the focus of intense controversy. Davis (1909) maintained that slopes become lower and divides more broadly rounded with time as relief diminishes during reduction of a landmass toward base level. However, Penck (1924) assumed that slopes were determined by relative rates of uplift of a landmass and that as slopes retreat, a constant angle is maintained. Others have suggested that slopes reach stability with angles controlled by rock resistance and rates of weathering and mass wasting, so that a stable slope persists through time without changes in form, receding parallel to itself indefinitely.

The contrasting views of Davis (1909) and Penck (1924) —that is, whether slopes gradually decrease with time or are controlled by tectonic uplift rates and retreat parallel to themselves —sparked lengthy debate in the geological literature for many years. Few North American geomorphologists accepted Penck's concepts, which were more popular in Europe. Johnson, commenting on Penck's ideas, wrote the following:

"Penck's conception that slope profiles are convex, plane, or concave, according to the circumstances of the uplifting action, is in my judgment one of the most fantastic errors ever introduced into geomorphology. One may in the same region, which has experienced the same history throughout, as for example in the Bad Lands of South Dakota or in certain portions of the Black Hills, photograph both convex and concave slopes within a short distance of each other." (Johnson, 1940)

Gilbert (1909) was among the first to recognize that the smooth, gently curving slopes of valley sides represent equi-

FIGURE 4–36
Distal portion of a 1983 debris
torrent in Smith Creek,
Washington.

librium systems, and he coined the phrase "graded slopes" for them. The position of any point in the profile of a graded slope depends on all other points of the system, so that any changes that take place have consequences both upslope and downslope, much in the same fashion as in a graded stream (Cotton, 1952). The form of the slope is stable because the particles are arranged such that they support one another. Friction prevents any particle from sliding past another.

The upper segments of slope profiles are typically convex upward, in contrast to the concave-upward profiles in the lower segments of the slope. Creep dominates the upper, convex part of the profile as a result of weathering of bedrock, which produces loose debris that is reduced to finer and finer sizes downslope. The thickening layer of weathered debris mantling the bedrock enhances chemical weathering because retention of water in the weathered material encourages chemical weathering and because the breaking up of rock particles produces an increase in surface area that facilitates chemical reactions. Creep is driven by gravity, so that the steeper the slope, the greater the creep rate. The rate of movement reaches a maximum near the ground surface and decreases to zero at the base of the weathered mantle.

The form of a slope profile depends on the rate at which debris is produced by weathering and the rate at which it is moved by creep. The amount of material creeping through any two planes normal to the slope is equal to the amount of material moved into the lower plane from the upslope plane by creep, plus the amount added between the two planes by weathering. Thus, in order to transport an increasing amount of material downslope, either creep must accelerate downslope, or else the weathering mantle between the two planes

must thicken. Gilbert (1909) observed that the weathered material making up the convex-upward surface of a slope is usually quite uniform in thickness, and he concluded that in a given interval of time, a constant thickness of the weathered material must be removed by creep from the divide area, implying that progressively larger quantities of weathered material must creep through successive cross sections downslope. The amount of material that creeps through any cross section is proportional to the distance from the divide. Because creep is driven by gravity, the slope angle must therefore increase away from the divide in order to move the progressively greater amounts of debris downslope, and the increasing slope makes the profile convex upward. Gilbert's conclusion that the convexity of hilltop slopes is primarily the result of soil creep has since been verified by others (for example, Schumm, 1956; Baulig, 1957; Carson and Kirkby, 1972).

Surface flow of water on the convex-upward surface of a slope is discontinuous and diffuse. However, flowing water assumes increasing importance downslope as the profile steepens along the convex-upward slope until water begins to flow over the surface instead of infiltrating the creeping material. The increase in fine, weathered material downslope, which decreases permeability, the cumulative amount of runoff from the upper slopes, and the steepening slopes, promote sheetwash and rillwash on the surface of the lower slopes. As runoff is gathered into rills, fine-grained particles are more easily moved downslope on the surface, and the rill channels begin to impose their own concave-upward profile on the slope. Below the junction of two rills, the increase in channel cross section relative to the friction-producing wetted perimeter enhances the velocity, allowing the flow to transport its sedi-

FIGURE 4–37
Multiple logging road failures leading to a debris torrent in the valley below, Sygitowitz Creek, Washington. (Photo by Washington Dept. of Natural Resources)

ment load on a more gentle slope, that is, on a concave-upward slope. At some point on a slope, surface wash becomes dominant over creep, and the slope profile changes from convex to concave. Schumm (1956, 1962) observed broadly rounded, convex slopes in badlands developed on permeable material that impeded runoff and promoted creep. He noted, on the other hand, that concave-upward slopes were formed on less permeable material where runoff was higher.

A question that has stimulated controversy through several generations of geomorphologists is whether slopes decrease with time as divides are progressively lowered or whether they maintain a constant angle as they retreat parallel to themselves. The declivity of graded slopes that have attained equilibrium on homogeneous rock depends on the rates of weathering, wash, and creep. An equilibrium slope, once achieved, will maintain the same angle as long as no changes in controlling variables take place. However, as stream valleys are eroded incessantly nearer to base level, their rate of downcutting must decrease and eventually approach zero, so that equilibrium conditions on hillslopes must also gradual-

ly shift. Downwasting of the slopes to a surface of low relief could be avoided under such conditions, and parallel slope retreat could take place, only if rates of weathering and erosion were matched by long-continued, uniform crustal uplift over vast areas.

Penck (1924) steadfastly contended that slopes maintained a constant angle by parallel retreat. Because slope evolution is slow, taking thousands of years to produce measurable changes, direct observation of slope retreat is not possible. Instead, a deductive approach is necessary. If parallel slope retreat occurs universally on all slopes, it should develop on rocks of homogeneous composition. However, examples are virtually impossible to demonstrate within a short, historic time frame, and over longer, geologic, time periods, evidence of the history of evolution of slopes is equally difficult to obtain. Parallel slope retreat can be proven only under special conditions where steep cliffs are maintained by resistant, flat-lying beds that overlie less-resistant beds susceptible to sapping and undercutting of slopes by streams or wave action (Johnson, 1940).

Modeling of the evolution of slopes recently generated by faulting or erosion in unconsolidated sediment has been attempted (Colman and Watson, 1983; Nash, 1984; Pierce and Colman, 1986). Newly created scarps steeper than the angle of repose are quickly modified as sediment eroded from the upper part of the slope is deposited at its base. Changes in the slope with time can be described by the diffusion equation.

The rate at which the elevation of any point on the initial slope changes (*dy*) with time *(dt)* is equal to a diffusion constant (*c*) multiplied by the slope curvature (d_2y/dx_2). If the diffusion constant for a specific sediment type within a climatic zone can be determined by measuring the profiles of scarps of known age, the diffusion equation can estimate the age of an unknown scarp by its shape.

$$\frac{dy}{dt} = c\,\frac{d^2y}{dx^2}$$

REFERENCES

Abrahams, A. D., ed., 1986, Hillslope processes: Allen and Unwin, Boston, MA.

Abrahams, A. D., and Parsons, A. J., Cooke, R. U., and Reeves, R. W., 1984, Stone movement on hillslopes in the Mojave Desert, California: a 16-year record: Earth Surface Processes and Landforms, v. 9, p. 365–370.

Akagi, Y., 1980, Relations between rock type and the slope form in the Sonora Desert, Arizona: Zeitschrift für Geomorphologie, v. 24, p. 129–140.

Alden, W. C., 1928, Landslide and flood at Gros Ventre, Wyoming: American Institute of Mining and Metallurgical Engineers Transactions, v. 76, p. 347–361.

Alger, C., and Brabb, E., 1985, Bibliography of U. S. landslide maps and reports: U. S. Geological Survey Open File Report 85–585.

Anderson, R., and Humphrey, N., 1990, Interaction of weathering and transport processes in the evolution of and landscapes: *in* Cross, T., ed., Quantitative Dynamic Stratigraphy, Prentice Hall, Englewood Cliffs, N.J., 625 p.

Andersson, J. G., 1906, Solifluction, a component of subaerial denudation: Journal of Geology, v. 14, p. 91–112.

Andrews, D., and Hanks, T., 1985, Scarp degraded by linear diffusion: Inverse solution for age: Journal of Geophysical Research, v. 90, p. 19193–19208.

Bagnold, R. A., 1954, Experiments on a gravity-free dispersion of large solid spheres in a Newtonian fluid under shear: Proceedings of the Royal Society of London, v. 225, p. 49–63.

Balk, R., 1939, Disintegration of glacial cliffs: Journal of Geomorphology, v. 2, p. 305–334.

Barton, N., 1976, The shear strength of rock and rock joints: International Journal of Rock Mechanics, Mineral. Science, and Geomechanics, v. 13, p. 255–279.

Barton, N., 1978, Suggested methods for the quantitative description of discontinuities in rock masses: ISRM Commission on Standardization of Laboratory and Field Tests, International Journal of Rock Mechanics, Mineral. Science, and Geomechanics, v. 15, p. 319–368.

Barton, N., and Choubey, V., 1977, The shear strength of rock joints in theory and practice: Rock Mechanics, v. 10, p. 1–54.

Baulig, H., 1957, Peneplains and pediplains: (translated C.A. Cotton): Geological Society of America Bulletin, v. 68, p. 913–930.

Behre, C. H., 1933, Talus behavior above timber in the Rocky Mts.: Journal of Geology, v. 41, p. 622–635.

Benedict, J. B., 1970, Downslope soil movement in a Colorado alpine region. Rates, processes, and climatic significance: Arctic and Alpine Research, v. 2, p. 165–226.

Bentley, S. P., and Smalley, I. J., 1984, Landslips in sensitive clays: *in* Brunsden, D., and Prior, D. B., eds., Slope instability, John Wiley and Sons, Chichester, England, p. 457–490.

Blackwelder, E., 1928, Mudflow as a geologic agent in semi-arid mountains: Geological Society of America Bulletin, v. 39, p. 465–480.

Blackwelder, E., 1942, The process of mountain sculpture by rolling debris: Journal of Geomorphology, v. 5, p, 324–328.

Bovis, M. J., 1993, Hillslope geomorphology and geotechnique: Progress in Physical Geography, v. 17, p. 173–189.

Bovis, M. J., and Jones, P, 1993, Holocene history of earthflow mass movements in south-central British Columbia: The influence of hydroclimatic changes: Canadian Journal of Earth Sciences, v. 29, p. 1746–1755.

Browning, J. M., 1973, Catastrophic rock slides, Mount Huascaran, north-central Peru, May 31, 1970: American Association of Petroleum Geologists Bulletin, v. 57, p. 1335–1341.

Brunsden, D., 1971, Slopes—form and process: Institute of British Geographers Special Publication, no. 3, 178 p.

Brunsden, D., 1984, Mudslides: *in* Brunsden, D., and Prior, D. B., eds., Slope instability: John Wiley and Sons, Chichester, England, p. 363–418.

Brunsden, D., 1993, Mass movement; the research frontier and beyond: a geomorphological approach: Geomorhology, v. 7, p. 85–128.

Brunsden, D., and Kesel, R. H., 1973, Slope development on a Mississippi River bluff in historic time: Journal of Geology, v. 81, p. 576–597.

Brunsdlen, D., and Prior, D. B., eds., 1984, Slope Instability: John Wiley and Sons, N.Y.

Bryan, K., 1934, Geomorphic processes at high altitudes: Geographical Reviews, v. 24, p. 655–656.

Bryan, K., 1940, The retreat of slopes: Association of American Geograph. Annals, v. 30, p. 254–268.

Buckman, R. C., and Anderson, R. E., 1979,. Estimation of fault-scarp ages from a scarp-height-slope-angle relationship: Geology, v. 7, p.11–14.

Bunting, B. T., 1961, The role of seepage moisture in soil formation, slope development, and stream initiation: American Journal of Science, v. 259, p. 503–518.

Caine, N., 1980, The rainfall-duration control of shallow landslides and debris flows: Geografiska Annaler, v. 62A, p. 23–27.

Caine, N., 1981, A source of bias in rates of surface soil movement as estimated from marked particles: Earth Surficial Processes and Landforms, v. 6, p. 69–75.

Caine, N., 1982, Toppling failures from alpine cliffs on Ben Lomond, Tasmania: Earth Surficial Processes and Landforms, v. 7, p. 133–152.

Campbell, R. H., 1975, Soil slips, debris flows, and rainstorms in the Santa Monica Mountains, southern California: U. S. Geological Survey Professional Paper 851.

Cairnes, D. D., 1912a, Wheaton district, Yukon Territory: Canadian Geological Survey Memoir 31, 153 p.

Cairnes, D. D., 1912b, Differential erosion and equiplanation in portions of Yukon and Alaska: Geological Society of America Bulletin, v. 23, p. 333–348.

Carson, M. A., 1969, Models of hillslope development under mass failure: Geographical Analysis, v. 1, p. 76–100.

Carson, M. A., 1976, Mass-wasting, slope development, and climate: in Derbyshire, E. D., ed., Geomorphology and climate: John Wiley and Sons, London, p. 101–136.

Carson, M. A., and Kirkby, M. J., 1972, Hillslope form and process: Cambridge University Press, Cambridge, England, 475 p.

Carter, R. M., 1975, A discussion and classification of subaqueous mass-transport with particular application to grain-flow, slurry-flow, and fluxoturbidites: Earth Science Reviews, v. 11, p. 145–177.

Cluff, L. S., 1971, Peru earthquake of May 31, 1970; engineering geology observations: Seismological Society of America Bulletin, v. 61, p. 511–521.

Coleman, J. M., 1988, Dynamic changes and processes in the Mississippi River delta: Geological Society of America Bulletin, v. 100, p. 999–1015.

Coleman, J. M., and Prior, D. B., 1988, Mass wasting on continental margins: Ann. Rev. Earth and Planetary Sciences, v. 16, P. 101–119.

Colman, S. M., 1982, Clay mineralogy of weathering rinds and possible implications concerning the sources of clay minerals in soils: Geology, v. 10, p. 370–375.

Colman, S. M., 1983, Progressive changes in the morphology of fluvial terraces and scarps along the Rappahannock River, Virginia: Earth Surface Processes and Landforms, v. 8, p. 201–212.

Colman, S. M., and Watson, K., 1983, Ages estimated from a diffusion model for scarp degradation: Science, v. 221, p. 263–265.

Costa, J. E., and Jarrett, R. D., 1981, Debris flows in small mountain stream channels of Colorado, and their hydrologic implications: Bulletin Association of Engineering Geologists, v. 18, p. 309–22.

Costa, J. E., and Wieczorek, G. F., eds., 1987, Debris Flows/ Avalanches: Process, Recognition, and Mitigation: Geological Society of America Reviews in Engineering Geology, v. 7.

Cotton, C. A., 1952, Erosional grading of convex and concave slopes: Geographical Journal, v. 118.

Coulomb, G. A., 1773, Sur une application des regles de maximum et minimus a quelques problemes de statique relatifs l'architecture: Acad. Roy. des Sci., Mem. de Math. et de Phys par divers Sovans, V. 7, p. 343–382.

Crandell, D. R., 1971, Postglacial lahars from Mount Rainier volcano, Washington: United States Geological Survey Professional Paper 677, 75 p.

Crandell, D. R., Miller, C. D., Glicken, H. X., Christiansen, R. L., and Newhall, C. G., 1984, Catastrophic debris avalanche from ancestral Mount Shasta volcano, California: Geology, v. 12, p. 143–146.

Crozier, M. J., 1969, Earthflow occurrence during high intensity rainfall in eastern Otago (New Zealand): Engineering Geology, v. 3, p. 325–334.

Crozier, M. J., 1984, Field assessment of slope instability: in Brunsden, D. and Prior, W., eds., Slope Instability, John Wiley and Sons, N.Y., p. 103–142.

Crozier, M. J., 1986, Landslides: causes, consequences, and environment: Croom Helm, London, 252 p.

Cruden, D. M., and Eaton, T. M., 1987, Reconnaissance of rockslide hazards in Kananaskis Country Alberta: Canadian Geotechnical Journal, v. 24, p. 414–429.

Culling, W. E. H., 1965, Theory of erosion on soil-covered slopes: Journal of Geology, v. 73, p. 230–254.

Cummans, J., 1981, Mudflows resulting from the May 18, 1980, eruption of Mount St. Helens, Washington: U. S. Geological Survey Circular 850–B.

Cunningham, F., and Griba, W., 1973, A model of slope development, and its applications to the Grand Canyon, Arizona: Zeitschrift für Geomorphologie, v. 17, p, 43–77.

Davis, W. M., 1892, The convex profiles of bad-land divides: Science, v. 20, p. 245.

Davis, W. M., 1909, Geographical Essays: Ginn and Company, Boston, MA (reprinted 1954 by Dover Publications, Inc., N.Y.).

De Frietas, M. H., and Walters, R. J., 1973, Some field examples of toppling failure: Geotechnique, v. 23, p. 495–513.

Dietrich, W. E., Wilson, C. J., and Reneau, S. L., 1986, Hollows, colluvium, and landslides in soil-mantled landscapes: in Ahrahams. A. D., ed., Hillslope processes: Allen and Unwin, Boston, MA, p. 361–388.

Dunkerley, D. L., 1980, The study of the evolution of slope form over long periods of time: a review of methodologies and some new observational data from Papua, New Guinea: Zeitschrift für Geomorphologie, v. 24, p. 52–67.

Dunne, T., 1978, Field studies of hillslope flow processes: in Kirby, M. J., ed., Hillslope Hydrology, John Wiley and Sons, N. Y., p. 227–94.

Dyrness, C. T., 1967, Mass soil movements in the H. J. Andrews experimental forest: U.S. Forest Service Research Paper PNW–42, 12 p.

Eakin, H. M., 1916, The Yukon-Koyukuk region, Alaska: U.S. Geological Survey Bulletin 631.

Easterbrook, D. J., 1983, Processes related to origin of debris torrents and debris chutes: Geological Society of America, Abstracts with Program, p. 564–565.

Easterbrook, D. J., 1993, Surface Processes and landforms: Macmillan Publishing Co., p. 84–87.

Eisbacher, G. H., 1982, Mountain torrents and debris flows: Episodes, v. 1982, p. 12–17.

Eisbacher, G. H., and Clague, J. J., 1984, Destructive mass movements in high mountains: hazard and management: Geological Survey of Canada Paper 84–16, 230 p.

Engebretson, D. C., Easterbrook, D. J., and Kovanen, D. J., 1995, Relationships of very large, deep-seated, bedrock landslides and concentrated shallow earthquakes: Geological Society of America Abstracts with Program, v. 27, p. A377.

Engebretson, D. C., Easterbrook, D. J., and Kovanen, D. J., 1996, Triggering of very large, deep-seated, bedrock landslides by concentrated, shallow earthquakes in the North Cascades, Washington: Geological Society of America Abstracts with Program, v. 28, p. 64.

Ericksen, G. E., Pflacker, G., and Fernandez, J. V., 1970, Preliminary report on the geological events associated with the May 31, 1970 earthquake: U.S. Geological Survey Circular 639.

Erismann, T. H., 1979, Mechanisms of large landslides: Rock Mechanics, v. 12, p 15–46.

Erismann, T. H., Heuberger, H., and Preuss, E., 1977, Der bimsstein von Kofels (Tirol), ein bergsturz-"friktionit": Tschermaks Mineralogische und Petrographische Mitteilungen, v. 24, p. 67–119.

Fairchild, L. H., 1987, The importance of lahar initiation processes: in Costa, J. E., and Wieczorek, G. F., eds., Debris flows/avalanches: process, recognition and mitigation, Geological Society of America Reviews in Engineering Geology, v. 7. p. 51–61.

Farres, P. J. 1987, The dynamics of rainsplash erosion and the role of soil aggregate stability. Catena, v. 14, p. 119–130.

Findlayson, B., 1981, Field measurements of soil creep: Earth Surface Processes and Landforms, v. 6, p. 35–48.

Fleming, R. W., and Johnson, A. M., 1975, Rates of seasonal creep of silty clay soil:

Quaternary Journal of Engineering Geology, v. 8, p. 1–29.

Fleming, R. W. Johnson, R. B. and Scuster, R. L., 1987, The reactivation of the Manti Landslide, Utah: U. S. Geological Survey Survey Prof. Paper 1311-A, 22p.

Gardner, J. S., 1979, The movement of material on debris slopes in the Canadian Rocky Mountains: Zeitschrift für Geomorphologie, v. 23, p. 45–57.

Garwood, N. C., Janos, D. P., and Brokaw, N., 1979, Earthquake-caused landslide: A major disturbance to tropical forests: Science, v. 205, p.997–999.

Gates, W. C. B., 1987, The fabric of rockslide avalanche deposits: Bulletin of Association of Engineering Geologists, v. 24, p. 389–402.

Gilbert, G. K., 1909, The convexity of hilltops: Journal of Geology, v. 17, p. 344–350.

Graham, J., 1984, Methods of stability analysis: in Brunsden, D., and Prior, D. B., ed., Slope Instability,.John Wiley and Sons, N.Y., p. 171–215.

Grainger, P, and Kalaugher, P. G., 1987, Intermittent surging movements of a coastal landslide: Earth Surface Processes and Landforms, v. 12, p. 597–603.

Hadley, J. B., 1964, Landslides and related phenomena accompanying the Hebgen Lake earthquake of August 17, 1959: U. S. Geological Survey Professional Paper 435, p. 107–138.

Haig, M., 1979, Ground retreat and slope evolution on regraded surface-mine dumps, Waunafon, Gwent: Earth Surface Processes and Landforms, v. 4, p. 183–189.

Haig, M. J., and Wallace, W. L., 1982, Erosion of strip-mine dumps in LaSalle County, Illinois: preliminary results: Earth Surface Processes and Landforms, v. 4, p. 183–189.

Hamblin, W. K., and Christiansen, E. H., 1995, Earth's Dynamic Systems: Prentice Hall, Englewood Cliffs, N. J., 710 p.

Hayden, E. W., 1956, The Gros Ventre Slide (1925) and the Kelly Flood (1927): Wyoming Geological Association 11th Annual Field Conference Guidebook, p. 20–22.

Heuberger, H., et al., 1984, Quaternary landslides and rock fusion in central Nepal and in the Tyrolean Alps: Mountain Research and Development, v. 4, p. 345–362.

Heuberger, H., 1989, The giant landslide of Kofels between Langenfeld and Unthausen, Otz Valley, Tyrol: in Ives, J. D., and Heuberger, H., eds., Symposium S03: Mountain Hazard Geomorphology, 2nd International Conference on Geomorphology, Frankfurt/Main, p. 36–42.

Hoek, N. B., 1971, The influence of structure on the stability of rock slopes: Proc. 1st Symposium on Stability in Open Pit Mining, Vancouver, AIME, p. 49–63.

Holmes, C. D., 1955, Geomorphic development in humid and arid regions—a synthesis: American Journal of Science, v. 253, p. 377–390.

Horton, R. E., 1933, The role of infiltration in the hydrological cycle: American Geophysical Union Transactions, v. 14, p. 446–460.

Hsu, K. J., 1975, Catastrophic debris streams (sturzstroms) generated by rockfalls: Geological Society of America Bulletin, v. 86, p. 129–140.

Hsu, K. J., 1978, Albert Heim: observations on landslides and relevance to modern interpretations: in Rockslides and avalanches, v. 1, Elsevier, Amsterdam, Netherlands, p. 71–93.

Hupp, C. R., 1983, Geo-botanical evidence of late Quaternary mass wasting in blockfield areas of Virginia: Earth Surface Processes and Landforms, v. 8, p. 439–450.

Hupp, C. R., 1984, Dendrogeomorphic evidence of debris flow frequency and magnitude at Mount Shasta, California: Env. Geol. and Water Science, v. 6, p. 21–28.

Jacobson, R. B., Cron, E. D., and McGeehin, J. P., 1989, Slope movements triggered by heavy rainfall, November 3–5, 1985: in Schultz, A. P., and Jibson, R. W., eds., Landslide processes of the eastern United States and Puerto Rico, Geological Society of America Special Paper 236, p. 1–13.

Jacobson, R. B., Miller, A. J., and Smith, J. A., 1989, The role of catastrophic geomorphic events in central Appalachian landscape evolution: in Gardner, T. W., and Sevon, W. D.,eds., Appalachian Geomorphology, Geomorphology, v. 2, p. 257–284.

Janda, R. J., Scott, K. M., and Martinson, H. A., 1981, Lahar movement, effects, and deposits: in Lipman, P. W., and Mullineaux, D. R., eds., The 1980 eruptions of Mount St. Helens, Washington: U.S. Geological Survey Professional Paper 1250, p. 461–478.

Johnson, D., 1940, Comments on the geomorphic ideas of Davis and Penck: Annals of Association of American Geographers, v. 30, p. 228–232.

Johnson, A. M., and Rahn, P. H., 1970, Mobilization of debris flows: Zeitschrift für Geomorphologie, v. 9, p. 168–186.

Johnson, A. M., and Rodine, J. R., 1984, Debris flow: in Brunsden, D., and Prior, D. B., eds., Slope instability: John Wiley and Sons, Chichester, England, p. 257–361.

Jones, F. O., Embody, D. R., and Peterson, W. L., 1961, Landslides along the Columbia River valley, northeastern Washington: U.S. Geological Survey Professional Paper 367, 98 p.

Karrow, P. F., 1972, Earthflows in the Grondines and Trois Rivieres areas, Quebec: Canadian Journal of Earth Sciences, v. 9, p. 561–573.

Keefer, D. K., and Johnson, A. M., 1983, Earth flows; morphology, mobilization, and movement: U.S. Geological Survey Professional Paper 1264, 56 p.

Keefer, W. D., and Love, J. D., 1956, Landslides along the Gros Ventre River, Teton County, Wyoming: Wyoming Geological Association Guidebook, 11th Annual Field Conference.

Kent, P. E., 1966, The transport mechanism in catastrophic rockfalls: Journal of Geology, v. 74, p. 79–83.

Kiersch, G. A., 1964, Vaiont reservoir disaster: Civil Engineering, v. 34, p. 32–39.

Kiersch, G. A., 1965, The Vaiont reservoir disaster: Mineral Information Service, v. 18, p. 129–138.

Kirkby, M. J., 1967, Measurement and theory of soil creep: Journal of Geology, v. 75, p. 359–378.

Kovanen, D. J., and Easterbrook, D. J., 1996, Extensive Readvance of Late Pleistocene (Y.D.?) Alpine Glaciers in the Nooksack River Valley, 10,000 to 12,000 Years Ago, Following Retreat of the Cordilleran Ice Sheet, North Cascades, Washington: Friends of the Pleistocene, Pacific Coast Cell Field Trip Guidebook, 74p.

Lamarche, V. C., 1968, Rates of slope degradation as determined from botanic evidence, White Mountains, California: U.S. Geological Survey Professional Paper 352–1, p. 341–377.

Masch, L., Wenk, H. T., and Preuss, E., 1985, Electron microscopy study of hyalomylonites—evidence for frictional melting in landslides: Tectonophysics, v. 115, p. 131–160.

McDowell, B., 1986, Eruption in Columbia: National Geographic, May 1986, p. 641–652.

McDowell, B., and Fletcher, J. E., 1962, Avalanche! 3,500 Peruvians perish in seven minutes: National Geographic, v. 121, p. 855–880.

Melosh, H. J., 1983, Acoustic fluidization: American Scientist, v. 71, p. 158–165.

Melosh, H. J., 1987, The mechanics of large rock avalanches: in Costa, J. E., and Wieczorek, G. F., eds., Debris flows/avalanches, process, recognition, and mitigation, Geological Society of America Reviews in Engineering Geology, v. 7, p. 41–49.

Milton, D., 1964, Fused rock from Kofels, Tyrol: Tschermaks Mineralogische und Petrographische Mittelungen, v. 9, p. 86–94.

Mitchell, J. K., 1976, Fundamentals of soil behavior: John Wiley and Sons, N.Y., 422 p.

Moon, B. R, 1986, Controls on the form and development of rock slopes in fold terrane: in Abrahams, A. D., ed., Hillslope processes: Allen and Unwin, Boston, MA, p. 225–243.

Moser, M., and Hohensinn, F., 1983, Geotechnical aspects of soil slips in alpine regions: Engineering. Geology, v. 19, p. 185–211.

Moyersons, J., 1988, The complex nature of creep movements on steeply sloping ground in southern Rwanda: Earth Surface Processes and Landforms, v. 13, p. 511–524.

Nash, D. B., 1980, Forms of bluffs degraded for different lengths of time in Emmet County, Michigan, U.S.A.: Earth Surface Processes and Landforms, v. 5, p. 331–345.

Nash, D. B., 1984, Morphologic dating of fluvial terrace scarps and fault scarps near West Yellowstone, Montana: Geological Society of America Bulletin, v. 95, p. 1413–1424.

Nash, D. F. T., 1987, Comparative review of limit equilibrium methods of stability analysis: in Anderson, M. G., and Richards, K. S., eds., Slope stability, geotechnical engineering and geomorphology, John Wiley and Sons, Chichester, England, p. 11–75.

National Academy of Sciences, 1978, Landslides: analysis and control: Transportation Research Board Special Report 176, Washington D. C.

Neary, D. G., and Swift, L. W., Jr., 1987, Rainfall thresholds for triggering a debris avalanching event in the southern Appalachian Mountains: in Costa, J. E., and Wieczorek, G. F., eds., Debris flows/avalanches, process, recognition, and mitigation, Geological Society of America Reviews in Engineering Geology, v. 7, p. 81–92.

Parsons, A. J., 1988, Hillslope form: Routledge, London.

Peltier, L., 1950, The geographical cycle in periglacial regions as it is related to climatic geomorphology: Annals of Association of American Geographers, v. 40, p. 214–236.

Penck, W., 1924, Die morphologische analyse: Englehorn's Nachfolger, Germany, Stuttgart.

Phipps, R. L., 1974, The soil creep-curved tree fallacy: U.S. Geological Survey Journal of Research, v. 2, p. 371–377.

Pierce, K. L., and Colman, S. M., 1986, Effect of height and orientation (microclimate) on geomorphic degradation rates and processes, late-glacial terrace scarps in central Idaho: Geological Society of America Bulletin, v. 97, p. 869–885.

Pierson, T. C., 1980, Erosion and deposition by debris flows at Mount Thomas, New Zealand: Earth Surface Processes and Landforms, v. 5, p. 227–247.

Pierson, T. C., and Costa, J. E., 1987, A rheologic classification of subaerial sediment-water flows: in Costa, J. E., and Wieczorek, G. F., eds., Debris flows/avalanches, process, recognition, and mitigation, Geological Society of America Reviews in Engineering Geology, v. 7, p. 1–12.

Plafker, G., and Ericksen, G. E., 1978, Nevados Huascaran avalanches, Peru: in Voight, B., ed., Rockslides and avalanches 1: natural phenomena: Developments in Geotechnical Engineering 14A, Elsevier Scientific Publishing Company, Amsterdam, p. 277–314.

Plafker, G., Ericksen, G. E., and Concha, J. F., 1971, Geological aspects of the May 31, 1970, Peru earthquake: Seismological Society of America Bulletin, v. 61, p. 543–578.

Pomeroy, J. S., 1980, Storm-induced debris avalanching and related phenomena in the Johnstown area, Pennsylvania, with references to other studies in Appalachians: U.S. Geological Survey Professional Paper 1191.

Porter, S. C., and Crombelli, G.,1981. Alpine rockfall hazards: American Scientist, v. 69, p. 67–77.

Preuss, E., 1974, Der birnsstein von Kofels im. Ortztal/Tirol, die reibungsschmelze eines bergsturzes: Jahrbuch des Vereins zum Schutze der Alpenpflanzen und Tiere, v. 39, p. 85–95.

Preuss, E., 1986, Gleitflachen und neue friktionitfunde im, bergsturz von Kofels im Otztal, Tirol: Material und Technik, v. 14, p. 169–174.

Prior, D. B., and Stephens, N., 1972, Some movement patterns of temperate mudflows: examples from northeastern Ireland: Geological Society of America Bulletin, v. 83, p. 2533–2543.

Qian, Y., Yang, W., Zhao, W., Cheng, X., Zhang, L., and Xu, W., 1980, Basic characteristics of flow with hyperconcentrations of sediment: in Proceedings of the International Symposium on River Sedimentation: Beijing, Chinese Society of Hydraulic Engineering, p. 175–184.

Rahm, D. A., 1962, The terracette problem: Northwest Science, v. 36, p. 65–80.

Rapp, A., 1960, Recent development of mountain slopes in Karkevagge and surroundings, northern Scandinavia: Geografiska Annaler, v. 42, p. 65–206.

Reneau, S. L., Dietrich, W. E., Dom, R. I., Berger, C. R., and Rubin, M., 1986, Geomorphic and paleoclimatic implications of latest Pleistocene radiocarbon dates from colluvium-mantled hollows California: Geology, v. 14, p. 655–658.

Richmond, G. M., 1949, Stone nets, stone stripes, and soil stripes in the Wind River Mts., Wyoming: Journal of Geology, v. 57, p. 143–153.

Robinson, E. S., 1970, Mechanical disintegration of the Navajo Sandstone in Zion Canyon, Utah: Geological Society of America Bulletin, v. 81, p. 2799–2805.

Rodine, J. D., 1974, Analysis of mobilization of debris flows [PhD thesis]: Stanford University, 226 p.

Russell, R. J., 1933, Alpine land forms of the western United States: Geological Society of America Bulletin, v. 44, p. 927–950.

Saunders, I., and Young, A., 1983, Rates of surface processes on slopes, slope retreat, and denudation: Earth Surface Processes and Landforms, v. 8, p. 473–501.

Scheffel, E. R., 1920, "Slides" in the Conernaugh formation near Morgantown, West Virginia: Journal of Geology, v. 28, p. 340–355.

Schumm, S. A., 1956, Evolution of drainage systems and slopes in badlands and slopes in badlands at Perth Amboy, New Jersey: Geological Society of America Bulletin, v. 67, p. 597–646.

Schumm, S. A., 1962, Erosion of miniature pediments in Badlands National Monument, South Dakota: Geological Society of America Bulletin, v. 73, p. 719–724.

Schumm, S. A., 1967, Rates of surficial rock creep on hillslopes in western Colorado: Science, v. 155, p. 560–561.

Schumm, S. A., and Chorley, R. J., 1964, The fall of Threatening Rock: American Journal of Science, v. 262, p. 1041–1054.

Schumm, S. A., and Chorley, R. J., 1966, Talus weathering and scarp recession in the Colorado Plateaus: Zeitschrift für Geomorphologie, v. 1, p. 11–36.

Schuster, R. L., and Krizak, R. L., 1978, Landslides: analysis and control: Transportation Research Board, National Academy of Science, National Research Council Special Report 176, 234 p.

Selby, M. J., 1966, Methods of measuring soil creep: Journal of Hydrology, v. 5, p. 54–63.

Selby, M. J., 1980, A rock mass strength classification for geomorphic purposes: with tests from Antarctica and New Zealand: Zeitschrift für Geomorphologie, v. 24, p. 31–51.

Selby, M. J., 1982, Hillslope materials and processes: Oxford University Press, N.Y.

Selby, M. J., 1987, Rock slopes: in Anderson, M. G., and Richards, K. S., eds., Slope stability, geotechnical engineering and geomorphology: John Wiley and Sons, Chichester, England, p. 475–504.

Sharp, R. P., 1942, Mudflow levees: Journal of Geomorphology, v. 5, p. 222–227.

Sharp, R. P., and Nobles, L. H., 1953, Mudflow of 1941 at Wrightwood, southern California: Geological Society of America Bulletin, v. 64, p. 547–560.

Sharpe, C. F. S., 1938, Landslides and related phenomena: Columbia University Press, N.Y., 137 p.

Sharpe, C. F. S., and Dosch, E. F., 1942, Relation of soil creep to earthflow in the Appalachian Plateaus: Journal of Geomorphology, v. 5, p. 312–324.

Shreve, R. L., 1966, Sherman landslide, Alaska: Science, v. 154, p. 1639–1643.

Shreve, R. L., 1968, The Blackhawk landslide: Geological Society of America Special Paper 108, 47 p.

Shroder, J. F., Jr., 1989, Hazards of the Himalaya: American Scientist, v. 77, p. 564–573.

Sidle, R. C., Pearce, A. J., and O'Loughin, C. L., 1985, Hillslope stability and land use: American Geophysical Union Monograph Series, v. 11, 140 p.

Sidle, R. C., and Swanston, D., 1982, Analysis of a small debris slide in coastal Alaska: Canadian Geotechnical Journal, v. 19, p. 167–174.

Skempton, A. W., 1953, Soils mechanics in relation to geology: Yorkshire Geological Society Proceedings, v. 29, p. 33–62.

Skempton, A. W., 1964, The long-term stability of clay slopes: Geotechnique, v. 2, p. 75–102.

Skempton, A. W., and Hutchinson, J. N., 1969, Stability of natural slopes and embankment foundations: Proceedings of 7th International Conference on Soil Mechanics, Foundation Engineering, p. 221–242.

Strahler, A. N., 1950, Equilibrium theory of erosional slopes approached by frequency distribution analysis: American Journal of Science, v. 248, p. 673–696, 800–814.

Strahler, A. N., 1956, Quantitative slope analysis: Geological Society of America Bulletin, v. 67, p. 571–596.

Surenian, R., 1988, Scanning electron microscope study of shock features in pumice and gneiss from Kofels (Tyrol, Austria): Innsbruck, Geolog. Palaontolog. Mitteilungen, v. 15, p. 135–143.

Swanson, F. J., and Dyrness, C. T., 1975, Impact of clear-cutting and road construction on soil erosion by landslides in the western Cascade Range, Oregon: Geology, p. 393–396.

Swanston, D. N., 1970, Principal mass movement processes influenced by logging, road building and fire: Proceedings of

Symposium Forest Land Uses and Stream Environment, Corvallis, Oregon, p. 29–117.

Swanston, D. N., 1974, Slope stability problems associated with timber harvesting in mountainous regions of the western United States: U.S. Forest Service General Technical Report PNW-21, 14 p.

Syverson, T., Easterbrook, D. J., and McCarten, C. A., 1985, Cause of debris torrents in the Cascade foothills of Washington: Geological Society of America, Abstracts with program, v. 17, p. 411.

Takahaski, T., 1978, Mechanical characteristics of debris flows: Journal of the Hydraulics Division, American Society of Civil Engineers, v. 104, p. 1153–1169.

Takahaski, T., 1981, Debris flow: Annual Reviews of Fluid Mechanics, v. 13, p. 57–77.

Terzaghi, K., 1936, The shearing resistance of saturated soils: Proceedings of the 1st international Conference on Soils Mechanics and Foundation Engineering, v. 1, p. 54–66.

Terzaghi, K., 1943. Theoretical soils mechanics: John Wiley and Sons, N.Y.

Terzaghi, K., 1950, Mechanisms of landslides: in Paige, S., ed., Application of geology to engineering practice: Geological Society of America, Berkey Volume, p. 83–123.

Terzaghi, K., 1962, Stability of steep slopes on hard unweathered rock: Geotechnique, v. 12, p. 251–270.

Thomas, D. G., 1963, Non-Newtonian suspensions: physical properties and laminar transport characteristics: Industrial and Engineering Chemistry, v. 55, p. 18–29.

Torrance, J. K., 1987, Slope stability: in Anderson, M. G., and Richards, K. S., eds., Slope stability, geotechnical engineering and geomorphology: John Wiley and Sons, Chichester, England, p. 447–473.

Toy, T. J., 1977, Hillslope form and climate: Geological Society of America Bulletin, v. 88, p. 16–22.

Twidale, C. R., and Campbell, E. M., 1986, Localised inversion on steep hillslopes: Gully gravure in weak and in resistant rocks: *Zeitschrift für Geomorphologie, v.* 30, p. 35–46.

Van Burkalow, A., 1945, Angle of repose and angle of sliding friction: an experimental study: Geological Society of America Bulletin, v. 56, p. 669–707.

VanDine, D. F., 1985, Debris flows and debris torrents in the southern Canadian Cordillera: Canadian Geotechnical Journal, v. 22, p. 44–68.

Varnes, D. J., 1958, Landslide types and processes: in Eckel, E., ed., Landslides and engineering practice: Washington D.C., Highway Research Board Special Report 29, p. 20–47.

Varnes, D. J., 1978, Slope movement types and processes: in Schuster, R., and Krizak, R., eds., Landslides: Washington, D.C., National Academy of Sciences Transportation Research Board, p. 11–33.

Voight, B., 1983, Nature and mechanics of the Mount St. Helens rockslide-avalanche of 18 May 1980: Geotechnique, v. 33, p. 243–273.

Ward, W. H., 1945, The stability of natural slopes: Geographical Journal, v. 105, p. 170–197.

Washburn, A. L., 1947, Reconnaissance geology of portions of Victoria Island and adjacent regions, Arctic Canada: Geological Society of America Memoir 22, 142 p.

Washburn, A. L., 1967, Instrumental observations of mass-wasting in the Mesters Vig District, northeast Greenland: Meddelser om Gronland, v. 166, 296 p.

Wasson, R. J., and Hall, G., 1982, A long record of mudslide movement at Waerenga-0-Kuri, New Zealand: Zeitschrift für Geomorphologie, v. 26, p. 73–85.

Watson, R. A., and Wright, H. E., Jr., 1969, The Saidmarreh landslide, Iran: in Schumm, S. A., and Bradley, W. C., eds., U.S.

contributions to Quaternary research, Geological Society of America Special Paper 123, p. 115–139.

Wells, J. T., Prior, D. B., and Coleman, J. M., 1980, Flowslides in muds on extremely low angle tidal flats, northeastern South America: Geology, v. 8, p. 272–275.

Williams, G. P., and Guy, H. P., 1959, Debris avalanches—a geomorphic hazard: in Environmental geomorphology: Binghamton, State University of New York, Publications in Geomorphology, p. 25–46.

Witkind, I. J., 1964, The Hebgen, Lake, Montana, earthquake of August 17, 1959, Events on the night of August 17, 1959: the human story: U. S. Geological Survey Professional Paper 435, 4 p.

Wood, A., 1942, Development of hillside slopes: Geologists Association Proceedings, v. 53, p. 128–140.

Young, A., 1960, Soil movement by denudational processes on slopes: Nature, v. 188, p. 120–122.

Young, A., 1961, Characteristic and limiting slope angles: Zeitschrift für Geomorphologie, v. 5, p. 126–131.

Young, A., 1972. Slopes: Oliver and Boyd, Edinburgh, U.K., 288 p.

Young, A., 1974, The rate of slope retreat: in Brown, E. H., and Waters, R. S., eds., Progress in geomorphology: Institute of British Geographers Special Publication no. 7, p. 65–77.

Young, A., and Saunders, L, 1986, Rates of surface processes and denudation: in Abrahams, A. D., ed., Hillslope processes: Allen and Unwin, Boston, MA, p. 3–27.

Ziemer, R. R., and Swanston, D. N., 1977, Root strength changes after logging in southeast Alaska: in USDA Forest Service Research Note, PNW-306, 10 p.

Fluvial Processes

Meandering stream with many cutoff meanders. (Photo from U.S. Dept. of Agriculture)

INTRODUCTION

Running water is the single most important surface process in shaping the configuration of the Earth's surface. Other processes may play more important roles locally, but on a world-wide basis, streams are responsible for more modification of the landscape than any other process. Plate tectonics and volcanism are responsible for creation of large mountain chains, but the sculpturing of mountains into their present forms is accomplished primarily by water.

In addition to carving the topography of the land, streams can also constitute catastrophic hazards. Virtually all streams flood from time to time, sometimes with devastating effects. In the United States alone, floods cause more than $1 billion in property damage annually, and an average of 85 people per year are killed by floods. Approximately 100,000 floods are estimated to occur worldwide in the next decade

VARIABLES OF STREAM PROCESSES

The Hydrologic Cycle

The quantity of water presently at the Earth's surface remains essentially constant over long periods of time, with occasional minor additions of "new" water from volcanic activity. The oceans contain more than 97 percent of the Earth's water, the remainder residing in lakes and streams and in the atmosphere.

Water moves in a continuous cycle from the oceans to the land and back again to the oceans. Water vapor enters the atmosphere by evaporation from the ocean surface and is carried over the continents by atmospheric circulation where precipitation occurs. Some of the precipitation is intercepted by vegetation, some evaporates back into the air, some is returned to the atmosphere by plants, some infiltrates into the ground, some is stored temporarily in glaciers, and some runs off in streams. Eventually, the water makes its way back into the oceans where the cycle starts all over again. The total amount of water moving through the hydrologic cycle is estimated to be more than 100 million billion gallons per year.

Discharge

The discharge of most streams is derived largely from precipitation, abetted by contributions from ground water and snowmelt. Of the total precipitation that falls on the earth, only about 25 to 40 percent runs off on the surface as stream discharge; the remainder falls on vegetation or other material that intercepts the moisture. This relationship may be written as

$$precipitation = runoff + interception + storage$$

where interception includes such factors as evapotranspiration of vegetation, evaporation, and infiltration.

$$interception = evapotranspiration + evaporation + infiltration$$

Storage is the temporary holding of water as snowpack or ground water before being released into streams as runoff.

Precipitation

Precipitation in a region is highly variable, in both long-term and short-term views. Annual amounts differ from year to year, and even in a single storm, the amount of rainfall may vary considerably from place to place. The characteristics of a given storm may be considered in terms of precipitation per unit time at any given recording station or total storm precipitation.

Most stations keep 24-hour records, recording the amount of precipitation daily, then resetting rain gauges to zero for the next 24-hour period. Some may also keep hourly records, which are very useful for calculating rainfall **intensity.** The magnitude of a storm at a station may be depicted by calculating a storm **recurrence interval** (RI). This interval is calculated by counting the number of times a storm of defined magnitude has been equaled or exceeded, divided by the period of record.

$$RI = \frac{\text{number of years of record}}{\text{number of storms beyond a given magnitude}}$$

For example, if storms of 4 inches or more of rainfall occurred (in 24 hours) five times in a 100-year interval, a storm of that magnitude would occur on an average of once every 20 years; that is, its recurrence interval would be 20 years. The precipitation from storms of various magnitude can then be plotted against their recurrence intervals on semilog paper, and a probability chart can then be derived by fitting a straight line through the points (Figure 5–1). Thus, even if a station has only 30 to 40 years of record, the magnitude of "100-year" storms or "50-year" storms can be read by extrapolation of the line from known points on the graph.

Intensity is the amount of precipitation that falls per unit time on the land surface. It is generally measured by recording the amount of precipitation hourly. The intensity of precipitation during a storm plays a significant role in determining runoff values. For example, if a given amount of rainfall is evenly distributed over a 24-hour period, much of

it may be intercepted; however, if the same amount of rainfall is concentrated in a very short period of time, it may overwhelm interception, and more of it may run off.

Most storms hold peak intensities for only a few hours, and thus stations that record only 24-hour precipitation do not accurately portray intensities. A graph of hourly precipitation intensity may be constructed by plotting time in hours versus amount of precipitation recorded per hour. From such data, intensities for various time periods may be calculated and their recurrence interval determined; 1-hour, 6-hour, and 12-hour intervals are commonly used. The recurrence interval for maximum 6-hour intensities can be found using the same techniques as for daily (24-hour) periods. The NOAA Atlas of Precipitation shows **isopluvial** maps for such short intervals using recurrence interval values of 1, 50, and 100 years.

Duration is the length of time that precipitation of a given intensity lasts. In general, storms of unusually high intensity are of short duration and have high recurrence intervals; that is, they do not occur as often as storms of low intensity. Thus, precipitation intensity is inversely proportional to duration and to recurrence interval.

The longer the duration of a high-intensity storm, the higher the amount of rainfall over short time periods and the greater the likelihood of significant geomorphic effects. Short thunderstorms in mountainous areas are examples of short-duration, high-intensity events, and hurricanes or typhoons are examples of longer-duration, high-intensity events. Both are capable of producing catastrophic results. In 1972, 35 centimeters (14 inches) of rain fell during the night of June 9 in the Black Hills of South Dakota, causing sudden severe flooding of Rapid City (Figure 5–2) and resulting in about 200 deaths (Rahn, 1975).

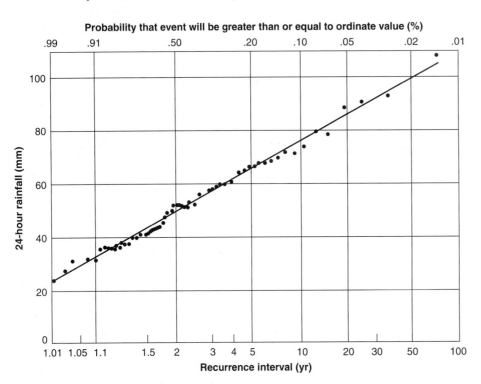

FIGURE 5–1

Recurrence interval of 24-hour precipitation, Buffalo, N.Y., 1891–1961. (From Dunne and Leopold, 1978)

FIGURE 5–2
House carried onto highway by 1972 Black Hills, South Dakota, flood.

Intensity-duration curves for areas can be derived for precipitation of various recurrence intervals. Families of curves such as those in Figure 5–3 can then be used to evaluate storms of different intensities and duration. For example, comparison of duration versus intensity curves for Seattle, Washington and Miami, Florida shows that the 100-year return period for any intensity/duration in Seattle corresponds to a less-than-2-year return period in Miami, i.e., high intensity storms for any given duration are much more common in Miami than Seattle.

Antecedent precipitation is the amount of precipitation preceding some event. In areas that experience pronounced rainy seasons, high-intensity storms may occur following extended periods of precipitation. The geomorphic importance of antecedent precipitation is that if soil and vegetational interceptors are already saturated by antecedent rainfall, an intense storm may produce higher runoff than would otherwise occur. Another factor may be the amount of snow on the ground during a warm rainstorm. Melting of the snow can add significantly to the total amount of runoff during a given storm.

Interception

Interception is the amount of precipitation that falls from the sky but never reaches the ground. Only a portion of precipitation reaches the ground and runs off in streams. Some of the precipitation is intercepted by vegetation or other ground cover before it reaches the ground and is evaporated back into the atmosphere or is taken in by plants (Figure 5–4).

The amount of precipitation that is intercepted by vegetation depends upon (1) the kind of plant cover, (2) how intense the precipitation is, (3) how long the precipitation lasts, (4) the air temperature and humidity, and, (5) the season of the year.

In heavily forested regions, most of the precipitation lands on leaves, stems, and branches where it remains until evaporated back into the atmosphere. If a rainstorm is prolonged, water temporarily stored on leaves and branches in the upper portions of a tree canopy drips down onto lower parts of the tree, and some may eventually reach brush, grass, or other low-growing vegetation beneath the tree where further interception occurs before the water reaches the ground. Some of the water is taken in by the plants and is later transpired back into the atmosphere.

The amount of precipitation intercepted by vegetation depends upon the nature of plant cover in an area, the rainfall intensity and duration, the air temperature and humidity, and, in some cases, the season of the year. Plant cover varies from species to species, including differences in the size and shape of leaves, leaf density between the canopy and the ground (including not only tree canopies but also lower-growing plants, brush, and grass), and spacing of individual plants. Differences between the broad leaves of deciduous trees and the needles of conifers affect the amount of interception, as do strong differences between forest cover and grassland.

Water filters through vegetation and onto the land surface as *throughfall* (dripping off leaves and branches) or stemflow (down the trunks of trees). Throughfall may be measured by collecting all of the water dripping from tree foliage or by placing random rain gauges under the tree canopy. *Stemflow* can be measured by placing a collar around the tree trunk to collect water flowing down the trunk. Water intercepted by grass and very small plants is much more difficult to measure. Generally, throughfall is much more significant than stemflow, which usually amounts to only a small percentage

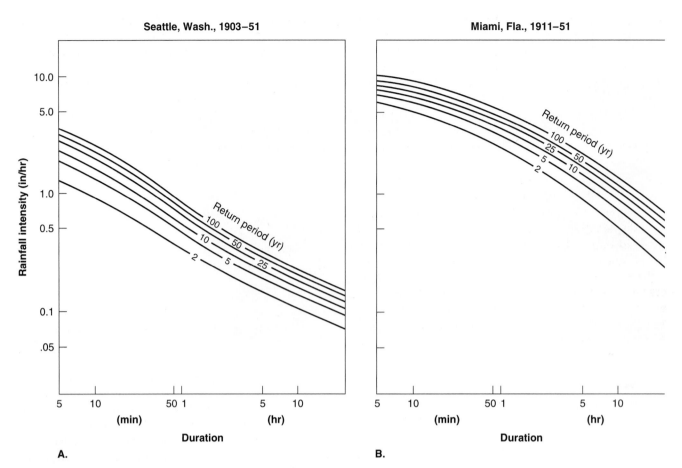

FIGURE 5–3
Intensity-duration recurrence interval curves for Seattle, Washington, and Miami, Florida. The family of curves for each station shows the recurrence interval for intensities of various durations. (From Dunne and Leopold, 1978)

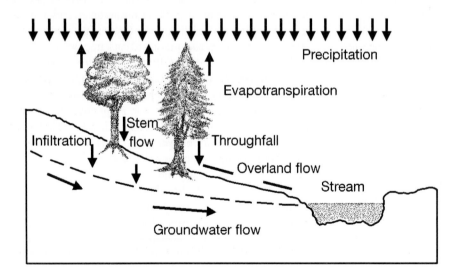

FIGURE 5–4
Components of water budget derived from precipitation.

of total precipitation. Coniferous forests typically intercept somewhat more precipitation than deciduous forests because they have greater density of foliage and because needles are capable of retaining more water than are broad leaves. In addition, deciduous trees lose their leaves during winter, allowing much more throughfall.

Another method of estimating interception is to measure total precipitation in a drainage basin and stream discharge from the mouth of the basin. The difference between these two gives an approximation of the total interception over a given time period. Compilations of such measurements suggest a wide variation from region to region, depending largely on the type of vegetation and the intensity and duration of precipitation. The values for total interception relative to precipitation shown in Figure 5–5 vary from about 50 to 90 percent. Much lower values for canopy interception relative to annual or seasonal precipitation, averaging about 13 percent for deciduous forests and 27 to 35 percent for coniferous forests, have been measured (Dunne and Leopold, 1978).

Largely because of the difficulties in measuring and generalizing the effects of all of these variables, opinions differ as to the effectiveness of vegetation in intercepting precipitation. Interception values measured for entire drainage basins are usually higher than those measured for individual trees.

Of the precipitation intercepted by vegetation, most is lost by transpiration (diffusion of water vapor into the atmosphere from plants) and evaporation. Because of the difficulties involved in measuring transpiration and evaporation separately, these two processes are typically combined under the term **evapotranspiration.** The rate of evaporation of water from the surface of leaves and branches is governed largely by air temperature and humidity. Water returned to the atmosphere by transpiration is replaced in plants by water taken into roots from the soil.

Infiltration   Some of the water that filters through vegetation interceptors to the ground runs off on the surface, but a significant portion soaks into the soil where it is temporarily stored. This process is known as **infiltration.**

The amount of water that infiltrates the soil depends largely on the following:

1. The soil's ability to accept water from the surface and permit the water to percolate through it, and

2. The rate of supply of water to the ground surface.

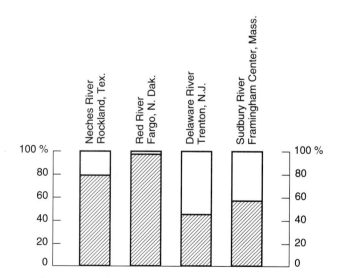

FIGURE 5–5
Ratio of evapotranspiration to runoff for four drainage systems. (Data from Williams et al., 1940; plot from Morisawa, 1968)

As long as rainfall input is equal to or less than the infiltration capacity, the maximum rate at which soil absorbs water, no runoff occurs. When rainfall input exceeds infiltration capacity, the excess water runs off on the land surface (Horton, 1945). The infiltrated water moves slowly through the soil and may eventually discharge into streams, but the time lag behind direct runoff is relatively long, and a portion of it is taken up by plant roots and is then returned to the atmosphere by evapotranspiration.

The factors that control infiltration rates include the following:

1. Vegetation

2. Soil permeability

3. Slope

4. Soil saturation

Vegetation plays an important role in infiltration. Organic material on the ground surface protects the soil from packing by direct raindrop impact; roots take up soil moisture; and root growth breaks up dense soils to make them more permeable.

Permeability, the ease with which water passes through the soil, depends on the amount, size, and distribution of pore spaces in the soil. Soils composed of sand-size material or larger have large pores that allow water to pass through rapidly, whereas fine-grained silt and clay have very small pores, which inhibit passage of water. Fine-grained material has greater total pore space (**porosity**) than coarse-grained material, but the pores are much smaller and are not well connected.

Infiltration rates on low slopes are greater than on steep slopes because of the time required for percolation of water into the soil. On steep slopes, the water falling on the surface may run off before it has time to soak into the ground. Thus, infiltration rates are generally inversely proportional to slope.

Soil saturation occurs when all available pore spaces in the soil are filled with water. If additional water is then added to the soil surface, with all pore spaces already filled, it has nowhere to percolate and will run off on the surface. When rainfall input exceeds infiltration capacity, runoff occurs. During severe storms when rainfall intensity is great, the soil may soon reach its infiltration capacity, and runoff results. Ground water may also make an important contribution to streamflow when it reaches the stream.

Runoff

On upper hillslopes, sheet runoff becomes concentrated into **rills,** which in turn feed larger gullies, which eventually run into larger and larger streams. In a drainage system such as the Mississippi, an estimated 22 trillion ft^3 (0.6 trillion m^3) of water flows every year from the land to the sea.

Velocity and turbulence are regulated by the boundaries of the flow and by discharge. **Discharge,** the volume of flow per unit time through a given cross section, is the product of average velocity and cross-sectional area. By substituting the product of average width and depth for cross-sectional area, the equation can be written as

$$Q = VA$$
$$Q = wdv$$

where
- Q = discharge
- w = width
- d = depth
- v = velocity

Thus, discharge is accommodated by some combination of width, depth, and velocity, and changes in discharge are compensated by adjustments in some or all of these dependent variables.

The cross-sectional area of a stream channel is approximately width times depth, and the wetted perimeter is width plus twice the depth (Figure 5–6). As discharge increases, the channel enlarges to accommodate the added water, and both

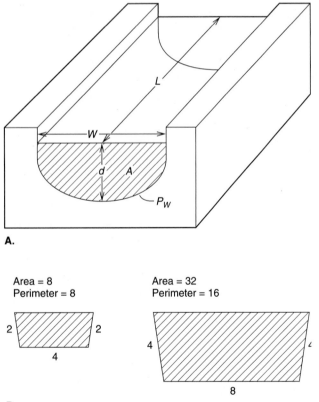

A.

Area = 8
Perimeter = 8

Area = 32
Perimeter = 16

B.

FIGURE 5–6
(A) Typical stream channel. *W* is channel width; *d* is depth; *L* is length af channel segment; *P$_w$* is wetted perimeter (portion of channel in contact with water); *A* is cross-sectional area; and *R* is the hydraulic radius, *A/P$_w$*. **(B)** Change in ratio of channel cross-sectional area to wetted perimeter with increase in discharge.

cross-sectional area and wetted perimeter increase. However, w × d increases more rapidly than w + 2d, so cross-sectional area increases more rapidly than wetted perimeter, resulting in a relative decrease in frictional retardation by the channel bed and banks. Thus, flow velocity increases with increasing discharge.

The cross-sectional area of the channel shown in Figure 5–6(B) is 2 × 4 = 8, and the wetted perimeter is 4 + 2(2) = 8. If discharge increases and the channel enlarges as shown in Figure 5–6(B), the cross-sectional area of the channel is 4 × 8 = 32, and the wetted perimeter is 8 + 2(4) = 16. Doubling the width and depth of the channel has only doubled the wetted perimeter while increasing cross-sectional area by four times. Thus, the increasing of cross-sectional area at a greater rate than wetted perimeter yields less frictional retardation of water relative to cross sectional area and flow velocity of increases. The general observation that streams flow faster during floods is a result of this relationship.

Empirical data show that changes in discharge produce changes in channel depth, width, and velocity (Leopold and Maddock, 1953; Leopold and Wolman, 1957; Leopold, et al., 1964). Figure 5–7 depicts data for a stream at a single gauging station and shows that as discharge Q increases, channel width w, depth d, and velocity v increase according to the following equations:

$$w = aQ^b$$
$$d = cQ^f$$
$$v = kQ^m$$

where the exponents b, f, and m express the slope of the lines shown in Figure 5–7, and the variables *a, c,* and *k* are numerical coefficients. Because $Q = wvd$, Q also equals $aQ^b \times cQ^f \times kQ^m$. Thus, the sum of the exponents, $b + f + m$, must equal 1.0, and the product $a \times c \times k$ must equal 1.0. Channels have large values for f and low values for b in resistant bank-forming material, such as cohesive silts, whereas values are high for b and low for f in weak bank-forming material, as in loose sand.

Floods

Stream channels are adjusted to carry the normal discharge of water from upstream and from tributaries. Most of the time, the water level remains within the confines of the stream banks, but periodically the flow of water is beyond the capacity of the channel to hold, and the water spills over the banks, causing flooding. Modest flooding commonly happens every two to three years, but periodically large volumes of water cause severe flooding.

Floods Caused by Heavy Precipitation   Most floods result from unusually heavy precipitation (rain or snow). However, large floods can also be caused by dam failure and glacial outbursts. Many of the worst historic floods have been produced

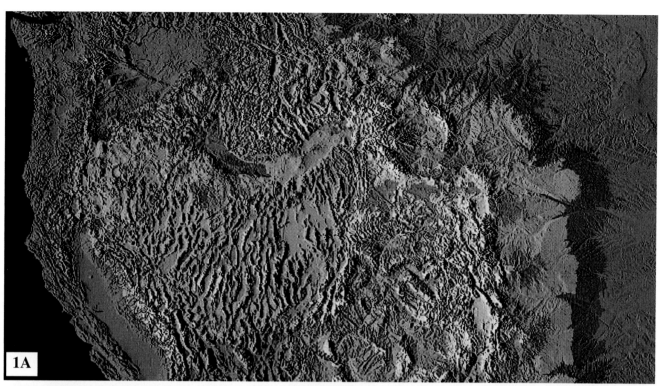

1A

1B

1C

1D

PLATE 1A
Computer-generated map of the western U. S.
(Courtesy of D. Simpson)
PLATE 1B
Balanced rock created by differential erosion of weak shale beneath
a more resistant sandstone, southern Utah. (W. K. Hamblin)
PLATE 1C
Scars made by shallow slides, Brazil.
(F. O. Jones, U. S. Geological Survey)
PLATE 1D
Rock slide along a bedding plane, Mendosa Valley, Argentine Andes.
(Photo by Author)

2A

2C

PLATE 2A
Delicate Arch, Utah. (W. K. Hamblin)
PLATE 2B
Grand Canyon of the Yellowstone River, Wyoming.
(W. K. Hamblin)
PLATE 2C
Meanders and point bar deposits in an Alaskan
stream. (Infrared photo by U. S. Dept. of
Agriculture)
PLATE 2D
Oxbow Lakes, Flathead River, Montana. (D. A.
Rahm)

2B

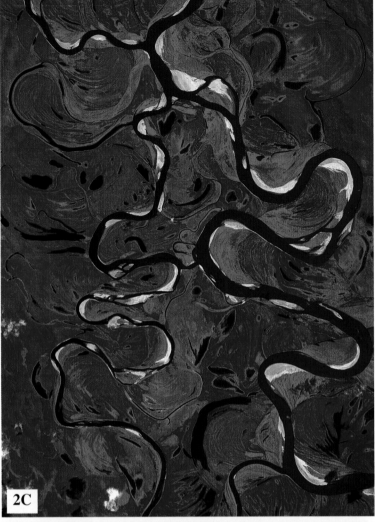

2D

3A

3C

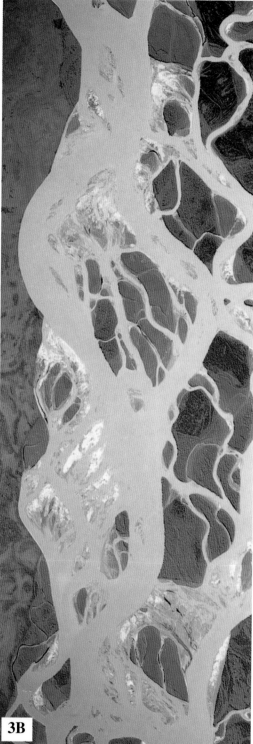

3B

PLATE 3A
Badlands, Petrified Forest National Monument, Arizona. (Photo by Author)
PLATE 3B
Braided stream, Alaska. (Infrared photo by Dept. of Agriculture)
PLATE 3C
Stream terraces incised by a braided stream, southeastern California. (D. A. Rahm)

PLATE 4A
Meanders of the Snake River, Idaho. (W. K. Hamblin)
PLATE 4B
Entrenched meanders, Green River, Utah. (W. K. Hamblin)
PLATE 4C
Subsummit erosion surface cutting across rock structures,
Rocky Mt. Font Range, Colorado. (D. A. Rahm)
PLATE 4D
Alluvial fan, Death Valley, California (W. K. Hamblin)
PLATE 4E
Mississippi birdsfoot delta. (NASA)

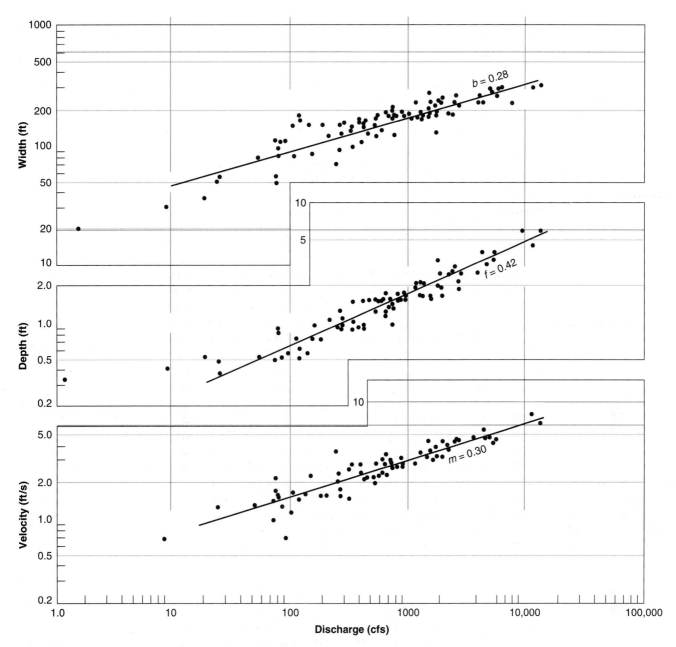

FIGURE 5–7
Relationship of discharge to width, depth, and velocity in a stream. (From Leopold and Maddock, 1953)

by exceptionally heavy precipitation associated with hurricanes or typhoons. Parts of Asia, especially China, have been hit with intense storms that have killed hundreds of thousands of people. For example, in one horrendous Asian storm, 200 cm (6 ft. 8 in) of rain fell in less than three days. That is more than twice the average annual rainfall for the United States. In a normal, rainy day, one to two inches of rain might fall, so this would be roughly equivalent to 2 1/2 months of continuous, normal rain compressed into three days. In a storm in the United States on May 31, 1935, 56 cm (22 in.) of rain fell in less than three hours. In 1972, 30 cm (12 in.) of rain fell in one hour.

The Discharge (Q) of a stream, the volume of flow per unit time through a channel cross section, is the product of average velocity (v) times the width (w) and depth (d) of the stream.

$$\text{Discharge} = \text{width} \times \text{depth} \times \text{velocity}$$

Thus, in order to calculate the discharge of a stream during flooding, the flow velocity must be measured, along with the width and depth of the stream. The velocity is usually measured by placing a flow meter in the stream. If the banks of the channel contain the flow at a constant width, then the only other thing that needs to be measured is the depth, or *stage height.*

Floods generated by heavy precipitation are preceded by rising water levels in the channel, until finally the stream channel can no longer contain the volume of water. *Flood stage* is reached when the stage height exceeds the height of the channel banks. The magnitude of a flood can be described either by calculating the maximum discharge or comparing the maximum stage height to those of previous floods. The stream is said to **crest** when the maximum stage height is reached.

Fluctuations in stream discharge (or stage height) can be plotted against time on a **hydrograph** (Figure 5–8). Hydrographs are very useful in developing a picture of how a particular stream typically behaves during a flood. The steepness of the ascending limb of the discharge curve is a measure of how rapidly a flood developed, and the steepness of the descending limb of the curve shows how quickly the flood waters dissipated. Some streams respond very quickly to precipitation events, others more slowly. For example, a flash flood in a desert stream would have a very steeply rising hydrograph with a sharp peak, whereas a large river with a vast drainage area would have a more gently rising hydrograph with a broader, gentler peak.

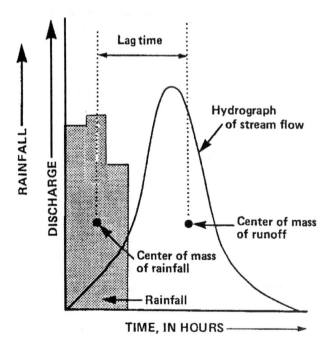

FIGURE 5–8
Flood hydrograph. The shaded histogram shows precipitation and the line curve shows stream discharge during a flood. The shape of the curve is determined by how rapidly flood waters rise and recede, which depends on runoff rates of a particular drainage system.

Stream discharge during a flood varies with time, and maximum flood discharges vary widely from flood to flood. However, these variations may be handled by statistical methods similar to those used for precipitation recurrence intervals. Measurements of maximum, average, and minimum discharge values at stream gauging stations provide the data for recurrence interval analysis. The recurrence interval (*RI*) for a flood of a given magnitude can then be calculated by dividing the number of times a given discharge has occurred into the number of years of record. For example, if three floods have exceeded a given discharge during 60 years of record, the *RI* would be 60/3 = 20 yrs. In other words, every 20 years you could expect a flood of this magnitude to occur. Plotting Q versus RI on log graph paper shows the frequency of recurrence for a discharge of any given magnitude (Figure 5–9). Records of flood discharge are seldom more than about 50–70 years, especially in the western United States. Thus, in order to determine the magnitude of "100-year flood," recurrence interval graphs must be extrapolated beyond the period of known discharges. Fortunately, that can be done with reasonable assurance. Flood-frequency curves are very useful in assessing flood hazards. For example, if the severity of the one-hundred-year flood can be estimated from a RI graph, then even if a flood of that discharge has not occurred within historic records, the area that would be flooded in a one-hundred-year flood can be determined, and flood-hazard maps can be constructed.

One of the largest floods in U.S. history from rainfall alone was the great 1993 Mississippi River flood. Weather conditions were abnormal in the upper Mississippi River basin in late 1992 and continued into 1993. Heavy rains fell in September 1992 and continued for eight months. A heavy snowmelt helped saturate the ground in the upper Mississippi River basin in the spring of 1993. In early June, a stationary high-pressure center that developed on the East Coast began drawing moist unstable air into the upper Mississippi River basin. The moist air resulted in unusually heavy rainstorms, and the high pressure center prevented storm systems from moving eastward. In places the rainfall was exceptionally intense. For example, 15 cm (6 in.) of rain fell in 6 minutes in Nebraska. By June 1993, some areas had experienced more than three times the normal annual rainfall. The ground over almost the entire region became saturated, and reservoirs were near their maximum capacity.

The excessive precipitation lasted through June and July, and with the ground already saturated, the water had no place to go but into the river. More than three fourths of the upper basin received more than twice its normal precipitation for July, and more than a fourth of the area received four times its normal precipitation for July. The summer of 1993 was the wettest on record for Iowa, Illinois, and Minnesota.

The flooding that occurred in the summer of 1993 lasted from late June to early August and inundated large areas in nine states. The flood exceeded the 100-year high, caused 50 deaths, and resulted in more than $10 billion in property damages.

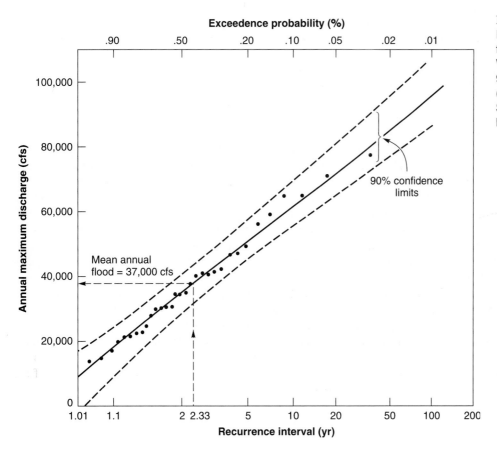

FIGURE 5–9
Flood frequency curve for annual
floods on the Skykomish River,
Washington. Dashed lines are
90-percent confidence limits.
(Data from U.S. Geological
Survey; plot from Dunne and
Leopold, 1978)

Most of the rain fell in the upper Mississippi basin, and flood crests migrated progressively downstream. The flood crested at Minneapolis/St. Paul, Minnesota on June 26, at Dubuque, Iowa on July 6, at Quincy, Illinois on July 13, and at St. Louis, Missouri on July 19 (Figure 5–10). Levees were topped, and about 10 million acres (55,000 km²) of flood-plain were inundated, including many towns and farmlands. Many roads were submerged, millions of acres of productive farmland remained under water for weeks during the growing season, the Missouri and Mississippi rivers were closed to shipping, thousands of rills and gullies were cut into farmlands, and huge volumes of topsoil were eroded.

Extreme flood events may fall well beyond discharges of historic record. For example, the 1976 flood in the Big Thompson River north of Boulder, Colorado was more than four times the previously recorded maximum discharge in the previous 88 years. Intense rain resulted from the pushing of a moist, unstable air mass upslope into the Front Range by strong, low-level, easterly winds associated with a polar front. The rain caused devastating floods in the Big Thompson River basin from July 31 to August 1, 1976, and resulted in 139 deaths, 5 missing persons, and more than $35 million in damages (Balog, 1978; Costa, 1978; Grozier et al., 1976; Jarrett and Costa, 1988; Miller et al., 1978; Shroba et al., 1979). During the storm 10 to 12 inches (25 to 30 cm) of rain fell (Figure 5–11). Precipitation intensities were extremely high; for example, 7.5 inches (19 cm) were reported from radar data between 7:30 and 8:40 p.m.

The peak measured discharge of the Big Thompson River flood was 30,100 cfs (850 m³/s). The flood crest passed through about 8 mi (13 km) of the canyon in approximately 30 minutes at an average travel rate of 15 mi/hr (25 km/hr).

Physical evidence of the flood remains long after its passing. Large boulders moved by the flood are common on gravel bars and along the channel (Figure 5–12A). One of the largest measured 11.8 by 12 by 22.9 ft (3.5 by 3.7 by 7 m). Boulders such as these are very useful in reconstructing the discharges of paleofloods. Scouring and erosion were widespread in the Big Thompson canyon. Sections of U.S. Highway 34 were obliterated; houses (Figure 5–12B), and many small bridges were washed away, and the distal ends of tributary fans, built by tributaries into the main canyon, were truncated by the Big Thompson River for the first time in thousands of years. Where sediments in the truncated fans can be radiocarbon dated, an idea of the return interval of the flood can be gained.

Many trees along the banks above the Big Thompson River were badly scarred from collision of boulders and other debris carried by the flood. Many years after the flood, the scars were still very obvious, and although they will eventually heal and disappear, the record of the flood will be retained in the tree ring history. Hundreds of years later, the tree rings will still show the scar made by this flood. Such evidence provides a very useful means of identifying pale-ofloods because not only is the event recorded in the tree

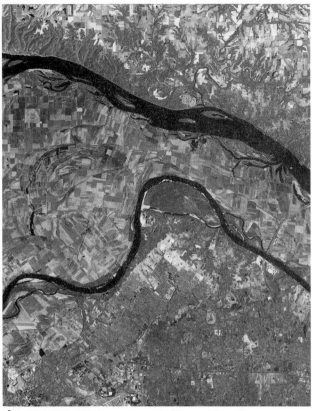

A.

B.

FIGURE 5–10
Landsat satellite images showing the extent of the June 1993 flooding of the Mississippi and Missouri Rivers near St. Louis. (Enhanced satellite images by Earth Satellite Corp.)
A. Image taken July 4, 1988 when the average stream gauge was 1.8 ft.
B. Image taken July 18, 1993 when the gauge reading was 46.5 ft.

rings, but the age of the flood responsible for the scar can also be determined by counting the number of tree rings since the scar occurred.

Stream terraces that were once part of ancient flood plains of the Big Thompson River are preserved in some parts of the canyon. The sediments making up the terraces contain many boulders similar to the ones moved by the 1976 flood. Radiocarbon dating of the terrace sediments can then provide a means of bringing together evidence for paleofloods and determining their age.

Floods from Dam Failures. Failure of dams and sudden release of large quantities in the reservoir cause disastrous floods downstream. One such dam was the St. Francis dam, built in 1926, in San Francisquito Canyon north of the San Fernando Valley, California. Filling of the reservoir was completed March 5, 1928, and one week later the dam failed, releasing a huge wall of water downstream that killed 450 people.

The St. Francis Dam was the type of dam that depends on the weight of the structure to hold back the water in its reservoir. Failure of the dam was not within the structure itself, but rather in the rock making the foundation of the dam, which had been examined by engineers, but not by geologists. The west abutment of the dam consisted of sandstone and con-

glomerate, normally a firm foundation material, but in this instance interbedded with gypsum, notable for its solubility in water and lack of strength. The east abutment was in mica schist, not noted for its strength. Even worse, the contact between the two rock types was an inactive fault zone containing crushed rock and gypsum about five feet (1.5 m) thick. Engineers tested samples of the rock for strength, but with no geologic input, did not recognize the undesirable geologic conditions.

Soon after the reservoir had filled, water began leaking from the sandstone abutment. A few days later, the damkeeper noticed a new leak at the sandstone abutment and alerted authorities responsible for the dam's construction, but they did not believe the leak was serious. That night, water under high pressure at the base of the reservoir burst through the ground-up rock in the fault zone between the sandstone and schist, and the dam collapsed, releasing a wall of water 185 feet (56 m) high. Huge sections of the dam, some weighing 10,000 tons, were carried more than a half-mile downstream by the floodwaters, leaving only the center section of the dam. Downstream, the floodwaters wiped out everything in its path. Where the water swept away a construction camp of tents, the air-filled tents of the workers quickly bobbed up to the surface and the men were able to ride the wave to safety.

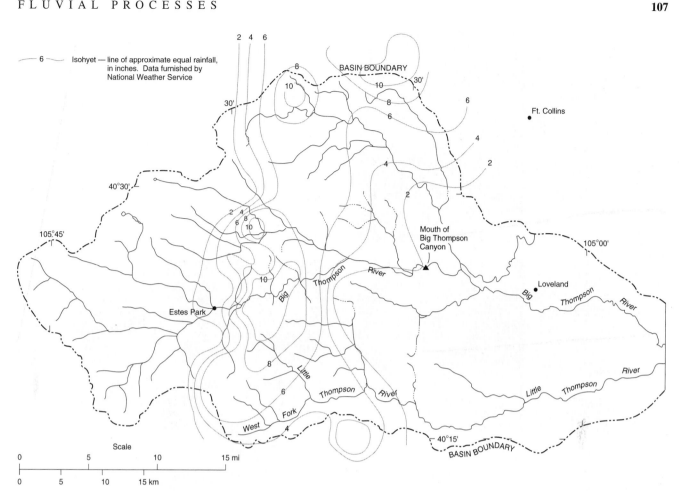

FIGURE 5–11
Precipitation values from the 1976 Big Thompson River flood. (From U.S. Geological Survey Professional Paper 1115)

On July 15, 1982, the 26-ft-high (8 m) earthen, Lawn Lake dam in Rocky Mountain National Park failed, sending an estimated peak discharge of 18,000 cfs (510 m³/s) down Roaring Fork to the Fall River where floodwaters overtopped the 17-ft-high (5 m), concrete Cascade Lake dam, which also failed and added more discharge to the flood (Jarrett and Costa, 1986). The flood then ripped through the town of Estes Park where it caused extensive damage. Three people were killed and damages totaled $31 million. Calculated peak discharges included 18,000 cfs (510 m³/s) from the Lawn Lake dam failure, 7210 cfs (205 m³/s) into Cascade Lake dam, 16,000 cfs (450 m³/s) from the failure of Cascade Lake dam, and 6650 cfs (188 m³/s) at Estes Park. Flood peaks were 2 to 30 times the 500-year flood.

The geomorphic effects of the flood from the dam failures were profound. At the junction of Roaring Fork with the Big Thompson drainage, impressive physical evidence from the Lawn Lake dam failure remains today. Stream channels were scoured 5 to 50 ft (1.5 to 15 m) and were widened tens of feet. The Roaring River channel below Lawn Lake dam was scoured by as much as 50 ft (15 m) and an alluvial fan of 364,600 yd³ (278,900 m³) of boulders and sediment was built in less than one hour, to a maximum thickness of 44 ft (13 m) and covering 42 acres (17 hectares).

Floods from Glacial Outbursts. Floods caused by sudden release of water held by glacial dams are relatively rare, but the results are spectacular when they do happen. They occur most frequently in Iceland, where volcanic heat melts large amounts of glacial ice, and sudden, large-scale floods occur about once every seven years.

Perhaps the greatest of all floods on Earth occurred on the Columbia Plateau of eastern Washington during the last Ice Age. A tongue of ice from the Cordilleran Ice Sheet blocked the Clark Fork River of western Montana, impounding a huge lake with water depths up to 700 m (2000 ft). The ice dam burst a number of times and released the most gigantic floods yet recognized anywhere in the world. Each time the ice dam failed, an estimated 2300 km³ (500 mi³) of water rushed across the Columbia Plateau, carving out long, deep, channels, cataracts, and waterfalls in the rock. The amount of floodwater was so great that even the deepest channels overflowed, and flood waters spilled across channel divides, cutting notches in ridge crests high above the valley floors. Waterfalls and cataracts migrated rapidly upstream, leaving deep, now-dry, canyons such as Grand Coulee and Moses Coulee. One such falls, Dry Falls, remains today as a 120 m (400 ft) precipice nearly 6 km (3.5 mi) wide. Huge gravel bars up to 100 m

FIGURE 5–12
(A) Boulders along the Big Thompson River carried by the 1976 flood. Eye witnesses said they saw boulders being tossed above the river surface during the height of the flood. (B) House carried away by the 1976 Big Thompson River flood and lodged on a bridge. (U.S. Geological Survey)

(~300 ft) high and 100 to 1000 m (~300 to 3000 ft) long were deposited, many with giant ripples having wave lengths of 100 m (300 ft) on their surface.

Reconstruction of Flood Discharges Using Paleoflood Analysis   Standard statistical hydrologic techniques of calculating the recurrence interval of rare, large-magnitude floods become increasingly unreliable as the recurrence interval of the flood exceeds the length of historical gauging records. Paleoflood analysis provides a much more reliable way of estimating the frequency of large floods.

Paleohydrologic analysis is used to reconstruct the magnitude and frequency of unusually large, rare floods. Several techniques may be used to estimate paleoflood peak discharges:

1. Calculation of hydraulic factors (velocity, hydraulic shear stress, and stream power) based on the sizes of flood-transported boulders

2. Slack-water sediments deposited by backwater of floods into tributaries

3. Effects on floodplain vegetation

4. Hydraulic modeling

5. Flood erosion

6. Dimensions of former channels

7. Truncation of tributary fans

Paleoflood reconstruction plays an important role in predicting probable maximum flood discharges for a river because it offers the opportunity to look at a much longer time span than is afforded by the historic record. A good example is the Pecos River in Texas, which had never before experienced a flood greater than 4000 m³/s (141,000 cfs) in gauging records dating back to 1900. In 1954, more than 100 centimeters (39 in.) of rain fell in 48 hours from Hurricane Alice, causing a flood of 27,400 m³/s (967,000 cfs) (Baker, 1977). Extrapolation of flood recurrence intervals (Figure 5–13) based on historic flood data would not have come anywhere near predicting the magnitude of such an event. Conventional flood frequency estimates would yield recurrence intervals ranging to more than 10 million years for this flood. Then in 1974, yet another hurricane produced another extreme flood of 16,000 m³/s (565,000 cfs), again unpredictable by normal methods.

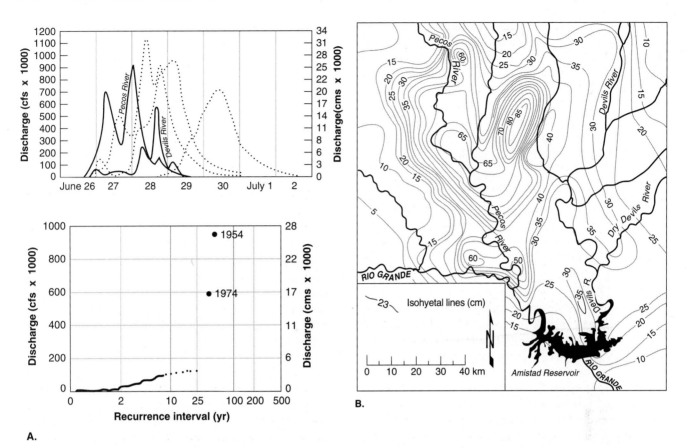

FIGURE 5–13

(A) Annual maximum peak discharges for the Pecos River near Langry, Texas, 1900–1977. The two outliers are from the severe floods of 1954 and 1974. The remainder of the data plot along a well-defined curve. (B) The flood of June 26–28, 1954, on the Pecos and Devil Rivers in southwest Texas. The isoheytal map shows the rainfall from Hurricane Alice. Resulting flood hydrographs on the Rio Grande and on the Pecos and Devil Rivers are shown in the accompanying graph (From Kochel, 1988).

The "paleodischarge" of the 1976 Big Thompson flood could be estimated from the evidence left by the flood. The stage height of the flood could be determined from the maximum elevation of flood-caused features in the canyon and used to back-calculate the flood discharge. Among the interesting implications of back-calculations is the notion that values of the roughness coefficients *n* used by many geologists and hydrologists are probably much too low and are in need of revision for high-gradient streams in the western United States.

How to treat such extreme flood events in estimating maximum flood discharges and their frequency can be a serious problem. However, stratigraphic principles, sediments, vegetation, and dating techniques can be used to determine the magnitude and frequency of large, prehistoric floods. For example, a fairly realistic paleohydraulic reconstruction of the mythical biblical deluge can be made from a 3.4-m (11-ft) thick layer of artifact-free mud between artifact-rich layers dating from 3500 B.C. in archaeological excavations of the ancient Sumerian city of Ur in the Tigris and Euphrates Valley. Woolley (1954) interpreted the sediment to be water deposited and believed that it represented the Old Testament flood of Noah. He estimated that a flood at least 7.6 m (25 ft) deep would be required to deposit 3.4 m (11 ft)

of sediment, and that such a flood would cover an area 161 km (100 mi) wide at Ur. Assuming a velocity of 3 m/s (10 ft/s) for such a large flood in the flat lowlands of Mesopotamia, and an average depth of 3 m (10 ft) across the valley, the discharge would have been about 1.5×10^6 m³/s (52.8×10^6 cfs).

In the early 1920s, Bretz presented compelling evidence that a colossal flood was responsible for the remarkable channeled scablands of the Columbia Plateau of eastern Washington (Bretz, 1923). The resulting scientific controversy created intense interest in paleoflood hydrology. Bretz proposed that the Channeled Scablands, covering an area of 40,000 km² (15,400 mi²), were formed by catastrophic flooding from the failure of a vast lake, Pleistocene Lake Missoula near Missoula, Montana, dammed by a lobe of the Pleistocene Cordilleran Ice Sheet. The lake was 7770 km² (3000 mi²) in area and had a volume of 2.0×10^{12} m³ (2.6×10^{12} yds³) of water, about half the size of Lake Michigan. When the ice dam failed, the lake drained catastrophically, sending enormous floods across the Columbia Plateau. Despite initial skepticism, Bretz was able to demonstrate that such gigantic floods had occurred, based on extensive sedimentologic and geomorphic evidence. He described slack-water sediments deposited in the mouths of tributaries that were inundated by the giant floods released from Lake Missoula. Later investigators analyzed Missoula flood slack-water sediments to estimate the number and chronology of Pleistocene floods (Baker, 1978; Bunker, 1982; Waitt, 1980, 1984, 1985; Baker and Bunker, 1985).

The discharge of the Missoula flood at Wallula Gap, Washington, was estimated by using erosional evidence on uplands for estimates of water depth and employing the Chezy formula.

$$v = C(RS)^{1/2}$$

where v = average velocity
            C = "smoothness" coefficient
            R = hydraulic radius
            S = slope

Hydraulic computations, made by an engineering colleague of Bretz, yielded velocity estimates of 6.4 to 9.1 m/s (21 to 30 ft/s) and conservative discharge estimates of 1,873,000 m³/s (66,000 cfs). Pardee (1942), who had earlier discovered the source of the Lake Missoula floodwaters, also estimated the discharge of the Missoula flood near its source in Montana, based on the Manning and Chezy equations. Hydraulic computations were made by Rubey and Langbein of the U.S. Geological Survey on field data reported by Pardee. Their calculations gave an estimated flood discharge of 10,946,000 m³/s (386,390,000 cfs) and water depths of hundreds of meters. A more comprehensive paleoflood study was made 30 years later by Baker (1973), who used eroded channels in loess at minor divide crossings as highwater marks and calculated a discharge of 21,300,000 m³/s (752,000,000 cfs) for the Missoula flood.

Slack-water sediments can provide the most abundant and accurate source of data for determining the history of multiple, large floods in a river system. Slack-water deposits usually consist of fine sand and coarse silt deposited by floods in areas sheltered from high-velocity flows. Slack-water sediments have been utilized in climatically diverse regions, including the relatively arid southwestern United States, the Mississippi Valley, the eastern U.S. Piedmont, the Columbia Plateau and Snake River Plain, and the Rocky Mountains.

Bretz (1929) found that the slack-water sediments formed in the mouths of tributaries backflooded by the Missoula floods ranged in size from coarse gravel to coarse silt and were deposited in backflooded tributaries extending tens of kilometers upstream from their main trunk channels (Bretz, 1929; Baker, 1973; Patton et al., 1979).

Most slack-water sediments are deposited from suspension where flood flow velocities are rapidly diminished. Four environments are especially suited for repeated accumulation of slack-water sediment:

1. Tributary mouths
2. Shallow caves along the bedrock channel walls

3. Downstream from major bedrock obstructions and where sudden channel widening occurs

4. Overbank deposits on high terraces

Figure 5–14 illustrates some of these environments.

Slack-water sedimentation occurs in virtually all river systems, but it is best preserved in narrow, deeply incised bedrock canyons where large increases in flood stage may be produced by small increases in flood discharge. Sediments transported during high flood flows are deposited high along canyon walls, in bedrock caves, and far upstream in tributary mouths. The highest such deposits provide a record of peak stage height.

Tributary-mouth slack-water sedimentation is favored when the peak flood stage of the main stream is not in phase with the flood stage of tributary streams, a common situation for tributaries with small drainage basins entering main streams with large drainage areas. Sediment-laden floodwater surges up into the tributaries, rapidly depositing slack-water sediments as backflood velocity decreases. These deposits can usually be traced uptributary as distally thinning wedges of sand and silt draped over the tributary channel and flood-plain surfaces.

When tributary flooding precedes mainstream backflooding, the two flood events produce two discrete sedimentary layers. For example, slack-water sediments from floods of Hurricane Agnes in tributary mouths along the Susquehanna River consist of basal gravel deposited by the tributaries, overlain by mud deposited by backflooding of the Susquehanna (Moss and Kochel, 1978). Pieces of coal in the mud indicate that it came from the Susquehanna because coal does not occur anywhere in the tributary drainages.

Tributary sediment is typically coarser grained than mainstream sediment because tributary gradients are steeper, and it contains only locally derived lithologies. Slack-water sediment deposited from mainstream backflooding contains sediment derived from distant sources, and paleocurrent directions are upstream into the tributary. For example, slack-water sediments of Pecos River floods are composed of fine-grained quartz sand derived from Permian sandstone outcrops hundreds of kilometers upstream, whereas tributary sediments are composed of coarse limestone gravel derived entirely from local Cretaceous limestone. Where streams cut across terrains of distinctly different rock types, mainstream and tributary sediment can usually be clearly distinguished from one another.

The length of time that waters from backflooding occupy tributary mouths varies among rivers, depending on the characteristics of mainstream floods and on tributary gradients. For example, slack-water inundation periods for major floods along the lower Pecos River last only a few hours to a day or two. However, Mississippi River backflooding of southern Illinois River tributaries typically lasts for weeks or months during major floods. Mississippi River floods of 1984–85 inundated tributary mouths for three to six weeks, depositing slack-water sediments several kilometers upstream in tributaries.

New slack-water sediments are deposited with each successive flood that is large enough to reach the previous level

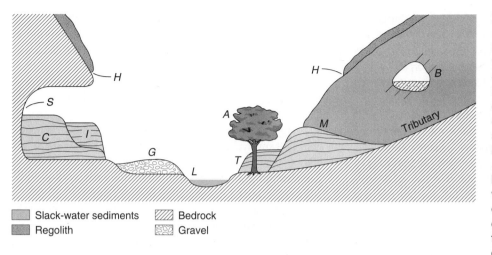

FIGURE 5–14
Idealized cross section of a bedrock canyon river. *A* is a tree with adventitious roots growing on flood deposits; *B* is a high-level, slack-water sediment in a cave on the canyon wall; *C* is a cave deposit of slack-water sediments; *G* is a gravel and boulder bar on the canyon floor; *H* is a high-water mark created by scour of soil on the canyon walls; *I* is inset slack-water deposits; *L* is the low-flow channel of the active river. *M* is tributary-mouth, slack-water deposits; *S* is the silt line of a paleoflood preserved in a cave; and *T* is a slack-water terrace. (After Baker, 1987)

Slack-water sediments   Bedrock
Regolith   Gravel

of slack-water sediment deposition. Because erosion of previously deposited sediments may completely remove preexisting slack-water sediments during subsequent floods, the slack-water sediment stratigraphy at any one site may be incomplete. Best results are obtained by investigating multiple sites to obtain a composite slack-water stratigraphy for an area.

Paleoflood reconstruction for long time intervals depends on the preservation of relatively complete stratigraphic sequences of slack-water sediments. These sequences occur where extensive slack-water sedimentation occurs during large floods and where the sedimentation site is protected from tributary and mainstream erosion from subsequent floods. The most important factors affecting the deposition and preservation of slack-water deposits include:

1. The tributary junction angle
2. Mainstream gradient
3. Tributary gradient
4. Tributary drainage basin morphology

To produce thick accumulations of slack-water sediments, the main stream must be able to backflood into tributary mouths. Field studies in southwest Texas and flume experiments have demonstrated that maximum backflood discharges occur when the angle between tributary junctions and the main stream is between 55° and 125° (looking downstream) (Baker and Kochel, 1988). Mainstream floodwaters are likely to bypass tributaries with particularly acute junction angles (less than 45°) with little backflooding. However, where tributaries join the main stream at angles greater than 130°, mainstream floodwaters run more easily up tributary channels at high velocity. Slack-water sediments are deposited during these powerful incursions from the main stream up tributary valleys, but these flows are typically so forceful that earlier-deposited slack-

water sediments may be destroyed by erosion. Complete slack-water stratigraphies are rare under high-angle conditions.

Flume experiments show that the gradient of the main stream is an important factor in determining the height of mainstream backflooding into tributary mouths (Baker and Kochel, 1988). Backflooding levels in tributary mouths decrease as the mainstream gradient increases because the high velocity of steep mainstream gradients causes floodwater to bypass tributary mouths, rather than flow up into them.

Steep tributary gradients are not favorable for the preservation of thick slack-water sediments because they produce high-velocity flows that are likely to destroy previously deposited slack-water sediments. Complete records of floods in slack-water sediments are more likely to be preserved where tributaries have low gradients. The most favorable sites occur along the insides of tributary meander bends a few tens of meters from the main stream and just downstream from major bedrock or talus obstructions to tributary channel flow where flood velocities are generally minimal.

Discrimination of individual flood events in slack-water sediment sequences is based on recognition of abrupt vertical grain size changes, buried soils, changes in sediment induration, color changes, buried mudcracks, colluvial horizons, and interbedded coarse tributary alluvium. Fine-grained organic detritus concentrated in the upper few meters of floodwaters may cap slack-water sediments. Stratigraphic relationship between individual flood units can usually be clearly defined in slack-water deposits in semiarid regions, but these relationships may be less distinct in humid regions because of increased bioturbation by roots, greater vegetation, and accelerated rates of soil formation.

Slack-water sediments thin away from the main stream and eventually pinch out up the tributary, marking the approximate peak stage associated with each flood. Slack-water sediments from a single flood become finer grained with distance from the main stream. The lateral variation is most pronounced in coarse slack-water sediments.

Compared to historic flood data, paleoflood discharge calculations might be assumed to have very large error factors. Although this is true of some paleohydrologic techniques, such as paleohydraulic estimates based on boulder transport (O'Connor and Webb, 1988) and in alluvial channels where substantial errors can occur, large errors are not typical of slack-water paleoflood procedures. With stable cross sections and well-preserved high-water indicators, slack-water paleoflood magnitude determinations can be superior to historic flood data.

Past floods can be identified based on the form and age of parts of trees growing on the flood plain. The date of deposition of sediment can be determined from the structure of wood in the buried part of living tree trunks, and the thickness of the deposit can be determined by measuring to the level of the original tree base. Anatomical characteristics of the wood of roots exposed by bank erosion provide data for determining the year that the roots were first exposed.

A significant advance in using vegetation to quantitatively reconstruct flood events was made by Sigafoos (1964), who established botanical techniques (tree ring dating) that can be used to reconstruct the paleoflood history for hundreds of years along channels bordered by trees. Sigafoos found that trees growing on flood plains are typically scarred, broken, or toppled by floodwaters but are rarely killed. New wood grows over the scars, but the scar remains easily distinguishable from the new wood, and the number of annual rings that have grown since the tree was damaged gives an accurate count of the number of years since the flood occurred. When trees are knocked over by floodwaters, new sprouts grow vertically upward from the inclined trunk, and their annual rings can be used to determine the length of time since the tree was damaged by a flood. If a tree survives partial burial by flood deposits, new wood that grows in the buried portion of the trunk resembles root wood more than stem wood. This difference allows the age of this change and the amount of sedimentation of a particular flood to be measured.

Scars on trees are especially useful in indicating floodwater levels. The height of scars on floodplain trees do not necessarily reflect the maximum stage of floodwaters, but they often serve as an approximation of minimum flood levels.

Evidence of past floods is provided by ages of trees on bottomland landforms and ages of scars and sprouts from flood-damaged stems (Sigafoos, 1964; Harrison and Reid, 1967; Helley and LaMarche, 1973; Hupp, 1988) and by differences in properties of wood anatomy related to flooding (Yanosky, 1982, 1983). Damage to trees during flooding commonly is severe: Floods typically scar tree bark, prune the tops and branches, or knock over trees on a flood plain, but the number of trees killed is often low. High water velocities, turbulence, and large quantities of transported debris are most damaging to bottomland vegetation (Sigafoos, 1964), but even relatively gentle inundation can cause datable anomalies in intraring structure (Yanosky, 1982, 1983). New wood and bark grow over the scars, and sprouts grow from inclined trunks and branches. Thus, the number of annual rings added since scarring of the bark or the beginning of new sprouts represents the number of growing seasons since the trees were damaged.

When a tree survives partial burial, the new wood formed in the buried part of the trunk is more like root wood than stem wood, so the number of growth rings since the change of wood type dates the year of burial. The ground surface upon which the trees were growing prior to burial can be identified by digging to roots growing from the flared base of the trunk.

Trees growing on the higher parts of a flood plain are usually large, straight, and single stemmed, whereas flood plain trees that survive floods show varying degrees of damage, including bending of small trees, removal of leaves, abrasion of bark, breaking of branches from trees, uprooting, and scarring. A line of scarring in vegetation may develop for substantial distances along the bank (Sigafoos, 1964). Although the tops of many trees may be killed, the roots of most survive. Small willow trees (Salix *interior*) may show deformation similar to that of "wind-trained" trees. Rare, severe floods may destroy all trees on parts of the flood plain, but some stands of trees may survive several major floods.

Trees continue to grow after they are damaged by floods, and subsequent growth preserves the scars of the injury. Scars on riparian trees and shrubs are the most conspicuous evidence of past flooding. Parts of the bark and underlying tissues that are killed are eventually covered by scar tissue over the damaged area. Sprouts grow from the residual stumps of flood-felled and decapitated trees, and the start of growth of these sprouts can be dated. Because of differences in wood produced during the early and late parts of the growth season, determination of the season when the scar was formed is often possible. Corrasion destroys the cambium (the wood-producing tissue) in the area of impact. Thus, growth ceases in the damaged tissues, and the event is recorded as undamaged tissue grows in annual increments around and over the scar. Evidence of the scar eventually disappears. A sharp blow to a tree trunk by debris carried by high water will kill the living tissues between the bark and the wood. Homogeneous tissue forms around the margin of the scar and across it if the scar is small. In time, a new layer of wood will cross the scar, and a complete annual ring will encircle the trunk. A scar can be dated by tracing the last ring formed prior to injury along its circumference to the sound part of the stem, and counting the number of rings that lie outside, giving the number of growing seasons elapsed since injury. Counting the annual rings in wood directly outside the scar is rarely possible, because the wood does not contain annual rings but generally shows an incomplete number of rings because the wood exposed by damage to the bark may rot and subsequently erode before the scar is healed.

Flood-felled trees typically consist of sprouts growing from stumps or inclined trunks. They consist of one or more straight trunks that are unbranched for about a third or more of their length, unlike multistemmed, multibranched shrubby plants. Sprouts start to grow from the stumps during the first growing season following the damage and, if not subsequently damaged by another flood, continue to grow as straight-stemmed trees. The year that a tree was felled can be dated by counting the rings of sprouts stemming from the inclined trunk.

Flood sediment surrounding the bases of trees gives the trees a characteristic appearance that can easily be recognized. If a tree remains buried through at least one growing season, ring counts and laboratory study of wood structure can accurately determine the year in which the burial took place.

Trees that grow on stable surfaces typically have flared bases that become buried by flood sediments. Digging around the base of a buried tree will expose the roots and may show the flared base that was once above the ground surface.

Roots commonly start to grow from the buried part of a trunk. Because wood formed in the buried part of a trunk structurally resembles the wood of roots, the year of burial can be determined.

Flood damage changes wood anatomy, forming flood rings grown when trees produce a second crop of leaves after having been stripped by an early-season flood. Typical spring growth in certain deciduous species is characterized by large-diameter cells and by smaller, thicker-walled, fibrous cells for the remainder of the growth year.

Abrupt tilting of a tree by floodwater results in subsequent rings that are wide on one side of the trunk and narrow on the opposite side. When this pattern follows concentric ring growth, the year of the onset of eccentric growth is usually within one year of a flood event.

Hydrodynamic interpretations based upon hydraulic formulas and sedimentary features may be based on such variables as the following:

1. Flow regime
2. Channel morphology, planimetry, and slope
3. Water velocity, energy, and/or depth

In one of the first reports devoted solely to paleohydraulic reconstructions from boulder sizes, Birkeland (1968) calculated the mean velocity and tractive force necessary to transport glacial outwash boulders in the Truckee River in California and Nevada. Paleoflood velocities of 9.1 m/s (30 ft/s) and tractive forces of 958 to 1437 N/m^2 were estimated using the Manning equation for flood-transported boulders 12.2 by 6.1 by 3.0 m (40 by 20 by 9.9 ft).

Particle size may be used as the independent variable to calculate average velocity and depth (Foley, 1980; Bradley and Mears, 1980). The basis for calculations rests on the assumption that the largest particles present in flood deposits represent the maximum competence of the stream during the flood. However, this may not necessarily be the case for the following reasons:

1. The largest boulders in a flood deposit may represent only the maximum size available for movement, and the flood might have been able to move still larger ones than those available for transport.

2. Large boulders may have been deposited in the channel not by a flood but by rafting, mass movements, or bank erosion (Rubey, 1938; Gage, 1953).

Krumbein and Lieblein (1956), however, showed that most of the anomalously large particles in gravel deposits are normal members of the stream particle population. Because of these uncertainties, the particle sizes used in computations are usually some average size of the largest rocks moved.

The first incipient motion studies probably date back to 1753, when Brahms formulated the "sixth power law," which states that at incipient motion, the weight of a particle is proportional to the sixth power of the critical bottom velocity (Leliavsky, 1955). The earliest investigations focused on the erosional velocity required to move particles. Assuming that weight is proportional to the third power of the particle diameter, Sternberg showed in 1875 that the velocity at which motion begins is proportional to the square root of the diameter. For particles coarser than 3 cm (1.2 in), Fahnestock (1963) concluded that the size of particles in motion varied with the 2.6 power of average velocity, consistent with results reported by Nevin (1946). Channel scour problems in the late nineteenth century led Kennedy in 1895 to propose an equation for critical average velocity for designing noneroding irrigation canals in India and Pakistan. Today, both velocity and tractive force are used to study particle movement in stream channels (Stelzer, 1981). Rubey (1938) showed that "bed velocity" is more significant than the tractive force (the depth-slope product) when measuring incipient motion of particles larger than 2.5 mm. Mean velocity is easier to measure, but it is not necessarily related to the velocity near the bed, where initiation of motion occurs. Many competence investigations utilize "bed" or "critical bottom" velocity estimated by measurements near the channel bottom or by extrapolation of velocity profile measurements to the bottom of the channel.

Costa (1983) used equations and graphs developed from boulder deposits to estimate single values of paleoflood velocity and depth. When these values are used in conjunction with valley cross sections near boulder deposits, a discrete paleoflood discharge value can be calculated.

Several methods may be used to reconstruct the average velocity of maximum floods, using boulder sizes transported by the floods (Novak, 1973; Bradley and Mears, 1980). The two theoretical methods used to compute average velocity are:

1. Balancing forces using turning moments (Helley, 1969)
2. Equating fluid drag *FD* and fluid lift *FL* against gravitational frictional resistance *FR*

The approach of both methods is similar—equating the forces of drag, lift, and resistance at incipient motion.

The velocity needed to move large particles is greater than the velocity needed to sustain transport (Hjulstrom, 1935; Sundborg, 1956). The 2.6 power law advocated by Nevin (1946) approximates average velocities needed to sustain large particle movement, but the sixth power law more closely approximates the average velocity needed to initiate motion.

The second empirical method for estimating average velocity from boulder sizes uses the extensive data of the U.S. Bureau of Reclamation on riprap stability. The limiting size needed for riprap stability is given by the equation

$$v = 5.9d^{1/2}$$

Particles from 0.01 m (0.03 ft) to more than 0.6 m (2 ft) were used to formulate this equation. Bed velocity (*v*) is multiplied by 1.2 to obtain an estimate of average velocity. The following guidelines are recommended in selecting cross-section sites for flood discharge estimates based on paleohydraulic reconstructions (Costa, 1983):

1. A straight reach is preferred, neither expanding nor contracting.

2. The site should not be an abnormally wide, narrow, steep, or flat part of the valley.

3. At least one, and preferably both, valley walls should be bedrock; thin colluvium over bedrock is acceptable.

4. Valley fill should be thin so that a reasonable estimate can be made of the elevation of the underlying bedrock surface.

5. The site should be close to the depositional site of boulder bars used to measure particle size and close to the point at which the discharge estimate is wanted. This criterion may require the selection of a less favorable site, but in small basins, where discharge is estimated, it is very important.

6. At least two cross sections, spaced about one valley width apart, should be measured.

7. No major tributaries should enter the main channel between the cross-section site and the point where the discharge estimate is desired.

If steady flow is assumed to have been attained, the Manning formula can be used to give a rough estimate of the mean velocity during a flood.

$$v = \frac{1.49}{n}R^{2/3}S^{1/2}$$

where v = the mean velocity in fps
         n = coefficient of roughness
         R = hydraulic radius (channel cross-sectional area divided by the wetted perimeter)
         S = slope

Most of the data needed to solve the equation are obtainable except for a value of n. A reasonable value of n, however, can be calculated and values can be obtained for flood velocities needed to move most boulders. For example, a boulder 12 by 6 by 3 m (40 by 20 by 10 ft) could be moved by a 12- to 24-m (40- to 80-ft) deep flood moving at a mean velocity near 9 m/s (30 ft/s) on a slope of 0.007.

Because of the problems inherent in extrapolating historic flood records to determine recurrence intervals for extreme floods, the concept of probable maximum precipitation (PMP) and probable maximum flood (PMF) were devised. PMP is the most severe precipitation that might occur from the most intense combination of meteorological and hydrological conditions that are reasonably probable in a region. This concept assumes some natural limit to precipitation intensity-duration conditions. Most spillways of dams in the United States are designed to handle floods resulting from PMPs.

The PMP approach has the advantage of producing numbers that are likely to represent the most severe conditions considered probable. The main disadvantages are that:

1. PMP values must be converted to PMF values that must include precipitation interception, storage, and runoff ratios

2. The method focuses on maximum severity of storms, not on such high-runoff events as rain-on-snow floods

Myers defined PMP as follows:

"The theoretical greatest depth of precipitation for a given duration that is physically possible over a particular drainage area at a certain time of year [and] that magnitude of rainfall over a particular basin which will yield the flood of which there is virtually no risk of being exceeded." (Myers, 1967)

The practical limitations on PMP include limits on humidity, the rate at which wind may carry humid air into a drainage basin, and the fraction of inflowing water vapor that can be precipitated.

The concept of PMF avoids the problem of evaluating the frequency of extreme events, but it assumes that the numerical values of all variables can be determined.

PMP/PMF analysis suffers from some of the same shortcomings as extrapolation of historic flood records to determine frequencies of extreme events, but various paleoflood techniques can be used to supplement both approaches to develop realistic estimates of flood levels.

Flow Velocity

The potential energy inherent to a mass of water poised above a base level becomes kinetic energy as the water flows downslope. The velocity of flowing water in a channel is a function of conversion of potential energy above a given point to kinetic energy of water as it drops under the influence of gravity. A mass of water on a sloping surface is impelled downslope by the component of gravity parallel to the slope.

$$F_p = F_g \sin \theta$$

where F_p = component of gravity parallel to slope
         F_g = force of gravity
         θ = angle of slope

The steeper the slope, the greater F_p becomes. Water is a Newtonian viscous fluid; that is, it has zero shear strength and undergoes continuous deformation under any stress. Thus, the potential energy PE of water on a slope is easily transformed into kinetic energy KE.

$$PE = KE$$

Once in motion, water acquires momentum M where

$$M = \text{mass} \times \text{velocity}$$

Momentum is a measure of the tendency of a moving mass of water to remain in motion. The velocity of flowing water, and thus its momentum, is limited by the friction between the water and the bed and banks of a channel and by internal friction between moving particles of water. Energy losses through friction prevent an infinite acceleration of flow. In fact, friction of water against water, and water against solids, either fixed or movable, determines the characteristics of stream flow and the ability of the stream to do geomorphic work.

Most of the kinetic energy expended by a stream, perhaps as much as 95 percent, is used in overcoming frictional forces.

$$KE = \begin{array}{l} \text{internal friction} + \text{external friction} \\ + \text{ transportation} \end{array}$$

The effect of friction is to reduce velocities in the vicinity of the bed and banks of a stream, where friction is greatest (Figure 5–15). Velocity decreases from a maximum just below the surface in mid-channel to zero at the bed and banks of a channel. The average velocity in a stream is normally 0.6 of the distance from the water surface to the channel bed (Leopold et al., 1964). The average flow velocity may be estimated by averaging the velocities at depths of 0.2 and 0.8. However, the rate of change of velocity of water flowing in a channel, known as the velocity gradient, is much more important than the average velocity because it determines the hydraulic shear against the channel walls and bed.

Velocity considerations are common to all of the factors that control fluvial processes, including discharge (volume of water per unit time), gradient of the channel (slope), load, and channel characteristics. Together they make up a system of mutually interdependent variables.

Hydraulic Shear

Because the velocity of water flowing in a channel increases away from the bed, hydraulic shear is developed. Figure 5–16 illustrates the basic concept of hydraulic shear. The moving water may be thought of as a series of thin, parallel layers moving with different velocities such that each layer of water moves over the one beneath it, causing hydraulic shearing stresses between them.

Hydraulic shearing stress is generated by the resistance with which each parallel layer of water moves over the one beneath it. It is proportional to the difference in velocity from one layer to the next. Thus, for laminar flow, the hydraulic shearing stress μ at any point y above the channel floor is

$$\tau = \mu \frac{dv}{dy}$$

where μ = viscosity
         dv = rate of change of velocity
         dy = rate of change of distance
         dv/dy = difference in velocity (velocity gradient) between layers

Flowage of water may be laminar or turbulent. **Laminar flow** is streamlined flow of water particles along parallel paths. In laminar flow, the streamlines remain distinct, and the direction of movement at every point remains unchanged with time. Laminar flow rarely exists for any distance in natural streams except for a thin layer at the channel bed.

Observation of natural flow in open channels reveals perturbations of flow related to the friction of any part of the current against its boundaries. This turbulence involves complex mixing of the flow in such a way that eddies appear at the surface and suspended particles tumble along erratic paths at varying speeds. A filament of dye injected into turbulent flow partakes of the mixing and spreads more rapidly than it would by molecular diffusion alone.

The Reynolds number Rn, a dimensionless parameter, is an approximate measure of conditions necessary for turbulent or laminar flow.

$$Rn = \frac{\rho dv}{\mu}$$

where ρ = density
         d = depth
         v = velocity
         μ = viscosity

Another frequently used form of this equation substitutes hydraulic radius (cross-sectional area divided by wetted perimeter). Reynolds numbers below 500 are usually indicative of laminar flow, whereas Reynolds numbers above 1200 are typical of turbulent flow.

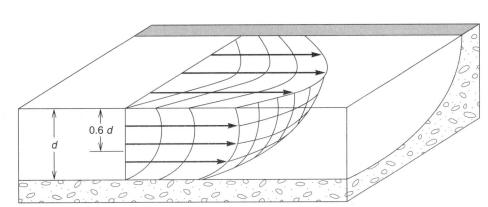

FIGURE 5–15
Three-dimensional, velocity-gradient curve for water flowing in a stream channel.

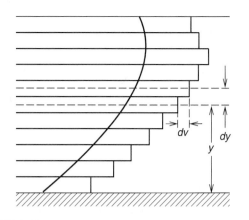

FIGURE 5–16
Hydraulic shear of flowing water, *dv*=horizontal displacement between layers, and *dy*=vertical displacement between layers.

Assuming that changes in water density are negligible, laminar flow is disrupted by deepening and/or acceleration of the current. On the other hand, increases in viscosity (more important than concomitant increases in density), such as might occur with introduction of suspended sediment, dampens turbulence. Laminar flow is effectively disrupted by protuberances along the bed and banks (channel roughness). In fact, completely laminar flow is seldom seen in natural channels and is rare even in sheetwash.

Under certain flow conditions, a thin, but well-defined, laminar layer occurs at the channel bed. Depending upon the perfection or degeneration of this laminar sublayer, three kinds of flow may be recognized:

1. *Hydrodynamically smooth flow* in which the laminar sublayer is thick enough and projections (such as grains) are small enough that all irregularities of the bed are enveloped by the laminar sublayer. Under these conditions the bed surface offers no more resistance to flow than a completely smooth surface.

2. *Transitional flow* in which the more exposed irregularities project through the laminar sublayer and offer resistance to the flow.

3. *Fully rough flow* in which a coherent laminar sublayer is no longer present.

All of the irregularities that project through the laminar zone offer a resistance proportional to the square of flow velocity (Sundborg, 1956).

The laminar sublayer is thickened by reduced temperatures, higher turbidity, and decreased velocity. It is thinned by acceleration of the current and reductions in density and viscosity. It is disrupted by bed roughness and channel irregularities.

Turbulence, represented by the magnitude and frequency of velocity fluctuations, is best developed near the wetted perimeter of the channel. Friction at the solid boundaries of the flow destroys the flow energy, but the turbulence is in turn exerted along the wetted perimeter

where geomorphic work (erosion, transportation, and deposition) is concentrated. Two types of turbulent flow are recognized. **Streaming flow** is the normal turbulence found in most streams. **Shooting flow** occurs at higher velocities. The difference between them is described by the **Froude number** *F*.

$$F = \frac{v}{\sqrt{gd}}$$

where v = velocity
            g = gravity
            d = depth of water

Froude numbers less than one are typical of streaming flow, and those greater than one are characteristic of shooting flow. Velocity increases rapidly during shooting flow and the water surface drops.

As water in open-channel flow moves downstream, the velocity has a positive relation to the slope or steepness of the bed and to the hydraulic radius (cross-sectional area divided by wetted perimeter) of the channel, which for most wide, relatively shallow channels is equivalent to depth. Obviously, gravity becomes more effective where channel slope steepens. In addition, perhaps not so obviously, the bulk of the water is freer from its wetted perimeter, and upper parts of the flow are transported by undercurrents (while they continue to spill forward internally), as hydraulic radius or depth is increased. Logically, increase in channel roughness, with attendant increases in friction, reduces the velocity of the flow.

The distribution of alluvial material along the length and across the wetted perimeter of a channel greatly affects the channel roughness. Protuberances above the channel bed, made by particles of different sizes and shapes, determine roughness of the bed and banks. Granular particles (sizes from fine sand to small pebbles) respond to the current by forming ripples, dunes (or oversized ripples), and antidunes, all of which create a kind of form roughness against the current. Ripples and dunes, to the extent that they migrate with the current, do not offer as much resistance to the flow as they would if they were fixed forms. Increases in velocity, however, can eradicate dunes and replace them with antidunes—forms that are characteristic of high velocity flow and that face steeply upstream. Antidunes, on the other hand, capture oncoming particles and gradually migrate upstream; their resistance is generally seen in steep standing waves on the surface of the water.

Relationships between velocity, slope, hydraulic radius, and roughness are summarized empirically in the Manning equation, which has been widely applied to open-channel flow.

$$v = \frac{1.49}{n} R^{2/3} S^{1/2}$$

where v = velocity
            R = hydraulic radius
            S = slope
            n = coefficient of roughness

Manning roughness is not directly measurable and is computed for a channel after the other terms in the equation have been determined. Photographs of stream beds prepared by the U.S. Geological Survey allow workable estimates of roughness (Barnes, 1967). However, these *n* values are commonly too low for high-gradient, mountain streams of the western United States.

Velocity increases at an exponential rate away from the wetted perimeter. Maximum velocities occur at the surface above the deepest sections of the channel, that is, toward the center of a straight reach of channel surface rather than along the banks. In a meandering section, on the other hand, the swiftest current swings toward the outsides of bends where the deepest pools are also located.

Both the magnitude and the rate of change of velocity are important in establishing turbulence and hydraulic shear, the shearing stress within the flow and against its wetted perimeter. Inasmuch as frictional drag is greatest against the wetted perimeter, this area is where the greatest rate of change of velocity, and hence turbulence and hydraulic shear, occurs. Local protuberances or increases in roughness, by imposing additional friction, increase turbulence and shear accordingly. By focusing additional kinetic energy upon themselves, they encourage their own destruction.

Because of the exponential increase in velocity from the bed upward, the velocity at or within the bed can be considered zero, and the maximum velocity is usually at the surface (although it may locally be deflected below the surface in the tee of some obstacles, and it may sometimes be depressed below the surface by the friction of upvalley winds). Depth-integrated velocity measurements, taken by lowering and raising a current meter at a constant rate, show this to be true. On this basis, for streams of the same cross-sectional area, average velocity, and (hence) discharge, the shape of the channel cross section plays an important part in regulating the rate of change of velocity. The shape of the cross section, and particularly the shape (even the detailed shape) of the wetted perimeter, makes its frictional demands on the energy of the flow and establishes the distribution of turbulence and the magnitude of hydraulic shear. A semicircular cross section, for example, transmits the flow at any given cross-sectional area past the least length of wetted perimeter. The hydraulic radius is proportional to the radius of the semicircle (1/2 the radius), and rate of change of velocity is nearly equivalent along all radii.

Natural channels do not approximate semicircles in cross section, even though this happens to be the shape that opposes the least friction to the discharge. Most channels are wider than their depth, and, with sloping sides, their cross section is roughly trapezoidal. Under a given discharge and with a given cross-sectional area, increases in width at the expense of depth impose a greater friction against the flow. Not only does wetted perimeter increase in length, but relative roughness also increases across the bed. The flow through the wide, shallow channel, on the other hand, opposes a greater hydraulic shear against the resistance of the bed. Analogous increases in depth at the expense of width, while reducing bed shear, apply additional shear against the banks in opposition to increased bank resistance.

The effective energy of a stream is not simply proportional to its average velocity. It depends also on the distribution of velocity or, in other words, on the way in which the resistance of the channel allows the discharge to develop turbulence and hydraulic shear, Wide, shallow channels exert maximum bed shear for traction of bed material.

Effect of Slope

As previously noted, as slope increases, the component of gravitational force parallel to the slope also increases. Thus, velocity is directly proportional to slope and increases as slope increases.

An increase in velocity due to an increase in slope has a greater effect on transportation of bed load than does an increase in velocity due to an increase in discharge (Gilbert, 1914). The reason is illustrated in Figure 5–17. Recall that the discharge of a stream is Q = wdv (width × depth × velocity). If Q increases, the channel width and depth increase, as shown in the figure. However, if velocity increases due to an increase in slope at *constant discharge,* depth must *decrease, so* that the product w × *d* × v remains the same. As shown in Figure 5–17, greater velocity is accompanied by shallower water. The higher velocity, coupled with the lower depth, produces a steeper velocity gradient and enhanced hydraulic shear, which intensifies transporting power along the channel bed.

The maximum velocity that can be developed without resulting in erosion of the channel floor is a function of depth. A deep channel with the same average velocity as a shallow channel will not erode its bed as much because as depth increases, the water above the bed merely drags ahead more water without affecting particles on the channel floor.

Effect of Load

Streams continually wear down the land and transport an immense amount of rock detritus to the oceans annually. For example, the Mississippi River system removes approximately 517 million tons of rock material from the land each year, enough to fill a train of freight cars extending six times around the equator.

The load supplied to a stream over a period of years is dependent upon the following:

1. Topographic relief

2. Lithology of slope-forming material

3. Climate

4. Vegetation

5. The nature of processes acting in a drainage basin

Relief determines gravitational forces and potential energy of slopes. Denudation rates increase exponentially with

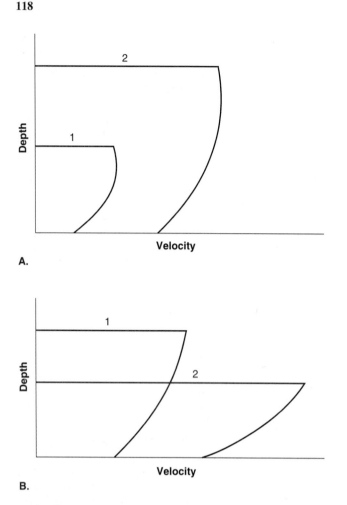

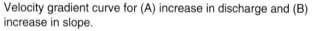

FIGURE 5–17
Velocity gradient curve for (A) increase in discharge and (B) increase in slope.

be important, as, for example, in the case of a glacier entering the headwaters of a system.

Total load is an important aspect of denudation of land masses, but for individual streams, the **caliber of load** is more significant with respect to the effect on stream processes. The mechanics of transportation of load include solution, flotation, suspension, saltation, and traction. Each of these mechanics is discussed in the following sections.

Dissolved Load

The largest percentage of dissolved load is contributed by chemical weathering of slopes, by tributary runoff, and by effluent ground water. The dissolved load reflects the weathering regime and solubilities of the drainage basin, both from the surface and underground. However, because ground-water divides may not coincide with the topographic divides of a drainage basin, dissolved load may not reflect solution in the drainage basin with complete fidelity.

Sampling of dissolved loads of the great rivers of the world suggests that about half the material carried is in solution. Although this is a tribute to the efficacy of chemical weathering and solution by ground water, the transportation of dissolved load is more continuous than the transport of clastic load. Fluctuations in the daily amount of dissolved load of a full order of magnitude or more reflect primarily the dissolving activities of runoff during floods and are not related to the transporting ability of the stream, as is the case with clastic load. Consequently, concentrations of dissolved load are slight, ranging in the United States from about 50 to 1000 parts per million according to samples from 170 stations examined by Langbein and Dawdy (Leopold et al., 1964). The important conclusions are that solution transport quietly accounts for a large share of total stream load, but that attendant changes in the density and viscosity of the water are negligible as far as their influence on stream flow and channel morphology is concerned.

Flotation

Flotation load depends upon rafting phenomena, which, although minor and perhaps inconsequential for most rivers, may locally be important. Rock fragments and mineral grains of appropriate sizes may be rafted by surface tension, bubbles, ice, and vegetation. Patches of sand, held up by surface tension, may travel indefinitely as long as they remain partially dry. The process requires a relatively quiet stream surface because strong surface turbulence wets the grains and sinks them. Transport of small particles by bubbles relies on similar forces capable of holding small grains to the surface of the bubble. River ice may account for the flotation of a variety of bed material wherever the water has frozen to the bed, as it may near the shallow margins of a stream. Breaking of the ice during floods causes sheets of ice, frozen into the gravel of the streambed, to lift free and float downstream. Vegetational rafting is accomplished primarily by the roots of fallen trees, which may clasp rock particles well enough for trans-

increase in relief (Figure 5–18). The lithology of slope-forming material establishes the physical nature of detritus produced by weathering and erosion. For example, erosion of columnar-jointed basalt typically yields hard, dense, physically durable boulders and cobbles, whereas shale weathers and erodes to form small, easily disintegrated particles. Climate is important for two reasons:

1. The amount and distribution of precipitation affect weathering and erosion rates.

2. Vegetation is highly dependent on climate.

A high level of correlation exists between vegetation and erosion rates. As shown in Figure 5–19, maximum sediment yields occur where little vegetation exists, and effective precipitation is about 10 in/yr (25 cm/yr). Sediment yield decreases with lower precipitation (because of low runoff rates) and with higher precipitation because of increased vegetative cover. Once vegetative cover reaches about 70 percent, any additional ground cover has little effect on sediment yield. Variation in local processes operating in the drainage system may

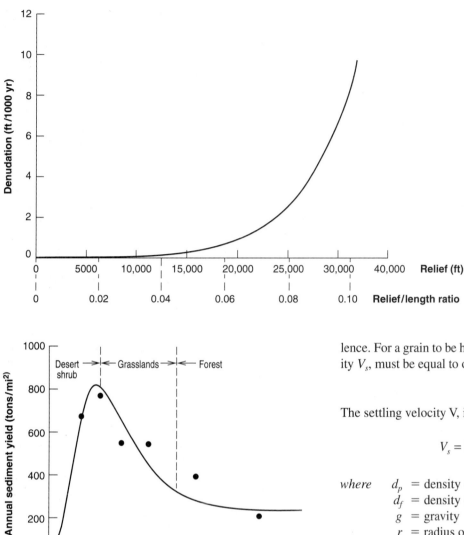

FIGURE 5–19
Relationship between sediment yield and
climate/vegetation. (From Langbein and Schumm, 1958)

port. The vegetation itself may be an important part of the sedimentary load.

Suspended Load

Fine-grained, clastic particles may be carried mechanically in suspension by streams. Smooth, streamlined, laminar flow is not maintained for appreciable distances in stream channels because water swirls and eddies in an irregular fashion, producing turbulent flow. Turbulent flow is critical to the transportation of suspended load because turbulence generates random movements with upward components.

The velocity with which silt and clay settle toward the bottom of a stream depends upon the strength of upward turbu-

lence. For a grain to be held in suspension, the settling velocity V_s, must be equal to or less than the turbulent velocity V_t

$$V_s \leq V_t$$

The settling velocity V, is governed by Stokes' law.

$$V_s = 2/9 \frac{(dp - df)gr^2}{\mu}$$

where d_p = density of particle
d_f = density of fluid
g = gravity
r = radius of particle
μ = viscosity

According to Stokes' law, settling velocity increases with grain density and the square of the radius of the particle. Because the density and viscosity of water are nearly constant relative to the other factors, and because the density of most rock particles varies only between about 2.6 and 3.3, the radius of a suspended particle is highly important. The square of the radius of a grain increases rapidly, producing a high settling velocity, which may overcome the turbulent velocity in a stream to the extent that it is unable to hold a particle in suspension.

As discharge increases, the suspended load carried by a stream increases at a more rapid rate than the discharge (Leopold and Maddock, 1953) (Figure 5–20). The enlarged concentration is a result not of scouring of the channel but rather of erosion of the drainage basin.

Bed Load: Saltation and Traction

Bed-load transport and suspended-load transport are so gradational into one another that a distinction is often difficult to make. Moreover, the proportions of each depend upon the characteristics of the streamflow at any time, and these may fluctuate over short intervals. With an increase in energy, parts of the bed load are thrown into suspension, and more of both suspended load and bed load is entrained. Deposition of bed material and reversion of some suspended sediment to bed-load transport accompanies energy losses.

Part of the gradation of bed load into suspended load is due to transport by **saltation,** a process of skipping or bouncing along the channel bed (Gilbert, 1914). Sand grains, momentarily lifted off the channel bed, move downstream before sinking to the bottom, only to be picked up again and the process repeated.

Traction, the rolling and sliding of particles along the bed, constitutes true bed transport. However, in many situations in which traction is established, most of the material moves by saltation. Particles involved in saltation roll upon the backs of their neighbors downstream and are tom forward by the current along low, extended trajectories. The storm of saltating particles concentrated near the bed has a definite boundary against the less heavily loaded water above. The impulse for saltation comes primarily from the water at the expense of fluid momentum. Consequently, a cloud of saltating grains acts as a brake on the flow, and the stress against the bed is due not so much to the (enfeebled) flow of water as to the impact of saltating grains (Bagnold, 1956).

Bed load may be defined as granular material supported directly by the channel bed, thus adding its immersed weight to the bed (Bagnold, 1956). Suspended load obtains its sup-

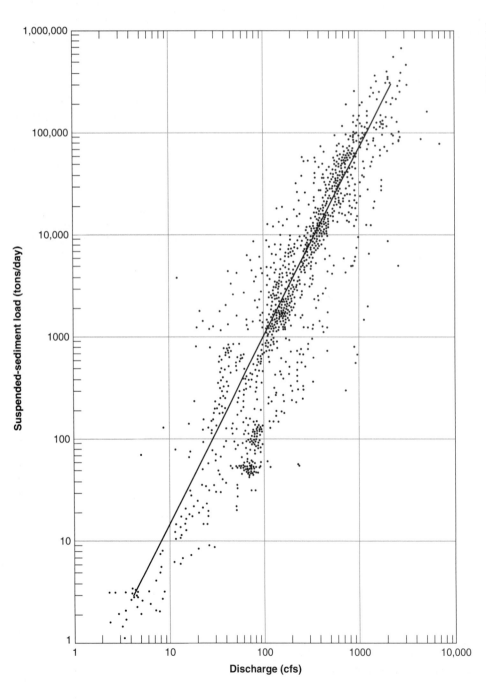

FIGURE 5–20
Increase in suspended load and discharge, Powder River at Arvada, Wyoming. (From Leopold and Maddock, 1953)

port from the water and thus adds its immersed weight to the hydrostatic column. In light of these definitions, saltation load is truly a hybrid form. For a particle to be maintained in suspension, its settling velocity must be exceeded by the upward velocity of water in turbulent transfer. However, turbulent exchange operates in all directions, so that while some particles are carried upward in the current, others are driven down onto the bed. The cloud of sediment in suspension is statistical: As some particles are dropped to the bed, others rise to take their place, and none of them is certain to remain indefinitely in suspension. Moreover, the concentration and average grain size of particles in suspension decrease with distance above the bed. Interchange of material in suspension and in the bed of the stream is suggested by a correspondence of at least part of the grain sizes in both places. On the other hand, some reaches of a stream may have bedrock or alluvium coarser than the suspended load at the boundaries of the flow, indicating that suspension is virtually 100 percent efficient and that residence of suspended particles on the bed is exceedingly temporary.

The hydraulic shear stresses required to move particles on a stream bed are complex and interactive. Figure 5–21 shows the relationship between critical shear stress needed to initiate movement of grains of varying diameters. At some critical shear stress, particles on the bed begin to move. The

transport rate increases with increasing shearing stress, as shown in Figure 5–22.

Rubey's study (1938) of hydraulic forces necessary to initiate grain movement in streams led to identification of the sixth power law, hydraulic lift, and critical tractive force as important factors. The relationship between the force of water pushing on the upstream side of a particle and the resistance of the particle to movement is described by the **sixth power law:** A particle on a stream bed is on the verge of motion when the force of the water against it equals the resistance of the particle to movement. The radius of the largest particle that can be set in motion by a given velocity is

$$r^3 = kv^6$$

where r = radius of the particle
k = a constant that includes gravity, grain density, and other variables which change very little
v = flow velocity

A small increase in velocity (taken to the sixth power) will therefore produce a very large increase in the size of particle that can be moved. The power of streams to transport surprisingly large material during floods (Figure 5–23) is thus more readily understood.

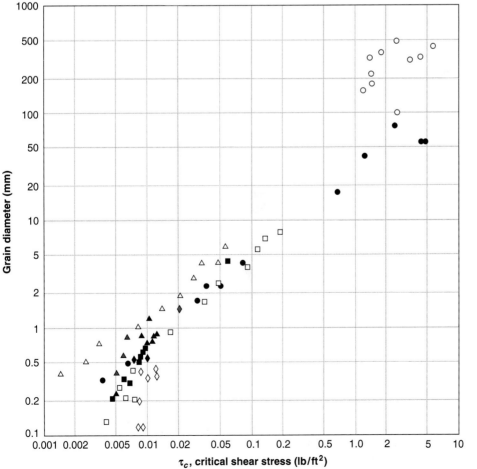

FIGURE 5–21
Critical shear stress required to initiate movement of particles. (From Leopold et al., 1964)

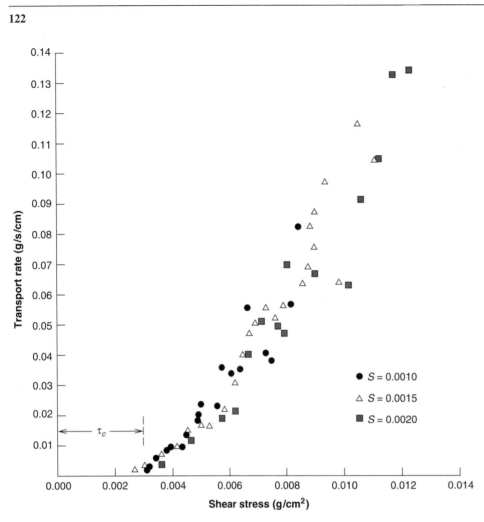

FIGURE 5–22
Relationship between transport
rate and shear stress. (From
Leopold et al., 1964)

Because of the steep rate of change of velocity near the bed of a stream, grains on a streambed in the area of steepest velocity gradient experience a lowering of pressure on the top of each particle surface, known as **hydraulic lift.** This phenomenon, which is analogous to an airplane wing, is more effective on small particles than on larger ones.

A particle on a sloping stream bed supports a column of water above which exerts a **critical tractive force** proportional to the depth of water and the channel slope.

$$Ft = \pi gds$$

where Ft = critical tractive force
 ρ = density of water
 g = gravity
 d = depth of water
 s = gradient of stream

The critical tractive force is significant for moving smaller particles, whereas the sixth power law is prominent in moving large particles (Rubey, 1938).

Before grains can be set in motion on a streambed, both gravitational and cohesive forces must be overcome. Cohesion between sand-size or larger grains is low and less important than gravitational forces. However, for silt and clay, cohesion plays a prominent part in entrainment of grains. Clay particles that can be transported by a given velocity are more difficult to move from a streambed because of cohesion between grains, which inhibits their motion. The higher velocities required to initiate movement of clay and silt than for larger grains was described by Hjulstrom (1935).

The threshold of erosion curve in Figure 5–24 indicates that the velocity required to move grains between about 0.1 and 1.0 mm decreases until about 0.1 mm and then increases again for particles less than 0.1 mm. The *increase* in velocities required to move smaller particles is due to greater cohesion between the smaller grains and the diminished surface resistance of the smoother channel bed. The threshold of deposition curve (shown in the figure) illustrates that deposition will take place for velocities below the curve, and that for velocities above the curve, grains already in motion will remain in motion.

Channel Relationships

Channel patterns in streams vary considerably, both in plan view and in cross section, and the volume of published literature on the subject is enormous. The pattern of a channel in

FIGURE 5–23
Two large boulders moved by Olsen Creek in 1983. That they were indeed moved is shown by green foliage beneath them. At the time the photo was taken during the following summer, the creek was about 1/2 meter wide and 6–8 cm deep.

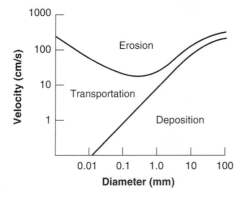

FIGURE 5–24
Relationship between flow velocity and particle size for erosion, transportation and deposition. (After Hjulstrom, 1935)

plan view (meandering, braided, or straight), channel cross-section form (the width/depth ratio), and bed roughness are important to the understanding of stream processes. However, many of these topics are controversial among geomorphologists, hydrologists, engineers, and geographers, and much diversity of opinion may be found in the published literature. The approach taken here is to discuss the relevant factors, point out some of the differences in philosophy, and, in a number of instances, await further research and more conclusive answers.

The pattern of stream courses varies from sinuous to straight to braided. Straight reaches of rivers are so rare as to be suspect of human intervention. Because straight streams are exceedingly rare, most of the discussion to follow deals with meandering and braided streams. Meandering streams typically have single, sinuous channels, few channel islands, and deep, narrow channels. Braided streams generally have low sinuosity, multiple channels that branch repeatedly into dividing and reuniting channels, numerous islands, and wide, shallow channels.

Meandering Streams

Meandering channels are by far the most common of all channel forms and have piqued the interest of geomorphologists for many years. Close to a thousand papers have been published on various aspects of meandering (see, for example, Ikeda and Parker, 1989).

The style of meandering of a stream may be measured quantitatively by its **sinuosity index** S, measured in one of the following ways:

$$S = \frac{\text{stream length}}{\text{valley length}}$$

$$S = \frac{\text{thalweg length}}{\text{valley length}}$$

$$S = \frac{\text{channel length}}{\text{length of meander belt axis}}$$

$$S = \frac{\text{channel length}}{\text{meander wavelength}}$$

Meander Wavelength and Radius

That the wavelength of meanders varies with the discharge of a stream has long been recognized; that is, large streams have large meanders and small streams have small meanders (Figure 5–25; Plate 2–C). The empirical relationship between meander wavelength and discharge may be quantified as follows:

$$\text{wavelength} = kQ^x$$

where Q = discharge
       k and x = constants

Values for discharge may be available from gauging stations, although some uncertainty exists about whether meanders are related more directly to average or to maximum discharge. In addition, meanders are sometimes inherited from an

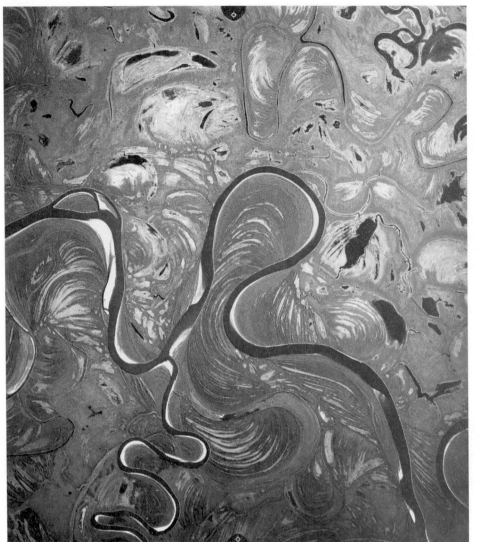

FIGURE 5–25
Difference in meander size from
streams of different discharge.
(Photo from Dept. of Agriculture)

ancestral stream having a discharge different from that of the present stream, thereby complicating the picture.

Friedkin (1945) demonstrated experimentally that the size of meanders enlarges with increased discharge (Figure 5–26). When the discharge of a laboratory stream was increased, the previously developed meander pattern was destroyed by a larger set of meanders.

> "To determine the effect of discharge on the size of bends, three tests were conducted in which all conditions were similar except the discharge. Each of these small scale rivers began with a straight channel except for an initiating bend. The bed and bank material, valley slope, initial angle of attack and duration were the same as the discharge increased, the size of bends increased both in length and width, that is, an increase in discharge resulted in an increase in radii of bends." (Friedkin, 1945)

Carlston (1965) studied 14 natural rivers and found a good correlation between meander wavelength L and mean annual discharge Qm expressed by the following equation:

$$L = 106Q_m^{0.46}$$

However, Schumm (1967) concluded that discharge alone did not fully account for meander wavelengths in the streams he studied, and he suggested that meander wavelength is also a function of the type of sediment load. That is,

$$L = \frac{1890Q_m^{0.34}}{M^{0.74}}$$

where Q_m = mean annual discharge
        M = percentage of silt/clay in channel perimeter

This raises the question, could factors other than Q be important in determining meander wavelength, and, if so, how are they related to causal mechanisms? The causal relationship between meander size and discharge appears to be largely a matter of inertia; that is, to move a large volume of water around a tight bend requires that a substantial amount of inertial energy be directed at the outside of the bend (because $I = ma$), and this, of course, allows accelerated erosion and enlargement of the bend to a more stable radius. What, then, determines the limiting size of the meander? At what radius does

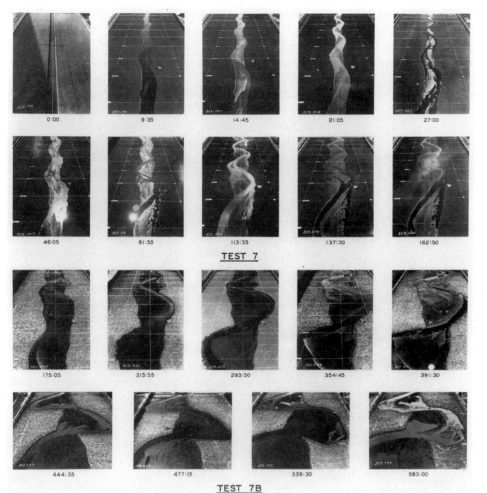

FIGURE 5–26
Development of a meandering channel from an initially straight channel in a flume. The numbers beneath each photo give the time since the beginning of the experiment. The lower two photos depict the change in meander radius with increased discharge. (From Friedkin, 1945)

TEST 7

TEST 7B

the meander cease to enlarge and achieve some sort of stable configuration? In natural streams, one limiting factor is the size to which meanders may grow before two meander loops intersect and a meander cutoff occurs. However, as shown by Friedkin's experiments, a given Q produced a characteristic meander radius without any meander cutoffs taking place. Conversely, reduction in discharge of natural streams produces decreased meander size and underfit streams, the implication being that with lower flow, the larger ancestral meanders are no longer stable. Clearly, discharge regulates meander size in both natural and experimental streams.

But how might sediment load play a role in influencing meander size? The effect of suspended load may be largely discounted because the viscosity of the water in most streams is not changed enough to affect the flow. However, the amount of silt and clay in sediment making the banks and floor of a channel plays an important role in determining the shape of the channel cross section (Schumm, 1960). Thus, silt/clay percentages in channel sediments may be indirectly, rather than directly, related to meander size in that the bank-forming sediment may determine channel width, which in turn affects meander radius.

Meanders size M can be shown empirically to vary with channel width W according to the equation

$$M = hW^x$$

where h and x = constants close to 1

Leopold and Langbein (1966) studied the ratio between meander wavelength Mw and radius of curvature Rc, and they concluded that Rc tends to be more constant. Values for Rc for looped meander are about 3:1, whereas for more open meanders, Rc is closer to 5:1. For meanders on different streams, Leopold and Langbein found that Rc averaged 4.7. A similar analysis by the author showed variations of Rc from 1.7 to 14.0. Thus, this ratio seems to have little significance other than in describing the geometry of meanders for a particular reach of a given stream.

Cause of Meandering Straight channels seldom persist long in either natural or flume courses. Even starting with accurately molded, straight channels in flumes, channels quickly evolve into sinuous patterns (Friedkin, 1945). Because perfectly laminar flow in channels is so difficult to maintain for any length of time, small perturbations in flow eventually deflect flow against the bank, reflecting flow toward the opposite bank, thus setting in motion a positive feedback system that leads to a meandering course. The deflected flow against a bank causes a small amount of bank erosion and initiates a more permanent bend in the channel. The force of inertia of the water as it flows around the bend causes the water to attempt to continue to flow in a straight line and to impinge against the outside of the bend, resulting in further erosion of

the bend and deepening of the outside of the channel (Figure 5–27). The deeper channel on the outside of the bend produces greater velocity and still more erosion (Figure 5–28).

At this point, the inertia of the moving water causes it to "pile up" on the outside of the bend and to slightly elevate the water surface. Inertia causes water near the center of channel to cross over toward the outside of the bend, and a cross-channel component of flow develops (Figure 5–27). Water near the floor of the channel is returned toward the inside of the bend, setting up a cross-channel, **helical flow pattern** that has the form of a spiral or helix (Figure 5–27) (Friedkin, 1945; Callander, 1978; Parker, 1976; Leopold, 1982; Ikeda and Parker, 1989). The force of the helical flow pattern can be considerable. Geomorphologist Dave Rahm once capsized a canoe on a river, found himself on the floor of the channel at the outside of a meander, and was thwarted in his attempts to swim upward to the surface by the downward direction of helical flow. Recognizing the problem, he simply walked along the bottom toward the inside of the bend until the downward motion of helical flow was no longer effective, and then he swam to the surface.

The effect of helical flow is to establish an orderly flow pattern that localizes deposition and erosion and further promotes development of the bend. Sediment entrained by bank

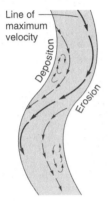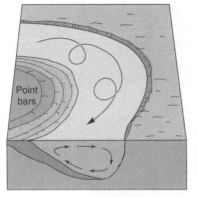

FIGURE 5–27
Helical flow in a meander.

FIGURE 5–28
Erosion on the outside bank of a meander, Nooksack River, Washington.

erosion on the outside of a bend, caused by the greater veloc-ity there, is carried downstream to the next bend in the chan-nel where it is deposited on **point bars** (Figures 5–29, 5–30; Plate 2–C) on the same side of the channel from which it was eroded (Matthes, 1941). Most of the sediment eroded from the outside of a bend is transported by helical flow toward the point-bar deposits on the inside of bends. Perpetuation of this process leads to further development of the meandering pattern.

However, as in most positive feedback systems, the process eventually ends in self-destruction of the system. As the cur-vature of a meander continues to increase as a result of erosion on the outside of the bend, the processes become self-accel-erating. The greater the curvature of a meander, the greater the erosion on the outside of the bend, and the greater the rate of increased curvature, until finally two bends intersect (Fig-ure 5–31; Plates 4A and 2D) to form a **meander cutoff.**

The abandoned channel of a cutoff meander remains as a horseshoe-shaped remnant of the former bend. If the slope of the channel is fairly steep, once water runs out the lower end, the abandoned channel is dry. If the slope is relatively gentle, the lower end of the cutoff channel may be dammed by sed-iment, and water is retained to form an **oxbow lake** (Plate 2–D). As time passes, the oxbow lake gradually fills with sediment and shallows; vegetation then encroaches, and the lake ultimately vanishes.

Braided Streams

Braided streams are characterized by multiple channels that branch again and again, with anastomosing, dividing and re-uniting, wide, shallow channels filled with numerous islands (Figures 5–33, 5–34, and Plate 3–B). Frequent lateral shifting of channels, bars, and islands can convey the impression that a braided stream is unable to carry all the sediment load, and

FIGURE 5–30
Crescentic point bar deposits left behind as a meander migrates toward the outside of the bend. (Photo by D. A. Rahm)

FIGURE 5–31
Two meanders about to intersect and produce a meander cutoff, Snake River near Rexburg, Idaho. (Photo by W. K. Hamblin)

FIGURE 5–29
Point bar deposits on the inside of meanders, White River, South Dakota. (Photo by D. A. Rahm)

this shifting has led to misleading statements in beginning geology textbooks that the cause of braiding is "overloading of a stream." However, Friedkin (1945), Rubey (1952), Mackin (1956), and others have shown that braided streams are not necessarily aggrading. Braided streams that are ac-tively incising their channels are not rare (Figure 5–35 and Plate 3–C). Streams with both braided and meandering reach-es also provide evidence that excess load cannot be the cause of braiding.

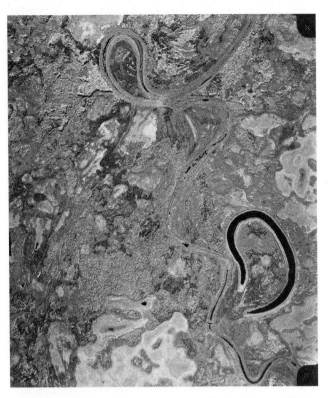

FIGURE 5–32
Oxbow lake (lower right) and abandoned channel with a cutoff meander (upper center). (Photo by Dept. of Agriculture)

The physical processes whereby some streams (reaches) braid while others meander are reasonably well known, but, as we shall see in later discussions of the graded stream, some confusion exists over cause-and-effect relationships involved in the roles of channel characteristics, slope, and adjustments of streams to externally imposed changes in load and discharge.

Friedkin's flume experiments (1945) were among the first to demonstrate the relationship between braiding and erodibility of bank-forming material. He found that narrow, deep channels developed in cohesive bank material produced meandering channel patterns, whereas broad, shallow channels developed in less cohesive bank material created braided channels for the same flow on the same initial slope. Channels formed in banks of high silt/clay content made meandering patterns, whereas channels formed in banks of sand with low silt/clay content made braided channels. Mackin (1956) suggested that braiding in a stream channel is directly controlled by erodibility of banks and its constraint on the cross-sectional form of the channel. If a channel has relatively strong, cohesive banks that can withstand the hydraulic shear of flowing water, a narrow, deep channel forms, and the inertial and helical flow effects lead to development of meanders. If a stream has relatively weak, low-cohesion banks that are incapable of withstanding the hydraulic shear of flowing water, a wide, shallow channel forms, inhibiting helical flow. The resulting channel disorder breaks up the current into rapidly shifting, self-baffling channels with many

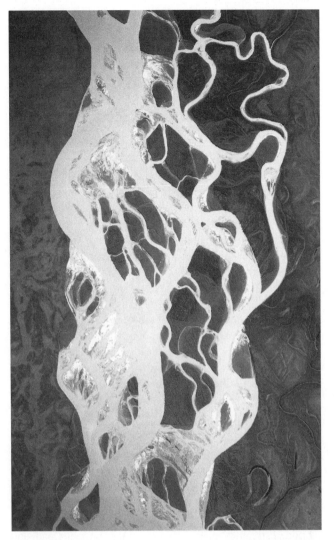

FIGURE 5–33
Braided stream, Alaska. (Photo by Dept. of Agriculture)

randomly placed bars and islands. Thus, the critical factor in braiding is the control of bank erodibility on channel width/depth ratio.

Schumm (1960) has shown that cross-channel form (width/depth ratio) is related primarily to the distribution of hydraulic shear and bank resistance. For example, a wide, shallow channel has a clustering of isovels (lines of equal velocity), a higher velocity gradient, and a greater hydraulic shear stress near the bed than higher in the channel (Figure 5–36). Thus, the ease with which banks and floor of the channel can be eroded under the hydraulic shear conditions permits the stream to develop a channel form consistent with the erodibility of the bank-forming material.

Schumm (1960) found that the type of sediment in the bank-forming material was the dominant factor in controlling cross-sectional form (Figure 5–37). Where banks consisted mostly of sand or gravel with low silt/clay content, the channel cross sections were wide and shallow, and the profile of the stream was relatively steep. Figure 5–38 shows a plot of

FIGURE 5–34
Braided stream, Nooksack River, Washington. (Photo by D. A. Rahm)

FIGURE 5–35
Deeply incised channel cut by a braided stream after 1980, Toutle River, Washington.

channel depth versus channel width for a variety of streams having different silt/clay percentages. As shown in the figure, Sand Creek has about 22 percent silt/ clay, whereas the Smoky Hill and Kansas River systems have silt/clay contents of only about 3–5 percent. Schumm found that for the Smoky Hill River, as the silt/ clay content decreased in the downstream direction, a large increase in channel width relative to depth occurred. Where a local increase in silt/clay occurred in the bank-forming material, depth increased greatly relative to width. From these studies, Schumm concluded that the width/depth ratio of a stream channel was directly related to the percentage of silt/clay in the bank-forming material, and that the higher amount of fine material characterized greater resistance of the banks through channel erosion. Sediment with high silt/ clay content is cohesive and resists hydraulic shear. A channel with higher silt/clay content will be narrower and deeper because it can withstand the higher hydraulic shear on its banks, whereas coarse sand, with less cohesion, is more easily eroded and thus does not resist lateral erosion as well, producing a wide, shallow cross section. As pointed out by Mackin (1956), bank vegetation may also increase bank resistance and thus influence channel cross-section form in a manner similar to banks of high silt/clay content. Mackin suggested that in streams that have braided reaches between meandering reaches, such as the Big Wood River in Idaho, bank vegetation may be the critical difference.

Rapid fluctuations in discharge are common to some braided streams, especially braided glacial meltwater streams. Some workers have suggested that the rapid discharge changes play an important part in the braiding process. However, in Friedkin's (1945) flume experiments (as well as many performed since then), braiding was produced and maintained from initially straight channels with constant discharge. Thus, although rapid channel fluctuations may be common in braided streams, they do not appear to be the cause of the braiding.

As noted by Friedkin (1945), Leopold and Wolman (1957), Schumm and Khan (1972), and many others, slopes of braided reaches are typically steeper than meandering reaches for streams with similar discharge (Figure 5–39). This relationship is especially well shown where a meandering stream braids for a distance and then resumes a meandering course —the slope of the braided reach is greater than that of the meandering reaches. The reason is shown by the Chezy equation, which tells us that the shape of the channel cross section, expressed by the hydraulic radius R (area/wetted perimeter), affects flow velocity.

$$v = C\sqrt{RS}$$

where v = velocity
C = smoothness coefficient
R = hydraulic radius
S = slope

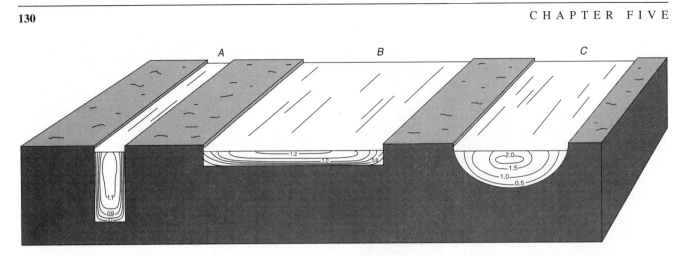

FIGURE 5–36
Variation of width/depth relationships in stream channels.

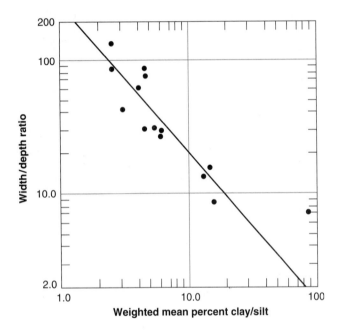

FIGURE 5–37
Relationship between silt/clay content of bank-forming material and width/depth of a channel. (From Schumm, 1960)

The hydraulic radius R for a wide, shallow channel of a braided stream is smaller than for a narrow, deep, meandering channel because the wetted perimeter Pw relative to the cross sectional area A is greater.

$$R = \frac{A}{Pw}$$

$$R_{braided} < R_{meandering}$$

Thus, in a channel reach of constant discharge, if the channel pattern changes from meandering to braided, R decreases, and, because velocity must remain constant in order to carry the load, slope S in the Chezy equation must increase.

If the channel pattern then changes from braided back to meandering, the situation reverses; that is, R increases, requiring a corresponding decrease in slope S. A very important point to remember here is that the change in slope is a *response* to changes in channel shape, not a cause of braiding. Increasing the slope of a stream does not *cause* it to braid.

Anastomosing Channels

Another type of stream pattern with multiple channels is known as **anastomosing.** These types of channels are not the same as braided channels because they typically consist of a network of low-gradient, narrow, deep channels with stable banks (Schumm, 1989).

EQUILIBRIUM IN STREAMS

Concept of the Graded Stream

Streams are not endowed with the ability to make a variety of decisions, but they expend their energy wisely because, automatically, they must. The terms reached between water and the solids with which it must deal are an expression of equilibrium. The idea that streams may reach some kind of equilibrium among controlling variables has been recognized by many authors. As early as the turn of the twentieth century, Gilbert (1877, 1914, and 1917), Powell (1876), Dutton (1882), and Davis (1909) recognized that streams behaved in an orderly manner, and the term *graded* was coined to describe streams that appeared to have achieved an equilibrium between the governing discharge, load, and slope conditions. The concept was further elaborated by Mackin who defined a **graded stream** as follows:

"A graded stream is one in which, over a period of years, slope is delicately adjusted to provide, with available discharge and with prevailing channel characteristics, just the velocity required for transportation of the load supplied from the drainage basin." (Mackin, 1948)

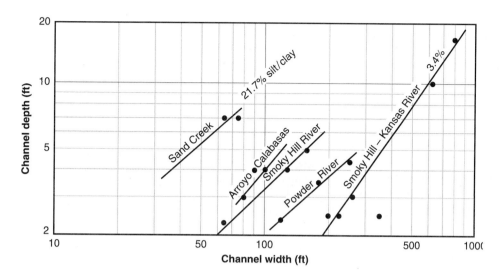

FIGURE 5–38
Relationship of width/depth ratios to silt/clay content in alluvial streams on the Great Plains. (From Schumm, 1960)

The four variables that control the equilibrium system of a graded stream are:

1. Slope
2. Discharge
3. Load
4. Channel characteristics

These variables are discussed in the following sections.

Slope of the Graded Profile

The slope of a graded stream is a concave-upward curve, which can be expressed mathematically. Different conditions of load and discharge cause the shape of the concave-upward curve to vary from one stream to another.

A stream operates within the constraints of the distance and the relief between its head and mouth. The potential energy of the water depends upon its height above the mouth of the stream **(base level).** However, the rate at which the potential will be expended or converted into kinetic energy (energy of motion) depends upon the steepness of slope. Within these constraints, few streams cover most of the distance at the altitude of their headwaters and finally plunge over waterfalls to a base level far below. Nor do many streams descend to the altitude of their mouths while almost at their headwaters and then wander most of the distance on a nearly flat gradient. Instead, most streams come to terms with the materials of the landscape by flowing down a slanting profile, which is typically steepest near the head and progressively gentler toward the mouth, making a concave-upward curve that connects the head and the mouth of the stream most efficiently. The concave-upward profile permits a growing discharge of water to expend its energy at a rate just sufficient to transport the load toward base level. Any bumps or shoulders in the profile would encourage the very concentration of energy that would eradicate them, and troughs or hollows, locally deficient in energy, would soon fill up. Irregularities along a pro-

file, then, should be suspected as clues to additional constraints of structure, lithology, or recent deformation of the profile by tectonic or mass movements.

Deepening affects the distribution of the flow velocity at the bed. Deepening, like widening, is a self-limiting process. Expansions in cross section prevent the stream from becoming indefinitely wide or deep because, by widening and deepening, the stream loses its ability to do more of the same.

The concave-upward curve of the long profile of a graded stream progressively decreases in slope downstream as a direct result of equilibrium in the stream. The stream may have minor irregularities below major tributary junctions because of changes in discharge or caliber of the load but may still be graded throughout, and short sections along the course may be ungraded because of local conditions.

The slope of a graded stream depends upon load-discharge relationships. High-gradient streams, such as many of the streams in the western United States, have gradients of 70 to 100 ft/mi (13 to 20 m/km), whereas others, such as the Illinois River, have gradients as low as 2 in/mi (3 cm/km). Both high- and low-gradient streams are graded, and their differences in slope are a result of differences in discharge and caliber of transported load.

Slope is the principal factor in the equilibrium system of a graded stream that is automatically adjustable *by the stream itself* to accommodate changes in the controlling factors of the system that would otherwise cause changes in velocity. The slope of a graded stream adjusts to provide just the velocity required to transport all the load supplied to the drainage system.

The slope of the long profile of a graded stream is essentially a profile of equilibrium:

"Its diagnostic characteristic is that any change in any of the controlling factors will cause a displacement of the equilibrium in a direction that will tend to absorb the effect of the change." (Mackin, 1948)

An example of this kind of adjustment is the response of Thompson Creek, Iowa, to changes that artificially increased the gradient. Thompson Creek responded by incising its channel 12 in (40 ft), with concomitant widening of the channel, resulting in the collapse of bridges, lowering of the water table, and intense gullying of land as small tributaries extended their channels headward.

Discharge and the Graded Profile

Discharge—volume of flow per unit time in a given area—is an important variable in the equilibrium system of a graded stream. We have already seen that as discharge increases, flow velocity increases because of the relationship between cross-sectional area and wetted perimeter. Thus, changes in discharge affect the velocity of a stream and call for an adjustment to the system to accommodate the changes. For example, an increase in the discharge of a graded stream produces an increase in velocity, which enhances the transporting power of the stream to the point that larger particles that previously could not be moved can now be set in motion. By setting these particles in motion, the stream channel is eroded, and the slope is thereby decreased. As the slope is lowered, velocity is also lowered, which limits the ability to further erode the channel and restricts continued lowering of the slope, and a new equilibrium between slope and discharge becomes established. Conversely, if the discharge of a stream is decreased, the accompanying decrease in velocity dictates that some of the larger particles, previously within the ability of the stream to move, now must be deposited, which, in turn, steepens the slope. As the slope is increased, velocity is also increased until finally a velocity is attained that is just sufficient to carry the load under the new discharge.

Short-term fluctuations in discharge, which may occur seasonally, daily, or even hourly, may cause local, temporary scouring of the channel during high-water stages or deposition during low-water stages. Because of the fluctuating nature of discharge, equilibrium of discharge with other variables in a graded stream refers only to the long-range picture. As Mackin pointed out:

> "Over a period of years sufficiently long to include all the vagaries of the stream, the two independent controls (discharge and supplied load) may be essentially constant. Scouring and filling with seasonal fluctuations in discharge and velocity occur in all streams; it is the peculiar and distinctive characteristic of the graded stream that after hundreds or thousands of such short-period fluctuations, entailing an enormous total footage of scouring and filling, the stream shows no change in altitude or declivity. In this long-term sense, there is an equivalence of opposed tendencies in the graded stream." (Mackin, 1948)

Load and the Graded Profile

The sediment load supplied to a graded stream is essentially independent of the stream itself and beyond the ability of the stream to regulate. The stream must, in some fashion, deal with whatever load the valley sides and tributary streams contribute to it. The nature of sediment load delivered to a stream depends on the following:

1. *The lithology of rocks being weathered and eroded from the valley sides.* For example, columnar-jointed lava will commonly produce large blocks of material, whereas shale will produce only small flakes of material.

2. *Slope and topographic relief.* A drainage basin with steep slopes and high relief sheds much coarse material into streams, whereas basins with low slopes and low relief shed mostly fine-grained material into streams.

3. *The weathering and erosional processes operating in the drainage basin.* Different weathering and erosional processes produce different detritus to be shed into streams. For example, a glacier in the headwaters of a drainage will generally deliver a coarse load of material for a stream to handle, whereas chemical weathering in a tropical or humid environment usually sheds only fine material.

4. *Vegetation.* Desert vegetation is notable for its sparseness and low percent of ground cover, whereas the opposite is true in humid climates. Thus, runoff in semi-arid, desert areas is much more concentrated and likely to deliver coarser sediment load to streams.

If conditions in a drainage system change so that coarse debris is introduced into the stream (that is, the caliber of the load increases), the particles that, in order to move require a velocity greater than the stream can provide will accumulate on the floor of the channel. The deposition of these particles raises the level of the streambed, thus increasing its slope. Velocity increases as the level of the bed rises, until the velocity is just sufficient to move the new coarser load, at which time further deposition and slope increase are limited. A new equilibrium is established under the new conditions of increased load and slope. Conversely, if the caliber of the load is decreased, the previously established velocity is then greater than that required to move the previous load of smaller particles. Thus, the excess velocity causes the stream to begin incising its channel until the decrease in slope produces a decrease in velocity to the point just necessary to carry the load.

Downvalley Changes in Load-Discharge Relationships

The profile of a graded stream is concave upward, with progressively decreasing slope downstream. This profile represents an equilibrium system in which two of the control variables—discharge and load—change progressively downstream.

The discharge of most streams increases downvalley as tributary streams contribute their flow to the main stream. As the discharge increases downstream, the channel becomes deeper and wider, so that the cross-sectional area of the channel increases relative to the wetted perimeter. If all else were to remain unchanged, the increase in discharge would cause an increase in the velocity of the stream. However, as the velocity rises, particles on the channel bed, which previously had been too large to transport, are now set in motion by the increased velocity, and the slope of the stream is lowered. As the slope diminishes, so does velocity, until a balance is reached between the new discharge and the slope. In other words, a downvalley segment with a larger discharge can maintain the velocity necessary to transport the load on a lower slope than an upvalley segment with a smaller discharge. The slope downstream from a tributary junction of two graded streams is often less than that of either of the tributaries because the cross-sectional area relative to the wetted perimeter increases below the junction as a result of the added discharge. The sum of the discharges of the two streams is equal to the discharge of the main stream, but the wetted perimeter of the main stream is less than the sum of wetted perimeters of the two streams above the junction. Thus, the slope required to produce the velocity necessary to transport the load is reduced.

The second factor that changes progressively in streams is the downvalley decrease in the caliber of load. The downstream decrease in the size of particles is due partly to the abrasion of particles as they bang against each other during transport and partly to selective transportation. If a truckload of gravel were dumped into the headwaters of a stream, by the time the material was transported to the mouth of the stream, the particles would be considerably reduced as a result of abrasion.

The smaller particles of transported load remain in motion longer and are more frequently moved by saltation or traction than larger particles. Thus, the smaller particles move downstream at a faster rate than larger particles. Smaller particles "outrun" their larger comrades, which lag behind, and thus the finer sediment is concentrated in the distal part of the stream. Distinguishing between the effects of abrasion and selective transportation in causing downstream decrease in particle size is virtually impossible, but because both result in decrease in particle size downstream, their effect on the stream is cumulative.

The downstream decrease in caliber of load means that less velocity is required for transportation, and thus, slope requirements to develop the necessary velocity are less. For example, the lower Greybull River in Wyoming derives a coarse sediment load from the volcanic rocks of the Absaroka Range and flows across the Big Horn Basin. It receives no tributaries in a 50-mi (80-km) segment, and the discharge remains nearly the same throughout this reach. However, the caliber of the load decreases progressively downstream, resulting in a decrease in slope from 60 to 90 ft/mi (11 to 17 m/km) at the upper end of the segment to 20 ft/mi (4 m/km) at the lower end (Mackin, 1948). Because the discharge is constant throughout the reach, the reduction in slope must be due directly to the decrease in caliber of the load.

This phenomenon also works in the opposite direction. If the caliber of the load increases relative to discharge, as, for example, by an influx of coarse debris from a glacier in the headwaters or by the influx of coarser material from a tributary, slope will increase. The profile of the Missouri River steepens below its junction with the Platte River because the Platte brings in coarser material than the Missouri. Both streams are graded, and the profile will remain steepened as long as the same equilibrium conditions prevail.

Responses of a graded stream to various changes are listed in the following table:

	Response*	Slope
Increase in load	aggradation	increases
Decrease in load	degradation	decreases
Increase in discharge	degradation	decreases
Decrease in discharge	aggradation	increases

*until a new equilibrium is reached

A number of factors, most of which bear directly or indirectly on velocity, determine whether a stream is capable of eroding its channel and adding new material to its load, whether it is able to transport just the load supplied, or whether it is able to transport the weathered debris shed from the valley sides. That which increases velocity increases carrying power; that which decreases velocity decreases carrying power.

Slope and Channel Morphology

We have previously seen that a stream may adjust its profile to changes in equilibrium conditions by eroding its channel to lower its slope (and decrease velocity) or by depositing material in its channel to increase its slope (and increase velocity). We now ask the question, What is the relationship between channel morphology and the slope of the graded profile? Does a stream adjust to new conditions by changing its channel morphology, either in cross section or in plan?

Schumm (1960) found a relationship between the pattern of streams and their slope. In general, meandering streams are characterized by relatively high silt/clay content in the bank-forming material and by deep, narrow channels. Meandering streams generally have lower gradients than streams in less cohesive bank-forming material, which generates wide, shallow channels. This poses an interesting question concerning cause-and-effect relationships: Is the channel form the cause of the difference in slope, or is the difference in slope the cause of the difference in channel form?

How are slope and meanders related in a cause-andeffect sense? If the slope of a stream is greater than that required to generate just the velocity to transport the load, then, of course, increased shearing stress will be developed on the outside of the bends of the meanders as well as on the floor of the channel; thus, a tendency toward increased erosion on the outside of bends will result. However, this does not produce new meanders but rather increases the rate of curvature of already developed meanders, thus increasing the probability of meander cutoff by intersection of two meanders. It is a positive feedback system in which the enlarging bends and asymmetry of the channel cross section feed upon themselves. The greater the curvature, the greater the rate and extent of the curvature until two meanders finally come together and the meander is destroyed (which further steepens rather than lowers the gradient). Thus, increasing the number of meanders in a particular reach of a river is not one of the options by which the stream may adjust to changes in controlling variables.

As previously discussed, the slope of braided channels is greater than that of meandering channels. We can ask the fundamental cause-and-effect question, Is slope the cause of the channel form (braided or meandering), or is the channel cross section the cause of the slope? The channel cross section is determined by the relationship between hydraulic shear and bank resistance. In Friedkin's flume experiments for bank material of low cohesion, a wide, shallow channel quickly formed, whereas with high silt/clay ratios and more resistant banks, the channel became relatively narrow and deep and developed a meandering pattern. A wide, shallow channel consumes a greater amount of energy in frictional shear at the bank/water interface, and therefore it requires a greater slope to develop the velocity necessary to transport the bed load. Thus, the nature of the bank-forming material is the critical factor in that it determines the nature of the channel cross section, which in turn determines the slope by dictating the velocity necessary for transportation of the load under the prevailing discharge.

The response of channel characteristics to changes in the controlling variables of graded streams is a controversial topic. In the graded-stream concept of Mackin (1948), channel characteristics are not considered one of the mechanisms by which a stream can adjust itself. Rather, channel characteristics are considered as an effect of the relationship between hydraulic shear and bank resistance. However, some geomorphologists (Morisawa, 1968; Bull, 1979; Leopold and Wolman, 1960) believe that streams adjust their channel characteristics to balance changes in the system. For example, if a stream were to undergo an increase in discharge, requiring a lowering of gradient, the Mackin concept would expect incision of the stream as the mechanism to reduce the gradient. Others, however, would endow the stream with the ability to lower its gradient by increasing its number of meanders, thus lengthening the course and reducing the slope, or by changing from a braided pattern to a meandering pattern.

"The river will adjust to the new conditions by changing its slope, cross section, bed roughness, its length, or channel pattern." (Morisawa, 1968)

In this sense, the stream is considered to have the ability to decide for itself how many meanders it will generate to control its length and gradient. What determines the number of meanders in a stream? According to the philosophy just enumerated, the number of meanders would be *caused* by the stream length needed for a particular gradient needed to produce the velocity to carry the load. However, much experi-

mental work in flumes and studies of natural streams indicate that the total population of meanders is a function of the rate at which new meanders are born, minus the rate at which they are destroyed by cutoffs. If streams had the ability to adjust to external changes by altering the number of meanders in a reach, they would have to have some way of altering the "birth rate" and "death rate" of meanders.

An interesting example of the response of a stream to an artificially changed gradient is the case of the Blackwater River in Missouri (Emerson, 1971). There, meanders were artificially straightened, thereby shortening the length of the stream through a particular reach, as a flood-control measure (Emerson, 1971). As a result of shortening the stream length by 24.6 km (15.3 mi) the gradient of the stream was increased from 1.67 m/km (9 ft/mi) to 3.1 m/km (17 ft/mi). The stream thus needed to somehow decrease its gradient to return to equilibrium. Instead of returning to its former state of meandering to accomplish the lowering of its gradient, as an option in the Morisawa-Leopold approach, the stream instead rapidly incised its channel to achieve the lower gradient, as predicted by the Mackin model. As the Blackwater River incised its channel, the width also increased, and the cross-sectional area changed from 38 m^2 (410 ft^2) to 484 m^2 (5210 ft^2). Tributary streams also deeply eroded their channels downward; for example, one tributary's cross section increased from 12 m^2 (130 ft^2) to 255 m^2 (2754 ft2). As a result of the channel widening that accompanied the incision, most of the bridges in the county had to be replaced. One bridge that was 15 m (49 ft) long had to be extended to 124.1 m (407 ft). The channel-straightening project ended at the county line, so discharge of floodwaters from the deepened and widened channel caused unprecedented flooding in the next county, causing burial of two generations of fence posts (~2 m) by overbank deposits.

Streams most certainly adjust their channel cross sections, change from meandering to braiding to meandering, and make other channel modifications. The critical question is the role of cause and effect in these changes: Are the changes a means of adjustment to external changes in load or discharge, or are they an incidental expression of the relationship between hydraulic shear and bank resistance?

Degradation

The hydraulic action of streamflow against its banks is due fundamentally to the stresses established by the onward motion of the fluid. Protuberance of the wetted perimeter or particles resting upon it are subject to hydraulic forces of impact, drag, and lift. The impact of oncoming water is taken by the maximum cross section of the obstacle that opposes it and is distributed over the surface of the obstacle upstream from this cross section according to its shape. Streamlined shapes, presenting the least resistance to flow, are least likely to be dislodged by impact.

Drag and lift depend on magnitude and rate of change of velocity. Because velocity increases from the base to the top of an obstacle, a shearing stress or couple is imposed. If the obstacle is a loose particle, the resultant torque may cause it to slide, roll, or flop according to its shape. Where flow lines are compressed above an obstacle, velocity increases locally as pressure is relaxed, creating a lift similar to that above an airplane wing.

Turbulent flow accentuates each of these effects through local spasmodic increases in velocity accompanied by turbulent exchange of fluid. Such eddying effects are especially concentrated at obstacles that, in their deflection of the current, focus the expenditure of additional hydraulic energy. Macroturbulent phenomena (Matthes, 1941) demonstrate how the pattern of a channel, the cross-sectional shape of the channel, and the irregularities of the wetted perimeter react to produce the very turbulence that either modifies these channel characteristics or eradicates them. Thus, the energy of the flow versus the friction of its boundaries is fundamental to a condition of equilibrium approached by the stream within its channel.

Entrainment of unconsolidated material within the channel is accomplished by hydraulic action, and, depending on the energy of the current versus the resistance of the banks, hydraulic action can accomplish quarrying. In addition to the forces described above, the turbulent driving of wedges of water into joints may pry slabs and blocks from the wetted perimeter, while perhaps other blocks above the surface of the stream are undermined until they too are captured. Buoyant forces, which effectively decrease the weight of solids underwater, facilitate these processes.

Another strictly hydraulic effect, called **cavitation,** depends upon the formation of small bubbles of water vapor in high-velocity flow. Rapid increases in velocity, as over the rim of a cascade or the downstream face of a large boulder, are accompanied by attenuation of the flow, depression of its surface, and a drop in internal pressure. Separation of the attenuated flow allows the local vaporization of the decompressed water and the formation of bubbles. Recompression of the flow, as at the base of the cascade or in zones of reduced velocity, causes the violent collapse (implosion) of the vapor-filled bubbles. Innumerable shock waves deliver sharp blows to the adjacent rock until a condition similar to metal fatigue develops and the rock begins to flake away. The origination and coalescence of depressions caused by flaking may give the rock a fretted appearance.

Not only is hydraulic action erosive in itself, but in most cases it also accounts for the transport of clastic load, which may be used in allied processes of abrasion while undergoing attrition. Corrasion of bedrock in the floor and walls of the channel first requires entrainment of any unconsolidated material, which, set in motion, becomes a natural source of abrasion (in addition to any material carried in from upstream). Consequently, the bedrock perimeter of a channel may be insulated during all but the most extreme discharges by an armor of gravel subject only to partial entrainment

and gradual attrition. Subaqueous sandblast and abrasion by gravel, although perhaps infrequent, are important processes in the erosion of bedrock, as is attested by observations of the rounded, more or less polished forms on the rocky floors of channels exposed during drought or in the construction of dams.

Pothole drilling is a particularly effective form of abrasion characteristic of swift, turbulent streams that carry coarse gravel over bedrock. Some of these potholes may evolve from shallow cavitation pits or small depressions hollowed by scour, but in any case they are localized by the formation of strong eddies with steep, nearly vertical axes capable of swirling pebbles or larger fragments in a spiral against the rock. A relatively deep, cylindrical hole with a round bottom is drilled into the rocky bed or bank of the channel by one or several generations of fragments. Vigorous abrasion accounts for the characteristic smoothness of the walls and floor and the excellent rounding of the grinding tools, which often remain in the pothole. The importance of pothole erosion to many streams is also illustrated by exposure of their rocky beds, which consist of a complex of potholes (Figure 5–40).

The rapid transport of coarse material, particularly cobbles and boulders, batters the wetted perimeter. Protuberances may be knocked away by the barrage, and impacts against unconsolidated bed material may either crush it or dislodge it and help to keep it in motion. This process is difficult to evaluate because of accompanying abrasion. Ragged edges and corners are rapidly rounded over in such a torrential regime. However, the presence of broken roundstones (abraded fragments cracked into parts) attests to the participation of this process in some situations. Many smaller grains of sand or silt size appear to be products of subaqueous crushing.

Ephemeral streams are affected by nonhydraulic processes during times of drought. Weathering may make the channel more vulnerable to erosion during the ensuing wet season. Mass movements, on the other hand, and perhaps the accumulation of windblown material, may impose additional work on subsequent flow until the channel is swept clean.

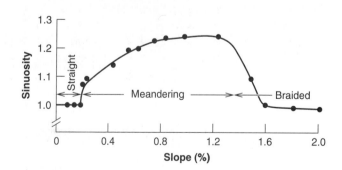

FIGURE 5–39
Stream gradients of meandering and braided streams. (From Schumm and Khan, 1972)

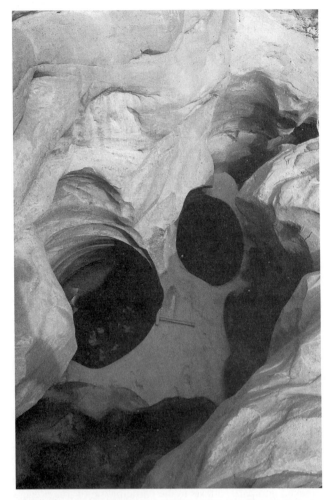

A.

B.

FIGURE 5–40
(A) Pothole erosion of bedrock in a stream channel, Mt. Fujiama, Japan. (Photo by W. K. Hamblin)
(B) Development of a bedrock channel by pothole erosion, New York. (Photo by P. J. Fleischer)

Aggradation

Fluvial deposition (**aggradation**) refers to long-term accumulation of sediments in a channel, in contrast to the temporary shifting of material within the depth of scour of a stream. Short-term accretion of sediment in a channel, which can be moved only during times of high discharge, is not considered aggradation.

Aggradation can be initiated either by changes in load/discharge relationships calling for an increase in slope or by the raising of base level. An example of the former occurs when a glacier introduces large amounts of high-caliber load into the headwaters of a drainage, and aggradation proceeds downvalley. Another example might be a decrease in discharge that calls for an increase in slope to produce the velocity necessary to carry the load.

In the case of a rise in base level, aggradation progresses headward rather than downvalley. The rise in base level may be due to a rise in sea level, but it can also be produced locally. For example, aggradation of a trunk valley caused by a glacier in the headwaters raises the local base level of tributary streams and results in aggradation, which progresses upvalley. Thus, the two causes may be complexly interrelated until a new equilibrium is reached.

If the cause of aggradation is brought about by load-discharge adjustments, aggradation will result in an increase in gradient of the stream. However, if the cause of the aggradation is a rise in base level, the finally adjusted profile will have the same slope as before the base-level change.

The rate of aggradation can be slow and barely noticeable over a human life span, or it can be sudden and dramatic if caused by an abrupt geologic event (Figure 5–41). However, since aggradation must be thought of as a long-term event, the effects of most short-lived geologic events are likely to be erased by subsequent adjustments to the system.

Among the characteristics of an aggrading stream that distinguish it from a graded stream is that the deposits lie below the depth of scour of the stream, even during large flood events. In this sense, the sediments are effectively removed from further participation in downstream transport. Ultimately, of course, as the stream gradient increases due to the accumulation of aggrading sediment, the velocity of the stream will increase sufficiently to allow transportation of even the largest particles supplied, and a new equilibrium will be established. But just how deeply can streams scour their channels? How can we tell whether or not a given thickness of sediment is below the depth of scour? The depth of scour is certainly different for almost every stream and in many cases may be surprisingly great. Gravel more than 30 m (100 ft) thick exposed during the building of Hoover Dam was thought to represent a Pleistocene valley fill, until a piece of lumber containing nails was found, indicating that the Colorado River could indeed scour to such depths. As yet, very little is known about scour depths of streams because when the sediment is in motion, it is not visible, and once it comes to rest, little evidence may remain to show that it has moved recently.

FIGURE 5–41
(A) Bridge over a stream before a flood.
(B) The same bridge buried by gravel from a flood. (Photos by U.S. Geological Survey)

REFERENCES

Ackers, P., 1970, The slope and resistance of small meandering channels: Proceedings of the Institution of Civil Engineers, Supplement (XV) Paper 7362S, p. 349–370.

Ackers, P., 1982, Meandering channels and bed material: *in* Hey, R. D., Bathurst, J. C., and Thorne, C. R., eds., Gravel-bed rivers, John Wiley and Sons, N.Y., p. 389–421.

Ackers, P., and Charlton, F. G., 1970, Meander geometry arising from varying flows: Journal of Hydrology, v. 11, p. 230–252.

Andrews, E. D., 1979, Scour and fill in a stream channel, East Fork River, western Wyoming: U.S. Geological Survey Professional Paper 1117, 49 p.

Andrews, E. D., 1981, Measurement and computations of bed-material in a shallow sand-bed stream, Muddy Creek, Wyoming: Water Resources Research, v. 17, p. 131–141.

Andrews, E. D., 1983, Entrainment of gravel from naturally sorted river-bed material: Geological Society of America Bulletin, v. 94, p. 1225–1231.

Ashmore, P. E. 1991. How do gravel-bed rivers braid? Canadian Journal of Earth Science 28:326–341.

Ashworth, P. J., and Ferguson, R. I., 1989, Size-selective entrainment of bed load in gravel bed streams: Water Resources Research 25:627–634.

Bagnold, R. A., 1956, The flow of cohesionless grains in fluids: Philosophical Transactions of the Royal Society, v. 249.

Bagnold, R. A., 1960, Some aspects of the shape of river meanders: U.S. Geological Survey Professional Paper 282-E, p. 135–144.

Bagnold, R. A., 1966, An approach to the sediment transport problem from general physics: U.S. Geology Survey Professional Paper 422-1, 37 p.

Bagnold, R. A., 1973, The nature of saltation and of "bed-load" transport in water: Proceedings of the Royal Society of London, 332A, p. 473–504.

Bagnold, R. A., 1977, Bed load transport by natural rivers: Water Resources Research, v. 13, no. 2, p. 303–312.

Baker, V. R., 1973, Paleohydrology and sedimentology of Lake Missoula flooding in eastern Washington: Geological Society of America Special Paper 144, 79 p.

Baker, V. R., 1974, Paleohydraulic interpretation of Quaternary alluvium near Golden, Colorado: Quaternary Research, v. 4, p. 94–112.

Baker, V. R., 1975, Competence of rivers to transport coarse bedload material: Geological Society of America Bulletin, v. 86, p. 975–978,

Baker, V. R., 1977, Stream-channel response to floods with examples from central Texas: Geological Society of America Bulletin, v. 88, p. 1057–1071.

Baker, V. R., 1978, Paleohydraulics and hydrodynamics of Scabland Floods: *in* Baker, V. R., and Nummedal, D., eds., The

Channeled Scabland, Washington, D.C., National Aeronautics and Space Administration, p. 59–79.

Baker, V. R., 1987, Paleoflood hydrology and extraordinary flood events: *in* Kirby, W. H., Hua, Shi-Quian, and Beard, L. R., eds., Analysis of extraordinary flood events, Journal of Hydrology, Elsevier Scientific Publishing Company, Amsterdam, Netherlands, v. 96, p. 77–99.

Baker, V. R., 1988, Flood erosion: *in* Baker, V. R., Kochel, R. C., and Patton, P. C., eds., Part II: flood processes: flood geomorphology, John Wiley and Sons, N.Y., p. 81–96.

Baker, V. R., and Bunker, R. C., 1985, Cataclysmic late Pleistocene flooding from Glacial Lake Missoula: a review: Quaternary Science Reviews, v. 4, p. 1–41.

Baker, V. R., and Costa, J. E. 1987. Flood Power: *in* Mayer, L., and Nash, D., eds., Catastrophic Flooding, Allen and Unwin, London, p. 1–24.

Baker, V. R., and Kochel, R. C., 1988, Flood sedimentation in bedrock fluvial systems: *in* Baker, V. R., Kochel, R. C., and Patton, R C., eds., Part II: flood processes: flood geomorphology, John Wiley and Sons, N.Y., p. 123–138.

Baker, V. R., and Komar, P. D., 1987, Cataclysmic flood processes and landforms: *in* Graf, W. L., ed., Geomorphic systems of North America, Geological Society of America Centennial Special Volume, p. 423–443.

Baker, V. R., Ely, L. L., O'Connor, J. E., and Partridge, J. B., 1987, Paleoflood hydrology: *in* Singh, V. P., ed., Regional flood frequency and analysis: Proceedings of the International Symposium on Flood Frequency and Risk Analysis, Reidel Publication, Dordrect, Netherlands, p. 325–338.

Baker, V. R., Kochel, R. C., Patton, P. C., and Pickup, G., 1983, Paleohydrologic analysis of Holocene slackwater sediments: International Association of Sedimentologists Special Publication, v. 6, p. 229–239.

Baker, V. R., and Nummedal, D., eds., 1978, The Channeled Scabland: N.A.S.A. Field Guidebook, Washington D. C., 186p.

Balog, J. D., 1978, Flooding in Big Thompson River, Colorado, tributaries: controls on channel erosion and estimates of recurrence intervals: Geology, v. 6, p. 200–204.

Barnes, H. H., 1967, Roughness characteristics of natural channels: U.S. Geological Water Supply Paper 1849, 213 p.

Bathurst, J. C., Thorne, C. R., and Hey, R. D., 1979, Secondary flow and shear stress at river bends: American Association of Civil Engineers, Journal of Hydraulics Division, v. 105, p. 1277–1295.

Bayazit, M., 1978, Scour of bed materials in very rough channels: American Society of Civil Engineers, Journal of Hydraulics Division, v. 104, p. 1345–1349.

Begin, Z. B., 1981, Stream curvature and bank erosion: a model based on the momentum equation: Journal of Geology, v.89, p. 497–504.

Begin, Z. B.; Meyer, D. F.; and Schumm, S. A. 1980. Knickpoint migration due to base-level lowering: American Society of Civil Engineers, Journal of Water, Port, Coastal and Ocean Division, v. 106, p. 369–87.

Begin, Z. B., Meyer, D. F., and Schumm, S. A., 1981, Development of longitudinal profiles of alluvial channels in response to base-level lowering: Earth Surface Processes and Landforms, v. 6, p. 49–68.

Benson, M. A., 1950, Use of historical data in flood-frequency analysis: Transactions of the American Geophysical Union, v. 31, p. 419–424.

Benson, M. A., 1964, Factors affecting the occurrence of floods in the Southwest: U.S. Geological Survey Water Supply Paper 1580-D, 72 p.

Birkeland, P. W., 1968, Mean velocities and boulder transport during Tahoe age floods of the Truckee River, California and Nevada: Geological Society of America Bulletin, v. 79, p. 137–142.

Boothroyd, J. C., and Ashley, G. M., 1975, Processes, bar morphology and sedimentary structure on braided outwash fans, northeastern Gulf of Alaska: *in* Jopling, A. V., and MacDonald, B. C., eds., Glaciofluvial and glaciolacustrine sedimentation, Society of Economic Paleontologists and Mineralogists Special Publication 23, p. 193–222.

Bradley, W. C., and Mears, A. I., 1980, Calculations of flows needed to transport coarse fraction of Boulder Creek alluvium at Boulder Creek Colorado: Geological Society of America Bulletin, v. 91, p. 1057–1090.

Bretz, J H., 1923, The Channeled Scablands of the Columbia Plateau: Journal of Geology, v. 31, p. 617–649.

Bretz, J H., 1929, Valley deposits immediately east of the Channeled Scablands of Washington: Journal of Geology, v. 37, p. 393–427.

Bretz, J H., 1969, The Lake Missoula floods and the Channeled Scabland: Journal of Geology, v. 77, p. 505–543.

Bretz, J H., Smith, H. T. U., and Neff, G. E., 1956, Channeled Scablands of Washington: New data and interpretations: Geological Society of America Bulletin, v. 67, p. 957–1049.

Brice, J. C., 1964, Channel patterns and terraces of the Loup River in Nebraska: U.S. Geological Survey Professional Paper 422-D, 41 p.

Brice, J. C., 1974, Evolution of meander loops: Geological Society of America Bulletin, v. 85, p. 581–586.

Brotherton, D. 1., 1979, On the origin and characteristics of river channel patterns: Journal of Hydrology, v. 44, p. 211–230.

Bull, W. B., 1979, Threshold of critical power in streams: Geological Society of America Bulletin, v. 90, p. 453–464.

Bunker, R. C., 1982, Evidence of multiple late-Wisconsin floods from Glacial Lake Missoula in Badger Coulee, Washington: Quaternary Research, v. 18, p. 17–31.

Callander, R. A., 1969, Instability and river channels. Journal of Fluid Mechanics, v. 36, p.465–80.

Callander, R., 1978, River meandering: Annual Review of Fluid Mechanics, v. 10, p. 129–158.

Carling, P. A., Kelsey, A., and Glaister, M. S., 1992, Effect of bed roughness, particle shape and orientation on initial motion criteria: *in* Billi, P., Hey, R. D., Thorne, C. R., and Tacconi, P., eds., Dynamics of Gravel-Bed Rivers, John Wiley and Sons, N.Y., p. 23–39.

Carlston, C. W., 1963, Drainage density and streamflow: U.S. Geological Survey Professional Paper 422-C.

Carlston, C. W., 1965, The relation of free meander geometry to stream discharge and its geomorphic implications: American Journal of Science, v. 263, p. 864–885.

Chang, H. H., 1979, Minimum stream power and river channel patterns: Journal of Hydrology, v. 41, p. 303–327.

Cheetham, G. H., 1979, Flow competence in relation to stream channel form and braiding: Geological Society of America Bulletin, v. 90, p. 877–886.

Church, M., 1972, Baffin Island sandurs: a study of arctic fluvial processes: Geological Survey of Canada Bulletin, v. 216, 208 p.

Church, M., 1978, Paleohydrological reconstructions from a Holocene valley fill: *in* Miall, A. D., ed., Fluvial sedimentology, Canadian Society of Petroleum Geologists, Memoir 5, p. 743–772.

Church, M., 1988, Floods in cold climates: *in* Baker, V. R., Kochel, R. C., and Patton, P. C., eds., Part III: floods, climate, landscape: flood geomorphology, John Wiley and Sons, N.Y., p. 205–230.

Church, M., and Jones, D., 1982, Channel bars in gravel-bed rivers: *in* Hey, R. D., Bathurst, J. C., and Thorne, C. R., eds., Gravel-bed rivers, John Wiley and Sons, N.Y., p. 291–338.

Colby, B., 1963, Fluvial sediments: a summary of source, transportation, deposition, and measurement of sediment discharge: U.S. Geological Survey Bulletin 1181-A, 21 p.

Colby, B., 1964, Scour and fill in sand bed streams: U. S. Geological Survey Professional Paper 462-D.

Costa, J. E., 1974, Stratigraphic, morphologic, and pedologic evidence of large floods in humid environments: Geology, v. 2, p. 301–303.

Costa, J. E., 1978, Colorado's Big Thompson flood: geologic evidence of a rare hydrologic event: Geology, v. 6, p. 617–620.

Costa, J. E., 1983, Paleohydraulic reconstruction of flash-flood peaks from boulder deposits in the Colorado Front Range: Geological Society of America Bulletin, v. 94, p. 986–1004.

Costa, J. E., 1986, A history of paleoflood hydrology in the United States, 1800–1970: *in* History of geophysics, American Geophysical Union, Washington, D.C., v. 3, p. 49–53.

Costa, J. E., 1987, Hydraulics and basin morphometry of the largest flash floods in the conterminous United States: Journal of Hydrology, v. 93, p. 313–338.

Costa, J. E., 1988, Floods from dam failures: *in* Baker, V. R., Kochel, R. C., and Patton, P. C., eds., Part V: environmental management: flood geomorphology, John Wiley and Sons, N.Y., p. 439–464.

Crowley, K. D., 1983, Large-scale bed configurations (macroforms), Platte River basin, Colorado and Nebraska: Primary structures and formative processes: Geological Society of America Bulletin, v. 94, p. 117–133.

Dalrymple, T., 1960, Flood frequency analysis. Manual of hydrology, part 3, Flood flow techniques: U.S. Geological Survey Water Supply Paper 1543-A.

Daniels, R. B., Entrenchment of the Willow Drainage Ditch, Harrison County, Iowa: American Journal of Science, v. 258, p. 161–176.

Davis, W M., 1902, Baselevel, grade, and peneplain: Journal of Geology, v. 10, p. 77–111.

Davis, W. M., 1909, Geographical Essays: Ginn and Company (reprinted in 1954 by Dover Publications, New York), Boston, MA, 777 p.

Dietrich, W E., 1987, Mechanics of flow and sediment transport in river bends: *in* Richards, K. ed., River Channels: Environment and Process, Basil Blackwell, Ltd., Oxford, England, p. 179–227.

Dunne, T., and Leopold, L. B., 1978, Water in environmental planning: W. H. Freeman and Co. , N.Y., 817 p.

Duty, G. H., 1976, Discharge prediction, present and former, from channel dimensions: Journal of Hydrology, v. 30, p. 219–245.

Dutton, C. E., 1882, Tertiary history of the Grand Canyon district, with atlas: U.S. Geological Survey Monograph 2, 264 p.

Einstein, H. A., 1942, Formulas for the transportation of bedload: Transactions of the American Society of Civil Engineers, v. 107, p. 561–597.

Einstein, H. A., 1950, The bed-load function for sediment transportation in open channel flows: U.S. Department of Agriculture Technical Bulletin 1926, 71 p.

Einstein, H. A., 1968, Deposition of suspended particles in a gravel bed: American Society of Civil Engineers, Journal of the Hydraulics Division, v. 94, no. HY5, p. 1197–1205.

Einstein, H. A., and Shen, H. W., 1964, A study of meandering in straight alluvial channels: Journal of Geophysical Research, v. 69, p. 5239–5247.

Ely, L. L., and Baker, V. R., 1985, Reconstructing paleoflood hydrology with slackwater deposits: Verde River, Arizona: in Physical geography, V. H. Winston and Sons, Silver Spring, MD, v. 6 p. 103–126.

Ely, L. and Baker, V. R., 1990, Large floods and climate change in the southwestern United States: in French, R. H., ed., Hydraulics/Hydrology of Arid Lands,. American Society of Civil Engineers, p. 361–366.

Emerson, J., 1971, Channelization: a case study: Science, v. 173, p. 61–63.

Ethridge, F. G., and Schumm, S. A., 1978, Reconstructing paleochannel morphologic and flow characteristics: methodology, limitations, and assessment: Memoirs of Canadian Society of Petroleum Geology, v. 5, p. 703–721.

Everitt, B. L., 1968, Use of cottonwood in an investigation of the recent history of a floodplain: American Journal of Science, v. 266, p. 417–439.

Fahnestock, R. K., 1963, Morphology and hydrology of a glacial stream—White River, Mt. Rainier, Washington: U.S. Geological Survey Professional Paper 422A.

Fahnestock, R. K., and Bradley, W. C., 1973, Knik and Matanuska rivers, Alaska: a contrast in braiding: in Morisawa, M. E., ed., Fluvial geomorphology, SUNY, Publications in Geomorphology, Binghamton, N.Y., p. 220–250.

Ferguson, R. I., 1975, Meander irregularity and wavelength estimation: Journal of Hydrology, v. 26, p. 315–333.

Ferguson, R., 1987, Hydraulic and sedimentary controls of channel pattern: in Richards, K., ed., River Channels: Environment and Process, Basil Blackwell, London, p. 129–58.

Florsheim, J. L., Keller, E. A., and Best, D. W., 1991, Fluvial sediment transport in response to moderate storm flows following chaparral wildfire, Ventura County, southern California: Geological Society of America, v. 103, p. 504–511.

Foley, M. G., 1978, Scour and fill in steep, sand-bed ephemeral streams: Geological Society of America Bulletin, v. 89, p. 559–70.

Foley, M. B., 1980, Bedrock incision by streams: Geological Society of America Bulletin, v. 91, p. 2189–2213.

Friedkin, J. F., 1945, A laboratory study of the meandering of alluvial rivers: Vicksburg, Miss., U.S. Waterways Experiment Station, 40 p.

Gage, M., 1953, Transport and rounding of large boulders in mountain streams: Journal of Sedimentary Petrology, v. 23, p. 60–61.

Gilbert, G. K., 1877, Report on the geology of the Henry Mts., Utah: U.S. Geographical and Geological Survey, Rocky Mt. Region, U.S. Govt. Printing Office. Washington, D.C., 160 p.

Gilbert, G. K., 1914, The transportation of debris by running water: U.S. Geological Survey Professional Paper 86, 263 p.

Gilbert, G. K., 1917, Hydraulic mining debris in the Sierra Nevada: U.S. Geological Survey Professional Paper 105, 154 p.

Gottsfield, A. S., and Gottsfield, L. M., 1990, Floodplain dynamics of a wandering river; dendrochronology of the Morice River, British Columbia, Canada: Geomorphology, v. 3. p. 159–79.

Graf, W. H., 1971, Hydraulics of sediment transport: McGraw-Hill, N.Y., 513 p.

Graf, J. B., Webb, R. H., and Hereford, R., 1991, Relation of sediment load and floodplain formation to climatic variability, Paria River drainage basin, Utah and Arizona: Geological Society of America Bulletin, v. 103, p. 1405–1415.

Gregory, K. J., 1974, Streamflow and building activity: in Gregory, K. J., and Walling, D. E., eds., Fluvial processes in instrumented watersheds: Institute of British Geographers, Special Publication 6, p. 107–122.

Gregory, K. J., 1977, River channel changes: John Wiley and Sons, N.Y., 450 p.

Gregory, K. J., and Walling, D. E., 1973, Drainage basin form and process: John Wiley and Sons, N.Y., 456 p.

Grozier, R. U., McCain, J. F., Lang, L. F., and Merriman, D. C., 1976, Big Thompson River flood, July 31–August 1, Larimer County: Denver, Colorado Water Conservation Board, 78 p.

Gupta, A., 1988, Large floods as geomorphic events in humid tropics: in Baker, V. R., Kochel, R. C., and Patton, P. C., eds., Part III: floods, climate, landscape: flood geomorphology, John Wiley and Sons, N.Y., p. 301–315.

Gupta, A., and Fox, H., 1974, Effects of high-magnitude floods on channel form: A case study in Maryland Piedmont: Water Resources Research, v. 10, p. 499–509.

Haan, C. T., and Johnson, H. P., 1966, Rapid determination of hypsometric curves: Geological Society of America Bulletin, v. 77, p. 123–25.

Hammad, H. Y., 1972, River-bed degradation after closure of dams: American Society of Civil Engineers, Journal of Hydraulics Division, v. 98, p. 591–607.

Harms, J. C., 1969, Hydraulic significance of some sand ripples: Geological Society of America Bulletin, v. 81, p. 363–396.

Harrison, S. S., and Reid, J. R., 1967, A flood-frequency curve based on tree scar data: Proceedings of the North Dakota Academy of Sciences, v. 21, p. 23–33.

Harvey, A. M., 1975, Some aspects of the relations between channel characteristics and riffle spacing in meandering streams: American Journal of Science, v. 275, p. 470–478.

Harvey, A. M., Hitchcock, D. H., and Hughes, D. J., 1979, Event frequency and morphological adjustment of fluvial systems: in Rhodes, D. D., and Williams, G. P., ed., Adjustments of the Fluvial System, Kendall Hunt, Dubuque, Iowa, p. 139–167.

Helley. E. J., 1969, Field measurements of the initiation of large bed particle motion in Blue Creek near Klamath, California: U.S. Geological Survey Professional Paper 562-G, 19 p.

Helley, E. J., and LaMarche, V. C., 1973, Historic flood information for northern California streams from geological and botanical evidence: U.S. Geological Survey Professional Paper 485-E, 16 p.

Hereford, R., 1984, Climate and ephemeral stream processes: Twentieth-century geomorphology and alluvial stratigraphy of the Little Colorado River, Arizona: Geological Society of America Bulletin, v. 95, p.654–668.

Hey, R. D., 1976, Geometry of river meanders: Nature, v. 262, p. 482–484.

Hey, R. D., Bathurst, J. C., and Theme, C. R., eds., Gravel-bed rivers: John Wiley and Sons, N.Y., p. 260–338.

Hickin, E. J., 1969, A newly-identified process of point bar formation in natural streams: American Journal of Science, v. 267, p. 999–1010.

Hickin, E. J., 1974, Development of meanders in natural river channels: American Journal of Science, v. 274, p. 414–442.

Hjulstrom, F., 1935, Studies on the morphological activity of rivers as illustrated by the river Fryis: University of Upsala Geological Institute Bulletin, v. 25, p. 221–527.

Hoey, T., 1992, Temporal variation in bedload transport rates and sediment storage in gravel river beds: Progress in Physical Geography, v. 16, p. 319–38.

Holmes, C. D., 1952, Stream competence and the graded stream profile: American Journal of Science, v. 250, p. 899–906.

Hong, Le Ba, and Davies, T. R. H., 1979, A study of stream braiding: Geological Society of America Bulletin, v. 90, p. 1839–1859.

Hooke, J. M., 1977, The distribution and nature of change in river channel patterns: the example of Devon: in Gregory, K. J., ed., River channel changes, John Wiley and Sons, N.Y., p. 265–280.

Hooke, J. M., 1980, Magnitude and distribution of rates of river bank erosion: Earth Surface Processes, v. 5, p. 143–157.

Hooke, J. M., 1987, Changes in meander morphology: *in* Gardiner, V., ed., International Geomorphology, Part I, 591–609 p.

Hooke, R. L., 1975, Distribution of sediment transport and shear stress in a meander bend: Journal of Geology, v. 83, p. 543–566.

Hornbeck, J. W., Pierce, R. S., and Federer, 1970, Streamflow changes after forest clearing in New England: Water Resources Research, v. 6, p. 1124–1131.

Horton, R. E., 1932, Drainage basin characteristics: Transactions of the American Geophysical Union, v. 13, p. 350–361.

Horton, R. E., 1945, Erosional development of streams and their drainage basins: hydrophysical approach to quantitative morphology: Geological Society of America Bulletin, v. 56, p. 275–370.

Howard, A. D., 1971, Simulation model of stream capture: Geological Society of America Bulletin, v. 82, p. 1355–1376.

Howard, A. D., and Kerby, G., 1983, Channel changes in badlands: Geological Society of America Bulletin, v. 94, p. 739–752.

Howard, Arthur D., and Dolan, R., 1981, Geomorphology of the Colorado River in the Grand Canyon: Journal of Geology, v. 89, p. 269–298.

Hupp, C. R., 1982, Extension of flood record for Passage Creek, Virginia, utilizing tree-ring analysis: Bulletin of Ecological Society of America, v. 63, no. 2.

Hupp, C. R., 1987, Botanical evidence of floods and paleoflood history: *in* Singh, V. P., ed., Regional flood frequency analysis: Proceedings of the International Symposium on Flood Frequency and Risk analysis, Reidel Publishing, Dordrect, Netherlands, p. 355–369.

Hupp, C. R., 1988, Plant ecological aspects of flood geomorphology and paleoflood history: *in* Baker V. R., Kochel, R. C., and Patton, R C., eds., Part IV: Paleofloods: flood geomorphology, John Wiley and Sons, N.Y., p. 335–356.

Ikeda, S., and Parker, G., eds., 1989, River meandering: Washington, D.C., American Geophysical Union, 485 p.

Jarrett, R. D., 1991, Paleohydrology and its value in analyzing floods and droughts: *in* Paulson, R. W., et al., ed., National Water Summary 1988–89; Hydrologic Events and Floods, U.S. Geological Survey Water Supply Paper.

Jarrett, R. D., and Costa, J. E., 1986, Hydrology, geomorphology, and dam-break of the July 15, 1982 Lawn Lake dam and Cascade dam failures, Latimer County, Colorado: U.S. Geological Survey Professional Paper 1369, 77 p.

Jarrett, R. D., and Costa, J. E., 1988, Evaluation of the flood hydrology in the Colorado Front Range using precipitation, streamflow, and paleoflood data for the Big Thompson River basin: Water Resources Investigations, Report No. WRI 87-4117, 37 p.

Jarrett, R. D., and Malde, H. E., 1987, Paleodischarge of the late Pleistocene Bonneville Flood, Snake River, Idaho:

Geological Society of America Bulletin, v. 99, p. 127–134.

Jefferson, M., 1902, Limiting widths of meander belts: National Geographic Magazine, v. 13, p. 373–384.

Johnson, D., 1932, Streams and their significance: Journal of Geology, v. 40, p. 481–497.

Johnson, D., 1933, Development of drainage systems and the dynamic cycle: Geographical Review, v. 23, p. 114–121.

Jopling, A. V., 1966, Some principles and techniques used in reconstructing the hydraulic parameters of a paleo-flow regime: Journal of Sedimentary Petrology, v. 36, p. 5–49.

Keller, E. A., 1971, Pools, riffles, and meanders: discussion: Geological Society of America Bulletin, v. 82, p. 279–280.

Keller, E. A., 1972, Development of alluvial stream channels: Geological Society of America Bulletin, v. 83, p. 1531–1536.

Keller, E. A., and Melhorn, W., 1978, Rhythmic spacing and origin of pools and riffles: Geological Society of America Bulletin, v. 89, p. 723–730.

Keller, E. A., and Swanson, F. J., 1979, Effects of large organic debris on channel form and fluvial process: Earth Surface Processes, v. 4, p. 361–380.

Keller, E. A., and Tally, T., 1979, Effects of large organic debris on channel form and fluvial processes in the coastal Redwood environment: *in* Rhodes, D. D., and Williams, G. P., eds., Adjustments of the fluvial system, Kendall/Hunt, Dubuque, Iowa, p. 169–197.

Kessel, R. D., Dunne, K. C., McDonald, R. C., Allison, K. R., and Kirby, M. J., 1969, Infiltration, throughflow, and overland flow: *in* Chorley, R. J., ed., Water, earth, and man, Methuen, London, p. 215–227.

Knighton, A. D., 1972, Changes in a braided reach: Geological Society of America Bulletin, v. 83, p. 3812–3822.

Knighton, A. D., 1974, Variations in width-discharge relation and some implications for hydraulic-geometry: Geological Society of America Bulletin, v. 85, p. 1059–1076.

Knighton, D., 1984, Fluvial Forms and Processes: Edward Arnold, London, 218 p.

Knighton, A., and Nanson, G., 1987, River channel adjustment the downstream dimension: *in* Richards, K.,ed., River Channels: Environment and Process, Oxford, Basil Blackwell, Ltd., p. 95–128.

Knighton, A., and Nanson, G., 1993, Anastomosis and the continuum of channel pattern: Earth Surface Processes and Landforms, v. 18, p. 613–625.

Knox, J. C., 1993, Large increases in flood magnitude in response to modest changes in climate: Nature, v. 361, p. 430–432.

Kochel, R. C., 1988, Geomorphic impact of large floods: review and new perspectives on magnitude and frequency: *in* Baker, V. R., Kochel, R. C., and Patton, P. C., eds., Part III: floods, climate, landscape: flood geomorphology, John Wiley and Sons, N.Y., p. 169–188.

Kochel, R. C., and Baker, V. R., 1988, Paleoflood analysis using slackwater deposits: *in* Baker, V. R., Kochel, R. C., and Patton, P. C., eds., Part IV: Paleofloods: flood geomorphology, John Wiley and Sons, N.Y., p. 357–376.

Komar, P. D., 1988, Sediment transport by floods: *in* Baker, V. R., Kochel, R. C., and Patton, P. C., eds., Part II: Flood processes: flood geomorphology, John Wiley and Sons, N.Y., p. 97–112.

Krumbein, W. C., and Lieblein, J., 1956, Geological application of extreme value methods to interpretation of cobbles and boulders in gravel deposits: American Geophysical Union Transactions, v. 37, p. 313–319.

Lane, E. W., 1937, Stable channels in erodible materials: Transactions of the American Society of Civil Engineers, v. 102, p. 123–194.

Lane, E. W., 1955, Design of stable channels: Transactions of the American Society of Civil Engineers, v. 120, p. 1234–1260.

Lane, E. W., 1957, A study of the shape of channels formed by natural streams in erodible material: Omaha, Nebraska, M. R. D. Sediment Series No. 9, U.S. Army Corps of Engineers, Engineering Division, Missouri River.

Lane, E. W., and Borland, W. M., 1954, River-bed scour during floods: American Society of Civil Engineers Transactions, v. 119, p. 1069–1079.

Langbein, W. B., and Schumm, S. A., 1958, Yield of sediment in relation to mean annual precipitation: Transactions of the American Geophysical Union, v. 39, p. 1076–1084.

Leliavsky, S., 1955, An introduction to fluvial hydraulics: Constable and Company, London.

Leopold, L. B., 1969, The rapids and pools—Grand Canyon: U.S. Geological Survey Professional Paper 669, p. 131–145.

Leopold, L. B., 1973, River channel change with time: an example: Geological Society of America Bulletin, v. 84, p. 1845–1860.

Leopold, L. B., 1982, Water surface topography and meander development: *in* Hey, R. D., Bathurst, J. C., and Thorne, C. R., eds., Gravel-bed rivers, John Wiley and Sons, N.Y., p. 355–388.

Leopold, L. B., and Bull, W. G., 1979, Base level, aggradation and grade: Proceedings of the American Philosophical Society, v. 123, p. 168–202.

Leopold, L. B., and Langbein, W. G., 1966, River meanders: Scientific American, v. 214, p. 60–70.

Leopold, L. B., and Maddock, T., 1953, The hydraulic geometry of stream channels and some physiographic implications: U.S. Geological Survey Professional Paper 252, 57 p.

Leopold, L. B., and Wolman, M. G., 1957, River channel patterns: braided, meandering, and straight: U.S. Geological Survey Professional Paper 282-B, p. 39–85.

Leopold, L. B., and Wolman, M. G., 1960, River meanders: Geological Society of America Bulletin, v. 71, p. 769–794.

Leopold, L. B., Wolman, M. G., and Miller, J. P., 1964, Fluvial processes in geomorphology: W. H. Freeman and Company, San Francisco, CA, 522 p.

Lewin, J., 1976, Initiation of bedforms and meanders in coarse-grained sediment: Geological Society of America Bulletin, v. 87, p. 281–285.

Mackin, J. H., 1936, The capture of the Greybull River: American Journal of Science, v. 31, p. 373–385.

Mackin, J. H., 1937, Erosional history of the Big Horn Basin, Wyoming: Geological Society of America Bulletin, v. 48, p. 813–894.

Mackin, J. H., 1948, Concept of the graded river: Geological Society of America Bulletin, v. 59, p. 463–512.

Mackin, J. H., 1956, Cause of braiding by a graded stream: Geological Society of America Bulletin, v. 67, p. 1717–1718.

Mackin, J. H., 1963, Rational and empirical methods of investigations in geology: in Albritton, C. C., ed., The fabric of geology, Addison-Wesley, N.Y., p. 135–163.

Maizels, J. K., 1983, Paleovelocity and paleodischarge determination for coarse gravel deposits: in Gregory, K. J., ed., Background to paleohydrology, John Wiley and Sons, N.Y., p. 10 1 – 139.

Malde, H. E., 1968, The catastrophic late Pleistocene flood in the Snake River Plain: U.S. Geological Survey Professional Paper 596, 52 p.

Manning, R., 1891, On the flow of water in open channels and pipes: Transactions of the Institution of Civil Engineers of Ireland, v. 20, p. 161–207.

Markham, A. J., and Thorne, C. R., 1992, Geomorphology of gravel-bed river bends: in Billi, P., Hey, R. D., Thorne, C. R., and Tacconi, P., eds., Dynamics of Gravel Bed Rivers, John Wiley and Sons, N.Y., p. 433–456.

Matthes, G. H., 1941, Basic aspects of stream meanders: American Geophysical Union Transactions, v. 22, p. 632–636.

Matthes, G. H., 1947, Macroturbulence in natural streams: Transactions of the American Geophysical Union, v. 28, p. 255–261.

Mayer, L., and Nash, D., eds., 1987, Catastrophic Flooding: Allen and Unwin, Boston, MA, 410p.

McGee, W. J., 1897, Sheetflood erosion: Geological Society of America Bulletin, v. 8, p. 87–112.

McQueen, K. C.; Vitek, J. D.; and Carter, B. J. 1993. Paleoflood analysis of an alluvial channel in the south-central Great Plains; Black Bear Creek, Oklahoma: Geomorphology, v. 8, p.131–146.

Meyer-Peter, E., and Muller, R., 1948, Formulas for bed-load transport: International Association for Hydrol. Structures Research Proceedings, Stockholm, p. 39–65.

Miall, A. D., ed., Fluvial sedimentology: Calgary, Memoirs of the Canadian Society of Petroleum Geologists, v. 5, p. 605–625.

Miller, D. L., Everson, C. E., Mumford, J. A., and Bertle, F. A., 1978, Peak discharge estimates used in the refinement of the Big Thompson storm analysis: American Meteorological Society Conference on Flash Floods, Boston, MA., p. 135–142.

Milliman, J. D., and Meade, R. H., 1983, World-wide delivery of river sediment to the oceans: Journal of Geology, v. 91, p. 1–22.

Morisawa, M. E., 1968, Streams, their dynamics and morphology: McGraw-Hill, N.Y., 175 p.

Moss, A. J., 1972, Bed-load sediments: Sedimentology, v. 18, p. 159–220.

Moss, J. H., and Bonini, W., 1961, Seismic evidence supporting a new interpretation of the Cody terrace near Cody, Wyoming: Geological Society of America Bulletin, v. 72, p. 547–556.

Moss, J. H., and Kochel, R. C., 1978, Unexpected geomorphic effects of the Hurricane Agnes storm and flood, Conestoga drainage basin, southwestern Pennsylvania: Journal of Geology, v. 86, p. 1 –11.

Myers, V. A., 1967, Meteorological estimation of extreme precipitation for spillway design floods: U.S. Department of Commerce, Weather Bureau Technical Memorandum YvBTM-HYDRO-5, 30 p.

Nevin, C., 1946, Competency of moving water to transport debris: Geological Society of America Bulletin, v. 57, p. 651–674.

Novak, I. D., 1973, Predicting coarse sediment transport: the Hjulstrom curve revisited: in Morisawa, M. E., ed., Fluvial geomorphology, SUNY, Publications in Geomorphology, Binghamton, N.Y., p. 13–25.

O'Connor, J. E., and Webb, R. H., 1988, Hydraulic modeling for paleoflood analysis,:in Baker, V. R., Kochel, C. R., and Patton, P. C., eds., Flood geomorphology, John Wiley and Sons, N.Y., p. 393–402.

O'Connor, J. E., Baker, V. R., and Webb, R. H., 1996, Paleo-hydrology of pool and riffle pattern development, Boulder Creek, Utah: Geological Society of America Bulletin, v. 97, p. 410–420.

O'Connor, J., Ely, L., Wohl, E., Stevens, L., Melis, T., Kale, V., and Baker, V., 1994, A 4500–year record of large floods on the Colorado River in the Grand Canyon, Arizona: Journal of Geology, v. 102,p 1–9.

Osterkamp, W. R., 1978, Gradient, discharge and particle-size relations of alluvial channels in Kansas, with observations on braiding: American Journal of Science, v. 278, p. 1253–1268.

Pardee, J. T., 1942, Unusual currents in Glacial Lake Missoula, Montana: Geological Society of America Bulletin, v. 53, p. 1569–1600.

Park, C., 1977, World-wide variations in hydraulic geometry exponents of streams channels: An analysis and some observations: Journal of Hydrology, v. 33, p. 133–146.

Parker, G., 1976, On the cause and characteristic scale of meandering and braiding in rivers: Journal of Fluid Mechanics, v. 76, p. 459–480.

Parker, G., 1978, Self-formed rivers with equilibrium banks and mobile bed: part II, the gravel-bed river: Journal of Fluid Mechanics, v. 89, p. 127–146.

Parker, G., 1979, Hydraulic geometry of active gravel rivers: Journal of Hydraulics Division, American Society of Civil Engineers, v. 105, no. HY9, p. 1185–1201.

Parker, G., Klingeman, P. C., and McLean, D. G., 1982, Bedload and size distribution in paved gravel-bed streams: American Society of Civil Engineers, Journal of Hydraulics Division, HY4,v. 108, p. 544–571.

Partridge, J. B., and Baker, V. C., 1987, Paleoflood hydrology of the Salt River, Arizona: Earth Surface Processes and Landforms, v. 12, p. 109–125.

Patton, P. C., and Baker, V. R., 1977, Geomorphic response of central Texas stream channels to catastrophic rainfall and runoff: in Doehring, D. O., ed., Geomorphology of Arid and Semiarid Regions, SUNY, Publications in Geomorphology, Binghamton, N.Y., p. 189–217.

Patton, P. C., and Dibble, D. S., 1982, Archaeologic and geomorphic evidence for the paleohydrologic record of the Pecos River in west Texas: American Journal of Science, v. 282, p. 97–121.

Patton, P. C., Baker, V. R., and Kochel, R. C., 1979, Slack-water deposits: a geomorphic technique for the interpretation of fluvial paleohydrology: in Rhodes, D. D., and Williams, G. P., eds., Adjustments of the fluvial system, Kendall/Hunt, Dubuque, Iowa, p. 225–253.

Petts, G., and Foster, I., 1985, Rivers and landscape: Edward Arnold, London, 274 p.

Pizzuto, J. E. 1984. Bank erodibility of shallow sandbed streams. Earth Surface Processes and Landforms, v. 9, p. 9:113–124.

Playfair, J., 1802, Illustrations of the Huttonian theory of the earth: London, Cadell and Davies.

Powell, J. W., 1875, Exploration of the Colorado River of the West and its tributaries: Smithsonian Institution, Washington, D.C., 291 p.

Powell, J. W., 1876, Report on the geology of the eastern portion of the Uinta Mts.: Geographical and geological survey of the Rocky Mt. Region, U.S. Geographical and Geological Survey.

Prestgaard, K. I., 1983a, Variables influencing water-surface slopes in gravel-bed streams at bankfull stage: Geological Society of America Bulletin, v. 94, p. 673–678.

Prestegaard, K. L., 1983b, Bar resistance in gravel bed streams at bankfull stage: Water Resources Research, v. 19, p. 472–476.

Quraishy, M. S., 1944, The origin of curves in rivers: Current Science. (India), v. 13, p. 36–39.

Rahn, P. H., 1975, Lessons learned from the June 9, 1972 flood in Rapid City, South Dakota: Bulletin of Association of Engineering Geologists, v. 12, p. 83–97.

Reid, L., Layman, J., and Frostick, L., 1980, The continuous measurement of bedload discharge: Journal of Hydraulic Research, v. 18, p. 243–249.

Richards, K. S., 1973, Hydraulic geometry and channel roughness—a nonlinear system: American Journal of Science, v. 273, p. 877–896.

Richards, K. S., 1976, The morphology of riffle-pool sequences: Earth Surface Processes, v. 1, p. 71–88.

Richards, K. S., 1977, Channel and flow geometry: a geomorphological perspective: Progress in Physical Geography, v. 1, p. 65–102.

Richards, K. S., 1982, Rivers: Methuen and Company, London, 358 p.

Rubey, W. W., 1933, Equilibrium conditions in debris-laden streams: American Geophysical Union Transactions, p. 497–505.

Rubey, W. W., 1938, The force required to move particles on a stream bed: U.S. Geological Survey Professional Paper 189-E, p. 121–141.

Rubey, W. W., 1952, Geology and mineral resources of the Hardin and Brussels quadrangles: U.S. Geological Survey Professional Paper 218, 179 p.

Rundle, A. S., 1985, Braid morphology and the formation of multiple channels, the Rakaia, New Zealand: Zeitschrift für Geomorphologie, Supplement Band, v., 55, p. 15–37.

Rust, B. R., 1972, Structure and processes in a braided river: Sedimentology, v. 18, p. 221–245.

Schumm, S. A., 1956, Evolution of drainage systems and slopes in badlands at Perth Amboy, N.J.: Geological Society of America Bulletin, v. 67, p. 597–646.

Schumm, S. A., 1960, The shape of alluvial channels in relation to sediment type: U.S. Geological Survey Professional Paper 352-B, p. 17–30.

Schumm, S. A., 1963b, Sinuosity of alluvial rivers in the Great Plains: Geological Society of America Bulletin, v. 74, p. 1089–1099.

Schumm, S. A., 1967, Meander wavelength of alluvial rivers: Science, v. 157, p. 1549–1550.

Schumm, S. A., 1971, Fluvial geomorphology: channel adjustments and river metamorphosis: *in* Shen, H. W., ed., River mechanics, v. 1, p. 1–22.

Schumm, S. A., 1972b, Fluvial paleochannels: Society of Economic Paleontologists and Mineralogists Special Publication 13, p. 98–107.

Schumm, S. A., 1977, The fluvial system: John Wiley and Sons, N.Y., 338 p.

Schumm, S. A., 1979, Geomorphic thresholds: the concept and its applications: Transactions of the Institute of British Geographers, N.S. 4, v. 4, p. 485–515.

Schumm, S. A., 1989, Anastomosing streams or anastomosing patterns? Geological Society of America Abstracts with Program, v. 21, p. 153.

Schumm, S. A., and Khan, H. R., 1972, Experimental study of channel patterns: Geological Society of America Bulletin, v. 83, p. 1755–1770.

Schumm, S. A., Mosley, M. P., and Weaver, W. E., 1987, Experimental Fluvial Geomorphology: John Wiley and Sons, N.Y., 413 p.

Shen, H. W., 1971, Stability of alluvial channels: *in* Shen, H. W., ed., Environmental Impact on Rivers, (river mechanics 111): Fort Collins, The Editor, v. 1, no. 16, 33 p.

Shreve, R. L., 1966, Statistical law of stream numbers: Journal of Geology, v. 74, p. 17–37.

Shroba, R. R., Schmidt, P. W., Crosby, E. J., and Hansen, W. A., 1979, Storm and flood of 31 July–1 August 1976 in the Big Thompson River and Cache la Poudre River basins, Larimez and Weld Counties, Colorado: U.S. Geological Survey Professional Paper 422-K.

Shroba, R. R., Schmidt, P. W., Crosby, E. J., Hansen, W R., and Soule, J., 1976, Storm and flood of July 31 –August 1, 1976, in the Big Thompson River and Cache la Poudre River basins, Latimer and Weld Counties Colorado: *in* Geologic and geomorphic effects in the Big Thompson, U.S. Geological Survey Professional Paper 1115, Part B, p. 87–152.

Sigafoos, R. S., 1961, Vegetation in relation to flood frequency near Washington, D.C.: U.S. Geological Survey Professional Paper 424-C, p. 248–249,

Sigafoos, R. S., 1964, Botanical evidence of floods and flood plain development: U.S. Geological Survey Professional Paper 485-A, 35 p.

Sigafoos, R. S., and Sigafoos, M. D., 1964, Flood-history told by tree growth rings: Natural History, v. 50, p. 50–55.

Simons, D. B., and Richardson, E. V., 1962,. Resistance to flow in alluvial channels: American Society of Civil Engineers Transactions, v. 127, p. 927–952.

Simons, D. B., Al-Shaikh-Ali, K. S., and Li, R. M., 1979, Flow resistance in cobble and boulder riverbeds: American Society of Civil Engineers Proceedings, Journal of the Hydraulics Division, v. 105, p. 477–488.

Smith, D. G., 1976, Effects of vegetation on lateral migration of anastomosed channels of a glacier meltwater: Geological Society of America Bulletin, v. 87, p. 857–869.

Smith, N. D., 1971, Transverse bars and braiding in the Lower Platte River, Nebraska: Geological Society of America Bulletin, v. 82, p. 3407–3419.

Smith, N. D., 1974, Sedimentology and bar formation in the upper Kicking Horse River, a braided outwash stream: Journal of Geology, v, 82, p. 205–223.

Stedinger, J. R., and Baker, V. R., 1987, Surface water hydrology: historical and paleoflood information: Reviews of Geophysics, v. 25, p. 119–124.

Stedinger, J. R., and Cohn, T. A., 1986, Flood frequency analysis with historical and paleoflood information: Water Resources Research, American Geophysical Union, v. 22, p. 785–793.

Stelzer, K., 1981, Bed-load transport: Water Resources Publications, 295 p.

Stewart, J. H., and LaMarche, V. C., 1967, Erosion and deposition produced by the floods of December 1964 on Coffee Creek Trinity County, California: U.S. Geological Survey Professional Paper 422-K.

Sundborg, A., 1956, The river Klaralven, a study of fluvial processes: Geogratisker Annaler, v. 38, p. 127–316.

Task Committee on Sedimentation, 1966, Sediment transport mechanics: initiation of motion: Journal Hydraulics Division, American Society of Civil Engineers, v. 92, no. HY2, Proceedings Paper 4738, p. 291–314.

Thomson, J., 1879, On the origin of windings of rivers in alluvial plains: Royal Society of London Proceedings, v. 25, p. 5–6.

Thorne, C. R., 1977, Channel changes in ephemeral streams—observations, problems and models: *in* Gregory, K. J., ed., River channel changes, John Wiley and Sons, London, p. 317–335.

Thorne, C. R., 1982, Processes and mechanisms of river bank erosion: *in* Gravel-Bed Rivers, Hey, R. Bathurst, J., and Thorne, C., John Wiley and Sons, N.Y., p. 227–272.

Thorne, C. R., and Lewin, J., 1979, Bank processes, bed material movement and planform development in a meandering river: *in* Rhodes, D. D., and Williams, G. P., eds., Adjustments of the fluvial system, Kendall/Hunt, Dubuque, Iowa, p. 117–137.

Thorne, C. R., and Tovey, N. K., 1981, Stability of composite river banks: Earth Surface Processes and Landforms, v. 6, p. 469–484.

Tinkler, K. J., 1982, Avoiding error when using the Manning equation: Journal of Geology, v. 90, p. 326–328.

Vanoni, V. A., 1946, Transportation of suspended sediment by water: American Society of Civil Engineers Transactions, v. 3, p. 67–133.

Waitt, R. B., 1980, About forty last-glacial Lake Missoula Jokulhlaups through southern Washington: Journal of Geology, v. 88, p. 653–679.

Werner, P. W., 1951, On the origin of river meanders: American Geophysical Union Transactions, v. 32, p. 898–902.

West, E. A., 1978, The equilibrium of natural streams: GeoBooks, Norwich, U.K., 205 p.

Wilcock, D. N., 1971, Investigation into the relations between bedload transport and channel shape: Geological Society of America Bulletin, v. 82, p. 2159–2176.

Williams, G. P., 1983, Paleohydrological methods and some examples from Swedish fluvial environments: I. Cobble and boulder deposits: Geografiska Annaler, v. 65A, p. 227–243.

Williams, G. P., 1984a, Paleohydraulic equations for rivers: *in* Costa, J. E., and Fleisher, P. S., eds., Developments and applications of geomorphology, Springer-Verlag, Berlin, p. 343–362~

Williams. G. P., 1984b, Paleohydrological methods and some examples from Swedish fluvial environments: II. River meanders: Geografiska Annaler, v. 66A, p. 89–102.

Williams, G. P., 1988, Paleofluvial estimates from dimensions of former channels and meanders: *in* Baker, V. R., Kochel, R. C., and Patton, P. C., eds., Part IV: Paleofloods: flood geomorphology, John Wiley and Sons, N.Y., p. 321–334.

Williams, G. P., and Costa, J. E., 1988, Geomorphic measurements after a flood: *in*

Baker, V. R., Kochel, R. C., and Patton, P. C., eds., Part I: external controls and geomorphic measurements: flood geomorphology, John Wiley and Sons, N.Y., p. 65–77.

Williams, G. P., and Rust, B. R., 1969, The sedimentology of a braided river: Journal of Sedimentary Petrology, v. 39, p. 649–679.

Williams, G. P., and Wolman, M. G., 1984, Downstream effects of dams on alluvial rivers: U.S. Geological Survey Professional Paper 1286, 83 p.

Winkley, B. R., 1977, Man-made cutoffs on the lower Mississippi River—conception, construction, and river response: Report No. P1300-2, U.S. Army Engineer District, Vicksburg, Corps of Engineers, Vicksburg, Miss.

Wolman, M. G., 1955, The natural channel of Brandywine Creek, Pennsylvania: U.S. Geological Survey Professional Paper 271.

Wolman, M. G., 1959, Factors influencing erosion of a cohesive rivet bank: American Journal of Science, v. 257, p. 204–216.

Wolman, M. G., and Brush, L. M., Jr., 1961, Factors controlling the size and shape of stream channels in coarse noncohesive sands: U.S. Geological Survey Professional Paper 282-G, p. 191–210.

Wolman, M. G., and Leopold, L. B., 1957, River flood plains: some observations on their formation: U.S. Geological Survey Professional Paper 282-C, p. 87–109.

Wolman, M. G., and Miller, J. P., 1960, Magnitude and frequency of forces in geomorphic processes: Journal of Geology, v. 68, p. 54–74.

Woolley, L., 1954, Excavation at Ur: Thomas Crowell Co., N.Y., 262 p.

Yang, C. T., 1971, On river meanders: Journal of Hydrology, v. 13, p. 231–253.

Yanosky, T. M., 1982, Hydrologic inferences from ring widths of flood-damaged trees, Potomac River, Maryland: Environmental Geology and Water Science, v. 4, p. 43–52.

Yanosky, T. M., 1983, Evidence of floods on the Potomac River from anatomical abnormalities in the wood of flood plain trees: U.S. Geological Survey Professional Paper 1296, 42 p.

Yen, B. C., 1975, Spiral motion and erosion in meanders: in Proceedings of the 16th Congress, International Association for Hydraulic Research, Sao Paulo, Brazil, v. 2, p. 338–346.

Fluvial Landforms

Meander cutoff of incised stream, Canyon Lands, Utah. (Photo by W. K. Hamblin)

DRAINAGE SYSTEMS

The signatures of fluvial processes are inscribed upon the landscape as branching patterns of stream channels and drainage divides. Leonardo da Vinci (1452–1519) and Demarest (1725–1815) were among the first to recognize that valleys were cut by the streams that occupied them, rather than having been made by some cataclysmic event. As Playfair wrote in 1802:

"If we proceed in our survey from the shores, inland, we meet at every step with the fullest evidence of the same truths, and particularly in the nature and economy of rivers. Every river appears to consist of a main trunk, fed from a variety of branches, each running in a valley proportioned to its size, and all of them together forming a system of valleys, communicating with one another, and having such a nice adjustment of their declivities, that none of them joins the principal valley, either on too high or too low a level; a circumstance that would be infinitely improbable, if each of these valleys were not the work of the stream that flows in it. If indeed a river consisted of a single stream without branches, running in a straight valley, it might be supposed that some great concussion or some powerful torrent, had opened at once the channel by which its waters are conducted to the ocean; but, when the usual form of a river is considered, the trunk divided into many branches, which rise at a great distance from one another, and these again subdivided into an infinity of smaller ramifications, it becomes strongly impressed upon the mind, that all these channels have been cut by the waters themselves; that they have been slowly dug out by the washing and erosion of the land; and that it is by the repeated touches of the same instrument, that this curious assemblage of lines has been engraved so deeply on the surface of the globe." (Playfair, 1802)

This statement, sometimes referred to as **Playfair's law,** established streams as the cause of drainage patterns, the systems of valleys they occupy, and, indeed, the encompassing drainage basins. Modern observations have amplified, extended, and refined Playfair's concept to the point where each drainage basin can be characterized by its stream-carved topography in at least as many ways as a signature characterizes the handwriting of its author.

Qualitative appraisals of drainage patterns—for example, dendritic drainage, trellis drainage, annular drainage, and radial drainage—can be characterized in terms of morphometric parameters—quantitative aspects of the geometry of the basin. Morphometric parameters of drainage basins, used with the maps and photos from which they are derived and with qualitative statements about drainage patterns, provide not only an elegant description but also a powerful means of comparison of the form and performance of drainage basins that may be widely separated in size and in space and that may have been separated in time.

Strahler (1952) developed the concept of drainage basins in equilibrium around a morphometric parameter known as the **hypsometric integral.** The idea—that a drainage basin may be an open system in a steady state—is based on the assumption that after an initial phase of removal of excess material, the form of the basin adjusts, like a well-designed funnel, to convey runoff and sedimentary load from the divide to the mouth.

An analysis is made in which height above the mouth of the basin (taken from a contour line) is plotted against the planimetric area of the part of the basin that lies above the chosen contour line. The resultant graph shows the distribution of basin area with altitude, and the curve drawn through the points is known as a **nondimensional hypsometric curve.** The **hypsometric integral,** a fraction, is the portion of the area that lies below the hypsometric curve. It represents the fraction of volume that remains after the erosion of a hypothetical reference figure whose base is the planimetric area of the basin and whose height is the relief of the basin.

According to Strahler (1952), equilibrium is attained (and maintained) after erosion has removed enough material that the hypsometric integral stabilizes. Stable hypsometric integrals are in the range 0.60–0.35. Hypsometric integrals larger than this are characteristic of basins in an early stage of sculpture where much of the area of the basin is as yet at high altitude. This period of adjustment is analogous to the early phases of the erosion cycle, and, indeed, basins with large hypsometric integrals show such characteristics.

The significance of the stable hypsometric integral is in the implied stabilization of basin form. Forms are stable to the extent that the pattern of channels, slopes, and divides is established, and from that time on, even though all these elements may lose material, the ratio of area to attitude is fixed. In other words, as absolute relief diminishes during the approach to base level, relative relief is maintained. To this extent the drainage basin is in dynamic equilibrium, systematically adjusting its form to maintain an open system in a steady state.

Other morphometric parameters require measurements of lengths, angles, and areas. Substitution of surface area for planimetric area and length of stream profile for planimetric length of channel is recommended for description of rugged basins where surface area is appreciably greater than planimetric area. A method of approximating surface areas is based on measuring the lengths of a sample of topographic profiles drawn orthogonal to contours in the basin. Surface area is greater than planimetric area (measured with a planimeter) by the same amount that the sum of profile lengths exceeds the planimetric lengths of the traverses.

One advantage of measuring the geometry of drainage basins is the amenability of the morphometric parameters to comparison by statistical analysis. Statistical tests of significance can determine the probable similarity or difference between basins according to their parameters. The breakdown of basin geometry into parameters shows the basis of similarities between basins and the areas where significant differences exist. Such an analysis focuses attention on aspects of the drainage that may be critical to its behavior. The analysis asks for the reasons why a drainage basin is the way it is.

The mathematically regular ramification of drainage, free to develop without structural control or derangement by external processes, was recognized in 1945 by Horton, who envisioned a hierarchy among streams in a drainage basin.

Horton's method was modified by Strahler (1952). If the smallest, unbranched tributaries are called **first-order streams,** then streams of the second order are formed by combination of the first-order streams, and these second-order streams can in turn combine to form streams of the third order, and so on, until the trunk stream, receiving the largest class of tributaries, is assigned the highest order number of all (Figure 6–1). This is also the order number of the entire drainage basin. Thus, basins of the fifth order are drained by fifth-order trunk streams; basins of the tenth order, by tenth-order streams; and so on. A fifth-order stream or some other stream of high order number can receive a first-order stream directly; in fact, this is often the case. The condition that determines that this particular stream is of the fifth order is that somewhere along its course it receives at least one fourth-order tributary, which in turn receives a third-order stream, which in turn receives a second-order stream, and so on.

If the quantities of streams of successive orders are plotted against their order numbers on a graph, an inverse geometric series appears. This series is best shown by plotting the data on semilogarithmic graph paper, so that the curve for number of streams versus order number (an exponential function) appears as a straight line.

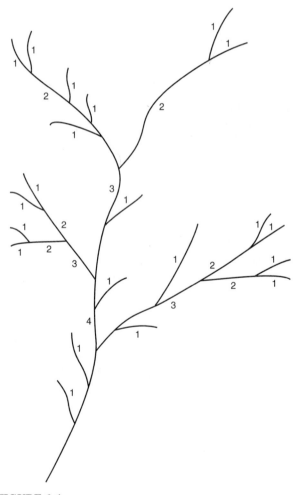

FIGURE 6–1
Stream ordering system. Numbers are stream orders for tributaries in the drainage.

The regular geometric increase of the number of streams of successively smaller order numbers can be seen in the hypothetical case of a sixth-order drainage basin with one master stream of the sixth order. If we now suppose three tributaries of the fifth order, then there should be 9 fourth-order streams, 27 third-order streams, 81 second-order streams, and 243 first-order streams. Because we specified three times as many fifth-order streams as sixth-order streams, the number of fourth-order streams should also be three times as many as fifth-order streams, and so on. That is, the ratio between the number of streams of successive orders— the bifurcation ratio—is constant.

The lengths of streams of successive order numbers are also related in a systematic way. However, in this case, the lengths of streams of successively higher orders approximate a direct geometric series. For example, the length of a sixth-order trunk stream could be 16 km (10 mi.), and the average length of fifth-order tributaries 8 km (5 mi). In this case, fourth-order streams would average 4 km (2.5 mi.) in length; third-order streams would average 2 km (1.2 km); second-order streams would average 1.6 km (1 mi.); and the numerous first-order tributaries would be, on average, 1 km (1/2 mi.) long. In other words, the stream length ratio is also a constant, although not necessarily the same constant as the bifurcation ratio of the drainage basin.

The slopes of streams of successively higher order numbers, where **slope** is defined as the difference in altitude between the head and mouth of a stream divided by stream length, approximate an inverse geometric series. Both of these distances can easily be measured on a contour map. If the slope of a sixth-order stream is 1/729, and if the average slope of fifth-order streams is 1/243, then slopes should be 1/81, 1/27, 1/9, and 1/3, respectively, for the average fourth-, third-, second-, and first-order streams. The successively smaller streams are steeper in this example by a constant factor of three.

The drainage areas of streams of successively higher order numbers are also related in a systematic way (Schumm, 1956). Just as length increases with order number, so also does drainage area approximate a direct geometric series with increasing order number. Drainage area can be outlined for each of the streams on a contour map and measured with a planimeter. Thus, a sixth-order stream could have a drainage area of 64 km^2 (25 mi^2), and fifth-order tributaries could have an average drainage area of 32 km^2 (12 mi^2), in which case the average areas of the successively smaller basins would be 16, 8, 4, and 2 km^2 (6.4, 3.2, 1.6, and 0.8 mi^2). In this example, the ratio between average areas of successive basins is always two.

The orderly composition of the drainage net serves equivalent parts of the drainage basin equally well, although a remote possibility exists that just below a drainage divide, channels are absent except for one corner of the drainage basin where they are extremely concentrated. If the basin is reasonably symmetrical about its length and composed of homogeneous material, this condition would be extremely unusual—just as unusual as if all the molecules of atmospheric gas in a room were concentrated in one corner near the ceiling. Both conditions are possible, but neither is very probable. More probably, runoff moves into the drainage basin with

reasonably orderly spacing, just as air in a room diffuses and fills all available space more or less uniformly.

As drainage is assembled in the systematic way described by Playfair and Horton, the numbers of stream channels that appear in a given area depend on:

1. The amount and erosive energy of the runoff
2. The erodibility of the surface

Every channel requires a certain amount of drainage surface from which it can collect its water in the form of unconcentrated runoff. Horton (1945) recognized this as a length of overland flow, that is, the average distance from divides that unconcentrated runoff flows before eroding channels or collecting into channels already available. Horton expressed the amount of stream channels per drainage area as **drainage density,** which he defined as the total length of streams in a drainage basin divided by the area of the basin.

$$D_d = \frac{L}{A}$$

where D_d = drainage density
L = sum of stream lengths
A = area of basin

Drainage density is a measure of how well or how poorly a watershed is drained by stream channels. It affords one of the best available quantitative expressions of drainage texture— an expression of the relative spacing of channels in fluvially dissected terrain. Thus, a well-drained basin of fine drainage texture (many, closely spaced channels) (Figure 6–2; Plate 3A) might have a drainage density of several hundred kilometers of channel length per square kilometer of drainage basin, whereas a poorly drained basin of coarse texture (relatively few channels, widely spaced) might have a drainage density of only a few kilometers per square kilometer. The quality of maps from which such measurements are taken must be such that all permanent drainage lines (both perennial and intermittent streams) are shown. For the United States, this generally limits computations to modern topographic maps at a scale of 1:24,000 or larger.

Texture ratio, although hampered by the variable quality of topographic drafting, is useful for assessing the texture shown on large-scale topographic maps. Texture ratio is found by dividing the number of crenulations along the most crenulated contour of a drainage basin by the perimeter of the drainage basin.

$$T = \frac{N}{P}$$

where T = texture ratio
N = number of crenulations of the most crenulated contour in basin
P = perimeter of basin in km

The crenulations are more or less V-shaped and point upstream to higher elevations. They show the presence of chan-

FIGURE 6–2
Fine-textured drainage system,
Petrified Forest, Arizona.

nels, even though symbols for all the streams may not be published on the map.

Because fluvial topography consists of an assemblage of drainage basins, a mean texture ratio can be found by averaging the texture ratios of the several basins that comprise a region. The various basins are given importance according to their size by the following equation.

$$T = \frac{AN/P}{A}$$

where T = weighted mean texture
A = basin area in km^2
P = basin perimeter in km
N = number of crenulations along the most crenulated contour of each basin

Regional topographic texture may be rated according to the mean texture ratio. Smith recommended the following textural categories:

$T < 4.0$ = coarse topographic texture
$4.0 - 10.0$ = medium topographic texture
> 10 = fine topographic texture
> 50 = ultrafine topographic texture

Badlands have a mean texture ratio of 50 or greater. Normally, coarse, medium, and fine textures can be detected on contour maps at a scale of 1:24,000. However, ultrafine textures usually require an even larger scale to show the many small channels and divides that occupy intricately dissected drainage basins.

A logarithmic relationship exists between texture ratio and drainage density. Inasmuch as drainage density appears as a power function of texture, estimated texture ratios can be computed indirectly from air photos, upon which only drainage density can be measured directly.

Topographic texture is an important indicator of the relative erodibility of the surface. The distribution of textures outlines the various rock bodies upon which the textures have developed and in this way helps to reveal geologic structures that might otherwise be hidden.

Three of the controls of topographic texture are slope, permeability of the surface material, and climate. Slope is important because it promotes runoff by inhibiting infiltration and provides the hydraulic shear necessary for effective erosion. Permeability, the ease with which water passes through material, determines whether precipitation runs off on the surface or descends into the ground. Low permeability of material generates high runoff rates. The permeability of a material is determined by the size, shape, and distribution of its constituent particles. Thus, fine-grained clay or shale has lower permeability than coarser material. Intensive gullying occurs where permeability is low and runoff rates are high, such as in areas underlain by clay or shale.

Climate determines the volume of runoff by controlling amount and type of precipitation, conditions of evaporation and transpiration, condition of the ground (moist, dry, frozen, or unfrozen), vegetation, and character of soils. Under comparable conditions of slope, permeability, and climate, erodibility of the surface is the critical factor that determines texture from place to place. Massive bodies of essentially unweathered sandstone or granitic rock may shed huge volumes of runoff in sheets, especially where a lack of constraining vegetation exists.

Badlands (Plate 3–A), having unusually high drainage density, occur where the following conditions prevail:

1. Clayey material of low permeability

2. Little vegetation to impede runoff

3. Rainfall concentrated in widely scattered showers

4. Downcutting of the drainage system

Coarse textures develop where the efficacy of runoff is reduced by infiltration, evapotranspiration, vegetal interference, gentle slopes, and strong resistance to erosion.

ORIGIN OF STREAM COURSES

The path taken by a stream is set by:

1. The initial slope of the land

2. Random headward erosion

3. Selective headward erosion

Once established, the course may be altered by conditions of rock resistance to form different configurations.

Consequent Streams

When precipitation first falls on a newly created land surface and water begins to run off, the direction of flow is determined by the direction of slope of the initial land surface (Figure 6–3), and the origin of the stream courses developed is said to be **consequent** upon the new surface (Powell, 1875).

Some familiar examples of newly formed land surfaces upon which consequent streams may form are:

1. Volcanic cones or lava fields

2. Recently deglaciated regions

3. Emergent marine areas

4. Uplifted domes, anticlines, or fault blocks

Insequent Streams

Insequent streams originate by extension of tributaries by random headward erosion. Sheet flow—downslope movement of water as a thin sheet over the land surface—converges in some places and diverges in others. Water is concentrated into small channels or rills where convergence of the flow occurs, and the deeper water, with its greater velocity and increased erosive power, extends and enlarges the heads of gullies (Figures 6–4 and 6–5) and deepens them by downward erosion. The deepened channel concentrates still more water from nearby slopes, and new rills develop as water flows into the deepening rill. The headward extension of insequent streams is not guided by weak rock zones, so they develop in random directions, usually on homogeneous rocks or on horizontally stratified sedimentary beds where structural control by underlying rocks is not evident.

Subsequent Streams

Subsequent streams originate by selective headward erosion or by adjustments of their courses to rocks of least resistance by differential erosion. They are especially common on folded or tilted sedimentary beds, on faulted or jointed rocks, and on other rocks of differing resistance.

Streams whose courses were originally consequent on a new land surface can be converted in time to subsequent streams as they adjust their channels to take advantage of weak rock zones (Figure 6–6). The mechanism of adjustment is usually by stream capture.

Obsequent and Resequent Streams

Over time, the consequent drainage of an area may become modified to the extent that streams flow in the direction opposite that of the original consequent drainage. Such streams are called **obsequent streams.** Alternatively, the streams may establish a direction that is the same as that of the consequent drainage, but developed at a lower erosional level than the initial slope. Such streams are called **resequent streams.**

Obsequent and resequent streams typically develop on tilted sedimentary beds. In Figure 6–6, a consequent stream flows down the initial slope of a newly uplifted sea floor. Selective headward erosion and stream capture along weak beds of a tilted sedimentary unit change the consequent drainage to subsequent drainage, and the stream at O, which flows down the scarp face of a dipping bed, is an obsequent stream because it flows in the direction opposite that of the original consequent drainage, C. The stream that flows down the dip of a resistant dipping bed at R is a resequent stream because it flows in the same direction as the original consequent stream, but at a lower level. Because consequent drainage is not always in the same direction as the dip of tilted beds, streams flowing down the dip of beds are not always resequent, and streams flowing in the opposite direction as the dip of the beds are not always obsequent.

DRAINAGE PATTERNS

The **pattern** of a fluvial system refers to the arrangement of streams in a drainage, which often reflects structural and/or lithologic control of underlying rocks. Because drainage patterns tell much about the substance of which the land surface is made, with a little practice, interpretation and identification of geologic structures and rock types in an area can be made from analysis of stream patterns on air photos or topographic maps.

Dendritic Patterns

Dendritic drainage patterns are an index of the homogeneous nature of underlying bedrock and the lack of structural control; they are characterized by the branching of stream valleys at acute angles (Figure 6–7) without systematic

FIGURE 6–3
Consequent drainage on the newly exposed, former floor of the Salton Sea (dark area at top of photo). (Photo by Wallace, U.S. Geological Survey)

arrangement. They are developed by random headward erosion of insequent streams on rocks that are of uniform resistance and that lack structural controls. The courses of tributary streams are determined by random headward erosion, implying that the underlying material is essentially homogeneous, without systematic differences in rock resistance. Dendritic patterns commonly form on horizontally bedded sedimentary rocks or massive igneous or metamorphic rocks.

Trellis Patterns

Trellis drainage patterns consist of nearly parallel streams that flow into main trunk streams at high angles (Figure 6–8). The streams making up a trellis pattern have etched their channels out of weaker rock and are subsequent in origin.

Tilted sedimentary beds of differing resistance typically form trellis drainage, with short tributary streams flowing from ridges of resistant beds into valleys of weak rock. Under certain circumstances, consequent streams may form less-well-developed trellis patterns. For example, drainage in recently deglaciated terrain may consist of roughly parallel streams consequent on the streamlined glacial terrain.

Rectangular and Angular Patterns

Rectangular drainage patterns consist of tributary streams that not only join trunk streams at right angles but also exhibit right-angle bends in their channels as a result of adjustment of subsequent streams to intersecting fault or joint systems (Figure 6–9). Rectangular drainage is usually formed on jointed igneous rock, flat-lying sedimentary beds with well-developed joint systems, or intersecting faults. Thus, these patterns serve as excellent indicators of fracture systems in rocks.

Both trellis and rectangular patterns have high-angle tributary junctions, but individual streams in rectangular patterns have many right-angle bends, generally controlled by joint or fault planes.

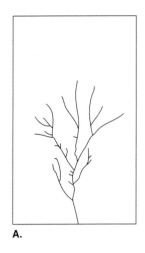

A.

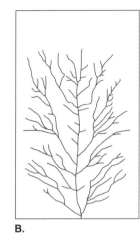

B.

C.

D.

E.

F.

FIGURE 6–5
Insequent drainage developing by random headward
erosion. (Photo by D. A. Rahm)

FIGURE 6–4
Progressive growth of insequent drainage system in a lab
experiment on a 9-by-14-m rectangle. (From Parker, 1976)

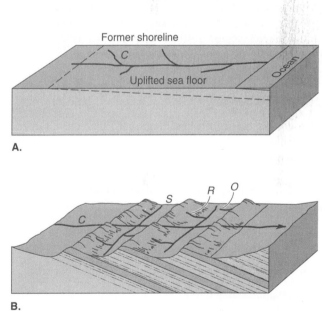

A.

B.

FIGURE 6–6
Evolution of subsequent drainage: (A) drainage consequent
on newly uplifted sea floor. (B) subsequent drainage after
tilting and selective headward erosion along weak rock
belts. *C* is a consequent stream, *S* is a subsequent stream,
R is a resequent stream, and *O* is an obsequent stream.

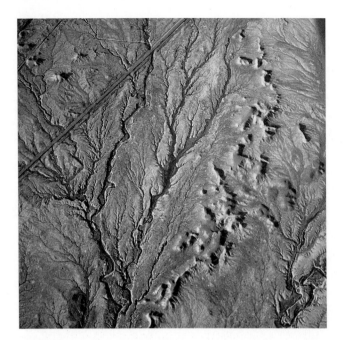

FIGURE 6–7
Dendritic drainage pattern developing by random headward erosion. (Photo by D. A. Rahm)

Because intersecting joint and fault systems may intersect at any angle, subsequent streams whose courses are controlled by joints or faults may show sharp channel bends that are not at right angles (Figure 6–10), as in rectangular patterns. Such patterns make up **angular drainage patterns.**

Radial Patterns

Consequent streams flowing from a central area in all directions make **radial drainage patterns.** They are usually developed on high, recently formed land surfaces, such as volcanic cones (Figure 6–11) or structural or intrusive domes. Radial drainage typically flows down an initial slope, which may be constructional, as in the case of volcanic cones, or structural, as in the case of uplifted domes.

Annular Patterns

Erosional breaching of domes of alternating weak and resistant sedimentary rock leads to formation of a circular outcrop pattern (resembling a bull's-eye) as subsequent drainage adjusts to weak rock beds. The result is that streams follow circular courses and develop an **annular drainage pattern** (Figure 6–12).

Centripetal Patterns

Streams converging into a central basin or depression form **centripetal drainage patterns**. Some typical examples are

consequent streams that flow into structural and topographic basins and streams that converge into sinkholes in limestone. Because the drainage is into a central depression, streams terminate there, either forming a lake or disappearing into the ground.

Parallel Patterns

Streams that flow in nearly parallel directions to one another form **parallel drainage patterns.** Their courses are usually controlled by some kind of unidirectional regional slope or by parallel to subparallel topographic features. They are not very common and are usually limited in their areal extent.

STREAM CAPTURE

An important mechanism for the adjustment of streams to the underlying material is **stream capture,** which may take place by abstraction or by intercision.

Abstraction

The position of divides that separate opposing drainage systems persists in the same place only if opposite sides of the divide are eroded at comparable rates. If rates of erosion of the divide between two drainages are unequal, the divide shifts toward the less energetic system. In so doing, the more vigorous tributaries intersect and divert portions of the less active drainage. The process of headward erosion of a stream until it undercuts and captures another stream is known as **abstraction.** Streams in the headwaters of a drainage system have an advantage over competing streams and are more likely to capture them if they have the following characteristics:

1. Steeper gradients
2. Shorter distances to base level
3. Less resistant rock to erode

In Figure 6–13, tributaries of the drainage in the foreground have eroded headward until they have intersected and captured portions of the drainage in the background.

Although capture of one stream by another is a fairly common event, capture within a single drainage system is rare. Tributaries almost always have steeper gradients than the main streams to which they are graded because of their lower discharge. As a result, even if a tributary were to erode headward in a direction that would eventually intersect a larger trunk stream, the tributary would be above it, not below it as required for capture. However, in rare instances, a tributary may capture another stream in the same drainage basin, or it may even capture a portion of the main stream to which it is graded. An example of this situation occurred in the Bighorn Basin of Wyoming, where a high-gradient main stream was

FIGURE 6–8
Trellis drainage pattern developed by selective headward erosion along joint planes in massive sandstone, Kane County, Utah. (Photo by U.S. Geological Survey)

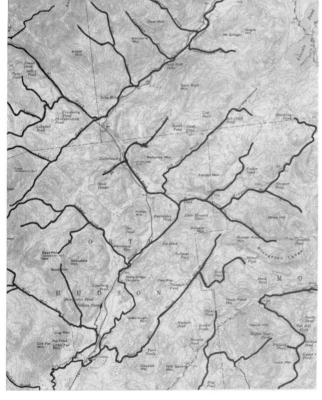

FIGURE 6–9
Rectangular drainage pattern, northeastern New York. (Elizabethtown quadrangle, U.S. Geological Survey)

captured by its own tributary with a lower gradient (Mackin, 1936). The Greybull River (Figure 6–14), a tributary to the Bighorn River, picks up a coarse load of volcanic cobbles and boulders from the Absaroka Mountains at its headwaters, thus requiring a high gradient as it crosses the Bighorn Basin. The capturing stream, also a tributary to the Bighorn River, originated in the Bighorn Basin, however, and carried only silt and sand derived locally from erosion of shale and sandstone. As a result, the capturing stream had a gradient of only a few feet per mile because of the small caliber of load it carried. Thus, at the point where headward erosion of the tributary stream caused it to intersect the Greybull River, the channel was significantly lower, and the Greybull was diverted into the tributary. This process is known as **autocapture** (Mackin, 1936).

Capture by abstraction is especially common on tilted sedimentary beds of differing resistance to erosion. Streams flowing transverse to resistant rock ridges are vulnerable to capture by streams flowing in less resistant shale. Abandoned former channels through resistant ridges are left as **wind gaps,** and streams in abandoned channels downvalley from the point of capture, supported only by local drainage, are underfit relative to the size of the channel. Bends in stream channels at the point of capture are known as **elbows of capture.**

FIGURE 6–10
Angular pattern of the Muddy
River, Utah. (Photo by U.S.
Department of Agriculture)

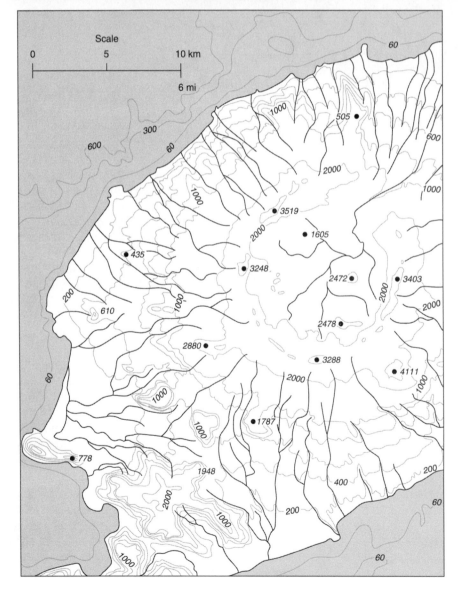

FIGURE 6–11
Radial drainage pattern on a
volcanic cone, Umnak, Alaska.
(Date from U.S. Geological
Survey Umnak quadrangle)

FIGURE 6–12
Annular drainage pattern on a breached structural dome, West Texas. (Photo by U.S. Department of Agriculture)

Intercision

Stream capture may also occur by **intercision,** whereby lateral movement of meanders causes two streams to intersect and one to be diverted into the other. This situation differs from a meander cutoff in that a meander cutoff involves only a single stream, whereas intercision takes place between two different streams.

For example, in Figure 6–15, the abandoned channel east of Frankfort around Thorn Hill was once occupied by the Kentucky River, and Benson Creek flowed in the channel between Bellepoint and Leestown, joining the Kentucky near Leestown. Migration of a meander of the Kentucky River and a meander of Benson Creek caused them to intersect near west Frankfort, and the Kentucky was captured by Benson Creek into the channel between Frankfort and Leestown, abandoning the old Kentucky channel around Thorn Hill.

FLOODPLAINS

Floodplains are portions of valley floors made by the streams that flow in them. They may consist largely of eroded bedrock,

thinly veneered with alluvium, or alluvial fills deeper than the depth of scour of the streams.

The landforms and sediments that make up a floodplain are often quite diverse, reflecting the numerous mini-environments found on most floodplains (Figures 6–16, 6–17, and 6–18). Floodplain sediments are formed by both lateral and vertical accretion of stable and aggrading channels (Table 6–1). Figures 6–16 and 6–17 show the geomorphic setting of floodplain deposits.

Most of the sediments on a floodplain are more or less continuously reworked as the main channel meanders back and forth across the floodplain. Erosion takes place on the outside of the meanders, while deposition occurs on the point bars on the inside of the meanders (Figure 6–18). These are known as **channel deposits,** in contrast to **overbank deposits** that settle out on the floodplain during floods. The mix of channel deposits relative to overbank deposits depends on the frequency with which the main channel meanders back and forth across the floodplain. The floodplain of a slowly meandering channel, which only infrequently reaches opposite sides of its valley, is likely to be characterized by mostly overbank silts and clays with smaller amounts of channel deposits. In contrast, the floodplain of a vigorously meandering channel will contain much more channel sediments.

FIGURE 6–13
Stream capture by abstraction.

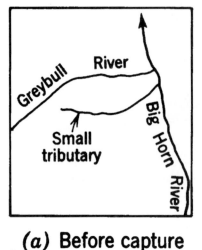

(a) Before capture

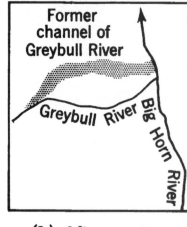

(b) After capture

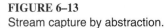

FIGURE 6–14
Autocapture of the Greybull
River, Bighorn Basin, Wyoming.
A small tributary of the Big Horn
river eroded its headwaters until
it intersected and captured the
Greybull River, diverting the
Greybull into a new course.

The manner of channel migration is often of much legal significance relative to land ownership. Many legal boundaries between nations, states or provinces, counties, and private property were established before the shifting nature of stream courses was fully appreciated. Thus, for example, the international boundary between the United States and Mexico, based on the position of the channel of the Rio Grande at the time of treaty signing, is now difficult to establish because of the rapid meandering of the Rio Grande. Similarly, the state boundary between Mississippi and Louisiana is, in places, indeterminant because the position of the Mississippi River at the time of establishment of the boundary is unknown. Private property boundaries may also be affected, even if the boundaries are not made by a river. In many states, as a stream meanders, point-bar deposits are considered to accrete to the adjacent property owner, so that as a stream meanders, property owners on the outside of a meander progressively lose their land while property owners on the inside of a bend gain property. However, if channel changes occur suddenly, as when a meander cutoff occurs, the change is considered **avulsion,** and property boundaries change only slightly. Usually, the state owns navigable channels of rivers; thus, when avulsion occurs, state ownership moves with the channel, and ownership of the abandoned channel is divided between adjacent property owners.

PEDIMENTS

Pediments are gently sloping, concave-upward, graded surfaces of erosional origin cut indiscriminately across rocks of varying resistance, mantled with a thin veneer of alluvium

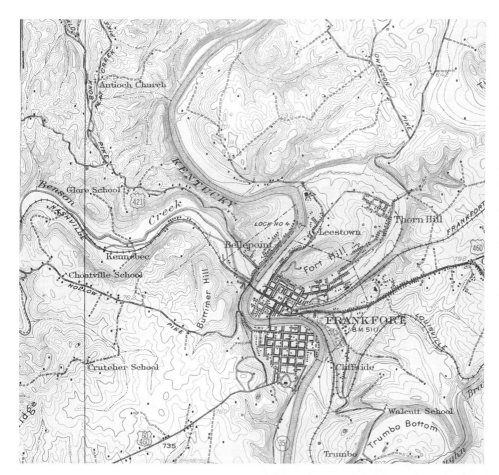

FIGURE 6–15
Stream capture by intercision. The Kentucky River once flowed around Thorn Hill, and Benson Creek flowed in the channel between Bellepoint and Leestown, joining the Kentucky near Leestown. Migration of meanders of the Kentucky River and Benson Creek caused them to intersect near west Frankfort, diverting the Kentucky into the lower Benson Creek channel. The former channel of the Kentucky is now an abandoned channel northeast of Frankfort.

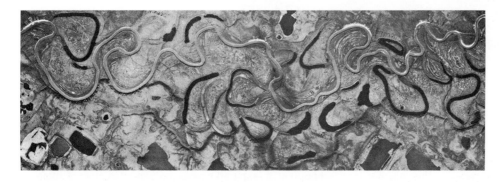

FIGURE 6–16
Floodplain of a meandering stream with many oxbow lakes. (Photo by U.S. Geological Survey)

(Figure 6–19). Pediments were first described by Gilbert (1877) as "hills of planation," cut across the upturned edges of tilted beds in the Henry Mountains of Utah. However, after more than 100 years of investigation, some aspects of their origin are still not entirely clear (McGee, 1897; Bryan, 1922; Blackwelder, 1931; Johnson, 1932; Rich, 1935; Gilluly, 1937).

Pediments typically slope gently away from desert mountainous areas with a concave-upward profile similar to those of graded streams, varying from a few hundred feet per mile (several tens of m/km) near the mountain front to about 40 ft/mi. (~10 m/km) in their distal portions. Pediment gradients are concave upward, and their declivities show a close correlation to the diameter of clasts (Gilluly, 1937; Akagi, 1980) and to the

size of streams on their surface (Bryan, 1922). Because of this correlation, and because pediments often cut across rocks of varying resistance (Figure 6–20), they are considered to be graded surfaces, similar to graded streams, and subject to regrading of their surfaces if load-discharge relationships change.

Origin of Pediments

The processes principally responsible for carving pediments include:

1. Lateral planation by streams
2. Sheetwash

TABLE 6–1
Classification of valley sediments.

Place of Deposition	Name	Characteristics
Channel	Transitory channel deposits	Primarily bed load temporarily at rest; part may be preserved in more durable channel fills or lateral accretions.
	Lag deposits	Segregations of larger or heavier particles, more persistent than transitory channel deposits, and including heavy mineral placers.
	Channel fills	Accumulations in abandoned or aggrading channel segments; ranging from relatively course bed load to fine-grained oxbow lake deposits.
Channel margin	Lateral accretion deposits	Point and marginal bars that may be preserved by channel shifting and added to overbank floodplain by vertical accretion deposits at top.
Overbank flood plain	Vertical accretion deposits	Fine-grained sediment deposited from suspended load of overbank floodwater; including natural levee and backland (backswamp) deposits.
	Splays	Local accumulations of bed-load materials, spread from channels onto adjacent floodplains.
Valley margin	Colluvium	Deposits derived chiefly from unconcentrated slope wash and soil creep on adjacent valley sides.
	Mass-movement deposits	Earthflow, debris avalanche, and landslide deposits commonly intermixed with marginal colluvium; mudflows usually follow channels but also spill overbank.

Source: Happ, 1971.

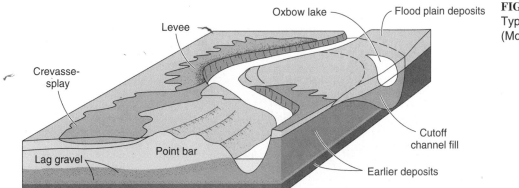

Levee · Oxbow lake · Flood plain deposits

Crevasse-splay

Lag gravel · Point bar

Cutoff channel fill

Earlier deposits

FIGURE 6–18
Typical floodplain landforms.
(Modified from Happ, 1971)

FIGURE 6–19
Pediments north of Nogales,
Arizona.

3. Rill wash

4. Weathering

However, much debate centers on which of these is most important. Gilbert (1877) attributed the origin of pediments in the Henry Mountains largely to stream planation, and the importance of active sideways erosion of desert streams in development of pediments was vigorously advocated by Paige (1912), Blackwelder (1931), and Johnson (1932). Johnson recognized three zones:

1. An inner zone composed of mountainous uplands in which vertical erosion was dominant

2. An intermediate zone of degradation, the pediment, beyond the mountain front

3. An outer zone of aggradation beyond the pediment where deposition prevailed

With time, the zone of degradation (the pediment) gradually encroaches upon the inner mountainous zone as the mountain front retreats from erosion, and the outer zone of aggradation encroaches upon the pediment, progressively covering it with alluvium beyond the thin veneer it normally carries. In support of the importance of lateral planation, Johnson (1932) pointed out that **rock fans**, having the same shape and slope as alluvial fans, were cut across bedrock structures by later-

FIGURE 6–20
Pediment cut across tilted
sedimentary beds near Sheep
Mt., Wyoming. (Photo by D. A.
Rahm)

al planation of streams issuing from canyon mouths, and he noted that such rock fans were transitional into pediments. Gilluly (1937) recognized that the gradient of pediments was directly related to the caliber of alluvium on them; that is, pediments with coarse debris had steeper slopes than those with finer load. Mackin (1937) showed that many pediments in the Bighorn Basin of Wyoming were cut by lateral planation of slowly degrading streams across truncated bedrock structures.

Many pediments show indisputable channels across pediment surfaces, providing clear evidence of stream involvement in their origin. However, unconcentrated flow of water as sheet flow or rill wash is also common on pediments. McGee (1897), who first used the term *pediment* for certain erosional surfaces in Arizona, observed the effects of sheet flow on pediments during desert storms, and he attributed the origin of pediments to such unconcentrated flows. Because the erosive power of unconcentrated flow on bedrock is limited, material must first be loosened by weathering in order for sheet flow or rill wash to be effective (Lawson, 1915; Rich, 1935). On some pediments, thick coatings of desert varnish on stones and other weathering phenomena indicate that a

substantial period of time has elapsed since that part of the pediment has been reached by vigorous stream flow. However, the question remains, Was the surface formed by weathering and sheet flow plus rill wash, or is it merely a dormant, stream-planed surface that will one day be revisited by the stream? How could weathering and unconcentrated flow develop a concave-upward surface on bedrock?

Gilluly (1937) recognized that lateral planation, sheet flow, rill wash, and weathering could all play a role, and he concluded that pedimentation could include "cooperation of lateral corrasion, rill wash, and weathering with subsequent removal of detritus by rills. One process may be dominant in one place and another elsewhere, but all are cooperative."

Mackin contended that lateral planation, weathering, and wash all took place on pediments and that the dominance of one process on any particular pediment was determined by the relationship between the rate of lateral planation and the rate of weathering and wash. The rate of lateral planation is controlled by discharge and load of streams issuing from the mountains, so those with high discharge and coarse load swing actively back and forth across the pediment, and lateral planation is most important in forming the pediment. How-

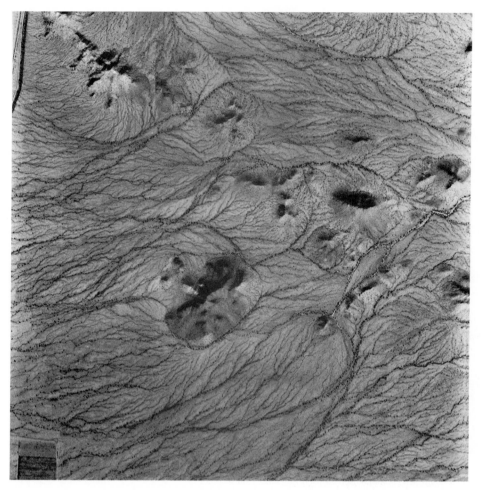

FIGURE 6–21
Pediment passes and inselbergs
in a mature stage of
pedimentation, Sacaton,
Arizona. (Photo by U.S.
Geological Survey)

ever, if discharge is feeble or infrequent across the pediment, and if the sediment load is fine-grained, lateral planation will not be effective, and weathering and wash will be more important. In either case, the general slope of the concave-upward profile of the pediment is due to the load-discharge relationships of streams crossing the pediment.

Most pediments meet the mountain front with an abrupt change in slope. In some cases, the sharp slope disjuncture may be structural, as when the mountain front is formed by a fault scarp or when a sharp change in rock resistance occurs. Where that is clearly not the case, the cause of the sharp change in slope has puzzled geomorphologists for many years and still remains controversial. Those who champion lateral planation as the dominant cause of pediments explain it simply as trimming of the mountain front by lateral cutting of streams as they swing against it from time to time. However, on pediments where lateral planation is not very active, this explanation loses much of its credibility. Bryan (1922) suggested that the break in slope was a result of the concentration of rill erosion at the base of the mountain front and of the difference in size of weathered material on the pediment and mountain front that results in strong slope differences.

The surface of pediments is often dotted with residual bedrock knobs, known as **inselbergs,** which stand above the general level of the pediment. Most are simply unconsumed bedrock remnants that have resisted pedimentation longer because of greater resistance to erosion (Kesel, 1977; Twidale, 1978). They become particularly numerous as mountain masses are reduced by erosion and as pediments intrude deeper and deeper into the mountains. Eventually, pediments on one side of a mountain range reach all the way through the mountain to pediments on the other side of the range to form **pediment passes** (Figure 6–21). The end result of long-continued pedimentation is total consumption of the mountain range, leaving a low, dome-like slope with pediments radiating from it.

Relict pediment surfaces are commonly found standing above modern pediments in somewhat the same fashion as stream terraces stand above the modern channel. These former pediment surfaces are progressively consumed by erosion on the modern pediment (Figure 6–22) and are thus related not to it but rather to some former cycle that is presently undergoing regrading. The cause of such regrading may be similar in nature to those that produce stream terraces, that

FIGURE 6–22
Dissected pediments, Book
Cliffs, Utah.

is, tectonic uplift, lowering of base level, or change in load-discharge relationships. In some cases, however, remnant pediment surfaces may be left above the modern surface without invoking any of these three causes. The slope of the upper portion of a pediment is generally quite a bit steeper than the distal portion. Thus, as a mountain front recedes in the path of an advancing pediment, that portion of the pediment surface becomes regraded and assumes a lower slope because its gradient is controlled by the caliber of load that must be carried across the pediment by the discharge of streams flowing across it; and because of the now-greater distance from the mountain front, the caliber of load is smaller and slope requirements are correspondingly less. The net result of such regrading is that the new, reduced slope leaves the old surface high and dry. Such pediments have been considered as "born dissected" (Gilluly, 1937).

Because pediments are graded slopes, they react to changes both in their source areas and in their base level. Changes in load-discharge relationships affect pediment gradients, and changes in local or regional base level also produce changes. Pediments may be graded either to a drainage system connected to the sea or to a closed basin. Through-flowing streams function as a stable, slowly degrading base level, whereas closed basins act as a slowly rising base level as the basin fills with sediment. As a result of these differences, pediments may respond in dissimilar fashions to changes. For example, a drop

in sea level might cause incision of pediments graded to through-flowing streams, but pediments in the same region graded to closed basins would remain unaffected.

Sediment transported across a pediment into a closed basin is deposited to form **bajadas**, which grade imperceptibly into pediments. In contrast to the erosional origin of pediments, bajadas are depositional in origin, often holding very shallow lakes that form **playas** (Figure 6–23). The exact boundary between pediments and bajadas is often difficult to determine because of the lack of any topographic break. Distinguishing the two based on the difference in depth of alluvium may also be unsatisfactory because of the difficulty in determining the depth of scour of streams flowing off the pediment.

ALLUVIAL FANS

Alluvial fans (Plate 4–D) are low, triangular-shaped deposits built from the accumulation of sediments deposited at the mouth of a valley that issues from a mountainous or upland area onto a piedmont into a larger mainstream valley or lowland. As such streams encounter reduced slopes, they deposit part of their sediment load in order to maintain an adequate gradient. The fan shape is made by the stream swinging back and forth from the fixed apex of the fan. Although alluvial

FIGURE 6–23
Playa (light-colored area near top), Racetrack Playa, California. (Photo by D. A. Rahm)

fans are more obvious in desert areas, they also are common in humid regions (Ryder, 1971; Boothroyd and Ashley, 1975).

Upon emerging from its valley at the head of the fan, a stream typically splits into multiple distributary channels on the fan (Figure 6–24). The high permeability of sand and gravel in alluvial fans permits loss of water by infiltration as water crosses the fan, so that between storms in arid climates, little or no water flows across the fan, and the fan surface is crossed by many dry washes separated by older surfaces of the fan mantled with desert pavement and stones coated with desert varnish. Thunderstorms in the source area of a fan produce surges in discharge, which lead to incorporation of much sediment and formation of mudflows capable of transporting large boulders and much coarse material onto the fan. Thus, fans typically consist of stratified fluvial sand and gravel interbedded with poorly sorted mudflows. The coarsest sediments, often cobbles and boulders, usually accumulate on the upper fan, grading downslope into finer sediments near the fan margins. All variations in type of sediment transport occur, from streamflow to watery mudflows to viscous debris flows. The type of sediment transport during any given time depends largely on the magnitude and intensity of precipitation in the drainage basin and the availability of detritus. In desert climates, fans typically grow intermittently, lying dry much of the time, then receiving considerable water and sediment during storms. Some fans consist almost entirely of fluvial ma-

terial, and others consist mostly of debris and mudflows composed of poorly sorted cobbles and boulders in a fine-grained matrix (Figure 6–24C).

Fan growth proceeds intermittently as deposition shifts from one part of a fan to another. As one part of a fan builds up by deposition, its surface is raised higher than other parts of the fan, so that periodically water diverts to the lower areas and builds them up.

The slope of fan surfaces is governed by load-discharge relationships; thus, slopes are typically steepest near the apex of the fan, declining toward the fan margins as coarseness of the load diminishes. The longitudinal profile of fans is concave upward from apex to margin, but individual channel segments may be relatively straight with progressively lower slopes downfan.

The size of an alluvial fan depends on the following:

1. Area of the drainage basin
2. Climate
3. Lithology of rocks in the source area
4. Tectonic activity
5. Space available for fan growth

The area of a fan is empirically related to the area of the drainage basin according to the following equation:

A.

B.

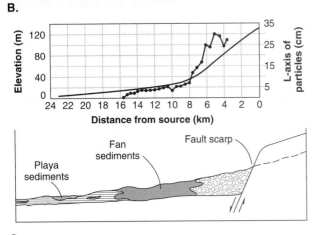

C.

FIGURE 6–24
(A) Alluvial fan, southeastern California. (Photo by W. K. Hamblin)
(B) Ground profile of an alluvial fan. (Photo by W. K. Hamblin)
(C) Relationship of sediment size to slope on an alluvial fan. (From Boothroyd and Ashley, 1975)

$$A_f = c(A_d)^n$$

where A_f = fan area
A_d = drainage basin area
c = constant varying with rock type and confinement
n = slope of regression line

Climate plays an important role in fan size. It determines the discharge of streams, the frequency of storm surges, and the weathering characteristics. The lithology of rocks in the source area determines the rate and character of sediment supply. Coarser sediment loads build steeper fans, and large sediment supplies build bigger fans.

Tectonic activity affects channel slopes in the drainage basin and thus load supplied. In Death Valley, California, eastward tilting of the valley resulted in larger fans on the west side than on the east side.

Fanhead Trenching

Many fans are incised by channels that cross them, especially near the heads of fans where fanhead trenching occurs (Figure 6–25). The depth of the trenches may reach 5 to 10 m (16 to 33 ft) below the fan surface, and the floors of the channels generally have lower gradients than fan surface, so that they are deeper near the fanhead than downfan. Some fanhead trenching is of a permanent nature and is related to climatic or tectonic changes, but others appear to be ephemeral, and the fan may undergo alternate trenching and filling.

Trenching and filling of fans may be caused by lateral swinging of the main stream to which the fan is graded. When the main stream swings to the fan side of the valley, the fan regrades by trenching, and the fan may be truncated in its distal portions by the main stream (Figure 6–26B). However, when the main stream swings to the far side of the valley, the

FIGURE 6–25
Fanhead trenching, southeastern California. (Photo by W. K. Hamblin)

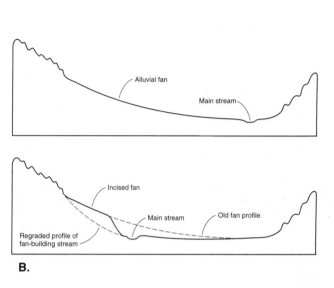

B.

FIGURE 6–26
(A) Truncated distal alluvial fan, Idaho. (Photo by D A. Rahm)
(B) Truncation of distal alluvial fan as a result of lateral swinging of the main stream to which the fan is graded. Above, initial fan profile; below, fan profile regarded to the closer main stream.

fan extends downstream and fills earlier trenches near the apex to provide the necessary gradient. Thus, fanhead trenches do not necessarily signal uplift or climatic changes.

DELTAS

Deltas are built by accumulation of sediment deposited by streams discharging into standing water. Herodotus first applied the name delta to sediments at the mouth of the Nile River because the triangular shape of the deposits resembled the Greek letter *delta*. However, other deltas have a variety of shapes. Deltas differ from fans in that they are built into bodies of water, whereas fans are constructed on land.

The distribution of water and sediment discharged into a body of water by a stream is determined largely by differences in density between the water of stream relative to that of the standing water, by the sediment load, and by the vigor of wave and current action in the lake or ocean. Multiple combinations of densities of water and water/sediment are possible. In general, fresh water is less dense than salt water, warm water is less dense than cold water, and clear water is less dense than muddy water. Thus, fresh water discharged by a stream into the sea will usually ride over the salt water because of its lower density, resulting in plumes of muddy water (Figure 6–27) extending for considerable distances from the mouth of the stream. However, if the stream water is unusually cold or has a high suspended load, it may be denser than the water it is flowing into, and it will quickly sink to the bot-

tom, leaving only a small sediment plume visible on top. The denser muddy water may flow for long distances along the bottom as a **turbidity current.** Such currents of muddy water occur in Lake Mead, the reservoir behind Hoover Dam in Nevada, where the cold, very muddy water of the Colorado River quickly disappears as it enters the upper end of the reservoir and flows as a turbidity current on the lake floor for many kilometers. Similar conditions occur where the Rhone River flows into Lake Geneva in Switzerland.

Fresh water flowing into the sea presents a different scenario. The fresh water is less dense than salt water and typically flows on top of the salt water, except when it is very muddy or very cold relative to the salt water, in which case it sinks to the sea floor and may result in turbidity currents. In addition, clay particles held in suspension by turbulence in fresh stream water are likely to flocculate in salt water, forming aggregates that become too large to remain in suspension.

When a stream discharges into standing water, its flow velocity is checked, causing deposition of sediment and extension of the delta. As the delta front progrades into the standing water, the surface of the delta is gradually raised by deposition of topset beds to provide the stream sufficient gradient to flow to the water's edge. Deposition in the main channel causes repeated splitting of the flow, and channels are repeatedly divided into distributary channels (Figures 6–27 and 6–28). However, most of the sediment load of the discharging stream is deposited on the delta front as foreset beds (Figures 6–29 and 6–30). Fine-grained silt and clay may be carried in sus-

FIGURE 6–27
Distributary channels on the delta of the Caetani River, Alaska. Note the sediment plume off the mouth of the delta. (Photo by U.S. Geological Survey)

pension beyond the delta front and may eventually settle out on the sea or lake floor as bottomset beds. However, delta forms may vary from this basic model, depending on the rate of sediment influx relative to the wave and current energy of the sea or lake. Much of the sediment influx into high-energy coastlines is distributed laterally along the shoreline by wave action.

The shape of a delta is determined largely by the rate of sediment influx relative to wave and current energy. Triangular-shaped deltas with broad, arcuate, seaward margins, such as the Nile Delta, are common forms, but many other types also occur. Some, such as the Mississippi Delta, are characterized by long extensions of distributary channels and are described as **bird-foot deltas** (Figure 6–31; Plate 4–E). Sediment discharged into high-energy shorelines is reworked by wave activity, which distributes the influx of sediment laterally along the shoreline to form smooth cuspate or arcuate shorelines with concentric bars and beach ridges (Figure 6–31). Where tidal currents are strong, tidal inlets and islands produce estuarine deltas (Figure 6–31).

THE CYCLE OF EROSION

Near the turn of the century, Gilbert, Dutton, and Powell formulated the idea that, over long periods of time, landforms pass through evolutionary stages of erosion and develop a systematic progression of landforms during erosion and downwasting of an elevated landmass. In conceiving a classification of stream systems based on their origin and relationship to geologic structure, Powell (1875) discerned that if fluvial erosion were to continue for a long enough period of time without renewed crustal uplift, the land would be reduced to a low plain only slightly above sea level. Dutton (1882) recognized a former period of "great denudation" in the Colorado Plateau of Utah and Arizona that had eroded the land to a plain of low relief prior to the present canyon-cutting phase. Dutton concluded:

> "All regions are tending to baselevels of erosion, and if the time be long enough, each region will, in its turn, approach nearer and nearer, and at last sensibly reach it." (Dutton, 1882)

Gilbert perceived the relationships between slope, discharge, and load in streams, and he noted the significance of lateral planation and stream-divide migration during fluvial erosion.

Davis (1909) integrated many of these ideas into the concept of the **cycle of erosion** in an attempt to devise a rational, systematic, genetic description of stream-sculptured landscapes, which would not only describe the present landforms but also present a clear scenario of the causative events that led to them. For many years, Davis's concept formed the basis for interpreting landforms, although not without controversy. In recent years, geomorphology has moved away from descriptive classifications, emphasizing instead the importance of understanding modern surficial processes. With much better understanding of tectonic processes, sea-level fluctuations, and surficial processes, less emphasis is now placed directly on the concept of the cycle of erosion. However, its underpinnings are implicit in

FIGURE 6–28
Two generations of delta growth. A now inactive delta lies to the right of the presently active delta. Note the distributaries on the active delta and the sediment plume offshore. (Her majesty the Queen in right of Canada, reproduced from the collection of the National Air Photo Library with the permission of Natural Resources Canada.)

FIGURE 6–29
Small delta built by stream flowing into standing water. Note the foreset beds forming beneath the water surface.

FIGURE 6–30
Foreset bedding in Pleistocene lake delta, Voges Mts. near Mellisy, France.

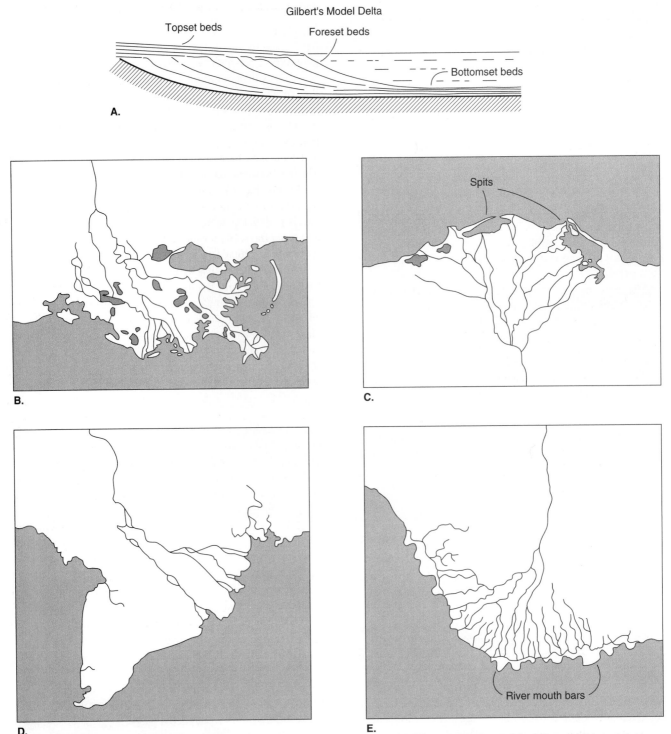

FIGURE 6–31

Characteristic delta forms: (A) delta cross section proposed by Gilbert; (B) Mississippi bird-foot delta; (C) Nile Delta whose distal margin is strongly influenced by wave processes; (D) Mekong Delta with many tidal tributaries; (E) Niger Delta made by a combination of fluvial deposition, wave processes, and tidal channels. (Modified from Hamblin, 1991)

understanding fluvial landforms. For example, without an appreciation of it, proper interpretation of fluvial terraces would be impossible.

Davis introduced the cycle of erosion with an ideal case in which rapid tectonic uplift was followed by stillstand of the land, allowing streams and valley-side processes to reduce the surface to a peneplain close to base level. The sequence of erosional events was depicted as irreversible, passing consecutively through evolutionary stages of development. Of the many possible categorizations of the continuum of stages, Davis chose three: youth, maturity, and old age, each of which was recognizable by its assemblage of landforms. Although Davis was himself well aware of a host of deviations from the ideal case, he emphasized the inevitability of old age to such an extent that critics assumed he recognized no other possibility. However, Davis clearly considered interruptions of the cycle to be far more common than completion.

The fundamental limitation to erosion on land, **base level**, is an extremely important consideration of the cycle of erosion. It had earlier been defined by Powell, the first person to navigate the Grand Canyon of the Colorado River:

"We may consider the level of the sea to be a grand base level, below which the dry lands cannot be eroded; but we may also have, for local and temporary purposes, other base levels of erosion, which are the levels of the beds of the principal streams which carry away the products of erosion . . . an imaginary surface, inclining slightly in all its parts toward the lower end of the principal streams draining the area." (Powell, 1875)

Davis adopted Powell's interpretation of base level and concluded the following:

1. Sea level is the ultimate base level for subaerial erosion.

2. Streams will always have at least some slope, however gentle, so that even when stream erosion has practically ceased, the land will always have a gentle slope.

3. Local, temporary base levels, such as inland lakes, will control points upstream from them.

However, at this time, neither the instability of tectonic processes nor the fluctuations of sea level were well known. Chamberlin (1930) showed that sea level could not be exactly still relative to land masses because of displacements of seawater by sediment and changes in ocean volume. He calculated that complete erosion of the continents with deposition of the sediment in the oceans would cause the sea level to rise about 670 ft (200 m), displacing low-level shorelines far inland. Chamberlin also estimated that melting of existing glaciers would raise sea level by about 80 ft (~25 m). (We now know that sea levels fluctuated through a range of about 300–450 ft (100–150 m) during Pleistocene glacial periods.) Sea level is thus a rather shifty entity, never in the same place for very long, and complicating long-term adjustments of streams to base level.

Sequence of Forms in the Cycle of Erosion

Landforms evolve with time through a sequence of forms that have characteristic features at successive stages, and although geomorphic research has generally moved away from landscape classification, an understanding of the ramifications of the concept of the cycle of erosion remains important.

Davis originally defined the stages of the cycle of erosion in terms of **youth, maturity, and old age** as a pedagogic tool. However, the evolution of landforms is a continuous process and could have been divided into any number of subdivisions. The line between stages is based on specific characteristics of landforms, but deciding where one stage ends and another begins is often somewhat arbitrary. The objective of this section is to develop an appreciation for the evolution of forms and to set the scene for the following sections on fluvial terraces and erosional surfaces.

As fluvial landscapes evolve with time, they pass through a sequence of forms. Beginning with newly initiated drainage on a consequent surface, stream systems are not well integrated, and lakes and swamps may be common. As streams vigorously cut their channels downward, drainage begins to become better integrated, and lakes and swamps disappear. Because streams have high gradients at this point, valley deepening predominates, cutting steep-sided canyons and V-shaped valleys. Rates of lateral stream erosion and valley widening are low relative to rates of incision, and the width of the channel may make up the entire valley floor (Figure 6–32). Erosional processes are not yet adjusted to rock resistance; thus, rapids and waterfalls are common (Figure 6–33; Plate 2–B), and large portions of the drainage system remain ungraded. Streams flow in irregular, steep-sided, V-shaped valleys, winding among spur ridges and cascading over irregularities in their channels.

Mass wasting and erosional processes have not yet attained equilibrium with rock resistance. Thus, just as the profiles of stream channels are irregular, so also are the valley sides with craggy outcrops of rock making cliffs. Headward erosion of tributaries extends the drainage system into the poorly defined divides and uplands.

As erosion proceeds, streams adjust their channels to load and discharge conditions; irregularities in stream profiles caused by differences in rock resistance become smoothed over; rapids and waterfalls are eliminated; and the profile becomes graded. At this point, the rate of downcutting, relative to lateral cutting, diminishes. The graded streams, although still cutting down, stabilize longer at any given level, allowing valley sides to begin to widen out and permitting lateral cutting on the outside of meanders to begin playing a more important role (Figure 6–34), widening floodplains as meanders cut more frequently into the valley sides. As streams meander more freely, floodplains widen (Figure 6–34), and the cross-valley profile loses its V-shape.

Concomitantly, drainage divides in the uplands become better defined, and a much greater proportion of the landscape consists of valley sides. Most of the original upland has

FIGURE 6–32
Vertical-walled canyon made by a stream early in the cycle of erosion when downcutting is dominant. (Photo by W. K. Hamblin)

FIGURE 6–33
Falls and rapids in V-shaped valley, Grand Canyon of the Yellowstone River, Yellowstone National Park, Wyoming. (Photo by W. K. Hamblin)

disappeared by now; the amount of the landscape consisting of divides has become minimal; and the divides consist mostly of sharp ridge crests (Figure 6–35). The adjustment of valley slopes provides an efficient system for transporting weathered material from valley sides and conducting the material to stream channels down smoothly linked profiles. As the divides are lowered by erosion, the caliber of debris supplied to the streams by mass wasting of the slopes gradually decreases. The decreased caliber of load reduces velocity requirements of the streams, and the streams respond by lowering their slopes. All of these changes are continuous and merge imperceptibly with one another.

The latter stages in the cycle of erosion are characterized by broad floodplains, wider than the meander belts. The valley sides are visited only infrequently by the slowly shifting channels of the streams, and **oxbow lakes** (Plate 2D) and swamps are numerous on the broad floodplains.

Adjustment of slopes reaches all the way to stream divides, and as the slopes diminish in steepness, they become mantled with an increasingly thick blanket of weathered soil. The thick soil mantle swallows more runoff, and stream divides become more and more broadly rounded and subdued until they stand only slightly above gently sloping valley sides. Uninterrupted continuation of these processes wears the landscape down to a plain of very low relief as the region ap-

FIGURE 6–34
Widening of a floodplain by lateral erosion of the valley sides. Note the cusp-shaped notches cut into the valley side by meanders of Ashley Creek, Utah. (Photo by D. A. Rahm)

INITIAL STAGE

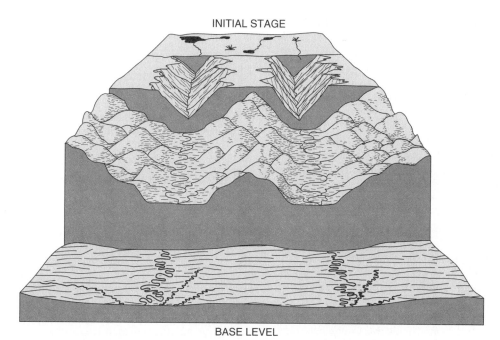

BASE LEVEL

A.

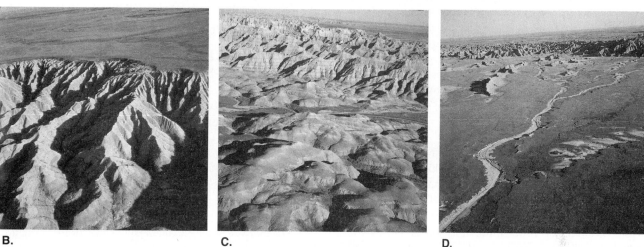

B.

C.

D.

FIGURE 6–35
(A) Characteristic features of landform evolution at Badlands National Monument, South Dakota.
(B) Early stage of landform development dominated by headward erosion and vertical incision of streams.
(C) Mature phase of landform evolution. Most of the land surface consists of valley slopes with very little surface area composed of stream divides. The slopes of the valley sides in the foreground have declined from the previous phase (background) and the divides have become rounded.
(D) Erosion has further reduced relief and developed a gently sloping plain at a new base level.

proaches base level. This phase would be equivalent to the old age stage of Davis, and the low-lying landscape produced would be a peneplain. The continuum of landscape evolution can be interrupted at any time and rarely proceeds to a peneplain stage because of tectonic activity.

These progressive changes with time can be observed on a miniature scale in Badlands National Monument, South Dakota (Figure 6–35). The initial stage is shown in Figure 6–35B, where headward erosion by streams tributary to the White River eat into the broad surface of poorly defined divides on

the High Plains. A later stage is seen in Figure 6–35C, where divides are well developed and most of the land surface consists of valley slopes. A still later stage is shown in Figure 6–35(D), where a broad, gently sloping plain close to base level (the White River) has developed by extensive erosion.

The end product of the cycle of erosion, **the peneplain** (literally, "almost a plain") has always been enigmatic because modern examples were difficult to find. Although the earth's surface includes many low-lying, nearly featureless plains, few can be unequivocally shown to be the result of long-continued fluvial erosion. Considering our present knowledge of plate tectonics and the restlessness of the earth's crust, this is not particularly surprising. Davis envisioned a peneplain as follows:

> "As old age comes on, the mounts and hills, already well subdivided and subdued, are reduced in area and are worn down to so moderate a relief that no sharp line of demarcation can be drawn between their slopes and the margin of the ever-widening and slowly lowering valley floors; but a zone of transition may be recognized where the convex profiles of the residual hills gradually become concave as they are continued down to the broad valley floors, which slant very faintly toward their streams. . . . Degradation may approach completion but a small difference of altitude must long remain between broad hill arches and broad valley floors, even though it is a diminishing difference. The main divides between the larger, opposing rivers of the region will be reduced to low and broadly convex swells, delicately diversified by wide-open and shallow valleys heads of branching streams. Minor divides will be of similar but less pronounced form. If a peneplain, thus worn down, extends a thousand km from its ocean border into a continental interior, its old rivers with a fall of 1 or 2 feet to a mile will not reduce its interior parts below an altitude of 1000 or 2000 feet; hence while a peneplain is a surface of low relief, it is not necessarily a lowland over all its extent.
>
> It is today extremely rare to find a peneplain, much less a plain of degradation, in a humid region still holding the altitude with respect to base level in which it was degraded; but if any such peneplains were found, we may be sure that they would have rock basements underlying their well-developed soil cover."
> (Davis, 1930)

With modern peneplains so rare, geomorphologists in the 1930s turned their attention to ancient erosional surfaces preserved as accordant summit levels in the landscape. Such remnant surfaces could possibly provide a datum for determining the timing of crustal uplift. The recognition of uplifted erosional surfaces as criteria for the age of mountain ranges was indeed tempting, and geomorphologists began wholesale identification of uplifted peneplains in many mountain ranges around the world. The result was development of general skepticism. However, some widespread low-relief surfaces cut across rocks of varying resistance in high mountainous areas, like the Rocky Mountains have quite clearly been sites of removal of thousands of meters of rock by erosion (Plate 4C). The evolution of many such surfaces remains controversial today, with avid proponents and opponents of cyclic erosion. Recognition of remnant cyclic erosional surfaces of any extent is complicated by development of surfaces that appear very similar to fluvial erosional surfaces by noncyclic processes. Among the most common are the following:

1. Stripped structural surfaces that owe their existence to stripping of weak rock from highly resistant rocks, often leaving a flat surface controlled not by erosion but by rock resistance

2. Marine abrasion surfaces

3. Various types of depositional surfaces, such as alluvial plains

4. Constructional surfaces not made by fluvial processes, such as lava

5. Exhumed surfaces of diverse origin

6. Altiplanation surfaces, developed at high altitudes by freeze-thaw processes

These surfaces are discussed further in a following section on stream terraces and erosional surfaces.

Interruptions to the Cycle of Erosion

The cycle of erosion may be interrupted at any time, and the graded surfaces may be incised by the streams that made them (Figure 6–36). Meanders that develop on a cyclic surface (Plate 4–A) may become deeply incised while retaining their sinuous loops (Figure 6–37, 6–38; Plate 4–B). Although the dominant mode of erosion in such cases is vertical incision, lateral erosion on the outside of meanders still occurs. Where an incised stream has a very sinuous course, the meanders enclose **meander cores.** If lateral erosion of the meander core cuts through the core, an arch (Figure 6–39) may be produced when a meander cutoff occurs.

CYCLIC STREAM TERRACES AND EROSIONAL SURFACES

Cyclic stream terraces and erosional surfaces are caused by erosion (or, in some cases, deposition) of stream channels that were once valley floors. They are remnants of previously formed valley floors that now stand above the active channels as a result of incision of the stream that produced them (Figure 6–40). Their slope represents the gradient of the stream that made them, and because they are graded surfaces of transportation, they truncate rocks of varying resistance (Figure 6–41; Plate 4–C).

The incision of the channel responsible for the formation of a fluvial terrace constitutes an interruption in the cycle of erosion—hence the term **cyclic stream terrace.** The causes of renewed downcutting of a stream may be:

1. Change of base level, typically eustatic sea-level change

2. Tectonic uplift

3. Climatic change that affects load-discharge relationships

Base-level changes may consist of eustatic sea-level changes or local base-level changes. **Eustatic sea level** is a raising or lowering of the level of the oceans as a result of withdrawal of large amounts of water from the sea during the ice ages or as a result of tectonic events that alter the volume of the ocean basins. Lowering of eustatic sea level leaves streams higher above base level than previously, and the increased velocity causes them to incise their channels to a new lower level, leaving their old floodplains high and dry along the valley sides as cyclic stream terraces. Eusta-

FIGURE 6–37
Stream channel incising into its former floodplain. (Photo by
W. K. Hamblin)

FIGURE 6–36
Stream beginning a new cycle of erosion as it incises its
channel into its previously formed floodplain. (Photo by D.
A. Rahm)

FIGURE 6–38
Incised meanders of the Green
River, Utah. The meanders were
inherited from an earlier cycle of
erosion. (Photo by W. K.
Hamblin)

FIGURE 6–39
Rainbow Bridge, Utah, formed by lateral erosion of a meander through the meander core of an incised stream. (Photo by D. A. Rahm)

FIGURE 6–40
Cut-in-bedrock, cyclic stream terrace, Huranui River, New Zealand.

FIGURE 6–41
YU-Emblem Benchs, cut in bedrock, cyclic stream terraces of the Greybull River, Wyoming (see also Figure 6–45). Note the truncation of the tilted strata by the upper terrace surface.

tic rise of sea level produces the opposite effect, drowning the lower portions of stream valleys and resulting in aggradation of streams to adjust to the new base level.

Tectonic uplift of the land increases the gradient of streams, just as a lowering of base level does, and the result is the same—incision of streams into their floodplains, leaving the former floodplains above the new channels. Tectonic downwarping of the land causes streams to aggrade their channels to adjust their gradients to an appropriate slope.

Climatic changes can cause streams to either incise or aggrade their channels by altering load-discharge relationships. If the caliber of load supplied to a stream changes, as, for example, in the case of a glacier entering the headwater of a stream and contributing large amounts of coarse detritus, the original velocity of the stream is no longer adequate to transport the new coarser material. Thus, the material accumulates on the floor of the channel, thereby steepening the gradient and producing an increase in velocity. The steepening effect continues until just the velocity that is needed to transport the load is attained. Climatic effects that increase the discharge of a drainage system generate just the opposite effect. With increased discharge, streams can generate the velocity needed to transport the load on a lower slope; that is, the greater discharge temporarily increases velocity, which scours the channel floor, lowering the channel gradient until the velocity becomes just adequate to carry the load. Thus, the effects of increased discharge are opposite those of an increased load, the former causing downcutting of streams and the latter causing aggradation. In some cases, both load and discharge may increase as a result of climatic changes. Then the effects may offset one another, or one may predominate over the other.

Lateral planation of meanders as they impinge against valley sides widens the valley floor, and the longer a stream erodes laterally at a given level, the wider the floodplain becomes. Incision of a stream into its floodplain may be caused by any of the changes just described, leaving remnants of the former valley floor as cyclic terraces above the new channel, gently inclined downvalley with the slope of the former stream gradient (Figures 6–42 and 6–43).

Types of Cyclic Terraces

The type of **cyclic stream terrace** developed depends on the nature and origin of the former floodplain that it represents; that is, it depends on whether the terrace surface was creat-

FIGURE 6–42
Cyclic stream terrace along Ashley Creek, Wyoming. (Photo by D. A. Rahm)

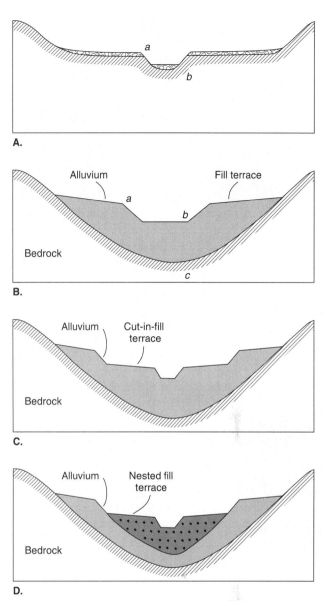

A.

B.

C.

D.

FIGURE 6–44
Types of stream terraces: (A) cut-in-bedrock terrace, (B) fill terrace, (C) cut-in-fill terrace, (D) nested fill terrace.

FIGURE 6–43
Multiple cyclic stream terraces, Pahsimeroi, Idaho. (Photo by W. K. Hamblin)

ed by stream erosion, by deposition, or by a combination of the two (Figure 6–44). Proper identification of the type of a particular cyclic stream terrace is essential to the interpretation of the sequence of events leading to the terrace. As will be seen in the discussion of terrace types that follows, each type of terrace owns a geomorphic history quite different from all others.

Cut-in-Bedrock Terraces

Floodplains carved by graded streams across rocks of differing resistance are floored with rock, mantled with a thin veneer of alluvium whose thickness does not exceed the depth of scour of the stream. Thus, when renewed incision of the stream channel leaves them as remnants above the active channel, the terraces consist of rock thinly veneered with alluvium (Figures 6–40, 6–41, and 6–44A). These terraces, known as **cut-in-bedrock terraces,** have the simplest geomorphic history of any of the terrace types.

Fill Terraces

Fill terraces are remnants of former valley floors that have been constructed by aggradation. Valleys are first filled with alluvium during aggradation; then incision of the stream channel into the fill follows, leaving terraces composed entirely of alluvium. The terrace surface in this instance is depositional in origin, in contrast to the cut-in-bedrock, which is erosional in origin. Fill terraces may have the same surface form as cut terraces and may have similar gradients, but they differ significantly in their geomorphic history. A cut terrace implies a period of floodplain development at a particular level, followed by incision of the channel (Figure 6–44B). On the other hand, a fill terrace implies downcutting, then aggradation to fill the valley, and finally renewed downcutting to leave the fill surface above the active

channel. Thus, distinguishing between these types of terraces is essential to proper interpretation of their geomorphic history.

Cut-in-Fill Terraces

Cut-in-fill terraces are remnants of former valley floors that have been cut in alluvium, followed by channel incision. They differ from fill terraces in that their surface is erosional in origin, whereas fill terrace surfaces are depositional in origin. Figure 6–44C illustrates the difference. A valley is first cut to level (*a*), followed by filling of the valley to level (*b*). The valley fill is then incised to level (*c*), and a new floodplain is widened out, followed by renewed incision to level (*d*), leaving the floodplain, which had been cut in fill, as a terrace. Thus, the terrace

FIGURE 6–45
Nested fill terrace, Rakaia River, New Zealand.

surface at *(c)* is erosional, and although it is somewhat analogous to a cut-in-bedrock terrace, it differs in being cut in alluvium rather than rock. Note that the highest terrace, level *(b)*, is a fill terrace because the origin of its surface was depositional.

Nested Fill Terraces

Nested fill terraces consist of fill terraces successively inset within one another (Figure 6–44D and 6–45). They are depositional in origin but are separated by periods of channel incision. For example, in Figure 6–44D, the valley is first cut down *to level (a),* then filled to level *(b),* followed by downcutting to level *(c),* filling back up to level (d), then downcutting *to level (e).* The sequence of terraces thus may resemble the cut-in-fill terraces in Figure 6–44C, but in this instance, all of the surfaces are depositional, rather than erosional, in origin, and the geomorphic history is considerably more complex.

Multiple terrace levels along a valley side may be highly complex because any given terrace surface may be of any of the four origins described above, and any combination of terrace origins is possible.

Correlation of Cyclic Erosional Surfaces

Cyclic erosional surfaces are relicts of former floodplains, so an important part of working with them is to correlate from remnant to remnant. If the remnants are fairly continuous, correlation may not be difficult. However, if remnants are widely scattered, or if correlations with terraces in different drainage basins are desired, then the longitudinal profile must be reconstructed by piecing together all of the terrace remnants and projecting them downvalley to the controlling base level. Once a longitudinal profile has been developed for a main trunk valley, terrace remnants in tributary valleys can be projected to establish correlation between terrace remnants in different streams of a drainage system. For

example, Figure 6–46 shows longitudinal profiles drawn from terrace remnants along the Greybull River in the Bighorn Basin of Wyoming (Mackin, 1937). Terrace remnants at Emblem Bench, YU Bench, Table Mountain, and Tatman Mountain cut across the upturned edges of interbedded sandstone and shale. Projection of the longitudinal profile, from YU Bench downvalley at a gradient of about 60 ft/mi. (11 m/km) clearly shows that Table Mountain, about 10 km (6 mi.) downstream, is a remnant of the same floodplain, and both are older than terrace remnants at the Emblem Bench level.

Cyclic surfaces are not restricted to narrow benches along stream valleys but may also include broad surfaces of low relief. When such surfaces are elevated by tectonic forces, remnants may persist as testimony to the former cycles of erosion (Figures 6–47 and 6–48; Plate 4–C).

NONCYCLIC SURFACES

Stripped Structural Surfaces

Stripped structural surfaces are formed by selective stripping of low-resistance rocks from high-resistance rocks, leaving a surface of low relief that may superficially resemble a cyclic surface. They are best-developed on resistant, nearly horizontal beds that temporarily act as a local base level for stripping of overlying softer beds. Erosion, effective in the soft beds, is retarded by the resistant layer, which stands in sharp relief by the etching out of the weaker strata. Figure 6–48 shows an example of a stripped structural surface adjacent to the Grand Canyon. Here, the Kaibab Plateau, an extensive, broad, nearly flat plain stretches for many kilometers in all directions; it developed on the upper surface of the resistant Permian Kaibab limestone. Even though several thousand meters of overlying weaker Mesozoic beds have been removed by erosion, the Kaibab Plateau is not a cyclic erosional surface because its origin is due to rock resistance rather than to long-continued erosion.

An important difference between stripped structural surfaces and cyclic erosional surfaces is that stripped structural surfaces correspond directly to the dip of the structure (Figure 6–49), whereas cyclic erosional surfaces truncate rocks of varying resistance (Figures 6–47, and Plate 4C).

The slope of a cyclic surface is the gradient of the stream that made it, but the slope of stripped structural surfaces is determined by the dip of the resistant bed that composes it. The dip of a resistant bed is unlikely to slope in the same direction as present streams; nor is it likely to be concave upward, as is the slope of a graded stream channel. Thus, the slope of a stripped structural surface is useful for identification of the origin of the surface.

Because of the importance of distinguishing stripped structural surfaces from cyclic erosional surfaces, several basic criteria are useful to keep in mind:

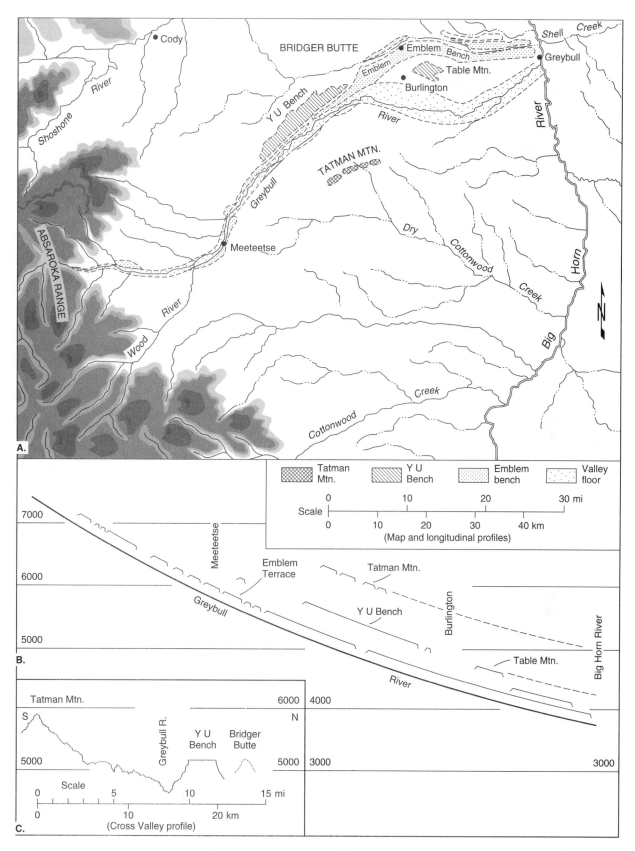

FIGURE 6–46
Correlation of cyclic terraces by reconstructing stream profiles. Each terrace profile is plotted against the present Greybull River profile and correlations can be made by projecting each terrace remnant downvalley. (From Mackin, 1937)

FIGURE 6–47
High altitude cyclic erosional surface cut across geologic structures and rocks of varying resistance, crest of the Rocky Mt. Front Range, Colorado. (Photo by D. A. Rahm)

1. *Truncation of geologic structure.* The slope of graded surfaces is cut across weak and resistant geologic units alike. Thus, if a surface truncates the upturned edges of rock units (Figure 6–47), it cannot be a stripped structural surface.

2. *Gradient of the surface.* Because the upper surface of a stripped structural surface conforms to the dip of the rock unit, the probability that it will parallel the concave-upward curve of the gradient of graded streams is extremely remote. For example, the Tonto Platform makes a flat bench along the Colorado River in the Grand Canyon (Figure 6–50) in much the same way as typical cyclic stream terraces do. However, the surface of the bench slopes gently upstream, corresponding to the dip of the resistant Tapeats sandstone.

3. *Shape of the slope of the surface.* Even if the surface of a resistant bed slopes downstream, the likelihood that the dip of the bed would change downdip at a rate exactly the same as that of a graded stream is highly remote.

4. *Cross-sectional shape.* The outcrop of a resistant bed making up a stripped structural surface is typically steep to vertical, conforming to the edge of the resistant bed (Figure 6–51). In contrast, the cross section of a cyclic stream terrace is likely to be more uniform.

FIGURE 6–48
Stripped structural surface on Kaibab limestone, Marble Canyon, Arizona. The broad bench along the Colorado River owes its existence solely to rock resistance so it is not a cyclic surface. (Photo by D. A. Rahm)

FIGURE 6–49
Stripped structural surface on resistant sandstone, Canyonlands, Utah. (Photo by W. K. Hamblin)

FIGURE 6–50
The Tonto platform, making the bench above the Colorado River in the Grand Canyon, is a noncyclic stripped structural surface developed on the resistant Tapeats sandstone.

Marine Erosional Surfaces

The action of ocean waves on an exposed coast has been compared to cutting with a horizontal saw. Waves cut into the land, undermining sea cliffs and extending an ever-expanding abrasion platform landward. If sea level remains stable relative to the land, the width of such an abrasion platform is lim-

ited because the wave energy is dissipated as the width of the platform grows and the ability of the waves to attack and remove material from sea cliffs declines. However, if sea level slowly rises, or if the land slowly sinks, extensive abrasion platforms may be cut, as suggested by the great marine transgressions of the stratigraphic record where truncated older rocks are unconformably overlain by marine conglomerate and sandstone over a wide area. Admittedly, the sea operates along a narrow front line of shore erosion, whereas fluvial erosion covers vast areas; however, under suitable conditions, marine erosion may account for the leveling of land masses. An example is the island of Helgoland, off the Atlantic Coast of Germany, which was reduced in area by marine erosion from about 300 mi^2 (780 km^2) in A.D. 800 to about 0.25 mi^2 (0.6 km^2) at present (von Engeln, 1942). Other examples of marine erosion are recorded in England and parts of coastal Europe where large areas, now being dissected by fluvial erosion, were thought to have been planed off by marine erosion. Ramsay made a clear statement of this concept as early as 1846 in his description of the truncation of folded strata in Wales. Perhaps because of the insular position of Great Britain, the interpretation of many erosional surfaces as marine planation surfaces has persisted in British geomorphology. However, the question always remains, were these surfaces reduced to a plain largely by fluvial erosion, followed by application of a finishing touch by marine submergence, or are they plains of marine abrasion now emerged from the sea and touched up by fluvial processes?

Nonpaired Fluvial Terraces

Under certain conditions, stream terraces may be produced by a continuously downcutting stream without interruptions in the cycle of erosion. As a stream swings back and forth across its floodplain, it cuts steadily downward. Each time the stream returns to the opposite side of its valley, it is cutting laterally at a lower level than the last time it was there, leaving portions of the former floodplain, which escape total destruction as terraces. The result is a set of terraces along the valley sides, but these terraces do not match up on opposite sides of the valley.

The degree to which an alluvial fill escapes destruction during downcutting of a stream may be random, or it may be controlled by the uncovering of bedrock at a lower level, which protects the terrace above from undercutting by the stream, leaving rock-defended terraces. A somewhat similar condition may result from undercutting of tributary alluvial fans by lateral swinging of the main stream, leaving terrace-like remnants that resemble cyclic stream terraces but that have no cyclic significance.

FIGURE 6–51
Edge of the Kaibab Plateau, a noncyclic stripped structural surface on resistant Kaibab limestone, Grand Canyon, Arizona. (Photo by D. A. Rahm)

REFERENCES

Akagi, Y., 1980, Relations between rock type and the slope form in the Sonora Desert, Arizona: Zeitschrift für Geomorphologie, v. 24, p. 129–140.

Bates, C. C., 1953, Rational theory of delta formation: American Association of Petroleum Geologists, v. 37, p. 2119–2162.

Bauer, B., 1980, Drainage density—An integrative measure of the dynamics and the quality of watersheds: Zeitschrift für Geomorphologie, v. 24:263–72.

Baulig, H., 1957, Peneplains and pediplains: Geological Society of America Bulletin, v. 68, p. 913–930.

Beaty, C. B., 1963, Origin of alluvial fans, White Mts., California and Nevada: Annals of the Association of American Geographers, v. 53, p. 516–535.

Beaty, C. B., 1970, Age and estimated rate of accumulation of an alluvial fan, White Mountains, California, USA: American Journal of Science, v. 268, p. 50–77.

Begin, Z. B., Meyer, D. F., and Schumm, S. A., 1980, Knickpoint migration due to base-level lowering: American Society of Civil Engineers, Journal of Water, Port, Coastal and Ocean Division, v. 106, p. 369–387.

Begin, Z. B., Meyer, D. F., and Schumm, S. A., 1981, Development of longitudinal profiles of alluvial channels in response to base-level lowering: Earth Surface Processes, v. 6, p. 49–68.

Blackwelder, E., 1931, The lowering of playas by deflation: American Journal of Science, v. 21, p. 140–144.

Blair, T., 1987, Sedimentary processes, vertical stratification sequences, and geomorphology at the Roaring River alluvial fan: Rocky Mountain National Park, Colorado: Journal of Sedimentary Petrology, v. 57, p. 1–18.

Blissenbach, E., 1954, Geology of alluvial fans in semiarid regions: Geological Society of America Bulletin, v. 65, p. 175–190.

Bluck, B. J., 1964, Sedimentation on an alluvial fan in southern Nevada: Journal of Sedimentary Petrology, v. 34, p. 395–400.

Bocco, G., 1991, Gully erosion; processes and models: Progress in Physical Geography, v. 15, p. 392–406.

Boothroyd, J. C., and Ashley, G. M., 1975, Processes, bar morphology, and sedimentary structure on braided outwash fans, northeastern Gulf of Alaska: in Jopling, A. V., and MacDonald, B. C., eds., Glaciofluvial and glaciolacustrine sedimentation, Society of Economic Paleontologists and Mineologists, Special Paper 23, p. 193–222.

Boothroyd, J. C., and Nummedal, D., 1978, Proglacial braided outwash: a model for humid alluvial-fan deposits: in Miall, A. D., ed., Fluvial sedimentology, Memoirs of the Canadian Society of Petroleum Geologists, v. 5, p. 597–604.

Born, S. M., and Ritter, D. F., 1970, Modern terrace development near Pyramid Lake, Nevada, and its geologic implications: Geological Society of America Bulletin, v. 81, p. 1233–1242.

Bridge, J. S., and Jarvis, R. S., 1977, The alluvial fan environment: Progress in Physical Geography, v. 1, p. 222–270.

Brush, L. M., and Wolman, M. G., 1960, Knickpoint behavior in noncohesive material: A laboratory study: Geological Society of America Bulletin, v. 71, p. 57–76.

Bryan, K., 1922, Erosion and sedimentation in the Papago country Arizona with a sketch of the geology: U.S. Geological Survey Bulletin 730, p. 19–90.

Bryan, K., 1940, Gully gravure, a method of slope retreat: Journal of Geomorphology, v. 3, p. 89–107.

Bryan, R. B., 1979, The influence of slope angle on soil entrainment by sheetwash and rainsplash: Earth Surface Processes and Landforms, v. 4, p. 43–58.

Bull, W. B., 1962, Relations of alluvial fan size and slope to drainage basin size and lithology in western Fresno County, California: U.S. Geological Survey Professional Paper 450-B.

Bull, W. B., 1964, Geomorphology of segmented alluvial fans in western Fresno County, California: U.S. Geological Survey Professional Paper 352-E, p. 89–129.

Bull, W. B., 1968, Alluvial fans: Journal of Geological Education, v. 16, p. 101–106.

Bull, W. B., 1970, The alluvial-fan environment: Progress in Physical Geography, v. 1, p. 222–270.

Bull, W. B., 1991, Geomorphic responses to climatic change: Oxford University Press, N.Y., 326 p.

Bull, W. B., and Knuepfer, P., 1987, Adjustments by the Charwell River, New Zealand, to uplift and climatic changes: Geomorphology, v. 1, p. 15–32.

Carlston, C. W., 1963, Drainage density and streamflow: U. S. Geological Survey Professional Paper 422-C.

Chamberlin, R. C., 1930, The level of base-level: Journal of Geology, v. 38, p. 166–173.

Chang, H. H., 1982, Fluvial hydraulics of deltas and alluvial fans: American Society of Civil Engineers, Journal of Hydraulics Division, HYII, p. 1282–1295.

Chorley, R. J., and Morley, L. S. D., 1959, A simplified approximation for the hypsometric integral: Journal of Geology, v. 67, p. 566–571.

Coleman, J. M., and Wright, L. D., 1975, Modern river deltas: variability of processes and sand bodies: in Broussard, M. L., ed., Deltas, Houston Geological Society, p. 99–150.

Cooke, R. U., and Reeves, A., 1972, Relations between debris size and the slope of mountain fronts and pediments in the Mohave Desert, California: Zeitschrift für Geomorphologie, v. 16, p. 76–82.

Cooke, R. U., and Warren, A., 1973, Geomorphology in deserts: Batsford Ltd., London.

Cotton, C. A., 1945, Significance of terraces due to climatic oscillation: Geological Magazine, v. 82, p. 10–16.

Daniel, J. R. K., 1981, Drainage density as an index of climatic geomorphology: Journal of Hydrology, v. 50, p.147–54.

Davis, W. M., 1902a, Baselevel, grade, and peneplain: Journal of Geology, v. 10, p. 77–111.

Davis, W. M., 1902b, River terraces in New England: Bulletin of the Harvard College Museum of Comparative Zoology, v. 38, p. 281–346.

Davis, W. M., 1909, Geographical essays: Boston, Ginn and Co., (reprinted by Dover Publications, Inc., N. Y.).

Denney, C. S., 1965, Alluvial fans in the Death Valley Region, California and Nevada: U.S. Geological Survey Professional Paper 466, 62 p.

Denny, C. S., 1967, Fans and pediments: American Journal of Science, v. 265, p. 81–105.

Dury, G. H., 1964, Principles of underfit streams: U.S. Geol. Survey Prof. Paper 452-A.

Dutton, C. E., 1882, Tertiary history of the Grand Canyon district: U.S. Geological Survey Monograph 2, 264 p.

Elliott, T., 1978, Deltas: in Reading, H. G., ed., Sedimentary environments and fans, Oxford University Press, London, p. 97–145.

Fisk, H. N. et al., 1944, Geological investigations of the alluvial valley of the lower Mississippi River: U.S. Army Corps of Engineers, Mississippi River Commission, Vicksburg, Mississippi, 78 p.

Fisk, H. N., McFarlan, E., Kilband, C. R., and Wilbert, L. J., 1954, Sedimentary framework of the modern Mississippi delta: Journal of Sedimentary Petrology, v. 24, p. 76–99.

Frye, J. C., and Leonard, A. R., 1954, Some problems of alluvial terrace mapping: American Journal of Science, v. 252, p. 242–251.

Gardiner, V., 1990, Drainage basin morphometry: in Goudie, A. S., ed., Geomorphological Techniques, Unwin Hyman, London, p. 71–81.

Gardner, T. W., 1983, Experimental study of knickpoint and longitudinal profile evolution in cohesive, homogeneous material: Geological Society of America Bulletin, v. 94, p. 664–72.

Gilbert, G. K., 1877, Report on the geology of the Henry Mts., Utah: U.S. Geographical and Geological Survey, Rocky Mt. Region, U.S. Government Printing Office.

Gilluly, J., 1937, Physiography of the Ajo Region, Arizona: Geological Society of America Bulletin, v. 43, p. 323–348.

Gregory, D. L, and Schumm, S. A., 1987, The effect of active tectonics on alluvial river morphology: in Richards, K., eds., River Channels: Environment and Process, Basil Blackwell, London, p. 41–68.

Hadley, R. F., 1967, Pediments and pediment-forming processes: Journal of Geological Education, v. 15, p. 83–89.

Hall, S. A., 1990, Channel trenching and climatic change in the southern U.S. Great Plains: Geology, v. 18, p. 342–345.

Hamblin, W. K., 1995, Earth's dynamic systems: Prentice Hall, Englewood Cliffs, N.J., 710 p.

Hammad, H. Y., 1972, River-bed degradation after closure of dams: American Society of Civil Engineers, Journal of Hydraulics Division, v. 98, p. 591–607.

Happ, S. C., 1971, Genetic classification of valley sediment deposits: American Society of Civil Engineers, Journal of Hydraulics Division, v. 97, p. 43–53.

Hereford, R., 1984, Climate and ephemeral stream processes: Twentieth-century geomorphology and alluvial stratigraphy of the Little Colorado River, Arizona: Geological Society of America Bulletin, v. 95, p. 654–68.

Hereford, R., 1986, Modern alluvial history of the Paria River drainage basin, southern Utah: Quaternary Research, v. 25, p. 293–311.

Holmes, C. D., 1952, Stream competence and the graded stream profile: American Journal of Science, v. 250, p. 899–906.

Holmes, C. D., 1964, Equilibrium in humid-climate physiographic processes: American Journal of Science, v. 262, p. 436–445.

Hooke, R. L., 1967, Processes on arid-region alluvial fans: Journal of Geology, v. 75, p. 438–460.

Hooke, R. L., 1968, Steady state relationships on arid-region alluvial fans in closed basins: American Journal of Science, v. 266, p. 609–629.

Hooke, R. L., and Rohrer, W. I., 1979, Geometry of alluvial fans: effect of discharge and sediment size: Earth Surface Processes, v. 4, p. 147–166.

Horton, R. E., 1933, The role of infiltration in the hydrological cycle: American Geophysical Union Transactions, v. 14, p. 446–60.

Horton, R. E., 1945, Erosional development of streams and their drainage basins. Hydrophysical approach to quantitative morphology: Geological Society of America Bulletin, v. 56, p. 275–370.

Howard, A. D., 1967, Drainage analysis in geologic interpretation: A summation: American Association Petroleum Geologists Bulletin, v. 51, p. 2246–2259.

Johnson, D., 1932, Rock fans of arid regions: American Journal of Science, v. 23, p. 389–416.

Johnson, D., 1938, Stream profiles as evidence of eustatic changes of sea level: Journal of Geomorphology, v. 1, p. 176–181.

Johnson, D., 1944, Problems of terrace correlation: Geological Society of America Bulletin, v. 55, p. 793–818.

Kenyon, P., and Turcotte, D. 1985. Morphology of a delta prograding by bulk sediment transport: Geological Society of America Bulletin, v. 96, p. 1457–1465.

Kesel, R. H., 1973, Inselberg landform elements: definition and synthesis: Revue, Geomorphologie Dynamique, v. 22, p. 97–108.

Kesel, R. H., 1977, Some aspects of the geomorphology of inselbergs in central Arizona, U.S.A.: Zeitschrift für Geomorphologie, v. 21, p. 119–146.

Knighton, A., and Nanson, G., 1987, River channel adjustment: the downstream dimension: in Richards, K., River Channels: Environment and Process, Basil Blackwell Ltd., Oxford, England, p. 95–128.

Knox, J. C., 1972, Valley alluviation in southwestern Wisconsin: Annals of the Association of American Geographers, v. 62, p. 401–410.

Knuepfer, P., 1988, Estimating ages of late Quaternary stream terraces from analysis of weathering rinds and soils: Geological Society of America Bulletin, v. 100, p. 1224–1236.

Kolb, C. R., and Van Lopik, J. R., 1958, Geology of the Mississippi River deltaic plain, southeastern Louisiana: U.S. Army Corps of Engineers, Waterway Experiment Station Report, 483 p.

Lawson, A. C., 1915, The epigene profiles of the desert: University of California Department of Geology Bulletin 9, p. 23–48.

Leopold, L. B., and Bull, W. B., 1979, Base level aggradation and grade: Proceedings of the American Philosphical Society, v. 123, p. 168–202.

Mabbutt, J. A., 1966, Mantle-controlled planation of pediments: American Journal of Science, v. 264, p. 78–91.

Mackin, J. H., 1936, The capture of the Greybull River: American Journal of Science, v. 31, p. 373–385.

Mackin, J. H., 1937, Erosional history of the Big Horn Basin, Wyoming: Geological Society of America Bulletin, v. 48, p. 813–894.

Mackin, J. H., 1948, Concept of the graded river: Geological Society of America Bulletin, v. 59, p. 463–512.

Mammerickx, J., 1964, Quantitative observations on pediments in the Mojave and Sonoran deserts (southwestern United States): American Journal of Science, v. 262, p. 417–435.

Martin, C. W., 1992, Late Holocene alluvial chronology and climate change in the central Great Plains: Quaternary Research, v. 37, p. 315–322.

McGee, W. J., 1897, Sheetflood erosion: Geological Society of America Bulletin, v. 8, p. 87–112.

McPherson, J., Sharunugam, G., and Moiola, R., 1987, Fan-deltas and braid deltas: Varieties of coarse-grained deltas: Geological Society of America Bulletin, v. 99, p. 331–340.

Morisawa, M. E., 1964, Development of drainage systems on an upraised lake floor: American Journal of Science, v. 262, p. 340–354.

Morisawa, M. E., 1968, Streams, their dynamics and morphology: McGraw-Hill, N.Y., 175 p.

Moss, J. H., and Bonini, W., 1961, Seismic evidence supporting a new interpretation of the Cody terrace near Cody, Wyoming: Geological Society of America Bulletin, v. 72, p. 547–556.

Nanson, G. C., and Young, R. W., 1981, Overbank deposition and floodplain formation on small coastal streams of New South Wales: Zeitschrift für Geomorphologie, v. 25, p. 332–345.

Nilson, T., 1982, Alluvial fan deposits: *in* Scholle, P., and Spearing, P., Sandstone Depositional Environments, American Association of Petroleum Geologists Memoir 31, p. 49–86.

Oberlander, T. M., 1974, Landscape inheritance and the pediment problem in the Mojave Desert of southern California: American Journal of Science, v. 274, p. 849–875.

Paige, S., 1912, Rock-cut surfaces in the desert ranges: Journal of Geology, v. 20, p. 442–450.

Parker, R. S., 1976, Experimental study of drainage system evolution: Unpublished report, Colorado State University.

Patton, P. C., and Schumm, S. A., 1975, Gully erosion, northwestern Colorado: a threshold phenomenon: Geology, v. 3, p. 88–90.

Petts, G., and Foster, I., 1985, Rivers and landscape: Edward Arnold, London, 274 p.

Playfair, J., 1802, Illustrations of the Huttonian theory the earth: William Creech, Edinburgh.

Powell, J. W., 1875, Exploration of the Colorado River of the West and its tributaries: Smithsonian Institution, Washington, D.C.

Rachocki, A., 1981, Alluvial fans: John Wiley and Sons, N.Y.

Rahn, P. H., 1966, Inselbergs and nickpoints in southwestern Arizona: Zeitschrift für Geomorphologie, v. 10, p. 217–225.

Rahn, P. H., 1967, Sheetfloods, stream floods, and the formation of pediments: *Annals of the Association of American Geographers*, v. 57, p. 593–604.

Rich, J. L., 1935, Origin and evolution of rock fans and pediments: Geological Society of America Bulletin, v. 46, p. 999–1024.

Rich, J. L., 1938, Recognition and significance of multiple erosion surfaces: Geological Society of America Bulletin, v. 49, p. 1695–1721

Russell, R., 1967, River and delta morphology: Louisiana State University Coastal Studies Institute Technical Rept. 52.

Ryder, J. M., 1971, The stratigraphy and morphology of paraglacial alluvial fans in south-central British Columbia: Canadian Journal of Earth Sciences, v. 8, p. 279–298.

Schlemon, R. J., 1975, Subaqueous delta formation—Atchafalaya Bay, Louisiana: *in* Broussard, M. L., ed., Deltas, Houston Geological Society, p. 99–150.

Schumm, S. A., 1956, Evolution of drainage systems and slopes in badlands at Perth Amboy, New Jersey: Geological Society of America Bulletin, v. 67, p. 597–646.

Schumm, S. A., 1962, Erosion of miniature pediments in Badlands National Monument, South Dakota: Geological Society of America Bulletin, v. 73, p. 719–724.

Schumm, S. A., 1976, Episodic erosion: a modification of the geomorphic cycle: *in* Melhom, W., and Flemal, R., eds., Theories of landform development, State University of New York, Binghamton, N.Y., p. 68–85.

Schumm, S. A., 1977, The fluvial system: John Wiley and Sons, N.Y., 338 p.

Schumm, S. A., and Hadley, R. F., 1957, Arroyos and the semiarid cycle of erosion: American Journal of Science, v. 255, p. 161–174.

Schumm, S. A., and Khan, H. R., 1972, Experimental study of channel patterns:

Geological Society of America Bulletin, v. 83, p. 1755–1770.

Scott, A. I., and Fisher, W. L., 1969, Delta systems and deltaic deposition: *in* Fisher, W., Brown, L., Scott, A., and McGowen, J., eds., Delta systems in the exploration for oil and gas, Bureau of Economic Geology, Austin, University of Texas.

Shepherd, R. G., and Schumm, S. A., 1974, Experimental study of river incision: Geological Society of America Bulletin, v. 85, p. 257–268.

Shlemon, R. J., 1975, Subaqueous delta formation—Atchafalaya Bay, Louisiana: *in* Deltas, Brousard, M., Houston Geologic Society. p. 209–221.

Shreve, R. L., 1966, Statistical law of stream numbers. Journal of Geology, v. 74, p. 17–37.

Strahler, A. N., 1952, Hypsometric (area altitude) analysis of erosional topography: Geological Society of America Bulletin, v. 63, p. 1117–1142.

Twidale, C. R., 1962, Steepened margins of inselbergs from northwestern Eyre Peninsula, South Australia: Zeitschrift für Geomorphologie, v. 6, p. 51–69.

Twidale, C. R., 1967, Origin of the piedmont angle as evidenced in South Australia: Journal of Geology, v. 75, p. 393–411.

Twidale, C. R., 1978, On the origin of pediments in different structural settings: American Journal of Science, v. 278, p. 1138–1176.

Twidale, C. R., 1982, Granite Landforms: Elsevier, Amsterdam, Netherlands, 372p.

Twidale, C. R., 1990, The origin and implications of some erosional landforms: Journal of Geology, v. 98, p. 343–64.

Twidale, C. R., and Bourne, J. A., 1975, Episodic exposure of inselbergs: Geological Society of America Bulletin, v. 86, p. 1473–1481.

Warnke, D. A., 1969, Pediment evolution in the Halloran Hills, central Mojave Desert, California: Zeitschrift für Geomorphologie, v. 13, p. 357–389.

Wells, S., 1977, Geomorphic controls of alluvial fan deposition in the Sonoran Desert, southwestern Arizona: *in* Doehring, D., ed., Geomorphology in and regions, 8th SUNY Geomorphology Symposium, Binghamton, N. Y., p. 27–50.

Wells, S., and Harvey, A., 1987, Sedimentologic and geomorphic variations in storm-generated alluvial fans, Howgill Fells, northwest England: Geological Society of America Bulletin, v. 98, p. 182–198.

Winkley, B. R., 1976, Response of the Mississippi River to the cutoffs: *in* Rivers, American Society of Civil Engineers Symposium on Inland Waterways Navigation, Flood Control, Water Diversions, v. 76, p. 1267–1284.

Wolman, M. G., and Leopold, L. B., 1957, River floodplains; some observations on their formation: U. S. Geological Survey Professional Paper 282-C.

Wright, L. D., 1977, Sediment transport and deposition at river mouths: a synthesis: Geological Society of America Bulletin, v. 88, p. 857–868.

CHAPTER SEVEN

Groundwater

Subterranean stream in a limestone cave, Krizna Jama, Slovenia. (Photo by A. N. Palmer)

INTRODUCTION

Although groundwater lies beneath the Earth's surface where it is not easily observable, it constitutes an important part of the hydrological system. In addition to feeding surface streams, groundwater is also capable of widespread solution of carbonate rocks.

We will first look at how subsurface water moves through various types of material, then consider how dissolving of subsurface carbonate rocks develops karst topography.

POROSITY AND PERMEABILITY

Precipitation that is not intercepted or does not run off on the surface infiltrates downward into the subsurface through pore spaces between mineral grains, cracks, solution cavities, or vesicles. The physical properties of subsurface material that control the volume and movement of groundwater are porosity—the percentage of the total volume of the rock consisting of voids—and permeability— the capacity of a rock to transmit fluids.

Porosity

Porosity is the percentage of rock or soil that consists of void space. It is defined by the equation

$$\text{porosity} = \frac{\text{volume of void space}}{\text{total volume}} \times 100$$

The nature of the void space varies considerably from one type of material to another. The most common types of pore spaces are:

1. Void spaces between mineral grains

2. Fractures

3. Solution cavities

4. Vesicles

Porosity in Unconsolidated Sediments

The porosity of unconsolidated sediments depends mostly on grain size distribution and grain packing. Surprisingly, fine-grained sediments (silt and clay) contain much more pore space than granular material (sand and gravel). Pore space in sand and gravel deposits may vary between about 12 to 45 percent, whereas pore space in clay/silt deposits may reach as high as 80 percent.

The particle size distribution is an important factor in determining the porosity of unconsolidated sediments. If many different grain sizes are present, the finer material can fit into some of the void space between larger grains, and the porosity is greatly reduced (Figure 7–1). The shape and packing of grains are also significant. The porosity of spherical grains packed together as in Figure 7–2 is about 47 percent, but if the grains are shifted by one-half diameter, the porosity drops to about 26 percent.

The filling of voids with cementing material greatly reduces porosity, as does compaction of unconsolidated sediments by the weight of overlying material. The porosity of

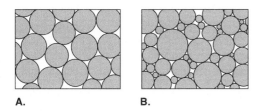

FIGURE 7–1
Comparison of porosities of (A) well-sorted sediments and (B) poorly sorted sediments.

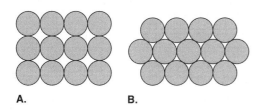

FIGURE 7–2
Effect of packing of spherical grains on porosity.

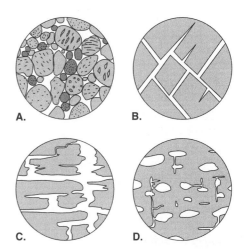

FIGURE 7–3
(A) Porosity resulting from spaces between grains.
(B) Porosity resulting from fracture occurs in most rocks.
(C) Porosity resulting from solution activity is common in limestone.
(D) Porosity resulting from vesicles is common in lava flows.

silt and clay overridden by continental glaciers during the Pleistocene is measurably less than similar sediment not compacted by the weight of overriding ice (Easterbrook, 1964).

Porosity of Rock

Virtually all rocks are cut by fractures, which may constitute the only significant pore space. Even dense granite may be made porous by sets of intersecting joint planes. If the fractures are numerous and interconnected, a surprisingly large volume of water may be contained in otherwise very dense rock (Figure 7–3). Some limestones have high porosity as a result of solution by water moving along joints and bedding planes. The moving water removes soluble material and progressively expands larger and larger conduits that may eventually become caves. Vesicles, formed by gas bubbles trapped in lava, can significantly affect the porosity of volcanic rocks (Figure 7–3). Vesicles are usually concentrated near the top of a lava flow, where they greatly increase porosity, and the vesicles may be interconnected with columnar joints or cinders and rubble at the top and base of the flow.

Porosity of Sedimentary Rocks   Groundwater occurs in the pore spaces between grains (**primary porosity**) of sedimentary rocks, as well as in fractures (**secondary porosity**). Sedimentary rocks are invariably fractured to some degree, ranging from widely spaced joints to intense fracturing that may almost completely shatter the rock. Some fine-grained, cohesive sediments (high in silt/clay) contain shrinkage cracks that develop when the sediment dries and contracts. The porosity of sedimentary rocks varies from 3 to 30 percent in clastic rocks and 1 to 30 percent in limestones and dolomites (Manger, 1963).

Limestone, dolomite, gypsum, and evaporite deposits are formed by chemical precipitation or biochemical accumulation of calcium carbonate, calcium-magnesium carbonate, calcium sulfate, or sodium chloride. All may be strongly affected by groundwater percolating through pore spaces, fractures, and bedding planes, dissolving and enlarging them to give the rock high secondary porosity.

Porosity of Crystalline Plutonic and Metamorphic Rocks
Interlocking crystals of plutonic rocks (formed by igneous intrusions) and metamorphic rocks typically give these rocks very low primary porosities, but intersection of fractures may allow significant secondary porosities. Joint sets in crystalline rock usually occur in three mutually perpendicular directions, which may increase the porosity of crystalline rocks by about 2 to 5 percent. Porosity due to jointing is concentrated along the fracture zones and increases with the width of the joints. Rocks in fault zones may be extensively fractured, giving the faults high porosity. Weathering of plutonic and metamorphic rocks can increase their porosities from 30 to 60 percent.

Porosity of Volcanic Rocks   Volcanic rocks cool and crystallize more rapidly than plutonic rocks; thus, volcanic rocks commonly have much higher porosity. Lava flows cool rapidly at the surface where escaping gases produce vesicles that may give the rock high porosity, although the vesicles may not be interconnected. The high vesicularity of pumice can produce porosities as high as 87 percent. More significantly, columnar joints, formed by shrinkage of the lava as it cools, can produce relatively high porosities with effective permeabilities. Additional porosity may be imparted by the following:

1. The rubbly tops of lava flows, formed by the breaking up of crusted-over lava as it continues to flow beneath

2. Lava tubes through which molten lava once poured

3. Stream gravels trapped between lava flows

Pyroclastic deposits, consisting of ash and cinders thrown into the air during eruptions, occur in loose, unconsolidated deposits that can have porosities ranging from 15 to 50 percent. Weathering of such volcanic deposits can increase the porosity to more than 60 percent.

Permeability

Permeability, the ease with which a material can transmit a fluid, varies with the fluid's viscosity and hydrostatic pressure, the size of void space, and the degree to which the open spaces are interconnected. Porosity and permeability are not synonymous —a material can have high porosity but low permeability. Porosity is a measure of how much water a rock can hold, whereas permeability is how easily water can be transmitted; thus, permeability is dependent on the number of connected voids in a material rather than on the total volume of voids.

Sediments or rocks with high porosity do not necessarily also have high permeability. For example, clay has very high porosity, typically 50 to 80 percent, but it has very low permeability because water has difficulty moving through the small pore spaces. Voids are very small and typically not connected, and the molecular attraction between particles and the thin films of water further inhibit movement. Sand has considerably lower porosity than clay, commonly less than 30 to 40 percent, but it is more permeable than clay because the pores are larger and are interconnected (Matsch and Denny, 1966).

Flow velocities of groundwater typically vary from about 1 m/day to 1 m/yr. The highest rate of percolation measured in the United States, in unusually permeable material, is only 250 m/day (820 ft/day). In special cases, such as the flow of water in caves, the velocity of groundwater may approach that of surface streams,

Specific Yield and Specific Retention

Specific yield is the ratio of the volume of water that drains by gravity from saturated sediment or rock to the total volume of the material. The surface tension of water molecules causes them to cling to grain surfaces. The gravitational force on a film of water surrounding a mineral grain will cause some of the water to pull away, in spite of the surface tension, and drip downward, leaving a thinner film of water with greater surface tension, until the gravitational force is equal to the surface tension. The moisture clinging to mineral grains by surface tension is known as **hygroscopic water.** If two samples have the same porosity but different grain sizes, the total grain surface area of the finer sample will be larger, allowing more water to be held as hygroscopic moisture.

The **specific retention** of a sediment or rock is the ratio of the volume of water a sample can retain against gravity drainage to the total volume of the rock. Specific yield Sy is the volume of water that a sediment or rock will yield by gravity drainage, and specific retention S_r is the remaining volume. Thus, the sum of the two is equal to porosity.

$$porosity = S_y + S_r$$

Because specific retention is greater for samples with small grain sizes, clay may have a porosity of 50 percent and a specific retention of 48 percent, yielding only a 2-percent specific yield. Maximum specific yield occurs in sediments having medium-to-coarse sand sizes (0.5–1.0 mm) (Johnson, 1967).

Hydraulic Conductivity

Darcy (1856) demonstrated that the discharge of a saturated system is proportional to the hydraulic head between two points, to the cross-sectional area, and to the permeability of the material. This relationship is now commonly expressed as **Darcy's law.**

$$Q = PIA$$

where Q = discharge
　　　　 P = a permeability coefficient sometimes called the **hydraulic conductivity**
　　　　 I = **hydraulic gradient,** represented by the change in hydraulic head between two points
　　　　 A = cross-sectional area

Permeability of Sediments

The most copious producers of groundwater are commonly coarse-grained, unconsolidated sediments that are noted for their high permeability. Conversely, clay has very low permeability. Mixtures of the two produce a wide-ranging continuum of permeability values for unconsolidated sediments.

The permeability of sediment depends largely on the size of pore spaces, which is directly related to grain size. The smaller the grain size the larger the overall grain surface area and the greater the resistance to flow. The relationship between permeability and grain size includes the following factors:

1. Permeability increases as the median grain size increases, due to larger pore openings that are more likely to be connected.

2. Permeability decreases for a given median grain diameter as the standard deviation of particle size increases (indicating a more poorly sorted sediment), because the finer material can occupy the voids between larger fragments.

3. Coarse-grained sediments show a greater decrease in permeability with an increase in standard deviation of particle size distribution than do fine-grained sediments.

4. Sediments having unimodal particle size distribution have greater permeability.

Hydraulic permeability may be measured in the field, based on grain-size analysis of sediments and hydraulic conductivity tests of monitoring wells. Aquifer pumping tests may be used to determine the hydraulic conductivity of sediments in the field, providing an integrated, average permeability over a large area.

Permeability of Rocks

The permeability of rocks depends on the primary void space generated when the rock was formed and the secondary void space created after the rock was formed. Clastic sedimentary rocks typically have primary permeability properties somewhat similar to those of unconsolidated sediments, only with reduced pore space caused by cementation and compaction, which can substantially reduce permeability even without a large change in primary porosity. Primary permeability may be enhanced by sedimentary structures, such as bedding. Conglomerate, sandstone, and some limestone are the sedimentary rocks most likely to have high permeability. Conglomerate and sandstone typically have high **primary** permeability, whereas permeability in limestone is usually **secondary,** from fractures and solution cavities. Shale has low permeability because of its small void spaces.

Crystalline igneous, metamorphic, and chemical sedimentary rocks typically have low porosity and low primary permeability because of tightly interlocking crystals, which allow very few openings. However, high secondary permeability can develop in crystalline rocks as a result of fracturing, and some volcanic rocks have high primary porosities because of large numbers of vesicles. If the vesicles are large and interconnected, high permeabilities result. Lava flows also may have high permeabilities because of extensive columnar jointing and because the tops of flows may consist of rubble from the broken crusts of flows.

THE WATER TABLE

As water infiltrates soil at the ground surface, it is impelled downward by gravity through pore spaces that are only partially saturated. As long as the water continues to encounter empty pore spaces, movement of the water is dominantly vertical (Figure 7–4). Eventually, however, all of the pore spaces become completely filled with water, leaving no more space for continued vertical percolation, and the material becomes totally saturated, producing a subsurface region known as the **zone of saturation.** The upper surface of the zone of saturation is the **water table.** Once water reaches the zone of saturation, it can no longer continue to move vertically downward because all of the pores beneath are already filled, so it moves laterally in the direction of the slope of the water table. The water table is not horizontal —it rises higher beneath hills and slopes in the direction of valleys where groundwater is discharged into streams, approximately mimicking the surface topography. The slope of the water table (*h/l*) represents the hydraulic gradient whose inclination depends on the permeability of material and the rate at which water is added to the zone of saturation.

The water table can be mapped from the level of water observed in wells and from the movement of water measured by injection of dyes and other tracers. It may be only a meter or so beneath the surface along flood plains in humid regions, but it may be hundreds of meters below the surface in deserts.

AQUIFERS

An **aquifer** is a geologic unit that can store and transmit water. Both unconsolidated sediments and rock have wide ranges of hydraulic conductivities. The permeability of aquifers is generally above about 10^{-2} darcy. Unconsolidated sand and gravel, sandstone, limestone, lava flows, and fractured plutonic and metamorphic crystalline rocks are examples of typical aquifers.

Unconfined Aquifers

Aquifers that extend continuously from the land surface downward through material of high permeability are known as **unconfined aquifers.** Recharge to the aquifer may be from percolation downward through the unsaturated zone, from lateral groundwater flow, or from upward seepage through underlying material.

FIGURE 7–4
Distribution and movement of subsurface water.

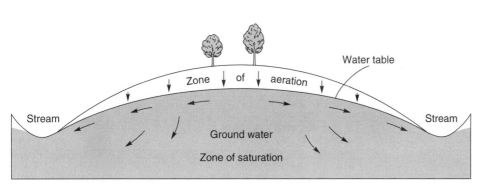

Aquifer **transmissivity** is a measure of the amount of water that can be transmitted horizontally by the full, saturated thickness of the aquifer under a hydraulic gradient of 1. The transmissivity T is the product of the hydraulic conductivity K and the saturated thickness t of the aquifer.

$$T = tK$$

Groundwater discharges wherever the water table intersects the land surface, usually along stream channels, marshes, and lakes where a link is established between groundwater reservoirs and surface elements of the hydrologic system. The discharge of groundwater into surface drainage systems provides a substantial amount of water for streamflow and keeps many streams from drying up between precipitation events.

Confined Aquifers

Subsurface material sometimes contains **confining layers,** having very low permeability, typically less than about 10^{-2} darcy. Groundwater may move through confining layers, but the rate of movement is usually very slow. Confining layers can be subdivided into aquitards, aquicludes, and aquifuges. An **aquitard** is a leaky confining layer of low permeability that can store groundwater and transmit it sluggishly from one aquifer to another. An **aquiclude** also has low permeability but is situated in a position to form the upper or lower bound-

ary of a groundwater flow system. An **aquifuge** is an almost totally impermeable body of rock or unconsolidated material.

Artesian water is confined groundwater that is under high hydrostatic pressure. The geologic conditions under which an artesian water system may develop include the following:

1. A permeable aquifer confined between impermeable layers
2. Surface infiltration to recharge the aquifer
3. Precipitation and infiltration adequate to fill the aquifer and maintain a hydrostatic head (Figure 7–5)

Artesian aquifers are overlain by an aquiclude. Recharge to them may occur where the aquifer is exposed at the surface, or by slow, downward percolation through an aquitard. Water from such an aquifer will rise above the top of the confined aquifer where it is intersected by a well or other conduit as a result of hydraulic head. The level to which water will rise in a well is the **potentiometric surface** (Figure 7–6). A potentiometric surface is inclined away from the recharge area because of frictional losses as water moves through the pores in an aquifer and because pressure is lost through leakage in fractures. If the potentiometric surface of an aquifer is above the land surface, a flowing artesian well is developed in which water flows from the well without pumping.

Where a layer of low-permeability material occurs as a lens within more permeable material, water moving downward through the unsaturated zone is intercepted and will collect on top of the impermeable lens. As saturation develops

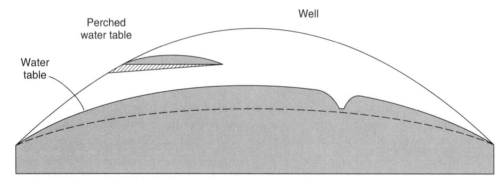

FIGURE 7–5
Regional and perched water tables.

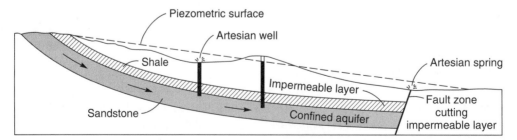

FIGURE 7–6
Artesian water resuslting from a confined aquifer.

above the impermeable lens, a **perched aquifer** is formed above the main water table (Figure 7–5). Water moves laterally above the low-permeability layer, much like a miniature expression of the regional water table, either until it spills downward over the edge of the lens toward the main water table or until it is intercepted by the ground surface where a spring will form (Figure 7–7).

A water well can withdraw groundwater by digging or drilling through the zone of aeration into the zone of saturation where water flows from pore spaces into the well. Without pumping the well, water will rise to the level of the water table. In order to extract the water, pumping is usually needed to bring it to the surface. However, under pumping, the removal of water causes the water table to be drawn down around the well, producing a **cone of depression** in the water table (Figure 7–5). If the rate of withdrawal of water by pumping is greater than the inflow of groundwater to the well, the cone of depression will grow increasingly large, flattening the hydraulic gradient in the vicinity of the well and progressively decreasing the rate at which water can be withdrawn, until ultimately the well goes dry.

THERMAL SPRINGS AND GEYSERS

In regions of unusually high geothermal gradients, especially those of recent volcanic activity, groundwater becomes heated to high temperatures, creating **thermal springs** (Figure 7–8) and **geysers** (Figures 7–9 and 7–10; Plate 5A) when the water discharges to the surface. If the water is hot enough, geysers may form when the water flashes into steam near the surface. In order for this to happen, rock temperatures just beneath the surface must be high, and groundwater must have access to the heated rocks through fracture systems. Geysers erupt when groundwater is heated to the boiling point, but because water near the base of the water column is under greater pressure than the water above, its temperature can rise above the boiling point of water at the surface. Boiling of water at the surface causes water to rise slightly, releasing pressure on the water column below. If the temperature of the water below is close to boiling, the water may flash into steam under the new reduced pressure, and the geyser will erupt (Figure 7–10).

GROUND SUBSIDENCE DUE TO WITHDRAWAL OF FLUIDS

Withdrawal of large quantities of groundwater or other fluids, such as petroleum, decreases the fluid pore pressure between grains. The decrease in pore pressure allows tighter grain-to-grain contact and leads to sediment compaction and subsidence of the ground surface. The amount of such subsidence can be measured by precision leveling relative to stable reference points. Such measurements have led to some startling results.

One of the most spectacular examples of ground subsidence due to groundwater withdrawal occurs in Mexico City, which is built on unconsolidated, water-saturated, alluvial and lacustrine sediments. Most of Mexico City's water supply is from wells located within the city. As groundwater is pumped for domestic and industrial use, pore pressures decrease, causing widespread, slow subsidence of the land surface. Between 1891 and 1959, the maximum subsidence was 7.5 m (25 ft), and all of the city except the western edge subsided at least 4 m (13 ft) (Figure 7–11). The opera house (weighing 54,000 metric tons) has settled more than 3 m (10 ft), and half of the first floor is now below ground level.

Another example of ground subsidence caused by withdrawal of subsurface fluids occurs in Long Beach, California, which has subsided about 9 m (27 ft) as the result of 40 years of oil production. This subsidence has caused almost $100 million worth of damage to wells, buildings, pipelines, transportation facilities, and harbor installations.

Compaction and subsidence also present serious problems in areas of recently deposited sediments. In New Orleans, for example, large areas of the city are now 4 m (13 ft) below sea level, a drop due largely to the pumping of groundwater. As a result, the Mississippi River flows some 5 m (16 ft) above parts of the city. Rainwater must be pumped out of the city at considerable cost, and water lines and sewers are damaged as the ground subsides.

KARST

Although groundwater is rarely seen, its effect on topography may be profound as a result of the dissolving of soluble rocks and the transportation of the dissolved material in solution. The effects of solution produce unique landforms that distinguish solution topography from other landforms.

Groundwater is typically slightly acidic and reacts chemically with subsurface rocks, especially those that are highly soluble, such as carbonate rocks (limestone and marble), dolomite (magnesium carbonate), gypsum (calcium sulfate), and some evaporites. Extensive solution of such rocks produces extraordinary landforms grouped under the general category of karst, characterized by numerous closed depressions, caves, various collapse features, and diversion of drainage underground.

The word karst stems from the German form of the Slavic word Krs or Kras and the Italian word carso, meaning "barren, stony ground," or "a bleak waterless place" (Monroe, 1970). Such features are typical of both the classical karst region on the high plateau near the Adriatic Sea between Italy and Yugoslavia and the Dinaric karst region to the south. This region is renowned for its large caves, irregular topography with many closed depressions, interrupted stream valleys, and

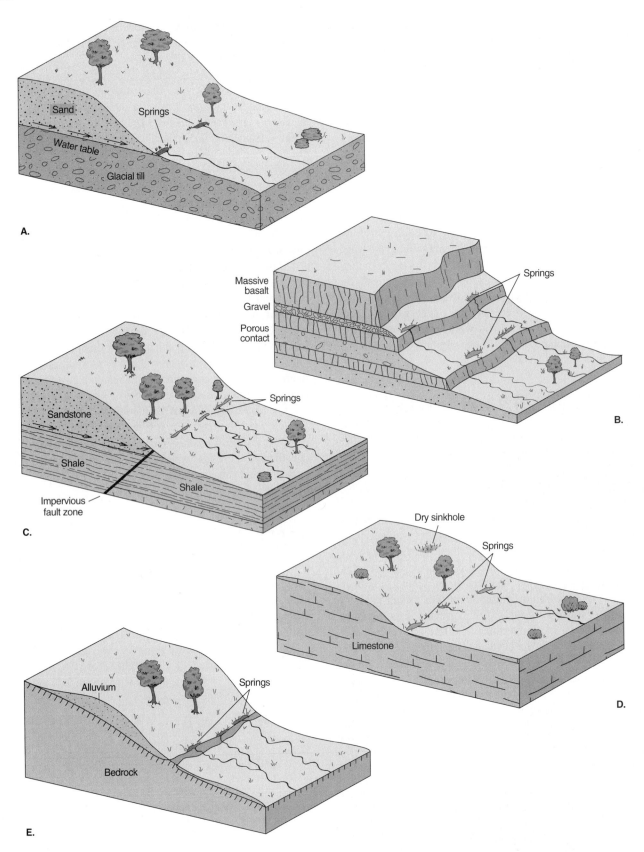

FIGURE 7–7
Springs (A) from the intersection of perched water and the land surface, (B) from interlayered permeable and impermeable rocks, (C) from water rising to the surface as a result of blockage of groundwater flow by an impermeable fault zone, (D) from intersection of solution cavities the surface, or (E) from rock protruding through alluvium.

FIGURE 7–8
Mammoth hotsprings,
Yellowstone National Park.
(Photo by W. K. Hamblin)

FIGURE 7–9
Thermal geysers, Yellowstone
National Park. (Photo by W. K.
Hamblin)

FIGURE 7–10
Old Faithful geyser, Yellowstone National Park.

an assortment of enclosed and subsurface-drained hollows (Figure 7–12).

Early studies of karst date back to the end of the eighteenth century (Roglic, 1972) when the collapse origin of solution depressions was recognized (Gruber, 1791; Hacquet, 1778; Virlet, 1834; Foumet, 1852; Prestwich, 1854). Owen (1856) and Cox (1876) first expressed the view that sinkholes in Kentucky and Indiana were formed by solution of limestone by infiltration of water.

Because the solution of limestone requires abundant water and dissolved carbon dioxide, the formation of karst landforms is affected by climate, especially the amount and distribution of rainfall. Variations in the calcium carbonate content of groundwater in karst terrains suggest that limestone solution may be different in various climates (Corbel, 1959). In arid climates, the water table is usually deep below the surface, greatly inhibiting solution activity. Some karst features do form in arid or semiarid climates, but the karst topography is quite subdued, and features such as collapsed sinkholes and deranged surface drainage are not as conspicuous. In some instances, arid karst features formed when the climate was more humid than at present, or they may have formed due to acidic fluids rising upward into the rocks from deep-seated sources such as at the famous Carlsbad Caverns (Egemeir, 1981; Hill, 1981). In high-latitude cold regions, water, when present, is often frozen, severely retarding free circulation and microbiologic activity. Karst is therefore quite rare in arctic and antarctic regions. Humid tropical climate, however, provides an optimal combination of temperature and precipitation to facilitate solution processes. Chemical reactions take place more rapidly at higher temperatures, and luxuriant vegetation, together with intense microbiotic activ-

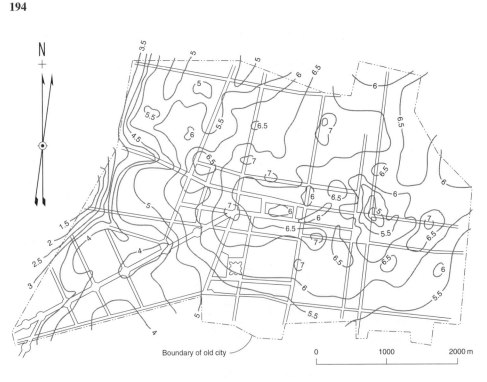

N
+

FIGURE 7–11
Subsidence of Mexico City,
1891–1959, shown by lines of
equal subsidence (meters).
(After Comision Hidrologica de la
Cuenca del Valle de Mexico,
1961)

Boundary of old city

0   1000   2000 m

FIGURE 7–12
Solution valley, Plitvice, Yugoslavia.

ity, imparts high partial pressures of carbon dioxide in tropical groundwater, making it exceptionally effective as an agent
of solution and producing strongly developed karst features
(Jennings and Bik, 1962; Monroe, 1976). Generation of bio-

genic CO_2 is perhaps the most important aspect of tropical
regions that enhances solution (Smith and Atkinson, 1976;
Trainer and Heath, 1976; Drake, 1980; Brook et al., 1983).

Karst Controls

Lithology and Structure

Although karst develops primarily on carbonate rocks (mostly limestones), not all carbonate rocks possess the proper
combination of physical and chemical properties that are con-
ducive to generation of karst topography. Geologic structure
is an important factor in karst development (Palmer, 1977).

Most of the world's karst regions are developed on limestone, which by definition consists of at least 50 percent calcite and/or aragonite ($CaCO_3$). Isomorphous substitution of
magnesium for calcium in the carbonate mineral structure
forms the mineral dolomite ($CaMgCO_3$). If more than 50 percent of the rock is composed of dolomite, the rock is a
dolomite (or dolostone). Because substitution of magnesium
for calcium in the mineral structure may occur in virtually
any amount, a considerable variation in composition of a carbonate is possible. In general, the purer the limestone is in
$CaCO_3$, the greater is its proclivity to form karst. Some evidence suggests that about 60 percent $CaCO_3$ is necessary to
form karst, and about 90 percent may be necessary to fully develop karst (Corbel, 1957). However, even pure limestones
may not produce karst because other important factors may
be lacking. Some karst features may form on dolomite, but
their permeability is typically lower than that of limestone
(Herman and White, 1985), so the occurrence of karst in
dolomites is usually relatively minor. Rock salt (halite, NaCl)

and gypsum ($CaSO_4$) are even more soluble than limestone, but they are not widely distributed.

Optimum conditions for karst development include more than just the mineralogy of carbonate rocks. Porosity and permeability also play important roles, especially the ease with which water moves through the rock. Primary porosity, the void spaces in the rock between mineral grains, commonly decreases with time because pore space is diminished by recrystallization of calcite and dolomite, precipitation of calcite/dolomite cement, and changes in mineralogy. The older the rock, the greater the tendency for increase in calcite grain size by recrystallization and replacement of calcite by dolomite, both of which decrease the primary porosity.

Primary porosity is generally considered much less important than secondary porosity and permeability produced by fractures and bedding plane partings. The type and distribution of secondary openings are commonly regarded as the most important factors in karst formation because they permit rocks to hold more water and facilitate groundwater circulation through the system by increasing permeability, which is more important than porosity in the formation of karst. Limestone with high primary porosity seldom develops karst without secondary permeability through fractures and bedding planes. Even rocks with high primary porosity may not develop strong karst features if circulation is limited. The most important structural feature of carbonate rocks consists of joints, which act as conduits for groundwater circulation. Joints in sedimentary rocks typically consist of sets that commonly intersect at angles of 70° to 90°, allowing free circulation between and among fractures. The spacing of joint sets is also important. If intersecting joint planes are spaced too far apart, circulation is impeded; but if they are too closely spaced, the rock may be too structurally weak to support karst features, even though the rock may be highly permeable.

Faults can also transmit water, but their spacing is usually much greater than that of joints, so they are not as effective in developing karst. Some fault zones that have fine-grained fault gouge or secondary mineralization along the fault plane have low permeability, which deters karst formation. If secondary crystallization along the fault plane includes sulfide minerals, oxidation may generate sulfuric acid, which aids the solution process (Pohl and White, 1965; Moorhouse, 1968).

Considering the above-mentioned factors, optimum lithologic and structural conditions for full karst development include the following:

1. Thick, pure calcite, crystalline limestone uninterrupted by insoluble beds

2. Intersecting joint sets that allow free circulation of groundwater along discrete flow paths with enough discharge to create or enhance significant solution openings

The Solution Process

The process of carbonate solution forms the basis for development of karst. The solution process consists of a series of steps:

1. Neither groundwater, nor the rainwater from which it was derived, consists of pure water (H_2O). Both contain carbon dioxide (CO_2), which is soluble in pure water, and the CO_2 reacts with the water to form carbonic acid.

$$CO_2 + H_2O \rightleftarrows H_2CO_3 \qquad (1)$$

The amount of CO_2 dissolved in water depends on the partial pressure of CO_2 at the air-water interface in the atmosphere or in pore space, and on the temperature of the water. Dissolved CO_2 in the water increases with increase in the partial pressure of CO_2 of the air and with decrease of water temperature. Cold water will hold more dissolved CO_2 than warm water at any given partial pressure of CO_2. Abnormally high dissolved CO_2 is commonly found in soil water (Jennings, 1971) as a result of biogenic CO_2 formed from the decomposition of organic material. This increased CO_2 becomes an important factor in the solution process.

The carbonic acid dissociates readily into its ionic state according to the equation

$$CO_2 \text{ (dissolved)} + H_2O \rightleftarrows H^+ + HCO_3^- \qquad (2)$$

3. Calcite dissociates into its ionic state as follows:

$$CaCO_3 \rightleftarrows Ca^{++} + CO_3^{--} \qquad (3)$$

4. The hydrogen atom of equation (2) combines with the carbonate ion to form another bicarbonate ion

$$CO_3^{--} + H^+ = HCO_3^- \qquad (4)$$

During this process, the Ca^{++} ion is released from the calcite crystal into the water.

These steps are illustrated in Figure 7–13. Although the net process can be written as

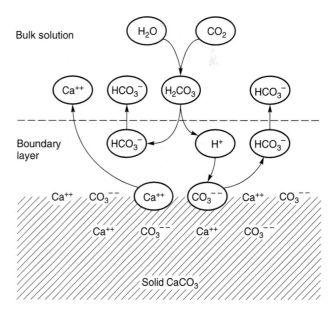

FIGURE 7–13
Steps in the solution of calcite in CO_2-bearing water. (From W. B. White, 1988, Geomorphology and Hydrology of Karst Terrains. Copyright 1988 by Oxford University Press, Inc.)

$$CaCO_3 + H_2O + CO_2 \text{ (dissolved)} \rightleftarrows Ca^{++} + 2HCO_3^- \quad \textbf{(5)}$$

the system is in reality exceedingly complicated, involving a number of mutually interdependent, reversible reactions. The bicarbonate ions (HCO_3^-), derived from the dissociation of $CaCO_3$ and the dissolving of CO_2 in water, introduce disequilibrium between the partial pressure of CO_2 in the air and the partial pressure of CO_2 in the water, which causes more CO_2 to dissolve in the water and drives more solution of calcite (Thrailkill, 1972; Thrailkill and Robi, 1981; Weyl, 1958). The amount of CO_2 dissolved in water is very important to the solution process. Where dissolved CO_2 is high, the water vigorously dissolves calcite. Soil water, with its high dissolved CO_2 from organic material, is effective in the solution of limestone (Drake and Wigley, 1975).

Effect of Climate and Vegetation on Solution

Because carbonate solution is the driving force of karst formation, factors that enhance solution also promote karst growth. Thus, climatic factors are important to karst, especially the amount and distribution of precipitation, the temperature, and the evaporation rate. The optimum combination of precipitation and temperature to promote solution processes occurs in tropical humid climates where chemical reactions proceed rapidly and luxurious vegetation combines with high levels of microbiotic activity to generate high partial pressures of carbon dioxide in groundwater. These conditions actively encourage solution and the proliferation of karst features (Monroe, 1976).

Perhaps an even more important aspect of climate than the amount of water supplied climatically is the high biogenic CO_2 values in groundwater from vegetation and biogenic activities that result from humid tropical climates (Smith and Atkinson, 1976; Trainer and Health, 1976; Drake, 1980; Brook et al., 1983). Areas of more intense solution within the same climatic zone occur as a result of vegetation differences (Woo and Marsh, 1977; Brook and Ford, 1982).

Arid or semiarid regions, where precipitation rates are low and temperature and evaporation rates are high, are less susceptible to karst development. Solution still occurs, but karst features are generally quite subdued, with fewer collapse features and fewer diversions of surface drainage. In cold arctic or subarctic climates (which may be arid to semiarid), karst development is further impeded by permafrost and by the lack of vegetation and biologic activity that is responsible for high biogenic CO_2 levels in groundwater (Smith, 1969). Much of the water that may be present is typically frozen, inhibiting free circulation necessary for solution. Karst topography is rare in such areas.

Karst Groundwater Hydrology

From the earliest days of research on karst phenomena, the nature of karst aquifers has been debated. Cvijic (1893), Grund (1903), and others argued that both surface and subsurface cave streams drained downward to a karst water table, rather than an integrated drainage system. On the other hand, Katzer (1909), Martel (1910), and others believed that underground cave streams were not necessarily connected and did not flow at common levels or to a karst water table. Thus, two schools of thought have evolved:

1. Groundwater in karst is considered to be similar to that found in nonlimestone terrains, with downward movement through the unsaturated zone to a defined water table below.

2. Groundwater in karst is believed to be confined to interconnected cavities with no defined water table (Baker, 1976; Brucker et al., 1972).

Hydrostatic levels in karst terrain are commonly quite different, with dry cavities close to others filled with water, some conduits crossing the paths of others, and some conduits flooding while others remain dry. White (1988) suggests that the water table controversy can be resolved by considering the water table to be very irregular, with rapid response of the conduit system to hydraulic changes relative to the diffuse system.

Karst Landforms

Sinkholes

Sinkholes are small, shallow, circular to oval, closed depressions formed by downward solution of limestone from the surface or by collapse of the roof of a solution cavity. A synonymous term, **doline,** meaning "small valley" in Slavic, was defined by Cvijic (1893) as a shallow, closed depression; this term was widely used in the early days of karst research in the Adriatic region. Sinkholes comprise the most widely distributed forms in karst terrains, where they occur by the tens of thousands, giving such regions much of their characteristic pitted topographic expression (Figure 7–14 Plate 5B).

Sinkholes vary considerably in size, typically ranging from 10 to 100 m (~30 to 300 ft) in diameter (average ~50 m, or 160 ft) and from 2 to 100 m (~7 to 330 ft) in depth (average ~10 m, or 33 ft), although they can be more than 1000 m (3300 ft) in diameter and hundreds of meters deep. Three principal forms, linked by transitional forms, occur in karst terrains:

1. Bowl-shaped, with very shallow depth relative to diameter, and side slopes of 10° to 12°.

2. Funnel-shaped, with diameter two to three times depth, and slope angles of 30° to 45°. They are much less common than the bowl form.

3. Cylinder-shaped, with depth greater than diameter and steep to vertical sides. They are rare relative to the bowl and funnel shapes.

Sinkholes can occur singly, but more typically they form in large numbers, commonly 40 to 50 in a square kilometer and

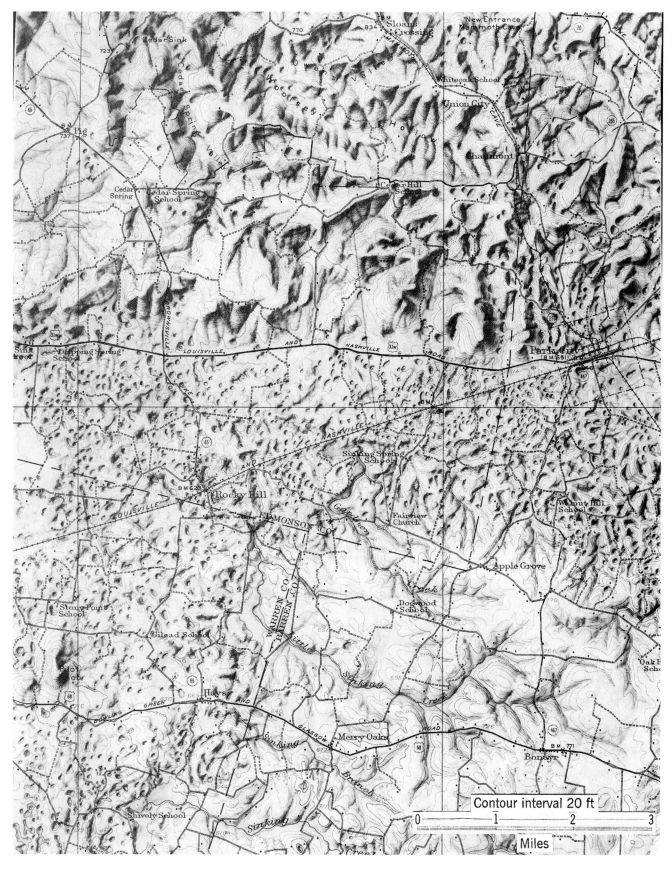

FIGURE 7–14
(A) Topographic map of karst features near Mammoth Cave, Kentucky (Mammoth Cave quadrangle, U.S. Geological Survey).

FIGURE 7–14 cont.
(B) air photo of same region (U.S. Geological Survey).

by the hundreds of thousands in a single karst terrain (Cvijic, 1893). The spatial distribution of sinkholes is generally relatively random, although in places sinkholes may develop in rows along the strike of a joint or lithologic contact.

Most of the early studies of karst features in the classic Adriatic areas emphasized the role of roof collapse of underground cavities in the formation of sinkholes, and caves were regarded as a precondition for formation. However, some noted that the collapse of roofs of solid rock should result in irregular and angular forms, rather than the round, funnel-shaped ones typical of most karst terrains, and that many small sinkholes were not connected with caves. This observation led to the conclusion that the great majority of the sinkholes could not possibly have originated from a collapsed roof.

Early English and American studies of karst differed from the generally prevailing collapse theory, finding instead that sinkholes terminated in small fissures attributed to solution of limestone along joints and cracks by atmospheric water (Prestwich, 1854). Cvijic (1893) argued vigorously against the collapse origin of all sinkholes on the basis that large sinkholes connected to caves by openings less than 1 meter could not have been formed by collapse.

Thus, two primary origins of sinkholes are now recognized: by solution from the surface downward or by collapse into solution cavities. The former are known as **solution sinkholes,** the latter as **collapse sinkholes.**

Solution Sinkholes.   Water seeping into joints and fissures in limestone dissolves the calcium carbonate from the surface downward, enlarging the conduits until surface water is diverted into funnel-shaped fissures or bowl-shaped, closed depressions known as **solution sinkholes,** which continue to enlarge about the central sink (Figure 7–15).

The most common form of solution sinkholes is approximately circular. Other forms of sinkholes are probably determined by variations in the progress of decomposition. Round sinkholes occur around central fissures where solution of the limestone proceeds from the center outward toward the periphery. Oval sinkholes form by solution along elongate fissures or major fractures.

Several factors play important roles in the development of solution sinkholes:

1. **Slope.** Sinkholes commonly occur in greatest numbers on level or gently sloping areas, and they are much less numerous on steep slopes. They are especially abundant on elevated plateaus where most of the precipitation seeps into solution fissures. On steep slopes, much of the water runs off on the surface.

2. **Lithology and fracturing.**   Dense, highly fractured limestones are more susceptible to solution than massive porous limestone. Limestones composed of pure calcite dissolve more readily than argillaceous limestone or dolomite.

3. **Soil and vegetation.**   Increased CO_2 saturation from organic material speeds up solution.

Collapse Sinkholes are produced by solution beneath the surface and enlargement of underground cavities until the roof collapses and a closed depression is formed on the surface. In the early days of karst investigation in Europe, nearly all sinkholes were thought to originate in this way, but later work has shown that although it is an important mechanism for solution formation, the collapse mode of origin is now considered less than ubiquitous.

In karst areas where limestone is overlain by weathered residual clay, loess, or other unconsolidated sediment, solution of the limestone enlarges subsurface cavities with roofs of unconsolidated material supported only by ever-decreasing limestone or, finally, only by natural arching in the sediment. When gravity overcomes the strength of the arching support, the roof collapses instantaneously, producing a depression at the surface, as shown in Figure 7–16. Another example of this type of collapse origin of sinkholes is shown in Figure 7–17 near Orlando, Florida, where a hole suddenly appeared at the surface and quickly developed into a major collapse feature.

Collapse may originate at varying depths below the surface and may include significant stoping of ceiling rock as well as direct solution. Stoping may play an especially important role in deep-seated collapse, where the initial solution cavity stopes its way upward to the surface, leaving the bottom of the cavity filled with rubble from the stoping, along with the collapsed roof rubble. Deep-seated collapse sinkholes are likely to have steep-sided walls, forming subsidence shafts. If groundwater circulates freely through the lower portions of collapse sinkholes or subsidence shafts, stoped blocks may be removed by solution or mechanical transport, thus facilitating further stoping. Deep-seated stoping may proceed from subsurface solution cavities through substantial thickness of overlying resistant rock. For example, Dante's Descent in

FIGURE 7–15
Sinkhole (doline) caused by subsidence into solution cavity in Cretaceous limestone near Planina, Yugoslavia. (Photo by A. N. Palmer)

FIGURE 7–16
Roof collapse into a solution cavity near Bloomington, Indiana.

northern Arizona is a vertical-sided pit 120 m (400 ft) deep made in Supai sandstone by upward stoping from the underlying Redwall limestone.

Evidence from sudden collapses in Florida and elsewhere suggests that collapsing may be brought about by lowering of the water table (Kemmerly, 1980 a, b). Ironically, some of the most spectacular collapses in the karst area around Orlando, Florida, have occurred following periods of drought when the water table has been substantially drawn down, allowing near-surface sand to collapse into solution cavities below.

Uvalas

Uvalas are essentially **compound sinkholes** made by the enlargement and coalescence of smaller, individual sinkholes.

They are typically shallow and somewhat irregular in shape as a result of the merging of multiple sinkholes of different diameter (Jennings, 1967). Their size may span several square kilometers in area, and their depth may vary from a few meters to 200 m.

Poljes

The term **polje** is Slovene for "field" and is used in karst literature to mean a large, closed depression with an alluvial flat floor, bounded by steep sides. The flat alluvial of poljes floors comprise one of their distinguishing characteristics. Poljes are perhaps best developed in the classic Adriatic karst region where they are typically 1 to 5 km (0.6 to 3 mi) wide and up to 60 km (36 mi) long. Their floors are flat and are usually filled with alluvium, making a sharp boundary with the surrounding limestone hills. Water enters the poljes from springs, flows across the flat floors as surface streams, and then disappears into sinkholes and caves at the opposite side. Water moves from one polje to another at successively lower elevations until it discharges as large springs along the Adriatic Coast. Poljes may flood annually from seasonally increased flow from springs, and the flat floors of the poljes may be entirely inundated to become shallow lakes, which later dry up.

The Adriatic poljes are formed in limestones that are part of a complex tectonic area, and the axes of elongation of the poljes seem to parallel structural trends in the region, suggesting that tectonism may play a significant role in their origin (Cvijic, 1960). Some of the poljes occur along faults, but where flat polje floors truncate bedding, factors other than structure must be important. Grund (1903) contended that the flat polje floors had been planed by solution to the level of the

FIGURE 7–17
Sinkhole that began in the backyard of a neighborhood in Winterpark, Florida and eventually swallowed up several new Porsches from a car dealership, several buildings, a swimming pool, a recreational vehicle, and a camper. (Photo by Geophoto)

water table. Louis (1956) and Roglic (1957) pointed out that the alluvium making up the polje floors comes from erosion of the adjacent mountains and that as the floors of the poljes become filled with sediment, local perching of groundwater occurs, which facilitates growth of the polje by lateral solution of the polje walls. Geophysical studies provide evidence to support both tectonic and karst origins for the poljes (Mijatovic, 1983). The alluvial fills of some poljes are quite shallow, simply mantling a thin veneer of sediment over solution-planated limestone floors. Other sediment fills are considerably deeper, extending below sea level in areas where active faults and earthquakes are common, suggesting a tectonic origin. The poljes of the Adriatic are thus a combination of intense solution processes and active block faulting.

Similar large, closed depressions with hydrologic features comparable to the Adriatic poljes occur in Canada (Brook and Ford, 1977), Cuba (Gradzinski and Radomski, 1965), Borneo (Sunartadirdja and Lehmann, 1960), Jamaica (Sweeting, 1958), and Ireland (Williams, 1970).

Sinkhole Ponds and Karst Lakes

Sinkhole ponds and karst lakes [Plate 5–B] are sinkholes that have filled with water because:

1. The central drain of the sinkhole has become plugged with clay and silt washed in from surface drainage or left behind as a residue during solution (Figure 7–18)

2. The sinkhole intersects the water table, and the level of the sinkhole pond or lake is the top of the water table (Figure 7–19)

The water surface of the lakes and ponds in Figure 7–19 is remarkably consistent in elevation, even though the surrounding topography has more than 30 m (100 ft) of relief, suggesting that the lakes and ponds represent the water table. Figure 7–20 is a cenote (sinkhole) in Yucatan, Mexico, which intersects the water table. The water in the sinkhole can be

FIGURE 7–18
Sinkhole pond.

observed to flow slowly in a consistent direction, down the slope of the regional water table.

Solution Chimneys and Vertical Shafts

Solution chimneys are formed by the dissolving of limestone walls along fissures or bedding planes, which are structurally controlled. Some are developed along a single fracture (Raines, 1972), whereas others follow steeply dipping bedding planes.

Vertical shafts are circular cylinders with vertical walls cutting across bedding. Their shape is generally controlled by groundwater hydraulics and is independent of structure and bedding (Pohl, 1955). They are most common where solution-resistant caprock prevents solution along vertical planes and solution is centered in restricted areas.

Vertical shafts are distinguished from solution chimneys primarily on the basis of hydraulic control versus structural control of solution morphology. Vertical shafts may be completely roofed over or may exit at the surface as a result of collapse of their roof. Shafts filled with rubble may appear as small, closed depressions. Other shapes of shafts and pits may act as subsurface drains, connecting closed depressions with cavities and caves below, or they may simply feather out at depth.

Solution along Fractures Solution of carbonates is usually enhanced along joint planes as water circulates through the fractures, leading to etching out of the joints (Figure 7–21). Cutters are planar notches in limestone made by vertical solution along joint planes (Howard, 1963). They are usually a few centimeters to several meters wide, a meter to tens of meters deep, and tens of meters or more long (Figure 7–22). Cutters are among the most common features in karst regions, but they generally do not have much surface expression because they taper downward and are typically filled with soil. They are most observed in road cuts or other vertical sections in horizontal to gently dipping limestone. Cutters occur in folded limestones as well as in horizontal beds. They are oriented along the strike of the bedding and make elongate, parallel limestone ledges. After prolonged solution, the limestone may have the appearance of Swiss cheese (Figure 7–23).

Cutters grade into larger and deeper features known as solution fissures, similar to cutters but with much greater depths than widths; **solution corridors,** having widths of tens of meters; and solution canyons, having widths and depths of tens to hundreds of meters and lengths of a kilometer or more (Monroe, 1976; Jennings and Sweeting, 1963). These forms typically have linear or rectilinear patterns as a result of structural control of fracture development.

Transition between Fluvial and Karst Drainage

Among the unique features of karst topography is the disruption of surface drainage and diversion underground

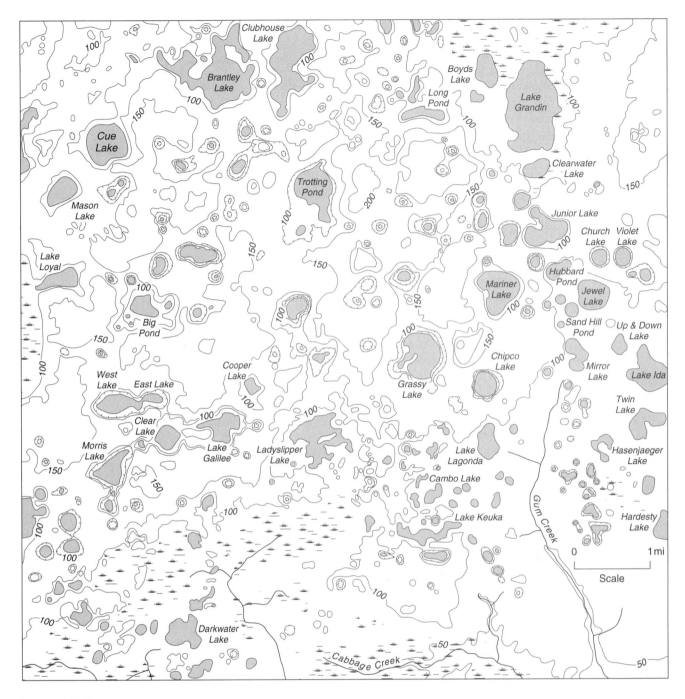

FIGURE 7–19
Sinkhole lakes and ponds representing the regional water table. (Data from Interlachen, Florida, quadrangle, U.S. Geological Survey)

through interconnected solution cavities. Surface streams end abruptly at sinkholes and swallow holes, and they emerge just as abruptly from karst springs as high-discharge streams. Thus, drainage in karst areas has both surface and subsurface components. Which of the two components predominates is a function of the thickness of limestone relative to the total thickness of exposed rocks and the surface area underlain by limestone relative to the area of the drainage basin.

In some of the classic Adriatic karst region, where thick limestones are exposed over extensive drainages, the drainage is entirely karstic, and although such areas are rare, their significance is such that they have become known as **holokarst** (Cvijic, 1960). The more common situation, with both karst and fluvial features, is referred to as **fluviokarst.** The transition from fluvial to karst drainage is best seen where normal fluvial drainage, developed on clastic beds overlying lime-

FIGURE 7–20
Cenote (sinkhole) in Yucatan, Mexico. The water level corresponds to the top of the regional water table, which can be observed to flow from right to left.

FIGURE 7–21
Joints in Carboniferous limestone widened by solution, Yorkshire, England. (Photo by A. N. Palmer)

stone, is let down onto the underlying limestone as the streams incised their channels. As the stream channels encounter the limestone, water becomes diverted underground through solution cavities, and the surface streams eventually disappear. As the clastic rocks are consumed by erosion, exposing more and more limestone, subterranean drainage becomes dominant over surface drainage, and the original fluvial drainage pattern ultimately vanishes. However, as in the case of the overlying clastic beds, solution and erosion eventually consume the limestone, and the topography returns once again to a normal fluvial system.

A variation of this scenario occurs when clastic segments of sedimentary sequences, whose weathering products must be mechanically transported, overwhelm subsurface streams, blocking underground channels, forcing streams to flow over a mantle of insoluble alluvium, and once again reverting to surface drainage.

Surface streams in a region of extensive sinkhole development rarely flow for long distances. Typically, surface streams that flow on noncarbonate rocks disappear into sinkholes or swallow holes once they reach carbonate rocks, and the drainage is diverted underground (Palmer, 1972). Such streams are known as **sinking creeks** or **disappearing streams** (Figures 7–24 and 7–25). The point where a sinking stream disappears underground is known as a **swallow hole (or swallet).** Swallow holes vary greatly in size and shape, and they may be sinkholes, pits, shafts, various forms of solution fissures, or cave entrances.

Karst Valleys

The transition from a fluvial valley to a **karst valley** takes place as the subterranean drainage system develops in the underlying limestone and surface flow becomes increasingly di-

FIGURE 7–23
Solution of limestone, Guilin, China. (Photo by W. K. Hamblin)

FIGURE 7–24
Sinking stream, West Kingsdale, Yorkshire, England. Water runs off peat-covered uplands and drops 72 m (235 ft) into a cave through a solutionally widened joint in Carboniferous limestone. (Photo by A. N. Palmer)

verted underground. The valley retains its fluvial characteristics (that is, shape, gradient, and surface channel) throughout the sequence of development as more and more surface flow disappears underground. When all of the surface flow is finally diverted underground, the valley becomes a **dry valley** or a **solution valley**, and surface stream flow occurs only during unusually high discharges. The former surface stream channels gradually become erased by further solution, weathering, and mass wasting. Sinkholes develop in the old channel floor, disrupting the former stream-valley profile, to be replaced by an irregular surface pitted with sinkholes. Woolsey Hollow in the upper part of Figure 7–14A shows topography developed on sandstone beds underlain by limestone. Incision of streams into the sandstone produced normal, fluvial, V-shaped valleys, but upon cutting into the underlying limestone, carbonate solution produced sinkholes in the valley floors into which surface drainage was diverted.

Sinking creeks that discharge into sinkholes over a long period of time erode their valley floors below the original level of the sinkholes, leaving the valley ending abruptly against the steep walls of the distal side of the sinkhole, forming a **blind valley** (Figure 7–25).

Streams emerging from springs in a karst area may also incise valleys and produce the opposite of a blind valley,

namely, a **pocket valley** in which the surface stream flows away from a blind headwall rather than into it. Pocket valleys typically have steep headwalls kept nearly vertical by spring sapping that undermines the overlying rock. Their size varies depending on the discharge of the springs and on the physical and chemical characteristics of the rocks. Some pocket valleys reach dimensions of 1 krn (0.6 mi) wide, 8 km (5 mi) long, and 300 to 400 m (990 to 1300 ft) deep (Sweeting, 1973). Some streams may even emerge from the blind headwall of a pocket valley and flow for some distance on the surface, only to disappear again into the sinkhole of a blind valley.

Caves

Caves are elongate cavities in limestone produced by solution, aided by mechanical erosion of subterranean flowing water (chapter opening photo). They form along paths of greatest groundwater solution and discharge where subsurface flow is great enough to remove dissolved limestone and keep undersaturated water in contact with the soluble cave walls. Cave formation usually requires a network of openings along which water can flow from recharge to discharge areas. Flow conduits that gain increasing discharge with time accelerate in growth, while the growth of others lags behind. As discharge increases, the maximum rate of cave wall retreat typically reaches about 0.01 to 0.1 cm/yr (0.004 to 0.04 in/yr), determined primarily by solution kinetics and nearly unaffected by further increase in discharge (Palmer, 1991). The time required to reach the maximum rate of enlargement varies directly with groundwater flow distance and temperature, and it varies inversely with initial fracture width, discharge, gradient, and partial pressure of CO_2 (Palmer, 1991). Most caves reach sizes large enough for humans to explore in about 104 to 105 years.

The morphology of caves is controlled by:

1. The spatial distribution of carbonate rocks,
2. The location of recharge and discharge points
3. The geologic structure
4. The distribution of vadose and phreatic flow
5. The geomorphic history of the region (Palmer, 1991)

The sizes and shapes of caves are usually best described by their diameter, length, and spatial pattern. The components of a cave system may be broken down into entrances, terminations, passages, and rooms.

Entrances. Subterranean cave systems are usually quite complex, consisting of both continuous and discontinuous cavities, conduits, and smaller openings in the limestone. Many cave systems are interconnected, serving as drainage ways for subsurface movement of water. Some are filled with water, others are dry; some are accessible from the surface, others are inaccessible because no openings reach to the sur-

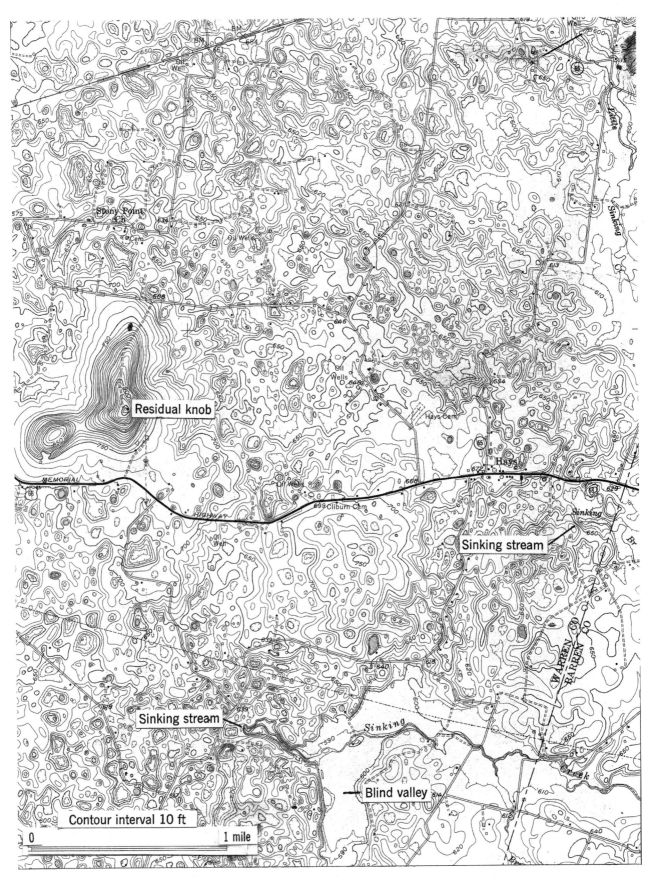

FIGURE 7–25
Topographic map of karst area. Note the numerous sinkholes, blind valley, sinking stream, and residual knob. (Smith Grove quadrangle, U.S. Geological Survey)

face. The number of accessible caves in most karst regions is only a small percentage of the total system.

The entrance to a cave may be largely fortuitous. The most common entrances are found as swallow holes, spring mouths, upward stoping of cave passages, sinkhole collapse, intersection of vertical shafts with cave passages, or simply incision of surface valleys that intersect underlying solutional cavities connected to the cave. However, the entrance need not be related to the sources and discharges of the water that formed the cave. Some cave entrances are formed by the downcutting of surface streams, which uncover subsurface cavities. However, truncation and dissection may not form entrances if thick soil covers or weathered bedrock slump into and obliterate the opening. Other entrances are formed by processes that randomly uncover subterranean passages. Most caves have no natural entrance at all and remain virtually undetected from the surface.

Terminations. Even the longest of caves, such as Mammoth Cave, Kentucky, with over 50 km (30 mi) of continuous subterranean passages, eventually end. They may be terminated by collapse of the ceiling, thus filling with debris, or by gradual tapering of the conduit to smaller and smaller sizes. Because virtually all karst caves are made by solution of carbonates, the water that created a cave must exit somewhere.

The courses of subsurface karst streams generate profiles and gradients generally similar to those of normal surface streams. The subsurface drainage may be controlled by structural and stratigraphic conditions, but underground streams seem to emerge from the karst areas at elevations not unlike those of surface streams in drainage basins.

Passages and Rooms. The conduits of a cave system are so intricate as to almost defy description. Many long, winding passageways branch into cavities that enlarge into rooms and chambers, or into a maze of anastomosing corridors and channels along intersecting joint and bedding systems. Such complex systems would be difficult enough to follow in two dimensions, but in karst areas, multiple levels of cave networks are common, substantially increasing the degree of complexity. Figure 7–26 shows some of the nomenclature used to distinguish various patterns.

Passages are segments of caves that are longer than they are wide or high. Thus, caves can be thought of as an assemblage of passages of different sizes and shapes. They can be divided into two broad categories—single-conduit passages and maze passages—each with further subdivisions (Palmer, 1975; White, 1960) (Figure 7–26):

Single-conduit Passages

Linear passages
Angulate passages
Sinuous passages

Maze Passages

Network mazes
Anastomosing mazes

Spongework mazes

The type of maze formed depends on the nature of fracture systems in the limestone and groundwater recharge (Palmer, 1975). Rooms are enlarged conduits of various size caused by

1. The intersection of several passages
2. The coalescence of cavities due to localized vigorous solution
3. The collapse of several cavities

Cave Patterns. Palmer (1991) describes caves as branchwork, network, anastomotic, spongework, and ramiform. The pattern formed depends primarily on the type of groundwater recharge.

Branchwork caves consist of passages that join downstream as tributaries, much like surface dendritic stream patterns. They form where groundwater recharge is from sinkholes, and they are by far the most common cave pattern (Palmer, 1991). Each branch serves as a channel for water from a separate recharge source, and each converges into successively higher-order channels downstream.

Maze caves, characterized by many closed passage loops, form where steep passage gradients and undersaturation of CO_2 allow many passages to enlarge at the same time.

Angular network caves are developed from angular systems of intersecting fissures shaped by the solution widening of principal fractures in carbonate rock. Cave passages are relatively straight, high, and narrow with widespread closed loops. Some simple angular networks consist of a set of angular, discontinuous fissures with infrequent closed loops.

Anastomotic caves consist of curving passages that coalesce and separate, much like the pattern of a braided stream, with many closed loops. They typically form along two-dimensional planes, usually parallel to bedding planes or low-angle fractures. Three-dimensional anastomotic cave systems sometimes form along more than one geologic structure, but they are relatively rare.

Spongework caves consist of three-dimensional, linked, solution cavities of diverse size in a more or less haphazard pattern, resembling a sponge. They typically form by the random joining of individual irregular solution cavities, commonly associated with mixing of chemically diverse water sources.

Ramiform caves consist of irregular, three-dimensional passages, rooms, and galleries, shaped like uneven splotches in plan (Palmer, 1991). Interconnected branches spread outward from the principal solution cavities, grading imperceptibly into spongework and network caves. Sequential outward branches of ramiform caves form by rising hydrothermal water enriched in H_2S.

Karst Windows

Caving in of portions of the roof of subterranean streams creates **karst windows** through which some of the underground stream is visible from the surface. When all but a few remnants of the roof of a cave have collapsed, the remaining portions form **natural bridges.**

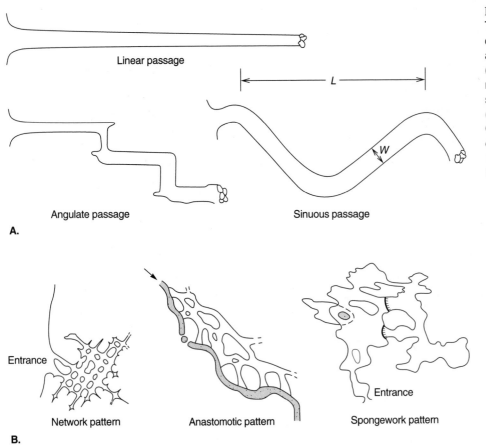

FIGURE 7–26
Types of caves: (A) single-conduit caves with linear, angulate, and sinuous patterns; (B) maze patterns, showing network, anastomotic, and spongework patterns. (From W. B. White, 1988, Geomorphology and Hydrology of Karst Terrains. Copyright 1988 by Oxford University Press, Inc.)

Origin of Karst Cave Systems

Theories of the origin of karst caves have evoked strong disagreements from the earliest days of karst investigation, and after more than 100 years of debate, the issue remains in doubt. Explanations for the origin of caves in the classic karst areas of Europe date back to the 1800s. Following the earlier work in the mid-1800s, great interest was displayed around the turn of the century (Cvijic, 1893; Penck, 1894, 1900, 1904; Grund, 1903; Katzer, 1909). Some of the controversy began to emerge even then (Roglic, 1972; Watson and White, 1985; White, 1988). Grund (1903) thought that caves were developed in the zone of active water circulation above the regional water table and that little if anything happened below the water table. A contrasting view was taken by Katzer (1909), who believed that cave development was independent of the water table, a view also favored by Martel (1921). Many of the earlier European ideas were applied to the karst areas of North America in the 1920s and 1930s (Martel, 1921; Malott, 1921; Davis, 1930; Piper, 1932; Gardner, 1935; Swinnerton, 1932), leading eventually to several contrasting models of cave evolution.

The principal theories of cave evolution in carbonate rocks all center on the relationship of circulating groundwater to the water table (Figure 7–27):

1. Caves form above the water table by solution by **vadose water.**

2. Caves form beneath the water table by deep circulation of **phreatic water.**

3. Caves form at the water table or in the shallow phreatic zone, often associated with fluctuations of the water table itself.

The concept of cave formation above the water table by vadose water arose in the early part of the twentieth century. Caves were thought to be created by subsurface streams flowing at or above the water table as infiltrating water descended downward toward the water table (Martel, 1921; Malott, 1921; Piper, 1932; Gardner, 1935). In addition to carbonate solution by vadose water, subterranean streams were thought to enlarge caves by abrasion (Martel, 1921). Malott (1929, 1932) and Woodward (1961) emphasized the role of solution by vadose water, and Gardner (1935) stressed the importance of water percolating down the dip of limestone beds.

The idea of the origin of caves by deep circulation of phreatic water dates back to the work of Cvijic (1893) and Grund (1903); it was later developed by Davis (1930, 1931) into a two-cycle process. Davis (1930) and later Bretz (1942, 1949, 1953) were impressed by the amount of travertine recently deposited in caves, and they inferred that such caves must have formed during an earlier phase of solution below the water table and had only recently risen above the water table where active carbonate deposition had begun. Davis envisioned that the solution of limestone caves occurred along curving flow lines deep below the surface in the phreatic zone

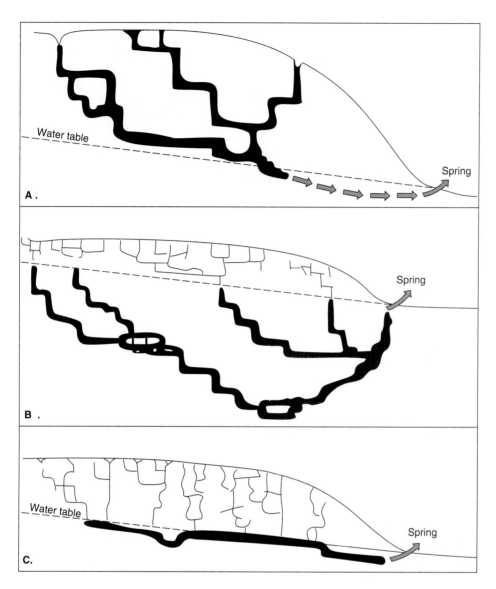

FIGURE 7–27
Zones of maximum cavern development, according to the (A) vadose, (B) deep phreatic, and (C) shallow phreatic theories. (From W. B. White, 1988, Geomorphology and Hydrology of Karst Terrains. Copyright 1988 by Oxford University Press, Inc.)

(Figure 7–27), with discharge emerging under the main river in the area. Subsequent rejuvenation of the region by uplift or base-level change causes incision of surface streams and lowering of the water table, leaving the caves high and dry to begin accumulating dripstone.

The shallow phreatic theory of cave formation calls for limestone solution within a zone where the water table fluctuates seasonally. Swinnerton (1932) noted that a great many caves are nearly horizontal and some seem to be stacked, one upon another. The common occurrence of horizontal caves in highly folded limestones suggests that they must be related to the water table. Swinnerton believed that cave development took place in the shallow phreatic zone between the high and low positions of the water table as it fluctuated with the season. Support for this origin of caves came from Sweeting (1950), White (1960), Wolfe (1964), and Thrailkill (1968). Davies (1960) proposed a four-stage, shallow phreatic origin of caves, beginning with random solution at depth, proceeding to integration of solution fissures and cavities

along the top of the water table, followed by deposition of clastic material and uplift and erosion to raise the cave above the water table.

Thrailkill (1968) noted that water table flow can be nearly horizontal and that water just beneath the water table flows along very shallow flow routes. Water near the water table surface mixes with vadose water and becomes under-saturated when it is cooled (dissolves more CO_2) or when new vadose water is added from back-flooding of nearby streams. Supporting evidence for this scenario came from White (1960), Davies (1960), White and White (1974), and Sweeting (1950), who showed that some caves seemed to be related to the elevation of nearby river terraces. White (1988) suggests that many caves show evidence of shallow phreatic solution with nearly horizontal caves even in highly folded limestones, and that three-dimensional cave networks typical of deep phreatic circulation are rare.

In contrast to the concept of cave solution in the phreatic zone and later enlarged by vadose water, Palmer (1991) pointed out that presently active solution in caves is well adjusted to surface recharge and is not dependent on an earlier phreatic stage. Cave development proceeds in the vadose and phreatic zones at the same time, rendering the arguments of Davis (1930), Swinnerton (1932), and Bretz (1942) irrelevant.

Cave Deposits. Calcium carbonate dissolved by groundwater can be precipitated in caves as travertine, often called **dripstone.** The solubility of calcium carbonate depends largely on the CO_2 content of the groundwater, which is controlled by climate, temperature, nature of the soil, vegetative cover, and other variables. As water enters the cave from fractures in the ceiling and drips toward the floor, some of the water evaporates, increasing the concentration of calcium carbonate and changing the partial pressure of CO_2. When supersaturation of calcium carbonate occurs, a small amount is precipitated. Each ensuing drop precipitates a little more calcium carbonate, which builds up as an icicle-shaped form, known as a **stalactite,** hanging down from the ceiling (Figure 7–28). The same thing happens to the water that drips off the ceiling to the floor directly below, precipitating additional calcium carbonate. Thus, stalactites are usually matched by similar forms, known as **stalagmites,** growing up from the floor. Given sufficient time, stalactites and stalagmites eventually meet and grow together to form **columns.** Pools of water that collect on the cave floor evaporate, depositing calcium carbonate as **travertine terraces.**

Tropical Karst

The common karst landforms associated with limestone terrains in humid temperate climates contrast rather sharply with the typical karst topography developed in humid tropical climates (Cooke, 1973; Day, 1976, 1978; Monroe, 1960, 1968). Although virtually all landforms recognized in karst regions of temperate climates also occur in tropical karst terrains, some notable differences exist. One of the more striking is that in tropical karst areas, residual hills, rather than the sinkholes so typical of temperate karst, dominate the topography. Residual steep-sided, cone-shaped hills are separated by cockpits, giving rise to **cone karst** and **tower karst** (Lehmann, 1936; Sweeting, 1958; Gerstenhauer, 1960; Wilford and Wall, 1965; Jennings, 1971; McDonald, 1976, 1979; Monroe, 1960, 1976; Silar, 1965; Williams, 1987). The slopes of towers may be steep-sided to vertical or overhanging, and the towers may stand several hundred meters above the adjacent lowland. The steepness of the residual hills of tower karst distinguishes it from **cockpit** karst (Figure 7–29 and 7–30), which is dominated by depressed cockpits.

FIGURE 7–28
Stalactitles hanging from the ceiling of a limestone cave in the Spanish Pyrenees.

Cone karst and tower karst are common in Central America, China, and parts of the South Pacific. The cone karst and tower karst of Cuba, Puerto Rico, and the Dominican Republic are characterized by short, stubby hills known as **mogotes** (Monroe, 1976). The exact mechanism for the development of mogotes rising above plains of blanket sand is not well understood. Differential solution of limestone, the collapse of caves, and stream erosion have all been suggested (Lehmann, 1954; Nunez-Jimenez, 1959; Panes and Stelcl, 1968; Monroe, 1976; Miotke, 1973). The mogotes of Puerto Rico are composed of the same limestone that underlies the blanket sands of the adjacent plains. However, the mogotes seem to have been more strongly indurated by calcium carbonate cementation.

Extensive tower karst occurs in southern China, where individual towers are relatively narrow with vertical sides. In southeast Asia, tower karst occurs in Java, the Celebes, Malaysia, New Guinea, and Borneo (Lehmann, 1936; Sunartadirdja and Lehmann, 1960; McDonald, 1976; Williams, 1971; Verstappen, 1964).

Cockpits are large, bowl-shaped sinkholes up to about 1 km (0.6 mi) in diameter that occur in thick limestones of tropical climates (Figure 7–30). The name comes from the Cockpit region of Jamaica where a 1000-m (3300-ft) thick limestone has been eroded into a karst belt 30 km (19 mi) wide and 90 km (56 mi) long (Sweeting, 1958). Cockpits are so large that secondary channel systems commonly develop on the cockpit walls, giving them a star-shaped form, in contrast to the circular or elliptical shapes of most sinkholes. Individual cockpits are typically separated by sharp ridgelines, and the rocks left at the intersection of several cockpits form pyramid-shaped hills.

FIGURE 7–29
Tower karst, Guilin, China. (Photo by W. K. Hamblin)

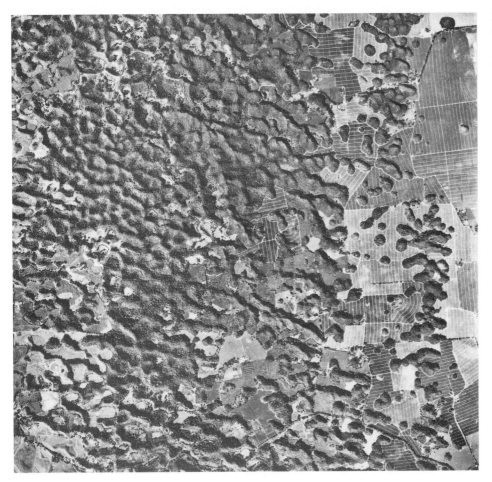

FIGURE 7–30
Cockpit karst, Puerto Rico.
(Photo by U.S. Geological
Survey)

REFERENCES

Baker, V. R., 1976, Hydrogeology of a cavernous limestone terrane and the hydrochemical mechanisms of its formation, Mohawk River basin, New York: Empire State Geogram 12, no. 2, p. 2–65.

Bosak, R., Ford, D., Jerzy, G., and Horacek, I., 1989, Paleokarst, a systematic and regional review: Developments in Earth Surface Processes, v. 1, 725 p.

Bretz, J H., 1942, Vadose and phreatic features of limestone caverns: Journal of Geology, v. 50, p. 675–811.

Bretz, J H., 1949, Carlsbad Caverns and other caves of the Guadalupe Block, New Mexico: Journal of Geology, v. 57, p. 447–463.

Bretz, J H., 1953, Genetic relations of caves to peneplains and Big Springs in the Ozarks: American Journal of Science, v. 251, p. 1–24.

Brook, G., Folkoff, M., and Box, E., 1983, A world model of soil carbon dioxide: Earth Surface Processes and Landforms, v. 8, p. 79–88.

Brook, G. A., and Ford, D. C., 1977, The sequential development of karst landforms in the Nahanni Region of northern Canada and a remarkable size hierarchy: Proceedings 7th International Congress of Speleology, Sheffield, p. 77–81.

Brook, G. A., and Ford, D. C., 1982, Hydrologic and geologic control of carbonate water chemistry in the subarctic Nahanni Karst, Canada: Earth Surface Processes and Landforms, v. 7, p. 1–16.

Brucker, R. W., Hess, J. W., and White, W. B., 1972, Role of vertical shafts in the movement of groundwater in carbonate aquifers: Ground Water, v. 10, p. 5–13.

Cooke, H. J., 1973, Tropical karst in northeast Tanzania: Zeitschrift für Geomorphologie, v. 17, p. 443–459.

Corbel, J., 1957, Karsts hauts-Alpins: Revue Geographic de Lyon, v. 32, p. 135–158.

Corbel, J., 1959, Vitesse de l'erosion: Zeitschrift für Geomorphologie, v. 3, p. 1–28.

Cox, E. T., 1876, Seventh annual report of the geological survey of Indiana made during the year 1875: Geological Survey of Indiana, Indianapolis, IN, 601 p.

Cvijic, J., 1893, Das Karstphanomen: Geographic Abhandlung, Ges., v. 5, p. 217–329.

Cvijic, J., 1960, La geographie des terrains calcaires: Monograph Acadamie Serbe, Science and Arts, v. 341, 212 p.

Darcy, H., 1856, Les fontaines publiques de la ville de Dijon: Victor Dalmont, Paris, 647 p.

Davies, W. E., 1960, Origin of caves in folded limestone: National Speleological Society Bulletin 22, p. 5–18.

Davis, S. N., 1969, Porosity and permeability in natural materials: in DeWiest, R. J. M, ed., Flow through porous media; Academic Press, N.Y., p. 53–89.

Davis, W. M., 1930, Origin of limestone caverns: Geological Society of America Bulletin, v. 41, p. 475–628.

Davis, W. M., 1931, The origin of limestone caverns: Science, v. 73, p. 327–331.

Day, M. J., 1976, The morphology and hydrology of some Jamaican karst depression: Earth Surface Processes, v. 1, p. 111–129.

Day, M. J., 1978, Morphology and distribution of residual limestone hills (mogotes) in the karst of northern Puerto Rico: Geological Society of America Bulletin, v. 89, p. 426–432.

Drake, J., 1980, The effect of soil activity on the chemistry of carbonate groundwater: Water Resource Research, v. 16, p. 381–386.

Drake, J. J., and Wigley, T. M. L., 1975, The effect of climate on the chemistry of carbonate groundwater: Water Resource Research, p. 958–962.

Dreybrodt, W., 1981, Kinetics of the dissolution of calcite and its application to karstification: Chemical Geology, v. 31, p. 245–269.

Dreybrodt, W., 1990, The role of dissolution kinetics in the development of karst aquifers in limestone: a model simulation of karst evolution: Journal of Geology, v. 98, p. 639–655.

Easterbrook, D. J., 1964, Void ratios and bulk densities as means of identifying Pleistocene tills: Geological Society of America Bulletin, v. 75, p. 745–750.

Egemeir, S. J., 1981, Cavern development by thermal waters: National Speleological Society Bulletin, v. 43, p. 1–52.

Ford, D. C., and Ewers R., 1978, The development of limestone cave systems in length and depth: Canadian Journal of Earth Sciences, v. 15, p. 1783–1798.

Ford, D. C., and Williams, P. W., 1989, Karst geomorphology and hydrology: Unwin Hyman, London, 601 p.

Fournet, J., 1852, Note sur les effondrements: Memoirs de Acadamerme de Lyon, p. 176–186.

Gardner, J. H., 1935, Origin and development of limestone caverns: Geological Society of America Bulletin, v. 46, p. 255–274.

Gerstenhauer, A., 1960, Der tropische Kegelkarst in Tabsco (Mexico): Zeitschrift für Geomorphologie Supplement, v. 2, p. 22–48.

Gradzinski, R., and Radomski, A., 1965, Origin and development of internal poljes "Hoyos" in the Sierra de los Organos: Poland, Bulletin of the Academy of Science, v. 13, p. 181–186.

Gruber, T., 1791, Briefe hydrographischen Imd physikalischen Inhalts aus Krain: Krauss, Wien, Austria, 162 p.

Grund, A., 1903, Die Karsthydrographie. Studien aus Westbosnien: Geographische Abhandlungen, v. 7, p. 1–200.

Grund, A., 1910, Beitrage zur Morphologie des dinarischen Gebirges: Geographische Abhandlungen, v. 9, p. 1–236.

Grund, A., 1914, Der geographische Zyklus im Karst: Zeitschrift für Geschiebeforschung Erdkunde, p. 621–640.

Gunn, J., 1981, Hydrological processes in karst depressions: Zeitschrift für Geomorphologie, v. 25, p. 313–331.

Gunn, J., 1983, Point recharge of limestone aquifers—A model from New Zealand karst: Journal of Hydrology, v. 61, p. 19–29.

Hacquet, B., 1778, Oryctographia Carniolica oder physikalische erd-beschreibung des herzogthums krain, istrien und zurn Tedder benachbarten lander: Gottlob und Breitkopf, Leipzig, Germany, v. 1, 162 p.; v. 2, 186 p.; v. 3, 184 p.

Herman, J. S., and White, W. B., 1985, Dissolution kinetics of dolomite: effects of lithology and fluid flow velocity: Geochimica et Cosmochemica Acta, v. 49, p. 2017–2026.

Higgens, C. G., and Coates, D. R., eds., 1990, Groundwater Geomorphology; The role of subsurface water in Earth-surface processes and landforms: Geological Society of America Special Paper 252, p. 177–209.

Hill, C. A., 1981, Speleogenesis of Carlsbad Caves and other caves of the Guadalupe Mountains: Proceedings of 8th International Congress of Speleology, Bowling Green, KY, p. 143–144.

Howard, A. D., 1963, The development of karst features: National Speleological Society Bulletin, v. 25, p. 45–65.

Jakucs, L., 1977, The morphogenesis of karsts: Halsted Press, N.Y., 284 p.

Jennings, J. N., 1967, Some karst areas of Australia: in Jennings, J. N., and Mabbutt, J. A., eds., Landform studies from Australia and New Guinea: Canberra, Australia, p. 256–292.

Jennings, J. N., 1971, Karst: M.I.T. Press, Cambridge, MA, 252 p.

Jennings, J. N., 1985, Karst geomorphology: Basil Blackwell, Oxford, England, 293 p.

Jennings, J. N., and Bik, M. J., 1962, Karst morphology in Australian New Guinea: Nature, v. 194, p. 1036–1038.

Jennings, J. N., and Sweeting, M. M., 1963, The limestone ranges of the Fitzroy Basin, Western Australia: Bonner Geographische Abhandlungen, no. 32.

Johnson, A. I., 1967, Specific yield—compilation of specific yields for various materials: U.S. Geological Survey Water Supply Paper 1662-D, 74 p.

Katzer, F., 1909, Karst und Karsthydrographie. Zur Kunde der Balkanhalbinsel: Sarajevo, Kajon, 94 p.

Kemmerly, P. R., 1980b, A time-distribution study of doline collapse: framework for prediction: Environmental Geology, v. 3, p. 123–130.

Kemmerly, P. R., 1982, Spatial analysis of a karst depression population: clues to genesis: Geological Society of America Bulletin, v. 93, p. 1078–1086.

LaFleur, R. G., ed., 1984, Groundwater as a geomorphic agent: Allan and Unwin, London, 390 p.

Larkin, R., and Sharp, J., 1992, On the relationship between river-basin geomorphology, aquifer hydraulics, and ground-water flow direction in alluvial aquifers: Geological Society of America Bulletin, v. 104, p. 1608–1620.

Lehmann, H., 1936, Morphologische Studien auf Java: Engelhorn, Stuttgart, Germany, 114 p.

Lehmann, H., 1954, Der tropische Kegelkarst auf den grossen Antillen: Erkunde, v. 8, p. 130–139.

Louis, H., 1956, Die Entstehung der Poljen und ihre Stellung in der Karstabtagung auf Grund von Beobachtungen im Taurus: Erkunde, v. 10, p. 33–53.

Malott, C. A., 1921, A subterranean cut-off and other subterranean phenomena along Indian Creek, Laurence Co., Indiana: Indiana Academy of Science Proceedings, v. 31, p. 203–210.

Malott, C. A., 1929, Three cavern pictures: Proceedings of Indiana Academy of Science, v. 38, p. 1–6.

Malott, C. A., 1932, Lost River at Wesley Chapel Gulf, Orange County, Indiana: Proceedings of the Indiana Academy of Science, v. 41, p. 285–316.

Manger, G. E., 1963, Porosity and bulk density of sedimentary rocks: U.S. Geological Survey Bulletin 1144-E.

Martel, E. A., 1910, La theorie de "Grundwasser" et les eaux souterraines du karst: Geographie, v. 21, p. 126–130.

Martel, E. A., 1921, Nouveau Traite des Eaux Souterraines: Doin, Paris, 840 p.

Matsch, F. E., and Denny, K. J., 1966, Grain-size distribution and its effect on the permeability of unconsolidated sands: Water Resources Research, v. 2, p. 665–677.

McDonald, R. C., 1976, Hillslope base depressions in tower karst topography of Belize: Zeitschrift für Geomorphologie Supplement, v. 26, p. 98–103.

McDonald, R. C., 1979, Tower karst geomorphology in Belize: Zeitschrift für Geomorphologie Supplement, v. 32, p. 35–45.

Mijatovic, B. F., 1983, Karst poljes in Dinarides: in Mijatovic, B. F., ed., Hydrogeology of Dinaric Karst, Geozavad, Belgrade, Yugoslavia.

Milanovic, P., 1981, Karst hydrology: Water Resources Publications, Littleton, CO.

Miotke, F. D., 1973, The subsidence of the surfaces between mogotes in Puerto Rico east of Arecibo: Caves and Karst, v. 15, p. 1– 12.

Monroe, W. H., 1960, Sinkholes and towers in the karst area of north-central Puerto Rico: U.S. Geological Survey Professional Paper 400-B, p. 356–360.

Monroe, W. H., 1968, The karst features of northern Puerto Rico: National Speleology Society Bulletin, v. 30, p. 75–86.

Monroe, W. H., 1970, A glossary of karst terminology: U.S. Geological Survey Water Supply Paper 1899K, 26 p.

Monroe, W. H., 1976, The karst landforms of Puerto Rico: U.S. Geological Survey Professional Paper 899, 69 p.

Moorhouse, D. E., 1968, Cave development via the sulfuric acid reaction: National Speleological Society Bulletin, v. 30, p. 1–10.

Nunez-Jimenez, A., 1959, Geografica de Cuba: Editorial Lex., Havana, Cuba.

Owen, D. D., 1856, Annual report of the Geological Survey in Kentucky: Frankfort, Ky., Geological Survey of Kentucky, p. 169–172.

Palmer, A. N., 1972, Dynamics of a sinking stream system, Onesquetaw Cave, N.Y.: National Speleological Society Bulletin, v. 34, p. 89–110.

Palmer, A. N., 1975, Origin of maze caves: National Speleology Society Bulletin 37, p. 57–76.

Palmer, A. N., 1977, Influence of geologic structure on groundwater flow and cave development in Mammoth Cave National Park, USA: in Tolson, J. S., and Doyle, F. L., eds., Karst hydrology: International Association of Hydrogeologists, 12th Memoir, p. 405–414.

Palmer, A. N., 1984a, Recent trends in karst geomorphology: Journal of Geological Education, v. 32, p. 247–253.

Palmer, A. N., 1984b, Geomorphic interpretation of karst features: in LaFleur, R. G., ed., Groundwater as a geomorphic agent, Allen and Unwin, London, p. 173–209.

Palmer, A. N., 1987, Cave levels and their interpretation: National Speleological Society Bulletin, v. 49, p. 50–66.

Palmer, A. N., 1990, Groundwater processes in karst terranes: in Higgens, C. G., and Coates, D. R., eds., Groundwater Geomorphology; The Role of Subsurface Water in Earth-Surface Processes and Landforms, Geological Society of America Special Paper 252, p. 177–209.

Palmer, A. N., 1991, Origin and morphology of limestone caves: Geological Society of America Bulletin, v. 103, p. 1–21.

Palmer, V. E., and Palmer, A. N., 1975, Landform development of the Mitchell Plain of southern Indiana: origin of a partially karstic plain: Zeitschrift für Geomorphologie, v. 19, p. 1–39.

Panes, V., and Stelcl, O., 1968, Physiographic and geologic control in development of Cuban mogotes: Zeitschrift für Geomorphologie, v. 12, p. 117–165.

Penck, A., 1894, Morphologie der Erdoberflache: Engelhorn, Stuttgart, Germany, p. 1–471; p. 2–696.

Penck, A., 1900, Geornorphologische Studien aus der Hercegovina: Zeitschrift Deutsche Osterrich. Alpenver, v. 31, p. 25–41.

Penck, A., 1904, Uber Karstphanomen: Vortrage Ver. Naturwiss, Kenntnisse, v. 44, p. 1–38.

Piper, C. D., 1932, Ground water in north-central Tennessee: U.S. Geological Survey Water Supply Paper 640.

Pohl, E. R., 1955, Vertical shafts in limestone caves: National Speleology Society Occasional Paper 2.

Pohl, E. R., and White, W. B., 1965, Sulfate minerals—their origin in the central Kentucky karst: American Mineralogist, v. 50, p. 1461–1465.

Prestwich, J., 1854, On some swallow holes on the chalk hills near Canterbury: Quarterly Journal, p. 222–224; p. 241.

Raines, T. W., 1972, Sotanito de Ahuacatlan: Association of Mexican Cave Studies Cave Report Series, no. 1.

Roglic, J., 1957, Quelques problemes fondmentaux du karst: L'Infonnation Geographique, v. 2 1, p. 1–12.

Roglic, J., 1972, Historical review of morphologic concepts: in Herak, M., and Stringfield, V. T., eds., Karst: important karst regions of the Northern Hemisphere: Elsevier, Amsterdam, Netherlands, p. 1–18.

Silar, J., 1965, Development of tower karst of China and North Vietnam: National Speleology Society Bulletin 27, p. 35–46.

Smith, D. I., 1969, The solution erosion of limestone in an arctic morphogenic region: in Stelcl, O., ed., Problems of the karst denudation, Brno, Czech., p. 99–110.

Smith, D. I., and Atkinson, T., 1976, Process landforms and climate in limestone regions: in Derbyshire, E., ed., Geomorphology and climate: John Wiley and Sons, N.Y., p. 367–409.

Sunartadirdja, M. A., and Lehmann, H., 1960, Der tropische karst von Maros und Nord-Bone in SW Celebes (Sulawesi): Zeitschrift für Geomorphologie Supplement, v. 2, p. 49–65.

Sweeting, M. M., 1950, Erosion cycles and limestone caverns in the Ingleborough District of Yorkshire: Geographical Journal, v. 113, p. 63–78.

Sweeting, M. M., 1958, The karstlands of Jamaica: Geographical Journal, v. 24, p. 184–199.

Sweeting, M. M., 1973, Karst landforms: Columbia University Press, N.Y., 362 p.

Sweeting, M. M., ed., 1981, Karst geomorphology: Academic Press, N. Y., 427 p.

Swinnerton, A. C., 1932, Origin of limestone caverns: Geological Society of America Bulletin, v. 43, p. 662–693.

Thrailkill, J., 1968, Chemical and hydrology factors in the excavation of limestone caves: Geological Society of America Bulletin, v. 79, p. 19–45.

Thrailkill, J., 1972, Carbonate chemistry of aquifer and stream water in Kentucky: Journal of Hydrology, v. 16, p. 93–104.

Thrailkill, J., and Robi, T., 1981, Carbonate geochemistry of vadose water recharging limestone aquifers: Journal of Hydrology, v. 54, p. 195–208.

Trainer, F., and Health, R., 1976, Bicarbonate content of groundwater in carbonate rock in eastern North America: Journal of Hydrology, v. 31, p. 37–55.

Troester, J. W., White, E. L., and White, W. B., 1984, A comparison of sinkhole depth frequency distributions in temperate and tropical karst regions: in Beck, B. R., Sinkholes, their geology, engineering, and environmental impact: A. A. Balkema, Rotterdam, Netherlands, p. 65–73.

Verstappen, H. T., 1964, Karst morphology of the Star Mountains (central New Guinea) and its relation to lithology and climate: Zeitschrift für Geomorphologie, v. 8, p. 40–49.

Virlet, J., 1834, Observations faites en Franche-Comte sur les caverns et la theorie de leurs formation: Bulletin Society Geologie France, v. 6, p. 148–163.

Watson, R. A., and White, W. D., 1985, The history of American theories of cave origin: Geological Society of America Centennial Special Volume 1, p. 109–123.

Weyl, P. K., 1958, The solution kinetics of calcite: Journal of Geology, v. 66, p. 163–176.

White, W. B., 1960, Termination of passages in Appalachian caves as evidence for a shallow phreatic origin: National Speleology Society Bulletin, v. 22, p. 43–53.

White, W. B., 1988, Geomorphology and hydrology of karst terrains: Oxford University Press, N.Y., 464 p.

White, W. C., and White, E. L., 1974, Base-level control of underground drainage in the Potomac River Basin: in Raunch, H. W., and Werner, E., eds., Proceedings of the 4th Conference on Karst Geology and Hydrology, West Virginia Geological Survey, p. 41–53.

Wilford, G. E., and Wall, J. R. D., 1965. Karst topography in Sarawak. Journal of Tropical Geography, v. 21, p. 44–70.

Williams, P. W., 1970, Limestone morphology in Ireland: Queens University, Irish Geographical Studies, Department of Geography, Belfast, Ireland, p. 105–124.

Williams, P. W., 1971, Illustrating morphometric analyses of karst with examples from New Guinea: Zeitschrift für Geomorphologie, v. 15, p. 40–61.

Williams, P. W., 1987, Geomorphic inheritance and the development of tower karst: Earth Surface Processes and Landforms, v. 12, p. 453–465.

Wolfe, T. E., 1964, Cavern development in the Greenbrier Series, West Virginia: National Speleology. Society Bulletin, v. 26, p. 37–60.

Woo, M., and Marsh, 1977, Effect of vegetation on limestone solution in a small High Arctic Basin: Canadian Journal of Earth Sciences, v. 14, p. 571–581.

Woodward, H. P., 1961, Limestone pavements with special reference to Western Iceland: Institute of British Geographers, Transaction Paper no. 40, p. 155–172.

Tectonic Landforms

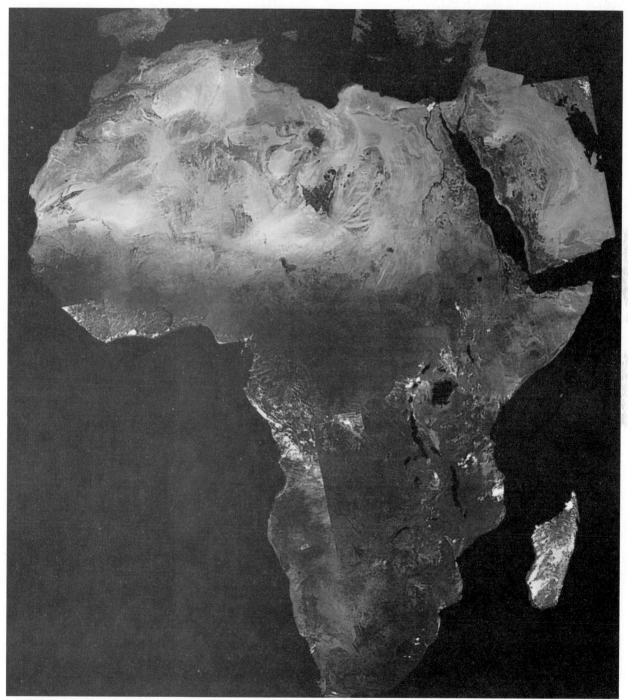

Satellite view of Africa, showing the Red Sea rift zone (right side). (NASA photo)

INTRODUCTION

The major surface features of the earth—the continents and ocean basins with their vast mountains and valleys—constitute the basic architecture upon which surface processes operate to carve the landscape as we see it today. A plot of elevations of the earth's surface (Figure 8–1) shows two prominent levels, corresponding to the average altitude of the continents and the average depth of the ocean basins. About two-thirds of the earth's surface consists of the floors of the oceans. The average altitude of the continents is 840 m (2770 ft) above sea level, and the average depth of the oceans is 3700 m (12,200 ft) below sea level. Only a small fraction of the earth's surface rises higher than the average altitude of the continent or lies above or below the mean ocean depth (Figure 8–1).

These two predominant topographic subdivisions of the earth differ in the nature of the rocks composing them and in chemical composition, density, geologic structure, age, and geologic history. Although largely unseen, except through various geophysical instruments, the ocean basins are not featureless plains, as once believed, but consist of impressive topographic relief made from extensive volcanic activity and crustal movements presently in operation. The continents stand above the ocean basins as great platforms, partially flooded by waters of the oceans. Today's shorelines do not mark the division between continents and oceans and have no real structural significance. Because the oceans spill out of their basins and lap up onto the **continental shelves**, shorelines fluctuate remarkably with relatively small deviations in sea level.

The altitude of the continents above the ocean basins is considerably more significant than the position of the shoreline and, unlike the shoreline, represents fundamental differences in rock density and geologic structure. The continents consist almost entirely of granitic rocks having an average density of about 2.6 gm/cc; these granitic rocks are less dense than the basaltic rocks of the ocean basins, which have a density of about 3.0 gm/cc. This difference in density is one of the main reasons that the continental crust "floats" above the denser oceanic crust. If these differences between the continents and ocean basins were not so profound, the total volume of water on the earth might well be sufficient to cover the entire surface of the earth with water, and the only life forms would be aquatic.

FIGURE 8–1
Major relief features of the Earth.

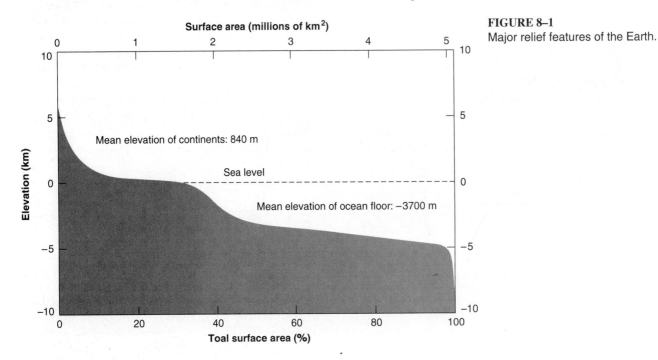

The oceanic crust also differs in other ways from the continental crust. The rocks of the ocean floor are surprisingly young, generally less than 150 million years old, whereas the ancient rocks of the continental shields are several billion years old. Compressional deformation is virtually absent in ocean floor rocks, in sharp contrast to the intricate deformation of continental rocks.

MAJOR FEATURES OF THE CONTINENTS

Although topographic relief on the continents often gives the impression of limitless diversity, the continents are really quite flat when viewed on a global scale. All of the continents are parts of crustal plates, and all have common components:

1 Cratons, composed of shields and stable platforms
2 Folded linear mountain belts

In addition, some continents also contain rift zones, linear volcanic chains, flood basalts, and tilted fault-block mountains.

Cratons

Parts of continents consist of **cratons**—expansive, stable regions of low relief. Those made of very old, complexly deformed, crystalline basement rocks are known as **shields**, whereas those buried beneath a relatively thin mantle of sedimentary rocks are called **stable platforms**. Both are characterized by long-term crustal stability, and both have remained relatively undisturbed for up to a billion years or so or have experienced only broad, gentle crustal warping.

Topographic relief on shields is typically quite low, often ranging within a hundred meters or less, as a result of prolonged erosion and low rates of deformation. Shields are usually made up of crystalline igneous and metamorphic rocks, some as old as 3.5 billion years; and although such areas have been very stable for long periods of time, the rocks exposed in them reveal great deformation and show evidence of having originated several kilometers below the earth's surface. The Canadian Shield, which covers a major portion of the North American continent, is a good example of a continental shield.

Extensive portions of cratons have been relatively stable since the late Precambrian or early Paleozoic eras and are covered with a thin veneer of sedimentary rocks. A good example is the east-central United States, which was covered by shallow Paleozoic seas, but its thin sedimentary cover has been only mildly warped since then.

Subduction Zone and Folded Linear Mountain Belts

The margins of most continents consist of young, linear mountain belts composed of intensely folded rocks (Figure 8–2) deformed by strong horizontal crustal stresses and intrud-

ed by igneous rocks. These belts typically occur where large crustal plates converge. Modern examples of such mountain belts are the circum-Pacific mountain chains of western North and South America and the Himalayan mountain belt of eastern Asia, all of which are characterized by very high rates of crustal deformation.

Crustal Spreading—Rift Zones

The topographic expression of oceanic ridges (discussed later in this chapter) marks the sites of spreading centers (**crustal divergence zones**). Where spreading centers extend under continents, **rift zones** are formed, such as the Red Sea rift (chapter-opening photo and Figure 8–3), which marks the landward extension of the Indian Ocean ridge. The pulling apart that occurs at divergent plate boundaries induces strong tensional stresses that cause block faulting, rifting (fracturing that creates open fissures), and eruption of basaltic lava. Extensive fissure eruptions are common. The vast rift valleys of eastern Africa (Figure 8–3), where mammoth down-dropped fault blocks make extensive elongate valleys, represent an early stage in this process. The opening of the Red Sea, separating the Arabian Peninsula from Africa, is an expression of a more advanced stage of rifting. The opening of the Atlantic Ocean along the mid-Atlantic ridge and the consequent separation of North and South America from Africa and Europe by thousands of kilometers reflect the advanced stage of sea-floor spreading.

Transform Faults—Large Strike-Slip Faults

Some crustal plates are bounded by great, steeply dipping strike-slip faults known as **transform faults** along which major plates slide past one another. One of the most dramatic examples is the 1000 km (600 mi) long San Andreas fault along which the North American and Pacific plates are presently sliding past one another, slicing California in a north-south direction and producing large earthquakes that threaten large metropolitan areas. Displacement along the San Andreas fault is thought to be several hundred kilometers. Another example is the Alpine fault which slices through New Zealand, separating the Pacific and Indian-Australian plates. Transform faults also cut oceanic ridges and trenches.

MAJOR FEATURES OF THE OCEANS

The topography of the ocean floors long remained hidden to view, and our knowledge of them was exceedingly sparse, limited to widely dispersed grab samples here and there. However, in the 1950s, the advent of precision, continuous, bottom profile recorders allowed construction of detailed maps of the ocean floor topography, and deep-sea drilling that began in the 1960s radically altered the prevailing view

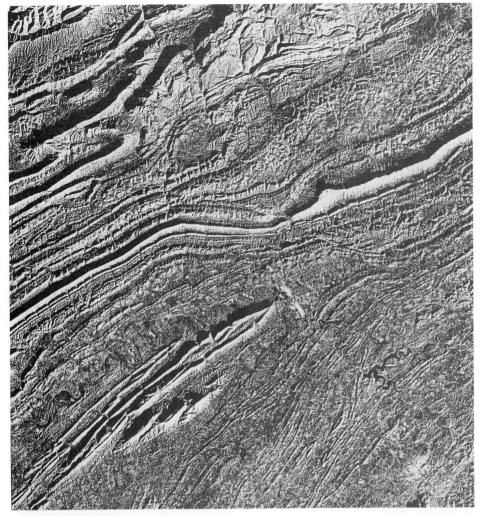

FIGURE 8–2
Folded mountain belt,
Appalacian Mts. (Radar mosaic,
U.S. Geological Survey)

FIGURE 8–3
East African rift system, Nile
delta on the left. (Landsat
satellite image)

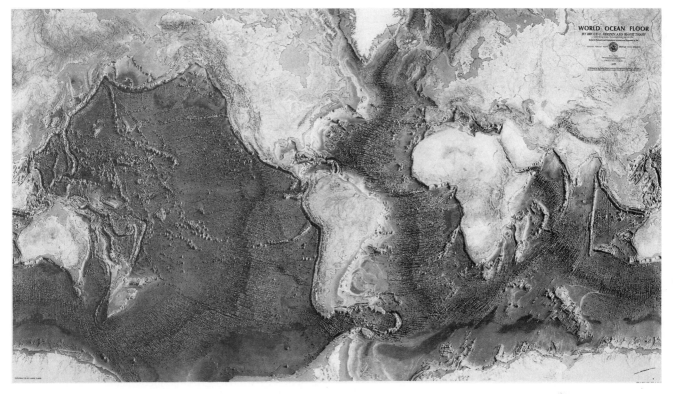

FIGURE 8–4
Topography of the ocean floors. (Bruce C. Heezen and Marie Tharp)

of the sea floors. The geophysical data obtained from the ocean floors led to formulation and confirmation of the plate tectonic concept.

Maps of the ocean floors showed that submarine topography is strongly controlled by crustal processes and structures (Figure 8–4). Among the most significant features discovered during investigation of the ocean floors were **oceanic ridges, rift zones, abyssal floor, deep-sea trenches**, and **volcanic arcs**. In addition, the discovery of magnetic anomaly patterns, transform faults, and lack of old ocean sediments verified the relationship of ocean floor topography to crustal plate motions.

Oceanic Ridges

The discovery of long **oceanic ridges** profoundly affected views of the origin of ocean floor topography. The mid-Atlantic ridge extends from the Arctic Ocean southward down the middle of the Atlantic Ocean for its entire length, and similar extensive oceanic ridges occur in the Pacific and Indian oceans. Oceanic ridges divide the Indian Ocean roughly in half and traverse the southern and eastern parts of the Pacific. The ridges are generally about 1500 km (900 mi) wide and rise up to 3000 m (9800 ft) above the ocean floor. A surprising aspect of the ridges was the detection of rift valleys, large linear valleys along the crest of the oceanic ridges. These are

now believed to be a result of pulling apart of the ridges by the process of sea-floor spreading.

Coring of deep-sea sediments has shown that deposits on the flanks of oceanic ridge spreading centers are geologically very young because new oceanic crust is being created along spreading centers by upwelling from below. Sediments lying directly above ocean floor basalt become progressively thicker and older farther away from the ridges. If modem sedimentation rates of about 1 cm/1000 yr (0.4 in/1000 yr) were projected back in time, the ocean floors would be mantled with about 5000 m (16,000 ft) of sediment. Instead, the greatest measured thickness of ocean floor sediments is about 300 m (980 ft). The reason for such a great discrepancy must be that the ocean basins are geologically quite young.

Abyssal Floor

Vast, low-relief, deep-ocean basins on both sides of oceanic ridges at depths of about 3000 m (9800 ft) are known as the **abyssal floor.** They extend from the oceanic ridges to the continental margins and can be subdivided into **abyssal hills** and **abyssal plains.** The abyssal hills cover about three-fourths of the Pacific Ocean floor, attaining relief as great as 900 m (3000 ft). The abyssal plains form flat, smooth, sediment-covered sea floors near the continental margins.

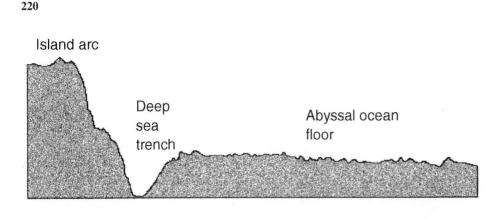

FIGURE 8–5
Profile of a typical deep sea trench.

Deep-sea Trenches and Volcanic Arcs

The deepest parts of the oceans are not out in the middle but rather occur in elongate **deep-sea trenches** close to continental margins. The trenches occur next to linear mountain ranges along continental margins and island arcs where they are typically associated with strong seismic activity and volcanic eruptions.

The Mariana Trench off the Philippine coast (Figure 8–5) is the greatest known ocean depth at $-11,000$ m ($-36,300$ ft). Depths in the other trenches are commonly -8000 m ($-26,400$ ft) or more.

RELATIONSHIP OF MAJOR TOPOGRAPHIC FEATURES TO PLATE TECTONICS

The plate tectonic concept is based on the division of the lithosphere into large, discrete, rigid crustal plates (Figures 8–6 and 8–7), which move about as a result of plastic flow in the underlying asthenosphere. As material flows plastically upward toward the surface due to thermal convection in the asthenosphere, it arches upward to form an oceanic ridge and flows laterally, pulling the overlying rigid crust apart at a spreading center. The crustal fragments form plates several thousand kilometers across. Molten rock rises into the rift zone as the plates move apart, creating new oceanic crust.

Where moving continental plates collide with oceanic plates, the lighter continental granitic rocks ride over the denser descending oceanic rocks to form a subduction zone. Deep-sea trenches form where the surface is dragged down by the descending plate, and linear fold mountains form by compression of the overriding plate.

Plate boundaries coincide with oceanic ridges, deep-sea trenches, and folded mountain belts, and they are zones of earthquakes and volcanic activity. Most plate boundaries are therefore not associated with coastlines between oceans and continents. The boundaries of the plates are recognizable today by linear patterns of earthquake activity (Figure 8–8) and volcanoes, oceanic ridges, deep-sea trenches, and rift zones.

Seven large crustal plates can be identified on the basis of their geophysical and topographic characteristics: the North American, South American, Pacific, Indian-Australian, African, Eurasian, and Antarctic plates (Figure 8–6). The Pacific plate consists almost entirely of oceanic crust; all of the other major plates contain both continental and oceanic crust. Continental crustal plates are thicker than oceanic plates, ranging in thickness up to about 150 km (90 mi) in contrast to only about 70 km (45 mi) for oceanic plates. Crustal plates are consumed as they move down subduction zones and new crust is generated over spreading centers. Thus, the size and the shape of plates are not fixed over long periods of geologic time.

Several smaller plates can also be distinguished: the Caribbean, Nazca, Cocos, Scotia, Arabian, Philippine, and Juan de Fuca plates (Figure 8–6). The smaller plates are usually found near convergence zones of major plates where continents are colliding, or between continents and island arcs. Rapid and complex motion is typical of the smaller plates.

The North American and South American plates are moving northwestward and are interacting with the Pacific, Juan de Fuca, Cocos, and Nazca plates along the west coast of the Americas. The mountain chains of western North America, from the Coast Ranges and Cascade Mountains to the Rocky Mountains and the coastal ranges of Alaska, are produced by collision of the North American plate with the Pacific and Juan de Fuca plates. The Andes Mountains of South America result from collision of the South American and Nazca plates.

The Pacific plate, composed almost entirely of oceanic crust and covering about 20 percent of the earth's surface, is moving northwestward from the East Pacific Rise oceanic ridge toward the Eurasian and Philippine Sea plates and parts of the Australian plate. The deep-sea trenches of the western Pacific are developed by collision and subduction of the Pacific plate with the adjacent plates. At its eastern margin, the Pacific plate slides past the North American plate more or less horizontally, causing a major fracture system between the two plates—the great San Andreas fault in California along which several hundred kilometers of movement has taken place.

Movement of the Australian plate northward from the oceanic ridge at the northern margin of the Antarctic plate

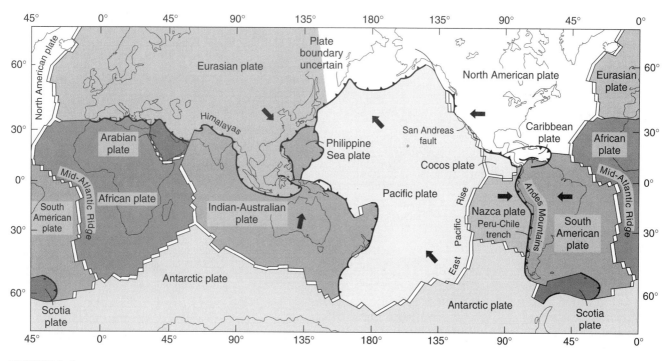

FIGURE 8–6
Crustal plates of the world.

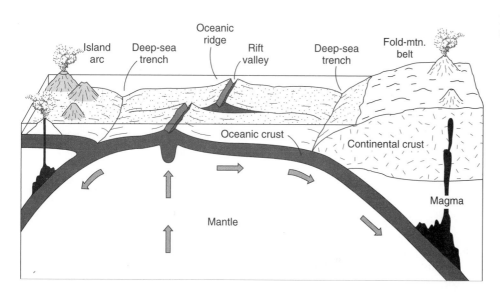

FIGURE 8–7
Cross section of crustal plates.

causes India to collide with the Eurasian plate, producing the Himalayan Mountains and the volcanic arc/deep-sea trenches of the East Indies.

The African plate is moving eastward from the southern mid-Atlantic ridge and northward from the oceanic ridge at the northern margin of the Antarctic plate. The Mediterranean Alpine Mountain belt is formed where it collides with the Eurasian plate. The Red Sea rift zone (Figure 8–3), which splits the Sinai and Arabian peninsulas from Africa, is an expression of the landward extension of the Indian Ocean spreading center.

The Eurasian plate, the largest of the continental plates, moves eastward from the northern mid-Atlantic ridge. It forms fold mountain belts along much of its southern margin where it collides with the African, Arabian, and Australian plates, and it forms volcanic arcs and deep-sea trenches where it collides with the Pacific plate. The Antarctic plate is bounded on almost all sides by oceanic ridges, which mark spreading centers.

Volcanic landforms are concentrated along plate margins, especially where plate convergence occurs. The great volcanic mountain belts of the continents all occur at converging plate boundaries. Volcanism is also extensive along spreading

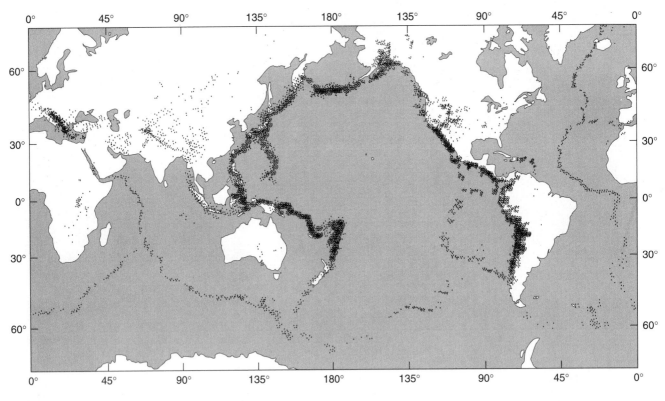

FIGURE 8–8
Earthquakes along plate margins.

centers, although it is much less obvious because it occurs along oceanic ridges below sea level.

Thus, the major topographic elements of the earth can be seen to be directly related to movement of crustal plates, and an understanding of plate tectonics is necessary to discern their origin and distribution. Because global tectonics plays such an important role in the arrangement of these large-scale features, the earth's crust can no longer be considered as a relatively stable place upon which surficial processes operate. The dynamic nature of plate movements is such that over periods of geologic time, the face of the earth will look quite different. In fact, if one projects the present directions and rates of movement into the future, the distribution of major landforms will change significantly.

REFERENCES

Bird, J. M., ed., 1980, Plate tectonics: American Geophysical Union, Washington, D.C.

Condie, K. C., 1989, Plate tectonics and crustal evolution: Pergamon Press, Elmsford, N.Y.

Gross, M. G., 1986, Oceanography: Prentice Hall, Englewood Cliffs, N. J.

Hamblin, W. K., and Christiansen, E. H., 1995, Earth's dynamic systems: Macmillan Publishing, N.Y., 710 p.

Hill, M. L., and Dibblee, T. W., 1953, San Andreas, Garlock, and Big Pine faults, California: Geological Society of America Bulletin, v. 64, p. 443–458.

Johnson, D. W., 1929, Geomorphic aspects of rift valleys: 15th International Geological Congress, South Africa, Comptes rendus, v. 2, p. 354–373.

McConnell, R. B., 1972, Geological development of the rift system of east Africa: Geological Society of America Bulletin, v. 83, p. 2549–2572.

McElhinny, M. W., and Valencio, D. A., 1981, Paleoreconstruction of the continents: Geological Society of America, Geodynamic Series, v. 2.

Oakshott, G. B., 1966, San Andreas fault: geologic and earthquake history: California Division of Mines and Geology, Mineral Information Service, v. 19, p. 159–165.

Orowan, E., 1969, The origin of the oceanic ridges: Scientific American, v. 221, p. 102–119.

Sieh, K. E., and Jahns, R. H., 1984, Holocene activity of the San Andreas fault at Wallace Creek, California: Geological Society of America Bulletin, v. 95, p. 883–896.

Van Andel, T. H., 1985, New Views on an old planet: Cambridge University Press, N.Y.

Vine, F. J., and Matthews, D. H., 1963, Magnetic anomalies over oceanic ridges: Nature, v. 199, p. 947–949.

Wilson, J. T., 1965, A new class of faults and their bearing on continental drift: Nature, v. 207, p. 343–347.

Wilson, J. T., 1976, ed., Continents adrift and continents aground: W. H. Freeman, San Francisco, CA.

CHAPTER NINE

Topographic Expression of Folded Strata

Topography developed on a plunging anticline at Sheep Mtn., Wyoming. (Photo by D. A. Rahm)

INTRODUCTION

The relationship between topography and geologic structure is from offsetting of the land surface by recent tectonic processes or by differential erosion of rocks whose spatial distribution is controlled by older geologic structures. The geomorphic expression of folded sedimentary beds is controlled by

1. The rate of crustal deformation
2. The three-dimensional geometry of the rock structure
3. Differential erosion of the rocks involved in the structure

The rate of crustal deformation relative to the rate of erosion in an area may be high enough and the folding geologically recent enough that weathering and erosion have not yet significantly modified the initial form of the folds. In such a case, the topographic expression of the folds stems from the configuration of the folds and their direct modification of the land surface; that is, anticlines make hills and synclines make valleys. However, the converse is not true—not all hills are anticlines and not all valleys are synclines.

Differential erosion of rocks of differing resistance plays an important role in development of topography on folded rocks. The form of the surface topography depends on the preferential etching out of weaker rocks, leaving the more resistant ones standing out in relief. The resulting surface forms then are controlled by relative rock resistance and by the three-dimensional geometry of the rock structure, which determines the relative position of the rocks in space. Because rates of weathering and erosion are high relative to rates of folding, most landforms developed on folded structures are related to differential erosion rather than to initial form caused by deformation. Most folds are quickly (in terms of geologic time) modified by erosion.

INITIAL FORMS

As folds grow in response to stresses in the earth's crust, the land surface is displaced. Initial forms may be generated so rapidly as to directly modify the land surface faster than erosion. If rates of erosion are less than rates of folding, folding of rocks into anticlines and synclines initially results in **anticlinal ridges** (the topographic high coincides with the structural high) and **synclinal valleys** (the topographic low coincides with the structural low). Where erosion has stripped away the uppermost beds from anticlines making topographic ridges, but the underlying beds are resistant enough to erosion to retain ridge topography, the term anticlinal *ridge is* still appropriate even though the landform is no longer the initial form of the fold. The key factor here is that the topographic crest of the ridge corresponds to the structural axis of the anticline. Similar relationships hold for synclinal valleys. Figure 9–1 shows examples of anticlinal ridges and synclinal valleys on the Columbia Plateau of eastern Washington, where folds in Miocene basalt flows have not yet been breached by erosion.

An idea of the rates of folding may be obtained from dating of newly developing folds such as the Ventura anticline in California (Figure 9–2). Dating of Quaternary terraces deformed across the Ventura anticline allows calculation of deformation rates for the past 200,000 years (Rockwell et al., 1988):

Interval (yr)	Deformation Rate (mm/yr)
30,000 to present	5
30,000–80,000/105,000	9
80,000/105,000–200,000	20

FIGURE 9–1
Young anticlinal ridge that owes its topographic form to recent uplift. The anticlinal ridge rose across the Yakima River, which was able to maintain its course by incising its channel as rapidly as the ridge rose.

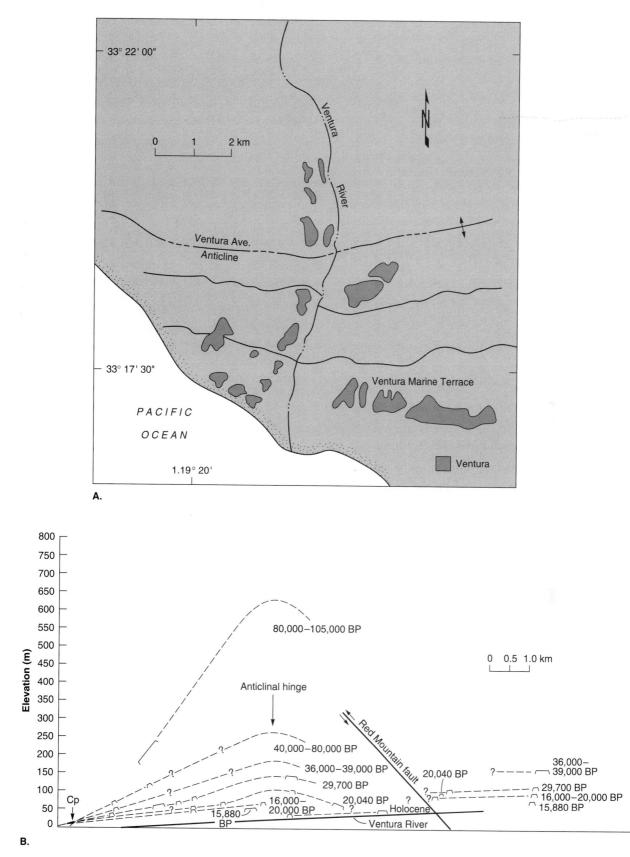

FIGURE 9–2
(A) Map of the Ventura anticline; (B) Cross section of the Ventura anticline showing rates of uplift. (From Rockwell et al., 1988)

DIFFERENTIAL EROSION

Variations in weathering and erosional attributes of sedimentary rocks cause etching out of weaker rocks, leaving the more resistant beds topographically higher than the weaker beds. Sandstone and conglomerate are more resistant to erosion than shale and make conspicuous ridges or mesas, whereas shale typically forms valleys or lowlands. The erosional resistance of carbonate rocks, such as limestone and dolomite, is greater in arid climates than in humid climates. Carbonate rocks are more susceptible to solution in humid regions, where precipitation supplies abundant groundwater, and carbonate rocks usually develop valleys or lowlands. However, in arid regions where solution is restrained by low precipitation, carbonate rocks are more resistant and typically produce ridges.

Gilbert (1877) recognized that topographic diversity is favored by geologic structures that set the stage for differential erosion. As the most erodible rocks are etched away in valleys, they collect the drainage while resistant rock masses are left as divides. In the course of their evolution, streams adjust their drainage pattern to the underlying rocks. Thus, if given enough time, streams will be guided by the weakest strata or by fracture zones in the rock.

FIGURE 9–3
Dendritic drainage pattern developed on flat-lying beds. (Photo by Dept. of Agriculture)

Geomorphic Expression of Flat-lying or Homogeneous Rocks

Where structural guidance is lacking, drainage branches in random fashion, like the branches of a tree. Such a dendritic drainage pattern is typical of flat-lying strata without any structural direction (Figure 9–3) and homogeneous rocks without any well-developed fracture patterns. Divides are randomly distributed between dendritic drainage basins. Three-dimensional structural homogeneity is typical of massive bodies of rock of uniform composition, as long as they are not cut by prominent joint, fault, bedding, or exfoliation planes. Thick, massive sandstone beds and large intrusive bodies develop similar dendritic drainage patterns if they are not highly jointed or faulted. Massive bodies of lava or pyroclastic material are similarly dissected, and even crystalline metamorphic rocks of complex structure, such as unfractured gneiss and schist, react to dissection as if they were homogeneous. Their very heterogeneity is so complex that none of the structures ranges far enough to effectively guide stream erosion.

In the examples given above, the topography is systematically uniform; that is, the slopes connecting divides with channels are smooth, continuous, and unbroken. However, many sequences of horizontal strata undergo a type of dendritic dissection, which is similar to that pattern, except that the slopes are broken into a staircase of cliffs and benches. Lack of homogeneity in the vertical dimension allows differential erosion that leads to **cliff-and-bench topography** where horizontally layered rocks characteristically reflect differences in rock type (Figure 9–4). Resistant sandstone, quartzite, chert,

conglomerate, limestone (in arid climates), lava flows, and sills are generally good cliff formers, whereas less resistant shale forms gentler slopes between cliffs. Cliff heights are a reflection of stratigraphic thickness so that on a topographic map such as Figure 9–4B both the outcrop distribution and bed thickness may be easily determined.

Differences in resistance may occur among rocks of the same composition. Such differences are well shown on the Columbia Plateau of the Pacific Northwest, where horizontal basalt flows of remarkably uniform composition are eroded differentially because of differences in jointing.

In humid regions, cliff and bench topography is more subdued, with all but the most resistant ledges masked by weathered material. The profile of slopes in a humid climate is not so much one of abrupt steps as one of reversals in curvature, the steeper segments marking the location of more resistant strata. Because of the rounding effects of weathering and creep, the heads of valleys are indistinct and are further obscured by a binding mat of vegetation, giving the staircase the appearance of wearing a thick carpet.

Under arid conditions, cliff and bench topography is chiseled into stark relief, where the cliff-forming strata cave away due to undermining as less-resistant beds are eroded. In this way, the cliffs maintain their ruggedness as they retreat by **basal sapping**.

Sapping of cliffs by rillwash, by sheetwash, and by erosion of underlying weak formations can cause long-distance retreat of cliffs to the extent that broad benches are left at the

A.

foot of the cliffs. Such structurally controlled benches may form on the slopes of canyons if the incision of the master stream does not outpace cliff retreat. They are prominent within the Grand Canyon where they are called **esplanades.** The formation of an esplanade can be the first stage in the development of a **stripped structural surface.**

Where the retreating escarpment is not uniformly linear, but crenulated with the buttresses and recesses created by many headward-eroding streams, detached outliers are cut off from the main line of cliffs to form **mesas.** The continued sapping of the caprock of a mesa causes it to cave away until only a relatively restricted summit area remains as a **butte** (Figure 9–5).

Geomorphic Expression of Tilted Strata

Tilted strata are eroded differentially in much the same way as horizontal strata. However, the definite strike and dip within the sequence provides direction to the drainage as high-re-

B.

FIGURE 9–4

(A) Cliff and bench topography, Grand Canyon, Arizona. (Photo by D. A. Rahm); (B) Topographic map of cliff and bench topography, Grand Canyon, Arizona. The thick dark lines are closely spaced contours on cliffs where the edges of resistant beds are exposed by erosion. The benches are developed on the tops of resistant beds. (Grand Canyon National Park map, U.S. Geological Survey)

FIGURE 9–5
Buttes of resistant sandstone capping weak shale,
Monument Valley, California. (Photo by D. A. Rahm)

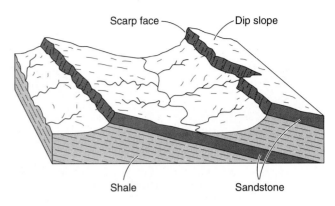

FIGURE 9–6
Adjustment of topography to geologic structure.

sistance and low-resistance beds are exposed in parallel strips. Ultimately, selective headward excavation of valleys along the strike of the least resistant strata leaves the resistant layers standing in relief and results in a topography of parallel ridges and valleys.

When sedimentary beds are tilted, drainage flows in the direction of dip. Such initial streams are called **consequent streams** because their courses are a consequence of the initial slope of the surface, cut gradually into the stratigraphic sequence in response to uplift. Tributaries to master consequent streams erode headward along weak beds of shale. As channels cut in easily erodible strata erode downward, their directly vertical incision through the dipping beds is temporarily interrupted when streams cut through the shale onto underlying resistant sandstone beds. Further excavation by the stream is arrested by the resistant bed, eventually causing the channel to shift laterally from the resistant beds to the more erodible adjacent shale beds. In this way, the resistant rock deflects streams laterally in a process called **homoclinal shifting** (Figure 9–6). Thus, the more easily erodible beds are exposed in belts as the floors of homoclinal valleys, as the more resistant formations become etched into higher relief and effectively rid themselves of streams.

As valleys are carved by the incision of **subsequent** streams into weak beds, the more resistant beds are left as homoclinal ridges (Figure 9–6). Such ridges are typically asymmetric in cross profile, with a steep **scarp face,** formed by basal sapping and accumulation of loose material at the angle of repose, and with a more gentle **dip slope,** corresponding to the dip of the resistant bed. Where the dip of the strata is gentle, homoclinal ridges are called cuestas (Figure 9–7A); where they are more steeply dipping (typically >30°– 40° they are hogbacks (Figure 9–7B). Homoclinal ridges become progressively less asymmetric with increase in dip, and

distinction of the dip slope and scarp face becomes more difficult on topographic maps. When the angle of repose of loose weathered material on the scarp face becomes approximately equal to the angle of dip, the ridge becomes symmetrical.

Because the scarp faces of most hogbacks and cuestas are steeper than the dip slopes, they are eroded more vigorously than the dip slopes. As a result, the divides are shifted in the direction of the dip—an effect that operates in alliance with the **homoclinal shifting** of streams at the base of the scarp slopes. The scarps of cuestas and hogbacks retreat in the direction of dip. If all **homoclinal valleys** are eroded at the same rate (both vertically and laterally), then the resistant ridges and strike valleys alike will migrate evenly in the dip direction, while maintaining an equivalent spacing and relative relief (Figure 9–6).

The intimate congruence of surface form to geologic structure permits determination of the direction of dip of beds. The asymmetry of homoclinal ridges, with gentle dip slopes and steep escarpments (Figure 9–7A, 9–7B) can be used to determine the dip of cuestas and hogbacks developed in strata dipping less than about 40°. The asymmetry is obscured on beds with greater dips because the angle of repose of weathered debris on the scarp face usually approximates 25° to 35°.

Because the surface slope of a homoclinal ridge is inclined in the direction of the dip of the bed, the angle of dip may be calculated if the dip slope is well defined. In Figure 9–8, the angle of dip (θ) of the bed may be calculated by determining the horizontal distance h relative to the vertical distance v.

$$\tan \theta = \frac{v}{h}$$

The value of v may be found from the number of contours intersected for a given horizontal distance h, which may be scaled directly from the map (Figures 9–8 and 9–9B). If v is 100 ft and h is 400 ft, then tan θ = (v/h) = 100/400 = 0.25, which is equivalent to an angle of 14° in a trigonometric table.

A.

B.

FIGURE 9–7
(A) Cuesta developed on tilted bed of resistant sandstone, Comb Ridge, Utah. Note that the ridge is asymmetric with a more gentle dip slope down the dip of the bed and steep scarp face made by the edge of the bed. (B) Hogback of more steeply dipping resistant sandstone. Note that with a steeper dip the ridge is more symmetric than the cuesta in (A). (Photos by W.K. Hamblin)

Where streams incise homoclinal ridges, **V-shaped notches** are cut into the resistant beds (Figure 9–9A and B). The apex of the V points in the direction of dip because of the decrease in height of the resistant bed above the stream in the direction of dip. Thus, the direction of dip of a homoclinal ridge may be determined by use V-shaped notches in ridges shown on topographic maps or air photos. Where two V-shaped notches are closely enough spaced, a triangular-shaped remnant of the bed between the notches forms **flatirons** (Figure 9–10).

Geomorphic Expression of Folds

As erosion attacks geologic structures, the increased elevation of anticlines stimulates more rapid erosion until initial forms become breached. As long as resistant beds are uncovered along the axial plane of the anticline, the form remains as an **anticlinal ridge** (Chapter opening figure). As erosion occurs along the crest of an anticline, more-easily erodible rocks uncovered by erosion near the axial plane of the leads to topographic inversion of the original relief to

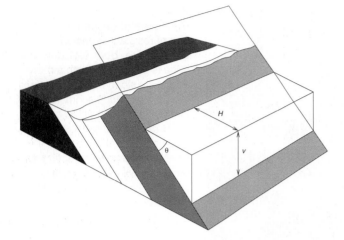

FIGURE 9–8
Calculation of the angle of dip of a bed making a homoclinal ridge.

form an **anticlinal valley** (Figure 9–11), with homoclinal ridges and valleys along the flanks marking the intersection of truncated beds with the surface. **Topographic inversion** may also affect a syncline where subsequent streams find more erodible beds along the flanks of the fold than in its trough, and where deep erosion of the flanks of the syncline leaves the more resistant rocks of the trough capping a **synclinal ridge.**

The escarpments of correlative homoclinal ridges on the flanks of breached anticlines face one another across the axial plane of the fold, whereas the escarpments developed on homoclinal ridges along the flanks of synclines face outward, away from the axial plane (Figure 9–12A). Parallel ridges and valleys are formed if the axes of the folds are horizontal (Figure 9–12A), but if the fold axes are not horizontal, a pattern of zig-zag ridges and valleys forms (Figure 9–12B).

Homoclinal ridges converge in the direction of plunge of anticlines (Figure 9–13; Plate 5C) but diverge in the direc-

FIGURE 9–9
(A) Photo of V-shaped notches cut in a homoclinal ridge, (B) Contour map of V-shaped notches. (Wolf Point quadrangle, WY, U.S. Geological Survey)

A.

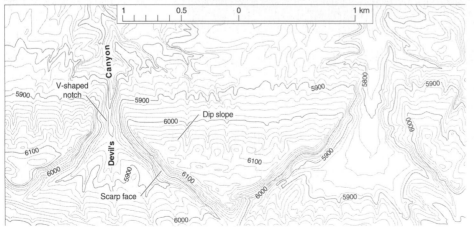

B.

FIGURE 9–10
Flatirons, triangular-shaped remnants of a bed between adjacent V-shaped notches, on a homoclinal ridge of the San Raphael Swell, Utah.

FIGURE 9–11
Breaching of the crest of an anticline has exposed underlying weaker beds that have been eroded into an anticlinal valley along the axial plane. Near the center of the photo, incomplete breaching of the overlying bed bridges the anticlinal valley and connects opposing homoclinal ridges along the flanks of the fold. (Side-looking radar photo of the Williamsport-Milton, Pennsylvania area, U.S. Geological Survey)

tion of plunge of synclines. A good place to determine whether a plunging fold is an anticline or a syncline is at the apex of the outcrop pattern, where the dip of the beds is the same as the angle of plunge of the structure. For example, the dip of homoclinal ridges along the flanks of a plunging fold may be too steep to show good asymmetry or V-shaped notches, but the plunge of the axes of folds is often less than the dips of beds along the flanks. Thus, steeply dipping hogbacks on the flanks of a fold typically pass into cuestas around the noses of plunging folds. Long, tapering dip slopes of beds facing in the direction of the convexity of the apex of a structure (Figures 9–14 and 9–15) are good indications of a plunging anticline. Conversely, a short, blunt escarpment facing in

the direction of convexity of the apex of the fold is typical of a plunging syncline (Figures 9–15, 9–16, 9–17, and 9–18). An easy rule to remember is that gently plunging anticlines have long, tapering noses, and plunging synclines have short, blunt noses.

Where notches are cut through an anticlinal ridge containing a resistant rock layer making a vertical cliff (Figure 9–19A), the spacing of contours is affected where they cross the vertical cliff. The contours "stack up" one upon another as they cross the vertical face, and thus they appear to be offset (Figure 9–19B). Any contour traced along the side of the ridge must make a sharp turn at the vertical cliff, cross the vertical face, and emerge on the far side with the appearance of hav-

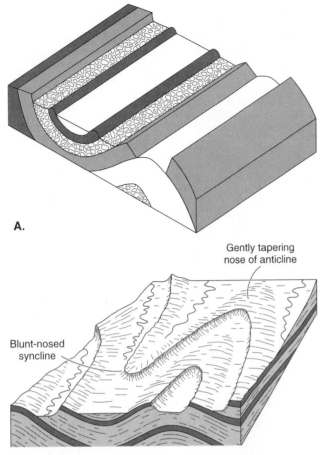

A.

B.

Gently tapering
nose of anticline

Blunt-nosed
syncline

FIGURE 9–12
Topographic expression of breached, nonplunging folds (A)
and plunging folds (B). The topographic expression of
eroded, nonplunging folds consists of parallel homoclinal
ridges and valleys, whereas the topographic expression of
eroded plunging folds consists of converging and diverging
homoclinal ridges and valleys.

FIGURE 9–13
The Virgin anticline, Utah, a breached, plunging anticline.
The plunge of the anticline is toward the top of the photo.
(Photo by W. K. Hamblin)

FIGURE 9–14
Gentle nose of a plunging
anticline, Split Mt., Utah.

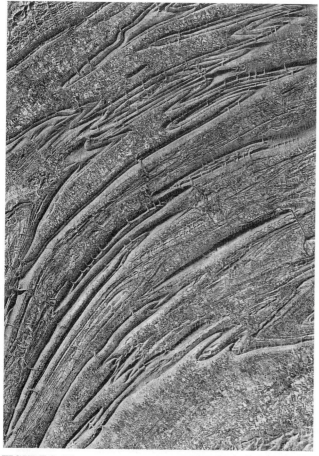

FIGURE 9–15
Plunging anticlines and synclines and homoclinal ridges and valleys in folded Paleozoic strata, Appalachian Mountains, Pennsylvania. (Side-looking radar photo mosaic, U.S. Geological Survey)

FIGURE 9–16
Blunt nose of a plunging syncline, Split Mt., Utah.

ing been offset. Thus, such offset contours on topographic maps can be used to determine the height of a vertical cliff made of resistant rock, and the inverted U shape provides a useful criterion for the recognition of an anticline.

Another method of establishing geologic structures is by **projection of known dip** along the strike of one part of a complexly folded area to permit determination of an entire set of folds. For example, Figure 9–15 shows a complexly folded area in the Appalachian Mountains. Analysis of the structure may be made by examining critical localities. The tapering noses of plunging anticlines and the short, blunt noses of plunging synclines allow determination of the strike and dip of the beds. The dip may then be extrapolated along homoclinal ridges. Projection of dips along the strike from the known points allows determination of dips for all of the beds making up the linear ridges.

The bowing up of **domes** may be accomplished either by tectonic warping or by igneous intrusion from below. In either case, the resulting landform in initial stages is an oval-shaped hill or mountain. Until erosion has stripped off the upper beds of a dome, intrusive igneous domes are difficult to distinguish from structural domes. The initial drainage on a newly up-

lifted dome is consequent in origin and forms a radial pattern. Erosional breaching of structural domes and basins leaves an elliptical or circular pattern of ridges (Figures 9–20 and 9–21). Easily eroded rocks discovered by incision near the crest of a dome lead subsequent streams to work headward along the weak beds to expose older rocks in the core of the structure.

Exposure of crystalline igneous or metamorphic rocks in the core of a breached dome results in a massive central area surrounded by concentric linear ridges of sedimentary rocks. However, the exposure of cores of granitic or other igneous rocks does not necessarily mean that the dome is intrusive. Deposition of sedimentary beds unconformably on a crystalline basement followed by structural arching into a dome yields the same general appearance as an intrusive dome, but in this case the crystalline rocks are older than the enclosing hogbacks.

Structural basins initially form oval topographic basins or large depressions. Consequent streams flowing down the dip of the beds toward the center of basins form centripetal drainage patterns.

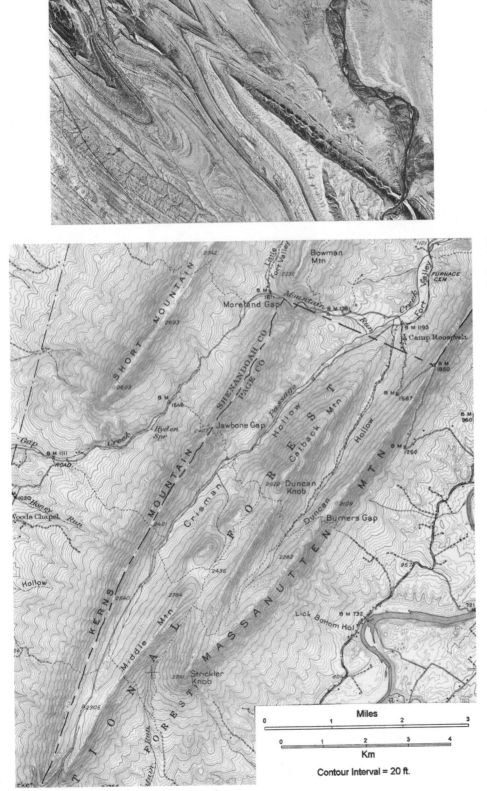

FIGURE 9–17
Gently tapering noses of plunging anticlines and short, blunt noses of plunging synclines, Sheep Mt., Wyoming. (Photo by Dept. of Agriculture)

FIGURE 9–18
Topographic map of plunging anticlines and synclines. Note the gently tapering nose of the plunging anticline that extends northeastward from Duncan Knob on Catback Mtn. and the short, blunt nose of the plunging syncline at Strickler Knob on Massanutten Mtn. (Mt. Jackson, Virginia, quadrangle, U.S. Geological Survey)

Contour Interval = 20 ft.

A.

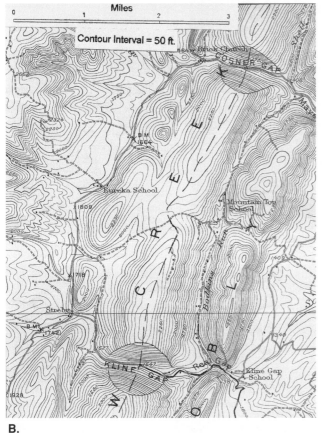

B.

FIGURE 9–19
Inverted U-shaped notches in anticlinal ridges: (A) Sheep Mt., Wyoming. (Photo by D.A. Rahm); (B) on a topographic map. (Greenland Gap, West Virginia, quadrangle, U.S. Geological Survey) The contours making the inverted U on the map at Cosner Gap and at Kline Gap appear to cut across other contours, but they really don't—the inverted U is made up of contours stacked one upon another on a vertical cliff that corresponds to the edge of a resistant bed folded into the anticline.

FIGURE 9–20
Upheaval dome, Utah, an eroded dome with annular drainage. (Photo by W. K. Hamblin)

Monoclines, structures in which the strata dip in one direction but with a local steepening of dip, form flat-topped upwarps with steep dips along their flanks (Figures 9–21A and B; Plate 5D).

Unconformities

Unconformities may sometimes be detected on the basis of their topographic expression. Discordances between superposed bodies of rock are the most immediate indication of such features (Figures 9–22 and 9–23).

Differential erosion of unconformable rocks shows striking differences in drainage and bedrock structure across the unconformity (Figure 9–24). If horizontal beds overlie folded or faulted strata, the discordance could be either an unconformity or a low-angle fault. If the contact is irregularly eroded and juts forward at interfluves, the possibility of a high-angle fault can be eliminated. An important rule that helps to differentiate a thrust sheet from an unconformable cover mass is that if the rocks of the cover mass strike into the contact with underlying rocks, the cover mass is the upper plate of a fault. If rock structures of the cover mass are essentially parallel to the contact (regardless of any discordance of the underlying rocks), the cover rocks may be either the upper plate of a low-angle fault or the overlying rocks of an unconformity.

Nonconformities may be detected by the erosional contrast between crystalline rocks (igneous or metamorphic) below and the layered rocks above. Again, care must be taken to differentiate this relation from an intrusive or fault contact. Differentially eroded dikes extending into the country rock beyond the general contact would be geomorphic evidence of an intrusion. However, the lack of such offshoots does not establish a nonconformity.

A.

FIGURE 9–21
San Raphael swell, Utah, a flat-topped dome bounded by monoclines. (A) satellite view (Enhanced NASA photo, Earth Satellite Corp.); (B) oblique air view. The nearly flat-lying bed in the upper left bends sharply down near the center of the photo, then flattens out again to the right, making a monocline. (Photo by W .K. Hamblin)

B.

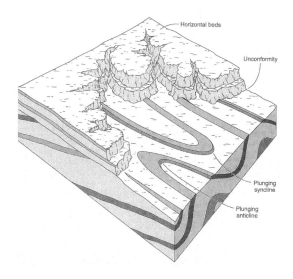

FIGURE 9–22
Topographic expression at an unconformity. Topography developed on the folded beds beneath the unconformity consists of homoclinal ridges and valleys, whereas the flat-lying beds above the unconformity develop cliff-and-bench topography. (Modified from Hamblin and Howard, 1995)

FIGURE 9–23
Unconformity between flat-lying sediments and underlying folded beds, Bighorn Basin, Wyoming. (Photo by D. A. Rahm)

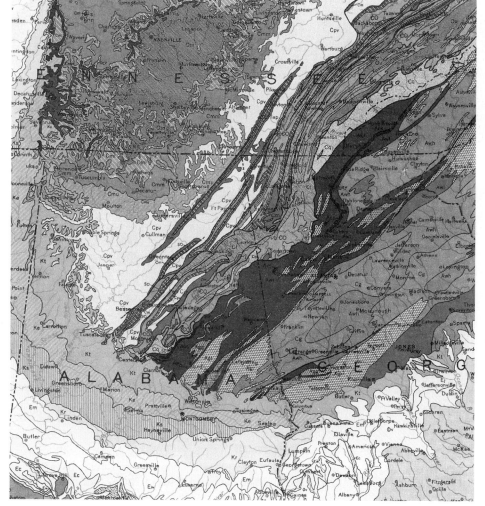

FIGURE 9–24
Unconformity between gently dipping Mesozoic and Cenozoic beds of the coastal plain (lower portion) and folded Paleozoic beds of the Appalachian Mts. Note the cross-cutting relationship of the coastal plain beds to the elongate trends of the Paleozoic beds.

RELATIONSHIP OF STREAMS TO GEOLOGIC STRUCTURES

Streams sometimes cut across anticlines and synclines, in many cases with a meandering pattern, as if entirely ignorant of the structures across their path (Figure 9–25). Such streams may be either antecedent or superposed in origin. The courses of antecedent streams were established prior to the growth of the structures they cross. As an anticline rises athwart the path of an antecedent stream, the stream is able to maintain its channel by cutting down at a rate greater than the rise of the anticline. In this way, the antecedent stream is not turned aside by the structural barrier.

Excellent examples of antecedent streams are the Ventura River of southern California, which cuts across the axis of the still-rising Ventura anticline (Figure 9–2) (Putnam, 1942); the Arun River as it cuts across the tectonically active Himalaya Mountains; and the Yakima River of Washington where it meanders through several anticlines of Pliocene-Quaternary age (Figure 9–1).

A superposed stream attains its course initially across geologic structures upon a cover mass, which initially buries the structures. At this time, the stream flows in complete independence of the concealed structures beneath. However, downcutting of the stream may incise through the thickness of the cover mass and discover the once-buried structure beneath. In this way, a superposed valley is cut across any anticline, syncline, fault block, or metamorphic terrain that happens to be below the unconformity. Spectacular examples of major superposition are abundant in the middle Rocky Mountains of Wyoming and adjacent states where faulted anticlinal structures of the late Cretaceous Laramide orogeny have been buried by the debris of their own erosion during the Cenozoic. Today, these structures are exhumed and stand as high ranges of the region. Remnants of the Tertiary cover rocks rest unconformably upon the Laramide structures. Major streams, which head along

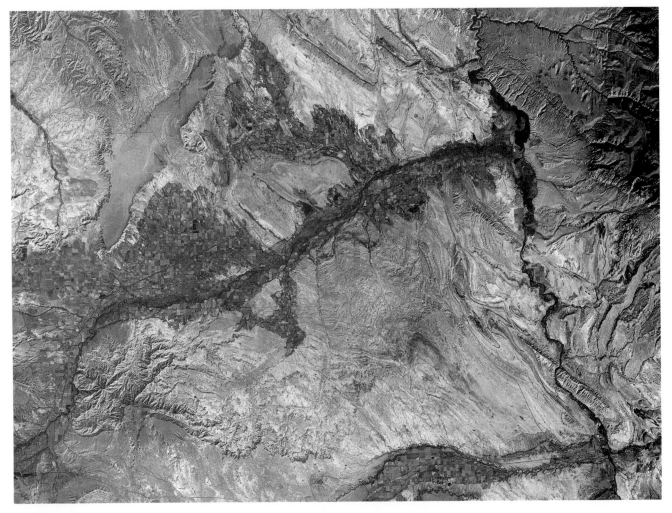

FIGURE 9–25
Superimposed drainage across the folds of the Bighorn Basin, Wyoming. The Bighorn River cuts across the Sheep Mt. Antieline (lower right), then traverses several other folds. The Shoshone River (center) cuts across a number of folds in the center of the basin. Both rivers were let down across the folded structures from an overlying cover of Tertiary sediments. (Enhanced NASA photo, Earth Satellite Corp.)

the Continental Divide, cut across other ranges in their effort to leave the middle Rockies. They do this in spite of many open basins, which seem to provide excellent detours around the mountains. Notable examples of such superposed rivers are the Bighorn, Green, and Yampa rivers that cut across mountain barriers rather than less encumbered routes.

The Bighorn River drainage flows southeast from its source in the Wind River Mountains into an open basin where the trunk stream turns abruptly north and flows directly across the Owl Creek Mountains. This range is transected by a deep, superposed canyon north of which the Bighorn meanders almost out of the open end of the Bighorn Basin. However, at the northern end of the Bighorn Basin, with gentle country ahead, the river turns again and cuts through both the Sheep

Mountain anticline and the northern Bighorn Mountains (Figure 9–25). The anticlinal structure of these latter two ranges is magnificently exposed in cross section on the canyon walls. The Green River behaves similarly as it wanders south from its source and cuts across the Uinta Range. A major tributary, the Yampa, whose incised meanders follow the trend of the Uinta Arch, joins the Green in the heart of the mountains. Both are excellent examples of superposed streams. A most convincing case can be made for superposition of the Sweetwater River. At Devil's Gate, the river flows through a granite gorge cut across the tip of a mountain ridge. Curiously, the river follows such a course when, by detouring a few hundred meters to one side, it could have easily circumvented the ridge.

REFERENCES

Billings, M. P, 1972, Structural geology: Prentice-Hall, Englewood Cliffs, N. J., 606 p.

Gilbert, G. K., 1877, Report on the geology of the Henry Mtns: U.S. Geographical and Geological Survey Report on the Rocky Mt. Region, p. 18–88.

Putnam, W. C., 1942, Geomorphology of the Ventura region, California: Geological Society of America Bulletin, v. 53, p. 691–754.

Rockwell, T. K., Keller, E. A., Clark, M. N., and Johnson, D. L., 1984, Chronology and rates of faulting of Ventura River terraces, California: Geological Society of America Bulletin, v. 95, p. 1466–1474.

Rockwell, T. K., Keller, E. A., and Dembroff, G. R., 1988, Quaternary rate of folding of the Ventura Avenue anticline, western Transverse Ranges, southern California:

Geological Society of America Bulletin, v. 100, p. 850–858.

Suppe, J., 1985, Principles of structural geology: Prentice-Hall, Englewood Cliffs, N. J., 537 p.

CHAPTER TEN

Topographic Expression of Joints and Faults

Wildrose graben, California, a down-dropped block on an alluvial fan. (Photo by W. K. Hamblin)

INTRODUCTION

Unique landforms are developed on joints and faults. Enhanced erosion along joint planes preferentially etches out linear ridges. Faults create landforms in two different ways; (1) direct tectonic offsetting of the land surface, and (2) differential erosion of rocks of different resistance that have been juxtaposed by offsetting along a fault.

JOINTING

Joints are fractures in rocks along which no significant movement has taken place on opposite sides of the plane of failure. Well-defined patterns reflect the response of brittle fracture in rocks resulting from crustal stresses or from shrinkage upon cooling.

Crustal stresses may produce joints in virtually any type of rock—igneous, sedimentary, or metamorphic—as a result of tensional, compressional, or shearing stresses. The joints may form parallel to or at some angle to the prevailing stress field. Under appropriate conditions, the stress field can be reconstructed on the basis of the joint patterns.

Joint patterns in rocks represent a brittle response to stress. Because the tensile strength of rocks is generally much lower than the compressive or shearing strength, even very strong rocks often show evidence of jointing. For example, the compressive strength of sandstone may be up to 2500 kg/cm^2 (35,000 lb/in^2) and the shearing strength 100 to 150 kg/cm^2 (1400 to 2100 lb/in^2), whereas the tensile strength may be only 10 to 30 kg/cm^2 (140 to 400 lb/in^2). Thus, brittle fracturing of seemingly resistant rocks is quite common. Tensional joints form at right angles to the direction of maximum stress. The origin of tensional stress can be either crustal extension or contraction due to cooling in igneous rocks.

Joints may also be caused by shearing stresses in rocks. The maximum shearing stresses in rocks occur in two planes that make angles of 30° to 45° with the maximum and minimum stress axes. Both tensional and shear stresses in rocks may originate as a secondary effect of crustal compression. For example, massive, resistant sandstone beds often show strong joint systems (Figure 10–1 and 10–2) as a result of gentle arching, which causes the top of the bed to elongate rel-

FIGURE 10–1
Vertical valley walls along joint planes in horizontal sandstone, Canyon de Chelly National Monument, Arizona.

FIGURE 10–2
Thin, vertical ridges developed along prominent joints in flat-lying sandstone, Arches National Park, Utah. (Photo by W. K. Hamblin)

ative to the base of the bed (Figure 10–2). In such cases, a single set of parallel joints aligned parallel to the strike of the bed is produced (Figure 10–3). If the axis of warping is itself slightly bent, then two sets of parallel joints are generated (Figure 10–2).

Topographic Expression of Joints

Joints can have a powerful effect on topography as a result of their influence on weathering and differential erosion. They are especially important because they provide planes of weakness in rock along which water, air, and plant roots can penetrate. Because joint planes provide avenues for percolation of surface water through rocks, they also become the sites of enhanced ionic exchange between water and the rock. This exchange greatly facilitates weathering processes, and weathering proceeds more rapidly along the planes than elsewhere in a rock body. Surface water finds runoff routes more easily in the open spaces of the joints, and the weathered material in the joint plane is easier to erode than the fresh rock. Thus, a prominent pattern of fractures in rocks may be emphasized by differential erosion to the point where steeply dipping joints appear as narrow, troughlike depressions. Further erosion may lead to isolation of large, vertical, slab-like walls of rock (Figure 10–4). In Figure 10–5, downward erosion of thick, highly jointed sandstone has been retarded by a thinner, very resistant caprock, but erosion has proceeded along vertical joint planes, eventually isolating a single fin of sandstone between adjacent joint planes. Differential erosion along intersecting joints pro-

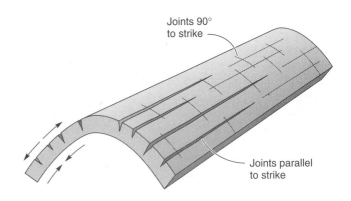

Joints 90° to strike

Joints parallel to strike

FIGURE 10–3
Jointing caused by flexing of a brittle sedimentary bed.

duces checkerboard-like columns of rock isolated between joints (Figure 10–6).

Shale is considerably less brittle than sandstone because the interlocking clay minerals are able to stretch in response to applied stress, whereas the grains of a sandstone cannot. The two rock types behave very differently under stress; thus, they respond differently to erosional processes and develop very different topographic expressions. Figure 10–7 and Plate 5E show dark shale beds overlying gently dipping, light-colored sandstone. The topography developed on the sandstone is strongly affected by a nearly vertical, single-direction joint system, but the overlying dark shale shows no evidence of jointing at all. The shale and the sandstone are both Jurassic in age, and both were gently warped by the late Cretaceous Laramide orogeny. Thus, the reason for such a sharp contrast

FIGURE 10–4
Rock fins developed along vertical joint faces in lying sandstone, Arches National Park, Utah. (Photo by D. A. Rahm)

FIGURE 10–6
Checkerboard topography developed on sandstone having joints that intersect at a right angle, southern Utah. (Photo by D. A. Rahm)

FIGURE 10–5
Single line of rock fins left by differential erosion of the beds shown in the background. The resistant cap rock on the fins in the left corner protected the underlying rocks from erosion except along the sides of vertical joint planes. (Photo by D. A. Rahm)

in topographic expression is related solely to the flexibility of the shale relative to the brittleness of the sandstone.

Jointing patterns in igneous and metamorphic crystalline rocks generally produce topographic expressions different from those of brittle sedimentary rocks. They are usually somewhat less regular and more widely spaced, and they lack the strong linearity of joints in sedimentary rocks (Figure 10–8). Jointing commonly has a strong effect on patterns of stream erosion. Rectangular and angular drainage patterns are common in areas of fractured rock.

FAULTING

Geomorphic Evidence of Faulting

As in the case of folds, the topographic expression of faults may be due to tectonic offsetting of the land surface or to differential erosion of rocks whose spatial position is determined by faulting. Scales may vary from small fault scarplets a few meters high to major rifting of crustal plates.

Louderback (1904, 1923) recognized that many mountain ranges in the Great Basin of the southwestern United States (Figure 10–9) must be fault blocks because late Tertiary or Quaternary lava flows rest unconformably on the summits of the ranges, high above remnants of the same flows on the downthrown block, indicating that such young

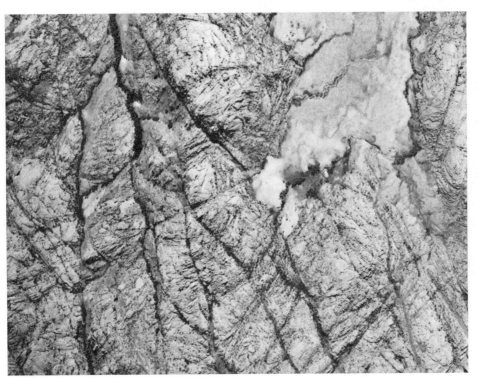

FIGURE 10–7
Well-developed vertical joints in gently dipping sandstone (light rock) overlain by unjointed shale (dark rock), southwest Utah. The shale is not as brittle as the sandstone and does not develop the prominent joints even though both beds were subjected to the same bending. (Photo by W. K. Hamblin)

FIGURE 10–8
Topographic expression of joints in igneous rocks. (Photo by U.S. Geological Survey)

flows could have been displaced only by very recent faulting. Davis (1913) later termed such dislocated lava flows *louderbacks*.

When V-shaped valleys are cut into scarps, the remainder of the fault face on spurs between valleys assumes a triangular shape known as a **triangular facet** (Figure 10–10A and B

and Plate 6A). Although triangular facets are commonly associated with faulting, other surface processes may create similar forms, so care must be used in their interpretation.

Displacement of ridges of tilted sedimentary beds and truncation of geologic structures both provide evidence of faulting. Resistant sedimentary beds are expressed topo-

FIGURE 10–9
Computer-generated view of
fault block ranges of the Basin
and Range Province, Nevada.
(Computer-generated map, D.
Simpson)

A.

B.

FIGURE 10–10
(A) Fault scarp cutting across alluvium. Triangular facets occur along the upthrown block as remnants of the fault scarp between adjacent V-shaped valleys. (Photo by D. A. Rahm)
(B) Triangular facets along the Wasatch fault, Utah. (Photo by W. K. Hamblin)

graphically as continuous ridges, so when a fault displaces the strata, the ridges are offset or terminate abruptly (Figure 10–11).

Determination of upthrown and downthrown blocks may be made by utilizing geometric relationships of displaced ridges. For example, in Figure 10–12, homoclinal ridges are displaced by fault *f-f*. Determination of the direction of fault movement may be simplified by application of the **downdip method of viewing structures** (Mackin, 1950). To use this method, rotate the map so that you are looking down the dip of the displaced bed. With the map in this position, the bed higher on the page will always be on the upthrown block, and the bed lower on the page will always be on the downthrown block.

FIGURE 10–11
Sedimentary beds offset along a fault across the center of the photo, Rainbow Gardens near Las Vegas, Nevada. (Photo by W. K. Hamblin)

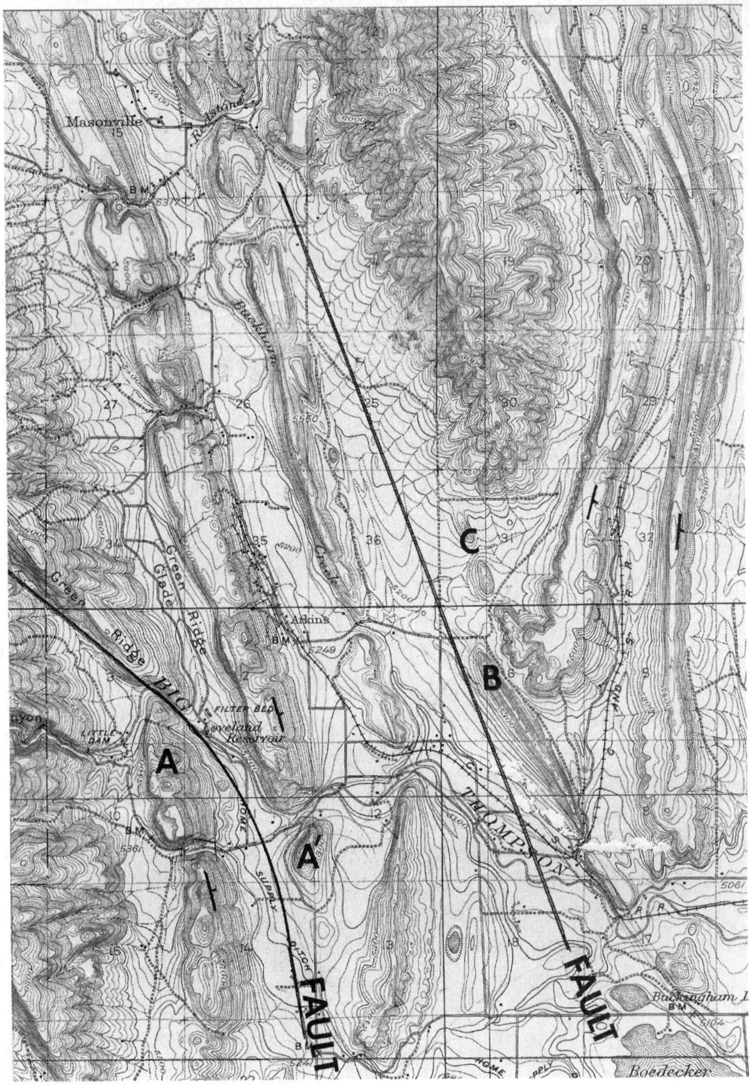

FIGURE 10–12
Homoclinal ridges offset by faults, Loveland, Colorado. The bed making the ridge at A has been displaced to A' and beds B and C are truncated abruptly at another fault. (Loveland quadrangle, U.S. Geological Survey)

PLATE 5A
Old Faithful Geyser, Yellowstone National Park, Wyoming. (W. K. Hamblin)
PLATE 5B
Sinkholes and ponds in a karst plain, Florida. (NASA)
PLATE 5C
Virgin anticline, southwest Utah. (W. K. Hamblin)
PLATE 5D
San Raphael swell, Utah. (Earth Satellite Corp.)
PLATE 5E
Vertical jointing in flat-lying sandstone overlain by unjointed shale, Canyonlands, Utah. (W. K. Hamblin)

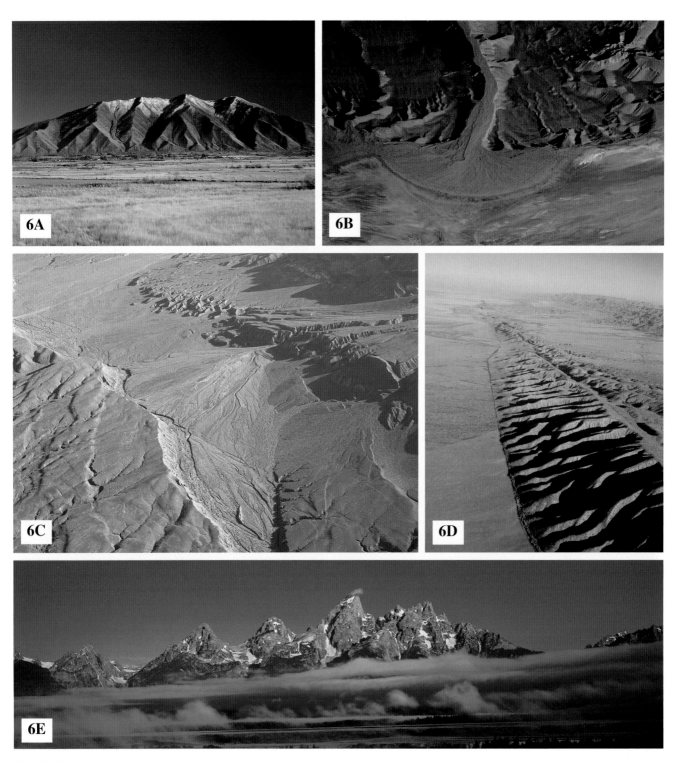

PLATE 6A
Triangular facets along the Wasatch fault, Utah. (W. K. Hamblin)
PLATE 6B
Faulted fan, Death Valley, California. (W. K. Hamblin)
PLATE 6C
Wildrose graben, California. (W. K. Hamblin)
PLATE 6D
San Andreas fault, California. (W. K. Hamblin)
PLATE 6E
Grand Teton Range, Wyoming, a tilted fault block. (W. K. Hamblin)

7A

7D

7B

7E

7C

7F

PLATE 7A
Mt. Rainier, Washington, a composite volcanic cone. (D. A. Rahm)

PLATE 7B
Eruption of pyroclastic flows from Mt. St. Augustine, Alaska. (M. E. Yount, U. S. Geological Survey)

PLATE 7C
Lava tube, pahoehoe lava, Hawaii. (W. K. Hamblin)

PLATE 7D
Mt. St. Helens, Washington, before 1980 eruption. (Photo by Author)

PLATE 7E
Mt. St. Helens, Washington, after 1980 eruption. (U. S. Geological Survey)

PLATE 7F
Volcanic neck and dike, Shiprock, New Mexico. (D. A. Rahm)

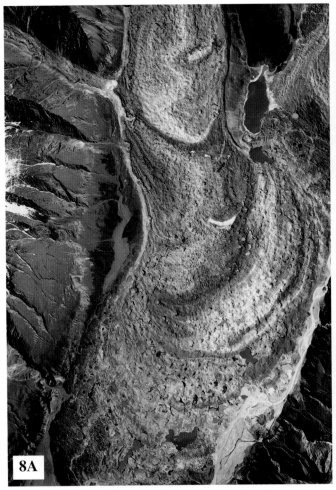

8A

8B

8C

PLATE 8A
Infrared photo of an Alaskan glacier. (Dept. of Agriculture)
PLATE 8B
Alaskan valley glacier with ogives below an icefall.
(D. A. Rahm)
PLATE 8C
Ice calving off a hanging glacier, The Monch, Switzerland.
(Photo by Author)
PLATE 8D
The Vatnajokull ice cap, Iceland. (NASA)
PLATE 8E
Piedmont glacier with ogives, LaPerouse glacier, Alaska.
(D. A. Rahm)

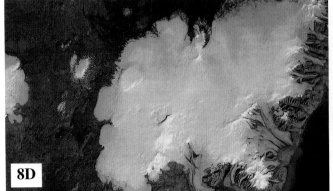

8D

8E

Often, beds displaced by a fault are bent in the vicinity of the fault as a result of frictional retardation of movement. In Figure 10–13, the dark bed has been bent back along the fault at 'A'. Such bending of beds near a fault is known as **fault drag** or **drag folding.**

Springs, especially if they occur in a linear zone, are suggestive of faulting, but they are not necessarily diagnostic. They are localized by faulting by:

1. Providing an impermeable barrier to groundwater movement, forcing the water to migrate upward to the surface
2. Providing a permeable zone of crushed rock through which water may rise to the surface

Hot springs aligned in a row are commonly associated with faulting. However, springs may originate in other ways and thus are not direct evidence in themselves.

Fault Scarps, Fault-Line Scarps, and Composite Scarps

Several types of scarps may be associated with faulting, each of which is characterized by certain distinctive features, depending on whether the scarp is due to direct displacement of the land surface or caused indirectly by differential erosion of rocks of differing resistance on opposite sides of the fault (Billings, 1972; Cotton, 1950; Johnson, 1939; Sharp, 1954).

Fault scarps originate by direct offset of the land surface by fault movement, so that the height of the scarp is equal to the vertical component of fault movement (Figures 10–14 and 10–15 and Plate 6B). The initial topographic expression may become deeply dissected by erosion of the upthrown block,

but as long as the scarp retains a recognizable degree of straightness, it is considered a fault scarp.

Fault-line scarps owe their topographic relief to differential erosion of rocks of contrasting resistance juxtaposed by movement of the fault. Less resistant rock on one side of the fault is eroded more rapidly than the more resistant rock on the other side of the fault, resulting in a scarp made out of the resistant rock. The height of the scarp is due only to the difference in rock resistance and does not reflect the amount of fault movement.

A **composite fault scarp** is one whose height is due partly to differential erosion and partly to direct movement on the fault. If a fault-line scarp that developed along a fault is offset by renewed movement on the fault, part of the height of the fault scarp would be due to erosion along a fault line, and part would be due to direct movement of the fault. In Figure 10–16, the total offset along the Hurricane fault is about 10,000 ft (~3000 m), but the height of the scarp is only about 1000 ft (300 m).

Fault Scarps — Tectonic Offsetting of the Land Surface

Fault scarps are steep, linear bluffs produced by tectonic offsetting of the land surface along the trace of faults. The height of a scarp is approximately equal to the amount of offset along the fault (Figure 10–17). Because fault movement is usually episodic, the height of large fault scarps is often the sum of multiple offsets, rather than a single major movement. Based on historic surface offsets accompanying earthquakes, offsets up to about 5 to 10 m (16 to 33 ft) seem to be fairly typical for single movements. During the Hebgen earthquake of West Yellowstone in 1959, a camping family was split by offset of 5.8 m (19 ft) in a matter of seconds (Figure 10–14). This off-

FIGURE 10–13
Drag of a sedimentary bed cut by a fault in Utah. The bed at 'A' bends as a result of frictional retardation along the fault. (Photo by U.S. Geological Survey)

FIGURE 10–14
Fault scarp, Klamath Lake,
Oregon.

FIGURE 10–15
Nineteen-foot fault scarp that
bisected a campground during
the 1959 Hebgen earthquake.

set is comparable to uplift along the Borah Peak fault near Chalis, Idaho, in 1983 where an eyewitness recounted seeing the base of the mountain "unzip" as movement propagated along the length of the fault (Scott et al., 1985).

The sudden increase in relief produced by the fault-offset surface leads to accelerated erosion of the scarp. The upper part of a newly formed scarp is soon worn back by the processes of weathering, mass wasting, and runoff, especially if the scarp is developed in unconsolidated sediments (Bucknam and Anderson, 1970; Colman and Watson, 1983; Pierce and Colman, 1986; Nash, 1984). The original fault plane becomes deeply notched by ravines and canyons, and the slope angle of the scarp is reduced to a lower angle (Figure 10–18). However, the position of the fault may re-

FIGURE 10–16
Composite scarp (across center of photo) along the
Hurricane fault, southwestern Utah. The scarp cuts Paleozoic
rocks that are displaced ~10,000 ft (~3000 m) on the down-
dropped block to the left. (Photo by W. K. Hamblin)

main discernible over long distances because the scarp main-
tains a generally straight base along the fault plane. Because
scarps can be created by other geomorphic processes in addition to faulting, such as from resistant beds, joint planes,
or ancient shorelines, a scarp is not necessarily proof of
faulting.

Although faults may occur as clean breaks on single
planes, they often break into thin, parallel slices, along which
the total displacement is distributed, or they may be splintered into several branching fault planes (Figures 10–19 and
10–20). Where **splintering,** also known as **splays,** occurs, the
height of the fault plane diminishes and is replaced by one or
more subparallel fault planes that increase in height. Between
the fault splinters, a narrow, sloping ramp typically connects
the upthrown and downthrown blocks (Figure 10–19). Because other processes do not generate these types of scarps,
they may be used as evidence of faulting.

Slumping is commonly induced by the sudden relief of
the upthrown block, resulting in multiple scarps parallel to
the original fault scarp. In such cases, the original scarp may
be difficult to identify. Scarps generated by movement along
parallel slices are especially difficult to distinguish from

FIGURE 10–17
Fault scarp across an alluvial fan
near Lone Pine, California.
(Photo by D. A. Rahm)

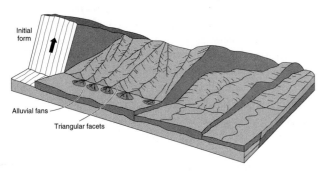

FIGURE 10–18
Progressive erosion of a fault scarp.

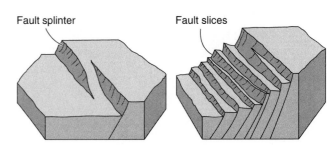

FIGURE 10–19
Fault splinters and slices.

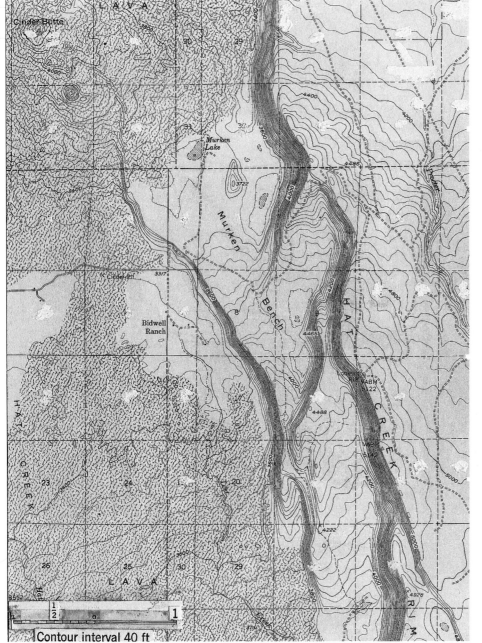

Contour interval 40 ft

FIGURE 10–20
Splintered fault scarps of varying height. Note that the scarps are not continuous but change height in both directions and splinter from the main scarps. The volcanic cone (Cinder Butte) in the upper left corner is aligned with the fault that makes the western margin of Murken Bench. (Jellico quadrangle, U.S. Geological Survey)

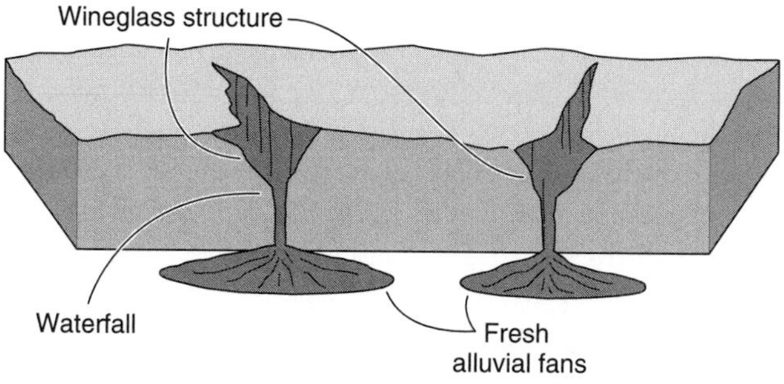

Wineglass structure

Waterfall

Fresh
alluvial fans

FIGURE 10–21
Geomorphic effects of faulting on drainage.

scarps made by slumping from the upthrown block. Although slumping can also occur on fault-line scarps, it is much more common on fault scarps because of oversteepening of slopes produced by the fault movement. Slumps are also common along thrust-fault scarps.

Stream profiles are commonly interrupted where they cross faults. Where streams flow toward the downthrown block, **hanging valleys** form at the fault scarp, and narrow gorges are incised vertically in the upthrown block by the stream. If the stream had a V-shaped valley profile before faulting, the combination of the V-shaped valley and the vigorous new incision produces a Y-shaped cross-channel profile known as **wineglass structure** (Figure 10–21).

Faults can provide avenues for lava to make its way to the surface, and volcanism is sometimes associated with faulting. The relationship of volcanism to faulting is especially noticeable where linear zones of cinder cones occur in alignment with faults (Figure 10–20, 10–22).

The courses of streams flowing toward the upthrown block are interrupted by offsetting of the land surface. Fault scarps are an effective dam across a stream, causing **ponding** (or aggradation) upstream from the scarp (Figure 10–23).

Crowding of streams against the base of a fault scarp results when impounded drainage ponds successively spill through their outlets to form a new stream. As a result, original drainage is shifted transverse to the fault into a new course at the base of the upthrown block, in some cases undercutting the fault scarp. In this way, erosional surfaces may be planed across the fault onto the upthrown block and preserved as a succession of terraces by recurrent faulting, which lifts them above the further influence of streams. In such cases, no terraces occur on the opposite side of the valley. Such a situation is most typical of humid regions where streams are relatively large and constant (Cotton, 1950). In dry regions, however, the downthrown block may become filled with alluvium to the extent that prograding alluvial fans force the piedmont streams well away from the scarp.

Frequent severe **earthquakes,** especially where the foci of earthquakes occur along known faults, indicate that escarpments are fault scarps or composite scarps. Although many historic earthquakes can be associated with movement along a particular fault, geologic evidence of past large-magnitude earthquakes has only recently been documented. Large-scale crustal subsidence and uplift in elongate zones accompanied the great 1960 Chilean and 1964 Alaska earthquakes approximately parallel to major plate boundaries. Subsidence of 1 to 5 m (3 to 17 ft) was observed along coastal regions during both quakes, and in places it coincided with **tsunamis (seismic sea waves)** that caused considerable destruction. During the 1960 Chilean earthquake, coastal tidal estuaries were dropped suddenly 1 to 3 m (3 to 10 ft), and

FIGURE 10–22
Volcanic cone erupted along the Toroweap fault, Utah. (Photo by W. K. Hamblin)

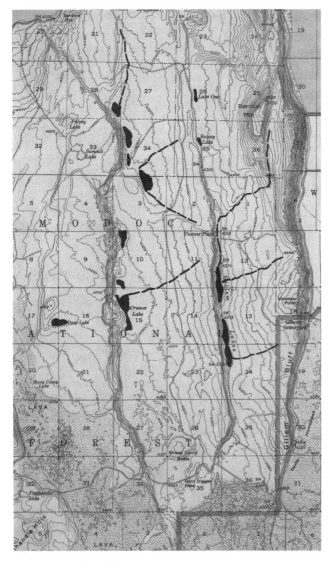

FIGURE 10–23
Ponding of drainage, splintering, and variable scarp height along fault scarps near Tule Lake, California. (Mt. Dome quadrangle, U.S. Geological Survey)

tsunamis swept several centimeters of marine sand over the area, drowning trees and marsh plants.

Studies of tidal estuaries in coastal areas in Washington, Oregon, and California have turned up remarkably similar records of sudden subsidence. Drowning of trees and marsh plants, and covering with thin sand layers, suggest that similar events have taken place periodically along the west coast of the United States (Atwater, 1987). Multiple sudden subsidence events are recorded in estuarine sediments, which suggest that slow uplift may occur between the subsidence events.

Disrupted and tilted terraces or strandlines are good evidence of faulting. Marine or lake terraces, which are horizontal at the time of their formation, may be offset or tilted by faulting. Good examples are found in the marine terraces near

Arcata, California (Carver, 1985; Kelsey and Carver, 1988). Offsetting of stream terraces can also be used to indicate faulting, but tilting of stream terraces is more difficult to work with than marine or lake terraces because the original gradient of stream terraces makes detection of tilting difficult.

Linear scarps in alluvium, especially across **alluvial fans** (Figures 10–17 and 10–24, Plates 6B and 6C), are good evidence of faulting because of the lack of any difference in erodibility of material on opposite sides of the fault. Considering that fault-line scarps owe their origin to differential erosion, if no difference in erodibility of material exists on opposite sides of the fault, the scarp could not be a fault-line scarp. Alluvial fault scarps usually do not survive very long (on a geologic time scale) because of the high erodibility of the material.

Material weathered and eroded from the scarp is deposited as fresh, new alluvial fans (Figure 10–25) and colluvium at the base of the scarp, burying the fault line. If renewed fault movements occur, small scarps are formed in the alluvium and colluvium. Small **alluvial fans** and dissected alluvial fans provide clues to recent faulting. Where dissection is related to linear alluvial scarps, small fans at the base of the scarp are especially significant. Where the pre-fault drainage was from the upthrown block to the downthrown block, the upthrown block becomes dissected by erosion, and new fan surfaces form on the downthrown block (Figure 10–26A). However, if the pre-fault drainage flows toward the upthrown block, new alluvial fans still form on the downthrown block, but the fault scarp blocks the drainage and either impounds it or re-routes it parallel to the scarp (Figure 10–26B).

Displacement of recent surface features is strong evidence of recent faulting. Examples include offset roads, fences, houses, orchards, and other human-made features (Figure 10–27) and once-continuous topographic surfaces such as terraces, streams, or ridges.

Sag ponds and fault-slice ridges sometimes form along fault zones where slices of rock are uplifted and depressed relative to one another. The uplifted slices, known as **fault-slice ridges** (Sharp, 1954), are squeezed up by pressure within the fault zone. **Sag ponds** form in closed depressions made by warping of the ground surface or over relatively down-dropped fault slices. Such isolated ridges and depressions could not have been carved by differential erosion, especially where the fault slices are composed of very resistant rock.

Sharp-crested anticlines in Quaternary sediments, and low ridges, small sinks, and discontinuous scarps, occur on the alluvial floor of a depressed portion of the San Andreas fault zone, produced by local compression within the active fault zone.

Fault-Line Scarps — Differential Erosion along Faults

Fault-line scarps are formed when a fault scarp is destroyed by erosion and a new scarp is produced by differential erosion of weak and resistant rocks on opposite sides of the fault. Thus, fault-line scarps have a geologic history quite different from that of fault scarps.

FIGURE 10–24
Fault scarp cutting across an alluvial fan, Death Valley, California. The upthrown block is being dissected by stream erosion, and the sediment is being deposited on the downthrown block. (Photo by W. K. Hamblin)

FIGURE 10–25
Fresh, new alluvial fans on the downthrown block of a fault, Death Valley, California. (Photo by D. A. Rahm)

Figure 10–28 shows the evolution of fault-line scarps from an original fault scarp. The height of the fault scarp is equal to the offset along the fault (Figure 10–28C). Where erosion has beveled the scarp to low relief and etched out a resistant bed across from a weak bed, a fault-line scarp along the trace of the resistant bed is formed (Figure 10–28C). At this point, the scarp faces in the opposite direction as the original fault scarp and is known as an **obsequent fault-line scarp**. Continued erosion once again bevels the scarp, as shown in Fig-

ure 10–28D, followed by development of another fault-line scarp along the outcrop of a resistant bed juxtaposed against a weak bed. This time the scarp faces in the same direction as the original fault scarp (but at a lower level), so it is a **resequent fault-line scarp** (Figure 10–28E).

Close correlation exists between fault-line scarps and rock resistance. A scarp on the downthrown block of a fault could not have formed by direct surface displacement and must therefore be a fault-line scarp caused by differential erosion. Scarps on the downthrown blocks of faults serve as an infallible criterion for identification of a fault-line scarp (as long as the downthrown side of the fault can be established by geologic evidence). Such scarps show a complete topographic inversion in which the upthrown block is eroded more deeply than the relatively resistant downthrown block.

Superposition of streams across a fault is evidence that a scarp along the fault is a fault-line scarp. This is also true if sedimentary rocks, unconsolidated deposits, or lava flows unconformably overlie any part of the fault, because they demonstrate that a scarp along the fault was formed after the cover was laid across the fault and without further movement of the fault. If a scarp is simply buried by later deposits or by lava, these materials will be much thicker on the downthrown block than on the upthrown block, allowing relatively easy distinction from the situation where the material was deposited when no relief existed across the fault. Figure 10–29 shows a scarp along a fault. At one point, the scarp leaves the fault and follows a resistant rock unit on the downthrown side of the fault. The resistant rock unit crosses the fault without displacement. Thus, even though the scarp follows the fault for some distance, no relief existed across the fault when the younger cover-mass unit was laid down, and, therefore, the entire scarp must be due to differential erosion and must be a fault-line scarp.

FIGURE 10–27
Offsetting of a sidewalk by fault creep along the San
Andreas fault, Hollister, California. (Photo by U.S.
Geological Survey)

A.

FIGURE 10–26
(A) Dissection of the upthrown
block on an alluvial and
deposition of new fans on the
downthrown block, Death Valley,
California. (Photo by D. A.
Rahm)
(B) Fault scarp across an alluvial
fan facing the drainage direction.
Note the new alluvial fans on the
downthrown block and the re-
rerouting of the drainage along
the base of the scarp. (Photo by
D. A. Rahm)

B.

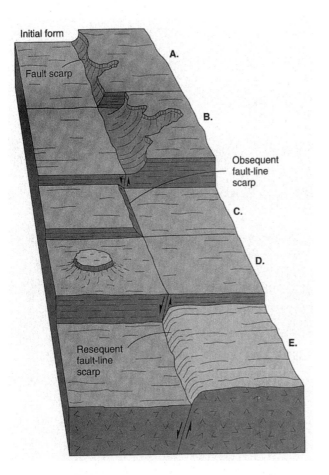

FIGURE 10–28
Evolution of a fault scarp into a fault-line scarp.

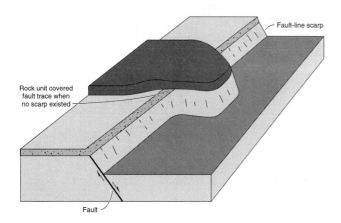

FIGURE 10–29
Fault-line scarp. The scarp was covered by a lava flow at a time when the earlier fault scarp had been eroded to no relief; then erosion of weak rock on the right-hand block etched the scarp into relief once more.

Fault Blocks

Tilted Fault-Block Mountains

Most of the mountain ranges in the Basin and Range Province of the southwestern United States (Figure 10–9) and such large mountain ranges as the Sierra Nevada (Figure 10–30) and the Grand Teton ranges (Davis, 1903; Gilbert, 1928) are **tilted fault-block mountains**. These mountains are asymmetric as a result of tilting of the blocks along range-front faults. Total displacement on the Sierra fault is more than 3300 m (10,000 ft) and more than 7500 m (25,000 ft) on the Grand Teton fault. Movement on large faults such as these is episodic, with individual events probably limited to 3 to 10 m (10 to 33 ft).

Tilted fault-block ranges in the Basin and Range Province of eastern Oregon are relatively young and undissected, whereas those in Nevada, Utah, and eastern California are more maturely eroded, and those in Arizona and New Mexico have been severely eroded. Original fault scarps commonly make up mountain fronts in the Oregon fault-block ranges (Figure 10–13), but fault scarps of ranges farther south are generally obscured by erosion (Fuller and Waters, 1929; Sharp, 1939).

Grabens and Horsts

Faults may occur in pairs, resulting in down-dropped blocks known as grabens and uplifted blocks known as horsts. Grabens often form broad, flat-floored valleys varying in size from a few meters to many kilometers (Figure 10–31). A good example is Death Valley, a graben measuring 200 km (120 mi) long and extending to 116 m (384 ft) below sea level.

The filling of grabens with sediments and the erosion of horsts may considerably alter the initial topography, so that grabens are not always topographic lows, and horsts are not always topographic highs.

Strike-Slip Faults

Strike-slip faults, in which the main fault movement is horizontal, are characterized by steeply dipping fault planes, **offset streams** and ridges, valleys etched out of zones of crushed rock, depressions, and **sag ponds.** Often strike-slip faults involve very large displacements; if the fault is extensive, it can consist of a fracture zone made up of multiple, parallel faults. Perhaps the best-known strike-slip fault is the San Andreas fault (Figure 10–32A; Plate 6D) which extends from Mexico to northern California where it bends westward into the Pacific Ocean. The San Francisco earthquake of 1906 was caused by up to 21 ft (6 m) of lateral movement along the San Andreas fault. Offsetting of roads, fences, orchards, sidewalks, drainage canals, and other surface features continues today at rates from $1/2$ to 2 in/yr (1 to

A.

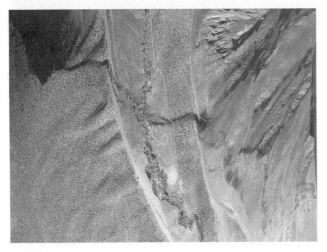

B.

FIGURE 10–30
A. The Sierra Nevada, a tilted fault block range. (Computer-generated map, D. Simpson)
B. Fault scarp cutting across a glacial moraine at the base of the Sierra Nevada Range, California. (Photo by W. K. Hamblin)

FIGURE 10–31
Grabens, Colorado Plateau. (Photo by W. K. Hamblin)

5 cm/yr). Sharp jogs of several hundred feet or more in the channels of streams crossing the fault (Figure 10–32B) suggest offsetting by recently active strike-slip movement (Sieh, 1984).

Ridges along the San Gabriel fault in southern California have been offset just enough to shut off or deflect drainage from valleys across the fault block, forming **shutter ridges** (Figure 10–33).

Thrust Faults

Thrust faults are low-angle, reverse faults that may achieve displacements of many kilometers (Figure 10–34). They usually do not produce impressive scarps, although they may produce large landslides, and often are characterized by very irregular outcrop patterns because of their gently dipping fault plane.

The Lewis thrust in Montana has pushed Precambrian sedimentary rocks many kilometers over weak Cretaceous shale (Figure 10–35). Erosion completely through the thrust fault has isolated remnants of the upper plate, known as **klippes,** such as Chief Mountain (Figure 10–36) (Billings, 1938).

A.

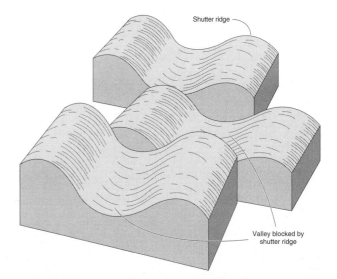

FIGURE 10–33
Shutter ridges caused by offset of a ridge across a valley.

B.

FIGURE 10–32
(A) The San Andreas fault, California, a strike-slip fault with displacement of hundreds of kilometers. (Photo by W. K. Hamblin)
(B) Offsetting of streams along the San Andreas fault, California. Note that as you look across the fault, the stream in the center of the photo has been displaced to the right, making a right-lateral fault. (Photo by R. Wallace, U.S. Geological Survey)

FIGURE 10–34
Thrust faults bringing Paleozoic sedimentary rocks (dark tone) over Jurassic sandstone (light tone), near Las Vegas, Nevada. Note the ragged outline of the thrust as a result of its gentle dip. (Photo by W. K. Hamblin)

FIGURE 10–35
Irregular mountain front developed on the Lewis thrust, Montana. Resistant Precambrian rocks on the left have been thrust more than 25 km (15 mi) eastward over Cretaceous shale. Chief Mt. Is a klippe, an erosional remnant of the upper plate of the thrust. (Glacier Peak quadrangle, U.S. Geological Survey)

FIGURE 10–36
Chief Mt., Montana, a klippe of Precambrian quartzite thrust over Cretaceous shale. (Photo by D. A. Rahm)

REFERENCES

Atwater, B. F., 1987, Evidence for great Holocene earthquakes along the outer coast of Washington state: Science, v. 236, p. 942–944.

Billings, M. P., 1938, Physiographic relations of the Lewis overthrust in northern Montana: American Journal of Science, v. 35, p. 260–272.

Blackwelder, E., 1928, The recognition of fault scarps: Journal of Geology, v. 36, p. 289–311.

Bucknam, R. C., and Anderson, R. E., 1970, Estimation of fault-scarp ages from a scarp-height-slope-angle relationship: Geology, v. 7, p. 11–14.

Buwalda, J. P., 1936, Shutter ridges, characteristic physiographic features of active faults: Geological Society of America Proceedings, p. 307.

Carver, G. A., 1985, Quaternary tectonics north of the Mendocino junction: the Mad River fault zone: Redwood Country American Geomorphological Field Group 1985 Field Trip Guidebook, p. 155–167.

Colman, S. M., and Watson, K., 1983, Age estimated from a diffusion equation model for scarp degradation: Science, v. 221, p. 263–265.

Cotton, C. A., 1950, Tectonic scarps and fault valleys: Geological Society of America Bulletin, v. 61, p. 717–758.

Davis, W. M., 1903, Mountain ranges of the Great Basin: Geographical Essays, Dover Publications, Inc. N.Y., p. 725–772.

Davis, W. M., 1913, Nomenclature of surface forms on faulted structures: Geological Society of America Bulletin, v. 24, p. 187–216.

Fuller, R. E., and Waters, A. C., 1929, The nature and origin of the horst and graben structure of southern Oregon: Journal of Geology, v. 37, p. 204–238.

Gilbert, G. K., 1928, Studies of Basin and Range structure: U.S. Geological Survey Professional Paper 153, 92 p.

Johnson, D., 1939, Fault scarps and fault-line scarps: Journal of Geomorphology, v. 2, p. 174–177.

Kelsey, H. M., and Carver, G. A., 1988, Late Neogene and Quaternary tectonics associated with northward growth of the San Andreas transform fault, northern California: Journal of Geophysical Research, v. 93, p. 4797–4819.

Louderback, G. D., 1904, Basin Range structure of the Humboldt region: Geological Society of America Bulletin, v. 15, p. 289–346.

Louderback, G. D., 1923, Basin range structure in the Great Basin: California University Publications in Geological Sciences, v. 14, p. 329–376.

Mackin, J. H., 1950, The down-structure method of viewing geologic maps: Journal of Geology, v. 58, p. 55–72.

Nash, D. B., 1984, Morphologic dating of fluvial terrace scarps and fault scarps near West Yellowstone, Montana: Geological Society of America Bulletin, v. 95, p. 1413–1424.

Pierce, K. L., and Colman, S. M., 1986, Effect of height and orientation (microclimate) on geomorphic degradation rates and processes, late glacial terrace scarps in central Idaho: Geological Society of America Bulletin, v. 97, p. 869–885.

Scott, W. E., Pierce, K. L., and Hait, M. T., 1985, Quaternary tectonic setting of the 1983 Borah Peak earthquake, central Idaho: Bulletin of the Seismological Society of America, v. 75, p. 1053–1066.

Sharp, R. P., 1939, Basin-range structure of the Ruby-East Humboldt Range, northeastern Nevada: Geological Society of America Bulletin, v. 50, p. 881–920.

Sharp, R. P., 1954, Physiographic features of faulting in southern California: California Department of Natural Resources, Division of Mines Bulletin 170, p. 21–28.

Sieh, K. E., 1984, Lateral offsets and revised dates of large prehistoric earthquakes at Pallett Creek, southern California: Journal of Geophysical Research, v. 89, p. 7641–7670.

Sieh, K. E., and Jahns, R. H., 1984, Holocene activity of the San Andreas fault at Wallace Creek, California: Geological Society of America Bulletin, v. 95, p. 883–896.

Wallace, R. E., 1977, Profiles and ages of young fault scarps, north-central Nevada: Geological Society of America Bulletin, v. 88, p. 1267–1291.

Willis, B., 1938, The San Andreas Rift in southwestern California: Journal of Geology, v. 46, p. 793–827, p. 1017–1057.

Landforms Developed on Igneous Rocks

Eruption of Mt. St. Augustine, Alaska. (Photo by M. E. Yount, U. S. Geological Survey)

INTRODUCTION

Igneous landforms result from both constructive and erosional processes. Eruptions of magma on the earth's surface build volcanic landforms. Magma intruded below the surface, however, can affect surface topography only by bowing the land surface upward during injection of magma or by exposure and differential erosion of the igneous body. Thus, volcanic landforms are primary features of accumulation of volcanic products on the land surface, and intrusive landforms are secondary erosional features. Volcanic landforms have characteristic features by which they may be identified, but weathering and erosion gradually modify or erase them with time.

VOLCANIC ERUPTIONS

The term **volcano** is derived from the Italian *Vulcano,* a volcano in the Lipari Islands just north of Sicily. Vulcano has been active in historic time, especially during the height of the Greek and Roman civilizations when Vulcan was believed to be the god of fire. The term **volcanic** now refers to all magmatic eruptive material.

The world's recently active volcanoes are not randomly distributed but are grouped in well-defined linear chains, the most striking of which is the circum-Pacific zone surrounding most of the Pacific Ocean (Figure 11–1). Volcanoes are generally considered to be **active, dormant,** or **extinct,** but the distinction between them is often blurred. Active volcanoes have been observed in eruption during historic time; dormant volcanoes have no historic record but show evidence of geologically recent activity; and extinct volcanoes appear to be geologically dead.

Relationship of Topographic Forms to Magma Characteristics

The nature of a volcanic eruption is governed largely by the physical and chemical characteristics of the erupting magma. The principal factors that determine the type of eruption, and thus the types of landforms produced, are magma composition, temperature, and gas content, all of which control lava viscosity:

1. *Composition of the magma.* The silica content of a magma plays an important role in determining its viscosity—the higher the silica content, the greater the viscosity. Most basaltic magmas have a silica content of about 50 percent, whereas rhyolitic (or dacitic) magmas contain about 70 percent silica. The reason that rhyolitic eruptions are typically more violent than basaltic ones is that the higher silica content makes the magma more viscous, often temporarily plugging up the vent until pressures become great enough to blow the plug out. Andesitic lavas are usually about 60 percent silica and are often intermediate in violence of eruption.

2. *Temperature.* The temperature of a lava may be established by use of an optical pyrometer to accurately measure the color of the glowing magma. Table 11–1 shows an approximate scale of temperatures associated with magma colors.

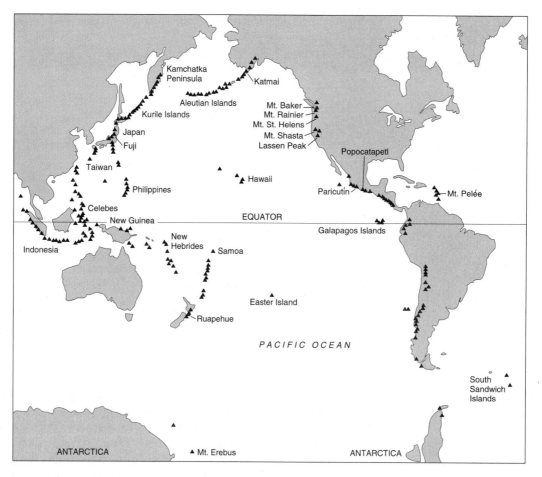

FIGURE 11-1
Distribution of active volcanoes. Note the concentration around the Pacific Rim at plate boundaries.

TABLE 11-1
Relationship between lava color and temperature.

Color	°C	°F
Incipient red heat	540	1000
Dark red heat	650	1200
Bright red heat	870	1600
Yellowish red heat	1100	2000
Incipient white heat	1260	2300
White heat	1480	2700

Hot lavas are more fluid than cooler ones. Because more heat is required to melt basalt (~1100°C, 2000°F) than rhyolite (~650°–700°C, 1200°F), basaltic lavas are much hotter and thus more fluid than rhyolitic lavas. Hence, basaltic eruptions are typically quieter and less violent because vents are less likely to clog up with viscous magma (Table 11-2).

3. *Gas content.* The explosiveness of a volcanic eruption depends to a large degree on the gas content of its magma. Often the explosiveness associated with rhyolitic/ dacitic eruptions is caused by the sudden release of high gas pressures in the magma. As magma with a high gas content rises toward the surface, the reduction in pressure allows expansion of gases that push the magma up the vent at a faster rate, which in turn accelerates pressure release and further rapid expansion of gases, until a violent eruption takes place.

Materials Erupted from Volcanoes

Volcanoes erupt three kinds of materials: solid, liquid, and gas. The relative amounts of each material depend largely on the violence of the eruption and thus on magma composition and viscosity.

Solid fragments blown from a volcano include rock torn from the sides of the vent during eruptions plus liquid magma blown into the air where it cools and solidifies before reaching the ground. All fragmental material ejected from a volcano is termed **pyroclastic** (from pyros, meaning "fire," and clastos, meaning "broken fragments"). Pyroclastic material ranges in size from microscopic dust to huge boulders weighing many tons. Although pyroclastic material is common to all eruptions, much greater volumes are emitted from siliceous volcanoes because of the higher viscosity of their magmas and their greater violence of eruption.

TABLE 11–2
Relationship between lava composition and eruption characteristics.

Composition	SiO_2	°C	Fluidity	Explosiveness	Cones
Rhyolite/ dacite	70%	~800	Low	High	Plug domes
Andesite	60%	~1000	Intermediate	Intermediate	Composite
Basalt	50%	~1200	High	Low	Shield

Liquid magma is emitted from volcanoes as lava that flows from the vent as streams or sheets of molten rock. The source of the lava may be a central vent, such as a crater, or a fissure.

Volcanologists have long recognized the importance of the release of gas as a magma approaches the surface and its role in determining the violence of an eruption. As magma nears the surface, the gas in it expands at an accelerating rate. The manner in which gas is released is related largely to the composition of the magma. If pressure builds up by clogging of the vent and pressure is released suddenly, a violent explosion occurs. However, as in the case of many basaltic eruptions, degassing may occur quietly. When pressure in the magma is released rapidly, the magma "froths," due to the rapidly expanding gases, and is ejected in the form of pumice and ash, consisting of vesicular glass.

Types of Eruptions

Although many variations in styles of eruptions exist, certain recurring characteristics allow grouping of eruptions by type, related mainly to explosiveness and dominant type of material extruded. Recognition of different types of eruptions dates back many years, based largely on observations of historic eruptions and concentrating on the violence of an event. The earliest attempts at classification were made in the 1800s, followed by those of Russell (1902) and Lacroix (1908). They used various combinations of the terms *explosive, intermediate,* and *quiet,* along with names of specific volcanoes that typified a particular type of eruption. Modified versions of Lacroix's types are commonly used today. They are of interest to geomorphologists because the landforms produced by each type of volcanic eruption are quite different.

Hawaiian Type

The **Hawaiian** type of eruption (Figure 11–2), typified by the outpourings of basaltic lava in the Hawaiian Islands, is characterized by relatively quiet emission of large volumes of fluid basalt from fissures a few kilometers long (Figure 11–3) or from central vents (Figure 11–4). The magma is hot (~1100°C, 2000°F) and fluid, allowing gases to be released readily without great violence. However, lava, thrown hundreds of feet into the air by the force of rapidly expanding gas, forms fountains known as "curtains of fire" that commonly occur along the length of a fissure, especially during the earlier stages of eruption (Figure 11–3). The highest fountain recorded in Hawaii occurred at Kilauea Iki during the 1959 eruption when a lava fountain reached a height of 575 m (1900 ft).

Great floods of hot, fluid lava issue from fissures and flow in "rivers of fire" for many kilometers down the mountainsides to the sea. The basalt is so hot and fluid that often flows may be exceptionally thin, some only a few centimeters thick. The 1969–71 eruptions along the South Rift zone of Kilauea near Mauna Ulu produced 185 million m^3 (240 million yd^3) of lava (Swanson et al., 1979), and the 1972–74 eruptions emitted 162 million m^3 (200 million yd^3) of lava (Tilling et al., 1987). More than 99 percent of the material making up the Hawaiian Islands consists of lava. The small amount of fragmental material forms cinder cones along fissures during late phases of eruptions.

Historic Hawaiian eruptions last from a few days to several years. Eruptions along the South Rift zone on Kilauea began in 1983 and continued to the present day.

Icelandic Type—Fissure Eruptions

In contrast to eruptions from central vents of volcanic cones, fissure eruptions originate along great fractures or rifts, pouring out great volumes of lava. The lava from fissure eruptions is basaltic, and the hot, fluid flows spread in sheets over great surface areas.

Great lava plateaus, consisting of hundreds of superimposed basalt flows covering thousands of square kilometers, have been built up in various parts of the world by fissure eruptions. The Columbia Plateau and the adjacent Snake River Plain of the northwestern United States were built up by innumerable basalt flows that covered an area of more than 500,000 km^2 (200,000 mi^2) in Washington, Oregon, and Idaho (Figure 11–5). The thickness of basalt flows in parts of the Columbia Plateau exceeds 4200 m (14,000 ft), although individual flows are generally only 10 to 30 m (30 to 100 ft) thick. Similar areas of basaltic fissure eruptions occur in the Deccan Plateau of India, in the Parana region of South America, and in parts of Africa, Australia, Mongolia, and Siberia.

Fissure eruptions differ from the other types of volcanic eruptions in that the magma issues from a long fracture or fissure, rather than from a central vent. Lava from a fissure eruption spreads out in great sheets that congeal into relatively flat-lying individual flows, often covering the fissure from which the magma was extruded.

FIGURE 11–2
Typical Hawaiian-type eruption of fluid basalt.

FIGURE 11–3
Lava fountains from a fissure making a "curtain of fire," Hawaii. (Photo by U.S. Geological Survey)

FIGURE 11–4
Lava fountain from a central vent, Hawaii. (Photo by U.S. Geological Survey)

FIGURE 11–5
Area covered by fissure eruptions of basalt on the Columbia Plateau, Washington, and the Snake River Plain, Idaho.

The only known historic fissure eruption occurred in Laki, Iceland, in 1783. Hence, these types of eruptions are often known as **Icelandic** types. On June 8, 1783, after a week of severe earthquakes, immense lava fountains began along the southern half of the 25 km (15 mi) long Laki fissure (Thorarinsson, 1970). From June 8 to July 29, large floods of basalt flowed from more than 22 vents along 16 km (10 mi) of the fissure, flowing as much as 14.5 km (9 mi) in one day. As the lava approached the coast, it spread out in a giant lobe 19 to 24 km (11 to 14 mi) across, and it terminated about 80 km (50 mi) from its source. On July 29, enormous quantities of lava began pouring out of the northeastern half of the fissure, and activity continued until the end of the eruption in early November.

During the first 50 days of the eruption, approximately 10 km³ (2 mi³) of lava flowed from the southwest half of the fissure and covered about 370 km² (140 mi²), corresponding to an average discharge of about 5000 m³ (175,000 ft³) of lava per second. During periods of maximum emission of lava, the discharge was perhaps 10 times the average. Thorarinsson (1970) estimated the total volume of lava and ash as about 12.5 km³ (3 mi³), covering an area of 565 km² (200 mi²).

The Laki eruption was a national disaster for Iceland. Damage to farms and livestock from lava flows and from a blue haze of gas emitted from the lava resulted in a famine that cost Iceland about one-fifth of its population, three-fourths of its sheep and horses, and about one-half of its cattle.

Strombolian Type

The **Strombolian** type of eruption takes its name from the volcano Stromboli, long known as the "lighthouse of the Mediterranean," off the coast of Sicily. Strombolian eruptions are typified by regular explosions of mild to modest intensity. The basaltic lava forms a crust over the vent, which is blown off when pressures become excessive, throwing pasty, incandescent lava and fragments of the crust into the air.

Vulcanian Type

Vulcanian eruptions are based on the characteristics of Vulcano, a volcanic island in the Mediterranean north of Sicily. Lava from Vulcano is basaltic to andesitic, somewhat more siliceous than Strombolian lava, and thus more viscous. It forms a thicker crust over the vent, and eruptions are less frequent. When magmatic pressures build up again, the vent plug is blown out by powerful explosions, and broken fragments of the crater plug—along with scoria, pumice, ash, and bombs from new lava—are ejected in great cauliflower-like clouds. After the vent cover is blown away, lava flows often emerge from the crater or from fissures on the sides of the cone. Because Vulcano has been inactive since 1889–90, Vesuvius is often used as:

1. A typical example of the Vulcanian type of eruption
2. A variety of Vulcanian eruption with a more protracted ejection of gas and pyroclastic material

Plinian Type

The violent eruption of Vesuvius in A.D. 79, which buried the Roman cities of Pompeii and Herculaneum, was described by Pliny the Younger in letters to the Roman historian Tacitus. Explosive eruptions of gas and pyroclastic material (tephra) accompanied by huge clouds that extend to altitudes of 50,000 to 60,000 m (165,000 to 200,000 ft) have become known as **Plinian eruptions.**

Pompeii was a long-established Roman city of about 20,000 people when Vesuvius erupted violently on August 24, A.D. 79, burying the city beneath 5 to 8 m (17 to 26 ft) of pumice and ash and killing thousands of inhabitants who were caught by surprise. Many of the victims were buried quickly by ash, and their forms were preserved as molds in the ash which could be filled with plaster of Paris to obtain rather detailed casts (Figure 11–6). Some 2000 such forms have been uncovered during excavation of about half the city, many with their hands or clothing over their mouths, apparently attempting (unsuccessfully) to prevent asphyxiation. In some houses, food laid out for meals suggests that the inhabitants were not expecting disaster. However, 18 bodies were found in one wine cellar, presumably victims seeking last-minute protection.

Forms of victims found about 60 to 70 cm (24 to 28 in.) above the base of the pumice layer indicate that they had sur-

FIGURE 11–6
Mold of a victim buried in ash at Pompeii, Italy, by the 79
A.D. eruption of Mt. Vesuvius. (Photo by W. K. Hamblin)

vived the early part of the eruption before perishing. Several
bodies were uncovered 3 m (10 ft) above the base of the
pumice layer and separated from the upper surface by a thin
layer of ash. One of the dead held a decomposed bag, which
had spilled 400 silver and bronze coins (Corti, 1951). Many
bodies were found near the sea, indicating that the victims
were attempting to flee. The greatest loss of life at Pompeii
probably occurred in the initial hours when the ash fall was
at its maximum. Pliny's writings suggest that the entire area
was covered with a black cloud, presumably containing much
ash, for several days.

The city of Herculaneum was closer to Vesuvius than Pompeii but seemed to have escaped the killing ash falls. Instead,
the city was covered by a 20-m (66-ft)-thick mudflow.

Pelean Type

At 7:52 a.m. on May 8, 1902, the city of St. Pierre on the island of Martinique in the West Indies was destroyed, along
with all but two of its 30,000 inhabitants, by a fiery cloud of
gas, lava, and pyroclastic material that swept down the slopes
of Mt. Pelee at high velocity (Figure 11–7A) (Anderson and
Flett, 1902; Heilprin, 1903; Jaggar, 1902; Lacroix, 1904,
1908; MacGregor, 1951; Perret, 1935; Russell, 1902). Eruptions of this type, known as **Pelean,** are characterized by **nuée
ardentes** and **pyroclastic flows,** glowing clouds of extremely hot gas and incandescent ash which rush down the slopes
of volcanoes at velocities close to 100 mi/hr (165 km/hr). The
nuée ardentes originate when highly viscous magma clogs
the vent, then breaks out during an explosive eruption as hot,
ash-charged gas (Davies et al., 1978).

Although St. Pierre was entirely destroyed by the nuée ardente from Mt. Pelee (Figure 11–7B), the event was witnessed
by survivors on some of the 18 ships in the harbor. At 7:50
a.m., Mt. Pelee exploded, and a black cloud shot upward and
laterally from the crater, sending out a fiery cloud of gas and
incandescent material, which swept down the slopes with hurricane speed, maintaining contact with the ground like an
avalanche. In two minutes, St. Pierre was overwhelmed by
the glowing cloud, which spread out in a fan shape into the
sea. The entire population of nearly 30,000, with the exception of two survivors, was almost instantaneously wiped out.
The clock on the tower of the military hospital was later found
stopped at 7:52.

Observers who witnessed the event described the upward
and lateral blasts as nearly identical, black, rolling clouds.
The blast clouds consisted of superheated gas, mostly steam,
and incandescent ash. The ash particles gave the cloud a density greater than that of air, which kept the cloud in contact
with the ground, and the gas content gave it great fluidity,
which accounted for the high velocity. The force of the rapidly moving cloud was hurricane-like. Stone walls 1 m (3 ft)
thick were demolished, roofs were ripped off houses, 6-inch
(15-cm) cannons were torn from their moorings, and a stat-
ue weighing 3 tons was carried 15 m (50 ft) from its base.
Heilprin (1904), a geologist who arrived in St. Pierre a few
weeks later, described "twisted bars of iron, great masses of
roof sheeting wrapped like cloth about posts upon which they
had been flung, and iron girders looped and festooned as if
they had been made of rope."

The hot gases instantly ignited any flammable material, and
the entire city was quickly in flames. More than three hours
after the incident, the heat was still so intense that boats could
not approach the shore. All but two of the ships in the harbor
capsized, and many of the crews and passengers were killed.

An idea of the nature of nuée ardente eruptions can be
gained from eyewitness accounts. The Vicar-General of Martinique, who had left St. Pierre the previous afternoon, returned on the afternoon of May 8 and gave the following
description:

> A little further out blazes a great American packet (the Roraima)
> which arrived on the scene just in time to be overwhelmed by
> the catastrophe. Nearer the shore, two other ships are in flames.
> The coast is strewn with wreckage, with the keels of the over-
> turned boats, all that remains of the twenty to thirty ships which
> lay at anchor here the day before. All along the quays, for a dis-
> tance of 200 meters, piles of lumber are burning . . . . St. Pierre,
> in the morning throbbing with life, thronged with people, is no
> more. Its ruins stretch before us, wrapped in their shroud of
> smoke and ashes, gloomy and silent, a city of the dead. Our eyes
> seek out the inhabitants, fleeing distracted, or returning to look
> for the dead. Nothing to be seen. No living soul appears in
> this desert of desolation, encompassed by appalling silence.
> (Jaggar, 1902)

Only two ships escaped—the British steamer *Roddoam* and
the American sailing ship *Roraima*. Twelve men on the *Rod-*

A.

B.

FIGURE 11–7
(A) The 1902 nuée ardente from Mt. Pelee on the island of Martinique
(B) Remains of the city of St. Pierre after the devastating nuée ardente of 1902. (Photos from Lacroix, 1904, courtesy of American Museum of Natural History)

doam were killed, and ten were badly burned. The entire superstructure of the *Roraima,* which had just arrived in the harbor an hour earlier, was swept away, and the remainder of the ship was on fire. Twenty-eight of the crew of 47 were killed, and all but two of the passengers, a small girl and her nurse, were killed. The nurse gave the following account:

> We had been watching the volcano sending up smoke. The captain (who was killed) said to my mistress, 'I am not going to stay any longer than I can help.' I went to the cabin and was assisting with dressing the children for breakfast when the steward (who was later killed by the blast) rushed past and shouted, 'Close the cabin door— the volcano is coming!' We closed the door and at the same moment came a terrible explosion, which nearly burst the eardrums. The vessel was lifted high into the air, and then seemed to be sinking down, down. We were all thrown off our feet by the shock and huddled crouching in one corner of the cabin. My mistress had the girl baby in her arms, the older girl leaned on my left arm, while I held little Eric in my right.
>
> The explosion seemed to have blown in the skylight over our heads, and before we could raise ourselves, hot moist ashes began to pour in on us; they came in boiling splattering splashes, like moist mud without any pieces of rock. In vain we tried to shield

ourselves. The cabin was pitch dark—we could see nothing. A sense of suffocation came next (but when the door burst open, air rushed in and we revived somewhat). When we could see each other's faces, they were all covered with black lava, the baby was dying, Rita, the older girl, was in great agony and every part of my body was paining me. A heap of hot mud had collected near us and as Rita put her hand down to raise herself up, it was plunged up to the elbow in the scalding stuff.

> The first engineer came now, and hearing our moans carried us to the forward deck and there we remained on the burning ship from 8:30 a.m. until 3:00 p.m. The crew was crowded forward, many in a dying condition. The whole city was one mass of roaring flames and the saloon aft as well as the forward part of the ship were burning fiercely; but they afterwards put out the fire. (Jaggar, 1945)

The assistant purser of the *Roraima* described the eruption as follows:

> I saw St. Pierre destroyed. It was blotted out by one great flash of fire. Nearly 40,000 people were killed at once. Of 18 vessels lying in the Roads, only one, the British steamship Roddam escaped and she, I hear, lost more than half on board.
>
> It was a dying crew that took her out. Our boat arrived at St. Pierre early Thursday morning. For hours before we entered the roadstead, we could see flames and smoke rising from Mt. Pelee. No one on board had any idea of danger. Capt. G. T. Muggah was on the bridge and all hands got on deck to see the show. The spectacle was magnificent. As we approached St. Pierre, we could distinguish the rolling and leaping red flames that belched from the mountain in huge volume and gushed high in the sky. Enormous clouds of black smoke hung over the volcano. The flames were then spurting straight up in the air, now and then waving to one side or the other a moment, and again leaping suddenly higher up. There was a constant muffled roar. It was like the biggest oil refinery in the world burning up on the mountain top. There

was a tremendous explosion about 7:45 soon after we got in. The mountain was blown to pieces. There was no warning. The side of the volcano was ripped out, and there hurled straight towards us a solid wall of flame. It sounded like a thousand cannon. The wave of fire was on us and over us like a lightning flash. It was like a hurricane of fire, which rolled en masse straight down on St. Pierre and the shipping. The town vanished before our eyes, and then the air grew stifling hot and we were in the thick of it. Wherever the mass of fire struck the sea, the water boiled and sent up great clouds of steam. I saved my life by running to my stateroom and burying myself in the bedding. The blast of fire from the volcano lasted only for a few minutes. It shriveled and set fire to everything it touched. Burning rum ran in streams down every street and out into the sea. Before the volcano burst, the landings at St. Pierre were crowded with people. After the explosion, not one living being was seen on land. Only 25 of those on the Roraima, out of 68, were left after the first flash. The fire swept off the ship's masts and smoke stack as if they had been cut by a knife. (Leet, 1948)

Only two people out of the entire population of St. Pierre survived. One was 25-year-old Auguste Ciparis, who was a prisoner in an underground cell at the time of the nuée ardente. When he was rescued three days later, he was badly burned on his back and legs. His cell had no openings to the outside except for a slot in the upper part of the door. According to Kennan (1902), who interviewed him a few days later, the cell suddenly became dark, and hot ashes almost immediately came through the door grating. Intense heat persisted only for an instant, but he was terribly burned. He did not remember hearing any noise accompanying the heat. The only other survivor was 28-year-old Leon Compere-Leandre, who was seated on the doorstep of his house at the time the nuée ardentes hit St. Pierre (Heilprin, 1903).

The first known description of a nuée ardente by experienced scientists was made by Anderson and Flett (1902). They were on a ship a short distance offshore on the evening of July 9, 1902, when they observed the eruption of a nuée ardente from Mt. Pelee. They portrayed it as follows:

As the darkness deepened, a dull red reflection was seen in the tradewind cloud which covered the mountain summit. This became brighter and brighter, and soon we saw red-hot stones projected from the crater, bowling down the mountain slopes and giving off glowing sparks. Suddenly the whole cloud was brightly illuminated and the sailors cried, 'The mountain bursts!' In an incredibly short span of time a red-hot avalanche swept down to the sea. We could not see the summit owing to the intervening veil of cloud, but the fissure and the lower parts of the mountain were clear, and the glowing cataract poured over them right down to the shore of the bay. It was dull red, with a billowing surface, reminding one of a snow avalanche. In it there were larger stones which stood out as streaks of bright red, tumbling down and emitting showers of sparks. In a few minutes it was over. A loud angry growl had burst from the mountain when this avalanche was launched from the crater. It is difficult to say how long an interval elapsed between the time when the great red glare shone on the summit and the incandescent avalanche reached the sea. Possibly it occupied a couple of minutes . . . . Had any building stood in its path they would have been utterly wiped out, and no living creature could have survived that blast.

Hardly had its red light faded when a rounded black cloud began to shape itself against the starlit sky, exactly where the avalanche had been. The pale moonlight shining on it showed us that it was globular, with a bulging surface, covered with rounded protuberant masses, which swelled and multiplied with a terrible energy. It rushed forward over the waters, directly towards us, boiling, and changing its form every instant . . . . The cloud itself was black as night, dense and solid, and the flickering lightnings gave it an indescribably venomous appearance.

The cloud still travelled forward, but now was mostly steam, and rose from the surface of the sea, passing over our heads in a great tongue-shaped mass . . . . Then stones, some as large as a chestnut, began to fall on the boat. They were followed by small pellets, which rattled on the deck like a shower of peas. In a minute or two fine grey ash, moist and clinging together in small globules, poured down upon us. After that for some time there was a rain of dry grey ashes.

There can be no doubt that the eruption we witnessed was a counterpart of that which destroyed St. Pierre . . . a mass of incandescent lava rises and rolls over the lip of the crater in the form of an avalanche of red-hot dust. It is a lava blown to pieces by the expansion of the gases it contains. It rushes down the slopes of the hill, carrying with it a terrific blast, which mows down everything in its path. The mixture of dust and gas behaves in many ways like a fluid. (Anderson and Flett, 1902).

These graphic descriptions provide a vivid view of the nature of nuée ardente eruptions. Attempts were made to estimate the temperature of the nuée ardentes from its effect on material where no fires occurred. Glass objects were softened (650°–700°C); fruit was carbonized; wooden decks of ships offshore in the harbor were set on fire; but copper (melting point 1058°C) was not melted.

From his studies of the many nuée ardentes emitted during the 1929–32 eruption of Mt. Pelee, Perret (1935) concluded that the expansion of gases given off by magmatic fragments is largely responsible for the great speed and power of the nuée ardentes. The rapid expansion of gases suspends solid particles in gas within the nuée ardente and causes fluidization (MacGregor, 1951).

Late in the summer and early fall, a domelike mass of exceptionally viscous magma appeared in the crater, and a spine with a diameter of 100 to 150 m (330 to 500 ft) rose to a height of 300 m (1000 ft) above the crater floor in a few months (Heilprin, 1904). It rose an average of about 10 m/day (30 ft/day), even though giant blocks continually broke away from the top, achieving its maximum height of 300 m (1000 ft) on May 30, 1903.

A phenomenon somewhat similar to nuées ardentes is a **base surge,** a basal cloud of gas and pyroclastic material that flows radially away from an eruptive source. Base surges are typical of nuclear bomb explosions, and the volcanic equivalent was described by Moore (1967) during the 1965 eruption of Taal Volcano in the Philippines.

During the 1980 eruption of Mount St. Helens, **pyroclastic flows** repeatedly swept down the mountain (Figure 11–8A), leaving trails of pumice the size of baseballs to volleyballs with pressure ridges and other flow structures (Figure 11–8B).

Whether an eruption behaves as a Plinian type or a Pelean type depends largely on whether degassing of the magma begins at relatively deep or shallow levels in the vent (Figure 11–9).

Krakatoan Type

The most violently explosive volcanic eruptions are known as **Krakatoan,** named after the 1883 explosion of Krakatoa in the Sunda Strait between Java and Sumatra, the most violent eruption in recorded history. Most of the accounts of what happened were collected by a Dutch fact-finding committee, which published a report six months later, and by a committee of the Royal Society in Great Britain, which published a lengthy volume in 1888.

The stirrings of Krakatoa began with relatively mild eruptions from the northernmost of three craters on the island on May 20, 1883. The early activity seemed so innocuous that one week later, people from nearby Djakarta (Batavia) chartered a steamer to visit the island. On August 26, the first of several colossal explosions of Krakatoa were heard by people 200 km (120 mi) away. More explosions, with increased intensity, continued, and black clouds rose 27 km (16 mi) above the island as the first gigantic sea wave occurred in Sunda Strait. Severe shock waves from explosions continued through the night and could be heard for 150 km (90 mi). Several more large sea waves were generated on August 27.

The last person known to see the island before it blew up was a Dutch surveyor who observed mild activity from three main vents on August 11. The explosion of Krakatoa on August 26 and 27 was vividly described in the log of the captain of the British ship *Charles Bal,* beginning on August 26:

At 2:30 P.M. noticed some agitation about the Point of Krakatoa; clouds or something being propelled with amazing velocity to the northeast. To us it looked like blinding rain, and had the appearance of a furious squall of ashen hue . . . . At 5 P.m. the roaring noise continued and increased, darkness spread over the sky and a hail of pumice stone fell on us, many pieces being of considerable size and quite warm. Had to cover up the skylights to save the glass, while feet and head had to be protected [by] boots and southwesters. About 6 P.m. the fall of larger stones ceased, but there continued a steady fall of a smaller kind, most blinding to the eyes, and covering the decks with 3 to 4 inches very speedily, while an intense blackness covered the sky and land and sea . . . . At 11 P.M. . . . . chains of fire appearing to ascend and descend between the sky and it, while on the southwest end there seemed to be a continued roll of balls of white fire; the wind though strong, was hot and choking, sulphurous, with a smell like burning cinders. From midnight to 4 A.M. the same impenetrable darkness continuing, the roaring of Krakatoa less continuous but more explosive in sound, the sky one second intense blackness and the next a blaze of fire; mastheads and yardarms are studded with electrical glows and a peculiar pinky flame coming from clouds which seemed to touch the mastheads and yardarms . . . . At 11:15 there was a fearful explosion in the direction of Krakatoa, now over 30 miles distant. We saw a wave rush right on to Button Island, apparently sweeping right over the south part and rising halfway up the north and east sides. This we saw repeated twice,

A.

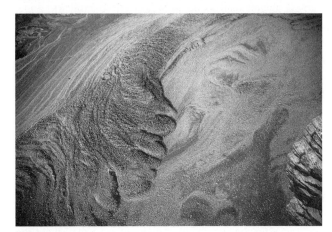

B.

FIGURE 11–8
(A) Nuée ardente from Mt. St. Helens, Washington, during the 1980 eruptive phase. (Photo by U.S. Geological Survey)
(B) Pyroclastic flow from 1980 Mt. St. Helens nuée ardente made up almost entirely of fresh pumice clasts about the size of baseballs. Note the pressure ridges and marginal flow lobes.

but the helmsman says he saw it once before we looked. The same waves seemed also to run right on to the Java shore. The sky rapidly covered in, by 11:30 A.M. we were inclosed in a darkness that might almost be felt. At the same time commenced a downpour of mud, sand, and I know not what . . . .

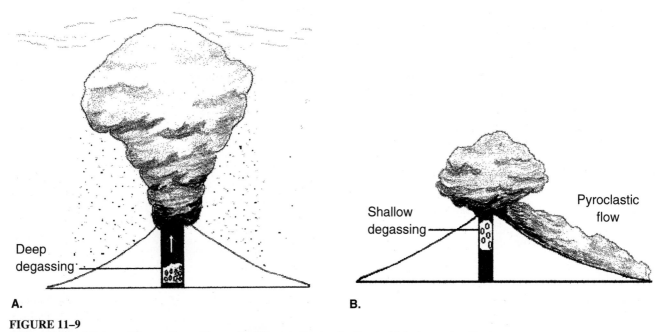

A.   **B.**

FIGURE 11–9
Plinian (A) and Pelean (B) eruptions. The main difference is the depth at which degassing begins to occur.

A second huge explosion occurred, but it did not generate another wave. Another loud explosion was followed by a small sea wave. Explosions of diminishing intensity continued during the night and the early morning hours of the next day, then essentially ended by noon. When eruptions finally ceased, an area of Krakatoa 5 by 8 km (3 to 5 mi) was gone, and an 800 m (2600 ft) high cliff remained at the south end of the island. Where the 450 m (1500 ft) high central peak of the island had been, the ocean was now 200 m (660 ft) deep. How much of the new huge caldera resulted from the explosion and how much formed by collapse of the former cones remains conjectural. In either case, the caldera owes its origin to the extreme explosiveness of the eruption and the expulsion of a huge volume of material.

The largest explosion ever recorded shot an ash cloud 80 km (48 mi) high and produced a large crater where Krakatoa had previously been. About 18 km³ (4 mi³) of material was blown out of Krakatoa, producing a caldera 6 km (3.5 mi) in diameter and 1 km (0.6 mi) deep (Figure 11–10). About half an hour after the cataclysm, a gigantic sea wave swept the coasts of Java and Sumatra, destroying 295 towns and killing an estimated 36,000 people. Because the sea wave had a very long wavelength in deep water, ships in the Sunda Strait simply rode up and down the wave. However, as the wave approached shore, it grew in height to about 30 m (100 ft). A Dutch warship was washed 1 km (0.6 mi) inland and stranded 10 m (30 ft) above sea level. Hours later, the noise of the explosion was heard 5000 km (3,000 mi) across the Indian Ocean, 3700 km (2200 mi) away in central Australia, and the shock wave was recorded on barometers around the world.

The immense amount of ash blown into the air by the eruption caused midday darkness over large areas, some as

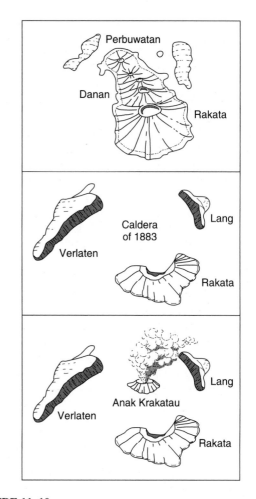

FIGURE 11–10
Map of the explosion of Krakatoa in 1883. (From Decker and Decker, 1981)

far away as 460 km (275 mi). The darkness persisted for 22 hours at a distance of 215 km (130 mi), and for 57 hours at a distance of 80 km (50 mi). Ash fell on the decks of ships 2700 km (1600 mi) away three days after the eruption, and fine ash remained in the atmosphere for months, traveling around the earth many times. Volcanic ash blown into the stratosphere completely encircled the earth, producing unusually colorful sunrises and sunsets and giving the sun and moon a blue-green appearance. The ash caused a decrease of 10 percent in the average solar radiation in Europe over the next three years, and average world temperatures dropped below normal.

Mount St. Helens.   Another example of a Krakatoan-type eruption is the 1980 explosion of Mt. St. Helens, Washington. On March 27, 1980, after 120 years of inactivity, the volcano began erupting steam laced with rock fragments ripped from the walls of the vent. During April and early May, swarms of earthquakes occurred every day as magma pushed its way toward the surface, steam eruptions continued intermittently, and a conspicuous bulge appeared on the north flank of the mountain. The bulge grew outward at a rate of up to 2.5 m/day (8 ft/day). By May 18, a 2 km (1.2 mi) area on the north flank had risen upward and outward 100 m (330 ft).

At 8:32 a.m. on May 18, 1980, a magnitude-5 earthquake north of the volcano triggered a gigantic landslide, and 2.5 km³ (0.6 mi³) of the north flank of the mountain slid away, including the entire bulge. The landslide released internal pressure under the bulge, and seconds later, a violent blast blew off the top 400 m (1313 ft) of Mt. St. Helen's cone (Figures 11–11 and 11–12; Plates 7D and E). Because pressure was released when the north side of the cone slid away, the main blast was directed laterally and surged outward at speeds up to 400 km/hr (250 mph), overtaking the fast-moving landslide and devastating an area of 600 km² (230 mi²) north of the volcano (Figure 11–11). The lateral blast wiped out entire forests at distances of 20 km (12 mi) from the mountain (Figure 11–13) leaving only splintered stumps close to the mountain (Figure 11–13A) and flattening an estimated 800 million board feet of timber. A Plinian cloud 20 km (12 mi) high rose rapidly above the summit. When the eruptive cloud had dissipated so the mountain could be seen, only a hollow crater with its floor 1000 m (3300 ft) below the original summit remained (Figure 11–12B) (Christiansen and Peterson, 1981; Lipman and Mullineaux, 1981; Lipman et al., 1981). The lateral blast, pyroclastic flows, mudflows, and floods killed 64 people, destroyed more than 100 homes, and caused an estimated damage of about $3 billion.

Solfataric Stage

Solfataric eruptions are restricted to the emission of gases, usually in the final phases of a volcanic eruption. A volcano may remain in the solfataric stage for many hundreds of years after its magmatic eruptive activity has ended.

VOLCANIC LANDFORMS

Volcanic cones are readily identified by their topographic form—a conical hill with a circular depression (crater) on the top. The form of a volcanic cone depends largely on the dominant type of the erupted material, whether mainly lava flows or pyroclastic material.

Shield Volcanoes

Eruption of hot, fluid, basaltic lava over a long period of time builds a **shield volcano,** with gently sloping sides and a low profile (Figure 11–14). Individual lava flows are typically only a few meters thick, but the accumulation of thousands of flows over long time periods builds large shield cones (so named because they resemble the shape of a Roman shield).

The Hawaiian Islands are good examples of shield volcanoes—they consist of broad shield cones rising from about 6000 m (20,000 ft) below sea level on the ocean floor to about 4000 m (13,000 ft) above sea level. Measured from its base on the ocean floor, Mauna Loa is about 10,000 meters (33,000 ft) high and 100 km in diameter at its base. The islands are the result of Hawaiian-type eruptions of basalt flows that emerged from central cones and from fissures along the flanks of the volcanoes. All eruptions were typically relatively quiet emissions of hot, fluid basalt with little explosive activity and little pyroclastic material. The vents that supply the lava change with time, so that eventually the accumulation of successive ribbons of individual flows meld into the broad shield form. Accumulation of basalt flows is fairly rapid, as shown by the absence of soil layers between successive flows.

Composite Cones

If the eruptive material from a vent consists of both lava flows and pyroclastic fragments, the resulting volcanic cone is a **stratovolcano** or **composite cone,** having slopes somewhat steeper than those of shield volcanoes (Figure 11–15; Plate 7A). Although andesite magmas are usually dominant, the composition of the material making up the cone may range from rhyolite to basalt. Pasty, gas-charged andesitic lavas intermittently erupt lava and pyroclastic material. Because andesitic volcanoes are so common at subduction zones, stratovolcanoes are the most common type of volcanic cone.

Plug Domes

Viscous dacitic/rhyolitic lavas ooze out on the ground surface like thick toothpaste and do not flow far from the vent, piling up steep-sided **plug domes** (Figure 11–16). They are

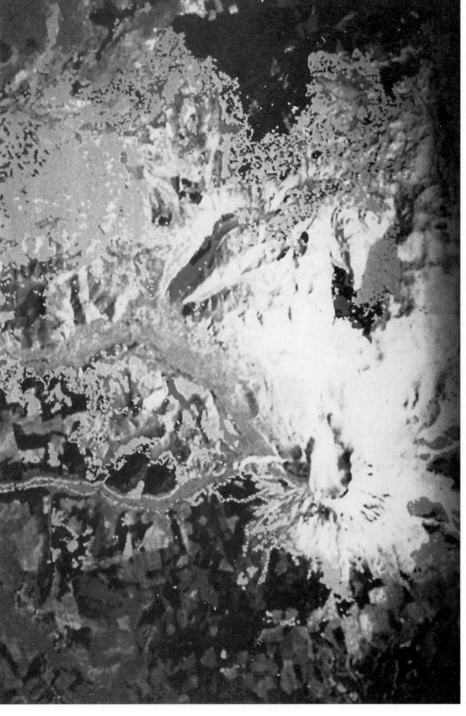

FIGURE 11–11
Devastation from the 1980 explosion of Mt. St. Helens, Washington. In the light gray area, entire forests were blown away; in the somewhat darker gray areas, entire forests were killed but left standing. Distance across the photo is about 20 km. (Photo by NASA)

typically composed of obsidian and pumice and have steep flow margins that may be tens or hundreds of meters high. The very viscous, pasty lava oozes up to the surface where it solidifies around the vent as a dome, usually with a very irregular summit surface rather than the distinct crater typical of most other volcanic cones (Williams, 1932).

Cinder Cones

Late in an eruptive cycle, vents may become clogged, and as a result, ejected pyroclastic fragments form steep-sided *cinder cones* (Figure 11–17), so called because they consist of loose cinders lying at the maximum possible angle without

FIGURE 11–12
(A) Mt. St. Helens before the
1980 explosion.
(B) Mt. St. Helens after the 1980
explosion. (Photo by U.S.
Geological Survey)

A.

B.

sliding, usually about 25° to 35°. Cinder cones may occur singly or in clusters aligned along fracture systems. Lava flows rarely emanate from the summit because the loose pyroclastic material making up the cone is not strong enough to sustain the fluid pressure of the magma high in the vent. Thus, lava flows typically break out through the base of a cinder cone (Figure 11–18), often carrying large blocks of the cone with it.

Spatter Cones

Spatter cones develop around small vents where molten lava is tossed into the air by the force of gases in the magma.

Blobs of airborne lava fall around the vent where they weld themselves to previously deposited material, building up a small cone.

Table Mountains

Table mountains are flat-topped, volcanic cones with steep sides composed of pillow lava, palagonite, and vitreous pyroclastic rocks. They form as a result of subglacial eruptions that melt the ice, creating a large pocket of steam and water beneath the ice. Eruption of lava into the water produces pillow lava, and interaction of the water and steam with the lava forms palagonite, which together build a flat-topped cone

A.

B.

FIGURE 11–13
(A) Shredded tree stumps, all that remains of a forest after the blast of May 18, 1980, Mt. St. Helens, Washington.
(B) Forest blown down by the blast.

FIGURE 11–14
Gently sloping profile of a typical shield volcano, Mauna Loa, Hawaii.

FIGURE 11–15
Composite volcanic cone of andesitic lava and pyroclastic material, Mt. Hood, Oregon.

(Kjartansson, 1967). When the enclosing glacier melts, a table mountain emerges from the ice. Good examples of table mountains occur in Iceland where volcanic eruptions have taken place beneath glaciers.

Craters and Calderas

Volcanic cones typically have craters at their summits, usually a few hundred to a thousand meters in diameter. However, the central depressions associated with some volcanoes are unusually large, having diameters measured in kilometers or tens of kilometers. Such large craters, known as **calderas,** may be produced by:

1. Explosion of the summit area

2. Collapse of the interior of the volcanic cone following ejection of immense volumes of pumice and ash

3. Subsidence due to draining of a magma chamber by flank eruptions

Although violent explosions such as Krakatoa and Santorini are associated with calderas, the deficiency of pyroclastic material around the calderas suggests a collapse origin, rather than explosion of an entire cone.

Crater Lake, Oregon (Figure 11–19), is a classic example of a caldera (Atwood, 1935; Diller and Patton, 1902; Williams, 1941a, 1941b, 1942). The circular caldera is about 9 km (5 mi) in diameter and filled with water as deep as 600 m (2000 ft). Nearly vertical cliffs above the lake rise to altitudes of about 2425 m (8000 ft) at the rim.

Diller and Patton (1902) recognized that the former summit of the volcano once extended more than 1800 m (6000 ft) above the present rim and that the present crater is only the stump of a former large volcanic cone. The summit cone must have been much higher than the present rim in order to support glaciers that carved deep glacial troughs into the lower flanks of the cone and that provided glacial debris interbedded with volcanic debris at the rim of the present caldera. Diller calculated that about 70 km³ (17 mi³) of material was missing from the former summit; he contended that if the destruction of the former summit were caused by a gigantic explosion, a great thickness of pyroclastic debris should surround the base of the volcano. He argued that because only a small amount of debris can be found nearby, the crater was not blown away but was formed principally by collapse of the volcano's sides. In order for collapse to occur, however, sufficient room must have been made to accommodate the collapsing material. Williams (1941a) concluded that about 1.5 mi³ (6 km³) of fragmental material could have been removed from the summit of the volcano by explosion and that another 23 km³ (5 mi³) could have collapsed into the cone to replace magma erupted during the last violent stages of activity. This still left about 44 km³ (10.5 mi³) of the former summit area to account for. Williams suggested that the missing 44 km³ (10.5 mi³) of material might have caved into the void left as magma withdrew from the area around the cone due to intrusion of the magma into adjacent country rock. Williams felt that the collapse probably followed shortly after a series of very violent eruptions that spread nuées ardentes down the side of the volcano and threw large amounts of ash into the air. Radiocarbon dating of wood associated with ash ejected during the final eruptive phases of Mt. Mazama indicates an age of 6700–7000 years.

A.

FIGURE 11–16
(A) Plug dome with steep-sided margins more than 100 m high, south of Mono Craters, California. (Photo by W. K. Hamblin)
(B) Plug domes and associated glass flows, Mono Craters, California. (Photo by D. A. Rahm)

B.

FIGURE 11–17
Cinder cone forming by eruption of pyroclastic material, Cerro Negro, Nicaragua. (Photo by U.S. Air Force)

FIGURE 11–18
Cinder cone formed by eruption of pyroclastic material, Lava Butte, Oregon. Note the lava flow that emanated from the base of the cone. (Photo by D. A. Rahm)

FIGURE 11–19
Caldera at Crater Lake, Oregon.

Santorini (also known as Thira), the largest island in a circular group of islands in the Agean Sea, marks the remains of a once-great volcano. Violent eruptions there about 1620 B.C., followed by collapse of the volcano into the sea, produced a caldera about 10 km in diameter (Figure 11–20B).

A series of explosive eruptions buried the Bronze Age city of Akrotiri beneath 70 m (230 ft) of ash and pumice. The first phase of eruptions buried buildings in Akrotiri with 6 m (20 ft) of pumice. This was followed by 56 m (180 ft) of ash, pumice, and large rock fragments that fell from the air, probably during collapse of the volcano. The last phase consisted of ash and small rock fragments deposited from hot gaseous clouds.

Some 30 km^3 of rhyolitic magma is estimated to have been blown out before the once-large island was virtually destroyed, leaving the 10 km-diameter caldera 390 m (1280 ft) below sea level. At that time, the shorelines around the Mediterranean Sea were populated by the Minoan civilization. The Minoans had a relatively advanced civilization, with a comfortable standard of living. Giant tsunamis, produced by the explosion and summit collapse of Santorini ravaged coastal populations with devastating effect.

About 1200 years later, Plato wrote of the vanishing of the island empire of Atlantis, which, after violent earthquakes and great floods, *"in a single day and night disappeared beneath the sea."* The legend of the sinking of Atlantis beneath the sea is believed to coincide approximately with the explosion and collapse of Santorini.

Even the greatest historical eruptions, such as Krakatoa and Santorini, are puny compared with some prehistoric ones. Yellowstone National Park has been the site of at least three mega-eruptions in the last 2 million years. Catastrophic erup-

tions occurred there 2 million, 1.2 million, and 600,000 years ago, creating three giant overlapping calderas (Figure 11–20A). Ash from each of these three giant eruptions extended 700–800 miles eastward into Nebraska, Iowa, and Kansas where thicknesses reach one meter. Imagine what would happen today if ash of this thickness were to blanket the central United States!

Yellowstone is famous for its geysers and hot springs, attesting to the presence of magma and hot rocks at shallow depths below the surface. Magma has been detected by subsurface geophysical methods not far below the surface, and the eruptive cycle of Yellowstone may not yet be finished. The mega-eruptions have recurred at intervals of about 600,000 years, and since the last one was 600,000 years ago, the possibility of yet another one is not out of the question.

Hawaiian calderas

Calderas can also be associated with quiet basaltic eruptions. A good example is Kilauea on the island of Hawaii (Figure 11–21), where the caldera is defined by a set of circular fractures. Subsidence of the floor of the caldera is thought to result from withdrawal of magma to flank eruptions.

Lava Flows

The topographic features of lava flows vary considerably with the composition of the magma. Basalt is typically fluid and forms thin, laterally extensive flows (Figure 11–22), whereas more siliceous, viscous rhyolite/dacite magmas form thick, pasty flows with steep margins (Figure 11–23).

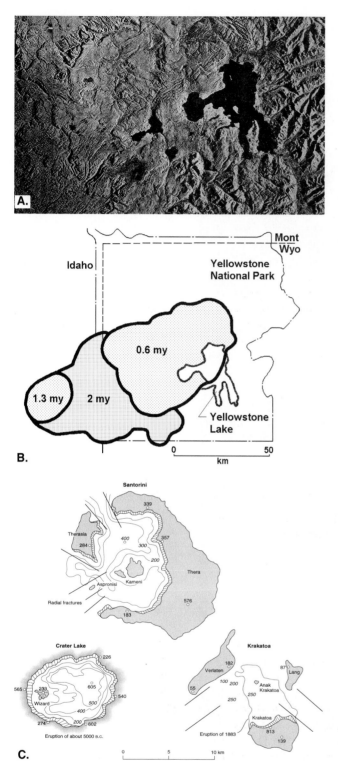

FIGURE 11–20

(A) Side-looking radar image of the giant calderas of Yellowstone National Park, Wyoming. (Photo by U.S. Geological Survey). (B) Map showing the calderas of Yellowstone National Park. (C) Maps of Santorini, Crater Lake, and Krakatoa calderas. Although all experienced violent eruptions, the calderas are generally believed to have formed by collapse of the summit cone, rather than by being blasted away. (Reprinted from Volcanoes of the Earth. 2nd revised edition, by Fred Bullard, copyright 1984. By permission of the author and the University of Texas Press)

Lava flows are characterized by lobate forms, lava levees, crescent-shaped pressure ridges, irregular surfaces, and lack of surface streams, all of which show up prominently on aerial photographs or topographic maps of recent lava flows (Figure 11–24). Contours on a topographic map of a lava flow similar to the one shown in Figure 11–25 would be highly crenulated because of the irregular surface.

As lava flows away from its vent area, the flow surface cools more rapidly than the interior, forming a darkening crust over the red-hot magma. As the still-molten interior of the flow continues to move, the crust breaks up into blocks, giving it a very rough, blocky surface known as aa. Solidified blocks of lava tumble over the front of the flow and are then overridden by the advancing flow, somewhat in the manner of a caterpillar tread. If the congealing surface of the lava is thin, it may stretch elastically to form smooth billowing or ropy forms known as **pahoehoe** (Figure 11–26). Both aa and pahoehoe may be erupted from the same vent (Figure 11–27), and pahoehoe may change to aa as viscosity of the lava increases due to cooling and loss of gases.

The molten interior of a flow often continues to move long after the surface and sides have solidified, forming **lava tubes** (Figure 11–28; Plate 7C). Basaltic eruptions in Hawaii typically begin high on the flanks of Kilauea or Mauna Loa, then flow unseen through lava tubes for distances of several kilometers, periodically breaking out through the front of the flow to initiate yet another set of lava tubes. The size of a lava tube is determined by the thickness of the flow, the viscosity of the lava, the rate of cooling, and the slope. Tubes are often 5 to 10 m (15 to 30 ft) in diameter and may be continuous for several kilometers. When the roofs of lava tubes collapse, elongate depressions are formed on the surface of the flow. Natural bridges are formed where all but a small portion has collapsed.

Pressure ridges, formed by movement of congealing lava while the interior is still mobile (Figure 11–24), are transverse to the direction of flow and are convex down slope. They are typically 1 m (3 ft) or so in height and 30 to 50 m (100 to 160 ft) long. Many have conspicuous cracks along the crests of ridges. **Lava blisters or squeezeups** are mounds formed by lava pushing up through the earlier-formed crust (Figure 11–29).

Fluid lava, usually basalt, may enclose trees that are in the path of the flow. The wood burns, but in the process the lava congeals around the trunk, leaving a cylinder-shaped **tree mold** where the tree formerly stood. For trees more than a few centimeters in diameter, burning of the base of the tree causes the portion of the tree above the flow surface to topple over onto the crusted-over top of the flow, which protects it from the incandescent lava beneath. Thus, leaves and small branches burn, but the central trunk usually remains on the flow surface.

Fluid basalt flowing into a forest congeals quickly around trees, and when the surface of the flow subsequently drops, lava trees remain standing above the surface (Figure 11–30). Lava trees can also be formed where spatter, thrown

FIGURE 11–21
Kilauea caldera, formed by subsidence along circular fractures around the rim of the caldera. (Photo by U.S. Geological Survey)

FIGURE 11–22
Thin, basaltic pahoehoe lava flow from the southwest rift zone, Hawaii.

FIGURE 11–23
Thick, pasty glass flow south of Mono Craters, California. Note the height and steepness of the flow margins relative to fully grown pine trees at the base. (Photo by W. K. Hamblin)

FIGURE 11–24
Lobate flows with pressure ridges and lava levees emanating from a central cone, Japan. (Photo by U.S. Geological Survey)

FIGURE 11–25
Irregular surface of a basalt flow near Craters of the Moon, Idaho. (Photo by W. K. Hamblin)

FIGURE 11–26
Pahoehoe lava formed by stretching of the plastic crust of a lava flow as the molten interior continues to advance. A few millimeters of glass on the surface of the lava make the flow surface resemble molten silver. Within a few years, the silvery crust weathers to a deep black.

into the air from lava fountains, lands on trees, coating them with lava.

The constructional features of lava flows become gradually erased by erosion with time and are replaced by new forms produced by differential erosion. Lava emitted from a vent flows down whatever slope it finds in its path, often preexisting valleys which give the flow an elongate form. The pre-flow drainage is destroyed as the lava fills the valley, and

water brought into the lava-filled valley by tributary streams is forced to flow between the valley sides and the margins of the lava flow because the flow is usually somewhat higher in the middle and the surface is rough with many depressions that inhibit flow of water. Thus, when drainage is reestablished, two streams occupy the valley, one along each side of the flow. As the two streams incise their channels downward between the lava flow and the valley sides, the flow, which

FIGURE 11–27
Aa lava on top of earlier pahoehoe flow, 1972–74 Mauna Ulu eruptions, Kilauea, Hawaii.

formerly occupied the topographic low in the valley, is left as a sinuous divide between the streams (Figure 11–31). This process, which results in conversion of a topographic low into a topographic high, is known as **topographic inversion.**

Rifts

Rifts are elongate cracks in the earth's surface. Some very large ones, like the great rift valleys of Africa, are caused by movement of crustal plates. Other much smaller ones, caused by local tensional forces in the rocks, are common in areas of basaltic eruptions (Figure 11–32A) where they may provide an avenue for the extrusion of lava, resulting in aligned cones along the rift (Figure 32B).

Predicting Volcanic Eruptions

Predicting volcanic eruptions is a lot like weather forecasting—long-term prediction is not very accurate, but short-

term forecasting may be reasonably precise. The shorter the time before an event, the more accurate the forecast. Volcanic eruptions are typically preceded by a number of observable physical phenomena that can help predict forthcoming eruptions. Eruptions of long-dormant volcanoes are more difficult to predict than forecasting the next phase of an actively erupting volcano because monitoring techniques can be employed for active volcanoes. The following phenomena commonly precede volcanic eruptions and thus can help predict their onset.

Seismic Activity

Volcanic eruptions are commonly preceded by swarms of earthquakes, sometimes several thousand in a single week. Although not all earthquake swarms are necessarily followed by volcanic eruptions, volcanoes seldom erupt without any antecedent earthquake activity. The frequency, magnitude, and location of earthquakes are meaningful indicators of impending volcanism.

Rising magma fractures the rock through which it travels and generates shallow earthquakes above the magma chamber, setting up nearly continuous seismic vibrations known as harmonic tremors (Figure 11–33). The tracing of such tremors on a seismogram differs from those of normal earthquakes, which are characterized by sharp, discontinuous jolts. Harmonic tremors typically immediately precede an eruption and provide one of the most useful methods of predicting an imminent eruption. This technique has been particularly effective in providing early warning of eruptions on Hawaii and at Mt. St. Helens.

Swelling and Tilting

A detectable upward bulging of the land surface occurs as magma rises in a vent beneath a volcano, and a corresponding deflation of the land surface takes place when magma is withdrawn elsewhere (Figure 11–34). The swelling beneath a volcano produces a tilting of the ground surface that can be measured by precise surveying of networks of points, such as those on the flanks of Kilauea in Hawaii.

The community around the popular ski area at Mammoth Lakes, California, lies in the Long Valley caldera, the site of a gigantic eruption 700,000 years ago that covered a large area with pyroclastic flows hundreds of meters thick. Thus, great concern was felt when, in 1980, the uplift rate in the area accelerated to about 25 cm (13.5 in.) in only 2 years, accompanied by swarms of earthquakes, some having magnitudes of 5 to 6 on the Richter scale. Careful monitoring of uplift rates and earthquake frequency suggested that a large body of magma had moved upward from a depth of 8 km in 1980 to 3.2 km in 1982. Increased concern led the U.S. Geological Survey to develop evacuation plans and issue a notice of potential volcanic hazard. Swarms of earthquakes continued until 1983, then abated, along with fears of a potentially catastrophic

A.

FIGURE 11–28
(A) Lava flowing in a lava tube,
Hawaii. (Photo by U.S.
Geological Survey)
(B) Thurston lava tube, Hawaii.
(Photo by U.S. Geological
Survey)

B.

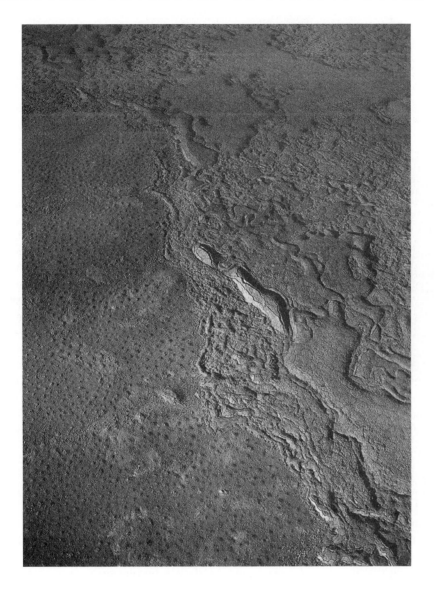

FIGURE 11–29
Squeezups and collapse features on the surface of a basalt flow, Snake River Plaine, Idaho. (Photo by W. K. Hamblin)

explosive eruption. Thus, even though the monitoring data suggested the potential for a dangerous eruption, subsurface movement of magma doesn't necessarily always lead to an eruption.

Monitoring of Volcanic Gases

Monitoring of gases emitted from a volcano may show changes in gas chemistry as an eruption become imminent. Changes in relative amounts of steam, carbon dioxide, and sulfur dioxide and changes in the emission rate of sulfur dioxide are related to upward movement of magma. In particular, substantial increases in the rate of sulfur emission from volcanoes have commonly preceded eruptions. During periods of low magmatic activity, Kilauea discharged a nearly constant flux of SO_2, whereas during high activity, the rate doubled (Casadevall et al., 1987).

Changes in Thermal Activity and Magnetic, Electrical, and Gravitational Properties

As magma rises upward in a vent prior to an eruption, the overlying rocks heat up, and the rise in temperature increases the activity of hot springs and steam vents. Periodic infrared remote sensing of volcanic terrain can disclose newly heated areas that may be the site of future volcanic activity.

When rocks are heated by proximity to magma, their magnetic and electrical properties change and the replacement of rocks by the rising, less-dense magma causes very slight, but detectable, changes in measurements of the gravitational force above.

Diversion of Lava Flows

Although volcanic eruptions can't be prevented, several methods can divert the direction of lava flows or impede their

FIGURE 11–30
Lava trees, southwest Kilauea, Hawaii. (Photo by W. K. Hamblin)

Iceland had long enjoyed the economic benefits of a prime fishing port, due to its excellent harbor made by ancient lava flows. On January 23, 1973, a fissure less than 1 km (3300 ft) from the town began erupting pyroclastic material, which quickly built a cinder cone and blanketed the town with a heavy fallout of cinders. The town was evacuated except for a volunteer workforce that stayed to shovel hot pyroclastic fragments from the roofs of buildings that would otherwise have collapsed from the heavy load. Despite their best efforts, pyroclastic fallout buried 70 buildings. The early pyroclastic phase was followed by eruption of basalt flows that buried about 300 buildings on the outskirts of town and began to fill in the harbor, threatening to block the entrance, the loss of which would have been an economic disaster. The Icelanders first attempted to bulldoze levees of pyroclastic material, then resorted to pumping large volumes of sea water onto the flow from boats and barges in the harbor so that evaporation of the water from the heat of the magma would cool the flow surface and cause the lava to congeal. The concept is similar to the cooling effect of the evaporation of perspiration from skin—the heat energy used to convert liquid into vapor results in a cooling effect on the surface. As the surface of the lava cooled, its velocity slowed and formed a barrier that impeded the flow of lava following behind. The effort succeeded and the harbor was saved. In fact, the remaining part of the harbor was partially closed off by the flows, improving protection from the sea as a natural breakwater.

IGNEOUS INTRUSIVE LANDFORMS

Although igneous intrusions rarely directly create landforms, subsurface injections of magma into the host rocks sometimes affect the surface by bowing up overlying rocks. However, the principal topographic effects of intrusions are due to later etching out of resistant intrusive rocks from less resistant host rocks by differential erosion. Igneous intrusions are distinguished on the basis of (1) their relationship to the host rocks (that is, whether concordant or discordant) and (2) their shape and size.

Discordant Plutons CUT THROUGH

Discordant plutons cut across the bedding or foliation of the host rocks. Discordant plutons are further subdivided on the basis of shape and size.

Volcanic Necks

Volcanic necks make cylindrical to spire-shaped landforms consisting of magma that has solidified in the conduit of a former volcano and has been exposed at the surface by dif-

progress. Bombing of flows to collapse lava tubes, hydraulic chilling to slow down flow progress, and construction of levees have been employed in Iceland and Hawaii to deflect or slow the progress of lava flows away from populated areas.

In Hawaii, attempts to stop or divert basalt flows by bulldozing levees along the path of oncoming lava have been only temporarily successful. Levees can divert small, thin, basalt flows, but if the rate of lava emission is high, the levees are soon overtopped.

Another approach to diversion of basalt flows that endangered the city of Hilo was attempted during the 1935 eruption on Mauna Loa. The flow vents originated high on the mountainside, and much of the lava flowed through long tubes to the lower slopes, where it periodically broke through the flow fronts and continued downslope toward Hilo. The director of the Hawaii Volcano Observatory, T. A. Jagger, proposed dropping bombs on the roofs of the lava tubes to force the lava onto the surface where cooling might slow the progress toward Hilo. When twenty 600-lb bombs were dropped on the flow surface to collapse the tubes that were delivering the basalt, the flow front, which had been advancing at 240 m/hr (800 ft/hr), slowed to 13 m/hr (44 ft/ hr) the next day and stopped entirely 30 hours after the bombing. Bombing of flows was also attempted in 1942 with similar results.

The 5300 inhabitants of the town of Vestmannaeyjar on the small island of Haimey just off the southern coast of

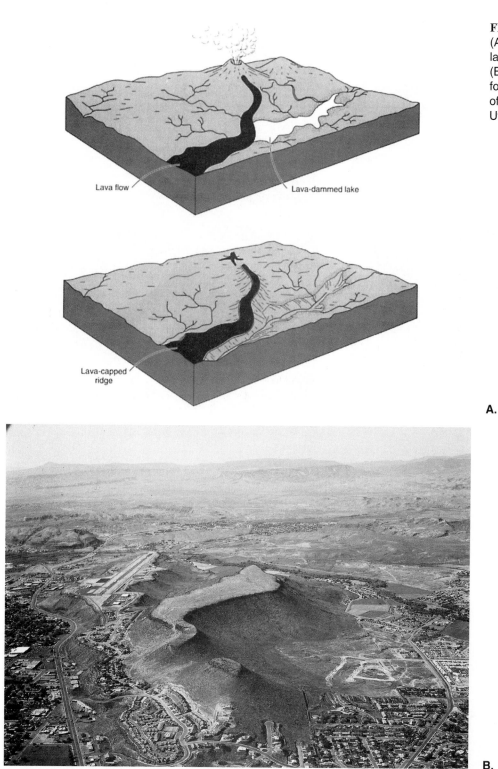

FIGURE 11–31
(A) Topographic inversion of a lava flow.
(B) Sinuous, lava-capped ridge formed by topographic inversion of a lava flow, near St. George, Utah. (Photo by W. K. Hamblin)

A.

B.

ferential erosion (Figure 11–35 and Plate 7F). They are etched into topographic relief because the igneous rock of the neck is more resistant to erosion than the host rock. They can be distinguished from volcanic cones by their steep-sided, spire-shaped forms and by the lack of a crater at their summit.

Dikes

Dikes are tabular plutons that cut across the structure of their host rocks, Their tabular form is usually a reflection of magma that has solidified in fissures. The thickness of dikes ranges from a few centimeters to hundreds of meters. The

A.   B.

FIGURE 11–32
(A) Southwest rift zone of Kilauea, Hawaii. (Photo by J. D. Griggs, U.S. Geological Survey)
(B) Rift associated with basaltic eruptions at Craters of the Moon, Idaho. (Photo by W. K. Hamblin)

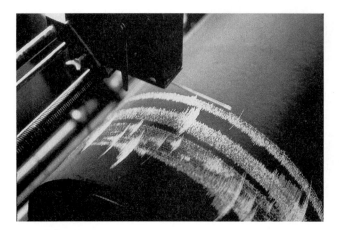

FIGURE 11–33
Harmonic tremor associated with a Hawaiian eruption.
(Photo by U.S. Geological Survey)

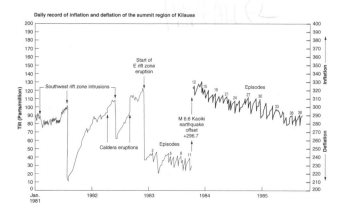

FIGURE 11–34
East-west component of ground tilt near Kilauea summit
1981–85. Large subsidence accompanied the beginning of
the southeast rift zone eruption in January, 1983, and
episodic inflation/deflation cycles characterize the
continuing eruption at Pu'u O'o. (From Heliker et al., 1986)

FIGURE 11–35
Volcanic neck with radiating dike, Shiprock, New Mexico. (Photo by D. A. Rahm)

FIGURE 11–36
Cross-cutting dike ridges. (Photo by U.S. Geological Survey)

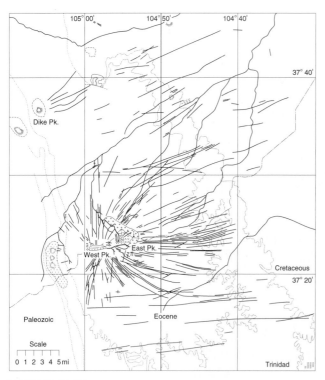

FIGURE 11–37
Dikes radiating from Spanish Peaks stock, Colorado. (From Billings, M. P., Structural Geology, p. 353, Copyright 1976 by Prentice Hall, Upper Saddle River, New Jersey)

FIGURE 11–38
Granite batholiths (light, oval-shaped masses) in western Australia. (NASA satellite photo)

Great Dike of Zimbabwe is 600 km (370 mi) long and 10 km (6 mi) wide.

When erosion exposes dikes at the surface, they commonly form narrow, elongate topographic ridges that may superficially resemble hogbacks of resistant sedimentary beds. However, dikes may be distinguished from hogbacks by the following:

1. Dikes sometimes cross one another (Figure 11–36) or cut across sedimentary beds.

2. Dikes may form a radial pattern, emanating from a central source (Figures 11–35 and 11–37).

3. Dikes may terminate abruptly.

Although the crystalline rock of dikes is usually more resistant than their host rocks, dikes need not always form ridges. Some consist of rock less resistant than their host rock and are etched out by differential erosion to form linear depressions.

Stocks and Batholiths   *IRREGULAR*

Stocks and **batholiths** are discordant, irregularly shaped plutons, generally larger than other types of intrusions. They differ from each other only by size—both are irregularly shaped, but, by definition, batholiths are greater than 100 km^2 (40 mi^2) in area and stocks are less than 100 km^2 (40 mi^2). Batholiths are often very large. For example, the Idaho batholith covers about 41,000 km^2 (16,000 mi^2) and the British Columbia batholith is more than 2000 km (1200 mi) long and 290 km (175 mi) wide. Differential erosion of stocks and batholiths typically develops topographic highs, especially in the cores of mountain ranges, and landforms are usually massive in character (Figure 11–38), lacking the distinct linear ridges and valleys formed by sedimentary beds.

Concordant Plutons   *PARRALLEL*

Concordant plutons are injections of magma parallel to the bedding or foliation of host rocks. They can be further subdivided on the basis of shape.

Sills

Sills are tabular plutons intruded parallel to the structure of enclosing rocks. They are commonly injected along bedding planes of sedimentary rocks where they produce **mesas** (Figure 11–39), **cuestas,** or **hogbacks,** depending on their dip, in a similar fashion as resistant sedimentary beds or buried lava flows. They vary considerably in size, ranging from a few centimeters in thickness to more than 300 m (1000 ft) thick and 80 km (50 mi) long.

FIGURE 11–39
Basaltic sill capping a mesa, southern Utah. (Photo by W. K. Hamblin)

Laccoliths

Laccoliths are mushroom-shaped plutons which, although injected parallel to the structure of the host rocks, dome up the overlying rocks (Figure 11–40). They differ from sills in the degree of doming, usually expressed as the ratio of their diameter to their thickness. Because laccoliths and sills grade into one another, a commonly used, arbitrary boundary is a ratio of 10:1 for diameter relative to thickness; that is, if the ratio is greater than 10, the body is a sill.

The irregular jointing of igneous plutons, in contrast to the strongly parallel joints of sedimentary beds (see Chapter 10), is useful in identifying plutons on air photos or topographic maps. Erosion of columnar jointing, formed by contraction upon cooling in lava flows, dikes, and sills, often results in vertical cliffs as the hexagonal columns break away during weathering and erosion.

FIGURE 11–40
Laccolith, Henry Mts. Utah. (Photo by D. A. Rahm)

REFERENCES

Anderson, T., and Flett, J. S., 1902, Preliminary report on the recent eruptions of the Soufriere in St. Vincent and of a visit to Mount Pelee, in Martinique: Proceedings of the Royal Society of London, v. 70, p. 423–445.

Atwood, W. W., Jr., 1935, The glacial history of an extinct volcano, Crater Lake National Park: Journal of Geology, v. 43, p. 142– 168.

Casadevall, T. J., et al., 1987, SO^2 and CO^2 emission rates at Kilauea volcano, 19791984: in Decker, R. W., et al., eds., Volcanism in Hawaii: U.S. Geological Survey Professional Paper 1350, v. 1, p. 771–780.

Christiansen, R. L., and Peterson, D. W., 1981, Chronology of the 1980 eruptive activity: in Lipman, P. W., and Mullineaux, D. R., eds., The 1980 eruptions of Mount St. Helens, Washington: U.S. Geological Survey Professional Paper 1250, p. 17–30.

Corti, E. C. C., 1951, The destruction and resurrection of Pompeii and Herculaneum: Roudedge and Kegan Paul.

Cotton, C. A., 1952, Volcanoes as landscape forms: John Wiley and Sons, N.Y., 415 p.

Davies, D. K., Quearry, M. W., and Bonis, S. B., 1978, Glowing avalanches from the 1974 eruption of the volcano Fuego, Guatemala: Geological Society of America Bulletin, v. 89, p. 369–384.

Decker, R. W., and Decker, B., 1981, Volcanoes: W. H. Freeman and Co., San Francisco, CA, 244 p.

Decker, R. W., Wright, T. L., Stauffer, P. H., eds., 1987a, Volcanism in Hawaii: U.S. Geological Survey Professional Paper 1350, v. 1, 839 p.

Decker, R. W., Wright, T. L., Stauffer, P. H., eds., 1987b, Volcanism in Hawaii: U.S. Geological Survey Professional Paper 1350, v. 2, p. 845–1667.

Diller, J. S., and Patton, H. B., 1902, The geology and petrography of Crater Lake National Park: U.S. Geological Survey Professional Paper 3, 167 p.

Fink, J. H., 1983, Structure and emplacement of a rhyolitic obsidian flow: Little Glass Mountain, Medicine Lake Highland, northern California: Geological Society of America Bulletin, v. 94, p. 362–380.

Gilbert, G. K., 1877, Report on the geology of the Henry Mountains: U.S. Geographical and Geological Survey, Rocky Mt. Region, 160 p.

Heilprin, A., 1903, Mount Pelee and the tragedy of Martinique: Philadelphia, Lippincott.

Heilprin, A., 1904, The tower of Pelee: Lippincott, Philadelphia, PA.

Heliker, C., Griggs, J. D., Takahashi, T. J., and Wright, T. L., 1986, Volcano monitoring at the U.S. Geological Society's Hawaiian Volcano Observatory: Earthquakes and Volcanoes, v. 18, p. 1–69.

Holcomb, R. T., 1987, Eruptive history and long-term behavior of Kilauea volcano: in Decker, R. W., et al., eds., Volcanism in Hawaii: U.S. Geological Survey Professional Paper 1350, p. 261–350.

Jaggar, T. A., 1902, Field notes of a geologist in Martinique and St. Vincent: Popular Science Monthly, v. 61, p. 352–368.

Kennan, G., 1902, The tragedy of Pelee, pt. 5: Outlook, v. 71, p. 769–777.

Kjartansson, G., 1967, Volcanic forms at the sea bottom in Iceland and mid-oceanic ridges: report of a symposium: Reykjavik, Societas Scientiarium, Islandica, v. 38, p. 53–64.

Lacroix, A., 1904, La Montague Pelee et ses eruptions: Ouvrage publie par I'Academie des Sciences sous les auspices des Ministeres de l'Instruction publique et des Colonies, Paris, Masson et Cie.

Lacroix, A., 1908, La Montagne Pelee apres ses eruptions, avec observations sur les eruptions du Vesuve en 79 et en 1906: Paris, Masson et Cie.

Leet, L. D., 1948, Causes of catastrophy: McGraw-Hill Book Co., N.Y., 232 p.

Lipman, P. W., and Mullineaux, D. R., eds., 1981, The 1980 eruptions of Mount St. Helens, Washington: U.S. Geological Survey Professional Paper 1250, 844 p.

Lipman, P. W., Moore, J. G., and Swanson, D. A., 1981, Bulging of the north flank before the May 18 eruption—geodetic data: in Lipman, P. W., and Mullineaux, D. R., eds., The 1980 eruptions of Mount St. Helens, Washington: U.S. Geological Survey Professional Paper 1250, p. 143–156.

MacGregor, A. G., 1951, Eruptive mechanisms: Mt. Pelee, the Soufriere of St. Vincent and the Valley of Ten Thousand Smokes: Bulletin Volcanologique, v. 12, p. 49–74.

Moore, J. G., 1967, Base surge in recent volcanic eruptions: Bulletein Volcanique, v. 30, p. 337–363.

Perret, F. A., 1935, Eruption of Mt. Pelee 1929–32: Washington, D.C., Carnegie Institute, Publication, 458 p.

Rowley, D. R., Kuntz, M. A., and MacLeod, N. S., 1981, Pyroclastic-flow deposits: in Lipman, P. W., and Mullineaux, D. R., eds., The 1980 eruptions of Mount St. Helens, Washington: U.S. Geological Survey Professional Paper 1250, p. 489512.

Russell, I. C., 1902, The recent volcanic eruptions in the West Indies: National Geographic Magazine, v. 13, p. 267–285.

Swanson, D. A., Duffield, W. A., Jackson, D. B., and Peterson, D. W., 1979, Chronological narrative of the 1969– 71 Mauna Ulu eruption of Kilauea Volcano, Hawaii: U.S. Geological Survey Professional Paper 1056, 55 p.

Takahashi, T. J., and Griggs, J. D., 1987, Hawaiian volcanic features: a photoglossary, in Lipman, P. W., and Mullineaux, D. R., eds., The 1980 eruptions of Mount St. Helens, Washington: U.S. Geological Survey Professional Paper 1250, p. 845902.

Tborarinsson, S., 1970, The Lakagigar eruption of 1783: Bulletin Volcanologique, v. 33, p. 910–929.

Tilling, R. I., et al., 1987, The 1972–1974 Mauna Ulu eruption, Kilauea Volcano: an example of quasi-steady-state magma transfer, in Volcanism in Hawaii, v. 1: U.S. Geological Survey Professional Paper 1350, p. 405–469.

Vereek, R. D. M., 1886, Krakatau: Batavia, Imprimerie de I'Etat.

Voight, B., 1981, Time scale for the first moments of the May 18 eruption: in Lipman, P. W., and Mullineaux, D. R., eds., The 1980 eruptions of Mt St. Helens, WA: U.S. Geological Survey Professional Paper 1250, p. 69–92.

Williams, H., 1932, The history and character of volcanic domes: Berkeley, University of California Department of Geological Sciences Bulletin, v. 21, p. 51–146.

Williams, H., 1941a, Crater Lake: the story of its origin: University of California Press, Berkeley, California, 97 p.

Williams, H., 1941b, Calderas and their origin: Berkeley, University of California Department of Geological Sciences Bulletin, v. 25, p. 239–346.

Williams, H., 1942, The geology of Crater Lake National Park, Oregon: Washington, D.C., Carnegie Institute, Publication 540, 162 p.

Glacial Processes

Alpine glacial system, Yentna, Alaska. (Photo by Austin Post, U. S. Geological Survey)

INTRODUCTION

Glaciers, both past and present, are responsible for sculpting the terrain of large areas of the earth's surface, especially in high latitudes and at high elevations. Spectacular scenery associated with mountain ranges is typically carved by glaciers, even where glaciers are absent or insignificant today. During the Pleistocene epoch, large sections of the earth's surface in temperate latitudes were covered with huge continental ice sheets, and mountain ranges were sites of intense alpine glaciation at altitudes 1000 m (3300 ft) below present snowlines. Although these glaciers melted away some 10,000 years ago, their impressions remain upon today's landscape.

Because glaciers are directly linked to climate, their deposits and landforms provide evidence for interpretation of climatic changes that have taken place during the past two million years. Glaciers also have important economic impacts in that huge volumes of sand and gravel from glacial outwash provide a substantial resource, and alpine glaciers store water in snow and ice that is released by melting throughout summer months to augment water supplies. The reading of geologic paleoclimate records requires an understanding of the glacial processes addressed in this chapter.

FORMATION OF GLACIERS

Glaciers are masses of ice and granular snow formed by compaction and recrystallization of snow, lying largely or wholly on land and showing evidence of past or present movement. The important parameters in this definition include the transformation of snow into ice in thicknesses great enough to promote motion on land. Note that sea ice, even the vast areas in polar regions, is not considered glacial because it does not lie on land and does not move by itself. Permanent snowfields that persist through the summer melt season are not glaciers because they lack motion.

Transformation of Snow into Glacial Ice

Glacial ice begins as light, fluffy snow with very high porosity and very low density, usually ranging between 0.07 and 0.18 g/cm^3 (Shumskii, 1964). The lowest known density of snow is 0.004 g/cm^3, the highest about 0.5 g/cm^3.

Snow that survives through the summer undergoes progressive transformation into more compact granules and finally into ice crystals. Snow (and ice) has a hexagonal crystal form, often with delicate hexagonal, prismatic, or star shapes. As earlier fallen snow is buried beneath a thickening mantle of snow, it becomes more tightly compacted, and the points of the lacy snow crystals become vulnerable to partial melting at points of grain-to-grain contact. The meltwater migrates to positions of lower stress where it recrystallizes, and it also percolates downward toward the base of the snowpack where it freezes in the pore spaces between grains or is added onto already existing grains, resulting in more spherical or hexagonal granules with a higher density, typically between 0.4 and 0.8 g/cm^3 (Anderson and Benson, 1963; Thorn, 1976). Such partially consolidated and recrystallized granules, known as **firn** or **neve**, constitute considerable portions of a glacier near its surface. The transition time from snow to firn varies, depending on climatic conditions and rates of accumulation of snow. In temperate climates, where oscillations of temperature back and forth across the freezing point facilitate melting and refreezing, the transformation may take only a few

PLATE 9A
Cirque on Mt. Aconcagua, Argentine Andes. (Photo by Author)
PLATE 9B
The Matterhorn, Switzerland. (Photo by Author)
PLATE 9C
Tarn in a cirque, North Cascades, British Columbia. (D. A. Rahm)
PLATE 9D
Glacial Lake Mascardi, Argentine Andes. (Photo by Author)
PLATE 9E
Half Dome, Yosemite National Park, California. (U. S. Geological Survey)

9A

9B

9C

9D

9E

PLATE 10A
Lateral moraines, Mt. Waddington, British Columbia.
(D. A. Rahm)
PLATE 10B
Moraines, Mt. Robson, Canada. (D. A. Rahm)
PLATE 10C
End moraine, arctic Canada. (J. Ives)
PLATE 10D
Lateral moraines, Freemont Lake near Pinedale,
Wyoming. (D. A. Rahm)

11A

11B

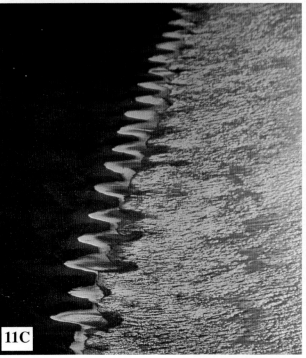

11C

11D

PLATE 11A
Giant ripple marks from Missoula flood, Washington. (D. A. Rahm)
PLATE 11B
End and lateral moraines, Mt. Aconcagua, Argentina. (Photo by Author)
PLATE 11C
Beach cusps. (Aspen Photographers)
PLATE 11D
Pleistocene shoreline of pluvial Lake Bonneville, near Provo, Utah. (D. A. Rahm)
PLATE 11E
Storm-breached washover of a barrier bar, North Carolina. (Dept. of Agriculture)

11E

12A

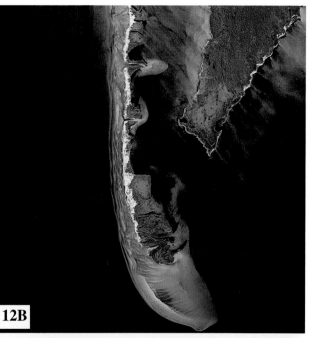

12B

12C

12D

12E

PLATE 12A
Effect of an offshore island on wave refraction and building of a tombolo. (D. A. Rahm)

PLATE 12B
Spit, North Carolina. (Dept. of Agriculture)

PLATE 12C
Marine terraces, northern California. (D. A. Rahm)

PLATE 12D
Transition between barchan dunes and parabolic dunes, Moses Lake, Washington. (D. A. Rahm)

PLATE 12E
Parabolic dunes, Moses Lake, Washington. (D. A. Rahm)

days, but in polar climates, where temperatures are well below freezing for substantial periods of time, the transformation may take years.

With further compaction and recrystallization, firn continues to become denser and is eventually transformed into ice having a density of about 0.8 to 0.9 g/cm^3. Because ice has 8-percent greater volume than the same weight of water, ice never reaches a density of 1.00 (the density of water), unless it contains appreciable dust or rock fragments. The time to make dense glacial ice may be only a few years in temperate glaciers, but in very cold polar regions, hundreds to thousands of years may be required. As shown in Figure 12–1, the firn/ice boundary (0.8 g/km^3) at Byrd Station, Antarctica, occurs at a depth of about 65 m corresponding to about 200 years (Gow, 1969, 1971). Elsewhere in Antarctica, the firn/ice boundary occurs at depths up to 160 m, equivalent to about 3500 years. The firn/ice boundary in the Greenland Ice Sheet occurs at a depth of about 70 m. In contrast, the firn/ice boundary of some Alaskan glaciers occurs at depths of only 13 m (Figure 12–1) (Sharp, 1951b; Behrendt, 1965). In general, for cold climates, the lower accumulation rate leads to longer compaction times and longer transformation times from snow to ice (Bader, 1963; Gow, 1969).

Because ice in a glacier is subject to gravitational stresses that cause movement, the original random orientation of crystals is not maintained (Kamb, 1972; Jones and Brunet, 1978; Duval, 1981). Measurements of the orientation of the C-axes of ice crystals in glaciers show that the C-axes become progressively more aligned with one another with depth, giving the ice a fabric. (Many such measurements have been made. For some examples, see Steinemann, 1954; Rigsby, 1960; Gow and Williamson, 1976; Kamb, 1972; Hook and Hudleston, 1980, 1981.)

For glaciers to develop, precipitation is needed in addition to low temperature. Some polar areas do not support substantial glaciers because even though the climate is cold, little snowfall occurs and optimum conditions needed to convert snow into ice occur infrequently.

Accumulation of dust and rock fragments on the surface of a glacier during the summer results in banded or stratified firn, representing annual snow layers. However, the process of converting firn to ice, accompanied by glacial motion, may obscure the layering. The banding often seen in glacial ice is caused by accumulation of rock debris along shear planes in the ice or by flow banding, which may be confused with annual layering.

Mechanical Properties of Ice

Ice is a crystalline mineral belonging to the rhombohedral division of the hexagonal system, characterized by an elongate c-axis at right angles to a **basal plane** containing three a-axes at angles of 60° to one another, as shown in Figure 12–2. A few of the many variations of shapes possible for snowflakes or ice crystals are also shown in the figure. The six points

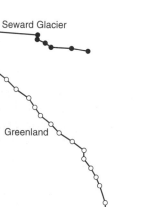

A.

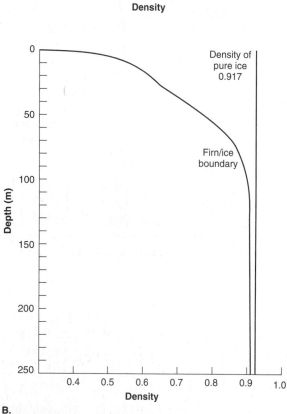

B.

FIGURE 12–1

Increase in density of snow/firn with depths in (A) the Greenland Ice Sheet and the Seward Glacier, Alaska, and (B) the Antarctic Ice Sheet at Byrd Station, Antarctica. (A) from Sharp a,b, 1951; Behrendt, 1965; (B) from Mellor, 1964)

common in stellar snowflakes correspond to each of the a-axes in the basal plane of the ice crystals.

Ice crystals are anisotropic and deform much more readily in the plane of the a-axes than along the c-axis. Disturbances in the crystal structure parallel to the basal plane disrupt only two atomic bonds per unit cell, whereas perpendicular to the basal plane, disruptions of four bonds per unit cell are necessary. The basal planes act as glide planes, which

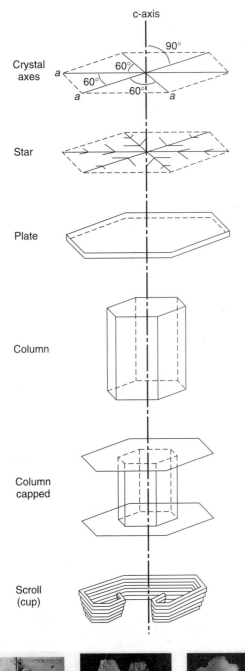

A.

B.

FIGURE 12–2
(A) The principal forms of snow crystals and their crystallographic axes. (After the Swiss Federal Institute for Snow and Avalanche Research)
(B) Electron microscope photos of a snow flake. (Photo by W. P. Wergin/E. F. Erbe; Electron Micxroscopy Unit; USDA-ARS; Beltsville, MD)

may slide past one another in a matter somewhat analogous to the sliding of individual cards past one another in a deck of playing cards. The crushing strength of ice parallel to the c-axis is 31 to 33 kg/cm^2 (440 to 470 lb/in^2), whereas parallel to the basal plane, crushing strengths are only 20 to 25 kg/cm^2 (285 to 355 lb/in^2).

Because ice has an 8-percent greater volume than liquid water, application of stress to ice near its melting point favors conversion from the solid state to water with its lower volume; that is, the melting point of ice decreases with increasing pressure, so that ice will melt at temperatures lower than 0°C under appropriate pressure (Bottomley, 1872). The term **pressure-melting point** is applied to the melting temperature at a particular pressure.

The melting point decreases at a rate of 1°C for every 140 bars of pressure (1 bar = 1.02 kg/cm^2). Therefore, the melting point of ice under confining pressure at the base of a glacier is usually slightly below 0°C. The pressure-melting point of ice at a depth of 2164 m (7100 ft) at Byrd Station in the Antarctic is −1.6°C (Gow, 1970 a,b). Pressure melting of ice plays an important role in many glacial processes, especially at points of greatest stress at the base of a glacier where pressure melting facilitates ice motion and the transfer of mass by melting and refreezing at points of lower stress (Drake and Shreve, 1973; Goodman et al., 1979; Morris, 1976). This process is known as **regelation.**

When stress is applied to ice, it may respond as an elastic, plastic, or brittle substance, depending upon the following:

1. The type of stress applied (tensional, compressional, shear, or hydrostatic)
2. The amount of stress applied
3. The rate at which the stress is applied
4. The temperature

When stress is applied to ice such that the amount of **strain** (deformation) is proportional to the stress applied, the response is said to be **elastic.** If, however, the amount of stress exceeds a certain limit, then strain is no longer directly proportional to stress, and a small increase in stress may produce a large increase in strain (Figure 12–3). Deformation is then said to be **plastic,** and the material no longer returns to its original shape upon release of stress. If enough stress is applied rapidly, so that the ultimate strength of the ice is exceeded, then the response is **brittle,** and fracturing or rupture occurs. A highly important characteristic of ice is that even though the crushing strength may be 30 kg/cm^2 (425 lb/in^2), prolonged application of stress as low as 1 kg/cm^2 may produce large amounts of deformation. The importance of this is that ice will behave plastically even under conditions of rather low stress. Figure 12–3 illustrates the relationship between stress and plastic deformation in ice.

In laboratory studies of strain rates in polycrystalline ice, Glen (1952, 1955) found that upon application of a constant stress beyond about 1 kg/cm^2 (14 lb/in^2), strain rates achieved a steady value and resulted in virtually continuous plastic de-

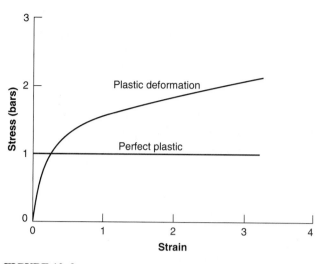

FIGURE 12–3
Stress-strain relationships in ice.

formation. The relationship between stress and strain in ice is given by the Glen-Nye equation:

$$\Sigma = k\tau^n$$

where Σ = strain rate
         τ = stress
n and k = constants varying from about 2 to 4
         (increasing with lower temperature)
     k = a function of temperature (Nye, 1957)

Temperature plays an important role in the behavior of ice (Hobbs, 1974; Jones and Brunet, 1978; Mellor and Testa, 1969). The warmer the ice, the more easily it deforms; that is, strain rates are extremely sensitive to increases in temperature as suggested by the Glen-Nye equation. For example, strain rates at 0°C may be 10 times greater than at −22°C (Paterson, 1981). Thus, temperate glaciers, which are near the pressure-melting point of ice, generally exhibit higher rates of plastic flow than do polar glaciers whose ice is well below the freezing point.

GLACIAL MOVEMENT

The continuous movement of alpine glaciers down their valleys has been recognized and measured since the eighteenth century. Early observations of the downvalley displacement of large boulders and other identifiable glacial debris were followed by more precise surveying techniques.

Glacial velocities vary from one glacier to another and from one point to another on a single glacier. Ice thickness, temperature, and slope all affect velocity. Alpine glaciers typically have surface velocities of 1 m or less per day, but polar glaciers, such as parts of the Antarctic Ice Sheet, may move only a few meters per year. The highest velocities are gener-

ally attained where glaciers descend very steep slopes and where outlet glaciers discharge from large accumulation areas. Velocities of 20 to 30 m/day (65 to 100 ft/day) have been recorded from some Greenland outflow glaciers.

Velocity measurements of glaciers indicate that the highest velocities occur near the midline of a glacier and diminish outward and downward toward the friction-producing surfaces at the base of the glacier (Figure 12–4A) and along the valley sides. Boreholes drilled vertically into glaciers to their base bend with time, demonstrating that the velocity of ice typically diminishes toward the base of a glacier (Figure 12–4B). Although strain rates might be expected to increase with depth as ice thickness increases, the highest velocity occurs at the surface because it represents the sum of the velocities of all layers beneath it.

Velocities also vary between **accumulation** and **ablation zones.** Maximum velocities occur at the equilibrium line (which corresponds approximately to the firn line) where the discharge of ice also reaches a maximum. Velocities decrease downvalley as ice discharge decreases. Maximum flow velocities of 32 cm/day (13 in/day) were measured at the firn line on the Saskatchewan Glacier, diminishing to less than 1 cm/day (0.4 in/day) at the terminus (Meier, 1960). Seasonal variations in velocities occur in most glaciers, with movement more rapid in the summer than in the winter in the ablation zone. Seasonal variations are much more pronounced in temperate glaciers where meltwater plays a significant role in the sliding of a glacier over its bed. Polar glaciers, whose bases are frozen to bedrock all year, do not show as much seasonal variation in flow rate. Even shorter-term variations (on a daily or hourly basis) are common in glaciers as a result of the very jerky motion of ice as it scrapes along its bed.

Basal Sliding

Basal sliding refers to the sliding of a glacier over its bed (Figure 12–5). Such motion is greatest in temperate glaciers and may be absent in polar glaciers frozen to their base. The contribution of basal sliding to total glacial motion varies considerably from one glacier to another, reaching maximums in thin glaciers resting on steep slopes (Boulton and Vivian, 1973; Boulton et al., 1979; Englehardt et al., 1978; Glen and Lewis, 1961; Hodge, 1974; Kamb, 1970; Kamb and LaChapelle, 1964; Lliboutry, 1968; Nye, 1970; Theakstone, 1967; Vivian, 1980; Weertman, 1957, 1964, 1967, 1971, 1979). Basal sliding is probably responsible for only a small fraction of the movement of thin glaciers on gentle slopes. Table 12–1 illustrates the amount of relative motion from basal sliding in a number of glaciers of varying thickness and slope (Paterson, 1981; Holdsworth and Bull, 1970).

Glacier velocity varies considerably, even over short time periods (Figure 12–6A), and motion between the glacier and its bed takes place in a very jerky fashion (Chamberlin,

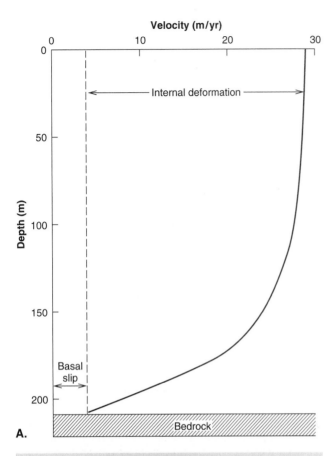

A.

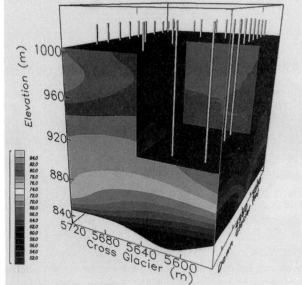

B.

FIGURE 12–4
(A) Vertical cross-section through a glacier showing the relative contributions of internal deformation and basal sliding to the movement of the Athabaska Glacier, Alberta. (From Savage and Peterson, 1963)
(B) Three-dimensional velocity components of the Worthington Glacier, Alaska. (From Harper, 1997)

A.

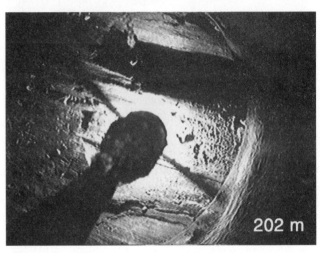

B.

FIGURE 12–5
(A) Basal sliding of the Breidamerkur Glacier, Iceland. Note the small grooves made by ice sliding over subglacial deposits.
(B) Borehole photo at the base of the Worthington Glacier, Alaska, 202 m below the surface of the glacier. (From Harper and Humphrey, 1995)

TABLE 12–1
Relative motion from basal sliding.

Glacier	Basal Sliding (%)	Ice Thickness (%)
Athabaska, Canada	75	322
Tuyuksu, USSR (formerly)	65	52
Aletsch, Switzerland	50	137
Salmon, Canada	45	495
Athabaska, Canada	10	209
Blue, United States	9	26
Skautbre, Norway	9	50
Meserve, Antarctica	0	80

Source: Paterson, 1981; Holdsworth and Bull, 1970.

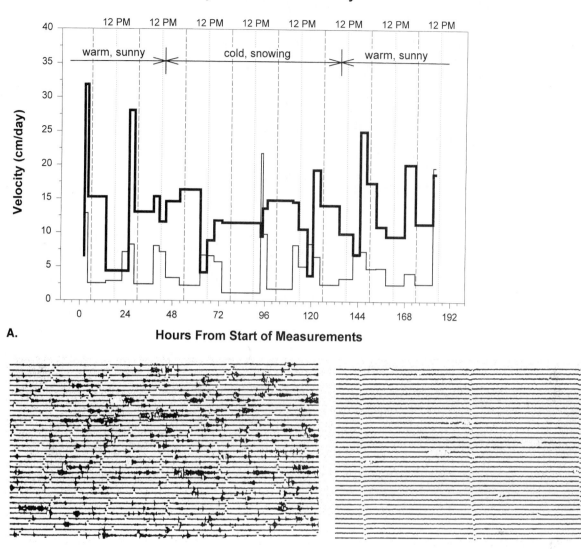

FIGURE 12–6
(A) Hourly variations in glacier velocity. (From Harper, 1997)
(B) Helicorder record from a station surrounded by active glaciers, August 13, 1973. Time pulses are at one-minute intervals. (From Weaver and Malone, 1979)
(C) Helicorder record from a nonglacial station for the same time period as in (B). Time pulses are at two-minute intervals. (From Weaver and Malone, 1979)

1928; Goldthwait, 1973; Hodge, 1974; Kamb and La-Chapelle, 1964; Meier et al., 1974) Weaver and Malone (1979) recorded numerous seismic events at the base of glaciers as a result of this type of motion and found many more events in the summer than in the winter (Figures 12–6B and 6C). The irregular movement of a glacier on its bed commonly leaves percussion marks in the form of crescentic fractures (Figure 12–7). Impressive evidence of basal sliding may also be found in glaciated regions where deep parallel grooves, striations, and glacial polish occur (see Chapter 13).

The increase in basal sliding of glaciers where subglacial meltwater is abundant suggests that water may play an important role in the slippage process (Hallet, 1979; Nye, 1973b; Shreve, 1972; Weertman, 1969). If the subglacial water is under hydrostatic pressure, it reduces the effective weight of the overlying ice, thus decreasing frictional retardation and leading to increased velocity (Bindschadler, 1983; Bind-

FIGURE 12–7
Crescent-shaped chattermarks from a glacier sliding over granite bedrock, Handegg, Switzerland.

schadler et al., 1987; Hodge, 1976; Iken et al., 1983; Weertman, 1964).

Deformation of water-saturated, low-strength, unconsolidated sediments beneath the basal ice of a glacier may also facilitate basal movement. Instead of the ice riding over a stationary subglacial surface, some of the shear strain is taken up within the deforming sediments. In some cases, this type of deformation may account for up to 90 percent of basal ice movement (Boulton, 1981; Boulton and Hindmarsh, 1987; Boulton et al., 1979).

Plastic Flow

Laboratory and field experiments show that ice subjected to low levels of stress over long periods of time will flow plastically while in the solid state (Barnes and Tabor, 1966; Bromer and Kingery, 1968; Chamberlin, 1928; Colbeck and Evans, 1973; Gerrard et al., 1952; Glen, 1956, 1958a, b; Hutter, 1982; Nye, 1951, 1953, 1959, 1965; Paterson, 1964, 1977; Perutz, 1940; Raymond, 1971; Robin, 1955; Russell-Head and Budd, 1979; Savage and Paterson, 1963; Sharp, 1953). Glaciers that spread out laterally into lobes as they move out

FIGURE 12–8
Spreading out of ice plastically as it emerges from the confinement of a valley, LaPerouse Glacier, Alaska. (Photo by A. Post, U.S. Geological Survey)

of the confinement of their valleys onto unrestrained plains are a reflection of the plasticity of ice (Figure 12–8).

As stress is applied to a substance such as ice, the initial deformation is elastic, but as stress exceeds the elastic limit, strain becomes plastic. For most materials, additional stress produces continued plastic strain. However, once the elastic limit of ice is surpassed, constant stress induces limitless permanent deformation (Figure 12–3) at a point known as *infinite yield stress*. For ice, the long-term infinite yield stress is considerably lower than the short-term crushing strength; that is, ice will yield plastically with long-term stress of only about 1 bar. The result is that ice will flow continuously under its own weight. The mechanisms by which ice flows plastically include:

1. Intergranular shifting
2. Intragranular shifting
3. Recrystallization

Differential movement between ice grains, known as **intergranular shifting,** takes place by rotation and sliding between individual ice crystals. It has been noted in firn, but because glacial ice commonly displays a strongly preferred orientation of crystal axes, movement between grains is believed to play a relatively minor role in glacial flow.

Ice undergoing plastic flow shows a consistent orientation of the crystallographic axes of the ice crystals as a result of yield by **basal plane gliding. Intragranular shifting** by gliding along the basal planes within crystals (the plane of the a-axes) appears to be a significant mechanism of glacial flow. The response of ice to stress exerted on a hexagonal prism of ice varies, depending on the direction of the force relative to the crystallographic axes. Slippage is easiest along planes parallel to the base of the crystals (**basal planes**) because, as noted earlier, fewer atomic bonds need to be broken for translation to occur. Basal plane gliding may be envisioned by imagining a force (shearing stress) applied to a deck of cards. If force is applied to a deck of cards at right angles to the flat surfaces, the cards will only bend. However, if the force is exerted parallel to the flat surfaces of the cards, they will slide easily past one another, deforming the original rectangular shape of the deck. Similar sliding takes place in ice along glide planes parallel to the basal plane of the ice crystal.

Recrystallization of ice facilitates downglacier transfer of material. Pressure at grain boundaries can melt ice at the pressure-melting temperature. The meltwater can then migrate to sites of lower pressure where it refreezes. Pressure melting at grain boundaries also diminishes friction between the grains and facilitates intergranular movement of the grains relative to each other.

A glacier flows plastically because the internal shearing stresses exceed the elastic limit of ice, and the infinite yield stress of ice is low. The source of the stress is gravitational, consisting of all-sided stress (**hydrostatic pressure**) and shear stress produced by the component of the gravity vector parallel to the surface slope of the glacier. Shear stresses in a glacier on an inclined surface are proportional to the thickness and surface slope of the ice. The thicker the ice, the greater the weight of ice; and the greater the slope, the higher the component of the gravitational force parallel to the slope. Shear stress in ice can be calculated by the equation

$$\tau = \rho g t \sin \alpha$$

where τ = shear stress
   ρ = density of ice (0.9 g/cm^3)
   g = acceleration of gravity (980 cm/s^2)
   t = thickness of the glacier
   $\sin \alpha$ = slope of the ice surface

As the equation reveals, high shear stresses (and hence plastic flow rates) are induced by thick ice or steep surface slopes. Ice flows plastically when the shear stress reaches about 10 times the elastic limit (Shumskii, 1964). Thus, a glacier having a 10° surface slope would undergo perceptible plastic flow under an ice thickness of slightly over 60 m (200 ft). If ice were perfectly plastic, it would undergo infinite deformation under a constant yield stress of about 1 bar (1.02 kg/cm^2). Observations suggest that shear stresses at the base of glaciers generally range between 0.5 and 1.5 bars (Nye, 1952a). Substituting a value of 1.0 in the general equation gives

$$\frac{\tau}{\rho g} = \tau \sin \alpha$$
$$\frac{1}{\rho g} = t \sin \alpha = \text{contstant}$$

Thus, a change in t (ice thickness) must be compensated for by a change in surface α. This explains why, in nature, glaciers become thinner on steep slopes and thicker on gentle slopes.

That glacial ice is not a perfect plastic is not surprising, considering that plastic flow in a glacier is an extremely complex process because of irregularities in the subglacial floor, changes in the slope of the ice surface, changes in ice thickness from place to place, and changes in ice temperature within the glacier. As long as ice thickness is significantly greater than relief on the subglacial floor, ice may flow up and over topographic obstacles because flow is governed by the surface slope of the glacier rather than by the slope of the subglacial floor.

Ice in a glacier thickens and thins in response to compressive and extending flow (Figure 12–9). **Compressive flow** results in a decrease in velocity and a "piling up" of ice. It occurs where the slope flattens or where loss of surficial ice occurs due to changes in accumulation or ablation rates. **Extending flow** results in increased velocity where slope increases or where ice is added to the surface of the glacier. Compressive flow is most common in the ablation zone where surface ice is being lost by ablation, and extending flow is

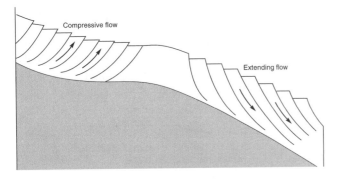

FIGURE 12–9
Compressing and extending flow in a glacier. Compressive flow takes place where the bed flattens, and extending flow takes place where the bed steepens. The surface features shown on the glacier are diagrammatic.

most common in the accumulation zone where ice is being added to the glacier surface.

Internal Shearing

In the upper 30 m (100 ft) of a glacier, especially near the terminus, ice may respond to stress as a brittle substance rather than flowing plastically, and ice within a glacier may move by internal fracturing along distinct shear planes, similar to thrust faulting in rocks. Movement occurs along distinct planes of slippage and is most common in thin ice near glacial margins (Figure 12–10). Concentrations of debris along the shearing planes emphasize these features when exposed in vertical ice cliffs. The shearing planes dip upvalley and provide an important mechanism for bringing debris in the lower portions of the glacier to the surface in the lower ablation zone. Although internal shearing may be of local importance in the

terminal zone of the glacier, few glaciologists consider it to be a significant contributor to overall glacial motion because to fracture along discreet shear planes, ice must be thin enough to act as a brittle, rather than a plastic, substance.

Crevasses

Tensional stress in the upper 50 m (165 ft) of a glacier causes crevasses, tensional fractures produced by differential motion in the ice. Crevasses rarely descend to depths beyond about 30 m (100 ft) below the surface because plastic flow at greater depths closes up the fractures. Figures 12–11 and 12–12 illustrate some of the common crevasse patterns found in glaciers. **Transverse crevasses** form at right angles to the direction of glacial movement where flexing of the ice surface occurs as the glacier moves over a subglacial steepening of the slope. **Chevron crevasses** form along the margins of a glacier where ice motion is restricted by the friction of the valley walls, making a chevron pattern of crevasses pointing upvalley. **Longitudinal crevasses** develop where glacial ice spreads laterally, setting up tensional stresses at right angles to the direction of movement. **Radial crevasses** are similar in origin to longitudinal crevasses except that the ice is spreading radially, so the crevasses make a radial, splaying pattern.

Ogives

Ogives are bands of alternating dark and light ice occurring as ridges and valleys on the surface of glaciers (Figure 12–13; Plates 8B and 8E) (Fisher, 1962; Haefeli, 1951; King and Lewis, 1961; Nye, 1958a; Post and LaChapelle, 1971). They invariably occur below icefalls and bend progressively downglacier in response to the greater velocity near the centerline of a glacier. The width of one dark and one light band

FIGURE 12–10
Shear planes at the terminus of the Breidamerkurjokull, Iceland.

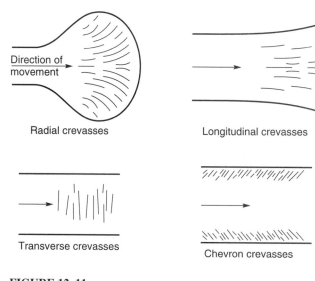

Radial crevasses

Longitudinal crevasses

Transverse crevasses

Chevron crevasses

FIGURE 12–11
Crevasse patterns on glaciers.

generally equals the annual movement of the glacier, suggesting that ogives are genetically linked to seasonal motion. Although not all icefalls have ogives below them, ogives occur only below icefalls, usually starting as alternating ridges and swales that gradually flatten out downglacier. Ice passing through an icefall is usually badly broken up, greatly increasing the ablation surface area in the summer and providing space for enhanced entrapment of snow during the winter. Thus, the winter ice passing through the icefall shows up as a bulge or ridge at the base of the icefall, and the summer ice becomes a swale.

TYPES OF GLACIERS

Glaciers can be classified on the basis of sets of similar physical characteristics. Ahlmann (1948) introduced a three-fold basis for glacier classification:

1. Thermal
2. Morphological
3. Dynamic

Thermal Classification of Glaciers

The temperature of a glacier plays an important role in its dynamic activity. Three subcategories are recognized:

1. **Temperate glaciers** in which the ice is at, or near, its pressure-melting temperature
2. **Polar or cold-base glaciers** in which the ice is below the pressure-melting temperature most of the time
3. **Subpolar glaciers**

A temperate glacier is at approximately the pressure-melting point throughout its thickness, except for the very upper few meters, which may be colder during the winter (Harrison, 1972, 1975). Ice temperature decreases downward at a rate of about 0.06°C/100 m (0.06°C/330 ft) as a result of the effect of increasing pressure on the melting temperature. Thus, ice may coexist with water at temperatures below 0°C near the base of a glacier. All heat goes into the melting of ice (as opposed to ice below the pressure-melting point where heat must be added to bring the ice to the melting point). Thus, meltwater is commonly found throughout

FIGURE 12–12
Longitudinal, chevron, and transverse crevasses, North Cascades, Washington.

FIGURE 12–13
Ogives at the base of an icefall, Yentna Glacier, Alaska. (Photo by A. Post, U.S. Geological Survey)

the thickness of the glacier. This fact allows more rapid conversion of snow and firn into ice and also facilitates plastic deformation of the ice. As a result, flow rates in temperate glaciers are greater than cold, polar glaciers. Temperate glaciers are also referred to as *wet-based glaciers* because of the common occurrence of meltwater at the base of the glacier, which facilitates sliding of ice over its bed. Thus, basal sliding is a more important factor in temperate glaciers than in polar glaciers. The production of much meltwater also leads to the building of large outwash plains in front of the glacier.

Ice in polar, cold-based glaciers is entirely below the pressure-melting point of ice; that is, much heat goes into raising the temperature of the ice (Hooke, 1977). In such glaciers, melting does not occur and the conversion of snow and firn to ice takes many times longer than in temperate glaciers. In addition, ablation takes place only by calving, wind erosion, or sublimation. Glacier velocities are slower than in temperate glaciers, usually only a few meters per year. Because ice in the basal portions of polar glaciers is below the pressure-melting point, meltwater is absent and the ice is frozen to the bedrock surface. Basal sliding is thus absent, and glacial motion is largely by plastic flow. Such

glaciers are not effective in eroding the subglacial bedrock floor by sliding. The lack of meltwater results in the absence of outwash plains.

Subpolar glaciers are frozen to their substrate, as with polar glaciers, but surface melting occurs in summers. Thus, they are intermediate between temperate glaciers and polar glaciers, possessing some characteristics of each.

Not all parts of all glaciers can be fitted neatly into these categories. For example, recent evidence has suggested that portions of polar glaciers may, in fact, be wet based, rather than frozen to their beds.

Morphological Classification of Glaciers

The morphology of glaciers is controlled largely by the relationship between glacier ice and topography. Some glaciers are confined by topography, while others are large enough to overwhelm the underlying topography, and the flow directions reflect the glacier regimen rather than the confining valley sides. Glaciers may be classified as one of the following, based on size, shape, and mode of occurrence:

1. Alpine

2. Piedmont

3. Ice sheet and ice cap

Alpine Glaciers

The configuration of glaciers in mountainous regions is governed largely by restrictions imposed by the topography. Two types of alpine glaciers are generally recognized:

1. Small **cirque glaciers** (Figure 12–14), limited to the area near the firn line

2. **Valley glaciers** (Figure 12–15), which flow well beyond their cirques in long, winding valleys (Raymond, 1980)

Cirque glaciers occupy amphitheater-like hollows (cirques) in alpine areas near the firn line (McCall, 1952, 1960). They are more frequent on northward- or eastward-facing slopes in the Northern Hemisphere, where solar energy is at a minimum. In mountainous terrain of high snow accumulation, ice may move downvalley out of cirques to form valley glaciers extending many kilometers below the source area. Contributions of ice from tributary valleys allow the ice to extend long distances downvalley. When this happens, the ice in a valley glacier consists of multiple ice streams having unique flow characteristics but all occupying the same valley. **Outlet glaciers** are similar to valley glaciers but are fed in their source areas by an ice cap or ice dome, rather than from individual cirque glaciers.

Where glaciers extend all the way to the sea, as, for example, along the Alaskan coast, the lower portions of the glacier are affected by buoyancy and may, in fact, partially float. Such glaciers are known as **tidewater glaciers.** Because of the typically rapid calving rates of floating ice, they are especially vulnerable to changes in water depth or regimen, and they may show spectacular rates of retreat or advance. For example, the tidewater glaciers in Glacier Bay were seen near the mouth of the bay by Vancouver around the turn of the nineteenth century, but they have since retreated many kilometers upvalley, largely as a result of extremely high calving rates.

Piedmont Glaciers

Piedmont glaciers form when valley glaciers discharge ice as broad, radially flowing lobes onto plains at the foot of mountains (Figure 12–16; Plate 8E). The Malaspina Glacier near Yakutat, Alaska (Figure 12–17) (Russell, 1893), is a classic example of a piedmont glacier. Ice in the St. Elias Mountains emanates from a central valley glacier, augmented by several smaller valley glaciers, and spreads out in a radial lobe about 50 km (30 mi) across and 40 km (25 mi) long. The principal distinction between a piedmont glacier and a valley glacier is that flow in a piedmont glacier is unconfined in its lower reaches.

Ice Sheets and Ice Caps

Ice sheets and **ice caps** are large enough not to be confined by topography, and their configuration reflects the size and shape of the glacier, rather than the underlying topography. The difference between an ice sheet and an ice cap is only size. Both are typified by ice flow in the direction of the slope

FIGURE 12–14
Cirque glacier, Eiger Glacier, Switzerland.

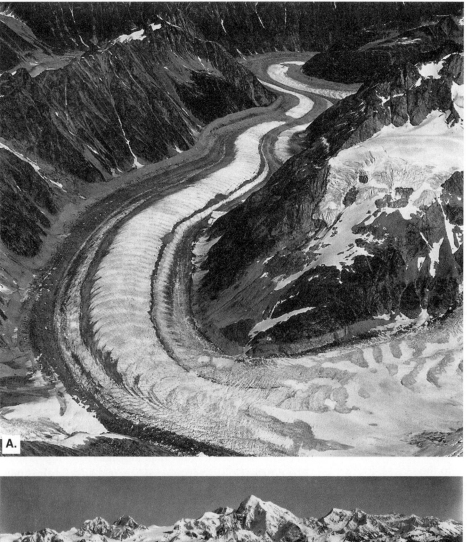

FIGURE 12–15
Alpine valley glaciers.
(A) Long, valley glacier,
Canadian Coast Range. (Photo
by D.A. Rahm)
(B) Complex alpine valley glacier
system fed by tributary glaciers,
Grand Plateau Glacier, Alaska.
(Photo by A. Post, U.S.
Geological Survey)

FIGURE 12–16
Piedmont glacier, Baffin Island, Canada. (Photo by Canadian Department of Energy, Mines, and Resources)

A.

B.

FIGURE 12–17
The Malaspina glacier, a typical piedmont glacier.
(A) View from space shuttle. (Photo by NASA)
(B) Low-angle oblique view. (Photo by A. Post, U.S. Geological Survey)

of the glacier surface, rather than flow in the direction of the underlying topography.

Ice caps are generally less than about 50,000 km² (~19,000 mi²). Modern examples include the Barnes Ice Cap on Baffin Island (Figure 12–18A) and the Vatnajokull in Iceland (Figure 12–18B; Plate 8D).

The Greenland and Antarctic glaciers are good examples of ice sheets. The Greenland Ice Sheet occupies about 1.7 million km² (0.67 million mi²) (Figure 12–19), and the Antarctic Ice Sheet about 13 million km² (5 million mi²). Both bury entire mountain ranges under ice about 3–4000 m (~10–13,000 ft) in thickness, and both depress the earth's crust under their own weight.

Dynamic Classification of Glaciers

The dynamic classification of glaciers is based on the nature of their mass balance, which determines whether they are advancing, neutral, or stagnating. Because conditions at the terminus of a glacier commonly reflect their mass balance, these conditions are useful in classifying glaciers on a dynamic basis.

GLACIER MASS BALANCE

The **mass balance** of a glacier is the relationship between accumulation and ablation of snow and ice. **Accumulation** includes all processes by which snow and ice are added to a glacier, and **ablation** refers to all processes by which snow and ice are lost from a glacier.

Glaciers are fed largely by precipitation as snow, although avalanching of snow from valley sides above a glacier and wind drifting of snow may also play a role. Most snowfall in alpine regions is orographic, caused by the cooling of moist air as it rises over the mountains. The orographic effect is also important in ice sheets and ice caps. Because moisture is as important as cold climate in generating snow, the rate of accumulation of snow and ice is generally great in mountains adjacent to oceans, which supply moisture-laden storms.

Most ablation of glacier ice takes place by melting, but evaporation, sublimation, calving, or wind erosion may be locally important. Heat for melting comes largely from solar radiation, with less significant contributions from warm air, rain, and geothermal heat.

Melting is facilitated by the occurrence of surface dark rock debris, which becomes warmed more rapidly than snow and ice because it absorbs radiation more efficiently. If rock debris is small enough to be warmed through, heat is transferred to the underlying ice where it helps melt the ice. However, if the rock debris is too thick, it acts as an insulating blanket that protects the underlying ice from solar radiation, and **debris cones** (Figure 12–20) or **rock tables** (Figure 12–21) are formed on the glacial surface. Warm rain failing on the glacial surface is effective in melting snow, but if refreezing

occurs, little snow/ice is lost. Some melting may also occur at the base of a glacier as a result of heat supplied by the geothermal gradient of rocks beneath the ice.

Evaporation and sublimation are not very significant and account for only a small amount of ablation, probably less than about one percent, largely because of the large latent heat of vaporization (540 cal/g) and because cold air is limited in its capacity to absorb moisture.

Calving of ice from the terminus of a glacier occurs where glaciers end in the sea or break off over cliffs (Plate 8C). Large pieces of ice may become separated from the glacier by calving and thus no longer be part of it (Figure 12–22). Calving rates of glaciers whose termini are partially floating may be exceptionally high, and large masses of ice may be lost to a glacier over short periods of time. Because of the importance of water depth to calving, glaciers that terminate in the sea where water depths vary considerably are subject to unusual rates of retreat or advance.

In the **zone of accumulation** at the head of a glacier (Figure 12–23), the accumulation of snow and ice is greater than the amount lost by ablation, whereas in the **zone of ablation** in the distal part of a glacier, more ice is lost than accumulated. The zone of accumulation is separated from the zone of ablation by the **equilibrium line,** along which the annual accumulation is equal to the annual ablation. The **firn line,** often marked by light new snow above and dense gray ice below (Figure 12–24), may be close to the equilibrium line, but dense ice below the snowline formed by refreezing of meltwater on the glacier surface may add mass (accumulation) below the firn line and thus cause the equilibrium line to differ from the firn line. However, in most instances, the firn line and equilibrium line closely approximate one another.

The annual difference in mass between accumulation and ablation on a glacier is the **net mass balance.** Figure 12–25 shows curves of accumulation and ablation over a balance year. Accumulation dominates in the winter, ablation in the summer. If the two are equal, then the net balance of the glacier is zero; if accumulation is greater, the net balance will be positive; and if ablation is greater, the net balance will be negative. Figure 12–26 illustrates the relationship between accumulation, ablation, and net mass balance.

The ice mass lost to the glacier in the ablation zone is replaced by the movement of ice from the accumulation zone. The mass of ice passing through any given cross section in the zone of accumulation is the volume of ice passing through a cross section upvalley plus the ice gained by net accumulation between the sections. Thus, ice discharge (average rate of flow times cross-sectional area) increases from the head of a glacier to the equilibrium line, where it reaches a maximum. Downglacier from the equilibrium line, the ice mass that passes through a given cross section in the zone of ablation equals the volume of ice passing through a cross section upvalley minus the ice lost by ablation between sections. Thus, ice discharge decreases from the equilibrium line to the terminus. Glacier ice in the accumulation zone has a downward component of movement within the ice, whereas ice in the accu-

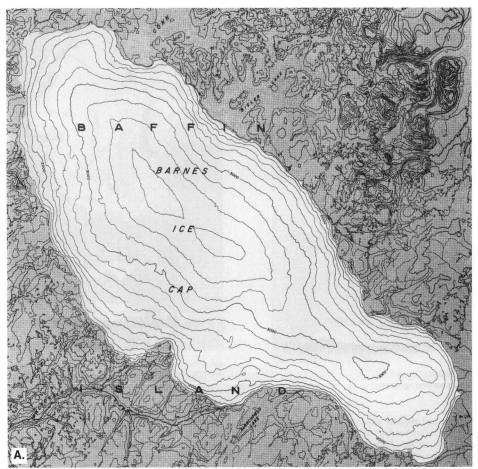

A.

B.

FIGURE 12–18
(A) Ice caps that overwhelm the underlying topography. Barnes Ice Cap, Baffin Island, Canada. (U.S. Geological Survey)
(B) Vatnajokull Ice Cap, Iceland. (Photo by NASA)

FIGURE 12–19
The Greenland Ice Sheet, about 2400 km long and 900 km wide.

FIGURE 12–20
Debris cones on the Breidamerkurjokull, Iceland. The cones are ice cored with only a few centimeters of debris on the surface.

FIGURE 12–21
Rock table, Palisade Glacier, California. (Photo by D.A. Rahm)

mulation zone has an upward component of movement (Figure 12–23).

The position and activity of the terminus of a glacier are determined by the ratio of accumulation and ablation. If accumulation exceeds ablation over a number of years, the glacier will have a positive net mass balance, the glacier front will advance, and the terminus typically will be steep and well defined with relatively clean ice (Figure 12–27). If accumulation and ablation are essentially equal over a substantial period of time, the terminus will be stable, and ice lost near the glacier terminus by ablation will be replaced by new ice transferred from the accumulation zone by plastic flow, basal sliding, and internal shearing. If ablation exceeds accumulation, the glacier will have a negative net mass balance, and the terminus will retreat or stagnate, usually giving the glacier a gently sloping, poorly defined front with much debris on the surface (Figure 12–28; Plate 8A). Note that even when its terminus is retreating upvalley, a glacier continues to flow downvalley (unless the ice is stagnant).

A glacier having a protracted negative net mass balance loses ice not only by retreat of the terminus but also by substantial thinning or downwasting of the glacier. Because rates of glacial movement are a function of ice thickness, thinning of the ice reduces the rate of ice transfer downglacier, and the rate of retreat and downwasting increases. Thus, acceleration of terminal retreat in a glacier does not necessarily indicate accelerated climatic warming.

FIGURE 12–22
Calving of the terminus of the
Columbia Glacier, Alaska.

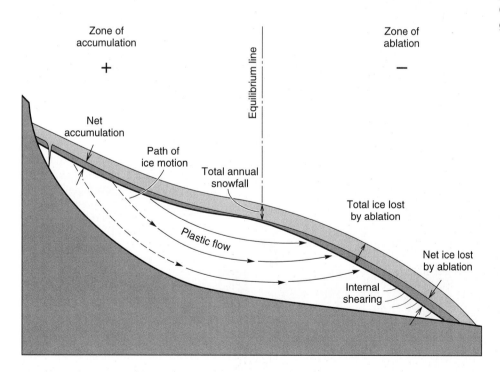

FIGURE 12–23
Cross-section through a typical
glacier.

FIGURE 12–24
Firn line, Klinaklini Glacier, British Columbia Coast Ranges, Canada. (Photo by D.A. Rahm)

from tributary glaciers was also a factor in the stagnation of the lower Wolf Glacier.

Because climatic factors control accumulation and ablation rates, glaciers are often used as indicators of past climatic conditions. Figure 12–29 illustrates the relationship between climate and glacial response (Nye, 1960; Meier, 1965). The time between a climatic perturbation and the resulting advance or retreat of the terminus of a glacier is known as the **response time.**

Glacier Surges

Startling advances of glacier termini occur periodically in some glaciers, often following stagnation in the lower ablation zone (Bindschadler et al., 1976; Clark, 1976; Clarke et al., 1986; Elliston, 1963; Field, 1969; Hewitt, 1969; Johnson, 1972; Liestol, 1969; McMeeking and Johnson, 1986; Meier and Post, 1969; Post, 1972; Robin, 1969; Robin and Barnes, 1969; Rutishauser, 1971; Schytt, 1969; Stanley,

Substantial thinning of a glacier with a negative economy may slow the rate of ice flow enough to cause stagnation of the lower part of the glacier. The amount of ablation on a stagnating glacier surface is often surprising. While the terminus of the lower 15 km (9 mi) of the Wolf Glacier in the St. Elias Mountains of Canada retreated only a few hundred meters, 100 to 150 meters (350 to 500 ft) of downmelting occurred as the glacier stagnated (Sharp, 1951). Loss of ice contributed

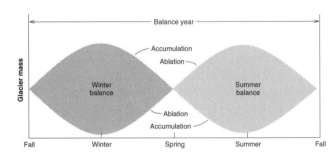

FIGURE 12–25
Relationship between accumulation, ablation, and mass balance on a glacier.

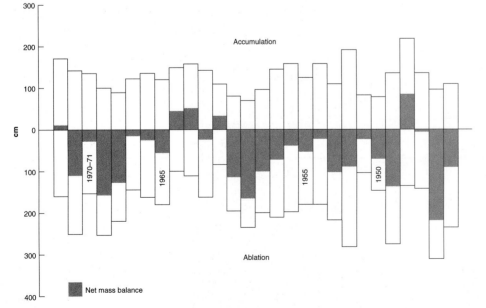

FIGURE 12–26
Mass balance of Storglaciaren, Sweden, from 1945 to 1973, showing total accumulation, total ablation, and net mass balance. (From Sugden and John, 1976; after Ostrem et al., 1973)

FIGURE 12–27
Clean, well-defined ice margin of a glacier having a positive mass balance, Easton Glacier, Mt. Baker, Washington. (Photo by A. Post, U.S. Geological Survey)

FIGURE 12–28
Poorly defined margin of a stagnant glacier, Fairweather, Alaska. Note the numerous depressions (ice-walled lakes) and forest growing on the glacier. (Photo by A. Post, U.S. Geological Survey)

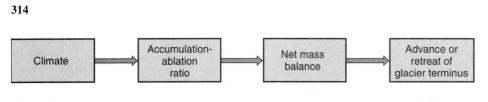

1969; Thorarinsonn, 1969; Weertman, 1962). After many years of recession of the glacier terminus, the Black Rapids Glacier in Alaska began to advance rapidly in 1936 and 1937, reaching a maximum rate of advance of about 60 m/day (200 ft/day). Soon thereafter, the glacier began to retreat again (Moffit, 1942). Post (1969) noted abrupt advances, known as **surges,** in 204 glaciers in western North America. Surging glaciers are characterized by brief, exceptionally rapid advances of their termini accompanied by dramatic thinning of ice in the upper segments of the glacier. Surges are most common in the ablation zone, somewhat downglacier from the equilibrium line. Stagnation of the glacier near its terminus typically precedes and follows surges (Post, 1960, 1966). During the surge, medial moraines become acutely contorted (Figure 12–30), and the glacier surface is extensively broken up into irregular blocks by extensional crevasses.

The ice in a surging glacier may move 10 to 100 times faster than normal and may transport large quantities of rock debris (Clapperton, 1975). Certain glaciers seem to surge periodically within a range of 15 to 100 years (Post, 1969), apparently as a result of physical conditions that lead to dynamic instability within the glacier.

Although glacial surges have been studied for many years, a unique cause for all surges remains to be demonstrated. However, observations suggest a cause-and-effect relationship between thickening of ice upglacier from a stagnating terminal zone and the generation of large volumes of meltwater. Stagnant ice near the terminus continues to ablate while ice in the upper ablation zone thickens behind the stagnant-ice plug. As the ice thickens, basal stresses increase until a critical limit is exceeded, and rapid acceleration of glacial movement takes place, often overriding the stagnant ice ahead of it. Coupled with the increase in basal shear stress, basal water pressure appears to play an important role in surges by reducing basal shear stress (Budd, 1975; Kamb et al., 1985; Robin and Weertman, 1973; Weertman, 1969). The association of unusual amounts of meltwater with surges suggests that water trapped beneath the glacier builds up pressure, which lowers frictional resistance at the glacier bed until a crucial limit is surpassed and basal sliding increases dramatically. Following a decade of thickening of the upper part of the glacier, the Variegated Glacier in Alaska began to surge in 1982 (Raymond and Harrison, 1988; Kamb et al., 1985). Velocity increased from 0.2 m/day (0.7 ft/day) to 40 to 60 m/day (130 to 200 ft/day). A borehole drilled to the base of the glacier showed that 95 percent of the movement was by

basal sliding and that the zone of water saturation in the ice was within 90 percent of the glacier thickness, that is, almost enough for the glacier to float. The water eventually drained out of the terminus in a dramatic flood, and the surge stopped (Kamb et al., 1985).

For this mechanism to be effective, drainage of subglacial water must be prevented in some way. Some possible modes of prevention include freezing of the glacier base to the bedrock (Jarvis and Clarke, 1975; Robin, 1976) and blocking of drainage by the stagnant ice plug, both of which entail some doubtful premises. Surges end with renewed stagnation of the glacier terminus.

Kinematic waves have also been used to explain glacial surges. They are masses of ice that move rapidly downglacier as waves having velocities of hundreds of meters a day (Harrison, 1964; Kamb, 1964; Lighthill and Whitham, 1955; Meier and Johnson, 1962; Nye, 1958a, b; Palmer, 1972; Weertman, 1958). The velocity of such waves is considerably greater than the velocity of transfer; that is, the waveform travels faster than the ice. Kinematic waves are discernible on glaciers as local bulges that travel rapidly downglacier, thickening the ice as the bulge arrives at a point, then thinning it as the wave passes. Although kinematic waves have been documented on glaciers, they have yet to be directly observed consistently on surging glaciers.

GLACIAL EROSION

Erosional Processes

Recognition that polished and striated rock surfaces (Figure 12–31) in alpine regions were made by abrasion of glaciers moving over them dates back to the latter part of the eighteenth century when Swiss alpinists and peasants observed polished and striated bedrock associated with active glaciers of the time. The first known written account apparently was by Saussure in 1779. In 1795, James Hutton, the famous Scottish geologist who is widely considered to be the "father" of modern geology, concluded that glaciers in the Alps must have been larger than the ones that he perceived in existence. In 1843, Forbes wrote about a glacier:

> Its stupendous unwieldy mass is dragged over the rocky surface, it first denudes it of every blade of grass and every fragment of soil, and then proceeds to wear down the solid granite, or slate, or limestone, and to leave most undeniable proofs of its actions upon these rocks. (Forbes, 1843)

FIGURE 12–30
Stagnation following a glacial surge, Yanert Glacier, Alaska. (Photo by A. Post, U.S. Geological Survey)

Two principal processes, **abrasion** and **plucking (quarrying),** are responsible for nearly all glacial erosion.

Abrasion

A number of factors affect the degree and rate of **glacial abrasion.** Among the most important is rock debris embedded in the base of a glacier, which acts as an effective scouring and polishing tool. On the basis of his study of striation patterns on bedrock, Chamberlin (1888) concluded that ice by itself is ineffective in abrading and striating rock and that rock debris held in basal ice provides the tools for abrasion and polishing. The common occurrence of rock flour—freshly broken fragments of minerals of silt/clay size—in glacial meltwater streams attests to the abrasion and grinding activity of glaciers (Collins, 1979; Vivian, 1970). Most rock flour consists of ground-up quartz and feldspar having hardnesses of 6 to 7, hard enough to polish and striate most bedrock. Among the factors that affect rates of glacial abrasion are the following

FIGURE 12–31
The surface of granite polished by glacial abrasion. In the grayer areas, the polish has been destroyed by post-glacial erosion. (Photo by G.K. Gilbert, U.S. Geological Survey)

(Hallet, 1979; Harbor et al., 1988; Nye and Martin, 1967; Sugden and John, 1976; Vivian, 1980):

1. The concentration of rock debris held in the basal ice. Clean ice is not hard enough to abrade rock. Thus, the greater the content of rock debris to act as cutting tools, the more effective the abrasion. However, if the concentration of rock debris becomes too great, it interferes with glacial motion and slows abrasion rates (Hallet, 1981; Rothlisberger, 1968).

2. A continuing supply of new rock debris brought to the ice-bedrock interface by basal melting. The cutting tools of the glacier (that is, the rock fragments embedded in the ice) become worn down by the abrasion, so a fresh supply facilitates active abrasion. Without replenishment of rock debris in the basal ice, abrasion becomes less and less effective. Rock debris is carried downward toward the glacier bed by melting of basal ice in temperate glaciers. In typical alpine glaciers, the downward movement may be 15 to 110 mm/yr (0.6 to 4 in/yr) but may reach 2 m/yr (7 ft/yr) (Rothlisberger, 1968). An obvious ramification of this is that renewal of basal debris occurs only in warm-based glaciers, not in polar glaciers where basal ice is frozen to the ground beneath. Thus, abrasion rates for cold-based glaciers must be lower than for warm-based glaciers.

3. Difference in hardness of bedrock relative to the rock fragments in the ice. For example, quartz and feldspar fragments (hardness 6 to 7) will quickly abrade limestone bedrock (hardness 3), but limestone fragments will not easily affect granite bedrock.

4. Basal meltwater to remove the "worn-out" grinding tools of a glacier. As a result, fresh fragments can work on the bedrock. Vivian (1970) observed a thin film of meltwater sufficient to remove rock flour less than 0.2 mm on the subglacial rock surface of the Glacier d'Argentiere. Removal of the rock flour then allows coarser grains in the basal ice to abrade the rock surface more effectively.

5. The velocity of basal sliding of the glacier over its bed. The faster the velocity, the greater the number of rock fragments that will be dragged over a given surface per unit time (Andrews, 1972a).

6. The thickness of the ice. The thicker the ice, the greater the pressure of particles on the underlying surface and the more effective the abrasion as particles are dragged across a rock surface (Boulton, 1974). However, because of the low yield stress for ice (2 kg/cm^2 at 0°C), the effect of ice thickness reaches an upper limit at which friction between large rock fragments in the basal ice and the subglacial surface becomes sufficiently great to retard their continued movement, and the ice flows around them rather than dragging them. Boulton (1974) proposed that abrasion rates increase with increasing ice thickness up to a critical thickness, beyond which additional increase in thickness causes the abrasion rate to decrease rapidly until rock debris becomes lodged (Figure 12–32). Direct measurements support this hypothesis (Boulton, 1979a).

7. Basal meltwater under high hydrostatic pressure, which reduces effective normal pressure. The reduced pressure has the effect of buoying up the glacier, with a twofold result: (a) because of the decreased normal pressure, basal abrasion rates are reduced; and (b) reduction in basal friction allows local increase in glacial velocity (Boulton, 1972).

8. Characteristics of rock fragments in the basal ice. Larger fragments exert more pressure on subglacial surfaces than smaller ones, and angular particles are more effective in abrading the surface.

Rates of abrasion have been measured on rock plates attached to bedrock surfaces beneath glaciers (Boulton, 1979a;

Boulton and Vivian, 1973; Boulton et al., 1979a) and by mechanical laboratory experiments (Mathews, 1979). Table 12–2 shows differences in abrasion rates as a function of lithology, ice velocity, and ice thickness. The effect of lithology on abrasion rates is obvious from Table 12–2. Marble, composed of calcite with a hardness of 3, abrades about three times as fast as basalt, composed largely of plagioclase, pyroxene, and olivine with hardness of 6 to 7. The effect of ice velocity and ice thickness on abrasion rates is also interesting; for example, where the ice velocity was doubled, the abrasion rate increased tenfold. As another example, if the average rate of abrasion for the whole of the Breidamerkurjokull drainage basin were 1 mm/yr, abrasion alone would yield approximately 975,000 m³ (1.26 million yd³) of debris per year (Boulton and Vivian, 1973).

Striations and **grooves** on flat bedrock surfaces (Figure 12–33) provide valuable evidence for directions of former glacial movement. However, some care must be exercised in their interpretation because local differences in flow directions may occur if ice is deflected by topographic obstacles. Striations are easily understood as merely scratches on

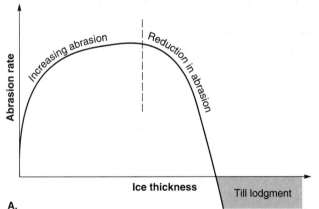

A.

FIGURE 12–32
Schematic diagram showing increasing friction accompanying increasing ice thickness until a critical ice thickness is reached. Then abrasion decreases and deposition of particles begins. The effect of basal water pressure is not considered in the diagram. (From Boulton, 1974)

TABLE 12–2
Abrasion rates.

Rock Plate Lithology	Ice Velocity (m/yr)	Ice Thickness (m)	Abrasion Rate (mm/yr)
Basalt	9.6	40	1
Marble	9.6	40	3
Basalt	19.5	15	9
Marble	19.5	15	34

bedrock made by rock fragments being dragged across the surface. They are most easily formed on fresh, relatively soft, fine-grained rocks, such as limestone, whose surface is gently inclined in the opposite direction as the ice movement. Polished rock surfaces, abraded by fine rock flour, are commonly associated with striations and grooves.

In general, the larger the fragment in contact with bedrock, the bigger the striae. However, large grooves, several meters deep and up to a hundred meters long (Figure 12–34), present a different problem. Some glacial grooves reach immense size, several kilometers long, 30 m (100 ft) deep, and 100 m (330 ft) wide (Goldthwait, 1979; Smith, 1948; White, 1972). They can hardly be made in a single pass of a large boulder, and many show evidence of long-continued abrasion. Why, then, should abrasion be concentrated along deep furrows, rather than appearing randomly across the bedrock? Linear concentrations of boulders, acting like the edges of a file, would help explain some deep grooves, but how common these might be in basal ice is uncertain.

The rock fragments that serve as the grinding tools of abrasion are themselves worn down as they are dragged over the subglacial surface. The results are **faceted, polished,** and **striated stones** commonly found in glacial deposits (Figure 12–35).

Plucking/Quarrying

Large amounts of rock debris can be eroded by **plucking** (also known as **quarrying**). Freezing of glacial ice to loosened blocks allows the glacier to extract the blocks as the ice moves. Plucking is most effective where rocks have previously been fractured. Jointed or fractured rocks are thus especially vulnerable to plucking, and large amounts of glacial erosion may take place in relatively short periods of time.

For plucking to be effective, fractures must be abundant in the bedrock, but whether glaciers can produce significant rock fractures themselves or whether the fractures are formed prior to the arrival of the glacier is a matter of some debate. The amount of prying apart of rocks by freezing and thawing beneath a glacier is not well established, in part because initial cracks and a supply of meltwater are required before frost shattering can occur effectively, and in part because of uncertainties about subglacial temperature fluctuations above and below the freezing point. Several lines of evidence suggest that unloading can produce fractures in rocks:

1. Rock bursts occur in quarries and tunnels where sufficient rock has been removed.

2. Fractures are common in rocks parallel to the walls of glacial troughs.

3. Fractures are typically more closely spaced near the surface.

The problem is whether significant dilatation occurs beneath a glacier or whether it can be caused by the glacier itself. In the Sondre Stromfjord of Norway, some 1000 meters of rock

A.

B.

have been removed by glacial erosion, enough to cause considerable unloading stress in the remaining rock (Sugden and John, 1976). Boulton (1974) has suggested that glaciers can cause rock fracturing where pressure differences on opposite sides of rock knobs exceed the strength of the rock (Boulton et al., 1974; Jahns, 1943).

The effectiveness of freeze/thaw processes in widening and extending fractures beneath glaciers is uncertain (Walder and Hallet, 1985a, 1985b). Measurements of temperatures of subglacial rock and the temperature of air in cavities suggest

that both are slightly above the freezing point (Souchez and deGroote, 1985; Vivian, 1970). However, regelation in the lee of rock knobs may be an important factor (Budd, 1970; Hallet, 1983; Lewis, 1954).

Friction Cracks. The episodic, jerky motion of glaciers as they slide over subglacial rock produces **friction cracks,** small fractures, usually 10 to 15 cm (4 to 6 in) long and 1 to 3 cm (0.4 to 1 in) deep, commonly found on glaciated rock surfaces. They occur as **crescentic fractures,** also known as

FIGURE 12–34
Glacial groove cut into
Precambrian tillite by a
Pleistocene continental ice
sheet, Lake Ontario, Canada.

FIGURE 12–35
Faceted, polished, and striated cobble from a glacial deposit
near Bellingham, Washington.

chattermarks, or **crescentic gouges** (Figure 12–7) (Dreimanis, 1953; Gilbert, 1906; Harris, 1943; Wintages, 1985). The
orientation and spacing of such features suggest intense local
stresses imparted by jerky glacial motion. Because rock is so
much stronger than ice, the participation of boulders pressing
on the bedrock surface is usually invoked as a mechanism.
The horns of crescent-shaped friction cracks may point either up-ice or down-ice. The direction of ice movement can
sometimes be determined by rubbing one's hand over the surface and noting which direction of movement feels smoothest.

Subglacial Erosion by Running Water.   Wet-based glaciers are known to be permeated by glacial meltwater that reaches the glacier bed through moulins or crevasses. Subglacial
meltwater facilitates the sliding of glaciers on their bed, and
channelized, subglacial flow of water has also been postulated as a significant subglacial erosional process (Fisher, 1973;
Hodge, 1976, 1979; Kamb et al., 1979; Mathews, 1964; Nye,
1973b, 1977; Rothlisberger, 1972; Rothlisberger and Spring,
1979; Shreve, 1972; Stenborg, 1969, 1973; Vivian and Zurn-
stein, 1973; Weertman, 1972, 1973a).

Much of what is known about subglacial water is based
on borehole data and on tunnels underneath the ice. Tunnels
driven through rock under glaciers to tap meltwater in the European Alps and in the mountains of Norway have revealed
the presence of much water in subglacial channels, confirming data from glacial boreholes (Rothlisberger, 1968, 1972;
Hooke et al., 1985). Natural subglacial water channels have
been observed from artificial ice-tunnels melted at the base of
the Bondjusbreen Glacier in Norway (Hooke et al., 1985).

Englacial and subglacial water channels close by plastic
deformation during the winter, so discharges are lowest in the
early spring. Channels enlarge by melting in the spring as
surface melting increases the discharge and raises hydrostatic water pressures in the system because meltwater output
increases faster than channel size can adjust (Hooke et al.,
1985; Iken et al., 1983). Cavities in the ice form in the lee of
bedrock knobs and provide space between the glacier and its
base for meltwater to flow; thus, basal water pressures are increased, helping to separate the glacier from its bed (Lliboutry,
1968; Iken, 1981). Although individual cavities may be small,
their cumulative volume can be considerable. The total cavity volume beneath part of Storglaciaren, Sweden, was found
to be equivalent to an average separation of 0.2 to 0.3 m (0.7
to 1 ft) between the glacier and its base (Hooke et al., 1983),

and the estimated depth of water in and under South Cascade Glacier, Washington, was found to be greater than 1 m (Tangborn et al., 1975). At times, water pressures at the base of a glacier may be sufficient to lift a glacier locally, causing extensive, audible icequakes (Holmlund and Hooke, 1983; Iken et al., 1983; Raymond and Malone, 1981).

Detailed mapping of erosional features on a recently deglaciated limestone surface formerly beneath the Blackfoot Glacier, Montana, led Walder and Hallet (1979) to conclude that at least 20 percent of the surface area had been separated from the base of the glacier by water-filled channels, with the remainder covered by a thin water film. They found a nearly continuous network of cavities and channels incised in bedrock that drained the base of the glacier. The subglacial channels were typically 5 to 25 cm (2 to 10 in) deep, 10 to 20 cm (4 to 8 in) wide, and 2 to 5 m (7 to 17 ft) long. Most were oriented nearly parallel to the former ice flow direction or parallel to the slope of the bedrock surface. Calculations of the melting rates of ice indicate that such a subglacial channel system could accommodate all meltwater, even during times of high rates of surface melting (Walder and Hallet, 1979).

Because of the close association with glacially polished surfaces, crescentic fractures, and other evidences of glacial abrasion, distinguishing how much erosion occurs by subglacial water relative to glacial erosion can be difficult. However, certain forms unique to flowing water—such as potholes, flow marks around obstacles, sinuous or swirling channels, narrow ridges in the lee of obstacles, and undercut faces—are helpful (Karcz, 1973). Unusually large subglacial channels, up to 1 km (0 .6 mi) in width and 30 km (19 mi) in length, known as **tunnel valleys,** have been identified near the margins of some late Pleistocene ice sheets (Mooers, 1989; Wright, 1973).

GLACIAL TRANSPORTATION AND DEPOSITION

Transportation of Material

Rock debris may be added to glaciers from either above or below the ice. As glaciers erode the land surface over which they move, material is entrained in the ice, and where the land surface stands higher than the glacier, mass wasting processes shed rock debris onto the glacier surface.

Glacial transportation and deposition include not only material directly in contact with glacial ice but also material transported and deposited by glacial meltwater; glacially derived sand, silt, and clay blown by the wind; and marine and lacustrine deposits transported by floating ice.

Because of the variety of ways in which glaciers transport and deposit material, much diversity exists in the nature of landforms produced and in the character of the deposits. For example, sediment deposited directly from basal ice differs from that deposited directly from ice along the glacial margin; sediment accumulated from stagnant ice differs from that laid down by active ice; and sediment winnowed by glacial meltwater differs from that derived directly from glacial ice. The details of these differences will be discussed as we look at various depositional landforms.

Deposition of Material

Few depositional environments are as diverse as those related to glaciers. Hence, the physical characteristics of glacial sediments vary widely. Those deposited directly from glacial ice, called **till,** are usually poorly sorted and not stratified, whereas those deposited from meltwater streams **(outwash)** or lakes **(glaciolacustrine** sediments) are sorted and stratified. Recognition of glacial imprints on sediments is important because of their implications about ancient climates.

Deposits whose source must have been far distant have been recognized for several centuries. For example, **erratic boulders** carried by glaciers (Figure 12–36A) that do not match the local bedrock or that are separated from bedrock by glacial deposits (Figure 12–36B) are commonly found in glaciated terrains. Early beliefs that such exotic deposits were related to a universal flood led to use of the term *drift* long before the glacial origin of such deposits was discovered. The term *drift* has persisted, but it is now used to pertain to all deposits associated with glacial processes. **Glacial drift** is traditionally subdivided on the basis of sediment sorting and stratification.

Nonstratified, Poorly Sorted Glacial Sediments (Diamictons)

Deposits that consist of nonstratified, poorly sorted clasts are known as **diamictons.** The most common glacial diamictons are various types of **till** and **glaciomarine** and **glaciolacustrine drift** deposited directly from glacial ice or dumped in marine or lake water upon melting of floating ice. These deposits typically contain a wide spectrum of particle sizes, ranging from clay to large boulders, mixed randomly together without noticeable stratification. Although the texture of these diamictons often resembles concrete, their textures and compositions range through great variations as a result of the many ways in which glaciers erode, transport, and deposit rock debris. A huge literature now exists on the texture, fabric, lithology, structure, compaction, particle size distribution, and other properties of glacial diamictons (Boulton and Dent, 1974; Boulton and Deynoux, 1981; Easterbrook, 1964; Evenson et al., 1983; Goldthwait, 1971; Goldthwait and Matsch, 1988). Only the most salient of these features is discussed here.

The stones in till or glaciomarine drift are commonly faceted, polished, and striated (Figure 12–35), reflecting the abrasion that they underwent during glacial transport. The glacial grinding process also seems to be responsible for the

A.

B.

FIGURE 12–36
(A) Large erratic boulder on a glacier. (Photo by J. Ives)
(B) Erratic boulder of basalt resting on glacial sediment, near Mansfield, Washington. (Photo by D.A. Rahm)

bimodal particle size distribution seen in many glacial diamictons. That the bimodality is due to crushing phenomena is suggested by laboratory crushing experiments (Gaudin, 1926; Lee, 1963) in which two dominant grain sizes emerge, one peak representing an optimum rock fragment size and a second peak representing an optimum mineral size. In other words, at some point in the crushing process, individual mineral grains are popped out of the rock intact, rather than continually reduced in size. Attempts have been made to use till texture as a measure of distance of glacier transport (Dreimanis and Vagners, 1971), but this is made difficult because glaciers may erode the land surface over such a vast area that new material is constantly added to the already long-traveled glacial debris.

Rock debris in modern glaciers is remarkably angular, yet till and glaciomarine drift often contain many well-rounded clasts. As suggested by the well-documented crushing activities of glaciers, long glacial transport should, if anything, increase the angularity of glacial debris. This apparent anomaly is probably due to the incorporation of stream-rounded clasts, either from glacial meltwater associated with the glacier or from older deposits (Evenson and Clinch, 1985).

Till. The physical characteristics of till that distinguish it from other sediments include the following (Easterbrook, 1982):

1. Poor sorting, commonly with bimodal particle size distribution

2. Absence of stratification

3. Faceted, polished, and striated clasts

4. Striated and polished bedrock surfaces under the deposit

5. Fabric of oriented, parallel, and elongate stones

6. High degree of compaction from the weight of overriding ice

7. Erratic lithologies of constituent rock and mineral debris

8. Deformation of the sediments, especially shear planes and overturned folds

Lodgment till consists of subglacially deposited, massive, poorly sorted glacial debris that has melted out of the basal ice and been plastered over the ground surface by the overlying moving ice (Figure 12–37A). Russell (1895) was among the first to note that as basal ice melts, debris becomes successively more concentrated at the base of the glacier until it begins to interfere with glacial motion. When this happens, the still-moving ice above shears away and smears the basal debris over the subglacial surface. More recent work (for example, Boulton, 1970a, 1970b, 1971, 1972a, 1972b, 1987; Mickelson, 1973) has amplified the early observations and led to a better understanding of subglacial processes. Melting of debris-laden basal ice is enhanced by geothermal heat, pressure melting, and friction. The shearing of still-plastic, moving ice over debris-rich basal ice produces deformation, thrusting, and fabric in the lodgment till being deposited. Shearing of debris and shear planes are common in lodgment till, as are folded and thrust sediments.

Basal-ice shearing commonly imparts a fabric to lodgment till in that the long axes of elongate particles are aligned parallel to one another. Two directions of alignment may be stable in shearing environments:

1. Elongate particles aligned parallel to the direction of ice motion

2. Elongate particles aligned at right angles to the direction of ice motion

As a result, measurement of the direction of elongate pebble orientation in lodgment till can allow reconstruction of the direction of glacial motion that deposited the till. Two directions of preferred orientation are typically found —the dominant one is usually parallel to the direction of ice motion, and the second, smaller mode is at right angles. This alignment is seen both in pebble-size clasts and in microscopic particles (Evenson, 1971; Harrison, 1957; Holmes, 1941). The orientation of elongate magnetic particles allows measurement of the magnetic microfabric with magnetometers (Easterbrook, 1983, 1988; Gravenor, 1975; Gravenor and Stupavsky, 1975; Stupavsky and Gravenor, 1975).

Shearing and the weight of overlying ice also compacts lodgment till more than in other sediments and provides a means of distinguishing lodgment till from other diamictons. The degree of compaction may be determined quantitatively by measurement of porosity and void ratios (Easterbrook, 1964), but it may also be apparent in the field.

Ablation till accumulates as debris is released from melting ice and is deposited on the land surface when the glacier melts away. Ablation till derived from supraglacial till typically consists of fairly coarse, angular rock debris (Figure 12–37B), and it is commonly coarser than lodgment till. It also generally has a random fabric and lacks the degree of compaction found in lodgment till. The term **melt-out till** is used for glacial deposits left by the melting of supraglacial or subglacial interstitial ice, leaving behind the englacial rock debris (Boulton, 1970b, 1972b). Ablation tills are often saturated during deposition and thus are prone to flowage and sliding on the downwasting ice surface (Boulton, 1971; Tarr, 1909; Tarr and Butler, 1909), resulting in **flow tills** (Figure 12–37C) (Boulton, 1968; Hartshorn, 1958; Marcusson, 1973). Flowtills may be thought of as small mudflows of water-saturated glacial detritus.

Glaciomarine and **Glaciolacustrine Drift.** Glaciers that terminate in the sea or in lakes produce many icebergs as a re-

A.

B.

C.

FIGURE 12–37
Types of till. (A) Lodgment till containing oriented clasts, Vatnajokull, Iceland. (B) Ablation till, near Uppsala, Sweden (C) Flow till, north of Stockholm, Sweden.

sult of calving, and if the water is deep enough, the glacial terminus may float. Melting of debris-laden, floating ice releases rock debris, which settles to the sea or lake floor. Sedimentation rates are often high, as much as 8 to 100 cm/yr (~3 to 40 in/yr) (Easterbrook, 1963; Powell, 1980). The deposits are typically poorly sorted and lack stratification, giving them a till-like appearance (Figure 12–38), but they are less compacted and show more random fabrics than lodgment till. Early studies sometimes referred to glaciomarine drifts as *marine* tills, but because they involve a marine interface between the melting ice and final deposition, the term **glaciomarine drift** is preferable.

The physical features of glaciomarine drift that distinguish it from other diamictons include the following (Easterbrook, 1982):

1. Poor sorting and general lack of stratification

2. Faceted, striated, and polished erratic clasts

3. Marine shells with unbroken, articulated valves

4. Mollusks buried in growth position

5. Preservation of delicate ornamentation of shells

6. Foraminifera and diatoms in matrix material

7. Absence of underlying fossiliferous deposits from which shells could be derived

8. Regional extent of deposits

9. High content of sodium and total exchange cations in clay-size fraction

Glaciolacustrine drift shows many of the same physical features as glaciomarine drift but lacks marine fossils. It is also likely to be more restricted in areal extent because of the limited size of most lakes.

Stratified, Sorted Glacial Sediments

Glaciofluvial Deposits

Sediments deposited from glacial meltwater streams may bear the imprint of glacial transportation. Large fluctuations in meltwater discharge that cause abrupt changes in particle size and sedimentary structures distinguish **glaciofluvial sediments** (also known as **glacial outwash**) from nonglacial fluvial sediments. Recognition of glaciofluvial sediments is easiest close to a glacier margin (Figure 12–39); downstream, the glaciofluvial sediments become indistinguishable from normal fluvial deposits. Near glacier termini, glaciofluvial sediments may be deposited on stagnant ice or may bury blocks of ice, producing collapse structures in sediments that identify its proximity to glacial ice.

Sediments deposited by meltwater streams exhibit better sorting and stratification than those deposited directly from ice. Because fine-grained particles are selectively winnowed out from the coarser fraction, curves of particle size distribution lose the bimodal distribution typical of glacial till and other diamictons. Larger particles round more quickly than smaller ones, so pebbles, cobbles, and boulders become rounded after only short distances of transport from their glacial source. Near the glacial margin or adjacent to stagnant ice, stratified sediments may become interbedded with flow till or ablation till.

Although mechanisms of transportation and deposition of glaciofluvial sediments are similar to those of other fluvial environments, sediments deposited from glacial meltwater may bear some imprints of their glacial source. Large daily, seasonal, or long-term fluctuations in discharge produce abrupt particle size changes and sedimentary structures that reflect the fluctuating discharge and proximity to glaciers

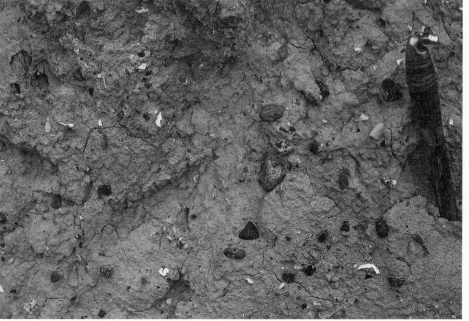

FIGURE 12–38
Poorly sorted Pleistocene glaciomarine drift containing marine mollusk shells, Whidbey Island, Washington.

FIGURE 12–39
Outwash sediments being deposited on stagnant ice, Breidmerkurjokull, Iceland.

(Church, 1972; Church and Ryder, 1972; Church and Gilbert, 1975; Davidson-Arnott, 1982; Fahnestock, 1963). Although measurements of many glaciofluvial characteristics have been made, recognition of a specific environment is commonly difficult because recycling of clasts through several fluvial and glacial environments obscures the effects of any single sedimentary process. For example, Price (1973) found no significant difference in roundness of particles between moraines and eskers at Breidarnerkurjokull, Iceland, presumably because much of the detritus in the moraines was reworked from glaciofluvial sediments.

As the glacial influence diminishes downstream from glacial termini, glaciofluvial sediments grade into normal fluvial deposits. Close to glacial margins, glaciofluvial clasts may exhibit less rounding than typical nonglacial fluvial sediments. However, rounding increases abruptly downstream.

Pebbles and cobbles in outwash deposits may be well imbricated, and upstream dips may provide good evidence of current direction. Long axes of large, elongate clasts are commonly oriented transverse to the principal current direction. Gravel bars near glacial margins commonly consist of poorly sorted, imbricated, stratified gravel. Downstream migration of megaripples leads to large-scale festoon cross-bedding, and migration of longitudinal and linguoid bars generates smaller-scale cross-bedding.

Outwash streams are commonly braided, with numerous anastomosing, wide, shallow channels filled with many mid-channel bars of sand or gravel that become emergent during periods of low to moderate discharges (Church and Jones, 1982; Hein and Walker, 1977). Although braided streams are not necessarily aggrading, many braided outwash streams aggrade rapidly during waning flood stages, filling their valleys with sediment.

Braided outwash streams commonly consist of coarse pebble to cobble gravel near their glacial source, grading to finer sandy gravel downstream. Channel gravel is typically flat-bedded with imbricated clasts. Tabular bars bounded by slip faces and single-clast gravel sheets may also be found (Hein and Walker, 1977; Smith, 1974). Large bars are typically complex forms made by accretion and coalescence of smaller bedforms of numerous high-discharge events. The morphology of gravel bars is usually complex because of recurrent erosion and deposition during frequent fluctuation of discharge.

Fine-grained overbank flood-plain deposits may be present in outwash streams, but they usually are not thick or extensive. Shallow lakes and swamps may occur on flood plains in abandoned channels or on swales between bars. They eventually fill with fine-grained sediment and organic material or may be inundated with sand and gravel as rapidly changing channels switch back and forth across the flood plain.

Outwash streams typically have easily erodible banks of sand and gravel banks characterized by low cohesion because of the lack of silt and clay. They also have high discharge peaks and steep gradients. Outwash deposits usually consist of horizontally bedded, clast-supported, imbricated gravel with the long axes of pebbles transverse to flow (Williams and Rust, 1969; Rust, 1972, 1975; Boothroyd and Ashley, 1975).

Glaciolacustrine Deposits

Glaciolacustrine drift consists of sediments that have been deposited in lakes and that owe their origin in some way to glaciers. They include ice-marginal lakes and various types of lakes produced by glacial erosion or deposition.

Gravel and sand carried into glacial lakes are deposited in deltas at the lake margin, whereas the suspended load of silt and clay is transported out into the lake in suspension or by turbidity currents along the lake floor (Kuenen, 1951). Clay that comes into a glacial lake in suspension settles out more or less continuously, but it becomes the principal deposit only during the winter when the lake may become frozen over and inflowing meltwater streams have reduced discharges and thus carry less coarse material.

Glaciolacustrine bottom sediments typically consist of fine-grained rhythmites composed of silt/clay couplets known as **varves** (Figure 12–40) that are believed to represent annual deposition. The term *rhythmite* is used for any kind of cyclic sedimentation, but varves are restricted to couplets deposited in one year (De Geer, 1912). Thus, all varves are rhythmites, but not all rhythmites are varves.

Sedimentation processes in glacial lakes have been studied for many years (Emerson, 1898; DeGeer, 1912; Antevs, 1922; Kuenen, 195 1; Mathews, 1956; Agterberg and Banerjee, 1969; Gustavson, 1972; Gustavson et al., 1975; Ashley,

1975). Many of these studies show that density currents, set up by differences in sediment concentration and water temperature, play an important role in glaciolacustrine sedimentation. Gustavson (1972) showed that concentration of suspended sediment is much more important than temperature differences in determining water density.

Ashley (1975) described a typical glaciolacustrine sedimentation environment for glacial Lake Hitchcock in Massachusetts and Connecticut. As inflowing streams, heavily laden with suspended sediment, entered the lake, underflow turbidity currents flowed down the prodelta slope and out onto the lake floor, depositing sediment on the lake floor. These flows were not single-pulse turbidity flows but were usually continuous because streamflow during the summer melt season was presumably fairly constant. Sedimentary structures in silt layers—such as erosional contacts, cross lamination, and multiple graded laminations—point to turbidity flows as the principal sedimentary process. Multiple graded units are caused by the variable nature of the flow or by fluctuations in the sediment concentration of the inflowing stream.

Ashley (1975) describes typical varve relationships in glacial Lake Hitchcock as follows:

1. Varves typically show a rhythmic pattern at any single locality, but relative thickness of individual layers, total couplet thickness, grain size, and sedimentary structures vary between localities.

2. Varve sediments are thickest in depressions and thinnest over high areas.

3. Rhythmic sedimentation occurred in deltas along the lake margin as well as in the lake deposits.

4. Near inflowing rivers, varved clays grade shoreward into varved deltaic deposits by gradual thickening of individual silt layers.

5. The thickness of silt layers varies considerably and appears to be directly related to proximity of inflowing rivers, whereas the thickness of clay layers is relatively constant throughout the lake.

6. A single varve is not a graded bed but consists of two texturally and genetically distinct layers, that is, a couplet that is not the result of a single sedimentation pulse.

7. Silt layers are composed of thin laminae that are commonly graded. A maximum of 40 graded beds was observed in one 5-cm (2-in) layer.

8. Silt layers do not always become finer-grained upward, whereas clay layers do.

9. The mean grain size of silt layers depends upon location in the lake, whereas grain size distribution of clay layers is much the same everywhere.

10. The range of grain sizes for the silt and clay layers is approximately the same, but each has a decidedly different mode. Positively skewed silt layers are coarser and better sorted than negatively skewed clay layers.

FIGURE 12–40
Varves, Stockholm, Sweden. The lighter layers are fine sand, the dark layers clay.

11. The silt/clay contact varies according to the environment of deposition. Fewer than 25 percent of the varves have gradational contacts.

12. Small-scale cross-lamination is common in varves of silt and fine sand but is rare in clay-dominated varves.

13. Sedimentary structures within the silt layer appear to be related to grain size: Multiple graded laminations occur in fine-grained units, whereas cross-bedding is found in the coarse-grained units.

14. Thin silt laminations sometimes occur in the clay layers, and thin clay laminations occur in the silt layers.

Glacioeolian Deposits

Glacioeolian sediments consist primarily of windblown sand, silt, and clay derived largely from glaciated areas during the Pleistocene. Among glacioeolian sand deposits, the Sand Hills of Nebraska make up the largest dune field in the Western Hemisphere, covering 56,890 km^2 (22,000 mi^2). The dunes, some up to 122 m (400 ft) high, are not presently active and are thought to be composed of sand picked up by the wind blowing across the outwash of Pleistocene glaciers.

The most widespread glacioeolian sediment is **loess,** (Figure 12–41) windblown silt and clay derived largely from glacial rock flour picked up by winds blowing across glaciofluvial plains and spread downwind over vast areas. Loess deposits measured in hundreds of meters occur in parts of China and the Mississippi Valley.

Loess usually consists of tan, unstratified silt with smaller amounts of clay and fine sand, forming nearly vertical bluffs. The name, meaning "loose" in German, originated in the Rhine Valley about 1821 and was used by Lyell in 1834. It is used in both a descriptive and a genetic sense, referring to a specific lithology (silt and clay) and genesis (glacial). Loess seems to be unequivocally a Pleistocene glacial sediment — pre-Pliocene loess is unknown. It covers broad areas downwind from areas glaciated during the Pleistocene and in areas marginal to some of the world's great deserts. Ice cores from Antarctica show large increases in loess content during times of Pleistocene glaciations.

Although loess is typically unstratified and fairly homogeneous, it commonly occurs as successive layers of silt and clay 1 to 5 m (3 to 16 ft) thick, separated by paleosols, solifluction material, or other sediment. Loess is usually moderately well sorted, consisting of 40 to 50 percent silt, 5 to 30 percent clay-size particles, and 5 to 10 percent sand.

The mineralogical composition of loess is fairly consistent on a global scale. Quartz is the most common mineral, ranging from about 40 to 80 percent, with a mean of about 65 percent. Other typical components include feldspar (10 to 20 percent) and calcium and magnesium carbonates (0 to 35 percent). The sand fraction above 0.25 mm usually consists largely of quartz, whereas the silt fraction is composed of quartz, feldspar, carbonates, and heavy minerals. Clay minerals can form authigenically in loess either concomitantly with deposition or subsequently. Heavy minerals reflect the source area of loess and can provide a means of evaluating where it originated. Loess is commonly calcareous, either as a primary depositional component or as secondary soil-forming processes, especially in areas of low precipitation.

FIGURE 12–41
Thick deposits of loess, China. (Photo by Ken Hamblin)

REFERENCES

Agterberg, F. P., and Banerjee, I., 1969, Stochastic model for the deposition of varves in glacial Lake Barlow-Ojibway, Canada: Canadian Journal of Earth Sciences, v. 6, p. 625–652.

Alley, R. B., 1991, Deforming-bed origin for southern Laurentide till sheets: Journal of Glaciology, v. 37, p. 67–76.

Anderson, D. L., and Benson, C. S., 1963, The densification and diagenesis of snow: *in* Kingery, ed., Ice and snow: MIT Press, Cambridge, MA, p. 391–411.

Anderson, R. S., Hallet, B., Walder, J., and Aubry, B. F., 1982, Observations in a cavity beneath Grinnell Glacier: Earth Surface Processes and Landforms, v. 7, p. 63–70.

Andrews, J. T., 1972, Englacial debris in glaciers: Journal of Glaciology, v. 11, p. 155–156.

Antevs, E., 1922, The recession of the last ice sheet in New England: American Geographic Society Research Series, no. 11, 120 p.

Ashley, G. M., 1975, Rhythmic sedimentation in glacial Lake Hitchcock, Massachusetts-Connecticut: *in* Jopling, A. V., and McDonald, B. C., eds., Glaciofluvial and glaciolacustrine sedimentation: Society of Economic Paleontologists and Mineralogists, Special Publication 23, p. 304–319.

Bader, H., 1963, Theory of densification of dry snow on high polar glaciers, II, *in* Kingery, ed., Ice and snow: MIT Press, Cambridge, MA, p. 351–376.

Barnes, P., and Tabor, D., 1966, Plastic flow and pressure melting in the deformation of ice: Nature, v. 210, p. 878–882.

Behrendt, J. C., 1965, Densification of snow on the ice sheet of Ellsworth Land and South Antarctica: Journal of Glaciology, v. 5, p. 451–460.

Bindschadler, R., 1983, The importance of pressurized subglacial water in separation and sliding at the glacier bed: Journal of Glaciology, v. 9, p. 3–19.

Bindschadler, R. A., Harrison, W. D., Raymond, C. F., and Gantet, C., 1976, Geometry and dynamics of a surge-type glacier: Journal of Glaciology, v. 18, p. 181–194.

Bindschadler, R. A., Stephenson, S. N., MacAyeal, D. R., Shabtaie, S., 1987, Ice dynamics at the mouth of ice stream B, Antarctica: Journal of Geophysical Research, v. 92, p. 8885–8894.

Boothroyd, J. C., and Ashley, G. M., 1975, Processes, bar morphology, and sedimentary structures on braided outwash fans, Northeastern Gulf of Alaska: *in* Jopling, A. V., and McDonald, B. C., eds., Glaciofluvial and glaciolacustrine sedimentation: Society of Economic Paleontologists and Mineralogists, Special Publication 23, p. 193–222.

Bottomley, J. T., 1872, Melting and regelation of ice: Nature, v. 5, p. 185.

Boulton, G. S., 1968, Flow tills and related deposits on some Vestspitsbergen glaciers: Journal of Glaciology, v. 7, p. 391–412.

Boulton, G. S., 1970a, On the origin and transport of englacial debris in Svalbard: Journal of Glaciology, v. 9, p. 213–229.

Boulton, G. S., 1970b, On the deposition of subglacial melt-out tills at the margins of certain Svalbard glaciers: Journal of Glaciology, v. 9, p. 231–245.

Boulton, G. S., 1971, Till genesis and fabric in Svalbard, Spitsbergen: *in* Goldthwait, R. P. ed., Till: a symposium: Ohio State University Press, Columbus, OH, p. 41–72.

Boulton, G. S., 1972a, Modern Arctic glaciers as depositional models for former ice sheets: Journal of Geological Society of London, v. 128, p. 361–393.

Boulton, G. S., 1972b, The role of thermal regime in glacial sedimentation—a general theory: *in* Price, K. I., and Sugden, G. P., eds., Polar geomorphology: Institute of British Geographers Special Paper 4, p. 1–19.

Boulton, G. S., 1974, Processes and patterns of glacial erosion: *in* Coates, D. R., ed., Glacial geomorphology: State University of New York, Publications in Geomorphology, 5th Annual Symposium, Binghamton, N.Y., p. 41–87.

Boulton, G. S., 1976, The origin of glacially-fluted surfaces—observation and theory: Journal of Glaciology, v. 17, p. 287–309.

Boulton, G. S., 1978, Boulder shapes and grain-size distribution of debris as indication of transport paths through a glacier and till genesis: Sedimentology, v. 25, p. 773–799.

Boulton, G. S., 1979a, Processes of glacier erosion on different substrata: Journal of Glaciology, v. 23, p. 15–38.

Boulton, G. S., 1979b, Processes of glacier erosion on different substrata: *in* Glen, J. W., Adie, R. J., Johnson, D. M., Homer, D. R., Macqueen, A. D., eds., Symposium on glacier beds: the ice-rock interface: Journal of Glaciology, v. 23, p. 15–38.

Boulton, G. S., 1981, Deformation of subglacial sediments and its implications: Proceedings of the Symposium on Processes of Glacier Erosion and Sedimentation, Annals of Glaciology, v. 2, p. 114.

Boulton, G. S., and Dent, D. L., 1974, The nature and rates of post-depositional changes in recently deposited till from southeast Iceland: Geographiska Annaler, Series A, 56 A (3–4), p. 121–134.

Boulton, G. S., and Deynoux, M., 1981, Sedimentation in glacial environments and the identification of tills and tillites in ancient sedimentary sequences: Precambrian Research, v. 15 (3–4), p. 397–422.

Boulton, G. S., and Eyles, N., 1979, Sedimentation by valley glaciers; a model and genetic classification: *in* Schluechter, C., ed., Moraines and varves; origin, genesis, classification: A. A. Balkema, Rotterdam, Netherlands, p. 11–23.

Boulton, G. S., and Hindmarsh, R. C. A., 1987, Sediment deformation beneath glaciers; theology and geological consequences: Journal of Geophysical Research, B, Solid Earth and Planets, v. 92, p. 9059–9082.

Boulton, G. S., and Jones, A. S., 1979, Stability of temperate ice caps and ice sheets resting on beds of deformable sediment: Journal of Glaciology, v. 24, p. 29–43.

Boulton, G. S., and Vivian, R., 1973, Underneath the glaciers: Geographical Magazine, v. 45, p. 311–316.

Boulton, G. S., Dent, D. L., and Morris, E. M., 1974, Subglacial shearing and crushing, and the role of water pressures in tills from south-east Iceland: Geographiska Annaler, Series A, 56 A (3–4), P. 135–145.

Boulton, G. S., Morris, E. M., Armstrong, A. A., and Thomas, A., 1979, Direct measurement of stress at the base of a glacier: Journal of Glaciology, v. 22, p. 3–24.

Bromer, D. J., and Kingery, W. D., 1968, Flow of polycrystalline ice at low stresses and small strains: Journal of Applied Physics, v. 39, p. 1688–1691.

Budd, W. F., 1970, Ice flow over bedrock perturbations: Journal of Glaciology, v. 9, p. 29–48.

Budd, W. F., 1975, A first simple model for periodically self-surging glaciers: Journal of Glaciology, v. 14, p. 3–21.

Chamberlin, R. T., 1928, Instrumental work on the nature of glacial motion: Journal of Geology, v. 36, p. 1–30.

Chamberlin, T. C., 1888, The rock-scorings of the great Ice Age: U.S. Geological Survey, 7th Annual Report, 254 p

Church, M., 1972, Baffin Island sandurs: a study of Arctic fluvial processes: Geological Survey of Canada Bulletin 216, 208 p.

Church, M., and Gilbert, R., 1975, Proglacial fluvial and lacustrine environments: *in* Jopling, A. V., and McDonald, B. C., Glaciofluvial and glaciolacustrine sedimentation: Society of Economic Paleontologists and Mineralogists, Special Publication 23, p. 22–100.

Church, M., and Jones, D., 1982, Channel bars in gravel-bed rivers: *in* Hey, R., Bathurst, J., and Thorne, C., eds., Gravel-bed rivers: John Wiley and Sons, N.Y., p. 291–338.

Church, M., and Ryder, J. M., 1972, Paraglacial sedimentation: a consideration of fluvial processes conditioned by glaciation: Geological Society of America Bulletin, v. 83, p. 3059–3072.

Clapperton, C. M., 1975, The debris content of surging glaciers in Svalbard and Iceland: Journal of Glaciology, v. 14, p. 395–406.

Clarke, G. K. C., 1976, Thermal regulation of glacier surging: Journal of Glaciology, v. 16, p. 231–250.

Clarke, G. K., Collins, S. G., and Thompson, D. E., 1984, Flow, thermal structure, and subglacial conditions of a surge-type glacier: Canadian Journal of Earth Science, v. 21, p. 232–240.

Clarke, G. K. C., Schmok, J. P., Ommanney, C. S. L., and Collins, S. G., 1986, Characteristics of surge-type glaciers: Journal of Geophysical Research, v. 91, p. 7165–7180.

Colbeck, S. C., ed., 1980, Dynamics of snow and ice masses: Academic Press, N.Y., 468 p.

Colbeck, S. C., and Evans, R. J., 1973, A flow law for temperate glacier ice: Journal of Glaciology, v. 12, p. 71–86.

Collins, D. N., 1979, Sediment concentration in melt waters as an indicator of erosion processes beneath an alpine glacier: Journal of Glaciology, v. 23, p. 247–257.

Davidson-Amott, R., Nickling, W., and Fahey, B. D., eds., 1982, Research in glacial, glacio-fluvial, and glacio-lacustrine systems: GeoBooks, Norwich, England, 318 p.

De Geer, G., 1912, A geochronology of the last 12,000 years: XI International Geological Congress, 1910, Compte Rendu, v. 1, Stockholm, Sweden, p. 241–258.

Drake, L. D., and Shreve, R. I., 1973, Pressure melting and regelation of ice by round wires: Proceedings of the Royal Society of London, v. 332, p. 51–83.

Dreimanis, A., 1953, Studies of friction cracks along shores of Cirrus Lake and Kasakokwog Lake, Ontario: American Journal of Science, v. 25, p. 769–783.

Dreimanis, A., 1988, Tills: their genetic terminology and classification: in Goldthwait, R. P., and Matsch, C. L., eds., Genetic classification of glacigenic deposits: A. A. Balkema, Rotterdam, Netherlands, p. 17–83.

Dreimanis, A., and Vagners, U. J., 1971, Bimodal distribution of rock and mineral fragments in basal tills: in Goldthwait, R. P., ed., Till: a symposiumOhio State University Press, : Columbus, OH, p. 237–250.

Drewery, D. J., 1986, Glacial geologic processes: Edward Arnold, London, 276 p.

Duval, P., 1981, Creep and fabrics of polycrystalline ice under shear and compression: Journal of Glaciology, v. 27, p. 129–140.

Easterbrook, D. J., 1963, Late Pleistocene events and relative sea level changes in the northern Puget Lowland, Washington: Geological Society of America Bulletin, v. 74, p. 1465–1484.

Easterbrook, D. J., 1964, Void ratios and bulk densities as means of identifying Pleistocene tills: Geological Society of America Bulletin, v. 75, p. 745–750.

Easterbrook, D. J., 1982, Characteristic features of glacial sediments: American Association of Petroleum Geologists, Memoir 31, p. 1–10.

Easterbrook, D. J., 1983, Remanent magnetism in glacial tills and related diamictons: in Evenson, E. B., Schluchter, C., and Rabassa, J., eds., Tills and related deposits; genesis, petrology, application, stratigraphy: A. A. Balkema, Rotterdam, Netherlands, p. 303–313.

Easterbrook, D. J., 1988, Paleomagnetism of Quaternary deposits: Geological Society of America Special Paper 227, p. 111 – 122.

Echelmeyer, K., and Zhongxiang, W., 1987, Direct observation of basal sliding deformation

of basal drift at subfreezing temperatures: Journal of Glaciology, v. 33, p. 83–98.

Elliston, G. R., 1963, Catastrophic glacier advances: International Association Sci. Hydrology Bulletin, v. 8, p. 65–66.

Embleton, C., and King, C. A. M., 1975, Glacial geomorphology: Edward Arnold, London, 583 p.

Englehardt, H. F., Harrison, W. D., and Kamb, B., 1978, Basal sliding and conditions at the glaciers bed as revealed by bore-hole photography: Journal of Glaciology, v. 20, p. 469–508.

Evenson, E. B., 1971, The relationship of macro- and microfabric of till and the genesis of glacial landforms in Jefferson County, Wisconsin: in Goldthwait, R. P., ed., Till: a symposium: Ohio State University Press, Columbus, OH, p. 345–364.

Evenson, E. B., and Clinch, J. M., 1985, Debris transport mechanisms at active alpine glacier margins: Alaskan case studies: in Kujansuu, R., and Saarnisto, M., eds., International Quaternary Association till symposium: Geological Survey of Finland, Special Paper 3, p. 111–136.

Evenson, E. B., Schluchter, C. H., and Rabassa, A., eds., 1983, Tills and related deposits: A. A. Balkema, Rotterdam, Netherlands, 454 p.

Fahnestock, R. K., 1963, Morphology and hydrology of a glacial stream—White River, Mount Rainier, Washington: U.S. Geological Survey Professional Paper 422A, 70 p.

Field, W. O., 1969, Current observations on three surges in Glacier Bay, Alaska, 1965–1968: Canadian Journal of Earth Sciences, v. 6, p. 831–843.

Fisher, J. E., 1962, Ogives of the Forbes type on alpine glaciers and a study of their origins: Journal of Glaciology, v. 4, p. 53–61.

Fisher, D., 1973, Subglacial leakage of Summit Lake, British Columbia, by dye determinations: Union Geodesique et Geophysique Internationale, Symposium on the Hydrology of Glaciers, Publication No. 95 de I'Association Internationale d'Hydrologie Scientifique, p. 111–116.

Forbes, J. D., 1843, Travels through the Alps of Savoy: Oliver and Boyd Edinburgh, U.K.

Gaudin, A. M., 1926, An investigation of crushing phenomena: Transactions of America Institute of Mining and Metallurgical Engineers, v. 73, p. 253–316.

Gerrard, J. A. F., Perutz, M. F., and Roch, A., 1952, Measurement of the velocity distribution along a vertical line through a glacier: Proceedings of the Royal Society of London, series A, v. 213, p. 546–558.

Gilbert, G. K., 1906, Crescentic gouges on glaciated surfaces: Geological Society of America Bulletin, v. 17, p. 303–316.

Glen, J. W., 1952, Experiments on the deformation of ice: Journal of Glaciology, v. 2, p. 111– 114.

Glen, J. W., 1955, The creep of polycrystalline ice. Proceedings of the Royal Society, v. 228, p. 519–538.

Glen, J. W., 1956, Measurement of the deformation of ice in a tunnel at the foot of an ice fall: Journal of Glaciology, v. 2, p. 735–745.

Glen, J. W., 1958a, Mechanical properties of ice: 1. The plastic properties of ice: Philosophical Magazine Supplement, v. 7, p. 254–265.

Glen, J. W., 1958b, The flow law of ice: A discussion of the assumptions made in glacier theory, their experimental foundations and consequences: International Association Sci. Hydrology Bulletin, v. 47, p. 171–183.

Glen, J. W., 1974, Physics of ice: U.S. Army Corps of Engineers, Cold Regions Research and Engineering Laboratory, Cold Regions Sci. and Engineering Monograph II-C2a, 81 p.

Glen, J. W., 1975, Physics of ice: U.S. Army Corps of Engineers, Cold Regions Research and Engineering Laboratory, Cold Regions Sci. and Engineering Monograph II-C2b, 43 p.

Glen, J. W., and Lewis, W. V., 1961, Measurements of side-slip at Austerdalsbreen, 1959: Journal of Glaciology, v. 3, p. 1109–1122.

Goldthwait, R. P., ed., 1971, Till: a symposium: Ohio State University Press, Columbus, OH, 402 p.

Goldthwait, R. P., 1973, Jerky glacier motion and meltwater: International Association Sci. Hydrology Bulletin 95, p. 183–188.

Goldthwait, R. P., 1979, Giant grooves made by concentrated basal ice streams: Journal of Glaciology, v. 23, p. 297–307.

Goldthwait, R. P., and Matsch, C. L., eds., 1988, Genetic classification of glacigenic deposits: A. A. Balkema, Rotterdam, Netherlands, 294 p.

Goodman, D. J., King, G. C. P., Millar, D. H. M., and Robin, G. Q., 1979, Pressure-melting effects in basal ice of temperate glaciers: laboratory studies and field observations under Glaciere D'Argentiere: Journal of Glaciology, v. 23, p. 259–270.

Gow, A. J., 1969, On the rates of growth of grains and crystals in South Polar firn: Journal of Glaciology, v. 8, p. 241–252.

Gow, A. J., 1970a, Deep core studies of the crystal structure and fabrics of antarctic glacier ice: U.S. Army Corps of Engineers, Cold Regions Research and Engineering Laboratory, Research Report 282, 21 p.

Gow, A. J., 1970b, Glaciological studies in Antarctica: Antarctic Journal, v. 5, p. 113–114.

Gow, A. J., 1971, Analysis of Antarctic ice cores: Antarctic Journal, v. 6, p. 205–206.

Gow, A. J., and Williamson, T., 1976, Rheological implications of the internal structure and crystal fabrics of the West Antarctic ice sheet as revealed by deep core drilling at Byrd Station: Geological Society of America Bulletin, v. 87, p. 1665–1677.

Gravenor, C. P., 1975, Erosion by continental ice sheets: American Journal of Science, v. 275, p. 594–604.

Gravenor, C. P., 1985, Magnetic and pebble fabrics of glaciomarine diamictons in the Champlain Sea, Ontario, Canada: Canadian Journal of Earth Sciences, v. 22, p. 422–434.

Gravenor, C. P., and Meneley, W. A., 1958, Glacial flutings in central and northern Alberta: American Journal of Science, v. 256, p. 715–728.

Gravenor, C. P., and Stupavsky, M., 1975, Convention for reporting magnetic anisotropy of till: Canadian Journal of Earth Sciences, v. 12, p. 1063–1069.

Gumell, A. M., and Clark, M. J., eds., 1987, Glacio-fluvial sediment transfer: an alpine perspective: John Wiley and Sons, Chichester, England, 524 p.

Gustavson, T. C., 1972, Sedimentation and physical limnology in proglacial Malaspina Lake, Alaska, in Jopling, A. V., and McDonald, B. C., Glaciofluvial and glaciolacustrine sedimentation: Society of Economic Paleontologists and Mineralogists, Special Publication 23, p. 249–263.

Gustavson, T. C., and Boothroyd, J. C., 1987, A depositional model for outwash, sediment sources, and hydrologic characteristics, Malaspina Glacier: a modem analog of the southeastern margin of the Laurentide Ice Sheet: Geological Society of America Bulletin, v. 99, p. 187–200.

Gustavson, T. C., Ashley, G. M., and Boothroyd, J. C., 1975, Depositional sequences in glaciolacustrine deltas: Society of Economic Paleontologists and Mineralogists, Special Publication 23, p. 264–280.

Haefeli, R., 1951, The formation of Forbe's bands: Journal of Glaciology, v. 1, p. 581–582.

Hallet, B., 1979a, A theoretical model of glacial abrasion: symposium on glacier beds; the ice-rock interface: Journal of Glaciology, v. 23, p. 39–50.

Hallet, B., 1979b, Subglacial regelation water film: in Symposium on Glacier Beds; the Ice-rock Interface, Journal of Glaciology, v. 23, p. 321–334.

Hallet, B., 1981, Glacial abrasion and sliding; their dependence on the debris concentration in basal ice, in Proceedings of the symposium on processes of glacier erosion and sedimentation: Annals of Glaciology, v. 2, p. 23–28.

Hallet, B., 1983, The breakdown of rock due to freezing: a theoretical model: Tempe, Fourth International Permafrost Conference, Proceedings, Arizona State University, p. 433–438.

Hambrey, M. and Alean, J., 1992, Glaciers. New York, Cambridge University Press, 208 p

Harbor, J. M., Hallet, B., and Raymond, C. F., 1988, A numerical model of landform development by glacial erosion: Nature, v. 333, p. 347–349.

Harper, J. T., 1977, The 3-dimensional structure and flow field of a temperate ice mass: surface and borehole deformation studies on Worthington glacier, Alaska: Ph.D Thesis, Dept. of Geology and Geophysics, University of Wyoming, Laramie, WY.

Harper, J. T., and Humphrey, N. F., 1995, Borehole video analysis of a temperate glacier's englacial and subglacial structure: Implications of glacier flow models: Geology, v. 23, n. 10, p. 901–904.

Harris, S. E., 1943, Friction cracks and the direction of glacial movement: Journal of Geology, v. 51, p. 244–258.

Harrison, A. E., 1964, Ice surges on the Muldrow glacier: Journal of Glaciology, v. 5, p. 265–268.

Harrison, P. W., 1957, A clay-till fabric: its character and origin: Journal of Geology, v. 65, p. 275–308.

Harrison, W. D., 1972, Temperature of a temperate glacier: Journal of Glaciology, v. 11, p. 15 –29.

Harrison, W. D., 1975, Temperature measurement in a temperate glacier: Journal of Glaciology, v. 14, p. 23–30.

Hartshorn, J. H., 1958, Flowtill in southeastern Massachusetts: Geological Society of America Bulletin, v. 69, p. 477–482.

Hartshorn, J.H., and Ashley, G.M., 1972, Glacial environment and processes in southeastern Alaska: Technical Report No. 4-CRC, Coastal Research Center, University of Massachusetts, Amherst, MA, 69 p.

Hein, F. J., and Walker, R. G., 1977, Bar evolution and development of stratification in the gravelly, braided, Kicking Horse River, British Columbia: Canadian Journal of Earth Science, v. 14, p. 562–570.

Hewitt, K., 1969, Glacier surges in the Karakoram Himalaya (Central Asia): Canadian Journal of Earth Sciences, v. 6, p. 1009–1018.

Hobbs, P. V., 1974, Ice physics: Clarendon Press, Oxford, England, 837 p.

Hodge, S. M., 1974, Variations in the sliding of a temperate glacier: Journal of Glaciology, v. 13, p. 349–369.

Hodge, S. M., 1976, Direct measurement of basal water pressure: a pilot study: Journal of Glaciology, v. 16, p. 205–217.

Hodge, S. M., 1979, Direct measurement of basal water pressures: progress and problems: Journal of Glaciology, v. 23, p. 309–319.

Holdsworth, G., and Bull, C., 1970, The flow law of cold ice; investigations on Meserve Glacier, Antarctica, in International symposium on Antarctic glaciological exploration, International Association of Sci. Hydrology Publication no. 86, p. 204–216.

Holmes, C. D., 1941, Till fabric: Geological Society of America Bulletin, v. 52, p. 1299–1354.

Holmlund, P., and Hooke, R. L., 1983, High water-pressure events in moulins, Storglaciaren, Sweden: Geografiska Annaler, Series A, v. 65, p. 19–25.

Hooke, R. B., 1977, Basal temperatures in polar ice sheets: a qualitative review: Quaternary Research, v. 7, p. 1–13.

Hooke, R. B., 1989, Englacial and subglacial hydrology: A qualitative review: Arctic and Alpine Research, v. 21, p. 221–233.

Hooke, R. B., and Hudleston, P. J., 1980, Ice fabrics in a vertical flow plane, Barnes Ice Cap, Canada: Journal of Glaciology, v. 25, p. 195–214.

Hooke, R. B., and Hudleston, P. J., 1981, Ice fabrics from a borehole at the top of the south dome, Barnes Ice Cap, Baffin Island: Geological Society of America Bulletin, v. 92, p. 274–281.

Hooke, R. L., Brzozowski, J., and Bronge, C., 1983, Seasonal variations in surface velocity, Storglaciaren, Sweden: Geografiska Annaler, Series A, v. 65, p. 263–277.

Hooke, R. L., Wold, B., and Hagen, J. 0., 1985, Subglacial hydrology and sediment transport at Bondjusbreen, southwest Norway: Geological Society of America Bulletin, v. 96, p. 388–397.

Hutter, K., 1982, Glacier flow: American Scientist, v. 70, p. 26–34.

Iken, A., 1981, The effect of the subglacial water pressure on the sliding velocity of a glacier in an idealized numerical model: Journal of Glaciology, v. 27, p. 407–423.

Iken, A., and Bindschadler, R. A., 1986, Combined measurements of subglacial water pressure and surface velocity of Findelenglescber, Switzerland: Conclusions about drainage system and sliding mechanism. Journal of Glaciology, v. 32, p. 101–119.

Iken, A., Rothlisberger, H., Flotron, A., and Haeberli, W., 1983, The uplift of Unteraargletscher at the beginning of the melt season—a consequence of water storage at the bed?: Journal of Glaciology, v. 19, p. 28–47.

Jahns, R. H., 1943, Sheet structure in granites: its origin and use as a measure of glacial erosion in New England: Journal of Geology, v. 51, p. 71–98.

Jarvis, G. T., and Clarke, G. K. C., 1975, The thermal regime of Trapridge Glacier and its relevance to glacier surging: Journal of Glaciology, v. 14, p. 235–249.

Johnson, P. G., 1972, The morphological effects of surges of the Donjek Glacier, St. Elias Mountains, Yukon Territory, Canada: Journal of Glaciology, v. 11, p. 227–234.

Jones, S. J., and Brunet, J. G., 1978, Deformation of single ice crystals close to the melting point: Journal of Glaciology, v. 21, p. 445–454.

Jopling, A. V., and McDonald, B. C., eds., 1975, Glaciofluvial and glaciolacustrine sedimentation: Society of Economic Paleontologists and Mineralogists, Special Publication 23, 320 p.

Kamb, B., 1964, Glacier geophysics: Science, v. 146, p. 353–365.

Kamb, B., 1970, Sliding motion of glaciers: theory and observation: Rev. Geophysics and Space Physics, v. 8, p. 673–728.

Kamb, B., 1972, Experimental recrystallization of ice under stress: Geophysical Monograph, American Geophysical Union, v. 18, p. 211–241.

Kamb, B., and LaChapelle, E., 1964, Direct observation of the mechanism of glacier sliding over bedrock: Journal of Glaciology, v. 5, p. 159–172.

Kamb, B., Engelhardt, H., and Harrison, W. D., 1979, The ice-rock interface and basal sliding process as revealed by direct observations in boreholes and tunnels: Journal of Glaciology, v. 23, p. 416–419.

Kamb, B., Raymond, C. F., Harrison, W. D., Engelhardt, H., Echelmeyer, K. A., Humphrey, N., Brugman, M. M., and Pfeffer, T., 1985, Glacier surge mechanism; 1982–1983 surge of Variegated Glacier, Alaska: Science, v. 227, p. 469–479.

Karcz, L, 1973, Reflections on the origin of small scale longitudinal stream-bed scours: in Morisawa, M., ed., Fluvial geomorphology, State University of New York, Binghamton, N.Y., p. 149–178.

King, C. A. M., and Lewis, W. V., 1961, A tentative theory of ogive formation: Journal of Glaciology, v. 3, p. 913–939.

Kuenen, P. H., 1951, Mechanics of varve formation and the action of turbidity currents: Stockholm, Geologiska Foreningen i Forhandlingar, v. 73, p. 69–84.

Lewis, W. V., 1954, Pressure release and glacial erosion: Journal of Glaciology, v. 2, p. 417–422.

Liestol, O., 1969, Glacier surges in West Spitsbergen: Canadian Journal of Earth Science, v. 6, p. 895–897.

Lighthill, M. J., and Whitham, G. B., 1955, On kinematic waves: Proceedings of the Royal Society of London, Series A, v. 229, p. 281–345.

Lliboutry, L., 1968, General theory of subglacial cavitation and sliding of temperate glaciers: Journal of Glaciology, v. 7, p. 21–58.

Marcusson, L, 1973, Studies on flow till in Denmark: Boreas, v. 2, p. 213–231.

Mathews, W. H., 1956, Physical limnology and sedimentation in a glacial lake: Geological Society of America Bulletin, v. 67, p. 537–552.

Mathews, W. H., 1964, Water pressure under a glacier: Journal of Glaciology, v. 5, p. 234–240.

Mathews, W. H., 1979, Simulated glacial abrasion: Journal of Glaciology, v. 23, p. 51–56.

McCall, J. G., 1952, The internal structure of a cirque glacier: report on studies of the englacial movements and temperatures: Journal of Glaciology, v. 2, p. 122–131.

McCall, J. G., 1960, The flow characteristics of a cirque glacier and their effect on glacial structure and cirque formation, in Lewis, W. V., ed., Norwegian cirque glaciers: Royal Geographical Society Research Series 4, p. 39–62.

McMeeking, R. M., and Johnson, R. E., 1986, On the mechanics of surging glaciers: Journal of Glaciology, v. 32, p. 120–122.

Meier, M. F., 1960, Mode of flow of the Saskatchewan Glacier, Alberta: U.S. Geological Survey Professional Paper 351, 70 p.

Meier, M. F., 1965, Glaciers and climate: in Wright, H. E., and Frey, D. G., eds., The Quaternary of the U.S.: Princeton University Press, Princeton, N.J., p. 795–805.

Meier, M. F., and Johnson, A., 1962, The kinematic wave on Nisqually Glacier, Washington: Journal of Geophysics, v. 67, p. 886.

Meier, M. F., and Post, A. S., 1969, What are glacier surges? Canadian Journal of Earth Science, v. 6, p. 807–817.

Meier, M. F., Kamb, B., Allen, C. R., and Sharp, R. P., 1974, Flow of Blue Glacier, Olympic Mountains, Washington, U.S.A.: Journal of Glaciology, v. 13, p. 187–212.

Mellor, M. N., 1964, Snow and ice on the earth's surface: U.S. Army Cold Region Research. Engineering Lab Research Report IT-CL, 163 p.

Mellor, M. N., and Testa, R., 1969, Effect of temperature on the creep of ice: Journal of Glaciology, v. 8, p. 131–145.

Mickelson, D. M., 1973, Nature and rate of basal till deposition in a stagnating ice mass, Burroughs Glacier, Alaska: Arctic and Alpine Research, v. 5, p. 17–27.

Moffit, F. H., 1942, Geology of the Gerstle River district, Alaska, with a report on the Black Rapids Glacier: U.S. Geological Survey Bulletin 926-B, p. 107–160.

Mooers, H. D., 1989, On the formation of the tunnel valleys of the Superior Lake, Central Minnesota: Quaternary Research, v. 32, p. 24–35.

Morris, E. M., 1976, An experimental study of the motion of ice past obstacles by the process of regelation: Journal of Glaciology, v. 17, p. 79–98.

Nye, J. F., 1951, The flow of glaciers and ice-sheets as a problem in plasticity: Proceedings of the Royal Society of London, Series A, v. 207, p. 554–572.

Nye, J. F., 1952a, The mechanics of glacier flow: Journal of Glaciology, v. 2, p. 82–93.

Nye, J. F., 1952b, A comparison between the theoretical and the measured long profiles of the Unteraar glacier: Journal of Glaciology, v. 2, p. 103–107.

Nye, J. F., 1953, The flow law of ice from measurements in glacier tunnels, laboratory experiments and the Jungfraufirn borehole experiment: Proceedings of the Royal Society of London, Series A, v. 219, p. 477–489.

Nye, J. F., 1957, The distribution of stress and velocity in glaciers and ice sheets: Proceedings of the Royal Society, 239, p. 113–133.

Nye, J. F., 1958a, A theory of wave formation on glaciers: International Association Sci. Hydrology Bulletin, v. 47, p. 139–154.

Nye, J. F., 1958b, Surges on glaciers: Nature, v. 181, p. 1450–1451.

Nye, J. F., 1959, The motion of ice sheets and glaciers: Journal of Glaciology, v. 3, p. 493–507.

Nye, J. F., 1960, The response of glaciers and ice sheets to seasonal and climatic changes: Proceedings of the Royal Society, v. 256, p. 559–584.

Nye, J. F., 1965a, The flow of a glacier in a channel of rectangular, elliptic, or parabolic cross-section: Journal of Glaciology, v. 5, p. 661–690.

Nye, J. F., 1965b, The frequency response of glaciers: Journal of Glaciology, v. 5, p. 567–587.

Nye, J. F., 1970, Glacier sliding without cavitation in a linear viscous approximation: Proceedings of the Royal Society, v. 315, p. 381–403.

Nye, J. F., 1973a, The motion of ice past obstacles: in Whalley, E., Jones, S. J., and Gold, L. W., eds., Physics and chemistry of ice: Ottawa, Royal Society of Canada, p. 387–395.

Nye, J. F., 1973b, Water at the bed of a glacier: Symposium on the Hydrology of Glaciers, International Association of Sci. Hydrology Publication no. 95, P. 189–194.

Nye, J. F., 1977, Water flow in glaciers, jokulhlaups, tunnels, and veins: Journal of Glaciology, v. 17, p. 181–207.

Nye, J. F., and Martin, P. C. S., 1967, Glacial erosion: Bern, Switzerland, International Association Scientific Hydrology Commission on Snow and Ice, p. 78–83.

Ostrom, G., Haakensen, N., and Melonder, O., 1973, Atlas over Breer i Nord-Scandinavia; Norges Vassdrags og electrisitetsjegen of Stockholm Univ., 315 p.

Palmer, A. C., 1972, A kinematic wave model of glacier surges: Journal of Glaciology, v. 11, p. 65–72.

Paterson, W. S. B., 1964, Variations in velocity of Athabasca Glacier with time: Journal of Glaciology, v. 5, p. 277–285.

Paterson, W. S. B., 1977, Secondary and tertiary creep of glacier ice as measured by borehole closure rates: Rev. Geophysics Space Physics, v. 15, p. 47–55.

Paterson, W. S. B., 1981, The physics of glaciers: Pergamon Press, Oxford, England, 380 p.

Perutz, M. F., 1940, Mechanism of glacier flow: Proceedings of the Royal Society of London, Series A, v. 52, p. 132–135.

Price, R. J., 1973, Glacial and fluvioglacial landforms. Oliver and Boyd Ltd., Edinburgh, U.K., 242 p.

Post, A., 1960, The exceptional advances of the Muldrow, Black Rapids, and Susitna glaciers: Journal of Geophysical Research, v. 65, p. 3703–3712.

Post, A., 1966, The recent surge of Walsh glacier, Yukon and Alaska: Journal of Glaciology, v. 6, p. 365–381.

Post, A., 1969, Distribution of surging glaciers in western North America: Journal of Glaciology, v. 8, p. 229–240.

Post, A., 1972, Periodic surge origin of folded medial moraines on Bering piedmont glacier, Alaska: Journal of Glaciology, v. 11, p. 219 – 226.

Post, A., and LaChapelle, E. R., 1971, Glacier ice: University of Washington Press, Seattle, WA, 110 p.

Powell, R. D., 1980, Holocene glaciomarine sediment deposition by tidewater glaciers in Glacier Bay, Alaska: Ph.D. thesis: Ohio State University, Columbus, OH, 472 p.

Raymond, C. F., 1971, Flow in a transverse section of Athabasca Glacier, Alberta, Canada: Journal of Glaciology, v. 10, p. 55–84.

Raymond, C. F., 1980, Valley glaciers: in Colbeck, S. C., ed., Dynamics of snow and ice masses: Academic Press, N.Y., p. 79–139.

Raymond, C. F., 1987, How do glaciers surge: A Review: Journal of Geophysical Research, v. 92, p. 9121–9134.

Raymond, C. F., and Harrison, W. D., 1988, Evolution of Variegated Glacier, Alaska, U.S.A., prior to its surge: Journal of Glaciology, v. 34, p. 154–169.

Raymond, C. F., and Malone, S. D., 1981, Relationship of directly observed basal sliding rate and conditions to variations in surface velocity, strain rates, and seismic activity: Progress Report for National Science Foundation Grant No. DPP-7903942, 34 p.

Rigsby, G. P, 1960, Crystal orientation in glacier and in experimentally deformed ice: Journal of Glaciology, v. 3, p. 589–606.

Robin, G. Q., 1955, Ice movement and temperature distribution in glaciers and ice sheets: Journal of Glaciology, v. 2, p. 523–532.

Robin, G. Q., 1969, Initiation of glacier surges: Canadian Journal of Earth Sciences, v. 6, p, 919–928.

Robin, G. Q., 1972, Polar ice sheets: a review: Polar Record, v. 16, p. 5–22.

Robin, G. Q., 1976, Is the basal ice of a temperate glacier at the pressure-melting point? Journal of Glaciology, v. 16, p. 183–195.

Robin, G. Q., and Barnes, P., 1969, Propagation of glacier surges: Canadian Journal of Earth Sciences, v. 6, p. 969–977.

Robin, G. Q., and Weertman, J., 1973, Cyclic surging of glaciers: Journal of Glaciology, v. 12, p. 3–18.

Rothlisberger, H., 1968, Erosive processes which are likely to accenctuate or reduce the bottom relief of valley glaciers: International Association Scientific Hydrology, v. 79, p. 87–97.

Rothlisberger, H., 1972, Water pressure in intra- and sub-glacial channels: Journal of Glaciology, v. 11, p. 177–203.

Rothlisberger, H., and Spring, U., 1979, Piezometric observations of water pressure at the bed of Swiss glaciers: Journal of Glaciology, v. 23, p. 429–430.

Russell, I. C., 1893, Malaspina Glacier: Journal of Geology, v. 1, p. 219–245.

Russell, I. C., 1895, The influence of debris on the flow of glaciers: Journal of Geology, v. 3, p. 823–832.

Russell-Head, D. S., and Budd, W. F., 1979, Ice-sheet flow properties derived from bore-hole shear measurements combined ice-core studies: Journal of Glaciology, v. 24, p. 117–130.

Rust, B. R., 1972, Pebble orientation in fluvial sediments: Journal of Sedimentary Petrology, v. 42, p. 384–388.

Rust, B. R., 1975, Fabric and structure in glaciofluvial gravels: Society of Economic Paleontologists and Mineralogists, Special Publication 23, p. 239–249.

Rutishauser, H., 1971, Observations on a surging glacier in East Greenland: Journal of Glaciology, v. 10, p. 227–236.

Savage, J. C., and Paterson, W. S. B., 1963, Borehole measurements in the Athabasca Glacier: Journal of Geophysical Research, v. 68, p. 4521–4536.

Schytt, V., 1969, Some comments on glacier surges in eastern Svalbard: Canadian Journal of Earth Sciences, v. 6, p. 867–873.

Sharp, R. P., 1951b, Features of the firn on upper Seward Glacier, St. Elias Mountains, Canada: Journal of Geology, v. 59, p. 599–621.

Sharp, R. P., 1989, Living ice: understanding glaciers and glaciation: Cambridge University Press, Cambridge, England, 225 p.

Sharp, R. R, 1951, Thermal regime of firn on Upper Seward Glacier, Yukon Territory, Canada: Journal of Glaciology, v. 1, p. 461–465.

Sharp, R. P., 1951b, Features of the firn on upper Seward Glacier, St. Elias Mountains, Canada: Journal of Geology, v. 59, p. 599–621.

Sharp, R. R, 1953, Deformation of a vertical bore hole in a piedmont glacier: Journal of Glaciology, v. 2, p. 182–184.

Shreve, R. L., 1972, Movement of water in glaciers: Journal of Glaciology, v. 11, p. 205–214

Shumskii, P. A., 1964, Principles of structural glaciology: Dover Publications, N.Y.

Smith, A., 1832, On the water courses and alluvial and rock formations of the Connecticut River Valley: American Journal of Science, v. 22, p. 205 – 23 1.

Smith, H. T. U., 1948, Giant glacial grooves in northwest Canada: American Journal of Science, v. 246, p. 503–514.

Smith, N. D., 1974, Sedimentology and bar formation in the upper Kicking Horse River, a braided meltwater stream: Journal of Geology, v. 82, p. 205 –223.

Souchez, R. A., and deGroote, J. M., 1985, Delta D-Delta-^{18}O relationships in ice formed by subglacial freezing; paleoclimatic implications: Journal of Glaciology, v. 31, p. 229–232.

Stanford, S. D., and Mickelson, P. M., 1985, Till fabric and deformational structures in drumlins near Waukesha, Wisconsin, U.S.A.: Journal of Glaciology, v. 31, p. 220–228.

Stanley, A. D., 1969, Observations of the surge of Steele glacier, Yukon Territory, Canada: Canadian Journal of Earth Sciences, v. 6, p. 819–830.

Stenborg, T., 1969, Studies of the internal drainage of glaciers: Geografiska Annaler, v. 51A, p. 13–41.

Stenborg, T., 1973, Some viewpoints on the internal drainage of glaciers: Union Geodesique et Geophysique Internationale, Publication No. 95 de I'Association Internationale d'Hydrologie Scientifique, p. 117–129.

Steinemann, S., 1954, Results of preliminary experiments on the plasticity of ice crystals: Journal of Glaciology, v. 2, p. 404–412.

Stupavsky, M., and Gravenor, C. P., 1975, Magnetic fabric around boulders in till: Geological Society of America Bulletin, v. 86, p. 1534–1536.

Sugden, D. E., and John, B. S., 1976, Glaciers and landscape: John Wiley and Sons, N.Y., 376 p.

Tangborn, W. V., Krimmel, R. M., and Meier, M. F., 1975, A comparison of glacier mass balance by glaciological, hydrological, and mapping methods, South Cascade Glacier, Washington: International Association of Scientific Hydrology Publication No. 104, p. 185–196.

Tarr, R. S., 1909, Some phenomena of the glacier margins in the Yakutat Bay region, Alaska: Zeitschrift für Gletscherkunde, v. 3, p. 81 –100.

Tarr, R. S., and Butler, R. S., 1909, The Yakutat Bay region, Alaska: U.S. Geological Survey Professional Paper 64, 178 p.

Tarr, R. S., and Martin, L., 1914, Alaskan glacier studies: Washington, D.C., National Geographic Society, 498 p.

Theakstone, W. H., 1967, Basal sliding and movement near the margin of the glacier Osterdalsiesen, Norway: Journal of Glaciology, v. 6, p. 805–816.

Thorn, C. E., 1976, Quantitative evaluation of nivation in the Colorado Front Range: Geological Society of America Bulletin, v. 87, p. 1169–1178.

Thorn, C. E., and Hall, K., 1980, Nivation: An arctic-alpine comparison and reappraisal: Journal of Glaciology, v.25, p. 109–124.

Thorarinsonn, 1969, Glacial surges in Iceland, with special reference to the surges of Bruarjokull: Canadian Journal of Earth Sciences, v. 6, p. 875–882.

Vivian, R., 1970, Hydologie et erosion sous-glaciares: Revue Geographie Alpine, v. 58, p. 241–264.

Vivian, R., 1980, The nature of the ice-rock interface: the results of investigation on 20,000 M2 of the rock bed of temperate glaciers: Journal of Glaciology, v. 25, p. 267–277.

Vivian, R., and Bouquet, G., 1973, Subglacial caviation phenomena under Glacier d'Argentiere, Mont Blanc, France: Journal of Glaciology, v. 12, p. 439–452.

Vivian, R., and Zumstein, J., 1973, Hydrologie sous-galciaire au galcier d'Argentiere (Mont-Blanc, France): Union Geodesique et Geophysique Internationale, Publication No. 95 de I'Association Internationale d'Hydrologie Scientifique, p. 53–64.

Walder, J., and Hallet, B., 1979, Geometry of former subglacial water channels and cavities: Journal of Glaciology, v. 23, p. 335–346.

Walder, J. S., and Hallet, B., 1985a, The physiucal basis of frost weathering; toward a more fundamental and unified perspective: Arctic and Alpine Research, v. 18, p. 27–32.

Walder, J. S., and Hallet, B., 1985b, A theoretical model of the fracture of rock during freezing: Geological Society of America Bulletin, v. 96, p. 336–346.

Weaver, C. S., and Malone, S. D., 1979, Seismic evidence for discrete glacier motion at the rock-ice interface: Journal of Glaciology, v. 23, p. 171–183.

Weertman, J., 1957, On the sliding of glaciers: Journal of Glaciology, v. 3, p. 33–38.

Weertman, J., 1958, Traveling waves on glaciers: International Association Sci. Hydrology, v. 47, p. 162–168.

Weertman, J., 1962, Catastrophic glacier advances: International Association Sci. Hydrology, v. 58, p. 31–39.

Weertman, J., 1964, The theory of glacier sliding: Journal of Glaciology, v. 5, p. 287–303.

Weertman, J., 1967, An examination of the Lliboutry theory of glacier sliding: Journal of Glaciology, v. 6, p. 489–494.

Weertman, J., 1969, Water lubrication mechanism of glacier surges: Canadian Journal of Earth Science, v. 6, p. 929–942.

Weertman, J., 1971, In defense of a simple model of glacier sliding: Journal of Geophysics Research, v. 76, p. 6485–6487.

Weertman, J., 1972, General theory of water flow at the base of a glacier or ice sheet: Reviews of Geophysics and Space Physics, v. 10, p. 287–333.

Weertman, J., 1973a, Can a water-filled crevasse reach the bottom surface of a glacier? Union Geodesique et Geophysique Internationale, Publication No. 95 de I'Association Internationale d'Hydrologie Scientifique, p. 139–145.

Weertman, J., 1973b, Creep of ice: in Whalley, E., Jones, S. J., and Gold, L. W., eds., Physics and chemistry of ice: Ottawa, Royal Society of Canada, p. 320–337.

Weertman, J., 1979, The unsolved general glacier sliding problem: Journal of Glaciology, vol. 23, p. 97–111.

Weertman, J., 1983, Creep deformation of ice: Annual Review of Earth and Planetary Sciences, vol. 11, p. 215–240.

Weertman, J., 1986, Basal water and high pressure ice: Journal of Glaciology, v. 32, p. 455–463.

White, W. A., 1972, Deep erosion by continental ice sheets: Geological Society of America Bulletin, v. 83, p. 1037–1056.

Williams, P. F., and Rust, B. R., 1969, The sedimentology of a braided river: Journal of Sedimentary Petrology, v. 39, p. 649–679.

Wintages, T., 1985, Studies on crescentric fractures and cresentic gouges with the help of close-range photogrammetry: Journal of Glaciology, v. 31, p. 340–349.

Wright, H. E., 1973, Tunnel valleys, glacial surges and subglacial hydrology of the Superior Lobe, Minnesota: in Black, R. F., Goldthwait, R. R, and Willman, H. B., eds., The Wisconsinan stage: Geological Society of America Memoir 136, p. 251–276.

Glacial Landforms

Alpine valley glacier and lateral moraines, Mt. Waddington, British Columbia. (Photo by D. A. Rahm)

INTRODUCTION

Glacial landforms are produced both by erosion and by deposition. Alpine glaciers have sculpted some of the most scenic topography in the world, and Pleistocene continental glaciers have created landforms over large portions of continents.

EROSIONAL LANDFORMS

Cirques

Because they make such striking landforms in alpine regions, **cirques** have long been the focus of attention of scientists and tourists alike. They consist of semicircular hollows set into mountain slopes and are characterized by steep headwalls (Plates 9–A, 9–C) that may vary from less than 100 m (330 ft) to many hundreds of meters high. The steepness of headwalls of cirques recently exposed by deglaciation is

often accentuated by the absence of talus at the base. Rock or morainal thresholds may separate the basins of cirques from downvalley slopes and commonly hold in **tarns,** glacial lakes on the cirque floor (Plate 9-C), after the ice has melted away.

The development of cirques from nonglacial topography begins with the initiation of permanent snowbanks above the firn line where **nivation,** a process consisting of freeze/thaw and mass wasting, occurs beneath the snowbanks (Thorn, 1976; Thorn and Hall, 1980). Rock beneath the snowbanks is broken apart by the freezing and expansion of meltwater in rock fractures, and the detached rock fragments move downslope by creep, solifluction, and rill wash. Thawing of frozen ground takes place from the surface downward, and thus water is prevented from percolating downward through the frozen soil and detritus beneath. The result is that the material becomes saturated and flows downslope as **solifluction lobes.** Continued excavation of material in this fashion develops shallow depressions known as **nivation basins** (Figure 13–1) beneath the snowbanks. A second set of processes comes into play when sufficient snow, firn, and ice accumulate in the nivation basin to cause motion. The mass then becomes a **cirque glacier,** and abrasion and plucking further modify the configuration of the basin (Fisher, 1948; White, 1970).

The behavior of a cirque glacier plays an important role in continued excavation of the cirque. Detailed studies of Norwegian cirque glaciers (Lewis, 1960) show that ice moves by rotational sliding on the bedrock floor. As noted in Chapter 12, downward motion of ice in the zone of accumulation

FIGURE 13–1
Nivation basins, Bighorn Mts., Wyoming. (Photo by D.A. Rahm)

and upward motion in the zone of ablation, combined with the short length of the glacier, produce a rotational component of movement about a horizontal axis. Figure 12–23 shows successive layers of snow added to the glacier in the accumulation zone. As each layer of snow moves downvalley and is buried under succeeding years of accumulation, the original downglacier slope becomes reversed until the snow layers reach the glacier terminus, where they dip steeply in the upglacier direction (Figure 13–2) (McCall, 1952, 1960). The net effect is as if the glacier were a rigid, rotating mass. Rotational sliding abrades the cirque floor, further excavating the basin and carrying the eroded material to the glacier margin. Because the erosive power of the glacier diminishes as the ice thins near its terminus, the basin is excavated more vigorously upglacier from the margin, and a rock threshold is produced.

As ice moves away from the motionless snowfields above, a deep crevasse, known as a **bergschrund** (Figure 13–3), is formed between the moving ice and the cirque **headwall.** Observers who have descended into deep bergschrunds have been impressed with the intense frost shattering and plucking

of the cirque headwall, leading to the widely held belief that the headwall is steepened and extended headward by this process. Bergschrunds are generally not more than 100 m (330 ft) deep, and because they do not extend all the way to the base of headwalls (which are commonly hundreds of meters high), plucking is not confined to the open bergschrund. Some water must melt its way between the glacier and headwall where freeze/thaw cycles and plucking keep the headwall steepened (Johnson, 1941; Lewis, 1938, 1940). However, active freeze/thaw cycles are difficult to imagine in Antarctica where steep cirque headwalls hundreds of meters high are common, in spite of the fact that air temperatures remain well below freezing for the entire year. This apparent anomaly may be explained by the observation that sunshine can cause snow to melt at air temperatures of $-17°C$. Although freeze/thaw cycles appear to be important in the steepening of cirque headwalls, the significance of bergschrunds has been challenged on the basis that temperature fluctuations are not severe enough to produce shattering of rock, and some bergschrunds do not extend deeply enough to intersect the headwall (Battle, 1960).

Although cirques are characterized by steep headwalls and basins (Figures 13–4 and 13–5), reflecting the processes described above, their sizes and shapes vary considerably. The rates at which cirques develop are difficult to ascertain because of differences in bedrock lithology, spacing and character of jointing, ice thickness, velocity of basal ice, temperature, amount and nature of basal debris, altitude, relief, and duration of glaciation (Anderson, 1978; Andrews and Lamasurier, 1973; Olyphant, 1981). Many cirques are the result of several periods of glaciation during the Pleis-

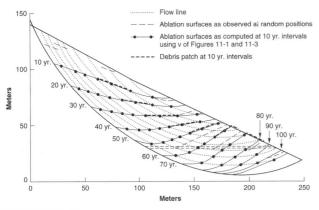

FIGURE 13–2
Flow lines in a cirque glacier. (From McCall, 1960)

FIGURE 13–3
Bergschrund at headwall of a cirque, Mt. Olympus, Washington.

FIGURE 13–4
Cirque with steep headwall and a tarn in the basin, Coast Range of British Columbia, Canada. (Photo by D. A. Rahm)

FIGURE 13–5
Multiple cirques with steep headwalls and tarns in their basins, Wind River Range, Wyoming. (Photo by A. Post, U.S. Geological Survey)

tocene, and others are formed by the coalescence of several cirques to make broad **compound cirques** with scalloped headwalls marking the position of intersection of individual cirques (Figures 13–6A and 13–6B).

As with many other geomorphic processes, landscapes of glacial erosion progress through evolutionary phases (Johnson, 1972; Matthes, 1900). Isolated individual cirques of the early phases of glacial erosion give way to coalescence into compound cirques, which continue to expand until much of the preglacial topography has been consumed, and only deeply scalloped, skeletal ridges and horns remain (Figure 13–7).

Cirques are commonly aligned, facing in a particular direction, reflecting differences in the amount of solar radiation received. For example, cirques may be more common on north-facing slopes because south-facing slopes receive more afternoon solar radiation and experience greater melting. In a similar fashion, east-facing cirques are sometimes more common than west-facing cirques because westfacing ones receive more direct afternoon solar radiation and thus have higher melt rates. The direction of prevailing storm tracks is also important.

Present altitudes of abandoned cirque floors provide useful evidence of rise and fall of composite snowlines during the Pleistocene. Equilibrium line altitudes (ELAs) are generally about three-fifths of the way up a glacier, and in the case of small cirque glaciers, they lie only a short distance above the cirque floors. Thus, contouring the altitude of abandoned cirque floors in a mountain range can give an idea of former

depressions of ELAs during the Pleistocene (Flint, 1971; Meierding, 1982; Porter, 1977; Trenhaile, 1977; Williams, 1983). Such data reflect a composite of all previous glaciations, not any single glaciation.

Aretes, Cols, and Horns

Aretes are formed when recession of cirque headwalls consumes much of a drainage divide, leaving only a narrow, knife-edged ridge between back-to-back cirques (Figure 13–8). In mountainous regions where long-continued glaciation has eroded away most of the preglacial topography, aretes may persist as skeletal ridges. Because cirque headwalls generally have curved outlines in plan view, their back-to-back confluences produce a low area between them, known as a **col**.

Back-to-back recession of three or more cirque headwalls leaves sharp, spire-shaped peaks (Figure 13–9), such as the Matterhorn in Switzerland (Figure 13–10; Plate 9–B). The number of faces on a **horn** depends on the number of cirque headwalls responsible for their formation, the most common being three or four.

Glacial Troughs

Glacial erosion of preglacial stream valleys greatly alters their topographic form by deepening and widening them and by planing off spurs to change the V-shaped cross sec-

A.

B.

FIGURE 13–6
A. Compound cirques, Uinta Range, Wyoming.
B. More advanced phase of compound cirques, Sierra Nevada, California.

FIGURE 13–7
Advanced phase of glacial erosion, Wind River Range, Wyoming. Only skeletal ridges (aretes) remain unconsumed by erosion.

FIGURE 13–8
Arete between two cirque headwalls, Twin Sisters Range, Washington.

tion of the original stream valley into a U-shaped cross section with steep sides and a wide valley floor (Figure 13–11). Because glaciers rarely create new valleys, their paths reflect the original preglacial stream courses in map view. However, because glaciers do not bend around sharp curves as easily as the running water of streams that initiated the valley, irregularities and sharp turns are pared away, straightening the valley somewhat and leaving **truncated spurs** (Plate 9–E).

Although glacial troughs are often characterized as having U-shaped cross profiles, a parabolic curve more closely approximates their shape. The evolution of a V-shaped stream valley profile to a parabolic form is undoubtedly related to the distribution of shear stresses exerted along the sides of a valley by a glacier attempting to fit into the V-shape. The size and shape of a glacial trough are thought to be in equilibrium with ice discharge. As noted by Penck (1905), although the

surfaces of tributary glaciers are accordant with main trunk glaciers, their valley floors hang discordantly above the floor of the larger trunk glaciers (Figure 13–11).

The cross-sectional shape that would afford the least frictional resistance to ice flowage would be a semicircle, but most glacial cross sections are parabolic, perhaps as a result of greater rates of erosion near the middle of valleys where ice velocities are higher. A good deal of variation exists between relatively narrow, U-shaped glacial troughs and wide, shallow troughs. At least some of these differences may be influenced by the physical properties of the bedrock. Narrow, deep troughs are often associated with resistant bedrock; wide, shallow troughs, with less resistant bedrock (Matthes, 1930; King, 1959). Some fjords have thresholds in their lower valley floors that have been interpreted both as submerged end moraines and as rock ridges left near the margin of the ice

FIGURE 13–10
The Matterhorn, Switzerland, a classic example of a horn peak.

FIGURE 13–11
Glacial trough and hanging valley, Yosemite National Park, California. Bridalveil Falls emerging from a hanging valley along the main valley side. (Photo by D.A. Rahm)

where the vigor of erosion was less than that upvalley (Shoemaker, 1986).

Although most glacial troughs are carved by glaciers from preexisting stream valleys, some are related to tectonic conditions. Dowdeswell and Andrews (1985) found that the intersection of many fjord segments on Baffin Island suggested structural control.

Where glaciers extend to coastal areas, glacial troughs may become submerged by the sea upon retreat of the ice to form **fjords** (Figure 13–12) (Crary, 1966; Holtedahl, 1967). Among the more spectacular fjord regions of the world are those in western Norway and western British Columbia where the sea extends many kilometers up into mountain ranges. Although some of the seaward incursions may be attributed to post-

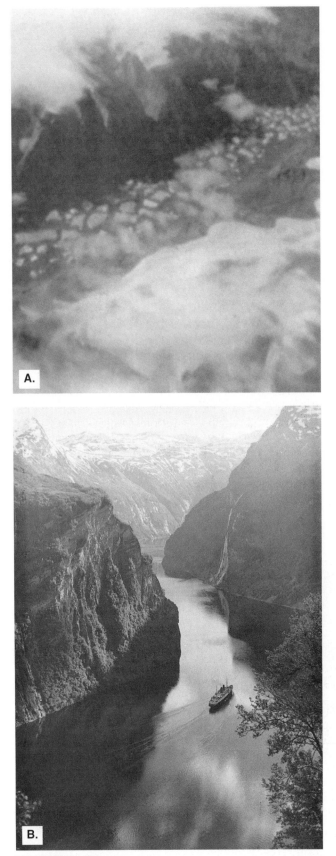

FIGURE 13–12
A. Fjord in eastern Greenland.
B. Fjord in western Norway.

Pleistocene sea-level rise following glacial erosion on land when sea levels were more than 100 m (330 ft) lower, some fjords are so deep that they must have been eroded below sea level. The Sogneflord and Hardangerflord fjords of western Norway extend to more than 1000 m (3300 ft) below sea level as ample evidence of the sometimes intense overdeepening of glacial troughs (Figure 13–12B).

The longitudinal profiles of glacial troughs are usually quite irregular, characterized by rock basins, steps, frequent breaks in slope, and **paternoster lakes,** chains of small lakes connected by inlet and outlet streams (Cotton, 1941). The steplike longitudinal profile of some glaciated valleys resembles a giant staircase, giving rise to the term **cyclopean stairs** (Figure 13–13) in which relatively flat treads or basins are separated by steep risers (Lewis, 1947; Matthes, 1930; Trenhaile, 1979). Cyclopean stairs may form in any of several different ways:

1. Differences in resistance to glacial erosion, especially with respect to frequency and spacing of joints. The subglacial floor may be differentially eroded because of these differences. For example, more rapid erosion in highly jointed granite susceptible to plucking can selectively etch out basins, whereas more massive granite is more resistant and forms steep risers (Matthes, 1930).

2. Discordance of subglacial floors at tributary junctions. Increased subglacial erosion due to glacial thickening below the confluence of tributary valleys deepens the valley floor successively below each junction (Bakker, 1965).

3. Backwasting of cirque headwalls at successively higher elevations at the heads of glaciated valleys. Progressively rising snowlines result in abandonment of lower cirques while headwall recession continues to steepen slopes at the heads of valleys.

As can be seen from the differences in processes related to these mechanisms of formation, not all cyclopean stairs can be attributed to a single type of development. Each occurrence must be evaluated on the basis of its own evidence.

Finger Lakes

Intense scouring and deepening of stream valleys by ice sheets flowing parallel to preglacial drainage are made possible by the greater thickness of ice in the valleys than over the adjacent divides. Under optimum conditions, the deepening of such valleys may be substantial. For example, the Finger Lakes of New York (Figure 13–14)—several elongate, north-south glacial troughs scoured out by Pleistocene ice sheets and named because of their resemblance to the fingers of a hand—vary in length from about 5 to 61 km (3 to 37 mi) and are up to 200 m (660 ft) deep. That they are essentially glacially scoured troughs is shown by the depth of their bedrock valley floors well below sea level. The bedrock floor of Seneca

FIGURE 13–13
Cyclopean stairs, Yosemite National Park, California.

Lake extends 304 m (1000 ft) below sea level, and the bedrock floor of Lake Cayuga is 249 m (817 ft) below sea level (Mullins and Hinchey, 1989). The troughs are filled with up to 275 m (900 ft) of unconsolidated Quaternary sediment. Similar deeply scoured glacial lakes found elsewhere in the world are referred to as **finger lakes.**

Streamlined Forms

Recently deglaciated regions commonly exhibit many types of elongate, streamlined forms of all sizes and gradations, from entirely bedrock to entirely unconsolidated deposits. The streamlining of material beneath a glacier minimizes the resistance of knobs and hills to glacial flow. Good examples of streamlined bedrock forms are found in southern Finland where thousands of smooth, elongate hills give the landscape a distinct linear aspect (Figure 13–15). Individual streamlined rock knobs are generally a few meters high and 10 to 20 m (~30 to 65 ft) long, usually with one end higher than the other. The term **whaleback** has been used to describe such forms because they resemble whales rising above water level. Whalebacks formed by intense abrasion of rock may be higher on the upglacial end, taper in the downglacial direction, and grade into **rock drumlins.** Others, known as **roches moutonnees** (Figure 13–16) or **stoss-and-lee topography,** have just the reverse symmetry, with a higher, more abrupt downglacial termination. The up-ice portions are usually highly abraded, whereas the down-ice ends are craggy and show the effects of steepening by plucking.

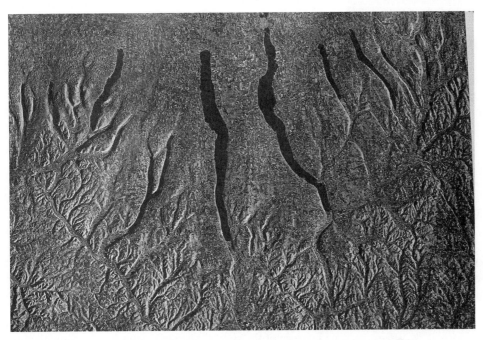

FIGURE 13–14
Finger Lakes, New York. (U.S. Geological Survey side-looking radar photo)

FIGURE 13–15
Glacially streamlined rock knobs, Helsinki, Finland.

FIGURE 13–16
Roches moutonnes, Mt. Baker, Washington.

DEPOSITIONAL LANDFORMS

Because of the wide diversity of glacial depositional environments, the variety of landforms produced by glacial deposition is almost endless, with many transitional forms. Depositional landforms may be grouped into categories related to the following:

1. Position of deposition from the glacier (subglacial, ice marginal, or proglacial)
2. Relative vigor of ice motion (stagnant ice or active ice)
3. Morphology and composition

Although most depositional glacial landforms are characterized by one or more of the sediment types discussed above, the landforms are named largely, but not completely, on the basis of their shape and size rather than on the material composing them.

Moraines

End Moraines

End moraines are constructed by the accumulation of rock debris at the terminus of a glacier. They are generally curvilinear in form, reflecting the shape of the glacier terminus (Plate 10–B, 10–C), and are typically composed mostly of till, although bulldozed material and lenses of sand and gravel are not unusual.

Several processes are important in the formation of end moraines. The state of activity of the ice— that is, whether the glacier terminus is advancing, stable, or stagnating plays a significant role in the determination of the shape and the internal character of an end moraine (Boulton, 1972b; Boulton and Eyles, 1979; Price, 1970).

If the accumulation/ablation ratio of a glacier is stable, then as long as equilibrium is maintained between the rate of ice movement and the rate of ablation, the glacier terminus remains in one position and the glacier acts like a giant conveyor belt, dumping an endless load of rock debris at the front of the glacier (Figure 13–17). Thus, the size of an end moraine depends more on the stability of the glacier terminus than on the size of the glacier making the moraine. Despite the proclivity of geologists to attach names to landforms, no term for moraines of such origin has been coined.

If the glacier terminus is advancing, material in front of the ice may be bulldozed up into a linear ridge to form a **push moraine** (Figure 13–18). However, push moraines are generally less prominent than the "conveyor-belt" type of moraine described above, because as soon as they become very large, they impede movement of the ice, which then shears over the top of the moraine and ultimately destroys it while beginning a new one at the current ice front. Because the shearing strength of ice is quite low relative to the resistance offered by a growing push moraine, the amount of material pushed by the glacier is quite limited before the ice shears over the top of the debris in front of it. In spite of this limitation, some large push moraines, consisting mostly of bulldozed sediment, do exist. Because of the strong shearing involved in the formation of

FIGURE 13–17
End moraine forming by release of debris from melting ice at the terminus of the Deming Glacier, Mt. Baker, Washington.

FIGURE 13–18
Push moraine made by an advance of an outlet glacier of the Vatnajokull, Iceland.

push moraines, they often show severe internal deformation (Boulton, 1981, 1986).

If the terminus of a glacier becomes stagnant, motion progressively decreases to zero, and as the ice thins by ablation, debris is concentrated on its surface (Russell, 1893; Boulton, 1967), often obscuring the ice beneath a mantle of rock debris thick enough to allow forests to grow on it. When the ice completely melts away, a broad, irregular end moraine results (Figures 13–19 and 13–20), differing in form from other types of end moraines in its breadth and irregularity. Such moraines are known as **dead-ice moraines.**

The topography developed on moraines varies a great deal, depending on the dominance of one or more of the three processes described above. Some moraines have very low relief and consist of indistinct mounds and depressions, where-

as others make prominent ridges 30 to 50 m (100 to 165 ft) high. Young alpine end moraines may consist of single, well-defined ridges with sharp crests (Figure 13–21), whereas dead-ice moraines of continental glaciers may occupy broad zones 1 to 30 km (0.5 to 18 mi) wide (Figure 13–22). Depressions are common on moraines as a result of the melting of buried blocks of ice or irregular deposition.

Multiple end moraines, built during successive retreats and stillstands of glacial margins, are known as **recessional moraines.** They may consist of any of the types of end moraines described above. However, not all multiple end moraines are necessarily recessional— some may be made by later readvances that did not reach as far as earlier ones. When evidence of significance age differences between moraines is lacking, whether a given end moraine is recessional or due to a readvance is very difficult to distinguish.

Lateral Moraines

Lateral moraines are linear ridges of till formed by the accumulation of rock debris along the sides of a glacier (Figure 13–23; Plate 10–A, 10–D). Whereas most of the material in an end moraine is delivered to the moraine by ice movement, much of the rock debris in a lateral moraine is derived from rockfalls and tributary streams from the valley sides above the glacier and is carried along the glacial margin until released by melting ice. As a glacier melts down and away from its lateral moraines, an ice core typically remains in the lateral moraine for some time.

Lateral moraines commonly consist of till with gravel lenses that are coarser grained than subglacial tills, lacking the silt/clay rock-flour component. As a result of their formation against valley sides, after a glacier melts away, lateral mor-

FIGURE 13–19
Broad end moraine forming at the margin of a stagnating glacial terminus, Alsek Glacier, Alaska. (Photo by A. Post, U.S. Geological Survey)

aines are usually preserved because they stand above the valley floor and thus escape postglacial stream erosion.

As glaciers wax and wane, multiple lateral moraines may be developed, some of which overtop earlier ones and some of which are nested inside earlier ones (Plate 10–D) (Rothlisberger and Schneebeli, 1979). Lateral moraines can be disproportionately large compared to their end moraines where multiple phases of lateral moraines are "piggy-backed" upon one another, but the end moraines are not.

Medial Moraines

When two tributary glaciers come together, the inside lateral moraines on each glacier no longer have a valley side against which to bank their lateral moraines, and they unite to form medial moraines that form ribbonlike bands downglacier (Figure 13–24A). Although some medial moraines form prominent ridges of debris on the glacier surface, their size is somewhat deceptive because as the ice moves downvalley and the surface ablates downward, rock debris of the medial moraine protects the underlying ice from ablation, so the moraine grows higher and wider downvalley. Medial moraines on some glaciers in Iceland and Alaska exceed 10 m (33 ft) in height as they near the terminus.

In spite of their sometimes impressive size on modern glaciers, medial moraines seldom persist as landforms once the glacier melts away, largely because most of the rock debris

making up the moraines is only a thin mantle on the ice surface. Figure 13–24B shows this rather dramatically. Two prominent lateral moraines merge to form a medial moraine, which continues for only a short distance downvalley before diminishing to an indistinct form.

Interlobate Moraines

The accumulation of rock debris between two adjacent lobes of ice builds an **interlobate moraine.** These moraines differ from lateral moraines in that the rock debris collects between two ice margins rather than between ice and a valley side. Like lateral moraines, interlobate moraines grade into end moraines around the terminus of each lobe.

Washboard Moraines

Groups of low, parallel ridges 1 to 2 m (3 to 7 ft) high and spaced evenly a few tens of meters apart occur in Iceland, Baffin Island (Andrews, 1963), South Dakota, southern Maine, and elsewhere. Because their appearance from the air resembles an old-fashioned washboard, they are known as **washboard moraines** (Elson, 1968). Their limited relief and even spacing suggested to some geologists that they represent a form of annual accumulation of rock debris; thus, they are sometimes also referred to as **annual moraines.** Washboard moraines in front of the Breidamerkurjokull in Iceland (Figure 13–25) were shown by Boulton to form as small push

FIGURE 13–20
Broad, irregular, dead-ice moraine, North Dakota. (Photo by
L. Clayton)

moraines caused by short advances of the glacier terminus
during the winter followed by somewhat greater recession during the summer.

Rogen Moraines

Transverse, ribbed, subglacial moraines near Lake Rogen
in Sweden and in Canada west of Hudson Bay have been
given the name **Rogen moraines** (Kujansuu, 1967;
Lundquist, 1969). These moraines consist of a series of
transverse ridges from 10 to 30 m (~30 to 100 ft) high and
up to 1 km (0.6 mi) long with ridge crests typically 100 to
300 m (~300 to 1000 ft) apart. They consist of till with a
fabric of preferred clast orientations transverse to the ridge
crests, which are commonly fluted. Although their origin
remains somewhat obscure, they seem to be formed subglacially in the vicinity of crevasses or shear planes that
reach to the subglacial floor (Figure 13–26).

Ground Moraine

Ground moraine consists of till-covered areas of low relief,
lacking transverse linear ridges. It consists of lodgment or
ablation till deposited as a blanket over glaciated regions,
often forming gently undulating plains covering large areas.

Most ground moraine is composed of lodgment till deposited from the base of glaciers, but some consists of ablation till with scattered lenses of gravel or sand. The irregular
topographic relief developed on ground moraine is usually
the result of nonuniform deposition of till, or it may reflect
topography buried beneath the till mantle, as seen in exposures through areas of ground moraine that display cores of
preglacial hills plastered with a mantle of till.

Drumlins

As glaciers slide along the ground surface, they often mold
streamlined forms with elongate axes oriented parallel to the direction of glacial movement. Sometimes the streamlined shapes
consist of **drumlins**—low, elliptical hills resembling inverted
teaspoons (Figures 13–27 and 13–28)—but these forms may
grade into narrow, elongate, parallel ridges known as **flutes.**

Although drumlins exist in a wide range of sizes and
shapes, most are elongate hills about 1 to 2 km (0.6 to 1.2
mi) long, 300 to 600 m (~0.25 mi) wide, and less than 50 m
(165 ft) high. The ratio of their length to their width is generally about 2:3.5. The proximal portions of drumlins are
usually steeper and higher than the distal portions, which
thin and narrow in the down-ice direction. The distribution
of shapes and spacing of drumlins has been extensively
studied (see, for example, Chorley, 1959; Reed et al., 1962;
Vernon, 1966; Gluckert, 1973; Mills, 1980; Menzies and
Rose, 1987).

Drumlins rarely occur singly, almost always developing
in large numbers covering many square kilometers. The extensive drumlin fields of southeastern Wisconsin (Alden,
1905), west-central New York, New England, Nova Scotia,
Ireland, England, and Germany consist of thousands of individual drumlins oriented with their long axes parallel to each
other. The spatial distribution of drumlins in such fields appears to be nonrandom (Reed et al., 1962; Vernon, 1966;
Smalley and Unwin, 1968; Trenhaile, 1971; Gluckert, 1973),
suggesting a relationship between distribution and origin.
Most drumlin fields occur in zones just behind end moraines
with their long axes at right angles to the end moraines. Where
the ice margin is lobate, the axes of drumlins diverge radially toward the curving arcs of the end moraines.

The material composing drumlins varies over such a wide
range that drumlins could be said to consist of almost anything. Many are composed of till corresponding to the ground
moraine between the drumlins, but others may consist almost
entirely of bedrock, sand, and gravel, or various combinations of material. Although the surface of many drumlins consists of till, the cores of the drumlins may consist of older
deposits of any kind or may be made entirely of rock. Where

A.

B.

FIGURE 13–21
Sharp, single-crested moraines.
A. Mt. Robson, Canada. (Photo by D.A. Rahm)
B. Schoolroom Glacier, Grand Tetons, Wyoming.

FIGURE 13–22
Broad end moraine of a
Pleistocene continental glacier,
near Withrow, Washington.
(Photo by D.A. Rahm)

FIGURE 13–23
Lateral moraines cut by a fault,
McGee Canyon, Sierra Nevada
Range, California. (Photo by Ken
Hamblin)

FIGURE 13–24
A. Medial moraines formed by joining of lateral moraines from merging valleys, Alaska. (Photo by D.A. Rahm)
B. Lateral moraines of two tributary glaciers converging to form a medial moraine, Mt. Rosa, Switzerland. Note how quickly the medial moraine dies out below the junction.

drumlins are composed entirely of rock, they are known as **rock drumlins.** Streamlined hills that consist of proximal bedrock knobs tapering to till at the distal end are known as **crag-and-tail** forms.

Although drumlins have been extensively studied for many years (Alden, 1905; Boulton, 1987; Menzies, 1979a, 1979b;

Menzies and Rose, 1987; Muller, 1974; Whittcar and Mickelson, 1979), the exact nature of the process of their formation remains obscure, in part because they cannot be observed forming in modern glaciers. Because of their streamlined shape and the parallel orientation of their long axes, drumlins are clearly shaped by moving ice, but the variability of

FIGURE 13–25
Washboard moraines, small push moraines made by short winter advances of the ice margin of a retreating glacier, Vatnajokull, Iceland.

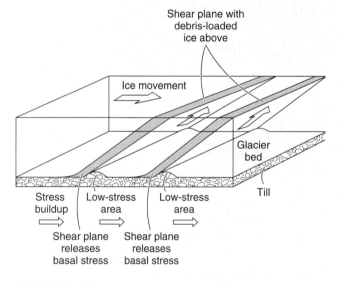

FIGURE 13–26
Rogen moraine formation. Debris accumulates at the base of shear planes in areas of compressive flow and migrates along pressure gradients. Pressure melting and lodgment of debris occur where shear planes leave the glacier bed. (From Sugden and John, 1976)

their internal composition indicates that they can be either depositional or erosional in origin. In considering the origin of drumlins, a clear distinction must be made between the origin of the shape of the drumlin and the origin of the material composing it. Bedrock, stratified sediment, or other older material in the cores of drumlins indicates reshaping of older, previously deposited material, thus indicating an erosional

origin of the shape (Gravenor, 1953; Kupsch, 1955; Lemke, 1958; Gravenor, 1974; Boulton, 1979).

Many drumlins are composed entirely of till. The till can be deposited as the drumlin is being shaped, or the till in the core of a drumlin could be from an earlier glaciation unrelated to the shape of the drumlin (Hill, 1971). As long ago as 1895, Russell suggested that drumlins could be formed by deposition of concentric layers of till at the base of a glacier by concentration of debris in basal ice. The debris decreases the rate of glacier flow until ice with less rock material overrides the debris-clogged ice and drift, shaping it into streamlined drumlins.

Drumlins composed of till of the same glaciation that shaped the drumlins and the surrounding drift between drumlins suggest a depositional origin. However, the fact that a drumlin is composed entirely of till does not prove that the shaping of the form was concomitant with the deposition; that is, it does not prove that the drumlins were shaped simultaneously with deposition of the till. Sometimes the relationship between the time of till deposition and the time of shaping can be shown by finding that the orientation of elongate pebbles in the till composing a drumlin is at an angle to the long axis of the drumlin, indicating that the glacier that deposited the till moved in a direction different from that of the glacier that shaped the drumlin (Andrews and King, 1968; Stanford and Mickelson, 1985).

Evenson (1971) suggested that shaping of drumlins might be related to low-pressure zones over the crest of a knob that encourage deposition from the flanks to the crest. However, this process would require some sort of preexisting knob under each drumlin in a drumlin field to generate conditions for drumlin formation. Boulton (1987) suggested that drumlins could form by subglacial sediment defor-

FIGURE 13–27
Drumlins, Waterville Plateau, Washington. Terminal moraine of the Cordilleran Ice Sheet in the background. (Photo by D.A. Rahm)

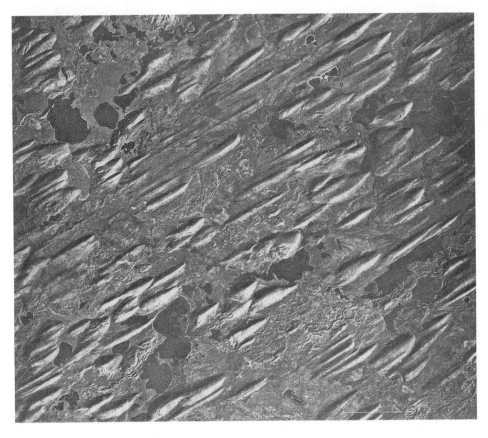

FIGURE 13–28
Drumlin field, northern Canada. (Photo by Canadian Dept. of Energy, Mines, and Resources)

mation, and Smalley and Unwin (1968) proposed that drumlin formation is related to dilatant properties of some tills under stress.

Obviously, no single process can be used to explain all drumlins. Some are clearly erosional and others are clearly depositional.

Flutes

Flutes are long, straight, narrow, parallel ridges that commonly are associated with drumlins but may also occur by themselves. They are typically only a few meters high, often a meter or less, but they may extend for hundreds of meters, giving the topography a fluted appearance. (Baranowski, 1970; Boulton, 1971, 1976, 1982; Boulton and Clark, 1990; Hoppe and Schytt, 1953). They are generally composed of till but may sometimes consist of sand or silt/clay. Some attain lengths up to several kilometers and heights of 1 to 10 m (~3 to 30 ft) in parts of North Dakota (Lemke, 1958) and in Alberta (Gravenor and Meneley, 1958). The long axes of the ridges are parallel to the direction of ice movement.

Studies of flute formation by modern glaciers in Iceland (Boulton, 1971, 1976) have demonstrated how they originate. Boulton (1976) described flutes in Iceland formed by squeezing up of water-soaked till into elongate cavities in the lee of boulders lodged on the subglacial floor (Figures 13–29 and 13–30). As boulders are gradually lowered to the subglacial floor by basal melting, they eventually become lodged in till and can no longer be moved by the ice. Small cavities are made in the wake of the boulders as they are

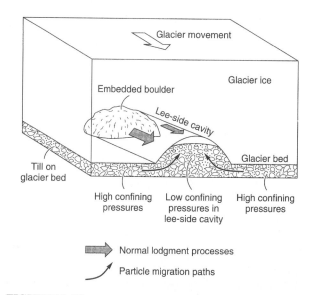

FIGURE 13–29
Flute formed by squeezing up of water-saturated sediment into the cavity in the lee of an embedded boulder. (From Sugden and John, 1976)

pushed along until finally emplaced. Once the boulders are emplaced, the ice must flow over and around them, producing elongate cavities in the ice parallel to the ice flow. Water-saturated till is then squeezed up into the cavities, making a long, low ridge. The most striking characteristic of the flutes in front of the Icelandic Breidamerkurjokull is that each of them can be traced to a single large boulder embedded in till (Figure 13–30).

FIGURE 13–30
Flute forming in the lee of a boulder embedded on the subglacial floor, Breidamerkurjokull, Iceland.

LANDFORMS ASSOCIATED WITH STAGNANT ICE

Sediments deposited on or against stagnant ice typically differ from sediments deposited from actively moving ice. Such sediments are commonly deposited in environments with much meltwater, and they typically lack the effects of deposition in the presence of subglacial shear. Thus, they usually have random clast fabrics and a low degree of compaction, and they may be deformed by collapse of the supporting ice upon melting.

Eskers

Eskers are elongate, sinuous ridges (Figure 3–31) composed primarily of sorted and stratified sand and gravel with water-worn, rounded clasts. They were recognized as early as 1866 by Close, and their origin has been discussed in many papers since then (for example, Crosby, 1902; Flint, 1928; Price, 1969; Cheel, 1982; Shreve, 1985).

Eskers may consist of a single, narrow, winding ridge, or they may unite with tributary eskers in complex branching patterns analogous to normal stream systems. They are commonly discontinuous, with many gaps separating sinuous segments of ridges. Some are closely associated with rounded hills, giving them a beaded appearance. The association of eskers with stagnant-ice features and observations of recently exposed eskers indicate that they form near the terminal zone of glaciers where ice is relatively thin and not vigorously moving.

Ridge crests are usually undulatory or knobby and are rarely level for very long. Shreve (1985) describes the crests of eskers of the Katahdin system, Maine, as sharp-crested, broad-crested, or multiple-crested. Sharp-crested eskers typically have a single, sharp, undulatory crest, approximately 20 m (65 ft) high and 150 m (490 ft) wide, with steep sides composed of sand, pebbles, cobbles, and boulders similar in composition to those in adjacent till, but sorted, rounded, and stratified. Broad-crested eskers are typically around 600 m (~2000 ft) wide and 10 m (~30 ft) high, but they may range up to 2 km (1.2 mi) wide. Multiple-crested eskers generally have from two to five or more parallel, sharp crests, making a system usually around 450 m (1500 ft) wide and 15 m (50 ft) high but sometimes up to 2 km (1.2 mi) wide.

Some eskers extend uphill, with reverse slopes in the direction of former water flow. A 500 km long (300 mi long) esker system west of Hudson Bay rises 275 m (900 ft) above its headwaters (Wilson, 1939; Shreve, 1985).

Eskers are typically composed of water-laid, coarse-grained, stratified, sorted sand and gravel, usually pebble/cobble-sized material, but with occasional boulders. Bedding is almost always present but may be irregular, and cross-bedding is common.

Although some eskers form by accumulation of sediment in supraglacier channels, in crevasses, in linear zones between stagnant ice blocks, or in narrow embayments at glacier margins, most eskers are believed to have been deposited in ice-walled tunnels at the base of glaciers on the basis of the nature of the water-washed sediment and their morphology. Water in superglacial channels, moulins, and englacial cavities collects in subglacial tunnels. The tunnels remain open only as long as inflow of the enclosing ice is balanced by melting of the ice making up the tunnel walls and by the hydrostatic pressure of the water filling the tunnels.

Shreve (1985) has suggested that the shape, composition, and structure of an esker are controlled by the size and shape of the subglacial tunnel, which in turn is determined by the rate of melting and plastic flow of the basal ice. The amount of sediment in an esker is related to the concentration of rock debris in the ice and the rate at which sediment is delivered to the tunnel by melting and from upstream transport.

Eskers may form in either confined or unconfined systems. Unconfined drainage may occur in supraglacial cavities or in some subglacial tunnels. Confined drainage occurs in subglacial tunnels that are hydrostatically closed, and the water may flow either uphill or downhill under hydrostatic pressure. Eskers that extend uphill and across divides may result from flow in hydrologically confined tunnels; or they may occur because the tunnels were partially floored by ice while they were being filled with sediment, and melting later dumped the material onto the land surface beneath the ice.

Kames

Kames include a whole family of stagnant-ice forms, many of which have more specific names. The term *kame* by itself is generally used for irregular mounds of sand, gravel, and till that have accumulated in depressions or cavities in or on a stagnating glacier (Figure 13–32A) and dumped helter-skelter on the land surface when the ice melts (Figures 13–32B and 13–33) (Clayton, 1964; Cook, 1946; Holmes, 1947).

Kames are usually found where a glacier has thinned from ablation so much that the ice has lost most or all of its ability to continue to move and dies in its tracks, eventually dumping its load of rock debris as irregular mounds and hills scattered among eskers (Figure 13–34).

Kettles

Kettles are depressions formed by the melting of buried ice (Figures 13–35 and 13–36). The size of a kettle depends on the size of the buried block of ice and may range from a few meters to more than a kilometer in diameter.

FIGURE 13–31
A. Sinuous subglacial tunnel at terminus of the Malaspina Glacier, Alaska. (Photo from Hartshorn and Ashley, 1972)

B. Sinuous esker formed by the Pleistocene Cordilleran Ice Sheet east of Mansfield, Washington. (Photo by D.A. Rahm)

B.

A.

FIGURE 13–32
A. Moulin being filled with rock debris, Breidamerkurjokull, Iceland.
B. Kame deposit on a Pleistocene moraine on the east flank of the Andes, Anfiteatro, Argentina.

FIGURE 13–33
A kame, an isolated mound of sediment deposited by the Malaspina Glacier, Alaska. (Photo from Hartshorn and Ashley, 1972)

FIGURE 13–34
Kames scattered among eskers on the Waterville Plateau, Washington. (Photo by D.A. Rahm)

FIGURE 13–35
Kettle forming from melting of a block of buried ice, Coleman Glacier, Mt. Baker, Washington.

FIGURE 13–36
Kettles in a kame terrace,
Columbia Plateau, Washington.
(Photo by D.A. Rahm)

Kame Terraces

Deposition of sediments in ice-marginal streams and lakes between stagnating ice and valley sides produces **kame terraces,** typically flat benches along valley sides, often with pitted surfaces made by numerous kettles (Figure 13–36). Collapse of the ice that supported one side during deposition commonly causes deformation of the sediment. Kame terraces usually slope downvalley with gradients similar to those of the glacier surface. They may be paired on opposite sides of a valley and may occur in successively lower sets, constructed as the ice surface downwasted.

Ice-walled Lake Plains

As motion diminishes toward zero on a glacier under a prolonged negative net mass balance, the ice stagnates and downwastes in its tracks. The glacier surface becomes mantled with an ever-thickening layer of rock debris, and the surface becomes very irregular, with many depressions and fissures (Figure 13–37A) (Reid, 1970; Hartshorn and Ashley, 1972). Meltwater lakes form in the ice-walled depressions and receive sediment from the melting ice. Crevasses

and meltwater channels (Figure 13–37B) in the ice also fill with rock debris from the melting ice. When the glacier melts completely away and this material is finally let down on the land surface, a collection of landforms results, which could reasonably be regarded as varieties of kames but which are given specific names, depending on their size, shape, and origin.

As supraglacial, ice-walled lakes on stagnant glaciers fill with sediment and the supporting ice melts away, inversion of topography takes place. The greater thickness of sediment on the lake floor than elsewhere on the glacier causes a hill to form when the ice melts away and leaves a thicker pile of sediment (Figure 13–38) (Gravenor and Kupsch, 1959; Clayton, 1967).

Disintegration Ridges and Trenches

A variation of the landforms created by accumulation of rock debris in ice-walled lake plains is the **circular disintegration ridge,** consisting of a circular ridge a few meters high and several tens of meters in diameter. Formation begins with accumulation of superglacial drift in a depression on

FIGURE 13–37
A. Ice-walled lakes on the
surface of the stagnating Bering
Glacier, Alaska. (Photo from
Hartshorn and Ashley, 1972)

FIGURE 13–37
B. Ice-walled lakes and streams flowing across the forest-
covered surface of the stagnating Malaspina Glacier,
Alaska. (Photo from Hartshorn and Ashley, 1972)

stagnant ice. As the ice melts away, the insulating effect of
the debris in the bottom of the depression produces an inversion of topography, leaving a circular, flat-topped hill of
material with an ice core (Figure 13–39). Melting of the ice
core then causes collapse of the middle of the hill, producing a central depression in the hill and making a circular
ridge somewhat resembling a doughnut (Figures 13–39 and
13–40) (Parizak, 1969).

Accumulation of rock debris in crevasses or superglacial
channels on a stagnant glacier leads to the development of
linear disintegration ridges a few meters high and a few tens
of meters long (Gravenor and Kupsch, 1959). A related form,
a **disintegration trench** (Figure 13–41), is formed when:

1. A superglacial channel or crevasse is filled with rock debris

2. The enclosing ice melts, and an ice-cored ridge remains

3. The ice-cored ridge is buried by sediment

4. Melting of the ice core causes collapse of the ridge, making a linear depression a meter or so deep and a few tens
   of meters long in the overlying sediment (Figure 13–41)

Kame Deltas and Fans

As the sizes of supraglacial lakes increase on a stagnating
glacier, sediment is carried into them by supraglacial streams,
building **kame deltas** on top of the ice (Figure 13–42). Melting away of the supporting ice then lowers the kame delta
onto the land surface, usually with accompanying sediment
deformation and pitting of the surface by kettles (Figure 13–
43). A variation of this situation occurs when a delta is built
into an ice-marginal lake. **Kame fans** are built by streams depositing material on or against stagnant ice rather than into a
lake (Figure 13–44).

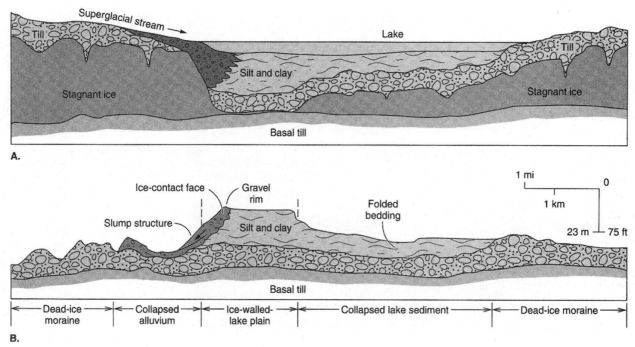

FIGURE 13–38
A. Cross sections illustrating the development of an ice-walled lake plain. Conditions at the time of deposition.
B. The resulting landforms. (From Clayton, 1967)

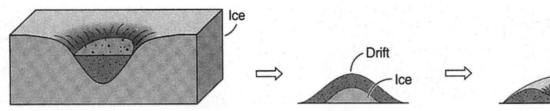

FIGURE 13–39
Formation of a circular disintegration ridge from an ice-walled lake. (After Clayton, 1967)

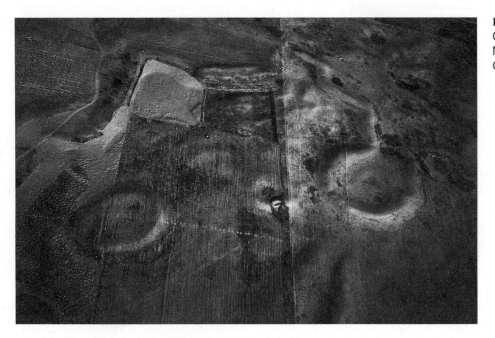

FIGURE 13–40
Circular disintegration ridges, North Dakota. (Photo by L. Clayton)

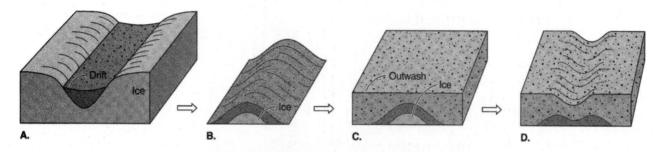

FIGURE 13–41
Formation of a disintegration trench from a buried crevasse filling. (After Clayton, 1967)

FIGURE 13–42
Kame delta being deposited against the Crillon Glacier, Alaska. (Photo from Hartshorn and Ashley, 1972)

FIGURE 13–43
Collapsed kame delta, Breidamerkurjokull, Iceland. The original flat surface of the kame delta has been almost completely disrupted by collapse of the underlying ice.

FIGURE 13–44
Esker ending at a kame fan at the margin of the Pleistocene Cordilleran Ice Sheet, near Sims Corner, Washington. (Photo by D.A. Rahm)

FIGURE 13–45
Outwash train from the Brady Glacier, Alaska. (Photo by A. Post, U.S. Geological Survey)

Outwash Plains

Deposition of sand and gravel by meltwater streams beyond the margin of a glacier produces flat, sloping **outwash plains** (Figure 13–45). The plains slope away from the glacier because of the gradient of the meltwater streams that deposit the sediments. Stream gradients can vary from a few meters per kilometer to 20 to 40 m/km (100 to 200 ft/mi), depending on the caliber of sediment transported and on the discharge of the meltwater streams (Gurnell and Clark, 1987; Gustavson and Boothroyd, 1987).

The surface of an outwash plain is commonly interrupted by kettles, especially near the terminus of the glacier, giving the outwash plain a pitted appearance. Unusually large kettles in an outwash plain occur when melting of stagnant ice detaches large blocks from the glacier, which are then surrounded by outwash sediments. Melting of the enclosed blocks can leave kettles that may be more than 1 km (0.6 mi) in diameter. Where outwash sediments are deposited over stagnant ice, wholesale collapse of the outwash plain may occur when the buried ice melts.

As a glacier retreats, the source of meltwater is lost, and the outwash streams may dry up completely or take a new channel, leaving abandoned channels on the outwash plain. Outwash channels cutting through previously deposited material may also be abandoned, leaving deeply incised, dry valleys.

REFERENCES

Alden, W. C., 1905, The drumlins of southeastern Wisconsin: U.S. Geological Survey Bulletin, v. 273, p. 9–46.

Anderson, L. W., 1978, Cirque glacier erosion rates and characteristics of neoglacial tills, Pangnirtung fjord area, Baffin Island, N.W.T., Canada: Arctic and Alpine Research, v. 10, p. 749–760.

Anderson, R. S., Hallet, B., and Walder, J., and Aubry, B. F., 1982, Observations in cavity beneath Grinnell Glacier: Earth Surfaces Processes and Landforms, v. 7, p. 63–70.

Andrews, J. T., 1963, Cross valley moraines of the Rimrock and Isotoq river valleys, Baffin Island: a descriptive analysis: Geographic Bulletin, v. 19, p. 49–77.

Andrews, J. T., 1972, Englacial debris in glaciers: Journal of Glaciology, v. 11, p. 155–156.

Andrews, J. T., and King, C. A. M., 1968, Comparative till fabric variability in a till sheet and a drumlin: a small-scale study: Proceedings of the Yorks Geological Society, v. 36, p. 435–461.

Andrews, J. T., and Lemasurier, W. E., 1973, Rates of Quaternary glacial erosion and corrie formation, Marie Byrd Land, Antarctica: Geology, v. 1, p. 75–80.

Aylsworth, J. M., and Shilts, W. W., 1989, Bedforms of the Keewatin Ice Sheet, Canada: in J. Menzies and J. Rose, eds., Subglacial Bedforms—Drumlins, Rogen Moraine and Associated Subglacial Bedforms: Sedimentary Geology, v. 62, p. 407–428.

Bakker, J. P., 1965, A forgotten factor in the interpretation of glacial stairways: Zeitschrift für Geomorphologie, v. 9, p. 18–34.

Baranowski, S., 1970, The origin of fluted moraine at the fronts of contemporary glaciers: Geographiska Annaler, v. 52, p. 68–75.

Battle, W. R. B., 1960, Temperature observation in bergschrunds and their relationship to frost shattering: in Lewis, W. V., ed., Norwegian cirque glaciers: Royal Geographic Society Research Series 4, p. 83–96.

Bouchard, M. A., 1989, Subglacial landforms and deposits in central and northern Quebec, Canada, with emphasis on Rogen moraines: in Menzies, J. and Rose, J. eds., Subglacial Bedforms Drumlins, Rogen Moraine and Associated Subglacial Bedforms, Sedimentary Geology, v. 62, p. 293–308.

Boulton, G. S., 1967, The development of a complex supraglacial moraine at the margin of Sorbreen, Ny Friesland, Vestspitsbergen: Journal of Glaciology, v. 6, p. 717–736.

Boulton, G. S., 1970a, On the origin and transport of englacial debris in Svalbard glaciers: Journal of Glaciology, v. 9, p. 213–229.

Boulton, G. S., 1970b, On the deposition of subglacial melt-out tills at the margins of certain Svalbard glaciers: Journal of Glaciology, v. 9, p. 231–245.

Boulton, G. S., 1971, Till genesis and fabric in Svalbard, Spitsbergen: in Goldthwait, R. P., ed., Till: a symposium: Ohio State University Press, Columbus, OH, p. 41–72.

Boulton, G. S., 1972a, Modem Arctic glaciers as depositional models for former ice sheets: Journal of Geological Society of London, v. 128, p. 361–393.

Boulton, G. S., 1972b, The role of thermal regime in glacial sedimentation—a general theory: in Price, K. I., and Sugden, G. P, eds., Polar geomorphology: Institute of British Geographers Special Paper 4, P. 1– 19.

Boulton, G. S., 1974, Processes and patterns of glacial erosion: in Coates, D. R., ed., Glacial geomorphology: State University of New York, Publications in Geomorphology, 5th Annual Symposium, Binghamton, N.Y., p. 41–87.

Boulton, G. S., 1976, The origin of glacially fluted surfaces—observation and theory: Journal of Glaciology, v. 17, p. 287–309.

Boulton, G. S., 1979, Processes of glacier erosion on different substrata: Journal of Glaciology, v. 23, p. 15–38.

Boulton, G. S., 1981, Deformation of subglacial sediments and its implications: Proceedings of the Symposium on Processes of Glacier Erosion and Sedimentation, Annals of Glaciology, v. 2, p. 114.

Boulton, G. S., 1982, Subglacial processes and the development of glacial bedforms: Proceedings, Guelph Symposium on Geomorphology, v. 6, p. 1–31.

Boulton, G. S., 1986, Push-moraines and glacier-contact fans in marine and terrestrial environments: Sedimentology, v. 33, p. 677–698.

Boulton, G. S., 1987, A theory of drumlin formation by subglacial sediment deformation: in Menzies, J., and Rose, J., eds., Drumlin symposium: A. A. Balkema, Rotterdam, Netherlands, p. 25–80.

Boulton, G. S., and Clark, C. D., 1990, A highly mobile Laurentide ice sheet revealed by satellite images of glacial lineations: Nature, v. 346, p. 813–817.

Boulton, G. S., and Deynoux, M., 1981, Sedimentation in glacial environments and the identification of tills and tillites in ancient sedimentary sequences: Precambrian Research, v. 15 (3–4), p. 397–422.

Boulton, G. S., and Eyles, N., 1979, Sedimentation by valley glaciers; a model and genetic classification: in Schluechter, C., ed., Moraines and varves; origin, genesis, classification: A. A. Balkema, Rotterdam, Netherlands, p. 11–23.

Boulton, G. S., and Vivian, R., 1973, Underneath the glaciers: Geographical Magazine, v. 45, p. 311–316.

Boulton, G. S., Dent, D. L., and Morris, E. M., 1974, Subglacial shearing and crushing, and the role of water pressures in tills from south-east Iceland: Geographiska Annaler, Series A, 56 A (3–4), P. 135–145.

Boyce, J. L, and Eyles, N., 1991, Drumlins carved by deforming till streams below the Laurentide ice sheet: Geology, v. 19, p. 787–790.

Cheel, R., 1982, The depositional history of an esker near Ottawa, Canada: Canadian Journal of Earth Sciences, v. 19, p. 1417–1427.

Chorley, R. J., 1959, The shape of drumlins: Journal of Glaciology, v. 3, p. 339–344.

Clayton, L., 1964, Karst topography on stagnant glaciers: Journal of Glaciology, v. 5, p. 107–112.

Clayton, L., 1967, Stagnant-glacial features of the Missouri Coteau in North Dakota: North Dakota Geological Survey Miscellaneous Series, v. 30, p. 25–46.

Cook, J. H., 1946, Kame complexes and perforation deposits: American Journal of Science, v. 244, p. 573–583.

Cotton, C. A., 1941, The longitudinal profiles of glaciated valleys: Journal of Geology, v. 49, p. 113–128.

Crary, A. P., 1966, Mechanism for fiord formation indicated by studies of an ice-covered inlet: Geological Society of America Bulletin, v. 77, p. 911–929.

Crosby, W. 0., 1902, Origin of eskers: American Geologist, v. 30, p. 1–38.

Dowdeswell, E.K., and Andrews, J. T., 1985, The fiords of Baffin Island: description and classification: in Andrews, J. T., ed., Quaternary environment: eastern Canadian Arctic, Baffin Bay and Western Greenland: Unwin and Allen, Boston, MA, p. 93–123.

Drewery, D., 1986, Glacial geologic processes: Edward Arnold, London, 276 p.

Elson, J. A., 1968, Washboard moraines and other minor moraine types: in Fairbridge, R. W., ed., Encyclopedia of geomorphology: Reinhold, N.Y., p. 1213–1219.

Embleton, C., and King, C. A. M., 1975, Glacial geomorphology: Edward Arnold, London, 583 p.

Emerson, B. K., 1898, Geology of old Hampshire County, Massachusetts: U.S. Geological Survey Monograph 29, p. 790.

Evenson, E. B., 1971, The relationship of macro- and microfabric of till and the genesis of glacial landforms in Jefferson County, Wisconsin: in Goldthwait, R. P., ed., Till: a symposium, Ohio State University Press, Columbus, OH, p. 345–364.

Fisher, J. E., 1948, The pressure-melting point of ice and the excavation of cirques and valley steps by glaciers: American Alpine Journal, v. 7, p. 62–72.

Fitzsimons, S. J. 1991. Supraglacial eskers in Antarctica: Geomorphology 4:293–99.

Flint, R. F., 1928, Eskers and crevasse fillings: American Journal of Science, v. 15, p. 410–416.

Flint, R. F., 1971, Glacial and Quaternary geology: John Wiley and Sons, N.Y., 892 p.

Gluckert, G., 1973, Two large drumlin fields in central Finland: Fennia, v. 120, 37 p.

Gravenor, C. P., 1953, The origin of drumlins: American Journal of Science, v. 251, p. 674–681.

Gravenor, C. P., 1974, The Yarmouth drumlin field, Nova Scotia, Canada: Journal of Glaciology, v. 13, p. 45–54.

Gravenor, C. P., and Kupsch, W. O., 1959, Ice disintegration features in western Canada: Journal of Geology, v. 67, p. 48–64.

Gravenor, C. P., and Meneley, W. A., 1958, Glacial flutings in central and northern Alberta: American Journal of Science, v. 256, p. 715–728.

Gurnell, A. M., and Clark, M. J., 1987, Glacio-fluvial sediment transfer: an alpine perspective: John Wiley and Sons, Chichester, England, 524 p.

Gustavson, T. C., and Boothroyd, J. C., 1987, A depositional model for outwash, sediment sources, and hydrologic characterists, Malaspina Glacier: a modern analog of the southeastern margin of the Laurentide Ice Sheet: Geological Society of America Bulletin, v. 99, p. 187–200.

Habbe, K. A., 1992, On the origin of the drumlins of the South German Alpine Foreland (II): the sediments underneath: Geomorphology, v. 6, p. 69–78.

Hanvey, P. M., 1992, Variable boulder concentrations in drumlins indicating diverse accretionary mechanisms examples from western Ireland: Geomorphology, v. 6, p. 69–78.

Hart, J. K., Hindmarsh, R. C. A., and Boulton, G. S., 1990, Styles of subglacial glaciotectonic deformation within the context of the Anglian Ice-Sheet: Earth Surface Processes and Landforms, v. 15, p. 227–241.

Hartshorn, J. H., and Ashley, G. M., 1972, Glacial environment and processes in southeastern Alaska: Technical Report No. 4-CRC, Coastal Research Center, University of Massachusetts, Amherst, MA, 69 p.

Hill, A. R., 1971, The internal composition and structure of drumlins in North Down and South Antrim, northern Ireland: Geographiska Annaler, v. 53, p. 14–31.

Holmes, C. D., 1947, Kames: American Journal of Science, v. 245, p. 240–249.

Holtedahl, H., 1967, Notes on the formation of fjords and fjord valleys: Geographiska Annaler, v. 49, p. 188–203.

Hoppe, G., and Schytt, V., 1953, Some observations on fluted moraine surfaces: Geographiska Annaler, v. 35, p. 105–115.

Johnson, D., 1941, The function of meltwater in cirque formation: Journal of Geomorphology, v. 4, p. 252–262.

Johnson, W. D., 1972, The profile of maturity in alpine glacial erosion: in Glaciers and glacial erosion: London, Macmillan, p. 70–78.

King, C. A. M., 1959, Geomorphology in Austerdalen, Norway: Journal of Geological Society of London, v. 125, p. 357–369.

Kujansuu, R., 1967, On the deglaciation of western Finnish Lapland: Finland, Commission Geologique (Geologinen Tutkimuslaitos) Bulletin No. 232, 98 p.

Kupsch, W. 0., 1955, Drumlins with jointed boulders near Dollard, Saskatchewan: Geological Society of America Bulletin, v. 66, p. 327–337.

Lee, H. A., 1963, Glacial fans in till from the Kirkland Lake fault: a method of gold exploration: Geological Survey Paper, 36 p.

Lemke, R. W., 1958, Narrow linear drumlins near Velva, North Dakota: American Journal of Science, v. 256, p. 270–283.

Lewis, W. V., 1938, The meltwater hypothesis of cirque formation: Geological Magazine, v. 75, p. 2249–2265.

Lewis, W. V., 1940, The function of meltwater in cirque formation: Geographical Review, v. 30, p. 64–83.

Lewis, W. V., 1947, Valley steps and glacial valley erosion: Institute of British Geographers Transactions, v. 14, p. 19–44.

Lewis, W. V., 1954, Pressure release and glacial erosion: Journal of Glaciology, v. 2, p. 417–422.

Lewis, W. V., ed., 1960, Norwegian cirque glaciers: Royal Geographical Society Research Series 4.

Loomis, S. B., 1970, Morphology and structure of an ice-cored medial moraine, Kaskawalsh Glacier, Yukon: Arctic Institute of North America Research Paper 57, 51 p.

Lundquist, J., 1969, Problems of the so-called Rogen moraine: Sveriges Geologiska Underssokning, C648, p. 1–32.

Matthes, F. E., 1900, Glacial sculpture of the Bighorn Mountains, Wyoming: U.S. Geological Survey 21st Annual Report, 1899–1900, Part 2, p. 167–190.

Matthes, F. E., 1930, Geologic history of the Yosemite Valley: U.S. Geological Survey Professional Paper 160, 137 p.

Matthes, F. E., 1972, Geologic history of the Yosemite Valley: Glaciers and Glacial Erosion, Macmillan, London, p. 92–118.

McCall, J. G., 1952, The internal structure of a cirque glacier; report on studies of the englacial movements and temperatures: Journal of Glaciology, v. 2, p. 122–130.

McCall, J. G., 1960, The flow characteristics of a cirque glacier and their effect on glacial structure and cirque formation: in Lewis, W. V., ed., Norwegian cirque glaciers: Royal Geographical Society Research Series 4, p. 39–62.

McKenzie, G. D., 1969, Observations on a collapsing kame terrace in Glacier Bay National Monument, S.E. Alaska: Journal of Glaciology, v. 8, p. 413–425.

Meierding, T. C., 1982, Late Pleistocene glacial equilibrium-line altitudes in the Colorado Front Range: a comparison of methods: Quaternary Research, v. 18, p. 289–310.

Menzies, J., 1979a, The mechanics of drumlin formation with particular reference to the change in pore-water content of the till: Journal of Glaciology, v. 22, p. 373–384.

Menzies, J., 1979b, A review of the literature on the formation of drumlins: Earth Science Reviews, v. 14, p. 315–359.

Menzies, J., 1989, Subglacial hydraulic conditions and their possible impact upon subglacial bed formation: in Menzies, J. and Rose, J., eds., Subglacial Bedforms—Drumlins, Rogen Moraine and Associated

Subglacial Bedforms, Sedimentary Geology, v. 62, p. 125–150.

Menzies, J., and Rose, J., 1987, Drumlins; trends and perspectives: Episodes, v. 10, p. 29–31.

Menzies, J., and Rose, J., 1989, Subglacial Bedforms—An introduction: Sedimentary Geology, v. 62, p. 117–122.

Mills, H. H., 1980, An analysis of drumlin forms in the northeastern and north-central United States: Geological Society of America Bulletin, v. 91, p. 2214–2289.

Moran, S., Clayton, L., Hooke, R. LeB., Fenton, M., and Andriashek, L., 1980, Glacier-bed landforms of the prairie region of North America. Journal of Glaciology, v. 25, p. 457–476.

Muller, E. H., 1974, Origin of drumlins: in Coates, D. R., ed., Glacial geomorphology: Binghampton, State University of New York, Binghamton Publications in Geomorphology, 5th Annual Symposium, p. 187–204.

Mullins, H. T., and Hinchey, E. J., 1989, Erosion and infill of New York Finger Lakes: implications for Laurentide Ice Sheet deglaciation: Geology, v. 17, p. 622–625.

Olyphant, G., 1981, Interaction among controls of cirque development: Sangre Cristo Mountain, Colorado: Journal of Glaciology, v. 27, p. 449–458.

Parizak, R., 1969, Glacial ice-contact rings and ridges: Geological Society of America Special Paper 123, p. 49–102.

Penck A., 1905, Climatic features in the land surface: American Journal of Science, v. 4, p. 165–174.

Porter, S. C., 1977, Present and past glaciation threshold in the Cascade Range, Washington, U.S.A.: topographic and climatic controls and paleoclimate implications: Journal of Glaciology, v. 18, p. 101–116.

Price, R. J., 1969, Moraines, sandar, kames, and eskers near Breidamerkurjokull, Iceland: Transactions of Institute of British Geography, v. 46, p. 17–43.

Price, R. J., 1970, Moraines at Fjallsjokull, Iceland: Arctic and Alpine Research, v. 2, p. 27–42.

Price, R. J., 1973, Glacial and fluvioglacial landforms: Oliver and Boyd Ltd., Edinburgh, U.K., 242 p.

Reed, B., Galvin, C. J., and Miller, J. P., 1962, Some aspects of drumlin geometry: American Journal of Science, v. 260, p. 200–210.

Reid, J. R., 1970, Geomorphology and glacial geology of the Martin River Glacier, Alaska: Arctic, v. 23, p. 254–267.

Roethlisberger, F., and Schneebeli, W., 1979, Genesis of lateral moraine complexes, demonstrated by fossil soils and trunks; indicators of postglacial climatic fluctuations: in Schlucter, C., ed., Moraines and varves: A. A. Balkema, Rotterdam, Netherlands, p. 387–419.

Russell, I. C., 1893, Malaspina Glacier: Journal of Geology, v. 1, p. 219[nds]245.

Russell, I. C., 1895, The influence of debris on the flow of glaciers: Journal of Geology, v. 3, 823–832.

Shoemaker, E. M., 1986, The formation of fjord thresholds: Journal of Glaciology, v. 32, p. 65–71.

Shreve, R. L., 1985, Esker characteristics in terms of glacier physics, Katahdin esker system, Maine: Geological Society of America Bulletin, v. 96, p. 639–646.

Smalley, I. J., 1966,Drumlin formation. A rheological model: Science, v. 151, p. 1379.

Smalley, I. J., 1981, Conjectures, hypotheses, and theories of drumlin formation: Journal of Glaciology, v. 27, p. 503–505.

Smalley, 1. J., and Unwin, D. J., 1968, The formation and shape of drumlins and their distribution and orientation in drumlin fields: Journal of Glaciology, v. 7, p. 377–390.

Souchez, R. A., and deGroote, J. M., 1985, Delta D-Delta-^{18}O relationships in ice formed by subglacial freezing; paleoclimatic implications: Journal of Glaciology, v. 31, p. 229–232.

Stanford, S. D., and Mickelson, P. M., 1985, Till fabric and deformational structures in drumlins near Waukesha Wisconsin, U. S. A.: Journal of Glaciology, v. 31, p. 220–228.

Sugden, D. E., and John, B. S., 1976, Glaciers and landscape: John Wiley and Sons, N.Y., 376 p.

Tarr, R. S., 1909, Some phenomena of the glacier margins in the Yakutat Bay region, Alaska: Zeitschrift für Gletscherkunde und Glazial Geologie, v. 3, p. 81 – 100.

Tarr, R. S., and Butler, R. S., 1909, The Yakutat Bay region, Alaska: U.S. Geological Survey Professional Paper 64, 178 p.

Tarr, R. S., and Martin, L., 1914, Alaskan glacier studies: National Geographic Society, Washington, D.C., 498 p.

Thorn, C. E., 1976, Quantitative evaluation of nivation in the Colorado Front Range: Geological Society of America Bulletin, v. 87, p. 1169–1178.

Thorn, C. E., and Hall, K., 1980, Nivation: an arctic-alpine comparison and reappraisal: Journal of Glaciology, v. 25, p. 109–124.

Trenhaile, A. S., 1971, Drumlins: their distribution, orientation and morphology: Canadian Geographer, v. 15, p. 113–126.

Trenhaile, A. S., 1977, Cirque elevation and Pleistocene snowlines: Zeitschrift für Geomorphologie, v. 21, p. 445–459.

Trenhaile, A. S., 1979, The morphology of valley steps in the Canadian Cordillera: Zeitschrift für Geomorphologie, v. 23, p. 27–44.

Vernon, P., 1966, Drumlins and Pleistocene ice flow over the Ards peninsula, Strangford Lough area, County Down, Ireland: Journal of Glaciology, v. 6, p. 401–409.

White, W. A., 1970, Erosion of cirques: Journal of Geology, v. 78, p. 123–126.

Whittcar, G. R., and Mickelson, D., 1979, Composition, internal structures, and a hypothesis of formation for drumlins, Waukesha County, Wisconsin, U.S.A.: Journal of Glaciology, v. 22, p. 357–371.

Williams, V. S., 1983, Present and former equilibrium-line altitudes near Mt. Everest, Nepal and Tibet: Arctic and Alpine Research, v. 15, p. 201–211.

Wilson, J. T., 1939, Eskers northeast of Great Slave Lake: Royal Society of Canada Transactions, Section IV, v. 33, p. 119–129.

Quaternary Climatic Changes and the Ice Ages

Late Pleistocene end moraine draped over a giant bar of the Missoula flood with a braided outwash terrace in the foreground, Moses Coulee, Washington. (Photo by D. A. Rahm)

364

INTRODUCTION

Recognition of the development of extensive mid- and high-latitude glaciers during ancient ice ages dates back to the early part of the nineteenth century, although the roots of such an idea began still earlier in the European Alps. The earliest recognition of previously more expansive glaciers was likely made by Swiss alpinists and peasants who could hardly escape the observation of polished and striated bedrock and boulders downvalley from identical features associated with active glaciers of the time. The first known written account seems to belong to Saussure, who observed granite boulders lying on limestone bedrock in the Swiss Alps in 1779. Saussure gave these boulders the name **erratic,** which we still use today, but he thought that they had been transported by water. In 1795, Hutton ascribed such boulders to transportation by glaciers that must have been much larger than modern glaciers. Venetz-Sitten, a Swiss engineer, presented a paper to the Helvetic Society of Natural History at Luzern in 1821 in which he proposed that the glaciers of the Alps had once been much more expansive. Soon thereafter, Esmark presented a similar idea for the alpine glaciers of Norway. Venetz-Sitten extended the concept of previously more extensive glaciers in the Alps to include the plains to the north. In 1832, a German professor, Bernhardi, carried the conclusions of Venetz-Sitten and Esmark even further by proposing that Scandinavian erratics in Germany had been carried there by a polar ice cap. Thus began the recognition of former ice ages.

In 1834, deCharpentier presented support for the views of Venetz-Sitten to the Helvetic Society, but a colleague, Louis Agassiz, a zoologist, remained unconvinced and set out in 1836 to the Rhone Valley with deCharpentier and Karl Schimper, a botanist, to disprove the idea. Schimper used the term *Eiszeit* (German for "ice age") for the glacier expansion. Instead of refuting the concept, Agassiz came away convinced and in 1837 proposed in a paper to the Helvetic Society that climatic changes had caused a great ice sheet to extend from the North Pole to the Alps during a great ice period. He published these ideas in 1840, shortly before more extensive evidence was published in a book by deCharpentier in 1841. The concept of an ice age spread to North America, where Hitchcock and Dana were instrumental in supporting the idea.

PLEISTOCENE GLACIATIONS

The Land Record

During the past 2.4 million years (m.y.), huge ice sheets have repeatedly built up in the high latitudes in response to worldwide climatic changes, and ice has covered up to 30 percent of the earth's land surface (Figure 14–1). The landforms of northern North America, northern Europe, Siberia, and many of the mountainous regions of the world are a direct result of glaciations caused by these climatic fluctuations. The Southern Hemisphere was also glaciated during the Pleistocene, but because most of the earth's surface there is oceanic, the effects of glaciation were not as widespread.

Although a detailed picture of the earliest glacial episodes is not yet completely clear, much is known now about the extent and chronology of the advance and retreat of continental ice sheets and alpine glaciers, based upon evidence from the deposits that they left behind. The extent of glacial advances is shown by end moraines, and the direction of ice movement is recorded in drumlins, flutes, grooves, and striations. Ice thickness can be reconstructed on the basis of maximum height of erratics, and the source of ice can often be deter-

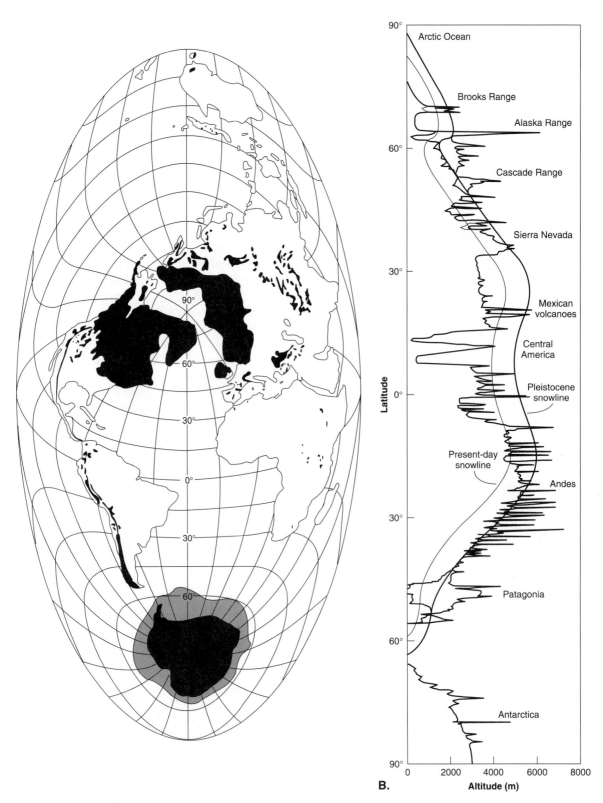

A.

B.

FIGURE 14–1

Distribution of Pleistocene ice sheets, alpine glaciers, and snowlines. Ice sheets and alpine glaciers expanded extensively in both hemispheres during the last ice age.

(A) Extent of land ice on the continents at the maximum of the last Pleistocene glaciation. Continental ice sheets extended seaward from some present coastlines because sea level was several hundred meters lower during glaciations.

(B) Average elevation of alpine snowlines in the mountains of North and South America during the last ice age. Pleistocene snowlines were about 1000 m (3300 ft) lower than present snowlines, regardless of latitude. (From Broecker and Denton, 1990)

mined from the provenance of the glacial sediments. Climatic changes related to these glacial episodes are recorded in oxygen isotope fluctuations in ice cores and deep-sea cores and in pollen extracted from sediments.

Great thicknesses of ice accumulated to form continental ice sheets greater than the present ice sheets in Antarctica and Greenland. One of the largest, the Laurentide Ice Sheet, grew to immense size in north-central Canada, extending from Hudson Bay into the United States (Figure 14–2) as far south as northern Missouri, Kentucky, Nebraska, Kansas, Illinois, Indiana, and Iowa (for example, see Eschman and Mickelson, 1986; Johnson, 1986; Matsch and Schneider, 1986; Richmond and Fullerton, 1986; Sibrava et al., 1986). For many years, glaciations that now bear the names of the states of Nebraska, Kansas, Illinois, and Wisconsin have been recognized throughout North America, and they have been used as a standard for correlations of glaciations elsewhere in the world. However, evidence regarding the two oldest, the Nebraskan and Kansan, has changed the picture significantly.

Although the term *Pleistocene* is intimately associated with ice-sheet glaciations, it was originally used by Lyell in 1839 on the basis of the fossil mollusk content of marine sediments in Europe. Because of the recognition of extensive glaciation during the Pleistocene (Penck and Bruckner, 1909; Kay and Apfel, 1929; Shimek, 1909; Chamberlin, 1895; Leverett, 1899), this epoch quickly became known as the Ice Age. Although ice ages are intimately associated with the Pleistocene, the definition of the Pleistocene is based not on glaciations but rather on a stratigraphic section in southern Italy, which dates back 1.65 m.y. However, evidence now suggests that the earliest late Cenozoic glaciations began about 2.2 to 2.4 m.y. ago (Boellstorff, 1978a, 1978b, 1978c; Zagwin, 1986).

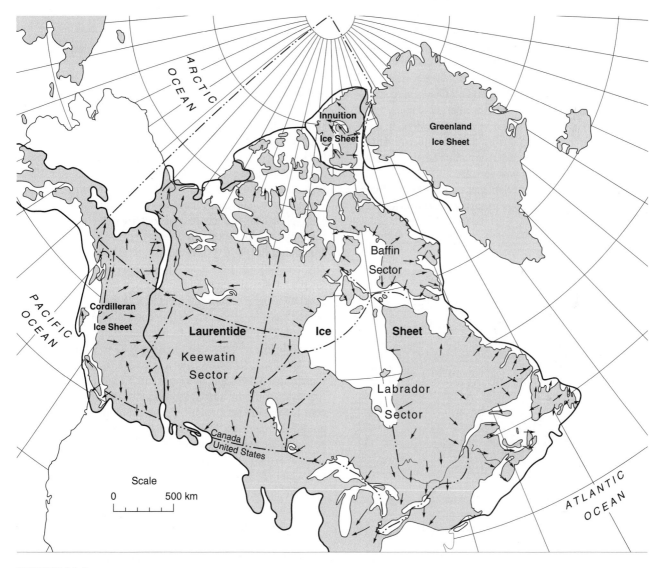

FIGURE 14–2
Approximate extent of the Laurentide, Cordelleran, and Greenland ice sheets during the last Pleistocene glaciation (Wisconsin). The arrows show directions of ice flow at the glacial maximum. (From Fulton, 1989)

The main centers of accumulations for the Pleistocene ice sheets were the broad uplands and lowlands around Hudson Bay in North America, central Scandinavia (Figure 14–3), and Siberia. Each of these great ice centers spread ice radially from areas of maximum ice buildup. Ironically, because the ice centers were south of the Arctic Ocean, the glaciers flowed in all directions, including northward toward the North Pole.

Early studies of glacial deposits in North America, beginning at the turn of century, suggested that four major periods of ice-sheet glaciation occurred during the Pleistocene, but recent work has shown that from 8 to 12 glaciations have, in fact, occurred (Boellstorff, 1976, 1978a, 1978b, 1978c; Easterbrook, 1986; Easterbrook and Boellstorff, 1981, 1984; Easterbrook et al., 1988; Westgate et al., 1987). Evidence from deep-sea cores suggests an even greater number of glacial events (Broecker and van Donk, 1970; Emiliani, 1955, 1966; Ericson et al., 1961; Ruddiman and McIntyre, 1976; Shackleton and Opdyke, 1973, 1976; Sibrava et al., 1986; van Donk, 1976).

The basis for subdivision of the Pleistocene is a composite effort utilizing different criteria for different parts of the Pleistocene because no single stratigraphic section encompasses deposits of all of the glacial and interglacial events. Consequently, establishment of a standard Quaternary stratigraphic nomenclature has evolved over a period of many years and has resulted in the grouping of glacial deposits into arbitrary time divisions, separated from those of other divisions by paleosols, nonglacial deposits, organic remains, unconformities, or magnetic polarity reversals.

The Pliocene/Pleistocene boundary, defined by the International Geological Correlation Program and the International Union for Quaternary Research, is at the Vrica section in southern Italy where the age of the lower Pleistocene boundary is 1.65 m.y. (Aguirre and Pasini, 1984).

The Matuyama-Brunhes magnetic polarity reversal has been used as the early Pleistocene/middle Pleistocene boundary (Richmond and Fullerton, 1986) at about 750,000 years. Middle Pleistocene Illinoian drift at its type locality in Illinois is used to mark the middle/late Pleistocene boundary. At least seven pre-Illinoian Pleistocene glaciations and two Pliocene glaciations have been recognized in the United States.

Early Pleistocene (1.65—0.7 m.y.)

The status of middle to early Pleistocene and late Pliocene glaciations (0.13 to 2.2 m.y.) in the Central Plains of the United States, where the early stages of glaciation were first defined in North America, is presently in somewhat of a state of confusion as a result of reevaluation of the chronology and correlation by Boellstorff (1978a, 1978b, 1978c). "Nebraskan" has long been used for the oldest glaciation in America, "Aftonian" for the first interglacial cycle, and "Kansan" for the second oldest glaciation, but this usage is now obsolete because tills in the classic areas of "Nebraskan-Aftonian-Kansan" deposits, where they were originally defined, have been shown to be younger than previously thought, underlain by older glacial deposits, and widely separated in age on the basis of fission-track dating of volcanic ash (Boellstorff, 1978a, 1978b, 1978c) and paleomagnetic measurements (Easterbrook, 1983, 1988; Easterbrook and Boellstorff, 1981, 1984). The chronology of glacial and paleomagnetic events is shown in Figure 14–4.

"Nebraskan" till, whose type locality in Nebraska was defined by Shimek (1909), lies beneath a thick section of loess containing Pearlette 0 volcanic ash, fission-track dated at 0.7 m.y. (Boellstorff, 1978a, 1978b, 1978c). The lithology of the till places it within a group of tills somewhat younger than Pearlette S ash, fission-track dated at 1.2 m.y. The till is reversely magnetized, which, in conjunction with the ash chronology, places its age in the late Matuyania Reversed Epoch (0.73 to .90 m.y. or more than 0.97 m.y.) (Easterbrook and Boellstorff, 1981, 1984). Thus, despite its long entrenchment in geologic literature for the earliest glaciation in North America, the term "Nebraskan" is clearly no longer usable for the oldest glaciation. The problem is further compounded because "Nebraskan" remains as a legitimate stratigraphic term for a younger glaciation.

A stratigraphic section near Afton, Iowa, described by Kay and Apfel (1929), consists of "Afton" soil underlain by "Nebraskan" till and overlain by "Kansan" till. Beneath this sequence, coring revealed sediments containing volcanic ash underlain by till lying on bedrock. The volcanic ash and underlying till are pre-"Nebraskan" because they lie beneath the classic Nebraskan-Aftonian-Kansan sequence. Glass from the ash yielded a fission-track age of about 2.2 m.y., so the underlying till predates type "Nebraskan" till by at least 1 m.y. (Boellstorff, 1978a, 1978b, 1978c). The pre-2.2 m.y. till is mostly

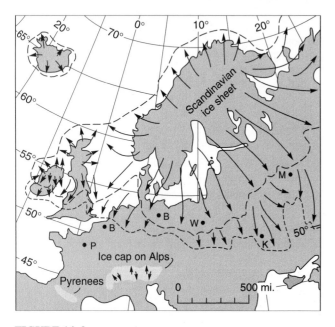

FIGURE 14–3
Approximate extent of the Scandinavian Ice Sheet during the last Pleistocene glaciation. The arrows show directions of ice flow at the glacial maximum. (From Strahler, 1963)

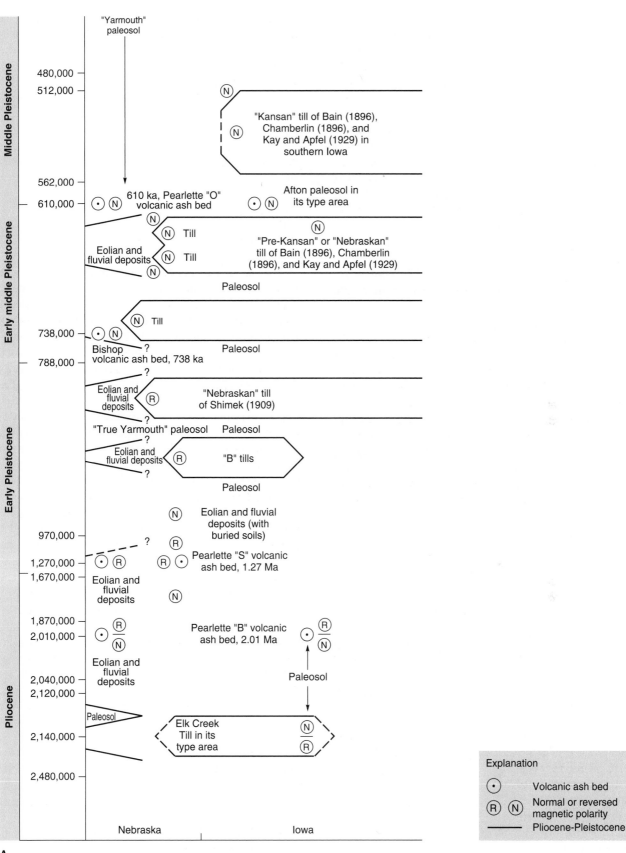

A.

FIGURE 14–4
Early and middle Pleistocene glaciations in the central United States. (From Richmond and Fullerton, 1986)

reversely magnetized (Easterbrook and Boellstorff, 1981, 1984). The earth's magnetic field was reversed during the lower Matuyama Epoch, so the till could be between 2.2 and 2.4 m.y.

Of six other tills in the area, the two oldest overlie Pearlette S ash (fission-track dated at 1.2 m.y.) and are reversely magnetized, placing them in the latest part of the Matuyarna Reversed Polarity Epoch; all of the four younger tills also lie above the Pearlette S ash, and one lies above the Pearlette O ash (fission-track dated at 0.6 m.y.); the three youngest of these tills are normally magnetized.

Based upon the fission-track and paleomagnetic dating of early and middle Pleistocene deposits in the Central Plains, we now know the following:

1. The early Pleistocene glacial sequence in the region is considerably more complex than previously thought.
2. "Nebraskan" till is not the oldest drift in the region.
3. "Nebraskan" tills at various classic sections are not the same age.
4. The oldest till is older than 2 m.y.
5. At least two post-Nebraskan tills are reversely magnetized and older than 0.73 m.y.
6. Two normally magnetized, still younger tills are 0.6–0.7 m.y. old, and another is younger than 0.6 m.y.

Three early Pleistocene tills dated at more than 1 m.y. are known from the Puget Lowland of Washington, where reversely magnetized glacial deposits lie beneath a volcanic ash, fission-track dated at 1.0 m.y. (Easterbrook et al., 1981, 1988; Westgate et al., 1987). Laser-argon dates of ash near the contact of the second oldest glacial drift and the underlying interglacial deposits indicate an age of 1.6 m.y. (Easterbrook, 1994a, 1994b).

Middle Pleistocene (0.7—0.13 m.y.)

The Matuyama-Brunhes magnetic polarity reversal marks the early Pleistocene/middle Pleistocene boundary at approximately 0.7 m.y. The early and middle portions of the middle Pleistocene are subdivided by the Pearlette O volcanic ash, derived from eruption of the Lava Creek Tuff in Yellowstone National Park; the ash was identified as a widespread marker horizon in stratigraphic sequences in many parts of the Rocky Mountains, the High Plains, and the Central Plains of the United States, and parts of southern Canada. Numerous fission-track ages of the ash range from 0.6 to 0.7 m.y. (Boellstorff, 1978a, 1978b, 1978c; Izett, 1981; Izett and Wilcox, 1982; Naeser et al., 1973), and the Lava Creek Tuff at its source is K-Ar dated at 0.61 m.y.

Illinoian drift deposited by the Lake Michigan lobe in Illinois (Leverett, 1899) is considered to represent the late middle Pleistocene in the United States. It is not directly dated; the 0.13 m.y. age for the end of the middle Pleistocene is derived from correlation of Illinoian glacial deposits with deep-sea isotope curves (which are also not directly dated). Illinoian

glacial sediments in the Puget Lowland of Washington have been directly dated by thermoluminescence at 167,000 to 185,000 years B.P. (Berger and Easterbrook, 1993) and amino-acid dated at 150,000 to 250,000 years B.P. (Blunt et al., 1987).

Late Pleistocene (130,000–10,000 Years)

The late Pleistocene was characterized by repeated advances and retreats of continental ice sheets. At least two major ice sheet advances, separated by a nonglacial climate, are well established by prominent moraines deposited during the Wisconsin Glaciation in North America, the Weichselian Glaciation in Europe, and the Zyryankan Glaciation in Siberia (for example, see Anderson, 1981; Sibrava et al., 1986).

The late Pleistocene began about 130,000 years ago with a prolonged interglacial period, during which the climate was similar to, or warmer than, the present. It is represented by widespread paleosols: the Sangamon in North America and the Eemian in western Europe (Dansgaard and Duplessy, 1981). Although the paleosols are not directly dated, sediments of the last interglacial in northwest Washington have been dated by thermoluminescence at 105,000 to 110,000 years B.P. (Easterbrook, 1994b). The Sangamon Interglacial is thought to correspond to marine oxygen isotope substage 5e, which, although not directly dated, is believed to have begun 130,000 years ago.

The term "Eowisconsin" has been applied to the time that followed the Sangamon and preceded widespread early Wisconsin glaciation (Richmond and Fullerton, 1986). In the United States, it is represented in places by deposits of two glacial advances that are separated by and also overlain by sediments containing pollen suggestive of climates nearly as warm as the present. Although "Eowisconsin" sediments have not been dated, they are thought to correspond to marine oxygen isotope substages 5d through 5a.

An early Wisconsin ice sheet advance formed in North America about 70,000 to 90,000 years ago. Amino-acid dates of 70,000 to 90,000 years B.P. have been obtained on deposits of this glaciation in the Puget Lowland (Easterbrook, 1986; Blunt et al., 1987). During the time that followed, the middle Wisconsin, the climate was characterized by alternating cool and warm episodes.

The last Pleistocene ice expansion occurred during the late Wisconsin, 10,000 to 20,000 years B.P., and is well dated by radiocarbon almost everywhere in the world. At about 17,000 to 18,000 years B.P., the Laurentide Ice Sheet extended to its maximum position in the region south of the Great Lakes where lobate end moraines mark the glacier terminus. As it retreated northward, the ice sheet built a series of looping end moraines (Figure 14–5).

A similar pattern of glaciation occurred in western Europe where the Scandinavian Ice Sheet moved southward into Denmark, northern Germany, and Poland, and adjacent areas during the late Weichselian Glaciation. A short period of climatic oscillation marked the late phases of deglaciation. A

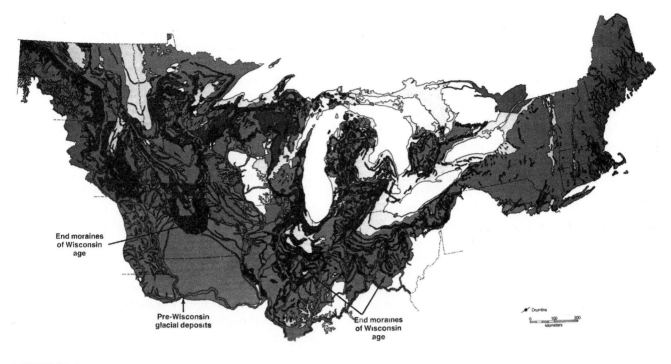

FIGURE 14–5
Map of late Pleistocene end moraines in the central United States. (From Geological Society of America map, 1959)

brief warm period, the Allerod, about 11,500 years B.P., was followed by a short glacial readvance, the Younger Dryas (Broecker et al., 1988; Dansgaard et al., 1989; Mercer, 1969, 1970, 1976; 1984; Paterson and Hammer, 1987; Siegenthaler et al., 1984). These short-term climatic oscillations are also found in the Southern Hemisphere and in western North America and have an especially important bearing on the Milankovitch theory of the cause of the ice ages (discussed later in the chapter). The Pleistocene ended abruptly with a dramatic climatic warming in both the Northern and the Southern hemispheres, beginning at 14,000 years B.P. By definition, the Pleistocene ended 10,000 years B.P., and the Holocene began.

Late Pleistocene Abrupt Climatic Changes:

The Problem of Abrupt Climatic Changes: At least three abrupt changes in the Earth's climate are known to have occurred near the end of the late Pleistocene: (1) sudden global warming 14–16,000 years ago that sent ice sheets into rapid retreat; (2) one or more abrupt climatic glacial readvances between ~11,000 and 13,000 years ago that interrupted retreat of the ice sheets after deglaciation was well underway, and (3) another abrupt climatic warming about 11,000 years ago that continued into the Holocene. Because dramatic climatic reversals such as these are not explained by Milankovitch orbital forcing, determining the extent, global or regional, and the causes of such events are critical to our understanding of the cause of climatic changes.

Although little doubt remains that rapid post-glacial reversals occurred, the exact timing and extent of these events around the world is one of the most important and strongly debated controversies in Quaternary geology today. One of the most precise records of late Pleistocene climate is found in the ice core stratigraphy of the Greenland Ice Sheet Project (GISP) and the Greenland Ice Core Project (GRIP) (Alley et al, 1993; Taylor et al., 1993). The GRIP ice core is especially important because the ages of the ice at various levels in the core have been determined by the counting down of annual layers in the ice, and climatic fluctuations have been determined by measurement of oxygen isotope ratios (see section that follows concerning oxygen isotopes).

The Younger Dryas Glacial Readvance: The Younger Dryas glacial readvance was characterized by a rapid ($<$ 100 year) cooling that persisted for about a thousand years, followed by a rapid ($<$ 100-year cooling) warming about 10–11,000 radiocarbon years ago (11,500–12,800 calendar years ago from Greenland ice stratigraphy) (Alley et al., 1993; Taylor et al., 1993). Evidence that mountain glaciers outside the North Atlantic region may have responded to regional climate changes at approximately the same time has been reported from western North America (Easterbrook and Kovanen, 1997; Kovanen and Easterbrook, 1996, 1997; Davis, 1994; Gosse et al., 1995a, 1995b; Osborn et al., 1995; Zielinski and Davis, 1987), the Southern Alps of New Zealand (Burrows, 1975, 1983, 1984, 1988; Mercer, 1984, 1988; Denton and Hendy, 1994; Lowell et al., 1995; McGlone, 1996), and southern South America (Heusser and Rabassa, 1987; Lowell et al., 1995). A few maintain that the late Pleistocene return to glacial conditions outside the North Atlantic preceded the Younger Dryas by more than 1000 years.

In an attempt to explain the rapid climatic reversal of the Younger Dryas, Broecker and Denton (1989, 1990) postulated that large amounts of fresh water were discharged into the north Atlantic about 11,500 years ago when retreat of the Laurentide ice sheet allowed drainage of glacial Lake Agassiz to spill eastward, rather than southward into the Mississippi River. They suggested that this large influx of fresh water might have stopped the formation of descending, higher-density water in the North Atlantic, thereby interrupting deepwater currents that distribute large amounts of heat globally and initiating a short-term return to glacial conditions. In this model, the cause of the sudden, Younger Dryas climatic reversal was initiated in the North Atlantic.

Oxygen isotope data from Greenland ice cores indicate that the climatic changes at the beginning and end of the Younger Dryas occurred within a few decades. Given the abruptness of climatic changes in the North Atlantic, if a global response was triggered via the ocean system, how much time elapsed before glaciers responded in the Southern Hemisphere, in the central part of North America, and on the northwest coast of North America? If glaciers responded much more quickly than the time required for oceanic distribution of thermal changes, then some other mechanism must be invoked to explain them. The possibility exists that an ocean/atmosphere system was responsible for triggering global cooling, but where the atmosphere is the dominant component, requiring essentially no lag time. The global timing of the Younger Dryas climatic event is of critical importance. As Denton and Hendy (1994) point out, if Younger Dryas events are synchronous in both hemispheres, then *the implication is that an atmospheric climatic signal was registered simultaneously in both polar hemispheres. This synchroneity is difficult to explain by a switch in North Atlantic deep-water production,* and *the source of Younger Dryas cooling* more probably *lies in the atmosphere, perhaps outside the North Atlantic region.*

If an abrupt change in the North Atlantic ocean-atmosphere system caused the late glacial (Younger Dryas) climatic reversal, a time lag should be noticeable in the Southern Hemisphere. If the Younger Dryas is globally synchronous, North Atlantic triggering of a global mode-switch in oceanic currents will be difficult to reconcile.

The ages of two of the abrupt climatic changes in the GISP2 core are 12,800 and 11,600 ($\pm$3%) years (~10,800 and 9,900 ^{14}C years B.P.) and are within 200 years of the GRIP ages. These ages are in close agreement with corrected radiocarbon ages of similar events in Europe. The dating of these events is largely undisputed because of the accurate methods of counting the annual snow accumulations in the ice cores.

Because the Younger Dryas event was first recorded and dated in Europe, its causes have been historically associated with changing physical conditions in the North Atlantic (Broecker et al., 1990; Weaver, 1994) or rapidly changing sea-ice cover (Johnsen, 1992; Dansgaard et al., 1989). Recent radiocarbon and cosmogenic dating of moraines in regions far removed from the North Atlantic suggests that the timing of major climate reversals in both hemispheres were essentially synchronous with one another. The apparent absence of evidence of the Younger Dryas in eastern and central North America had earlier led to a generally held belief that such an event did not affect North America. However, evidence for a late-glacial readvance of the Cordilleran Ice Sheet in western North America (Easterbrook, 1963, 1992; Easterbrook and Kovanen, 1997; Kovanen and Easterbrook, 1997), in the North Cascades Range of Washington (Kovanen and Easterbrook, 1996; Kovanen et al., 1996), in the Wind River Mountains of Wyoming (Gosse et al., 1995a, 1995b), in the British Columbia Coast Range (Mathews, 1993; Mathews and Heusser, 1993), and in the Canadian Rocky Mts. (Reasoner et al., 1994) indicate that the Younger Dryas event did indeed affect North America. Stable isotope and planktonic faunal records from northwest Pacific sediment cores (Duplessy et al., 1992; Kallel et al., 1988), pollen records from Alaska (Engstrom et al., 1990), and ^{14}C-constrained marine isotope records and terrestrial glacial deposits in both hemispheres (Kudrass et al., 1991; Mathewes, 1993; Mathewes et al., 1993; Heusser and Rabassa, 1987; Heusser, 1993) have suggested that the YD cooling was most likely a global event, although this conclusion remains hotly debated (Rind et al., 1986; Markgraf, 1989, 1991).

Rapid Glacial Responses to Abrupt Climatic Changes in the Northern Hemisphere: *The Late-Glacial Readvance of the Cordilleran Ice Sheet.* A brief readvance of the Cordilleran Ice Sheet into northern Washington during the Sumas Stade reversed the wholesale deterioration of the ice sheet that began shortly before ~14,500 ^{14}C-yrs. (Easterbrook, 1963, 1992; Armstrong, 1981; Armstrong et al., 1965). During the Sumas readvance, the Cordilleran Ice Sheet built two end moraines near the U.S.-Canadian border (Figure 14–6). Morphostratigraphic and radiocarbon evidence indicates that the ice sheet readvanced twice into the U.S. after 11,300 ^{14}C-yrs ago and did not retreat from the area until 10,000 ^{14}C-yrs ago, thus clearly overlapping with the Younger Dryas (Easterbrook and Kovanen, 1997; Kovanen and Easterbrook, 1997).

The older of the two Sumas moraines rests on sediments radiocarbon dated at 11,300$\pm$80 ^{14}C-yrs B.P. A 7 km (4.2 mi) long, outwash channel filled with ~10 m (33 ft) of peat cuts through and parallels the distal moraine margin. Radiocarbon ages of 11,000–11,400 ^{14}C-yrs B.P. from basal peat resting on the floor of the channel indicates that meltwater from the older Sumas advance must have ceased close to that time. [14]C ages of 10,400–11,100 ^{14}C-yrs B.P. from basal peat in an older Sumas kettle and outwash channel confirms the limiting age from the channel bog and ^{14}C ages of 11,500 from underlying sediments confirm the older limiting age from the

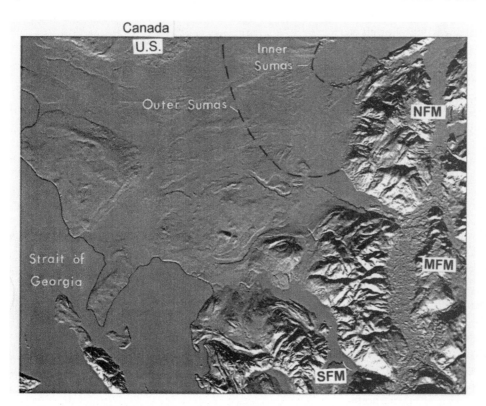

FIGURE 14–6
Glacial limits of two advances of the late Pleistocene Cordilleran Ice Sheet during the Sumas Stade and their relationship to alpine post-ice-sheet moraines in northwestern Washington. The outer Sumas glacial limit is bracketed by radiocarbon dates of 11,300 and 10,500 ^{14}C-yrs. B.P., and the younger, inner Sumas is bracketed by radiocarbon dates of 10,500 and 10,000 ^{14}C-yrs. B.P. NFM is the location of two alpine moraines in the North Cascades, one of which is the same age as the Sumas; MFM is the location of two alpine moraines, the older of which is radiocarbon dated at 12,000 ^{14}C-yrs. B.P.; SFM is the location of another alpine moraine. (From Kovanen and Easterbrook, 1996)

moraine. Thus, the age of the outer Sumas moraine must be younger than 11,300±80 ^{14}C-yrs and older than 10,500±80 ^{14}C-yrs., partially overlapping the age of the Younger Dryas.

The younger Sumas advance is represented by a second, well-defined moraine (Easterbrook, 1963) (Figure 14–6). This moraine is younger than the 10,500 ^{14}C-yr. kettle of the older Sumas advance and older than ~10,000 ^{14}C-yr. dates from basal peat in meltwater channels originating from the younger Sumas moraine. Thus, the age of the younger Sumas advance is between 10,500 and 10,000 ^{14}C-yrs. B.P., clearly overlapping the age of the Younger Dryas (Kovanen and Easterbrook, 1996, 1997; Easterbrook and Kovanen, 1997).

Late-Glacial Readvance of Alpine Glaciers in the North Cascades, Washington. Evidence for an extensive system of post-Cordilleran Ice-Sheet alpine glaciers not connected to the ice sheet has been found in the Nooksack Valley of the North Cascades of Washington immediately south of the Sumas readvance (Figure 14–6) (Kovanen and Easterbrook, 1996). This evidence consists of distinctive moraines and ice-contact deposits derived from local sources in the North Cascades by valley glaciers 23–45 km long that readvanced after disappearance of the Cordilleran Ice Sheet from the area. The presence of two moraines suggests a multiple response of the alpine glaciers to late Pleistocene climatic changes. The age of outermost moraine is limited by ^{14}C dates of 12,100±95 and 12,300±125 ^{14}C-yrs. from rootlets and wood in gravel at the base of a 20-m-deep peat bog in a kettle (Figure 14–7). Diatoms just above the organic material show that a deep lake formed in the kettle immediately following melting out of the ice, so the rootlets had to

be growing in rock debris on stagnant ice before the underlying ice melted out to make the kettle. Thus, the age of this moraine must be close to 12,000 ^{14}C-yrs. A large, younger, lateral moraine drapes across the kettled ice-contact deposits. In another fork of the Nooksack drainage, outwash, ^{14}C-dated at 10,600 ^{14}C-yrs., forms a terrace whose surface merges with Sumas outwash that rests on 11,500-yr.-old glaciomarine drift. Meltwater channels in the Sumas outwash were abandoned 10,000 ^{14}C-yrs. ago, indicating that the glacier had disappeared by then.

The Crowfoot glacial event, widely recognized in the Rocky Mountains of Canada, has been correlated with the Younger Dryas by Reasoner et al. (1994). The type locality of the Crowfoot moraine in the Rocky Mountains is bracketed by ^{14}C dates of 11,300 and 10,100 from lakes adjacent to the moraine. Pollen evidence of Younger Dryas climatic changes has also been found in British Columbia (Mathews et al., 1993).

However, late Pleistocene expansion of glaciers in the Sierra Nevada of California and in parts of the southern Cascade Range of Washington seems to have occurred somewhat earlier. Thus, the picture of the Younger Dryas and other late glacial advances in west-coast North America is not yet completely known (Gillespie and Molnar, 1995; Clark and Gillespie, 1997).

Late-Glacial Readvance of Alpine Glaciers in the Wind River Range, Wyoming. In the Wind River Range of Wyoming, [10]Be ages on bedrock surfaces associated with recessional moraines support the interpretation that ice had receded 35 km to cirque basins by 12,100±600 years B.P.

FIGURE 14–7
Two late Pleistocene moraines of the Nooksack alpine glacial system in the North Cascades, Washington. The irregular topography above KM is a kame-kettle moraine; P is a kettle containing a 20-m-deep peat bog whose base is radiocarbon dated at 12,300 ^{14}C-yrs. B.P.; LM is a younger lateral moraine that drapes across the kame-kettle moraine.

Cirque glaciers hesitated or re-expanded between 12,300 and 10,600 years ago and deposited two moraines 4 km from the cirque. Nine of ten boulder exposure ages plot between these dates (Table 14–1). A comparison with [10]Be measurements on a classic Swiss Alps Younger Dryas moraine (Ivy-Ochs et al.) shows excellent agreement of the [10]Be ages of the moraines (Table 14–1).

Rapid Glacial Responses to Climatic Changes in the Southern Hemisphere: *Late-Glacial Readvance of Alpine Glaciers in the Southern Alps of New Zealand.* Late-glacial moraines 20–40 km upvalley from terminal moraines of the last glaciation have been reported from a number of localities in New Zealand. The best known example, the Waiho Loop moraine, deposited by the Franz Josef Glacier about 20 km behind the moraine of the late-glacial maximum, represents a striking readvance of Franz Josef Glacier following large-scale recession of the glacier margin (Denton and Hendy, 1994; Lowell et al., 1995; McGlone, 1996; Mercer, 1988; Suggate, 1965; Wardle, 1978). Numerous ^{14}C dates have been obtained from wood preserved in glacial sediments 1.6 km upvalley from the Waiho Loop moraine (Denton and Hendy, 1994; Mercer, 1988; Wardle, 1978). Radiocarbon dates reported by Denton and Hendy (1994) from basal organic silt

TABLE 14–1.
[10]Be ages of the inner Titcomb moraines and the Swiss Alps.

[10]Be ages of moraines in the Wind River Range, Wyoming (Gosse et al., 1995a, 1995b)	[10]Be ages of moraines in the Swiss Alps (Ivy-Ochs et al.)
11,300	11,700
10,600	12,100
8,800	10,400
10,600	10,100
11,200	10,300
11,400	
11,700	
11,700	
11,300	
Mean=11,300	Mean=10,900

and 36 dates from three laboratories on 25 battered and broken pieces of wood from a diamicton yield a mean age of 11,200 ^{14}C yrs. Whether or not the Waiho Loop moraine is Younger Dryas or whether it predates the YD is still debated (Mabin, 1996). McGlone (1995) points out that a Cropp Valley advance (Lowell et al., 1995) is approximately the same age as the Younger Dryas.

Elsewhere in the New Zealand Alps, possible equivalents of YD-age moraines occur well upvalley from terminal moraines, but these moraines are not yet well dated. Prominent among such moraines are the Birch Hill and Prospect Hill moraines on the east side of the Southern Alps. The Birch Hill moraine (Figure 14–8) lies about 40 km upvalley from the terminal moraine, as does the Prospect Hill moraine (Figure 14–9).

Late-Glacial Readvance of Alpine Glaciers in the Southern Andes. Whether or not a cold, late-glacial, climatic change equivalent to the Younger Dryas occured in South America is still controversial. Caldenius (1932) mapped moraines several kilometers from the present ice front and well back from the terminal moraines of the last glaciation. Van der Hammen and Gonzalez (1960) inferred from pollen data that vegetation had fluctuated from wet and cold, to dry and warmer, to dry and cold after the last glacial maximum, and although not well constrained by ^{14}C, the changes were correlated with the late Pleistocene European climatic events. However, Mercer (1976) could find no morainal evidence of such fluctuations in the Southern Andes, and ^{14}C dates obtained from close to the Patagonian icefields, led him to conclude that the former ice cap had largely retreated by ~13,000 ^{14}C-yrs. and that only Neoglacial moraines were preserved since.

Evidence of late-glacial climatic fluctuations based on pollen and beetle studies has led to conflicting opinions. Pollen studies by Heusser and Streeter (1980) and Heusser and Rabassa (1987) suggested a cool and wet climatic interval at 11,0000–10,000 ^{14}C-ka, but other pollen studies (Markgraf,

FIGURE 14–8
Late Pleistocene end moraine 40 km (24 mi) upvalley from the late Pleistocene maximum stand, Birch Hill, New Zealand.

FIGURE 14–9
Late Pleistocene end moraine 40 km (24 mi) upvalley from the late Pleistocene maximum stand, Prospect Hill, New Zealand.

1989; Markgraf, 1991; Markgraf, 1993) suggested that the climate has remained consistently warm since 12,800 ^{14}C-ka. Studies of fossil beetles and pollen by Ashworth and Hoganson (1984), Ashworth et al. (1991), and Hoganson and Ashworth (1992) have not revealed any evidence of significant climatic deterioration between 12,000 and 10,000 ^{14}C-yrs.

Clapperton (1993) noted possible late-glacial moraines upvalley from terminal moraines of the late glacial maximum and younger than 14,000 years, but well downvalley from Neoglacial moraines, at Lago Nahuel Huapi in northern Patagonia. Mercer (1976) suggested that prominent moraines near Lago Argentina, well upvalley from terminal moraines of the last glaciation and ~20 km downvalley from Neoglacial moraines, could be late-glacial in age, but thought they could have been made by glaciers somewhat earlier (~14,000 yrs.). Geyh and Rothlisberger (1986) found a moraine older than 8,500 yrs. overlying a paleosol dated at ~13,000 yrs. that may represent a late-glacial advance of the Jose Glacier.

Moraines and outwash deltas 20 km from the present fronts of the Tyndall and Grey glaciers at the southern margin of the South Patagonian Icefield are older than 9,200 and may be late-glacial (Clapperton, 1993). Upper layers of the outwash contain reworked pumice clasts <10 cm in diameter that probably accumulated on the upper part of the glaciers during a large eruption of Reclus volcano ~10,330 yrs., implying that outlets of the South Patagonian Icefield were then 20 km downvalley from their present limits.

Clapperton (1993) and Osborn et al. (1995) note that in the entire Andes, of the few well-dated late glacial advances, the only ones documented between 12,500–10,000 ^{14}C-yrs. are Chimborazo (Ecuador) and Quelccaya ice cap (Peru) (Thompson et al., 1995). Elsewhere, ages of suspected late-glacial moraines have only been inferred on the basis of minimal limiting dates that are commonly younger than ~13,000 ^{14}C-yrs. and older than ~10,000 ^{14}C-yrs.. Morainal, palynological, and stratigraphic evidence of a clearly defined climatic reversal in the Andes between 12,500–10,000 ^{14}C-yrs. is *suggestive,* but not conclusive, and whether or not such an event was synchronous throughout the Andes remains debatable.

The Holocene

Climatic conditions also oscillated during the Holocene, but with far less amplitude than during the Pleistocene. A period of climatic warming, known as the *Hypsithermal,* has been widely documented in pollen profiles about 8500 years B.P., but recent data from ice cores in Greenland suggest that the early Holocene climate was by no means simple, including abrupt climatic changes (Alley et al., 1997; Obrien et al., 1995: Stuiver et al., 1995). Climatic cooling of 6±2°C is recorded in Greenland ice cores about 8200 years ago (~7500 ^{14}C yrs. ago) (Alley et al., 1993; Alley et al., 1997; Johnsen et al., 1992: Mayewski et al., 1994). The effect of this climatic cooling seems to be recorded in modest alpine glacial advances in several areas.

Oxygen isotope data and pollen records suggest that most of the remainder of the Holocene was not further interrupted by cooling until 2–3000 years ago when cirque glaciers showed modest expansion. Matthes (1900, 1930) informally referred to this cooling period as the "little ice age," and Moss (1951) later modified this to "Neoglaciation." More formal

designations were later made (Porter and Denton, 1967; Denton and Porter, 1970; Flint, 1971) to include the rebirth and/or growth of alpine glaciers following maximum shrinkage of ice during the Hypsithermal.

During the past several centuries, alpine glaciers have advanced and retreated over distances of hundreds of meters many times, displaying a well-defined pattern that also seems to be present in many parts of the world.

ISOSTATIC ADJUSTMENTS

That regions beneath the large continental glaciers were depressed by the weight of the ice during glaciations is now well established by several lines of evidence from North America and Scandinavia. The evidence includes the following (Andrews, 1970, 1987; Andrews and Barnett, 1972; Balling, 1980; Gutenberg, 1941; Farrand, 1962; Farrand and Gajda, 1962; Peltier, 1987):

1. The greatest amount of postglacial rebound coincides with areas of greatest ice thickness
2. The outer limits of uplift conform to the margins of the last glaciation (Figure 14–10)

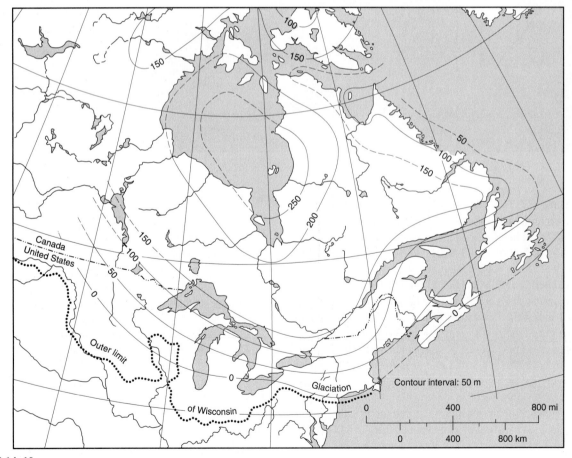

FIGURE 14–10
Isostatic rebound following retreat of the Laurentide Ice Sheet. (From King, 1965)

3. Lines of equal uplift are concentric around regions of greatest ice thickness and correspond with lines of equal ice thickness (Figure 14–10)

4. Negative gravity anomalies indicate deficiencies of mass below the surface, showing that isostatic equilibrium has not yet been attained

5. The areas occupied by the Laurentide and Scandinavian ice sheets are continuing to rise today at approximately the same rates

6. Marine terraces and sediments indicate that both areas were previously below sea level and have recently been uplifted

The area around Hudson Bay, depressed by the mass of the Laurentide Ice Sheet, is now rimmed by terraces up to altitudes of 250 m (825 ft) (Figures 14–10 and 14–11), and the area continues to rise. Estimates of the amount of uplift remaining to achieve isostatic balance suggest that another 100 to 150 meters (~300 to 500 feet) of uplift can be expected in the future.

The Scandinavian Ice Sheet similarly depressed Norway, Sweden, and Finland, whose coastlines are now rimmed with raised marine terraces, tilted southward due to the greater uplift in the north where the ice was thickest. As shown in Figure 14–12, Scandinavia is presently continuing to rise at a rate of about 1 m (3 ft) per century at the head of the Gulf of Bothnia, decreasing to zero in the southern Baltic area. The isostatic uplift has had some interesting effects on land-sea relationships in Sweden. The old Viking city of Stockholm was originally built on the Baltic where Viking sailing ships could access the sea. However, due to postglacial uplift, the city rose above the level of the Baltic, and the original embayment of the Baltic is now a lake. North of Stockholm along the Baltic, old castles, which used to be surrounded by moats connected to the Baltic, are now high and dry above the sea (Figure 14–13). The rate of continuing uplift is rapid enough that a historically documented mark chiseled in rock a meter above the Baltic in 1704 is now about 3 meters above the sea.

EUSTATIC SEA-LEVEL CHANGES

During expansion of the great Pleistocene ice sheets, enough water was tied up in the glacial ice to cause sea level to lower by about 120 m (400 ft) (Birkeland, 1972; Bull, 1985; Chapell, 1974, 1983; Chapell and Shackleton, 1986; Curray,

FIGURE 14–11
Marine terraces uplifted by isostatic rebound from the Laurentide Ice Sheet, Hudson Bay, Canada.

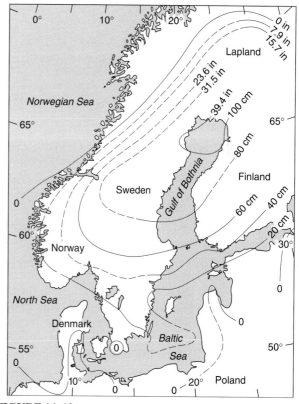

FIGURE 14–12
Present rate of uplift of the Baltic region shown by lines of equal uplift (cm/century/inches/century). (From Strahler, 1963)

FIGURE 14–13
Moat from an A.D. 1300 Swedish castle now 8 m (26 ft) above the Baltic as a result of continuing isostatic uplift following recession of the Scandinavian Ice Sheet near Sundsvall, Sweden. Water in the moat was originally connected to the water below.

1965; Fairbridge, 1961; Guilcher, 1969; Hopkins, 1967; Milliman and Emery, 1968; Morner, 1969, 1971). Intuitively, one would think that the best place to record **eustatic sea level** fluctuations would be a coastline very stable over long time periods. Not only are such coasts difficult to find, but, even if they exist, the effect of multiple shoreline changes would be superimposed upon one another in a fashion that would be almost impossible to decipher. The optimum condition for recording multiple sea-level changes is along a constantly rising shoreline where evidence for sea-level fluctuations during glacial and interglacial periods may be found in multiple terraces. However, because tectonic rates are known to be episodic rather than constant, separating out the tectonic from the eustatic components in a marine terrace sequence poses serious problems usually involving assumptions about uplift rates over thousands of years. High eustatic stands of sea level are much more likely to be preserved than low sea level stands, but subaerial organic material has been found below present sea level on continental shelves.

Figure 14–14 shows an example of sea level lowering on the Atlantic continental shelf of the United States during the last glaciation. Sea level began dropping about 25,000 to 30,000 years ago as the great ice sheets were building, reaching a minimum level some 120 m (400 ft) below present sea level 15,000 years ago. Rapid deglaciation beginning 14,000 years ago caused sea level to rise rapidly.

Calculations of the total ice volume during glacial episodes suggest that sea level should have dropped about 100 to 140 m (~300 to 450 ft) below present sea level. Some evidence suggests that during interglacials warmer than the present climate, sea level may have been up to 20 m (65 ft) above the present sea level. Establishment of worldwide, eustatic sea-level highstands has been attempted on the basis of uranium series dating of corals from tectonically deformed marine terraces in New Guinea and Barbados (Chappell, 1983; Bull, 1985). If uplift rates are high enough, terraces cut at former sea-level highstands are preserved above the present sea level, regardless of the elevation of those highstands relative to today's mean sea level. Most terraces cut at lower sea-level stands are now under water offshore.

Coral reefs on raised terraces on Barbados in the Caribbean have been dated by the uranium series method at 230,000, 170,000, 125,000, 105,000, 82,000, and 60,000 years. The ~80,000, ~100,000, and ~120,000 terraces correspond approximately with calculated insolation maxima (see the discussion of Milankovitch curves which follows) (Broecker, 1968; Mesolella et al., 1969).

Global, eustatic sea level interpretations are based on the following assumptions:

1. Eustatic sea-level fluctuations are essentially in phase.
2. The uplifted reef sequence on the Huon Peninsula of New Guinea (a) is correctly dated and (b) reflects the true nature of the timing and magnitude of eustatic events.
3. The local uplift rate, in New Guinea and elsewhere, is uniform over a geologically significant period of time.

Ice-volume, sea-level curves have been used to attempt eustatic sea-level correlations on a global scale. Terrace age assignments are based on the best matches between altitude sequences of local terraces and unique terrace altitude sets produced by applying uniform average uplift rates to known ages and altitudes of formation of New Guinea terraces.

PLUVIAL LAKES

Not long after the concept of great expansion of ice sheets during the Pleistocene was accepted, the effect of glacial climates on desert areas was recognized. In 1863, Jamieson wrote:

"Now this head and dryness being much lessened during the glacial period, there must have resulted a much smaller evaporation, which would no longer balance the inflow. These lakes would therefore swell and rise in level."

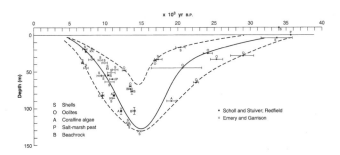

FIGURE 14–14
Late Pleistocene eustatic sea-level changes on the U.S. Atlantic continental shelf based on depths of sea-level indicators. The solid line is the inferred sea-level curve for the past 35,000 years; the dashed line is the envelope of values. (From Milliman and Emery, 1968)

Changes in evaporation rates and precipitation brought about by worldwide temperature lowering of several degrees Celsius during glaciations altered the delicate balance between evaporation and inflow of desert basins in many parts of the world and resulted in the development of **pluvial lakes** (Meinzer, 1922; Morrison and Frye, 1965; Gilbert, 1890; Russell, 1885, 1889). During interglacials, pluvial lake levels dropped as the evaporation/inflow ratio shifted back to negative with rising temperature. Evidence for former greater wetness in the Sahara Desert has been found in NASA X-ray photos showing ancient drainage channels now buried by sand.

The desert basins of the Great Basin in the southwestern United States held many pluvial lakes, the largest of which was Lake Bonneville, covering about 50,000 km^2 (19,000 mi^2) in western Utah (Figure 14–15). At its maximum, Lake Bonneville rose 300 m (1000 ft) above the level of Great Salt Lake (Gilbert, 1890) and spilled over a pass into the Snake River where outflow quickly incised through unconsolidated sediments, causing huge floods down the Snake River (Malde, 1968). The lake stabilized at a lower level and cut prominent shorelines along the margins of the lake basin (Plate 11–D and Figure 14–16).

Sediments that accumulate in pluvial lakes also record their fluctuations. For example, at Searles Lake in eastern California, glacial episodes are recorded as layers of mud deposited when the lake was high, whereas interglacial periods are represented by salt deposits left as the lake desiccated (Figure 14–17) (Smith et al., 1967). The mud contains freshwater mollusks that suggest temperatures about 4¡C lower than present and pollen that indicates forests in the vicinity of the lake. Radiocarbon dates suggest that deposition of the youngest mud began about 24,000 years ago and ended about 12,000 years ago, coinciding with the well-dated late Wisconsin glaciation.

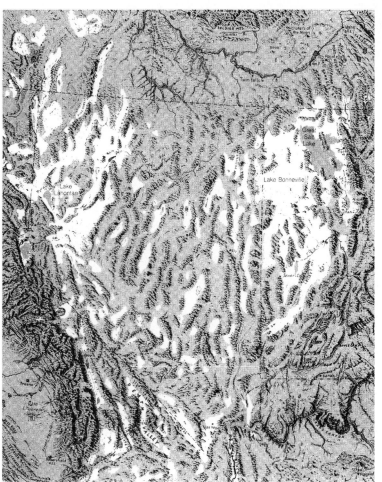

FIGURE 14–15
Pleistocene pluvial lakes of the southwestern United States. (From Hamblin, 1995)

FIGURE 14–16
Terraces of Pleistocene pluvial
Lake Bonneville against the
Wasatch Range south of Provo,
Utah. (Photo by D. A. Rahm)

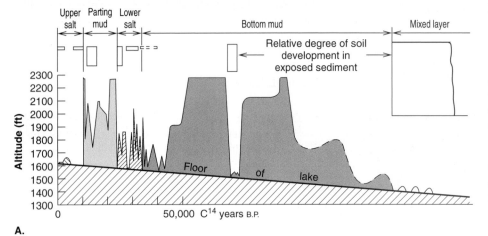

A.

FIGURE 14–17
Fluctuations of pluvial Searles
Lake, eastern California.
(A) Inferred fluctuation of the
level of Searles Lake. The
chronology of the past 45,000
years is based on radiocarbon
dates. Earlier events are
extrapolated.
(B) Subsurface stratigraphy of
Searles Lake. The age of the
base of the Bottom Mud is
estimated to be 130,000 years,
based on sedimentation rates.
(From Smith et al., 1967)

Unit	Depth (ft)		Lithology	Correlation	Dates (C^{14} yr B.P. approx.)
Overburden mud	0		Interbedded halite and mud	Late recent	
					6000
Upper salt	50		Salines, mostly halite, trona, hanksite, and borax	Early recent	
					10,000
Parting mud			Mud with gaylussite and pirssonite	Late Wisconsin	
					24,000
Lower salt	100		Saline layers, composed of trona, halite, burkeite, and borax, interbedded with mud containing gaylussite and pirssonite	Middle Wisconsin	
					33,000
Bottom mud	150 200		Mud containing gaylussite and a few thin beds of salines	Early Wisconsin	
Mixed layer	250 875		Salines composed of halite, trona, and nahcolite, grading down to mud	Sangamon (?) and Illinoian (?)	

B.

MISSOULA FLOODS AND THE CHANNEL SCABLANDS

Among the more impressive effects of Pleistocene glaciations was the scouring of the Channeled Scablands of the Columbia Plateau in eastern Washington, a network of anastomosing channels, cataracts, and waterfalls made by immense, catastrophic floods from the bursting of an ice-dammed lake in Montana (Bretz, 1923, 1969; Bretz et al., 1956; Baker, 1973, 1978, 1981; Baker and Bunker, 1985; Baker and Komar, 1987). A tongue of ice from the Cordilleran Ice Sheet of western Canada extended across the Clark Fork River of western Montana, blocking a major drainage system and impounding glacial Lake Missoula (Figure 14–18) with water depths up to 700 m (2300 ft). With each melting back of the ice sheet, the ice dam burst and released the most gigantic floods yet recognized anywhere in the world.

An estimated 2300 cubic kilometers (500 cubic miles) of water rushed across the Columbia Plateau west of the lake, carving out long, deep coulees and anastomosing channels in basalt (Figure 14–19), giving the land a scablike appearance. Many of even the deepest channels were filled to overflowing, and the floodwaters spilled across divides, cutting notches in ridge crests high above the valley floors. Waterfalls and cataracts formed quickly by intense quarrying of the columnar-jointed basalt, and these migrated rapidly upstream to leave large canyons such as Grand Coulee and Moses Coulee (Figure 14–20). One such falls, Dry Falls, remains today as a 120-m (400-ft) precipice nearly 6 km (3.5 mi) wide.

Huge gravel bars up to 100 m (~300 ft) high and 100 to 1000 m (~300 to 3000 ft) long occur on the inside of bends in the channel (Figures 14–20 and 14–21), in mid-channel, and across tributary mouths. Giant ripples with wavelengths of 100 m (300 ft) (Plate 3D and Figure 14–22) remain on the surface of some of the flood gravel. The relationship between the last giant flood that reamed out Moses Coulee and landforms of the last glaciation is shown in Figures 14–20 and 14–21. The end moraine of the last glaciation loops down into the coulee from the plateau above and drapes across a giant flood bar. An outwash terrace with braided channels still preserved extends down the coulee from the end moraine.

The gargantuan size of the floods, with so much greater erosion than any other event in recorded history, at first stretched the imagination. Bretz, who first wrote about these remarkable features in the 1920s, had a difficult time convincing people that it had indeed happened, in spite of the imprint that it left on the landscape. Perhaps the most convincing pieces of evidence were the giant ripple marks in cobble/boulder gravel (Figure 14–22) and divide crossings, which implied that multiple channels had to have been occupied with floodwaters simultaneously.

THE DEEP-SEA RECORD

Sediments of the deep-ocean basins carry a record of the earth's climatic changes during the Pleistocene (Broecker, 1981, 1982a, 1982b, 1986; Broecker and van Donk, 1970; Broecker et al., 1968, 1988; Emiliani, 1978; Emiliani and Shackleton, 1974;

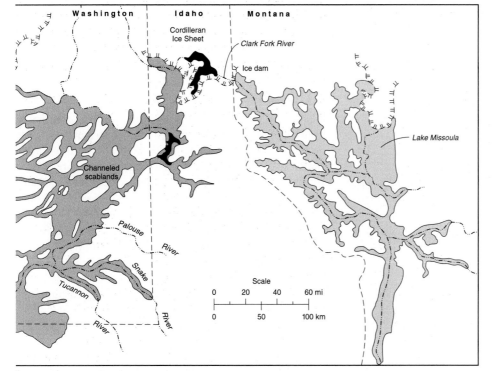

FIGURE 14–18
Map of glacial Lake Missoula and the Channeled Scablands, western Montana, northern Idaho, and eastern Washington. (Modified from Richmond et al., 1965)

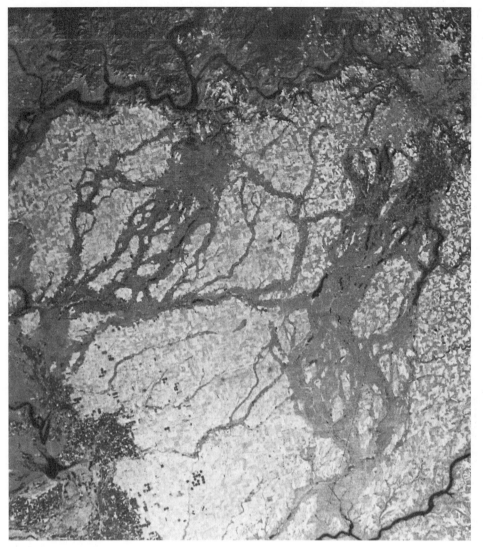

FIGURE 14–19
Satellite photo of the Channel Scablands of eastern Washington. Dark areas are anastomosing channels simultaneously occupied by floodwater from the bursting of the ice dam holding in glacial Lake Missoula. (Photo by NASA)

Epstein et al., 1951; Ericson and Wollin, 1968; Ericson et al., 1961; Hays et al., 1976; Imbrie et al., 1984; Imbrie and Kipp, 1971; Ruddiman, 1987a, 1987b; Ruddiman and McIntyre, 1976, 1981, 1984; Shackleton, 1977, 1987; Shackleton and Opdyke, 1973, 1976; Shackleton and Pisias, 1985; van Donk, 1976). This record is contained in deep-sea oozes, composed largely of foraminifera, radiolarians, diatoms, and coccoliths with various amounts of terrigenous mud. The material accumulated on the sea floor at rates of a few centimeters per year, adding from 6 to 11 billion metric tons of sediment annually.

Deep-sea sediments consist of both biogenic and terrigenous components. Biogenic sediments contain the microscopic skeletons of near-surface-dwelling (planktonic) and deep-water (benthic) organisms that record past climate and oceanic circulation. Although a great deal of research has been done on ocean cores in attempts to reconstruct past climates, and although the results indicate that the oceans underwent warming and cooling during Pleistocene climatic changes, some aspects of paleoclimatic research still remain equivocal to controversial. Most deep-sea sediments

contain biogenic material that serves as the basis for many paleoclimatic interpretations. However, the skeletal remains of organisms may not be truly representative of the assemblages of organisms living in the overlying water column, because differential scouring of easily transported species by bottom currents, selective dissolution of thin-walled shells at depth, and occasional contamination by exotic species transported from far away by large-scale ocean currents all introduce elements of uncertainty about the significance of organisms in ocean cores. Some organisms change the water depths at which they live during different times in their life cycles, a factor of special importance to isotopic measurements of calcareous shells because the oxygen isotope composition depends on the water temperature at which the carbonate is secreted; thus, if carbonate is added at varying depths through the life span of an organism, interpretations of water temperatures and salinities may be erroneous.

Sedimentation on the ocean floors was once thought to be continuous and to hold a history of the entire Quater-

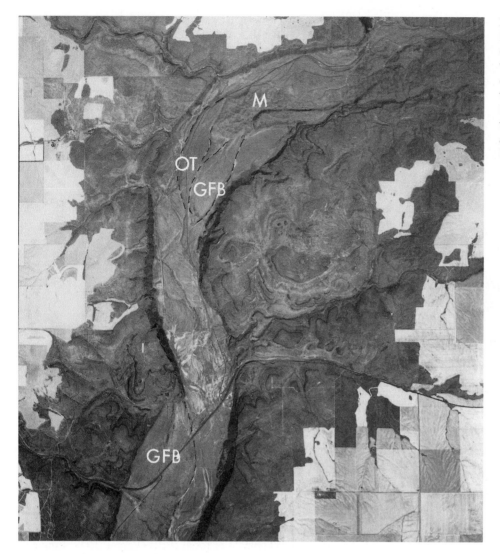

FIGURE 14–20
Moses Coulee, a large, sinuous, dry canyon cut by floodwaters from glacial Lake Missoula. GFB consists of giant flood bars from the Missoula floods; M is a late Pleistocene moraine deposited by the Cordilleran Ice Sheet; OT is an outwash terrace. (Photo by U.S. Dept. of Agriculture)

nary climatic changes, but the discovery of turbidity currents casts some doubts on exactly how uninterrupted deposition has been. Nonetheless, deep-sea cores have produced some extraordinary evidence of climatic fluctuations during the Quaternary. The climatic changes are best shown by the following:

1. Coiling directions of certain species of foraminifera (some warm-water forms coil in one direction, whereas some cold-water species coil in the opposite direction)

2. The relative abundance of warm- and cold-water species

3. Oxygen isotope ratios in shells, reflecting water temperature and fresh-water dilution differences

4. Morphological variations in a particular species resulting from environmental factors

The magnitude of Pleistocene temperature changes may be estimated from measurements of the oxygen isotope ratios of calcareous microfauna, largely foraminifera. When calcium carbonate ($CaCO_3$) is crystallized in water, ^{18}O is slightly con-

centrated in the $CaCO_3$ relative to that in the water. The ratio of ^{18}O to ^{16}O in fossil shells depends upon ocean water temperature at the time of formation (Urey, 1947, 1948). Fractionation of the two isotopes results from the higher vapor pressure of H_2O, which allows it to evaporate more readily from the ocean surface than the heavier $H_2^{18}O$. Calcium carbonate in shells is relatively enriched in ^{18}O in cooler ocean temperatures and when much ocean water is stored in ice sheets during glacial episodes. The lighter $H_2^{16}O$ molecule is preferentially evaporated from the ocean, transported as water vapor over land, and held in the large Pleistocene ice sheets, leaving a greater concentration of the heavier isotope in the ocean. The $^{18}O/^{16}O$ ratio in fossil shells can be measured with a mass spectrometer as a means of detecting such fractionation.

The oxygen isotopic composition of a sample is expressed as a departure of the $^{18}O/^{16}O$ ratio from an arbitrary standard, SMOW (standard mean ocean water).

$$S = \frac{(^{18}O/^{16}O)_{\text{sample}} - (^{18}O/^{16}O) \times 10^3}{(^{18}O/^{16}O)_{\text{standard}}}$$

FIGURE 14–21
Late Pleistocene moraine of the Cordilleran Ice Sheet draped across a giant flood bar in Moses Coulee. Note the well-preserved braided channels on the outwash terrace in the foreground. (Photo by D.A. Rahm)

where the ratio is expressed in per mil (0/00) units, negative values represent isotopically lighter samples with less ^{18}O than ^{16}O, and positive values are isotopically heavier with more ^{18}O than ^{16}O.

The relationship of the isotopic composition of $CaCO_3$, in marine shells to the temperature at the time of $CaCO_3$ deposition is given by the following equation:

$$T = 16.9 - 4.2(\delta_c - \delta_w) + 0.13(\delta_c - \delta_w)^2$$

where   T   = water temperature in degrees Celsius,
         δ_c = per mil difference between the sample carbonate and the SMOW standard
         δ_w = per mil difference between the δ ^{18}O of water in which the sample was precipitated and the SMOW standard (Epstein et al., 1951)

Although δ_w can be measured directly in ocean water for modern samples, the isotopic composition of the water from which fossil shells were precipitated is not necessarily the same as it is today. During glaciations, the removal of isotopically light water ($H_2{}^{16}O$) from the oceans to form continental ice sheets led to an increase in the $^{18}O/^{16}O$ ratio of the oceans. As the great ice sheets melted and returned their water to the oceans, the $^{18}O/^{16}O$ was again diluted. Thus, the effects of

both temperature and dilution/concentration are contained in the $^{18}O/^{16}O$ content of fossil shells (Berger, 1978; Emiliani, 1955, 1966). Because they both operate in the same direction —that is, they serve to reinforce one another—the fractionation between glacial and interglacial climates is enhanced. However, what proportion of the fractionation is attributable to temperature change and what to isotopic concentration/dilution is difficult to separate, making calculation of temperature differences uncertain. Emiliani (1955, 1966) estimated that about *70* percent of the $^{18}O/^{16}O$ ratio measured in shells from deep-sea cores was due to temperature and the rest to the dilution effect. This would correspond to about a 5°C to 6°C lowering of temperature during glacial episodes. However, Dansgaard and Tauber (1969) argue that the present isotopic composition of precipitation in the world, and the measured isotopic composition of ice-age precipitation in ice cores, point to low ^{18}O concentrations in continental and polar ice sheets; and they estimate the isotopic composition of ice-age ocean water to be due about 70 percent to dilution and 30 percent to temperature change. Similar conclusions were reached by Shackleton (1977).

In any event, the measurable $^{18}O/^{16}O$ ratio of foraminifera shells from deep-ocean cores reflects glacial and interglacial climates quite well; this ratio has led to the recognition of

FIGURE 14–22
Giant ripple marks left by the Missoula flood. (Photo by D.A. Rahm)

and further compounds the difficult problem of determining isotopic paleotemperatures.

Despite these problems, oxygen isotope analyses have been made on planktonic and benthic species from cores in all of the world's major oceans, and similar isotopic (also) variations have been found in them all, leading to the conclusion that they are all caused by the same glacial and interglacial conditions. If the variations in $^{18}O/^{16}O$ are indeed synchronous, then they will possibly provide global horizons that could be correlated worldwide. To do this, however, requires some kind of independent dating method for at least some of the cores. Radiocarbon is limited to only the upper part of the cores (Duplessy et al., 1986), and no reliable dating method is easily applicable to the older parts of the cores. To get around this problem, paleomagnetic measurements (discussed further in Chapter 18) have been made to identify the last major magnetic reversal in the sediment and to arbitrarily assume that it represents the Brunhes/Matuyama Polarity Epoch boundary, dated elsewhere at about 730,000 years B.P. The depth of sediment overlying this horizon is then divided by 730,000 years to determine the average sedimentation rate for the core. If a constant sedimentation rate is then assumed, the age of any level of the core can be calculated. If either of the two assumptions is incorrect—that is, the identity of the first magnetic reversal in the core or the constancy of the sedimentation rate—the age of oxygen isotope changes will be wrong. To check the validity of ages assigned to oxygen isotope changes, isotope stages, representing glacial and interglacial intervals, were established from the core data and compared with dated glacial and interglacial phase on land. For example, the interglacial sea-level highstands recorded on Barbados, dated at 82,000, 103,000, and 125,000 years B.P. by uranium series analyses of uplifted corals (Mesolella et al., 1969), correlate with interglacial oxygen isotope values in foraminifera from deep-sea cores (Shackleton and Opdyke, 1973; Shackleton, 1977).

Warmer interglacial and nonglacial periods were assigned odd numbers, starting with the present interglacial as 1, and colder, glacial intervals were assigned even numbers (Figure 14–23). Five marine oxygen isotope stages have been recognized in the past approximately 130,000 years. Stage 5 is divided into several substages and is assigned letter designations 5a to 5e. Substage 5e marks the peak of the last interglacial warm phase (Shackleton, 1977). Older marine oxygen isotope stages are more difficult to compare with glaciations recorded on land, and precise confirmations are not firmly grounded with independent age checks. Notwithstanding the lack of independent, accurate dating, comparison of oxygen isotope records in sediment cores from the Pacific and Atlantic oceans and the Caribbean Sea has led to the establishment of a marine oxygen isotope time scale (Emiliani and Shackleton, 1974; Shackleton and Opdyke 1973, 1976). Although the system stands on the sometimes shaky ground of its built-in assumptions, it serves as a highly useful composite standard for comparison. Like the original geologic time scale, it will undoubtedly continue to evolve as additional data and age checks are established.

glacial and interglacial periods during the last 425,000 years, and perhaps as many as 21 glaciations during the entire Pleistocene (van Donk, 1976).

Another complicating factor in calculating water temperatures from the isotopic composition of carbonate shells is the variation in depth habitat of foraminifera from glacial to interglacial times. Water temperatures change rapidly with depth in the upper few hundred meters of the ocean, so small changes in depth habitat can be the same as a temperature change of several degrees Celsius, perhaps equivalent to the glacial-to-interglacial difference at the ocean surface. To complicate matters even further, shells of living foraminifera contain $CaCO_3$ that has been produced by organisms in isotopic equilibrium with the upper water column. Measurements show that foraminifera shells on the sea floor are significantly enriched with ^{18}O compared to living foraminifera. (Duplessy et al., 1981), apparently from calcification of shells at depths greater than 300 m. This enriched ^{18}O content affects the isotopic composition of foraminifera shells as a whole

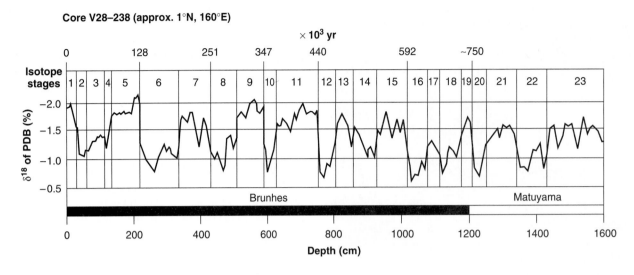

FIGURE 14–23
Oxygen isotope and paleomagnetic record of the last 1.6 million years in a deep-sea core from the equatorial Pacific. (After Shakleton and Opdyke, 1976)

A particularly interesting pattern of periods of gradual increase in $\delta^{18}O$, separated by shorter, relatively abrupt episodes of increased $^{18}O/^{16}O$ values, gives the marine oxygen isotope curves an asymmetric, saw-toothed appearance in which the gradual increase in ^{18}O apparently results from the slow buildup of ice sheets on the continents, followed by periods of very rapid deglaciation when isotopically light water is rather quickly returned to the oceans (Broecker and van Donk, 1970). These rapid decreases in $\delta^{18}O$ have been referred to as *terminations* and assigned numbers by Broecker and van Donk (1970), beginning with the end of the most recent deglaciation (Termination 1).

The marine oxygen isotope record is based on assumed continuous and constant sedimentation rates and appears to represent global ice volume changes. Thus, proposals have been made to use it as a standard reference for both marine and terrestrial deposits, in part because the land record of glacial sediments is rarely continuous. So far this has not been done, largely because of the lack of methods to connect local stratigraphies with the marine isotopic record and because of the difficulties in establishing independent ages for the deep-sea cores.

ICE CORES

The great ice sheets of Antarctica and Greenland contain a record of past climatic changes in the thick layers of snow and ice accumulated over thousands of years. The $^{18}O/^{16}O$ ratio in the ice can be used to identify past climatic changes; analysis of trapped gases in the ice allows measurement of the ancient CO_2 content of the atmosphere; and fluctuations in the rate of eolian dust influx and other atmospheric pa-

rameters can be determined from the ice (Barnola et al., 1987; Beer et al., 1983; Budd and Morgan, 1977; Dansgaard, and Tauber, 1969; Dansgaard et al., 1969, 1970, 1971, 1982, 1984, 1989; Delmas et al., 1980; Epstein et al., 1970; Hammer et al., 1985; Johnsen et al., 1972; Jouzel et al., 1987a, 1987b, 1989; Langway, 1970; Langway et al., 1985; Lorius et al., 1979, 1984, 1985; Mix, 1987; Mix and Ruddiman, 1984, 1985; Neftel et al., 1982, 1988; Oeschger et al., 1983; Paterson and Hammer, 1987; Raynaud et al., 1988; Robin, 1977, 1983; Stauffer et al., 1984; Cuffey et al., 1994, 1995; Grootes et al., 1993; Stuiver et al., 1995; Alley et al., 1997).

Many ice cores a few hundred meters long have been extracted from the Antarctic and Greenland ice sheets, and several long ice cores have been drilled. The longest ice cores include the Greenland Camp Century (1387 m), and the Antarctic Byrd (2164 m), Vostock (2083 m), and Dome Circle (905 m) cores.

Deep-ice cores from the Greenland and Antarctic ice sheets have yielded much valuable information about paleoclimates of the earth. Measurements of the $^{18}O/^{16}O$ content of the Camp Century ice core in Greenland revealed a remarkable record of Pleistocene climatic fluctuations (Figure 14–24). The $^{18}O/^{16}O$ measurements clearly show a sharp climatic change at 13,000 to 14,000 years that corresponds to the sudden retreat of the great ice sheets, as well as a number of other climatic oscillations that match reasonably well with known glacial advances and retreats (Figures 14–25 and 14–26). The Greenland core $^{18}O/^{16}O$ curve of the Byrd Station ice core in the Antarctic shows that climatic changes were synchronous in both the Northern and the Southern hemispheres (Figure 14–24).

Measurement of CO_2 and dust content of the Vostock and Dome Circle ice cores provide additional indications of paleoclimatic changes (Figure 14–25). Carbon dioxide content decreases during glaciations and increases during inter-glaciations (Figure 14–25), mirroring deuterium levels and dust content changes.

Although these measurements seem to accurately reflect climatic changes, establishing their age is fraught with difficulty because of the lack of a direct method of dating the ice.

Radiocarbon dating can be applied to the upper parts of an ice core, but the procedure is laborious and best suited to ice less than 10,000 years old. In the absence of a direct means of dating older ice, attempts have been made to utilize theoretical ice-flow models to estimate ages (Dansgaard et al., 1969). Unfortunately, for the ice-flow model approach to work, sev-

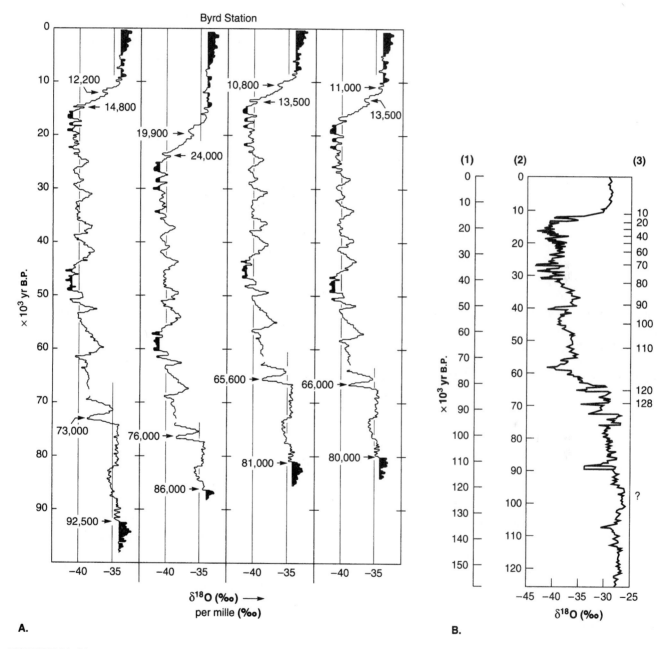

FIGURE 14–24
Oxygen isotope records from ice cores. (A)δ^{18}O record from Byrd Station, Antarctica. A flow model with different assumptions about annual accumulation thickness of ice and divergence of ice flow was used. Changing the initial assumptions has the effect of stretching or shrinking the age scale at the left, as shown by distinctive features of the record. (B)δ^{18}O record from Camp Century, Greenland, showing several revisions of the time scale: (1) the original time scale developed by Dansgaard et al. (1969), based on a theoretical model of ice flow; (2) based on periodic signals in δ^{18}O; and (3) the most recent interpretation of ice age with depth, based on correlations with the isotopic record in deep-sea sediments. (A, after Johnsen et al., 1972; B, from Dansgaard et al., 1982)

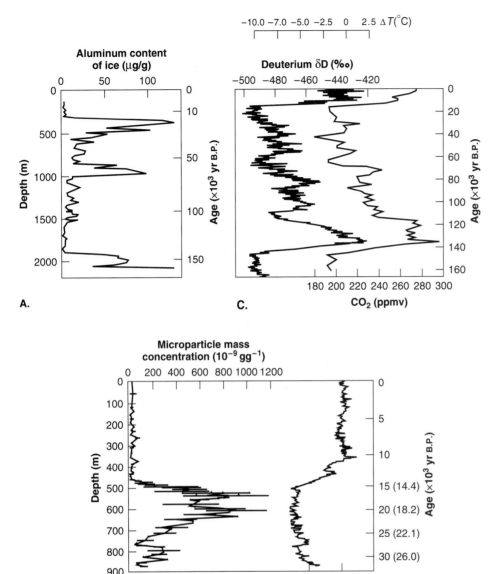

FIGURE 14–25
Antarctic ice cores: (A) CO_2 and $\delta^{18}O$ isotope record from the Vostock ice core; (B) dust content from the Vostock ice core; and (C) dust content and $\delta^{18}O$ isotope record from the Dome Circle ice core. The chronologies are based on ice-flow models. (From Broecker and Denton, 1990; A and B, after Jouzel et al., 1987a, and C, after Royer et al., 1983)

eral assumptions must be made about the constancy of the following factors throughout the time span of the core (including through times of global glaciations):

1. The rate of snow accumulation must remain unchanged.

2. The thickness must have remained constant.

3. The flow pattern must not have varied.

4. Ice in all sections of the core must have originated from places having the same accumulation rate, atmospheric temperature, and ice thickness.

The accuracy of age estimates from ice-flow models is thus highly speculative and has led to large variations in age approximations (in one case, from 60,000 to 115,000 years). In the search for a more accurate means of age determination, a number of other approaches have been attempted, such as matching ice-core $^{18}O/^{16}O$ curves with dated glacial events on land or in deep-sea cores.

Despite the problems associated with establishing the age of ice cores, they have produced dramatic results in interpreting Quaternary climatic changes.

CAUSE OF PLEISTOCENE GLACIATIONS

More than 170 years have elapsed since the first recognition of past ice ages. Yet, despite intense investigation of the Quaternary stratigraphic and geomorphic records, oxygen isotope analyses of deep-sea cores and ice cores, and astronomical calculations, demonstration of the driving force that causes ice

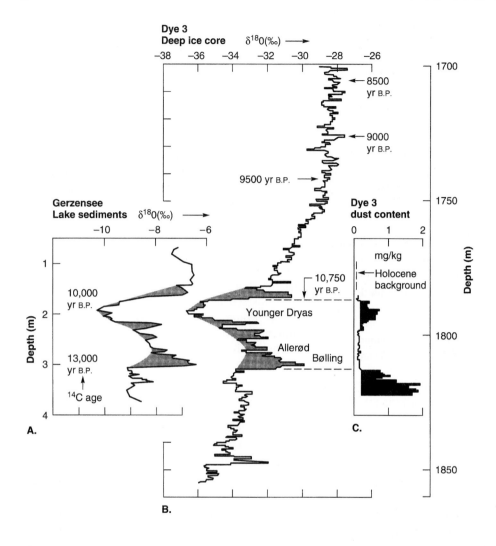

FIGURE 14–26
Depth variation of (A) $\delta^{18}O$ in lake sediments, Gerzensee, Switzerland; (B) $\delta^{18}O$ at Dye 3, Greenland, and (C) dust concentration at Dye 3. (From Paterson and Hammer, 1987; A, after Siegenthaler et al., 1984)

ages remains elusive. Although many theories have been proposed to explain the cause of global glacial/interglacial climatic changes, none has yet emerged as completely conclusive, and the mechanism of climatic changes continues to be a mystery. The problem is not only of academic interest, but also of considerable practical importance because of the distinct possibility of human interference in affecting climatic change.

Several theories of the cause of ice ages are reviewed here, but the reader will quickly recognize that none is conclusively proven and none of them satisfactorily explains all of the following features of ice ages:

1. The recurring ice ages of the Pleistocene are geologically recent events, not common throughout geologic time. Although evidence of glaciations in parts of Africa, South America, and North America exists in some Precambrian and Paleozoic sedimentary sequences, ice sheet glaciations were absent for several hundred million years prior to the Pleistocene ice ages.

2. Global temperatures fluctuated through a range of 2°C to 6°C between glacial and interglacial episodes.

3. The Pleistocene ice ages were preceded by a very gradual global cooling during the previous 65 m.y. of the Cenozoic, culminating in glacial/interglacial fluctuations beginning at about 2.5 m.y.

4. Snowlines in alpine regions of both hemispheres fluctuated through a range of about 1000 m (3300 ft) during the Pleistocene.

5. Glacial and interglacial climates were synchronous in both the Northern and the Southern hemispheres.

6. Desert areas experienced pluvial conditions that corresponded to glacial periods.

Changes in Solar Activity

Short-term fluctuations in solar activity have been well documented, especially those associated with sunspots and solar flares on the sun's surface (Eddy et al., 1983). Historic records document sunspot cycles with a period of about 11 years, and other, longer-term cycles may exist. During periods of low sunspot activity, anticyclonic polar winds increase and mid-latitude storm tracks shift toward the equator, resulting in steeper temperature gradients, intensified atmospheric circulation, increased evaporation, greater storminess, and generally colder and wetter climates. During times of greater sunspot activity, anticyclonic polar winds decrease and mid-latitude storm tracks shift poleward, resulting in lower temperature gradients, diminished atmospheric circulation, decreased evaporation, less storminess, and generally warmer and drier climates. However, whether or not such phenomena could produce 5°C to 6°C variations in global temperature or occur over long-period cycles is unknown.

Changes in Earth-Sun Relationships: The Astronomical Theory of Ice Ages

That the position and orientation of the earth relative to the sun is not constant has been known for centuries, but the possible climatic significance of such variations was not realized until 1842 when mathematician Adhemar published a book suggesting that the primary cause of the ice ages might be variations in the earth's movements relative to the sun. Adhemar knew that the path of the earth's orbit around the sun was an ellipse with the sun at one focal point, and he knew that the earth's axis of rotation is presently tilted 23° from a line vertical to the plane of the orbit. He hypothesized that the following systematic changes in these parameters were the fundamental causes of the ice ages (Figure 14–27):

1. At present, the earth's axis of rotation points toward the star Polaris, so that as the earth rotates on its axis, all stars except Polaris describe a circular path in the sky every 24 hours. However, in 2000 B.C. the axis pointed to a place midway between the Little Dipper and the Big Dipper, and in 4000 B.C. it pointed to the tip of the handle of the Big Dipper. By plotting the progressive positions of the stars, astronomers were able to show that the axis of rotation wobbles, describing a circle in space rather than a single point. Every 26,000 years the axis returns to the same point on the circle. This wobble, called axial *precession,* is caused by the gravitational pull of the sun and moon on the earth's equatorial bulge. Cyclic variation in the angle of tilt of the earth's axis relative to its plane of orbit varies from 21.8° to 24.4° every 41,000 years, the last maximum occurring about 100,000 years ago. At high latitudes, as the angle of tilt increases, seasonal differences become more pronounced: Summer solar radiation increases, causing hotter summers, and winter solar radiation decreases, causing colder winters.

2. Axial precession and rotation of the elliptical orbit cause the equinoxes and solstices to shift slowly along the orbital path, so that at times the summer solstice occurs when the earth is closest to the sun in its orbit, and at other times (as at present) the summer solstice is farthest from the sun. This slow, continual shifting, known as precession of the equinoxes, makes a complete cycle every 21,000 years. Precession of the equinoxes, periodic shift in the position of perihelion relative to the equinoxes and solstices, alters the distribution of solar energy over the earth's surface. Precession governs whether summer falls at a close or distant point in the earth's orbit; that is, it determines whether tilt seasonality is enhanced or weakened by distance seasonality. These orbital variations do not cause any long-term change in solar radiation received by the earth; rather, they affect only the seasonal distribution of energy.

3. The shape of the earth's orbit varies. At times the earth's orbital path is a circle, but at other times it is an ellipse. The orbit varies from nearly circular to maximum eccentricity with an average period of about 95,800 years. The more elliptical the orbit, the greater the distance from the sun to the earth's closest and farthest points, causing intensification of seasons in one hemisphere and moderating the seasons in the other. At present, the earth is farther away from the sun during Northern Hemisphere summers (Southern Hemisphere winters) than during winters (the earth is closest to the sun on January 4), resulting in slightly colder winters and slightly warmer summers in the Southern Hemisphere. Because of the earth's elliptical orbit [the earth is closest to the sun (perihelion) around January 3 and is farthest away from the sun (aphelion) around July 5], the distance traveled in its orbit from September 21–22 (fall equinox) to March 20–21 (spring equinox) is shorter than the distance traveled from March 20–21 to September 21–22. Eccentricity variations affect the relative intensities of the seasons, having an opposite effect in the Northern and Southern hemispheres.

Seasonality changes have a direct effect on glaciers of the Northern Hemisphere. A decrease in summer sunshine permits glaciers to increase in size, and an increase in summer sunshine causes glaciers to melt. The combined effects of all three changes on the amount of solar energy received at the earth's surface can be calculated (Figure 14–27C), giving a solar energy curve with highs and lows as a result of times when all three effects are in phase or out of phase, due to their different periods.

The growth or retreat of large ice sheets affects other parts of the earth's climatic system. Variations in the volume of global ice directly affects sea level, and changes in the surface area of ice sheets cause changes in global radiation balance. As the surface area of ice sheets increases, more solar heat is lost by reflection, encouraging global temperatures to decrease and more ice to form. Conversely, when ice sheets contract, less solar energy is reflected, temperatures rise, and additional contraction results. Such positive feedback is important to several theories of ice-sheet generation because small, initial changes can be amplified into significant changes in the size of ice sheets.

The effects of precession of the equinoxes on solar radiation received by the earth are moderated by variations in eccentricity. When the orbit is close to circular, the timing of the seasons at perihelion is relatively unimportant. However, at maximum eccentricity, timing of the seasons is much more important, and solar radiation may vary by up to 30 percent. Solar radiation at

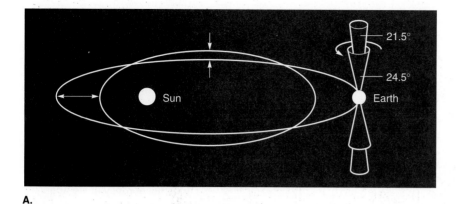

A.

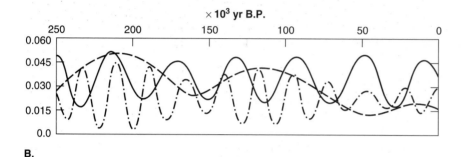

B.

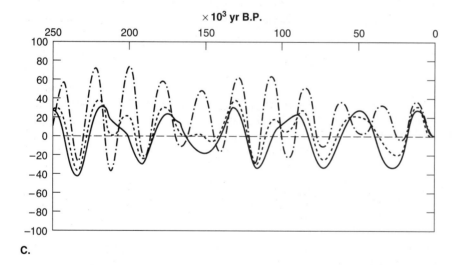

C.

FIGURE 14–27
(A) Cyclic change in earth-sun relationships; (B) variation of eccentricity procession (—), and obliquity (- - -) over the last 250,000 years; (C) Northern Hemisphere summer solar radiation at 80° N (- - -), 65° N (—), and 10° N (– · – ·) latitude (expressed as departures from A.D. 1950 values). The radiation signal at high latitudes is dominated by the ~41,000-year obliquity, whereas the ~23,000-year precessional cycle is dominant at lower latitudes. (B and C, from Bradley, 1985; after Berger, 1978)

low latitudes is affected primarily by variations in eccentricity and precession of the equinoxes, whereas solar radiation at high latitudes is affected principally by variations in obliquity.

Like Adhemar, Croll, a Scottish natural historian, was much impressed with the possible climatic effects of changes in orbital parameters, especially seasonal ones. He proposed that a decrease in winter solar radiation resulted in increased accumulation of snow, which led to further reduction of solar energy by intensifying reflection of solar energy back into space, setting up a positive feedback situation (Croll, 1864, 1867a, 1867b, 1875). Croll suggested that times of greater eccentricity of the earth's orbit provided optimum conditions for the growth of ice sheets. He considered both the preces-

sional cycle and variations in the shape of the earth's orbit together and hypothesized that either the Northern Hemisphere or the Southern Hemisphere experienced an ice age when the two factors reached their optimum conditions concurrently—that is, a maximum elongate orbit and the occurrence of the winter solstice at the maximum distance from the sun (Figure 14–28).

Croll's concepts were elaborated upon by Milankovitch, a Yugoslavian engineer, who calculated solar insolation curves based on the effects of the eccentricity of the earth's orbit, the tilt of the axis of rotation, and the position of the equinoxes in their precessional cycle (Milankovitch, 1920, 1941). He considered solar insolation variations at high northern lati-

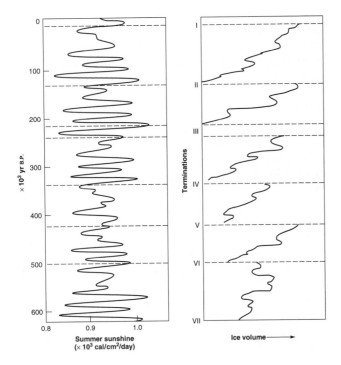

FIGURE 14–28
The eccentricity of the earth's orbit, the orientation of its spin axis, and the tilt of its axis affect the intensity of summer sunshine at high northern latitudes (shown on the left). The curve at the right shows the volume of glacial ice determined from isotopic studies of sea-floor sediments. Ice volumes increase gradually for about 100,000 years and then fall abruptly at ice-age terminations that correspond to episodes of increasing summer sunshine at northern latitudes. (From Broecker and Denton, 1990)

tudes (60°–70° N) to be critical for the growth of large ice sheets because at these latitudes, times of lower summer solar radiation would favor the survival of winter snow longer into summer months and eventually throughout the entire year. His calculations showed that optimum conditions for glaciers to form—minimum obliquity, high eccentricity, and the Northern Hemisphere summer coinciding with aphelion—should have occurred 185,000, 115,000, and 70,000 years B.P. The coincidence of Northern Hemisphere winters with perihelion during these times would have favored increased evaporation from subtropic oceans, which would enhance moisture for precipitation as snowfall at high latitudes. Thus, stronger equator-pole temperature gradients would have intensified circulation and moisture transportation to high latitudes (Milankovitch, 1941; Berger, 1977, 1978, 1980, 1981).

For many years the Croll-Milankovitch theory of the cause of climatic changes that produced Quaternary icesheet glaciations lacked reliable field evidence to corroborate or refute the concept, and a number of serious problems confronted the theory:

1. Evidence that the variations in solar radiations at high latitudes are large enough to produce gigantic ice sheets is lacking. Some kind of positive feedback system is probably required, perhaps related to increased reflectivity of solar energy as the ice sheets grow in size. However, this introduces the problem of what phenomenon is capable of reversing such a positive feedback system.

2. Strong evidence exists that glaciations in the Northern and Southern hemispheres were not only synchronous but, at least for the last glaciation, almost exactly simultaneous. Although a number of possible explanations have been put

forth to explain this vexing anomaly, so far none has proven satisfactory, and, as Broecker (1968) and Mercer (1984) put it, the globally synchronous climatic change "remains the fly in the ointment of the Milankovitch theory: the last interglacial and the last glaciation both appear to have affected the Northern and Southern hemispheres simultaneously, and with comparable severity."

3. Temperatures near the equator should be slightly increased during glacial episodes, in conflict with strong evidence that equatorial snowlines were substantially lower during glacial periods.

4. The sudden terminations of glaciations in the saw-tooth-like pattern that appears in the marine oxygen isotope curves is difficult to explain by orbital forcing.

5. The abrupt late Pleistocene climatic reversals between 13,000 and 10,000 years B.P. at the end of the last glaciation is impossible to explain using only orbital forcing. Attempts to explain these with deep ocean current changes in the North Atlantic seems doomed to failure in light of new evidence that these abrupt climatic changes were essentially simultaneous in both the Northern and Southern Hemispheres and on opposite sides of North America.

6. The timing of changes in the marine oxygen isotope curves from deep-sea cores and from polar ice cores depends on assumptions that may or may not be correct. Although oxygen isotope curves can be correlated quite well from core to core over wide distances, the age of terminations is based on first assuming a constant sedimentation rate, then "tuning" of the core ages by calculating the "astronomic" age of the Milankovitch curve that seems to best fit the isotope data. Unfortunately, the end

result is self-proving; that is, the ages of glacial terminations in the core are based on the assumed correlations with Milankovitch astronomical events, which in turn are used to prove correlations of deep-sea isotope curves with Milankovitch curves. A direct, independent, reliable dating method for dating of the marine oxygen isotope curves is badly needed but at present not available.

7. Ice ages were unknown during the preceding 65 m.y. What was different enough during the last 2.5 m.y. to cause recurring ice ages?

Despite these very serious problems with the Milankovitch theory and the uncertainty concerning the exact mechanism of how orbital forcing is translated into climate responses, the concept seems to be favored by many Quaternary geologists and climatologists as the basic cause of Pleistocene glaciations. Among the more persuasive arguments favoring the concept is that the Milankovitch curves predict recurrences of glacial/nonglacial conditions with periodicities of about 100,000, 41,000, and 21,000 years B.P., and spectral analyses of marine oxygen isotope curves suggest similar recurrences in deep-sea cores. However, these correlations are subject to the limitations of dating deep-sea cores and ice cores as discussed above, so final judgment awaits a suitable, direct dating technique for the cores.

Influence of the Oceans

In an effort to overcome the deficiencies in the Milankovitch theory, listed above, Broecker and Denton (1990) proposed that glacial/interglacial transitions are produced by major reorganizations of the ocean/atmospheric system, which oscillates between two modes, governed by changes in salinity, temperature, and density of seawater.

The concentration of CO_2 trapped in glacial ice in Greenland and Antarctica has fluctuated dramatically from glacial to interglacial climates (Figure 14–25). During the last glaciation,

the CO_2 concentration was about 67 percent of its interglacial level (Barnola et al., 1987; Delmas et al., 1980; Stauffer et al., 1984). Only a major shift in the oceans could produce such a drastic change in the CO_2 concentration in the atmosphere. The oceans hold 60 times as much CO_2 as the atmosphere, and the ocean CO_2 controls the atmospheric concentration (Hansen et al., 1981; Manabe, 1983; Manabe and Stouffer, 1980).

Changes in the microfossil content of deep-sea sediments suggest that the circulation of the oceans differed between glacial and nonglacial times. Northward-flowing, high-salinity water in the Atlantic Ocean cools from about 10° to 2°C, and the resulting increase in density causes the water to sink to the ocean floor. The amount of water involved is about 20 times the flow of all the earth's rivers combined. The dense, cold, high-salinity water flows southward the length of the Atlantic and curves around the southern tip of Africa to join the deep current that circles Antarctica. The effects are to release large quantities of heat in the Northern Hemisphere and to send huge volumes of water into the Southern Hemisphere. As orbital forcing warms the Northern Hemisphere, fresh water is added to the north Atlantic, diluting it until the density contrast is weakened to the point that the "Atlantic conveyor" ceases to function (Broeker et al., 1985). The Broeker-Denton hypothesis calls for orbital forcing to drive glacial-interglacial climates but joined with oceanic circulation changes to produce synchronism of climates in both hemispheres. However, the abrupt late Pleistocene climatic reversals (especially the Younger Dryas) that seem to be contemporaneous in both hemispheres are almost impossible to explain with this hypothesis. Broecker and Denton (1990) suggest that these short-term, dramatic climatic changes could be caused by the sudden release of much glacial meltwater into the Atlantic as the Laurentide Ice Sheet retreated, opening up eastern drainage in southeastern Canada and temporarily upsetting the oceanic circulation in the North Atlantic (Broecker et al., 1988). However, under this scenario, a time lag in climatic changes should exist between the Northern Hemisphere and the Southern Hemisphere.

REFERENCES

Aguirre, E., and Pasini, G., 1984, Proposal of the I. C. S. Working Group on the Pliocene/Pleistocene boundary concerning the definition of the Pliocene/Pleistocene boundary stratotype, 6 p.

Alley, R. B., Mayewski, P. A., Sowers, T., Stuiver, M., Taylor, K. C., Clark, P. U., 1997, Holocene climatic instability: A prominent, widespread event 8200 yr ago: Geology, v. 25, p. 483–486.

Alley, R.B., Meese, D.A., Shuman, C.A., Gow, A.J., Taylor, K.C., Grootes, P.M., White, J.W.C., Ram, M., Waddington, E.D., Mayewski, P.A., and Zielinski, G.A., 1993,

Abrupt increase in Greenland snow accumulation at the end of the Younger Dryas event: Nature, v. 362, p. 527–529.

Anderson, B. G., 1981, Late Weichselian ice sheets in Eurasia and Greenland: in Denton, G. H., and Hughes, T. J., eds., The last great ice sheets: John Wiley and Sons, N.Y., p. 1–65.

Andrews, J. T., 1970, A geomorphological study of postglacial uplift with particular reference to Arctic Canada: Institute of British Geographers Special Publication 2, 156 p.

Andrews, J. T., 1987, The late Wisconsin glaciation and deglaciation of the

Laurentide Ice Sheet: in Ruddiman, W. F., and Wright, H. E., eds., The geology of North America, North American and Adjacent Oceans during the Last Deglaciation, v. K-3, p. 13–38.

Andrews, J. T., and Barnett, D. M., 1972, Analysis of strandline tilt directions in relation to ice centres and post-glacial crustal deformation, Laurentide Ice Sheet: Geographiska Annaler, v. 54, p. 1–11.

Armstrong, J. E., 1981, Post-Vashon Wisconsin glaciation, Fraser Lowland, British Columbia: Geological Survey of Canada Bulletin, v. 322, p. 34.

Armstrong, J. E., Crandell, D.R., Easterbrook, D.J., and Noble, J.B., 1965, Late Pleistocene stratigraphy and chronology in southwestern British Columbia and northwestern Washington: Geological Society of America Bulletin, v. 76, p. 321–330.

Ashworth, A. C., et al., 1991, Late Quaternary climatic history of the Chilean Channels based on fossil pollen and beetle analysis, with an analysis of modern vegetation and pollen rain: Journal of Quaternary Science, v. 6, p. 279–291.

Ashworth, A. C., and Hoganson, J. W., 1984, Testing the Late Quaternary climate record of southern Chile with evidence from fossil Coleoptera: in Vogel, J.C., ed., Late Cainozoic Paleoclimates of the Southern Hemisphere: Balkema, Rotterdam, Netherlands, p. 85–102.

Baker, V. R., 1973, Paleohydrology and sedimentology of Lake Missoula flooding in eastern Washington: Geological Society of America Special Paper 144, 79 p.

Baker, V. R., 1978, Paleohydraulics and hydrodynamics of Scabland Floods: in Baker, V. R., and Nummedal, D., eds., The Channeled Scablands: Washington, D.C., National Aeronautics and Space Administration, p. 59–79.

Baker, V. R., ed., 1981, Catastrophic flooding: the origin of the Channeled Scablands: Benchmark Papers in Geology No. 55, Dowden, Hutchinson and Ross, Stroudsburg, PA, 360 p.

Baker, V. R., and Bunker, R. C., 1985, Cataclysmic late Pleistocene flooding from glacial Lake Missoula: a review: Quaternary Science Reviews, v. 4, no. 1, p. 1–41.

Baker, V. R., and Komar, P. D., 1987, Cataclysmic flood processes and landforms: in Graf, W. L., ed., Geomorphic systems of North America: Geological Society of America Centennial Special Volume, p. 423–443.

Balling, N., 1980, The land uplift in Fennoscandia, gravity field anomalies and isostasy: in Momer, N. A., ed., Earth rheology, isostasy and eustasy: John Wiley and Sons, N.Y., p. 297–321.

Barnola, J. M., Raynaud, D., Korotkevich, Y. S., and Lorius, C., 1987, Vostock ice core provides 160,000-year record of atmospheric CO_2: Nature, v. 329, p. 408–414.

Beer, J., Oeschger, H., Andree, M., Bonani, G., Suter, M., Wolfli, W., and Langway, C., 1983, Temporal variations in the ^{18}O concentration levels found in the Dye 3 ice core, Greenland: in Proceedings, Symposium on Ice and Climate Modeling, Evanston, Illinois: Annals of Glaciology, v. 5, p. 16–17.

Berger, A. L., 1977, Long-term variation of the earth's orbital elements: Celestial Mechanics, v. 15, p. 53–74.

Berger, A. L., 1980, The Milankovitch astronomical theory of paleoclimates: a modern review: Vistas Astronomical, v. 24, p. 103–122.

Berger, A. L., 1981, The astronomical theory of paleoclimates: in Berger, A. L., ed., Climatic variations and variability: facts and theories: Reidel Publishing Company, Dordrecht, Netherlands, p. 501–525.

Berger, A. L., Imbrie, J., Hays, J., Kukia, G., and Saltzman, B., eds., 1984, Milankovitch and climate, Part 1: Reidel Publishing Company, Dordrecht, Netherlands, 510 p.

Berger, W. H., 1978, Oxygen-18 stratigraphy in deep-sea sediments: additional evidence for the deglacial meltwater effect: Deep-Sea Research, v. 25, p. 473–480.

Birkeland, P. W., 1972, Late Quaternary eustatic sea level changes along the Malibu coast, Los Angeles County, California: Journal of Geology, v. 80, p. 432–448.

Blunt, D., Easterbrook, D. J., and Rutter, N, A., 1987, Chronology of Pleistocene sediments in the Puget Lowland, Washington: Washington Division of Geology and Earth Resources Bulletin 77, p. 321–353.

Boellstorff, J., 1976, The succession of late Cenozoic volcanic ashes in the Great Plains: a progress report: Guidebook Series 1, Kansas Geological Survey, p. 37–71.

Boellstorff, J. D., 1978a, North American Pleistocene stages reconsidered in light of probable Pliocene-Pleistocene continental glaciation: Science, v. 202, p. 305–307.

Boellstorff, J. D., 1978b, Chronology of some late Cenozoic deposits from the central U.S. and the ice ages: Transactions of Nebraska Academy of Science, v. 6., p. 35–49.

Boellstorff, J. D., 1978c, A need for redefinition of North American Pleistocene stages: Gulf Coast Association of Geological Societies Transactions, v. 28, p. 65–74.

Bradley, R. S., 1985, Quaternary paleoclimatology: methods of paleoclimatic reconstruction: Allen and Unwin, Boston, MA, 472 p.

Bretz, J H., 1923, The Channeled Scablands of the Columbia Plateau: Journal of Geology, v. 31, p. 617.

Bretz, J H., 1969, The Lake Missoula floods and the Channeled Scablands: Journal of Geology, v. 77, p. 505–543.

Bretz, J H., Smith, H. T. U., and Neff, G. E., 1956, Channeled Scablands of Washington: New data and interpretations: Geological Society of America Bulletin, v. 67, p. 957–1049.

Broecker, W. S., 1968, In defense of the astronomical theory of glaciation: in Mitchell, J. M., ed., Causes of climatic change: Meteorological Monograph 8, p. 112–125.

Broecker, W. S., 1981, Glacial to interglacial changes in ocean and atmosphere chemistry: in Berger, A. L., ed., Climatic variations and variability: facts and theories: Reidel Publishing Co., Dordrecht, Netherlands, p. 109–120.

Broecker, W. S., 1982a, Glacial to interglacial changes in ocean chemistry: Progress in Oceanography, v. 11, p. 151 –197.

Broecker, W. S., 1982b, Ocean chemistry during glacial time: Geochimica et Cosmochimica Acta, v. 46, p. 1689–1705.

Broecker, W. S., 1986, Oxygen isotope constraints on surface ocean temperatures: Quaternary Research, v. 26, p. 121–134.

Broecker, W. S., and Denton, G., 1990, What drives glacial cycles? Scientific American, v. 262, p. 49–56.

Broecker, W. S., and Peng T. H., 1984, The climate-chemistry connection: in Hansen, J. E., and Takahashi, T., eds., Climate processes and climate sensitivity: Geophysical Monograph 29, Maurice Ewing, v. 5, p. 327–336.

Broecker, W. S., and van Donk, J., 1970, Insolation changes, ice volumes and the ^{18}O record in deep-sea cores: Reviews Geophysics and Space Physics, v. 8, p. 169–197.

Broecker, W. S., Thurber, D. L., Goddard, J., Ku, T. H., Mathews, R. K., and Mesolella, K. J., 1968, Milankovitch hypothesis supported by precise dating of coral reefs and deep-sea sediments: Science, v. 159, p. 297–300.

Broecker, W. S., Peteet, D., and Rind, D., 1985, Does the ocean-atmosphere system have more than one stable mode of operation? Nature, v. 315, p. 21–25.

Broecker, W.S., Andree, M., Klas, M., Bonani, G., Wolfli, W., and Oeschger, H., 1988, New evidence from the South China Sea for an abrupt termination of the last glacial period: Nature, v. 333, p. 156–158.

Broecker, W. S., Andree, M., Wolfli, W, Oeschger, H., Bonani, C., Kennet, J., and Peteet, D., 1988, The chronology of the last deglaciation: implications to the cause of the Younger Dryas event: Paleoceanography, v. 3, p.1–19.

Broecker, W. S., Oppo, D., Peng, T. H., Curry, W., Andree, M., Wolfli, W., and Bonam, G., 1988, Radiocarbon-based chronology for the $^{18}O/^{16}O$ record for the last glaciation: Paleoceanography, v. 3, p. 509–515.

Budd, W. F., and Morgan, 1977, Isotopes climate and ice sheet dynamics from core studies on Law Dome, Antarctica: in Isotopes and impurities in snow and ice: International Association of Scientific Hydrology Publication, v. 118, p. 312–321.

Bull, W. B., 1985, Correlation of flights of global marine terraces: in Tectonic geomorphology, Morisawa, M., and Hack, J., eds., Proceedings of the l5th Annual Geomorphology Symposium, State University of New York at Binghamton: Allen and Unwin, Boston, MA, p. 129– 152.

Burrows, C.J., 1975, Late Pleistocene and Holocene moraines of the Cameron Valley, Arrowsmith Range, Canterbury, New Zealand: Arctic and Alpine Research, v. 7.

Burrows, C.J., 1983, Radiocarbon dates from late Quaternary deposits in the Cass District, Canterbury, New Zealand: New Zealand Journal of Botany, v. 21, p. 443–454.

Burrows, C.J., 1984, Problems of dating, correlation and environmental interpretation of the New Zealand Quaternary: in Mahaney, W.C., ed., Correlation of Quaternary Chronologies: Geo Books, Norwich, England.

Burrows, C.J., 1988, Late Otiran and early Aranuian radiocarbon dates from South Island localities: New Zealand Natural Sciences, v. 15, p. 25–36.

Caldenius, C., 1932, Las glaciaciones cuaternarlas en la Patagonia y Tierra del Fuego: Geografiska Annaler, v. 14, p. 164.

Chamberlin, T. C., 1895, The classification of American glacial deposits: Journal of Geology, v. 3, p. 270–277.

Chapell, J., 1974, Geology of coral terraces, Huon Peninsula, New Guinea: a story of Quaternary tectonic movements and sea level changes: Geologic Society of America Bulletin, v. 85, p. 553–570.

Chappell, J., 1983, A revised sea-level record for the last 300,000 years on Papua, New Guinea: Search, v. 14, p. 99–101.

Chappell, J., and Shackleton, J. J., 1986, Pleistocene sea levels and the oxygen isotope record: a reconciliation: Nature, v. 324, p. 137–140.

Clapperton, C., 1993, Quaternary Geology and Geomorphology of South America: Elsevier, Amsterdam, Netherlands, 779 p.

Clapperton, C.M., 1995, Fluctuations of local glaciers at the termination of the Pleistocene: 18-8 ka B.P.: Quaternary International, v. 28, p. 41–50.

Clapperton, C.M., and Sugden, D.E., 1988, Holocene glacier fluctuations in South America and Antarctica: Quaternary Science Reviews, v. 7, p. 185–198.

Clark, D. H., and Gillespie, A. R., 1997, Timing and significance of late-glacial and Holocene glaciation in the Sierra Nevada, California: Quaternary International, v. 38/39, p. 21–38.

Croll, J., 1864, On the physical cause of the change of climate during geological epochs: Philosophical Magazine, v. 28, p. 121–137.

Croll, J., 1867a, On the eccentricity of the earth's orbit, and its physical relations to the glacial epoch: Philosophical Magazine, v, 33, p. 119–131.

Croll, J., 1867b, On the change in the obliquity of the ecliptic, its influence on the climate of the polar regions and on the level of the sea: Philosophical Magazine, v. 33, p. 426–445.

Croll, J., 1875, Climate and time in the geological relations: Appleton and Co., N.Y., 388 p.

Cuffey, K. M., Clow, G. D., Alley, R. B., Stuiver, M., Waddington, E. D., and Saltus, R. W., 1995, Large Arctic temperature change at the Wisconsin-Holocene glacial transition: Science, v. 270, p. 455–458.

Curray, J. R., 1965, Late Quaternary history, continental shelves of the United States: in Wright, H. E., and Frey, D. G., eds., Quaternary of the U.S.: Princeton University Press, Princeton, N. J., p. 723–736.

Dansgaard, W., 1987, Ice core evidence of abrupt climatic changes: in Berger, W. J., and Labeyrie, L. D., eds., Abrupt climatic change: Evidence and implications, Reidel, Dordrecht, Netherlands, p. 223–233.

Dansgaard, W., and Duplessy, J. C., 1981, The Eemian interglacial and its termination: Boreas, v. 10, p. 219–228.

Dansgaard, W., and Oeschger, H., 1989, The environmental record in glaciers and ice sheets: Oeschger, H., and Langway, C. C.,

eds., Dahlem Workshop, John Wiley and Sons, N.Y., v. 8, p. 287–317.

Dansgaard, W, and Tauber, H., 1969, Glacier oxygen-18 content and Pleistocene ocean temperatures: Science, v. 166, p. 499–502.

Dansgaard, W., Johnsen, S. J., Moeller, J., and Langway, C. C., 1969, One thousand centuries of climatic record from Camp Century on the Greenland Ice Sheet: Science, v. 166, p. 377–381.

Dansgaard, W., Johnsen, S. J., Clausen, H. B., and Langway, C. C., Jr., 1970, Ice cores and paleoclimatology, in Olsson, U., ed., Twelfth Nobel Symposium, Radiocarbon variations and absolute chronology, John Wiley and Sons, N.Y., p. 337–351.

Dansgaard, W., Johnsen, S. J., Clausen, H. B., and Langway, C. C., 1971, Climatic record revealed by the Camp Century ice core: in Turekian, K. K., ed., Late Cenozoic glacial ages: Yale University Press, New Haven, CT, p. 37–56.

Dansgaard, W., Clausen, H. B., Gundestrup, N., Hammer, C. U., Johnsen, S. J., Kristinsdottir, P. M., and Reeh, N., 1982, A new Greenland deep ice core: Science, v. 218, p. 1273–1277.

Dansgaard, W., Johnsen, S. J., Clausen, H. B., Dahl-Jensen, D., Gundestrup, N., Hammer, C. U., and Oeschger, H., 1984, North Atlantic climatic oscillations revealed by deep Greenland ice cores: in Hansen, J. E., and Takahashi, T., eds., Climate processes and climate sensitivity: Geophysical Monograph 29, Maurice Ewing, v. 5, p. 288–298.

Dansgaard, W., White, J. W. C., and Johnsen, S. J., 1989, The abrupt termination of the Younger Dryas climate event: Nature, v. 339, p. 532–533.

Delmas, R. J., Ascencio, J. M., and Legrand, M., 1980, Polar ice evidence that atmospheric CO_2 29,000 yr. B.P. was 50% of the present: Nature, v. 284, p. 155–157.

Denton, G.H., and Hendy, C.H., 1994, Younger Dryas age advance of Franz Josef Glacier in the Southern Alps of New Zealand: Science, v. 264, p. 1434–1437.

Denton, G. H., and Hughes, T. J., 1981, The last great ice sheets: Wiley-Interscience, N.Y., 484 p.

Denton, G. H., and Hughes, T. J., 1983, Milankovitch theory of ice ages: hypothesis of ice-sheet linkage between regional insolation and global climate: Quaternary Research, v. 20, p. 125–144.

Denton, G. H., and Karlen, W., 1973, Holocene climatic variations—their pattern and possible cause: Quaternary Research. v. 3, p. 155–205.

Denton, G. H., and Porter, S. C., 1970, Neoglaciation: Scientific American, v. 222, p. 101–110.

Denton, G. H., Hughes, T. J., and Karlen, W., 1986, Global ice-sheet system interlocked by sea level: Quaternary Research, v. 26, p. 3–26.

Duplessy, J. C., Blanc, P. L., and Be, A. W. H., 1981, Oxygen-18 enrichment of planktonic foraminifera due to gametogenic calcification below the euphotic zone: Science, v. 213, p. 1247–1250.

Duplessy, J. C., Arnold, M., Maurice, P., Bard, E., Duprat, J., and Moyes, J., 1986, Direct dating of the oxygen-isotope record of the last deglaciation by ^{14}C mass spectrometry: Nature. v. 320, p. 350–352.

Duplessy, J. C., Labeyrie, L., Arnold, M., Paterne, M., Duprat, J., and van Weering, T. C. E., 1992, Changes in surface salinity of the North Atlantic Ocean during the last deglaciation: Nature, v. 358, p. 485–488.

Easterbrook, D.J., 1963, Late Pleistocene glacial events and relative sea-level changes in the northern Puget Lowland, Washington: Geological Society of America Bulletin, v. 74, p. 1465–1483.

Easterbrook, D. J., 1983, Remanent magnetism in glacial tills and related diamictons: in Tills and related deposits: Balkema Publishers, Rotterdam, Netherlands, p. 303–313.

Easterbrook, D. J., 1986, Stratigraphy and chronology of Quaternary deposits of the Puget Lowland and Olympic Mountains of Washington and the Cascade Mountains of Washington and Oregon: in Sibrava, V., Bowen, D. Q., and Richmond, G. M., eds., Quaternary glaciations in the Northern Hemisphere: Quaternary Science Reviews, v. 5, p. 145–159.

Easterbrook, D. J., 1988, Paleomagnetism of Quaternary sediments: in Dating Quaternary sediments: Geological Society of America Special Paper 227, p. 111–122.

Easterbrook, D.J., 1992, Advance and retreat of Cordilleran Ice Sheets in Washington, U.S.A.: Géographie physique et Quaternaire, v. 46, p. 51–68.

Easterbrook, D.J., 1994a, Chronology of pre-late Wisconsin Pleistocene sediments in the Puget Lowland, Washington: in Lasmanis, R., and Cheney, E.S., eds., Regional geology of Washington State, Washington Division of Geology and Earth Resources, Volume Bulletin 80, p. 191–206.

Easterbrook, D. J., 1994b, Stratigraphy and chronology of early to late Pleistocene glacial and interglacial sediments in the Puget Lowland, Washington: in Swanson, D. A., and Haugerud, R. A., Geologic Field Trips in the Pacific Northwest, Geological Society of America, p.1J-23–38

Easterbrook, D. J., and Boellstorff, J. D., 1981, Paleomagnetic chronology of "Nebraskan-Kansan" tills in the midwestern U.S., in Quaternary glaciations in the Northern Hemisphere: International Geological Correlation Program Project 73/1/24, Report No. 6, Prague, Czechoslovakia, p. 72–82.

Easterbrook, D. J., and Boellstorff, J. D., 1984, Paleomagnetism and chronology of early Pleistocene tills in the central United States: in Mahaney, W. C., ed., Correlation of Quaternary chronologies: Geobooks, Norwich, England, p. 73–90.

Easterbrook, D. J., and Kovanen, D. J., 1997, New evidence for relative sea level fluctuations during the Everson Interstade and multiple stillstands of ice during the Sumas Stade in the northern Puget Lowland: Geological Society of America Abstracts with Program.

Easterbrook, D. J., Briggs, N. D., Westgate, J. A., and Gorton, M., 1981, Age of the Salmon Springs Glaciation in Washington: Geology, v. 9, p. 87–93.

Easterbrook, D. J., Naeser, N. A., and Roland, J., and Carson, R. J., 1988, Application of paleomagnetism, fission-track dating, and tephra chronology to Lower Pleistocene sediments in the Puget Lowland, Washington: in Easterbrook, D. J., ed., Dating Quaternary sediments: Geological Society of America Special Paper 227, p. 139– 165.

Eddy, J. A., Gilliland, R. L., and Hoyt, D. V., 1983, Changes in the solar constant and climatic effects: Nature, v. 300, p. 689–693.

Emiliani, C., 1955, Pleistocene temperature: Journal of Geology, v. 63, p. 538–578.

Emiliani, C., 1966, Paleotemperature analysis of Caribbean cores P6304-8 and P6304-9 and a generalized temperature curve for the past 425,000 years: Journal of Geology, v. 74, p. 109–126.

Emiliani, C., 1978, The cause of the ice ages: Earth and Planetary Science Letters, v. 37, p. 349–352.

Emiliani, C., and Shackleton, N. J., 1974, The Brunhes epoch; isotopic paleotemperatures and geochronology: Science, v. 183, p. 511–514.

Engstrom, D. R., Hansen, B. S., and Wright, H. E. Jr., 1990, A possible Younger Dryas record in southeastern Alaska: Science, v. 250, p. 1383–1385.

Epstein, S., Buchsbaum, R., Lawenstam, H., Urey, H. C., 1951, Carbonate-water isotopic temperature scale: Geological Society of America Bulletin, v. 62, p. 417–425.

Epstein, S., Sharp, R. P., and Gow, A. J., 1970, Antarctic Ice Sheet: stable isotope analyses of Byrd station cores and interhemispheric climatic implications: Science, v. 168, p. 1570–1572.

Ericson, D. B., Ewing, M., Wollin, G., and Heezen, B. C., 1961, Atlantic deep-sea sediment cores: Geological Society of America Bulletin, v. 72, p. 193–286.

Ericson, D. B., and Wollin, G., 1968, Pleistocene climates and chronology in deep-sea sediments: Science, v. 162, p. 1227–1234.

Eschman, D. F., and Mickelson, D. M., 1986, Correlation of glacial deposits of the Huron, Lake Michigan, and Green Bay lobes in Michigan and Wisconsin: in Sibrava, V., Bowen, D. Q., and Richmond, G. M., eds., Quaternary glaciations in the Northern Hemisphere: Quaternary Science Reviews, v. 5, p. 53–58.

Espizua, L., 1993, Quaternary glaciations in the Río Mendoza Valley, Argentine Andes: Quaternary Research, v. 40, p. 150–162.

Fairbanks, R. G., 1990, The age and origin of the "Younger Dryas climate event" in Greenland ice cores: Paleogeography, v. 5, p. 937–948.

Fairbanks, R. G., and Mathews, R-K., 1978, The marine oxygen isotope record in Pleistocene coral, Barbados, West Indies: Quaternary Research, v. 10, p. 181–196.

Fairbridge, R. W., 1961, Eustatic changes in sea level: in Physics and Chemistry of the Earth, v. 5: Pergamon Press, London, p. 99–185.

Farrand, W. R., 1962, Postglacial uplift in North America: American Journal of Science, v. 260, p. 181–199.

Farrand, W. R., and Gajda, R. T., 1962, Isobases on the Wisconsin marine limit in Canada: Geographical Bulletin, v. 17, p. 5–22.

Fawcett, P. J., Agustsdottir A. M., Alley, R. B., and Shuman, C. A., 1997, The Younger Dryas termination and North Atlantic deepwater formation: Insights from climate model simulations and Greenland ice core data: Paleogeography, v. 12, p. 23–38.

Fisher, D. A., Koerner, R. M., and Reeh, N., 1995, Holocene climatic records from Agassiz ice cap, Ellesmere island, NWT, Canada: Holocene, v. 5, p. 19–24.

Flint, R. F., 1971, Glacial and Pleistocene geology: John Wiley and Sons, N.Y., 892 p.

Geyh, M., and Rothlisberger, F., 1986, Gletscherschwankungen der letzen 10,000 Jahre. Ein Vergleich zwischen Nord- und Sudhemisphere (Alpen, Himalaya, Alaska, Sudamerika, Neussland).: Verlag Sauerländer, Aarau, Switzerland, 417 p.

Gilbert, 1890, Lake Bonneville: U.S. Geological Survey Monograph 1, 438 p.

Gillespie, A., and Molnar, P., 1995, Asynchronous maximum advances of mountain and continental glaciers: Reviews of Geophysics, v. 33, p. 311–364.

Gosse, J.C., Klein, J., Evenson, E.B., Lawn, B., and Middleton, R., 1995a, Beryllium-10 dating of the duration and retreat of the last Pinedale glacial sequence: Science, v. 268, p. 1329–1333.

Gosse, J.C., Evenson, E.B., Klein, J., Lawn, B., and Middleton, R., 1995b, Precise cosmogenic ^{10}Be measurements in western North America: Support for a global Younger Dryas cooling event: Geology, v. 23, p. 877–880.

Grootes, P. M., Stuiver, M., White, J. W. C., Johnson, J., and Jouzel, J., 1993, Comparison of oxygen isotope records from the GISP2 and GRIP Greenland ice cores: Nature, v. 366, p. 552–554.

Guilcher, A., 1969, Pleistocene and Holocene sea level changes: Earth Science Reviews, v. 5, p. 69–97.

Gutenberg, B., 1941, Changes in sea level, postglacial uplift, and mobility of the earth's interior: Geological Society of America Bulletin, v. 52, p. 721–772.

Hamblin, W. K., 1995, Earth's dynamic systems: 7th ed: Prentice Hall, Englewood Cliffs, N. J., 647 p.

Hammer, C. U., Clausen, H. B., Dansgaard, W., Neftel, A., Kristinsdottir, P., and Johnson, E., 1985, Continuous impurity analysis along the Dye 3 deep core: in Langway, C., et al., eds., Greenland ice core, geophysics, geochemistry, and the environment: American Geophysical Union Monograph 33, p. 90–94.

Hansen, J., Johnson, D., Lacis, A., Lebedeff, S., Lee, P., Rind, D., and Russell, G., 1981, Climatic impact of increasing atmospheric carbon dioxide: Science, v. 213, p. 957–966.

Hays, J. D., Imbrie, J., and Shackelton, N. J., 1976, Variations in the Earth's orbit:

pacemaker of the ice ages: Science, v. 194. p. 1121–1132.

Heusser, C.J., and Rabassa, J., 1987, Cold climatic episode of Younger Dryas age in Tierra del Fuego: Nature, v. 328, p. 609–611.

Heusser, C.J., and Streeter, S.S., 1980, A temperature and precipitation record of the past 16,000 years in Southern Chile: Science, v. 210, p. 1345–1347.

Hoganson, J.W., and Ashworth, A.C., 1992, Fossil beetle evidence for climatic change 18,000 to 10,000 years B.P. in South-Central Chile: Quaternary Research, v. 37, p. 101–116.

Hopkins, D. M., 1967, Quaternary marine transgression in Alaska: in Hopkins, D. M., ed., The Bering land bridge: Stanford University Press, Stanford, CA, p. 47–90.

Hughen, K., Overpeck, J. T., Peterson, L. C. and Trumbore, S., 1996, Rapid climatic changes in the tropical Atlantic region during the last deglaciation: Nature, v. 380, p. 51–54.

Imbrie, J., and Kipp, N. G., 1971, A micropaleontological method for quantitative paleoclimatology: application to a late Pleistocene Caribbean core: in Turekian, K. K., ed., The late Cenozoic glacial ages: Yale University Press, New Haven, CT, p. 71–18 1.

Imbrie, J., et al., 1984, The orbital theory of Pleistocene climate: support from a revised chronology of marine ^{18}O record: in Berger, A. L., et al., eds., Milankovitch and Climate: Reidel Publishing Company, Dordrecht, Netherlands, p. 269–305.

Ivy-Ochs, S., Schluchter, C., Kubik, P.W., Synal, H.-A., Beer, J., and Kerschner, H., submitted, The exposure age of Egesen moraine at Julier Pass, Switzerland, measured with the cosmogenic radionuclides 10Be, 26Al, and 36Cl: Eclogae Geologica Helvetica.

Izett, G. A., 1981, Volcanic ash beds: recorders of Upper Cenozoic silicid pyroclastic volcanism in the western U.S.: Journal of Geophysical Research, v. 86, p. 10,200–10,222.

Izett, G. A., and Wilcox, R. E., 1982, Map showing localities and inferred distributions of the Huckleberry Ridge, Mesa Falls, and Lava Creek ash beds (Pearlette family ash beds) of the Pliocene and Pleistocene age in the western U.S. and southern Canada: U.S. Geological Survey Miscellaneous Investigations Map, I–1325.

Johnsen, S. J., Dansgaard, W. S., Clausen, H. B., and Langway, C. C., Jr., 1972, Oxygen isotope profiles through the Antarctic and Greenland ice sheets: Nature, v. 235, p. 429–434.

Johnsen, S. J., et al., 1992, Irregular glacial interstadials recorded in a new Greenland ice core. Nature, v. 359, p. 311–313.

Johnson, W. H., 1986, Stratigraphy and correlation of the glacial deposits of the Lake Michigan lobe prior to 14 ka B.P.: in Sibrava, V., Bowen, D. Q., and Richmond, G. M., eds., Quaternary glaciations in the Northern Hemisphere: Quaternary Science Reviews, v. 5, p. 17–22.

Jouzel, J., Lorius, C., Petit, J. R., Genthon, C., Barkov, N. I., Kotlyakov, V. M., and Petrov, V. M., 1987a, Vostock ice core: a continuous isotope temperature record over the last climatic cycle (160,000 years): Nature, v. 329, p. 403–408.

Jouzel, J., Lorius, C., Merlivat, L., and Petit, J. R., 1987b, Abrupt climatic changes: the Antarctic ice record during the late Pleistocene: in Berger, W. H., and Labeyrie, L. D., eds., Abrupt climatic change: evidence and implications: Reidel Publishing Company, Dordrecht, Netherlands, p. 235–245.

Jouzel, J., et al., 1989, A comparison of deep Antarctic ice cores and their implications for climate between 65,000 and 15,000 years ago: Quaternary Research, v. 31, p. 135–150.

Kallel, N., Labeyrie, L. D., Arnold, M., Okada, H., Dudley, W. C., and Duplessy, J. C., 1988, Evidence of cooling during the Younger Dryas in the western Pacific: Oceanologica Acta, v. 11, p. 369–375.

Kay, G. F., and Apfel, E. T., 1929, The pre-Illinoian Pleistocene geology of Iowa: 34th Annual Report, Iowa Geological Survey, 304 p.

Keigwin, L. D., and Jones, G. A., 1995, The marine record of deglaciation from the continental margin off Nova Scotia: Paleoceanography, v. 10, p. 973–985.

King, P. B., Tectonics of Quaternary time in middle North America: in Wright, H. E., and Frey, D. G., eds., The Quaternary of the United States, Princeton University Press, Princeton, N.J.

Kovanen, D. J., and Easterbrook, D. J., 1996, Extensive Readvance of Late Pleistocene (Y.D.?) Alpine Glaciers in the Nooksack River Valley, 10,000 to 12,000 Years Ago, Following Retreat of the Cordilleran Ice Sheet, North Cascades, Washington: Friends of the Pleistocene, Pacific Coast Cell Field Trip Guidebook, 74 p.

Kovanen, D. J., and Easterbrook, D. J., 1997, Two advances of the Cordilleran Ice sheet in the northern Puget Lowland, WA, during the Sumas Stade between 10,000 and 11,500 ^{14}C-yrs. B.P.: Geological Society of America Abstracts with Program.

Kovanen, D. J., Easterbrook, D. J., Evenson, E. B., and Olsen, O., Evidence for two readvances of long, post-Cordilleran-Ice-Sheet, alpine glaciers between 12,000 and 14,000 ^{14}C-yrs. B.P. in the Nooksack Middle Fork, North Cascades, WA: Abstracts with Program, Geological Society of America, v. 28, no. 27, p. A-434.

Kudrass, H. R., Erienkeuser, H., Volbrecht, R., and Weiss, W., 1991, Global nature of the Younger Dryas cooling event inferred from oxygen isotope data from Sulu Sea cores: Nature, v. 349, p. 406–409.

Langway, C. C., Jr., 1970, Stratigraphic analysis of a deep core from Greenland: Geological Society of America Special Paper 125, 186 p.

Langway, C. C., Jr., Oeschger, H., and Dansgaard, W., eds., 1985, Greenland ice core: American Geophysical Union, Geophysical Monograph, v. 33, 118 p.

Larcome, P., Carter, R. M., Dye, J. Gagan, M. K., and Johnson, D. P., 1995, New evidence for episodic post-glacial sea-level rise, central Great Barrier Reef, Australia: Marine Geology, v. 127, p. 1–44.

Legrand, M. R., Lorius, C., Barkov, N. I., and Petrov, V. N., 1988, Vostok (Antarctic ice core): atmospheric chemistry changes over the last climactic cycle (160,000 years): Atmospheric Environment, v. 22, p. 317–331.

Leverett, F., 1899, The Illinois glacial lobe: U.S. Geological Survey Monograph 38, 817 p.

Linsley, B.K., 1996, Oxygen-isotope record of sea level and climate variations in the Sulu Sea over the past 150,000 years: Nature, v. 380, p. 234–237.

Lorius, C., Merlivat, L., Jouzel, J., and Pourchet, M., 1979, A 30,000 year isotope climatic record from Antarctic ice: Nature, v. 290, p. 644–648.

Lorius, C., Raynaud, D., Petit, J. R., Jouzel, J., and Merlivat, L., 1984, Late glacial maximum-Holocene atmospheric and ice thickness changes form Antarctic ice core studies: Annals of Glaciology, v. 5. p. 88–94.

Lorius, C., Jouzel, J., Ritz, C., Merlivat, L., Barkov, N, L, Korotkevich, Y. S., and Kotlyakov, V. M., 1985, A 150,000 year climatic record from Antarctic ice: Nature, v. 316, p. 591–596.

Lowell, T.V., Heusser, C.J., Andersen, B.G., Moreno, P.I., Hauser, A., Heusser, L.E., Schlüchter, C., Marchant, D.R., and Denton, G.H., 1995, Interhemispheric correlation of Late Pleistocene glacial events: Science, v. 269, p. 1541–1549.

Lyell, C., 1830–1833, Principles of geology: J. Murray, London.

Mabin, M.C.G., 1996, The age of the Waiho Loop glacial event: Science, v. 271, p. 668.

Malde, H. E., 1968, Evidence in the Snake River Plain, Idaho, of a catastrophic flood from Pleistocene Lake Bonneville: U.S. Geological Survey Professional Paper 400-B, p. 295–297.

Manabe, S., 1983, Carbon dioxide and climatic change: Advances in Geophysics, v. 25, p. 39–82.

Manabe, S., and Stouffer, R. J., 1980, Sensitivity of a global climate model to an increase of CO_2 concentration in the atmosphere: Journal of Geophysical Research, v. 85, p. 5529–5554.

Markgraf, V., 1989, Reply to C.J. Heusser's "Southern Westerlies during the Last Glacial Maximum": Quaternary Research, v. 31, p. 426–432.

Markgraf, V., 1991, Younger Dryas in southern South America? Boreas, v. 20, p. 63–69.

Markgraf, V., 1993, Younger Dryas in southernmost South America—an update: Quaternary Science Reviews, v. 12, p. 351–355.

Mathewes, R. W., 1993, Evidence for Younger Dryas-age cooling on the north Pacific coast of America: Quaternary Science Reviews, v. 12, p. 321–331.

Mathewes, R. W., Heusser, L. E., and Patterson, R. T., 1993, Evidence for a Younger Dryas-like cooling event on the British Columbia Coast: Geology, v. 21, p. 101–104.

Matsch, S. L., and Schneider, A. F., 1986, Stratigraphy and correlation of the glacial deposits of the glacial lobe complex in Minnesota and northwestern Wisconsin, in Sibrava, V., Bowen, D. Q., and Richmond, G. M., eds., Quaternary glaciations in the Northern Hemisphere: Quaternary Science Reviews, v. 5, p. 59–64.

Matthes, F. E., 1900, Glacial sculpture of the Bighorn Mts., Wyoming: U. S. Geological Survey, 21st Annual Report, p. 167–190.

Matthes, F. E., 1930, Geologic history of the Yosemite Valley: U. S. Geological Survey Professional Paper 160, 137 p.

Mayewski, P. A., Meeker, L. D., Whitlow, S., Twickler, M. S., Morrison, M C., Alley, R. B., Bloomfield, P., and Taylor, K., 1994, The atmosphere during the Younger Dryas: Science, v. 263, p. 1747–1751.

McGlone, M.S., 1996, Lateglacial landscape and vegetation change and the Younger Dryas climatic oscillation in New Zealand: Quaternary Science Reviews, v. 14, p. 867–882.

Meinzer, D. E., 1922, Map of Pleistocene lakes of the Basin and Range Province and its significance: Geological Society of America Bulletin, v. 33, p. 541–542.

Mercer, J. H., 1969, The Allerod oscillation: a European climatic anomaly? Arctic Alpine Research, v. 1, p. 227–234.

Mercer, J. H., 1970, Variations of some Patagonian glaciers since the Late-Glacial: II: American Journal of Science, v. 269, p. 1–25.

Mercer, J. H., 1976, Glacial history of southernmost South America: Quaternary Research, v. 6, p. 125–166.

Mercer, J.H., 1982, Simultaneous climatic change in both hemispheres and similar bipolar interglacial warming: evidence and implications: Geophysical Monograph, v. 29, p. 307–313.

Mercer, J. H., 1984, Simultaneous climatic change in both hemispheres and similar bipolar interglacial warming: evidence and implications: in Hansen, H., and Takahash, R., eds., Climate processes and climate sensitivity: Geophysical Monograph 29, Maurice Ewing, v. 5, p. 307–313.

Mercer, J.H., 1988, The age of the Waiho Loop terminal moraine, Franz Josef Glacier, Westland, New Zealand: New Zealand Journal of Geology and Geophysics, v. 31, p. 95–99.

Mesolella, J. K., et al., 1969, The astronomical theory of climatic change: Barbados data: Journal of Geology, v. 77, p. 250–274.

Milankovitch, M., 1920, Theorie Mathematique des phenomenesthermiques procluits per la radiation solaire: Paris, Gauthier-Villars.

Milankovich, M., 1941, Kanon der Erdbestrahlung und seine Andwenclung auf das Eiszeitenproblem: Royal Serbian Academy Special Publication 133, Belgrade, Yugoslavia, 633 p.

Milliman, and Emery, 1968, Sea levels during the past 35,000 years: Science. v. 162, no. 3858, p, 1121–1123.

Mix, A. C., 1987, The oxygen-isotope record of glaciation: in Ruddiman, W. F., and Wright, H. E., eds., The geology of North America, v. K-3: North American and Adjacent Oceans during the Last Deglaciation, p. 111–135.

Mix, A. C., and Ruddiman, W. F., 1984, Oxygen-isotope analyses and Pleistocene ice volumes: Quaternary Research, v. 21, p. 1–20.

Mix, A. C., and Ruddiman, W. F., 1985, Structure and timing of the last deglaciation, oxygen isotope evidence: Quaternary Science Reviews, v. 4, p. 59–108.

Morner, N. A., 1969, The late Quaternary history of the Kattegat Sea and the Swedish west coast: Sveriges Geoliska Underssökning, Series C, No. 640, Arsbok 63, no. 3, 487 p.

Morner, N. A., 1971, The position of the ocean level during the interstadial at about 30,000 B.P.: a discussion from a climatic-glacialogical point of view: Canadian Journal of Earth Sciences, v. 8, p. 132–143.

Morrison, R. B., and Frye, J. C., 1965, Correlation of the middle and late Quaternary successions of the Lake Lahonton, Lake Bonneville, Rocky Mountain (Wasatch Range), southern Great Plains, and eastern mid-west areas: Nevada Bureau of Mines Report 9, 45 p.

Moss, J. H., 1951, Late glacial advances in the southern Wind River Mountains, Wyoming: American Journal of Science, v. 249, p. 865–883.

Naeser, C. W., Izett, G. A., and Wilcox, R. E., 1973, Zircon fission-track ages of Pearlette family ash beds in Meade County, Kansas: Geology, v. 1, p. 187–189.

Neftel, A., Oeschger, H., Schwander, J., Stauffer, B., and Zumbrunn, R., 1982, Ice core sample measurements give atmospheric CO_2 content during the past 40,000 years: Nature, v. 295, p. 220–223.

Neftel, A., Oechgerr, H., Staffelbach, T., and Stauffer, B., 1988, CO_2 record in the Byrd ice core 50,000–5,000 years B.P.: Nature, v. 331, p. 609–611.

Nelson, C. S., Hendy, C. H., Jarrett, G. R., and Cuthbertson, A. M., 1985, Near-synchromety of New Zealand alpine glaciations and Northern Hemisphere continental glaciations during the past 750 kyr:: Nature, v. 318, p. 361–363.

O'Brien, S. R., Mayewski, P. A., Meeker, L. D., Meese, D. A., Twickler, M. S., and Whitlow, S. I., 1995, Complexity of Holocene climate as reconstructed from a Greenland ice core: Science, v. 270, p. 1962–1964.

Oeschger, H., Beer, J., Siegenthaler, U., Stauffer, B., Dansgaard, W., and Langway, C. C., Jr., 1983, Late-glacial climate history from ice cores: in Hansen, J. E., and Takahashi, T., eds., Climate processes and climate sensitivity: Geophysical Monograph 29, Maurice Ewing, v. 5, p. 199–306.

Osborn, G., Clapperton, C., Davis, P.T., Reasoner, M., Rodbell, D.T., Seltzer, G.O., and Zielinski, G., 1995, Potential glacial evidence for the Younger Dryas Event in the Cordillera of North and South America: Quaternary Science Reviews, v. 14, p. 823.

Paterson, W. S. B., and Hammer, C. U., 1987, Ice core and other glaciological data: in Ruddiman, W. F., and Wright, H. E., eds., The geology of North America, v. K-3: North American and Adjacent Oceans during the Last Deglaciation: Geological Society of America, p. 91–110.

Peltier, W. R., 1987, Glacial isostasy, mantle viscosity, and Pleistocene climatic change: in Ruddiman, W. F., and Wright, H. E., eds., The geology of North America, v. K-3: North American and Adjacent Oceans during the Last Deglaciation, p. 155–182.

Penck, A., and Bruckner, E., 1909, Die Alpen im Eiszeitalter, Tauchmtz, Leipzig, Germany.

Peteet, D., 1995, Global Younger Dryas? Quaternary International, v. 28, p. 93–103.

Porter, S. C., and Denton, G. H., 1967, Chronology of the Neoglaciaton in the North American Cordillera: American Journal of Science, v. 265, p. 177–210.

Raynaud, D., Chappelaz, J., Barnola, J., Korotkevich, Y., and Lorius, C., 1988, Climatic and CH_4 cycle implications of glacial-interglacial change in the Vostock ice core: Nature, v. 333, p. 655–657.

Reasoner, M. A., Osborn, G., and Rutter, N. W., 1994, Age of the crowfoot advance in the Canadian Rocky Mts.: a glacial event coeval with the Younger Dryas oscillation: Geology, v. 22, p. 439–442.

Richmond, G. M., and Fullerton, D. S., 1986, Introduction to Quaternary glaciations in the U.S.A.: in Sibrava, V., Bowen, D. Q., and Richmond, G. M., eds., Quaternary glaciations in the Northern Hemisphere, Quaternary Science Reviews, v. 5, p. 3–10.

Richmond, G. M., Fryxell, R., Neff, G. E., and Weiss, P. L., 1965, The Cordilleran Ice Sheet of the northern Rocky Mts., and related Quaternary history of the Columbia Plateau: in Wright, H. E., and Frey, D. G., eds., The Quaternary of the United States: Princeton University Press, Princeton, N.J., p. 231 – 242.

Rind, D., Peteet, D., Broecker, W., McIntyre, A., and Ruddiman, W., 1986, Comparison of the observed geographic distribution of Younger Dryas impacts with that predicted by a G. C. M. model which assumes the CLIMAP-based temperature change for the northern Atlantic Ocean: Climate dynamics, v. 1, p. 3—33.

Robin, G. de Q., 1977. Ice cores and climatic changes: Philosophical Transactions of the Royal Society of London, B280, p. 143–168.

Robin, G. de Q., 1983, The climatic record from the ice cores: in Robin, G. de Q., ed., The climatic record in polar ice sheets: Cambridge University Press, Cambridge, England, p. 180–195.

Ruddiman, W. F., 1987a, Northern oceans: in Ruddiman, W. F., and Wright, H. E., eds., The geology of North America, v. K-3: North American and Adjacent Oceans during the Last Deglaciation, p. 137–154.

Ruddiman, W. F., 1987b, Synthesis~ the ocean/ice sheet record: in Ruddiman, W. F., and Wright, H. E., eds., The geology of North America, v. K-3: North American and Adjacent Oceans during the Last Deglaciation, p. 463–478.

Ruddiman, W. F., and McIntyre, A., 1976, Northeast Atlantic paleoclimatic changes over the past 600,000 years: in Cline, R. M., and Hays, J. D., eds., Investigation of late Quaternary paleoceanography and paleoclimatology: Geological Society of America Memoir 145, p. 111–146.

Ruddiman, W. F., and McIntyre, A., 1981, The North Atlantic Ocean during the last deglaciation: Paleogeography, Paleoclimatology, Paleoecology, v. 35, p. 145–214.

Ruddiman, W. F., and McIntyre, A., 1984, Ice-age thermal response and climatic of the surface Atlantic Ocean. 40°N to 63°N: Geological Society of America Bulletin, v. 95, p. 381–396.

Russell, I. C., 1885, Geological history of Lake Lahontan, a Quaternary lake of northwestern Nevada: U.S. Geological Survey Monograph 11, 288 p.

Russell, I. C., 1889, Quaternary history of Mono Valley, California: U.S. Geological Survey 8th Annual Report. p. 261–394.

Shackleton, N. J., 1977, The oxygen isotope record of the late Pleistocene: Philosophical Transactions of the Royal Society of London, B280, p. 169–182.

Shackleton, N. J., 1987, Oxygen isotopes, ice volume and sea level: Quaternary Science Reviews, v. 6, p. 183–190.

Shackleton, N. J., and Opdyke, N. D., 1973, Oxygen isotope and paleomagnetic stratigraphy of equatorial Pacific core V28–238: oxygen isotope temperatures and ice volumes on a 105-year time scale: Quaternary Research, v. 3, p. 39–55.

Shackleton, N. J., and Opdyke, N. D., 1976, Oxygen-isotope and paleomagnetic stratigraphy of Pacific core V 29-239 . Late Pliocene to latest Pleistocene: in Cline, R. M., and Hays, J. D., eds., Investigation of late Quaternary paleoceanography and paleoclimatology, Geological Society of America Memoir 145, p. 449–464.

Shackleton, N. J., and Pisias, N. G., 1985, Atmospheric carbon dioxide, orbital forcing, and climate: in Sundquist, E., and Broecker, W., eds., The carbon cycle and atmospheric CO_2: natural variations archean to present: Geophysical Monograph 32, p. 303–317.

Shimek, B., 1909, Aftonian sands and gravels in western Iowa: Geologic Society of America Bulletin, v. 20, p. 399–408.

Sibrava, V., Bowen, D. Q., and Richmond, G. M., eds., 1986, Quaternary glaciations in the Northern Hemisphere: Quaternary Science Reviews, v. 5, 514 p.

Siegenthaler, U., Eicher, U., Oeschger, H., and Dansgaard, W., 1984, Lake sediments as continental ^{18}O records from the glacial/post-glacial transition: Annals of Glaciology, v. 5, p. 149–152.

Smith, G. L., et al., 1967, Pleistocene geology and palynology, Searles Valley, California: Guidebook to Friends of the Pleistocene Field Excursion, 66 p.

Soons, J.M., and Burrows, C.J., 1978, Dates for Otiran deposits, including plant microfossils and macrofossils, from Rakaia Valley: New Zealand Journal of Geology and Geophysics, v. 21, p. 607–615.

Soons, J.M., and Gullentops, F.W., 1972, Glacial advances in the Rakaia Valley, New Zealand: New Zealand Journal of Geology and Geophysics, v. 16, p. 425–438.

Sowers, T., and Bender, M., 1995, Climate records covering the last deglaciation: Science, v. 269, p. 210–213.

Steig, E. J., Grootes, P. M., and Stuiver, M., 1994, Seasonal precipitation timing and ice core records: Science, v. 266, p. 1885–1886.

Strahler, A. N., 1963, The earth sciences: Harper and Row, N.Y., 680 p.

Stauffer, B., Hofer, H., Oeschger, H., Schwander, J., and Siegenthaler, U., 1984, Atmospheric CO_2 concentration during the last glaciation: Annals of Glaciology, v. 5. p. 160–164.

Stuiver, M., Grootes, P.M., and Braziunas, T.F., 1995, The GISP2 ^{18}O climate record of the past 16,500 years and the role of the sun, ocean, and volcanoes: Quaternary Research, v. 44, p. 341–354.

Suggate, R.P., 1965, Late Pleistocene geology of the northern part of the South Island, New Zealand: New Zealand Geological Bulletin, v. 77, p. 91.

Suggate, R.P., 1985, The glacial sequence of North Westland, New Zealand: New Zealand Geological Survey Record, v. 7, p. 22.

Taylor, K.C., Lamorey, G.W., Doyle, G.A., Alley, R.B., Grootes, P.M., Mayewski, P.A., White, J.W.C., and Barlow, L.K., 1993, The 'flickering switch' of late Pleistocene climate change: Nature, v. 361, p. 432–436.

Thompson, L. G., and al., e., 1995, Late Glacial Stage and Holocene tropical ice core records from Huascar‡n, Peru: Science, v. 269, p. 46–50.

Urey, H. C., 1947, The thermodynamic properties of isotopic-substances: Journal of Chemical Society, v. 152, p. 190–219.

Urey, H. C., 1948, Oxygen isotopes in nature and in the laboratory: Science, v. 108, p. 489–496.

Van der Hammen, T., and Gonzalez, E., 1960, A pollen diagram from Laguna de al Herrera (Sabana de Bogota): Leidse Geol. Meded, v. 32, p. 183–191.

Van Donk, J., 1976, An ^{18}O record of the Atlantic Ocean for the entire Pleistocene: in Cline, R. M., and Hays, J. D., eds., Investigation of late Quaternary paleoceanography and paleoclimatology: Geological Society of America Memoir 145, p. 145–163.

Wardle, P., 1978, Further radiocarbon dates from Westland National Park and the Omoeroa River mouth, New Zealand: New Zealand Journal of Botany, v. 16, p. 147–152.

Warrick, R. A., LeProvost, C., Meier, M. F., Oerlemans, J., and Woodsworth, P. L., 1995, Changes in sea level: in Houghton, J. T. et al., eds., Climate change 1995,: The science of climate change: Cambridge University Press, Cambridge, England, p. 359–405.

Weaver, A.J., and Hughes, T.M.C., 1994, Rapid interglacial climate fluctuations driven by North Atlantic ocean circulation: Nature, v. 367, p. 447–450.

Westgate, J. A., Easterbrook, D. J., Naeser, N. A., and Carson, R., 1987, The Lake Tapps tephra: an early Pleistocene stratigraphic marker in the Puget Lowland, Washington: Quaternary Research, v. 28, p. 340–355.

Zagwin, W. H., 1986, The Pleistocene of the Netherlands with special reference to glaciation and terrace formation: in Sibrava, V., Bowen, D. Q., and Richmond, G. M., eds., Quaternary glaciations in the Northern Hemisphere: Quaternary Science Reviews, v. 5, p. 341–346.

Zielinski, G.A., and Davis, P.T., 1987, Late Pleistocene age of the Type Temple Lake Moraine, Wind River Range, Wyoming, U.S.A.: Géographie Physique et Quaternaire, v. XLI, p. 397–401.

CHAPTER FIFTEEN

Periglacial Processes and Landforms

Ice wedges, MacKenzie Delta region, Northwest territories. (Photo by H. M. French)

and as the heat of the sun melts the ice, the trees frequently fall into the river."

Other early observations of frozen ground were made in northern North America by employees of the Hudson's Bay company (Richardson, 1839), in Yakutsk, Siberia by von Middendorf in 1844, and by search parties of the Franklin Expedition (Lefroy, 1889).

Interest in frozen ground phenomena was stimulated during the construction of parts of the Trans-Siberian Railway because of the many difficulties caused by frozen ground. Many casual observations of frozen ground phenomena were made during the Alaska and Yukon Klondike Gold Rush of 1896–8, and the subsequent expansion into Alaska in the early 1900s. Similar impacts of frozen ground were experienced during construction of arctic military bases and highways during World War II (Muller, 1945).

Chief among the early involvement of scientists in the processes and landforms of cold nonglacial environments was the International Geological Congress field excursion to Svalbard in 1910. Regional geological reconnaissance studies by Leffingwell (1915, 1919) and Capps (1910, 1919) in northern Alaska provided some of the first integrated observations of periglacial landform processes.

In 1909, Lozinski proposed the term *periglacial* to describe frost weathering conditions associated with the production of rock rubble in the Carpathian Mountains of eastern Europe. He later introduced the term *periglacial zone* for climatic and geomorphic conditions in areas marginal to Pleistocene ice sheets (Lozinski, 1912). However, frost-action phenomena are known to occur at great distances from ice margins and unrelated to ice-marginal conditions. They owe their distinctive nature not to their proximity to ice sheets but rather to low mean annual temperatures and a variety of processes that require neither an ice-marginal location nor an especially cold climate. Thus, most scientists who work in cold areas use periglacial to include all nonglacial, cold-climate phenomena related to intense frost activity, whether or not glaciers are now nearby or have been nearby in the past (Edelman et al., 1936; Cailleux, 1942; Budel, 1944, 1953; Dylik 1953, 1964; Jahn, 1975; Washburn 1979; French, 1996). Attempts have been made to define 'periglacial' on the basis of perennially frozen ground (Tricart 1970) or on the basis of climatic features, but certain periglacial processes occur in the absence of permafrost, and periglacial environments are highly diverse. Relict permafrost unrelated to present climatic conditions underlies extensive areas of forest in Siberia and northern North America, complicating any simple boundary of periglacial environments and extending periglacial areas beyond present-day, frost activity limits.

INTRODUCTION

Periglacial refers to nonglacial processes and landforms associated with cold climates, particularly with various aspects of frozen ground. The effect of cold climates on geomorphic processes has been recognized for more than two centuries. Frozen ground in the Arctic was reported by early explorers in various parts of the world, and news of frozen carcasses of mammoth and woolly rhinoceros in Siberia reached the western world in the seventeenth century. In 1789, Alexander Mackenzie wrote about the Mackenzie River in *Voyages to the Arctic*:

"…banks, which are about six feet above the surface of the water, display a face of solid ice, intermixed with veins of black earth,

PERIGLACIAL PROCESSES

Periglacial processes, primarily freezing and thawing, are similar to those in temperate climates, but differ in their frequency and intensity. Periglacial regions are distinctive because of the importance of intensive freeze-thaw processes and the growth and decay of subsurface ground ice that produce a distinctive suite of landforms.

The effects of periglacial processes on the landscape are not restricted to polar regions. Diurnal and seasonal freeze-thaw occurs in areas well beyond permafrost regions, perhaps as much as one-third of the world. During the Pleistocene, intensification of periglacial processes in mid-latitude regions accelerated hillslope development.

Freezing and Thawing

Freezing and thawing processes are discussed in Chapter 3 and are not repeated here. Instead, we will consider some aspects of freezing and thawing that are especially important in periglacial regions.

Ice Segregation and Frost Heave

As soil freezes, water in the soil may segregate into ice lenses. Early investigations into the mechanism of segregated ice formation by Taber (1929, 1930) led him to conclude that:

1. Ice segregation is favored in fine-grained sediments having a grain size of 0.01 mm (0.0004 in) or less diameter.

2. Ice crystals grow normal to the surface from which heat is being most rapidly conducted away. These conclusions have been verified by later studies.

When the ground freezes, tongues of ice may descend through the soil pores, or the freezing plane may remain in a constant position within the soil. If the freezing plane remains stationary within the soil, water will move upward through the soil pores toward the freezing plane from the unfrozen ground beneath. Ice crystals form when the water reaches the freezing plane, and ice will continue to be added as long as the supply of water is maintained, forming an ice lens. As the ice lens grows, it heaves the overlying sediment upward.

The supply of water to the freezing plane controls whether the freezing level remains stationary or descends deeper into the soil. Progressive *in situ* freezing of pore water causes ice to move downward through the soil, resulting in formation of pore ice rather than segregated ice. Formation of segregated ice is favored in fine-grained sediments that possess small interstices, whereas pore ice typically develops in coarser-grained material, such as sand, in which the pore spaces are larger.

Two kinds of frost heave have been recognized—primary and secondary (Miller, 1972; Mackay et al., 1979; Smith, 1985a). Primary heave occurs near the frost line and is related to the expansion of water as it freezes, combined with the growth of ice-segregation lenses. Secondary heave occurs within frozen layers at various temperatures. Large secondary heaving pressures may develop over time. Annual ground displacements of several centimeters are common.

The frost heaving may also occur in bedrock and is widespread in permafrost regions. Depending upon the geomorphic expression of bedrock and its fracture characteristics and groundwater conditions, the heave can vary from single ejected blocks to dome-shaped accumulations of blocks up to several meters in diameter. Annual movements up to 5 cm (2 in) may occur horizontally and vertically (Dyke, 1984).

Upfreezing of Stones

The progressive upward movement of stone and objects during frost heaving is called **upfreezing**. It occurs most commonly in sediments that contain both fine-grained material and coarser particles. Although the mechanics of upfreezing are still not completely understood, at least two hypotheses have been suggested. The *frost-pull* hypothesis assumes that as the ground freezes from the top downward, the top of a pebble or coarser particle is gripped by the advancing freezing plane and rises upward along with other heaved material. Upon thawing, the pebble cannot return to its original position because lateral frost heaving during the freezing process compresses the cavity vacated by the pebble, and during thawing, material slumps into the hollow.

According to the *frost-push* hypothesis, upfreezing of stones results from the greater thermal conductivity of stones that causes ice to form around and beneath the stones, and the stones are forced upward by the ice. Upon thawing of the ice, the pebble cannot return to its original position because fine-grained sediment has filled in the area beneath the stone. This process has been confirmed experimentally (Corte, 1966). Mackay (1984) suggested that the unfreezing of stones is governed by:

1. The direction of freezing, either upward or downward

2. The degree of frost susceptibility of the enclosing sediment

3. The degree of frost-susceptibility of the stone

Tilted stones, often standing on end, (Figure 15–1) are common in many areas subject to frost action. They result from differential frost heave at the top and bottom of the stone, causing rotation and tilting of the long axis of the stone. Upon repeated freezing and thawing cycles, a progressive increase in the angle of the stone axis takes place, along with progressive upward movement of the stone. The upward tilting of pebbles is a characteristic of cold-climate slope deposits.

The formation of needle ice is a small-scale, upfreezing process caused by diurnal one-sided freezing at, or just beneath, the ground surface. Delicate, vertical ice crystals, ranging in length from a few millimeters to several centimeters, grow upward in the direction of heat loss and may lift small particles. Needle ice formation is particularly common in wet, silty, frost-susceptible soils. The thawing and collapse of needle ice is significant in frost sorting, frost creep, differential downslope movement of fine and coarse material, and the origin of certain small-scale, patterned ground forms (Pissart, 1977).

FIGURE 15–1
Tilted frost debris in a blockfield with patterned ground, northern Yukon Territory. (Photo by H. M. French)

Frost Sorting

Frost sorting is a complex process by which migrating particles are sorted into uniform particle sizes by freezing and thawing. Early frost-sorting laboratory experiments showed that the movement of particles depends on the amount of moisture present, the rate of freezing, the particle size distribution and the orientation of the freeze-thaw plane Corte (1966). Three types of sorting mechanisms can be simulated under laboratory conditions.

1. Sorting by uplift of particles by frost heave when freezing and thawing occurs from the top. Vertical sorting takes place as the larger particles move upward and the smaller particles move downward.

2. Sorting by movement in front of a moving, freezing plane as freezing and thawing occurs from either the top or the sides. The finer particles migrate away from the advancing freezing plane, leaving the coarser particles nearest the cooling side and causing lateral sorting.

3. Mechanical sorting occurs as mounds and frost-heaved structures form. Coarser particles migrate from higher central areas of finer sediment to form borders of coarser material.

Experiments show that fine particles migrate more readily under a wider range of freezing rates than do coarser particles. Thus, a heterogeneous material becomes sorted by freezing (Corte, 1971). Several factors govern the mechanism of frost sorting.

1. Sorting decreases as the moisture supply decreases.

2. Sorting develops best under conditions of slow freezing rates and saturated soils. Fast freezing of heterogeneous sediments produces less sorting than the same sediment subject to very slow freezing.

3. Fine particles can be sorted under a wider range of freezing rates than coarser particles.

4. Vertical sorting can be changed to lateral sorting by changing the orientation of the freeze-thaw plane.

5. Horizontal layers of fine particles produced by vertical sorting can become dome-shaped when freezing takes place from the sides.

Periglacial Mass Movements and Slope Deposits

Frost-Creep

Frost creep is the downslope movement of particles resulting from frost heaving of the ground normal to the slope and subsequent nearly vertical settling upon thawing (Davison, 1889; Taber, 1929, 1930, 1943; Washburn, 1967, 1979; Benedict, 1970b; Jahn, 1975; French, 1976). Taber's classic experiments (1929; 1930, 1943) showed that the direction of ice-crystal growth and heaving is normal to the cooling surface. Upon freezing, the ground heaves at right angles to the slope, since this is the primary cooling surface.

As shown in Figure 15–1, upon freezing, a particle at *"a"* moves to *"b";* and the downslope component of movement is the distance *bc*, which represents the maximum possible movement by frost creep during a single freeze-thaw cycle. If the particle at *"b"* dropped vertically upon thawing of the ground, it would move to *"d"*. However, in fact, the movement during thawing is some point *"e"* between *"a"* and *"d"* as a result of particle cohesion (Davison, 1889), interference

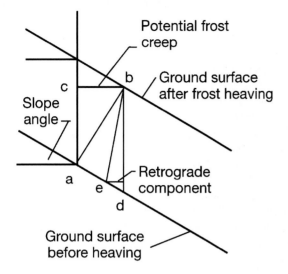

FIGURE 15–2
Frost creep caused by heaving of the ground during a single freeze/thaw cycle.

of particles with one another, or capillary pressures (Washburn, 1967). The rate of frost creep is quite variable, depending on the number of freeze-thaw cycles, the slope angle, the amount of moisture in the soil, and the susceptibility of the soil to frost activity.

Solifluction

Solifluction (solum = soil; fluere = to flow) was described by Andersson (1906), who recognized its great importance as a gradational agent in cold climatic regions where the soil is periodically saturated by the thawing of ice in the ground. Andersson defined the process as "the flowing from higher to lower ground of masses of waste saturated with water (this may come from snow-melting or rain)." The material on active solifluction slopes in cold regions reminds one of viscous concrete or mortar.

Sven Hedin, an early explorer of central Asia, battled solifluction slopes for weeks in the Tibetan Plateau and observed:

"I could not help thinking that the whole bunch of hills would, like a viscous fluid or thick porridge, gradually flatten themselves out to a uniform level...The slope consisted of a gigantic sheet of mire, of the consistency of porridge, and to judge from the cracks which ran across it and all round it, the entire mass was slipping slowly, though imperceptibly, down the mountain side."

Another early observer of this process was Edward Belcher on Buckingham Island (77°N), quoted by Geikie in 1874.

"As noon passed, the soil in all the hollows or small watercourses became semifluid and very uncomfortable to walk on or sink into. The entire slope, in consequence of the thaw, had become a fluid moving chute of debris for at least one foot in depth."

Solifluction is not limited to regions of frozen ground, but it finds its best expression there because of the ready source of meltwater, the impermeability of the frozen sublayer, the lack of constraining vegetation, and the absence of processes powerful enough to destroy the landforms of solifluction or to subordinate solifluction to a minor role.

A particular type of solifluction, *gelifluction*, has been defined as solifluction associated with frozen ground. The distinction between the two terms, however, is not universally agreed upon among geomorphologists. Two major types of solifluction processes have been distinguished: (1) the downslope movement of water-saturated debris (mud flows) and (2) downslope movement induced by alternate freezing and thawing of debris on slopes (frost creep).

Taber (1943) noted that some frozen silt in Alaska contained more than 80 percent ice that converted to fluid mud upon thawing. Another source of soil moisture is snow melt. The effect of water saturation in the upper soil is to reduce soil shear strength. The shear strength of a material depends on the internal friction between adjacent grains plus cohesion. Increased pore-water pressure at the contacts between individual grains tends to push them apart thus decreases the internal shear strength of the material. Saturated material may flow on slopes as low as 1°. Grain size plays an important role because in well-drained, permeable sand and gravel, little moisture remains in the soil to freeze. Fine-grained sediments have higher porosities and are much more like to remain wet longer, thus making them more vulnerable to flowage.

Flowage of saturated soil over frozen ground commonly results in formation of solifluction lobes, large tongue-like masses of material 30–50 m (~100–165 ft) wide (Figure 15–3). The lobes typically have a well-defined bulge 1–5 m (~3–16 ft) high at their distal margins. Successive piling up of lobes produces stone-banked terraces having a stair-step-like profile (Figure 15–3).

From a distance, a slope festooned with solifluction lobes appears to be melting down much like an overheated waxen figure. As their shape suggests, the lobes move most rapidly through their centers and more slowly near their lateral margins where they are retarded by friction (Washburn, 1947). The flows are attenuated into solifluction streams where they are guided by valleys. Concentrations of boulders may be built up along the leading edge of a lobe or sheet until they dam the flow and impound material upslope.

Solifluction is permitted primarily by the loss of cohesion due to filling of openings in the regolith with water. The weakened material flows like a viscous fluid. Accompanying frost heave is of minor importance, although it may dilate the solifluction flow and provide additional openings for meltwater. Measurements of porewater pressure in solifluction slopes show that pore-water pressures are not excessive; in fact, they fluctuate from slightly above atmospheric pressure to less than atmospheric pressure—a circumstance attributed to dilation of the flow by frost heave. Rates of flow appear to be seasonal in proportion to the degree of saturation of the re-

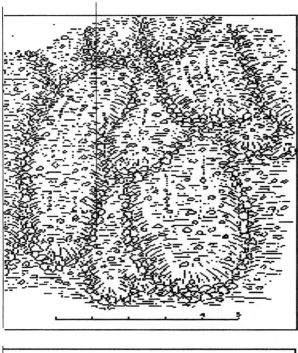

FIGURE 15–3
Solifluction lobes making stone-banked terraces. (Sharp, 1942)

Rock Glaciers

golith. The rates are essentially imperceptible but may range from centimeters per week to centimeters per melt season, for example, 1 to 10 cm (0.4 to 4 in) (Washburn, 1947).

Rock glaciers are large, lobate, tongue-shaped masses of angular rock debris frozen in interstitial ice (Figure 15–4). They move downslope by deformation of the ice contained within them, causing their surface to resemble those of glaciers.

Capps (1910) studied about 30 rock glaciers in Alaska and found that they consisted of masses of angular blocks that, in many cases, extended up to cirque headwalls, but showed no ice at the surface. Their lobate form and surfaces that exhibited concentric wrinkles and ridges (Figure 15–5) gave them distinct impressions of movement. Excavations showed that the interstices between the blocks were filled with ice at depths within a meter of the surface, and in some cases, they graded into true glaciers at their upper ends. Some large rock glaciers reach lengths of 3 km (~2 mi) with terminal embankments 60 m (~200 ft) high. Many of the blocks on the surface can be quite large, some up to 8 m (~25 ft) in diameter.

Measurements of the movements of rock glaciers have been made by in Switzerland (Chaix, 1923, 1943), in the Front Range of Colorado (Ives, 1940), in Alaska (Wahrhaftig and Cox, 1959), and in the French Alps. Chaix (1923, 1943) measured surface velocities of 1–1.5 m/year (3–5 ft) over a 24-year period, and Wahrhaftig and Cox (1959) measured the surface velocity of one Alaskan rock glacier to be 0.76 m/year (2.5 ft) over an 8-year period.

A unique opportunity to observe the internal constituents of a rock glacier arose in Colorado where a mining tunnel was dug from the lower end of a rock glacier. The tunnel first passed through loose rock, then rock with interstitial ice for some distance, and finally through a small amount of glacial ice before encountering solid rock (Brown, 1925). Thus, two rather different relationships of ice to blocks exist in rock glaciers:

1. Rock with interstitial ice
2. Ice carrying rock debris

One of the principal problems of the origin of rock glaciers is whether or not the ice within a rock glacier originated as glacial ice or represents freezing of pore water (Ostrem, 1963); Outcalt and Benedict, 1965; Barsch, 1971; Potter, 1972; Johnson, 1974; White, 1976; Washburn, 1980; Whalley, 1983; Haeberli, 1985; Martin and Whalley, 1987; Giardino and Vitek, 1988). The gradation of many rock glaciers upvalley into true glaciers suggests that tongue-shaped rock glaciers heading in cirques originated as stagnating glaciers. Flow features on the surface of rock glaciers may develop from

1. Deformation of the ice core
2. Movement of the debris cover along the debris-ice interface
3. Deformation from a period of glacial advance
4. Changes in the hydrologic balance

Some relic rock glaciers formed in unglaciated areas support a periglacial origin (Barsch, 1978, 1988, 1992; Haeberli, 1985). These types of rock glaciers form where an adequate supply of debris is combined with percolating snowmelt that freezes to form an ice-debris matrix.

Block Fields, Block Slopes, and Block Streams

Block fields, also known as *felsenmeer,* are broad, relatively level, mountainous areas covered with moderate-to-large-sized angular blocks of rock produced *in situ* from frost wedging, typically 350 m to 1.3 km (0.2 to 0.8 mi) long and 60 to 120 m (~200 to 400 ft) wide. They generally rest on slopes of 3–12° (Caine, 1963a; White, 1976). Similar accumulations of blocky debris on steeper slopes are known as block slopes (Figure 15–6), and accumulations confined to valleys make block streams (Andersson, 1906).

FIGURE 15–4
Rock glacier, Sierra Nevada Range, California. (Photo by W. K., Hamblin)

Accumulation of blocks on a slope typically shows an internal fabric (individual blocks oriented downslope) indicating downslope movement. Near the distal end of the deposit, block orientation may become transverse to the slope. Active block fields, slopes, and streams are common in polar regions (Nichols, 1966) and highland areas.

PERMAFROST AND FROZEN GROUND

Characteristics of Permafrost

All variations exist between seasonally frozen ground and perennially frozen ground *(permafrost)*. Seasonally frozen ground is ground frozen by low air temperatures only during the winter (Muller, 1945). Perennially frozen ground *(permafrost)* has been defined by Muller (1945) as ground:

"…at a variable depth beneath the surface of the earth in which a temperature below freezing has existed continually for a long time (from two to tens of thousands of years). Permanently frozen ground is defined exclusively on the basis of temperature, irrespective of texture, degree of induration, water content, or lithologic character."

This definition, based solely on ground temperature, does not distinguish between dry *permafrost*, in which insufficient ground moisture does not allow interstitial ice to form, and *frozen ground*, in which moisture allows ground ice to form.

FIGURE 15–5
Rock glacier, Sierra Nevada Range, California. (Photo by Ken Hamblin)

FIGURE 15–6
Blockfield of angular rocks broken by frost activity. (Photo by Ken Hamblin)

Because of the importance of subsurface ice, Stearns (1966) proposed an alternate definition of permafrost:

The term *permafrost* is defined as a condition existing below ground surface, irrespective of texture, water content, or geological character, in which;

1. The temperature in the material has remained below 0° continuously for more than 2 years

2. If pore water is present in the material a sufficiently high percentage is frozen to cement the mineral organic particles

A temperature definition alone is not considered sufficient, for often a geothermal situation exists in which a frozen or cemented state is not obtained even though the temperature of the material is well below 0°....

The definition of permafrost given above includes both dry-frozen and wet-frozen ground. In the dry-frozen or dry frost, condition there is very little or no water contained in the pores so that temperature becomes the only criterion. In the wet-frozen condition some cementing ice must be present.

Other definitions have also been proposed. Although workers do not agree upon any single definition, most agree that the term permafrost is useful for ground below the freezing point, whether it be wet or dry. Because of the significant impact of ground ice on geomorphic forms, attention will be focused here on wet frozen ground.

The distinction between seasonally frozen ground and permafrost becomes blurred where the upper zone of permafrost thaws seasonally. This zone is usually considered to be the **active permafrost layer.** The active layer thaws out when temperatures rise sufficiently, and refreezes in winter or during cold spells. Rooted plants may grow in the active layer, and mass movements typically occur while in the thawed state. The thickness of the active layer (typically 15 cm to 5 m; 6 in to 16 in) depends on the ambient air temperature, amount of exposure to solar insolation, conductivity of the soil, and amount of snow, water, or vegetative cover (Embleton and King, 1968; Owens and Harper 1977; Repelewska-Pekabwa and Gluza 1988). It is generally thickest in the subarctic region where temperatures seasonally rise above freezing, and thins toward both the pole and the mid-latitudes where temperatures are generally above or below freezing most of the time (Figures 15–7, 15–8). The active layer is usually thicker in sand and gravel than in fine-grained sediments.

The active layer usually rests directly upon the upper surface of permafrost (the permafrost table), but in places, unfrozen layers (**taliks**) may occur between the active layer and permafrost or within the permafrost (Figure 15–8). Temperatures in permafrost are generally lowest near the ground surface where they are influenced by the ambient air temperature. Daily temperature fluctuations generally affect only the upper meter, and in the active layer, temperatures, by definition, repeatedly cross the freezing point. Below the permafrost table, seasonal temperature fluctuations occur entirely below freezing point and gradually die out with depth, usually 10 to 30 m (~30 to 100 ft), until they vanish at the level of *zero annual amplitude*. At greater depths, temperatures do not fluctuate and rise steadily with depth to unfrozen ground. The geothermal gradient below the level of zero annual amplitude is about 1°C. per 30 to 40 m (100 to 130 ft). Beneath all permafrost is unfrozen ground caused by outflow of heat from the earth's interior. The depth of permafrost depends on a number of variables:

1. Ambient air temperature at the ground surface
2. Rate of escape of geothermal heat
3. Thermal properties of the ground material
4. Moisture content of the ground
5. Changes in thermal properties (thermal conductivity) when ground ice begins to form
6. Lack of protection of the ground from cover of vegetation, water, or snow

Distribution and Thickness

Three permafrost zones have been recognized (Ray, 1951):

- **Continuous permafrost**   In this zone, the only unfrozen areas lie beneath lakes, rivers, or the sea.
- **Discontinuous permafrost**   Small, scattered, unfrozen areas occur.
- **Sporadic permafrost**   Small islands of permafrost occur in a generally unfrozen area, sometimes as relics of a former colder climate.

Continuous or discontinuous permafrost underlies about one-fifth of the earth's surface, including about half of Russia, half of Canada, 80 percent of Alaska, and almost all of Greenland and Antarctica (Stearns 1966; Washurn, 1979; Brown 1970).

The depth of penetration of permafrost is known to reach at least 1500 m (~5000 ft) in the upper Markha River in Siberia, 900 m (~3000 ft) in Udokan, Russia, more than 600 m (~2000 ft) in Bakhynay and Isksi, Russia, Nordvik, Siberia, and Prudhoe Bay, Alaska (Budel, 1960; Ferrians, 1965; Brown, 1967; Washburn, 1979).

Origin of Permafrost

Whether permafrost is the result of present climatic conditions or originated during colder Pleistocene climates has long been discussed. Although most of the world's permafrost is in approximate balance with the present climate, and a general relationship exists between present-day mean annual air temperatures and permafrost zones, discrepancies do occur in some areas (Brown, 1960).

Washburn (1979, 1980) concluded that most permafrost may have originated during the Pleistocene and is now slowly melting with the climatic amelioration since the last glaciation on the basis of the following:

1. Tissue of wooly mammoths and other Pleistocene animals have been preserved in permafrost, indicating presence of permafrost at the time of their death (Gerasimov and Markov, 1968).
2. The upper boundary of some permafrost is considerably deeper than the present depth of winter freezing (Gerasimov and Markov, 1968).
3. In some places, the temperature of permafrost decreases with depth below the surface, suggesting that the residual cold temperature is a relict of colder times (Gerasimov and Markov, 1968).
4. The thickest permafrost is commonly found in unglaciated areas that were not insulated from air temperatures by glacial ice.
5. A permafrost area in Russia was covered by transgression of the Kara Sea, showing that the permafrost predated the marine transgression.

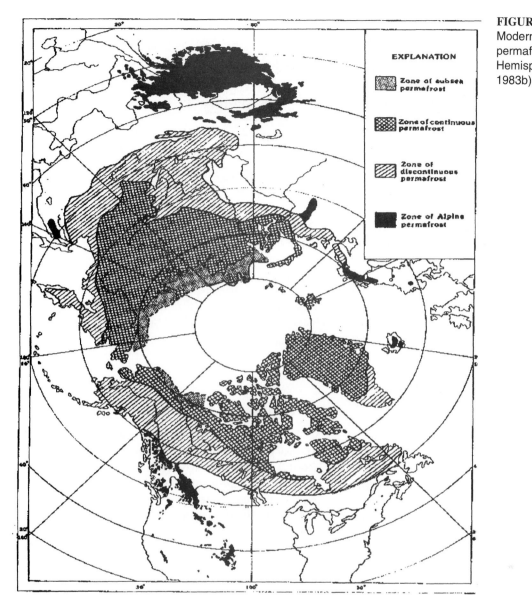

FIGURE 15–7
Modern distribution of permafrost in the northern Hemisphere. (From Pewe, 1983b)

However, permafrost is forming today in polar regions where Pleistocene glaciers formerly covered the land, suggesting that the permafrost formed after the ice disappeared. In Russia, a zone of melting permafrost in the south is separated from a zone of actively forming permafrost by a transition zone in which permafrost is stable under the present climatic conditions. Permafrost commonly occurs in recent floodplain silt (Taber, 1943). Draining of a swamp at Port Nelson, Manitoba in 1929 revealed that, although its bed was then unfrozen, within three winters permafrost had extended to a depth of 10 m (33 ft), the same as the base of permafrost in adjacent areas. The new permafrost extended no deeper, suggesting that the permafrost had reached equilibrium under the present climate. Temperature profiles from northern Alaska show a bending over of the upper profiles (Figure 15–9), indicating a more recent warming response to atmospheric temperature following

the original freezing of the permafrost. Permafrost is affected not only by air temperature, but also by other factors, such as the nature of vegetative cover, whether or not the area has been covered by glacial ice, and the original temperature of formation, so a direct relationship between modern air temperature and permafrost may not necessarily be expected.

Ground Ice

Subsurface frozen ground consists not only of ice filling pore spaces, but also includes masses of clear ice (Figure 15–9) as much as 30 m (~100 ft) across. When the volume of subsurface ice exceeds the pore space in material, discrete forms of solid ice occur.

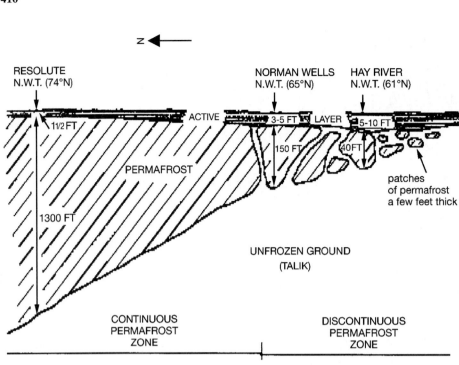

FIGURE 15–8
North-south profile of permafrost in Canada. Permafrost thickness decreases southward and becomes discontinuous.(Brown, 1970)

Ice Lenses

Bands, layers, or wedges of clear ice (segregated ice) form where ice crystals grow in the direction of most rapid conduction of heat away. Experiments by Taber (1929, 1930) showed that ground ice grew freely in silt, less readily in clay, and almost never in sand. Taber found that quartz dust, having a maximum particle size of 0.07 mm (0.003 in) was too coarse for ice to segregate, but in grains 0.01 mm (0.0004 in) or less, ice segregation occurred easily.

Other factors that affect the process of ice segregation include:

1. The shapes of particles (tabular particles are most favorable)

2. Very small voids may delay freezing due to supercooling of the water

3. The rate of cooling (very rapid cooling is unfavorable for ice segregation

4. Surface loading (lenses of ice are smaller and less common with depth below the ground surface because the greater overburden pressure lowers the freezing point

Movement of groundwater toward a subsurface freezing plane is complicated by many factors.

Subsurface ice lenses parallel to the ground surface are the most common forms of segregated ice. Under favorable conditions, ice lenses may grow to large sizes, up to 20 m (65 ft) thick (Taber, 1943; Mackay, 1966; Cressey, 1939). Where ice segregation is well developed, ground heaving occurs, and where thick ice lenses melt, collapse occurs (Harris, 1990).

Ice Wedges and Ice-Wedge Polygons

Near-vertical, downward tapering, ice wedges (Figure 15–10) may extend 10 m (33 ft) or more below the ground surface. Large ice wedges may reach thicknesses of 10 m (33 ft) at the top. They are commonly parts of a polygonal network of ground ice enclosing polygons of frozen ground from 1 to 30 m (~3 to 100 ft) in diameter.

Ice wedges begin as small fractures, ranging from 1 cm to 3 m (0.4 in to 10 ft) wide and from 1 to 10 m (~30 to 30 ft) deep, that penetrate the upper several meters of permafrost and the active layer. They commonly occur where the climate is characterized by intense winter cold and a large temperature range, usually between January and March when ground temperatures are lowest (Mackay 1974, 1992), and the rate of temperature decline is the highest. Ice wedges are usually three to six times deeper than they are wide and are best developed in fine-grained soils with a high ice content (Black 1976).

Growth of an ice wedge proceeds by increments, not necessarily on an annual basis. During the maximum cold of the winter, the active layer freezes and contracts, forming a ground crack that penetrates both the active layer and the permafrost, sometimes accompanied by an audible sound (Leffingwell, 1915). The crack may then fill with snow and water during the summer. Freezing of the water in the crack below the permafrost table creates a small ice wedge or ice vein. Subsequent cracking along the initial fracture during cold winters and filling and freezing of the water in the crack adds new ice that enlarges the ice wedge. The ice wedge does not grow every year. Mackay (1974, 1975a) found that only about 40 percent of the ice wedges on Carry Island in the Northwest Territories, Canada, cracked

FIGURE 15–9
Massive, deformed ground ice, Nicholson Peninsula, MacKensie Delta, Northwest territories. (Photo by H. M. French)

in any given year. Growth rates and the ultimate size of an ice wedge depend on the timing between opening and closing of the winter cracks and the arrival of meltwater in the summer (Mackay 1975a). The maximum width of cracking in any one winter observed by Leffingwell (1915) was 8 to 10 mm (0.3 to 0.4 in). The thickness of an annual increment is usually less than the size of the winter crack. Thus, a large ice wedge may take hundreds of years to form. Where the climate is particularly dry, wind-blown sand may fill the crack and form a sand wedge, rather than an ice wedge (Pewe, 1959; Carter, 1983).

Vertical foliation in an ice wedge is due to fine particles that are brought into the contraction crack by meltwater from above. The successive evolution of an ice wedge is shown in (Figure 15–11) (Lachenbruch, 1962).

Ice wedges cease to form when thermal contraction cracking ceases. This may be the result of a climatic change or burial by snow, water, or mass wasting debris that insulate the ground from the large changes of temperature necessary for thermal contraction. As ice wedges melt, sediment collapses into the space left by the melted ice, leaving a wedge-shaped mass of debris known as an **ice-wedge cast**. Although ice-wedge casts provide good evidence of an earlier permafrost condition, not all wedge-like features are ice wedge casts. For example, **soil wedges** are common in subarctic regions where a zone of deep seasonal frost is frequently present. They originate by frost cracking in this zone and filling of the crack with sediment.

Ice-wedge casts are common in areas that were once subjected to a severely cold climate. Features useful in their recognition include:

1. The source of the sediment filling a former ice wedge must have come from above or the sides
2. The former ice wedge is nearly vertical
3. The sides converge with depth
4. Elongate stones stand on end in the wedge where they have slid from above (Johnsson, 1959, 1962)

FIGURE 15–10
Ice wedge, MacKensie Delta region. (Photo by H. M. French)

cussed later) that are usually not larger than about 3 m (10 ft) and nearly always consist of sorted material. The perimeters of some ice-wedge polygons consist of minor ridges, slightly above the center of the polygon, whereas others have higher central cores than the perimeters, which are depressed by melting of the ice wedges (Pewe 1966a, 1966b; Black 1974, 1976).

Ice-wedge polygons may consist of tetragonal, rectangular, or five or six-sided polygons (Figure 15–12), almost always formed in sediments, but rarely occur on bedrock where intense frost activity has worked along polygonal jointing. The best-developed polygons occur on flat, nearly horizontal, surfaces.

PERIGLACIAL LANDFORMS

Pingos

Pingo, an Inuit term for an isolated, dome-shaped hill, was first used by Porsild (1938) to describe large ice-cored mounds found in the Mackenzie Delta region, Canada , (Figure 15–13). They are infrequent, but typical, features of the permafrost regions of Alaska, Canada, Greenland, and Siberia, and they have also been found in sub-arctic forests in the discontinuous permafrost zone of North America and Siberia (Holmes et al., 1968). On the Mackenzie Delta and on the northern coastal plain of Alaska, 1000–1400 pingos occur (Muller, 1962; Ferrians, 1988); ~70 occur in the Brooks Range, and more than 300 in the discontinuous permafrost zone of central Alaska and the Arctic Islands.

Pingos range in height from a few meters to more than 40 m (130 ft) and from a few meters to 1000 m (3300 ft) in diameter (Leffingwell, 1915; Mackay 1962, 1973; Washburn, 1980; Ferrians, 1988). Small pingos typically have rounded tops, but larger ones are commonly broken open at the top where melting of the ice core forms a crater resembling a volcanic cone (Figure 15–13). Where they occur in stratified silt or sand, the beds commonly dip outward from the center, much like those adjacent to an intrusive body. Some pingos that develop in bedrock show similar deformation (Cruickshank and Colhoun, 1965). The ice in the core of a pingo is typically massive and of segregation/injection origin. Tension fractures are common at the summit of the mound, but expansion of pingo ice is rare and short-lived. Ice up to 7 m (23 ft) thick has been found in pingos of Sweden (Lagerback and Rodhe, 1986). As the ice core melts, a small freshwater lake may occupy the summit crater that forms.

Open System Pingos

Open-system pingos form where groundwater under artesian pressure beneath thin permafrost forces its way upward and freezes as it approaches the surface where it forms an ice core that heaves the surface upward (Muller, 1959, 1963). Although the initial growth of these types of pingos may occur where ice lenses lie above the water table, their continued

Modern, actively growing ice wedges are found only in areas of continuous permafrost where the mean annual temperature is −6° to −8° C (Pewe 1973). They are generally absent or inactive in the discontinuous permafrost zone, although some have been reported in the Yukon at the southern limit of discontinuous permafrost. However, frost cracking is known from seasonally frozen ground in non-arctic regions. For example, frost-cracking occurred on a golf course in New Hampshire during an exceptionally cold December in 1958, during which the mean temperature was −15°C. Little or no snow cover insulated the ground, which froze to depths as great as 2 m (~7 ft), and a network of intersecting cracks appeared, some enclosing polygons 6 to 30 m (20 to 100 ft) across and 5 mm (0.2 in) wide at the surface (Washburn et al., 1963).

Ice wedges commonly form in a connected, polygonal pattern, somewhat similar in shape to polygonal mud cracks, and known as **ice-wedge polygons.** They may be more than 10 m (33 ft) in diameter, and some Pleistocene relicts in the Avon valley in England (Shotton, 1960) are up to 60 m (200 ft), in contrast to the smaller patterns produced by frost sorting (dis-

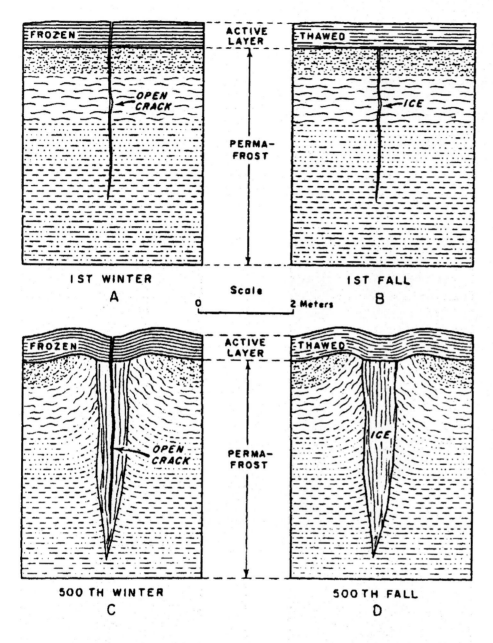

FIGURE 15–11
Evolution of ice wedges. (From Lachenbach, 1962)

growth requires a particular combination of hydrostatic pressure and soil permeability (Cruickshank and Colhoun, 1965; Washburn, 1969; Mackay, 1973). Thin, discontinuous permafrost and artesian water pressure play important roles in the development of open system pingos. The role of artesian pressure is not to force the overlying sediments upwards but rather to provide a slow, regular supply of groundwater to the growing ice core.

Most open-system pingos are oval or oblong in shape and typically occur as isolated mounds or in small groups developed in either soil or bedrock. Rupturing near their top is common. Concentrations of open-system pingos occur in the northern interior of the Yukon (Hughes, 1969) and central

Alaska (Holmes et al., 1968). They also occur in northern Alaska (Hamilton and Curtis, 1982); Spitsbergen, Norway; and central Yakutia, Siberia (Soloviev, 1973). Why nearly all open system pingos occur in unglaciated terrain is not clearly understood.

Closed System Pingos

When a lake in a permafrost environment is progressively drained and covered by encroachment of vegetation from the margins, the permafrost table progressively rises to the level of the former lake floor (Figure 15–14). The rising permafrost table expels pore water ahead of the freezing front, and when

A.

B.

FIGURE 15–12
A. Ice-wedge polygons and thaw lakes, southern Banks Island, Northwest Territories, Canada. (Photo by H. M. French)
B. Ice-wedge polygons near Amundson's 1924 winter camp. (Photo by R. Schmidt)

the pore-water pressure exceeds the overburden strength, upward heaving of the frozen ground occurs as the ice core grows progressively. The size and shape of the resulting pingo typically reflects that of the original body of water.

Closed system pingos vary in height from a few meters to over 60 m (~200 ft) and up to 300 m (~1000 ft) in diameter, ranging from symmetrical conical domes to asymmetric and elongate hills. The tops of the pingos are commonly ruptured to form small, star-like craters that eventually form shallow-rimmed depressions as the ice core melts.

Numerous pingos of this origin occur in the Mackenzie Delta of Canada. Of the 1380 pingos mapped in the Macken-

zie Delta (Mackay 1973), 98 percent are located at or near present or former lake basins (Porsild, 1938; Muller, 1959, 1962; Mackay, 1962, 1966, 1972 Pissart, 1970; Flemal, 1976). Others occur on the Yukon coastal plain, western Victoria Island, Banks Island, Baffin Island and other areas (French, 1996).

Figure 15–14 illustrates the mechanism of pingo formation in a closed system (Mackay, 1962). A deep, ice-covered lake is surrounded by permafrost. The lake inhibits the development of permafrost beneath it, and the ground remains unfrozen. As the lake slowly drains or is filled with sediment, at some point the lake ice freezes to the bottom, and the bottom sediments begin to freeze. As the layer of ice and per-

A.

B.

FIGURE 15–13
A. Pingo, Thompson River,
northern Banks Island,
Northwest Territories (Photo by
H. M. French)
B. Collapsed, thawed pingo,
southern Banks Island,
Northwest Territories. (Photo by
H. M. French)

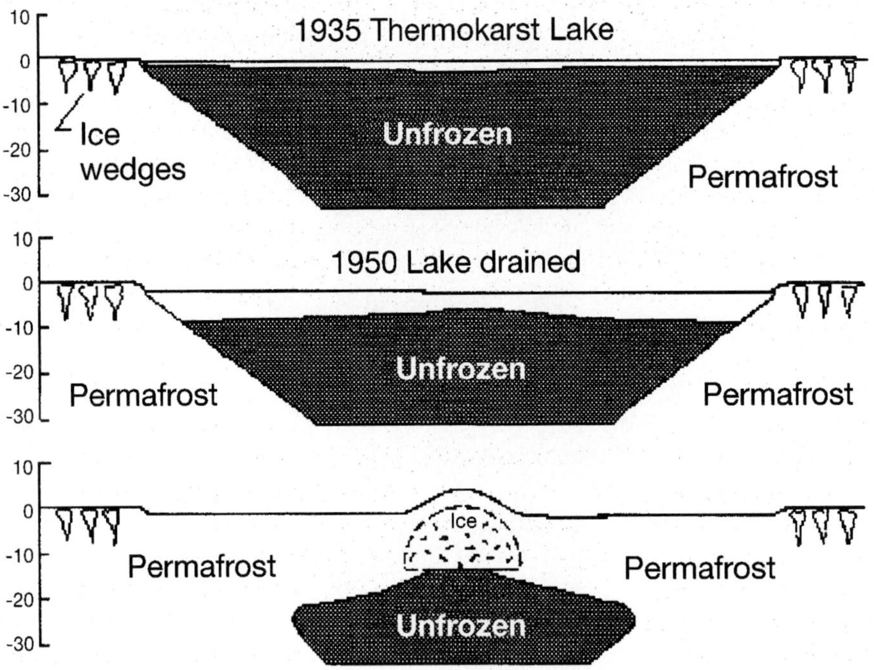

FIGURE 15–14
Growth of a closed-system pingo
in the Mackenzie delta area.
(Modified after MacKay, 1962)

mafrost covers the former lake floor, a closed-system is set up in the still-unfrozen ground beneath because the permafrost cap prevents the escape of groundwater. As permafrost continues inward growth around the unfrozen core, water pressure increases. Pore water is expelled from the unfrozen sediment by the advancing permafrost, and to relieve the pressure, the surface bulges upward. Eventually, all of the water in the enclosed system groundwater mass becomes frozen, and the excess water forms a core of clear ice under the bulge.

Growth Rate of Pingos

The birth and growth of a small pingo studied by Mackay (1988) is representative of more than 2000 closed system pingos of the western Canadian Arctic and adjacent Alaska. The pingo appeared on the former floor of a lake that drained suddenly about 1900. Small frost mounds began appearing between 1920 and 1930. Arctic botanist A. Porsild photographed one 3.7 m high (12 ft) frost mound in 1935 and later described it as part of a paper on earth mounds (Porsild, 1938). The pingo grew steadily until 1976, but the growth rate decreased after that. Mackay (1972, 1973, 1979b) also monitored the growth of other small pingos in a lake in the Mackenzie Delta region that drained between 1935 and 1950. The pingos grew rapidly in the initial years, commonly 1.5 m/yr (5 ft/yr), then decreased. Some of the largest pingos in the Mackenzie Delta are growing only 2.3 cm/yr (1 in/yr) and are probably more than a thousand years old (Mackay, 1986). Mackay suggests that about 15 new pingos per century appear in the Mackenzie Delta region, and only about 50 seem to be actively growing. Similar conclusions have been reached by Russian investigators in Siberia (Soloviev, 1973).

Mackay noted the growth rate of a pingo at about 15 cm/yr (6 in) between 1969 and 1971, but growth rates of more than a meter a year have also been reported (Washburn, 1980; Mackay, 1973, 1990).

Palsas

Palsas are low, permafrost mounds with cores of interlayered segregated ice and peat or soil. They are typically 1 to 7 m (3 to 23 ft) high, 10 to 30 m (33 to 100 ft) wide, and 15 to 150 m (50 to 500 ft) long (Forsgren, 1968) and less than 100 m (330 ft) in diameter. The term palsa comes from Scandinavia where it means "a hummock rising out of a bog with a core of ice" (Seppala, 1972). They originate by ice segregation in peat and soil and are distinct from seasonal frost mounds. Palsas are regarded reliable surface indicators of permafrost in the discontinuous zone.

The origin of palsas is clearly related to the unique thermal properties of peat that govern ground freezings and ice segregation, but exactly how these elements interact remains to be conclusively demonstrated (French, 1996; van Everdingen, 1978; Akerman, 1982; Seppala, 1982, 1986; Brown et al., 1983; Nelson et al., 1992).

Seppala (1982) suggested that palsa formation is triggered when thinning of the snow cover on some parts of a peat bog allows frost to penetrate more deeply, causing the surface to heave. Once formed, the mound becomes more snow-free each winter and permafrost continues to develop. In a three-year field experiment in Finnish Lapland, snow was removed from a plot of ground several times each winter. Permafrost formed and a small palsa grew ~30 cm (12 in) high (Seppala, 1982).

Palsas occur in greatest numbers at the southern margin of the discontinuous permafrost zone in the subarctic regions of Canada, Iceland, Sweden, and Russia but also occur wherever suitable climatic conditions combine with bog terrain (Kershaw and Gill, 1979). The southern limit of palsas in Canada has been mapped as coincident with the 0°C mean annual air temperature isotherm (Zoltai, 1971) south of the limit of sporadic discontinuous permafrost.

Thermokarst

Thermokarst includes a diverse assemblage of collapsed landforms resulting from thawing of ground ice. Thermokarst is best developed where numerous, thick ground-ice masses occur in the upper parts of the permafrost (Dylik, 1964; Kachurin, 1962; Higgens et al., 1990). Among the common forms created are surface depressions and basins, funnel-shaped sinks, dry valleys, ravines, and conical hillocks. They somewhat resemble landforms of karst regions, except that the basic process is melting, rather than solution. Among the many individual forms are linear and polygonal troughs, collapsed pingos, thaw lakes, beaded drainage, and alases (Kachurin, 1962).

The thawing of ice wedges commonly leaves troughs, and the thawing of ice wedge polygons may result in prominent inter-trough mounds or low, conical hills (Czudek and Demek, 1970). Areas of mound topography beyond the margin of Pleistocene glaciations have long defied conclusive explanation, but may be related to the thawing of polygonal ice wedges. The Mima Mounds of southwest Washington (Figure 15–15), first observed about 1845, consist of fairly symmetrical mounds 1 to 2 m (3 to 7 ft) high and 3 to 20 m (10 to 65 ft) in diameter developed on an outwash plain of sand and gravel. Although their origin remains a mystery, Ritchie (1953) suggested the following sequence of events:

1. Freezing of the outwash deposits and the formation of ice-wedge polygons

2. Thawing of the ice in the ice-wedges, leaving a rounded, frozen core

3. Removal of sediment during draining of the thawed ice wedges

However, in a review of possible origins of the mounds, Washburn (1988) concluded that none seemed to explain all of the observed features of the mounds.

The thawing of ice at junctions of ice-wedge polygons creates small ponds that may then be connected by short, straight streams flowing along thawing ice wedges (Figure 15–16). On the north slope of Alaska and in other wet tundra terrain,

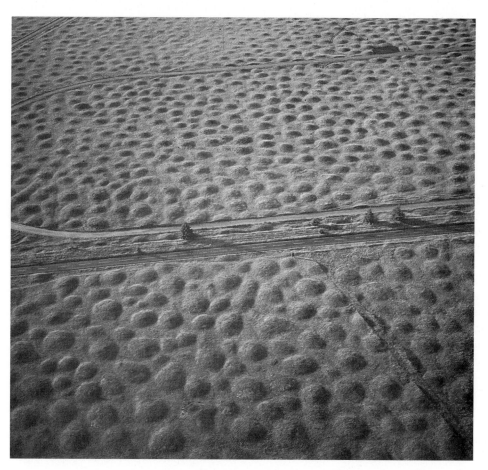

FIGURE 15–15
Mima Mounds near Tenino, Washington. (Photo by D. A. Rahm)

they give the appearance of a string of beads—hence the name, **beaded drainage** (Hopkins *et al.,* 1955). The ponds are typically 1 to 3 m (3 to10 ft) deep and up to 30 m (100 ft) in diameter.

Depressions formed by thawing of frozen ground and filled with water form **thaw lakes**, also known as thaw depressions, thermokarst lakes, and tundra ponds (Wallace, 1948; Hopkins, 1949; Black and Barksdale, 1949; Mackay, 1963; Hussey and Michelson, 1966; Black, 1969; Tedrow, 1969; Ferrians et al., 1969; Sellman et al., 1975). As ground ice melts, the water in the resulting lakes is warmer than the frozen ground, so the lakes expand along their margins (Burn 1992). Thus, bank erosion is rapid along lake margins and along river banks (Walker, 1988; Burn, 1992; Carter, 1987). Some thaw lakes may have originated as collapsed pingos, forming shallow, oval, depressions with a slightly higher rim. Such depressions are commonly difficult to distinguish from other thermokarst depressions or glacial kettles. Collapsed pingos have been found in the former periglacial regions in Europe and Asia (Mullenders and Gullentops, 1969; Pissart, 1956, 1965; Wiegand, 1965).

Elongate, oval, oriented lakes with long axes parallel to others having a similar alignment may develop in permafrost areas. Exceptionally good examples of oriented lakes have been described in the Point Barrow region (Carson and Hussey, 1962, 1963), as well as along the Arctic coastal plain (Black and Barksdale, 1949; Mackay, 1963) in the interior Yukon, northwest Banks Island (Harry and French, 1983), the alluvial lowlands of northern Siberia (Tomirdiaro and Ryabchun, 1978), and in other parts of Arctic Canada (Dunbar and Greenaway, 1956; Bird, 1967). Typical oriented lakes may have a variety of shapes: elliptical, rectangular, oval,

triangular, or D-shaped (Mackay, 1963; Bird, 1967; Harry and French, 1983), ranging in size from small ponds to lakes 15 km (9 mi) long and 6 km (3.7 mi) wide (Black and Barksdale, 1949). Some have a shallow shelf surrounding a deeper central part 6 to 10 m (20 to 33 ft) deep, while others have a uniform saucer-shaped cross profile with depths of less than 2 m (7 ft). Despite extensive study, and the observation that long axes of the lakes are typically at right angles to the present-day prevailing winds, the reason such lakes have parallel orientations remains unexplained (Mackay, 1963; French, 1996).

Alases are a specific type of closed thermokarst depression found in central Yakutia. Typically, they have steep sides and flat floors, formed by subsidence following the melting of ground ice (Czudek and Demek, 1970). They are commonly oval and may enclose shallow lakes.

Patterned Ground

The land surface in periglacial regions is commonly characterized by ground material having distinct, symmetrical, geometric shapes, known collectively as **patterned ground**. The most common shapes found in patterned ground are polygons, circles, stripes, nets, and steps. They are common in permafrost regions, but can be found anywhere within the periglacial environment where freezing and thawing alternate.

An early description of patterned ground was given in 1884 by von Bunge.

"When one steps upon the tundra almost anywhere in the (Lena) delta, it appears divided into countless irregular polygons of differing size, whose edges are higher than the middle. Between the edges of two such polygons, a small furrow is found, which is used as a pathway by lemming."

Since then, similar phenomena have been observed in many periglacial regions (e.g., Washburn, 1956a, 1956b, 1980; Shotton, 1960; Pewe, 1963; Corte, 1963; Goldthwait 1976; among many others).

In an attempt to classify various types of patterned ground, Washburn (1956a) suggested using a combination of geometric shape and determining whether the material composing the patterned ground is sorted or unsorted. Sorting of coarse and fine fractions of sediments typically results in patterns made of fine particles in the center, rimmed with stones.

Polygons

Polygons are a common form of patterned ground. They can form either in permafrost areas or in areas having seasonal frost. Some types of polygons form only in permafrost, but others have been shown to be actively forming under present climatic conditions in Britain and Scotland (Miller et al., 1954).

Small, sorted polygons typically consist of a central core of fine material, bounded by relatively straight lines of stones (Figure 15–17). Tabular stones making the sides of the polygon are commonly oriented parallel to the direction of the stone line. The stones typically decrease in size with depth. Small, nonsorted polygons differ in from sorted polygons in

that stones are generally absent along their borders and are marked by furrows instead.

Circles

Circles are more widespread forms ranging from a few centimeters to several meters in diameter. They typically reach dimensions of 1 to 3 m (3 to 10 ft) in diameter and occur singly or in groups. Sorted circles consist of a central area of finer material surrounded by a border of stones (Figure 15–18). The size of stones in the border appears to be bigger for larger-sized circles and to decrease in size with depth below the surface. Although common in both polar and high mountain areas, they are not restricted to permafrost areas.

Nonsorted circles consist of a central core of finer material, but lack a stony rim, being bounded instead by a margin of vegetation (Figure 15–19). They are commonly found in polar, subpolar, and high alpine environments.

Nets

Nets (Figure 15–20) consist of patterned ground that is neither circular nor polygonal, typically 1 to 2 m (3 to 7 ft) in diameter. Although they are most common in sub-arctic and Alpine areas, permafrost is not necessary for their development.

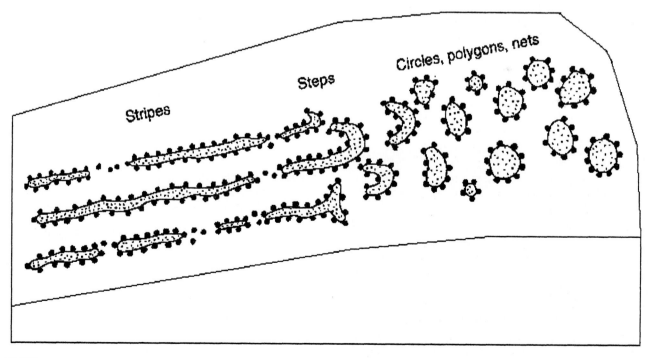

FIGURE 15–17
Pattern ground classification.

FIGURE 15–18
Sorted stone circles, Alaska Range, Alaska. (Photo by O. J. Ferrians, U.S. Geological Survey)

FIGURE 15–19
Nonsorted stone circle with vegetation rim, southern Banks island, Northwest Territories. (Photo by H. M. French)

Steps

Steps (Figure 15–20) are terrace-like forms of patterned ground with a downslope border of stones or vegetation making a low riser that fronts a tread of relatively bare ground upslope. They are developed from circles, polygons, or nets, rather than developing independently, and like their parent forms, can be nonsorted or sorted, depending on whether the riser is characterized by vegetation only or by stones (Washburn, 1973). The longest dimension of steps is typically transverse to the steepest slope, usually 5–15° (Sharp, 1942a; Washburn, 1969). Some sorted steps represent an intermediate stage between sorted polygons and sorted stripes (Sharp, 1942a,).

Stripes

The rather irregular step forms merge into **stripes** with increasing slopes, usually at angles between 2° and 7°. Stripes are lines of stones, vegetation, or soil on slopes Figures 15–21 and 15–22). As with other types of patterned ground, stripes may be either sorted or nonsorted.

Sorted stripes are elongate lines of stones separated by intervening zones of smaller stones, fine sediment, or vegetation. They are generally several centimeters to several meters wide and may be more than 100 m long. As in other patterned ground, big stripes usually contain larger stones and the largest stones occur at the surface. The long axes of stones are commonly oriented parallel to the slope. Stripes are the most common form of patterned ground on steep slopes and are rarely found on slopes of less than 3° (Evans 1976).

Nonsorted stripes consist of vegetation or soil with intervening zones of bare ground. They are usually not as long as the sorted variety and are sometimes discontinuous. Both large and small forms occur (Washburn, 1947, 1969). Smaller varieties of stripes have been found forming under present climatic conditions in areas of seasonal frost activity (Miller et al., 1954).

Origin of Patterned Ground

The conditions responsible for creating these diverse forms of patterned ground are varied enough that no single process can explain them all. Some forms undoubtedly result from a combination of processes and some forms are transitional into other forms.

Sorting of sediments in patterned ground takes place by:

1. Frost heaving where freezing and thawing occur from the surface downward. Coarse particles move upward and fine particles move downward to the bottom of the freeze-thaw layer, resulting in vertical sorting.

2. Frost-pushing of fine particles in front of a freezing plane. Freezing and thawing proceed either from the top or sides and particles move away from the cooling front, leaving the coarsest particles on the cooling side. This results in lateral sorting.

A number of recent studies suggest a circular or convection-like movement to be involved in patterned ground formation (Hallet and Prestrud 1986; Hallet et al., 1988). A circular motion of particles produced during freezing and

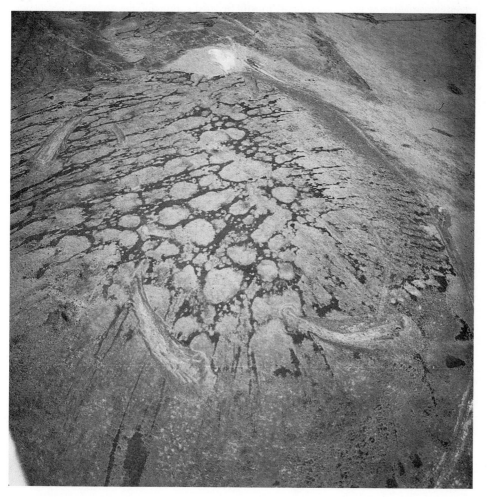

FIGURE 15–20
Stone circles grading downslope into nets and stone stripes, near Klamath Falls, Oregon. (Photo by D. A. Rahm)

FIGURE 15–21
Sorted stone stripes, Prince Patrick island, Northwest Territories. (Photo by H. M. French)

FIGURE 15–22
Non-sorted stone stripes, northern Banks Island, Northwest Territories. (Photo by H. M. French)

thawing concentrates fine-grained sediments in the centers and coarse sediments along the rims, resulting in a geometric pattern with regular spacing. Differences in soil texture, moisture content, and degree of compaction result in differences in thermal properties that influence the shape of the underlying ice front, which in turn determine the relative effectiveness and pattern of sorting mechanisms. The presence of relict patterned ground is commonly assumed to be an

indicator of former periglacial climatic conditions (Clark, 1968; Clark and Ciolkosz, 1988; Johnson 1990).

The rate of lateral movement of large stones (>1.3 cm (0.5 in) in length) ranges from 0.45 to 0.89 cm/yr (0.2 to 0.35 in/yr) (Vitek, 1983). Hallet et al. (1988) suggest that convective activity can displace stones at rates up to 10 mm/yr (0.4 in/yr). Larger stones move more slowly than smaller ones, probably mostly by frost action and the growth of needle ice.

REFERENCES

Akerman, H. J., 1982, Observations of palsas within the continuous permafrost zone in eastern Siberia and in Svalbard: Geografiska Tidsskrift, v. 82, p. 45–51.

Akerman, H. J., 1993, Solifluction and creep rates 1972–1991, Kapp Line, West Spitzbergen: in Frenzel, B., Matthews, J. A., and Glasere, B., eds., Solifluction and climatic variation in the Holocene, Gustav Fischer Verlag, Stuttgart, p. 225–250.

Anderson, S. R., 1988, Upfreezing in sorted circles, western Spitzbergen: in Permafrost, 5th International Conference Proceedings, K. Senneset, ed., Tapir Publishers, Trondheim, Norway, p. 666–670.

Andersson, J. G., 1906, Solifluction, a component of sub-aerial denudation: Journal of Geology, v. 14, p. 91–112.

Associate Committee on Geotechnical Research, 1988, Glossary of permafrost and related ground ice terms: Permafrost Subcommittee, National Research Council of Canada, Ottawa, Canada, Technical Memorandum 142.

Ballantyne, C. K., 1987, The present-day periglaciation of upland Britain: in Boardman, J. (ed.), Periglacial processes and Landforms in Britain and Ireland:

Cambridge University Press, Cambridge, England, p. 113–126.

Ballantyne, C. K., and Harris, C., 1994, The Periglaciation of Great Britain: Cambridge University Press, Cambridge, England, 330 p.

Barsch, D., 1971, Rock glaciers and ice-cored moraines: Geografiska Annaler, v. 53A, p. 203–206.

Barsch, D., 1977, Nature and impedance of mass wasting by rock glaciers in alpine permafrost environments: Earth Surface Processes, v. 2, p. 231–245.

Barsch, D., 1978, Active rock glaciers as indicators of discontinuous permafrost. Art example from the Swiss Alps: in Proceedings of the 3rd International Conference on Permafrost, National Research Council, Ottawa, Canada v. 1, p. 349–352.

Barsch, D., 1988, Rockglaciers: in Clark, M. I., ed., Advances in periglacial geomorphology, John Wiley and Sons, Chichester, England, p. 69–90.

Barsch, D., 1992, Permafrost creep and rock glaciers: Permafrost and Periglacial Processes, v. 3, p. 175–188.

Barsch, D., 1993, Periglacial geomorphology in the 21st century: Geomorphology, v. 7, p. 141 163.

Barsch, D., Fierz, H., and Haeberli, J.W., 1979, Shallow core drilling and bore-hole measurements in permafrost of an active rock glacier near the Grubengletscher, Wallis, Swiss Alps: Arctic and Alpine Research, v. 11, p. 215–228.

Benedict, J. B., 1970a, Downslope soil movement in a Colorado alpine region. Rates, processes, and climatic significance: Arctic and Alpine Research, v. 2 p. 165–226.

Benedict, J. B., 1970b, Frost cracking in the Colorado Front Range: Geografiska Annaler, v. 52A, p. 87–93.

Benedict, J. B., Benedict, R. J., and Sanville, D., 1986, Arapaho rock glacier, Front Range, Colorado; a 25-year resurvey: Arctic and Alpine Research, v. 18, p. 349–352.

Bennet, L. R., and French, H. M., 1991, Solifluction and the role of permafrost creep, eastern Melville Island. N.W.T., Canada: Permafrost and Periglacial Proceedings, v. 2, p. 95–102.

Bird, J. B., 1967, The physiography of Arctic Canada: Johns Hopkins Press, Baltimore, MD, 336 p.

Black, R. F., 1969, Thaw depressions and thaw lakes: A review: Biuletyn Peryglacjalny, v. 19, p. 131–150.

Black, R. F., 1974, Ice-wedge polygons of northern Alaska: *in* Coates, D. R., ed., Glacial geomorphology, Publications in geomorphology, State University of New York, Binghamton, N. Y., p. 247–275.

Black, R. F., 1976, Periglacial features indicative of permafrost: Ice and soil wedges: Quaternary Research, v. 6, p. 3–26.

Black, R. F., and Barksdale, W. L., 1949, Oriented lakes of northern Alaska: Journal of Geology, v. 57, p. 105–118.

Boardman, J., 1992, Periglacial Geomorphology: Progress in Physical Geography, v. 16, p. 339–345.

Bose, M., 1991, A palaeoclimatic interpretation of frost-wedge casts and aeolian sand deposits in the lowlands between Rhine and Vistula in the Upper Pleniglacial and Late Glacial: Zeitschrift für Geomorphologie, N.F. Supplement-Band 90, p. 15–28.

Brown, R. J. E., 1960, The distribution of permafrost and its relation to air temperature in Canada and the U.S.S.R: Arctic, v. 13, p. 163–177.

Brown, R. J. E., 1967, Permafrost in Canada: Canada Geological Survey Map 1246A.

Brown, R. J. E., 1970, Permafrost in Canada: Toronto University Press, Toronto, Canada.

Brown, J., Nelson, F., Brockett, B., Outcalt, S. T.; and Everett, K. R., 1983, Observations on ice-cored mounds at Sukakpak Mountain, South-Central Brooks Range, Alaska: Proceedings of the 4th International Conference on Permafrost, National Academy Press, Washington, D.C., v. 1, p. 91–96.

Brown, W. H., 1925, A probable fossil glacier: Journal of Geology, v. 33, p. 464–466.

Budel, J., 1944, Die morphologischen Wirkungen des Eiszeitklimas im geltscherfreien Gebiet: Geologische Rundschau, v. 34, p. 482–519.

Budel, J., 1953, Die 'periglazial' morphologischen Wirkungen des Eiszeitklimas auf der Ganzen Erde: Erdkunde, v. 7, p. 249–266.

Budel, J., 1960, Die Frostschott-zone Sudorst Spitzbergen: Collogquium Geographica, Bonn, v. 6, 105 p.

Burn, C. R., 1992, Thermokarst lakes: The Canadian Geographer, v 36, p. 81–85.

Burn, C. R., and Smith, M. W., 1990, Development of thermokarst lakes during the Holocene at sites near Mayo Yukon Territory, Canada: Permafrost and Periglacial Proc., v. 1, p. 161–176.

Caine, T. N., 1963a, The origin of sorted stripes in the Lake District, northern England: Geografiska Annaler, v. 45, p. 172–179.

Caine, T. N., 1963b, Movement of low angle scree slopes in the Lake District, northern England: Revue de Geomorphologie Dynamique, v. 14, p. 171–177.

Cailleux, A., 1942, Les actions eoliennes periglaciaires en Europe: Societe Geologique, Memoire 46, 176 p.

Capps, S. R., 1910, Rock glaciers in Alaska: Journal of Geology, v. 18, p. 359–375.

Capps, S. R., 1919, The Kantishna region, Alaska: U. S. Geological Survey Bulletin, v. 687, 116 p.

Carson, C. E., and Hussey, K. M., 1962, The oriented lakes of Arctic Alaska: Journal of Geology, v. 70, p. 417–439.

Carson, C. E., and Hussey, K. M., 1963, The oriented lakes of Arctic Alaska: a reply: Journal of Geology, v. 71, p. 532–533.

Carter, L. D., 1983, Fossil sand wedges on the Alaskan arctic coastal plain and their paleoenvironmental significance: *in* Proceedings of the 4th International Permafrost Conference, National Academy Press, Washington, D.C., p. 109–114.

Carter, L. D., 1987, Oriented lakes: *in* Graf, W., ed., Geomorphic Systems of North America, Geological Society of America, p. 615–619.

Chaix, A., 1923, Les coulees de blocs du Parc National Suisse dEngadine, Le Gloloe, Geneve, Switzerland, v. 62, p. 1–38.

Chaix, A., 1943, Les coulees de blocs du Parc National Suisse—nouvelles mesures et comparison avec les "rock streams" de la Sierra Nevada de Californie, Le Gloloe, Geneve, Switzerland, v. 82, p. 121–128.

Clark, M. J., 1988, Periglacial hydrology: *in* Clark, M. J., ed., Advances in periglacial geomorphology, John Wiley and Sons, Chichester, England, p. 415–462.

Clark, M. J., and Ciolkosz, E. J., 1988, Periglacial geomorphology of the Appalachian Highlands and Interior Highlands south of the glacial border—a review: Geomorphology, v. 10, p. 475–477.

Corte, A. E., 1962a, The frost behavior of soils: laboratory and field data for a new concept—II, Horizontal sorting: U.S. Army Cold Regions Research Engineering Laboratory, Rep. 85, 20 p.

Corte, A. E., 1962b, Vertical migration of particles in front of a moving freezing plane: Journal of Geophysical Research, v. 67, p. 1085–1090.

Corte, A. E., 1963, Relationship between four ground patterns, structure of the active layer, and type and distribution of ice in the permafrost: Biutetyn Peryglacjalny, v. 12, p. 7–90.

Corte, A. E., 1966, Particle sorting by repeated freezing and thawing: Biutetyn Peryglacjalny, v. 15, p. 175–240.

Corte, A. E., 1971, Laboratory formation of extrusion features by multicyclic freeze-thaw in soils *in* Etude des phenomenes periglaciaires en laboratoire, Colloque International de Geomorphologie, Liege-Caen, Centre de Geomorphologie a Caen, Bulletin 13-14-15, p. 117–131.

Cressey, G. B., 1939, Frozen ground in Siberia: Journal of Geology, v. 47, p. 472–488.

Cruickshank, J. G., and Colhoun, E. A., 1965, Observations on pingos and other landforms in Schuchertdal, north-east Greenland: Geografiska Annaler, v. 47 p. 224–236.

Czudek, T., and Demek, J., 1970, Thermokarst in Siberia and its influence on the development of lowland relief: Quaternary Research, v. 1, p. 103–120.

Davison, C., 1889, On the creeping of the soil-cap through the action of frost: Geological Magazine, v. 6, p. 255–261.

Dawson, A. G., 1992, Ice Age Earth. Late Quaternary Geology and Climate: Routledge, London, 293 p.

Dixon, J. C., and Abrahams, A. D., 1992, Periglacial geomorphology: John Wiley and Sons, Chichester, England.

Dunbar, M., and Greenaway, K. R., 1956, Arctic Canada from the air: Queen's Printer, Ottawa, Canada, 541 p.

Dybeck, M. W., 1957, An investigation into soil polygons in central Iceland: Journal of Glaciology, v. 3, p. 143–146.

Dyke, L. S., 1984, Frost heaving of bedrock in permafrost regions: Bulletin of the Association of Engineering Geologists, XXXI(4), p. 389–405.

Dylik, J., 1953, Periglacial investigations in Poland: Bulletin Societe des Sciences et des Lettres de Lodz, v. 4, p. 1–16.

Dylik, J., 1964, Elements essentiels de la notion de 'periglaciaire': Biuletyn Peryglacjalny, v. 14, p. 111–132.

Edelman, C. H., Florschutz, F., and Jeswiet, J., 1936, Uber spatpleistozane und fruhholozane kryoturbate Ablagerungen in den ostlichen Niederlanden: Verhandelingen: Geologish-Mijnbouwkundig Genootschap voor Nederland en Kolonien, Geologish Series, v. 11, p. 301–360.

Elton, C. S., 1927, The nature and origin of soil polygons in Spitsbergen: Quarterly Journal of the Geological Society of London, v. 83, p. 163–194.

Embleton, C., and King, C. A. M., 1968, Glacial and Periglacial Geomorphology: Edward Arnold Ltd., London, 608 p.

Embleton, C., and King, C. A. M., 1975, Periglacial geomorphology: Halsted Press, N.Y.

Etzelmuller, B., and Sollid, J. L. 1991, The role of weathering and pedological processes for the development of sorted circles on Kvadehuksletta, Svalbard - a short report: Polar Research, v. 9, p. 181–191.

Evans, R., 1976, Observations on a stripe pattern: Biuletyn Peryglacjalny, v. 25, p. 9–22.

Ferrians, O. J., 1965, Permafrost map of Alaska: U. S. Geological Survey Miscellaneous Geological Investigations Map I-445.

Ferrians, O. J., 1988, Pingos in Alaska: A review: 5th International Permafrost Conference, Senneset, K. ed., Tapir Publishers, Trondheim, Norway.

Ferrians, O. J., Jr., 1994, Permafrost in Alaska: *in* Plafker, G. and Berg, H. C., eds., The Geology of Alaska, The Geology of North America, Geological Society of America, v. G-1, p 845–854.

Ferrians, O. J., Kachadoorian, R., and Greene, G. W., 1969, Permafrost and related engineering problems in Alaska: U. S. Geological Survey Professional Paper 678.

Flemal, R. C., 1976, Pingos and pingo scars: Their characteristics, distribution, and utility in reconstructing former permafrost environments: Quaternary Research, v. 6, p 37–53.

Forland, K. S., Forland, T., and Ratkje, S. K., 1988, Frost heave: *in* Proceedings of the 5th International Permafrost Conference, Tapir Publishers, Trondheim, Norway, p. 344–348.

Forsgen, B., 1968, Studies of palsas in Finland, Norway, and Sweden, 1964–1966: Biutetyn Peryglacjalny, v. 17, p. 117–123.

French, H. M., 1971, Ice cored mounds and patterned ground, southern Banks Island, western Canadian Arctic: Geografiska Annaler, v. 53A, p. 32–38.

French, H. M., 1974, Active thermokarst processes, eastern Banks Island, Western Canadian Arctic: Canadian Journal of Earth Sciences, v. 11, p. 785–794.

French, H. M., 1976, The periglacial environment: Longman, London and N.Y., 308 p.

French, H. M., 1986, Periglacial involutions and mass displacement structures, Banks Island, Canada: Geografiska Annaler, v. 68 A, p. 167–174.

French, H. M., 1987a, Periglacial geomorphology in North America: current research and future trends: Progress in Physical Geography, v. 11, p. 569–587.

French, H. M., 1987b, Periglacial processes and landforms in the Western Canadian Arctic: *in* Boardman, J., ed., Periglacial processes and landform is in Britain and Ireland. Cambridge University Press: Cambridge, England, p. 27–43.

French, H. M., 1987c Permafrost and ground ice: *in* Gregory, K. J., and Walling, D. E., eds., Human activity and environmental processes: John Wiley and Sons, Chichester, England, p. 237–269.

French, H. M., 1988, Active layer processes: *in* Clark, M. J. ed., Advances in periglacial geomorphology: John Wiley and Sons, Chichester, England, p. 151–177.

French, H. M., 1996, The Periglacial Environment: Longman, London, 341 p.

French, H. M., and Harry, D. G., 1988, Nature and origin of ground ice, Sandals Moraine, southwest Banks Island, Western Canadian Arctic: Journal of Quaternary Science, v. 3, p. 19–30.

French, H. M., and Harry, D. G., 1990, Observations on buried glacier ice and massive segregated ice, western Arctic coast, Canada: Permafrost and Periglacial Processes, v. 1, p. 31–43.

Gerasimov, I. P., and Markov, K. K., 1968, Permafrost and ancient glaciation: Defence Research Board Translation T499R, Ottawa, Canada, p. 11–19.

Giardino, J. R., and Vitek, J. D, 1988, Rock glacier rheology: A preliminary assessment: *in* Proceedings of the 5th International Permafrost Conference, Tapir Publishers, Trondheim, Norway, p. 714–718.

Giardino, J. R., Shroder, J. F. Jr., and Vitek, J. D., eds., 1987, Rock glaciers: Allen and Unwin, Boston, MA.

Giardino, J. R., Vitek, J. D., and DeMorett, J. L., 1992, A model of water movement in rock glaciers and associated water characteristics: *in* Periglacial Geomorphology, Dixon, J. C.

and Abrahams, A. D. eds., John Wiley and Sons, Chichester, England, p. 159–184.

Gold, L. W., and Lachenbruch, A., 1973, Thermal conditions in permafrost—a review of North American literature: *in* Permafrost, North American Contribution, 2nd International Conference, Yakutsk, USSR, National Academy of Sciences, Publication 2115, Washington, D. C., p. 3–26.

Goldthwait, R. P., 1976, Frost-sorted patterned ground: A review: Quaternary Research, v. 6, p. 27–35.

Haeberli, W., 1985, Creep of mountain permafrost: Internal structure and flow of alpine rock glaciers: Mitteilungen der Versuchsanstalt fur Wasserbau, Hydrologie und Glaziologie, ETH, Zurich, Switzerland, v. 77, 142 p.

Haeberli, W., ed., 1990, Pilot analysis of permafrost cores from the active rock glacier Murtel I, Piz Corvatsch, Eastern Swiss Alps. Arbeitshaft 9: Versuchsanstalt fur Wasserbau: Hydrologie und Glaziologie, ETH, Zurich, Switzerland, 38 p.

Hallet, B., and Prestrud, S., 1986, Dynamics of periglacial sorted circles in Western Spitsbergen: Quaternary Research, v. 26, p. 81–99.

Hallet, B., and Waddington, E. D., 1992, Buoyancy forces induced by freeze thaw in the active layer: Implications for diapirism and soil circulation: *in* Periglacial Geomorphology, Dixon, J. C., and Abrahams, A. D., eds., John Wiley and Sons, Chichester, England, p. 251–280.

Hallet, B., J.S. Walder and C.W Stubbs (1991) Weathering by segregation ice growth in microcracks at sustained subzero temperatures: verification from an experimental study using acoustic emissions: Permafrost and Periglacial Processes, v. 2, p. 283–300.

Hallet, B., Anderson, S. R., Stubbs, C. W., and Gregory, E. C., 1988, Surface soil displacements in sorted circles, western Spitzbergen: *in* Proceedings of the 5th International Permafrost Conference, Tapir Publishers, Trondheim, Norway, p. 779–785.

Hamilton, T. D., and Curtis, M. O., 1982, Pingos in the Brooks Range, northern Alaska, U.S.A.: Arctic and Alpine Research, v. 14, p. 13–20.

Hamilton, T. D., Craig, J., and Sellman, P., 1988, The Fox permafrost tunnel: A late Quaternary geologic record in central Alaska: Geological Society of America Bulletin, v. 100, p. 948–969.

Harris, C., 1987, Solifluction and related periglacial deposits in England and Wales: *in* Periglacial Processes and Landforms in Britain and Ireland, Boardman, J., ed., Cambridge University Press, Cambridge, England, p. 209–223.

Harris, C., 1990, Periglacial landforms: *in* Natural Landscapes of Britain from the Air: Stephens, N., ed., Cambridge University Press, Cambridge, England, p. 43–48.

Harry, D. G., and French, H. M., 1983, The orientation and evolution of thaw lakes, southwest Banks Island, Canadian Arctic: *in*

4th International Conference on Permafrost, Proceedings, National Academy Press, Washington, D. C., v. 1, p. 456–461.

Hay, T., 1936, Stone stripes: Geographical Journal, v. 87, p. 47–50.

Higgins, C. G., et al., 1990, Permafrost and thermokarst; Geomorphic effects of subsurface water on landforms of cold regions: *in* Higgins, C. G., and Coates, D. R., eds., Groundwater geomorphology; the role of subsurface water in earth-surface processes and landforms, Geological Society of America Special Paper 252, p. 211–218.

Hinkel, K.M., and Outcalt, S. I., 1994, Identification of heat-transfer processes during soil cooling, freezing, and thaw in central Alaska: Permafrost and Periglacial Processes, v. 5, p. 217–235.

Hinkel, K. M., Nelson, F. E., and Outcalt, S. I., 1987, Frost mounds at Toolik Lake, Alaska: Physical Geography, v. 8, p. 148–159.

Hinkel, K.M., Outcalt, S. I., and Nelson, F. E., 1990, Temperature variation and apparent thermal diffusivity in the refreezing active layer, Toolik Lake, Alaska: Permafrost and Periglacial Processes, v. 1, p. 265–274.

Holmes, G. W., Hopkins, D. M., and Foster, H. L., 1968, Pingos in central Alaska: U.S. Geological Survey Bulletin 1241-H, 40 p.

Hopkins, D. M., 1949, Thaw lakes and thaw sinks in the Imuruk Lake area, Seward Peninsula, Alaska: Journal of Geology, v. 57, p. 119–131.

Hopkins, D. M., et al., 1955, Permafrost and ground water in Alaska: U. S. Geological Survey Professional Paper 264-F, p. 113–146.

Hughes, O. L., 1969, Distribution of open-system pingos in central Yukon Territory with respect to glacial limits: Geological Survey of Canada Paper 69-34, 8 p.

Hussey, K. M., and Michelson, 1966, Tundra relief features near Point Barrow, Alaska: Arctic, v. 19, p. 162–184.

Ives, R. L., 1940, Rock glaciers in the Colorado Front Range: Geological Society of America Bulletin, v. 51, p. 1271–1294.

Jahn, A. 1975, Problems of the periglacial zone: PWN Polish Scientific Publishers, Warsaw, Poland, 219 p.

Johnson, P. G., 1974, Mass movement of ablation complexes and their relationship to rock glaciers: Geographiska Annaler, v. 565A, p. 93–101.

Johnson, W. H., 1990, Ice-wedge casts and relict patterned ground in Central Illinois and their environmental significance: Quaternary Research, v. 33, p. 51–72.

Johnsson, G., 1959, True and false ice-wedges in southern Sweden: Geografiska Annaler, v. 41, p. 15–33.

Johnsson, G., 1962, Periglacial phenomena in southern Sweden: Geografiska Annaler, v. 44, p. 378–404.

Kachenbruch, A. H., 1962, Mechanics of thermal contraction cracks and ice-wedge polygons in permafrost: Geological Society of America Special Paper 70, 69 p.

Kachurin, S. P., 1962, Thermokarst within the territory of the U.S.S.R.: Biuletyn Peryglacjalny, v. 11, p. 49–55.

Kane, D. L., Hinzman, L. D., and Zarling, J. P., 1991, Thermal response of the active layer to climatic warming in a permafrost environment: Cold Regions Science and Technology, v. 19, p. 111–122.

Karte, J., and Liedtke, H., 1981, The theoretical and practical definition of the term 'periglacial' in its geographical and geological meaning: Biuletyn Peryglacjalny, v. 28, p. 123–135.

Kershaw, G. P., and Gill, D., 1979, Growth and decay of palsas and peat plateaux in the Macmillan Pass-Tsichu River area, Northwest Territories, Canada: Canadian Journal of Earth Sciences, v. 16, p. 1362–1367.

Kondratjeva, K. A., Khrutzky, S. F., and Romanovsky, N. N., 1993, Changes in the extent of permafrost during the Late Quaternary Period in the territory of the former Soviet Union: Permafrost and Periglacial Processes, v. 4, p. 113–119.

Lachenbruch, A. H., 1962, Mechanics of thermal contraction cracks and ice wedge polygons in permafrost: Geological Society of America Special Paper 70, 69 p.

Lagerback, R., and Rodhe, L., 1986, Pingos and palsas in northernmost Sweden; preliminary notes on recent investigations: Geografiska Annaler, v. 68A, p. 149–154.

Leffingwell, E. K., 1915, Ground ice wedges; the dominant form of ground-ice on the north coast of Alaska: Journal of Geology, v. 23, p. 635–654.

Leffingwell, E. K., 1919, The Canning River region, northern Alaska: U. S. Geological Survey Professional Paper, v. 109, 251 p.

Lefroy, General Sir J. H., 1889, Report upon the depth of permanently frozen soil in the polar regions, its geographical limits and relations to the present poles of greatest cold: Proceeding of the Geographical Section, Royal Geographical Society, London, p. 740–746.

Lewkowitz, A. G., 1992, A solifluction meter for permafrost sites: Permafrost and Periglacial Processes, v. 3, p. 11–18.

Lozinski, W. von, 1909, Uber die mechanishe Verwitterung der Sandsteine im gemassigten Klima: International Bulletin of the Academy of Science Cracovie. v. 1, p. 1–25

Lozinski, W. von, 1912, Die periglaziale Fazies der mechanischen Verwitterun: Coimptes Rndus, XI Congres Internationale Geologie, Stockholm, Sweden, p. 1039–1053.

Mackay, J. R., 1962, Pingos of the Pleistocene Mackenzie River delta: Geographical Bulletin, v. 18, p. 21–63.

Mackay, J. R., 1963, The Mackenzie Delta area: Geographical Branch Memoir 8, 202 p.

Mackay, J. R., 1966, Segregated epigenetic ice and slumps in permafrost: Mackenzie delta area, N.W.T.: Geographical Bulletin, v. 8, p. 59 80.

Mackay, J. R., 1972, Offshore permafrost and ground ice, southern Beaufort Sea, Canada: Canadian Journal of Earth Sciences, v. 9, p. 1550–1561.

Mackay, J. R., 1973, The growth of pingos, western Arctic coast, Canada: Canadian Journal of Earth Sciences, v. 10, p. 979–1004.

Mackay, J. R., 1974, Ice-wedge cracks, Garry Island, Northwest Territories: Canadian Journal of Earth Sciences, v. 11, p. 1366–1383.

Mackay, J. R., 1975a, The closing of ice-wedge cracks in permafrost, Garry Island, Northwest Territories: Canadian Journal of Earth Sciences, v. 12, p. 1668–1674.

Mackay, J. R., 1975b, The stability of permafrost and recent climatic change in the Mackenzie Valley, NWT: Canada Geological Survey Paper 75-18, p. 173–176.

Mackay, J. R., 1978, Contemporary pingos: A Discussion: Biuletyn Peryglacjalny, v. 27, p. 133–154.

Mackay, J. R., 1979a, An equilibrium model for hummocks (non-sorted circles), Garry Island, Northwest Territories: Geological Survey of Canada Paper 79-1A, p. 165–167.

Mackay, J. R., 1979b, Pingos of the Tuktoyaktuk Peninsula area, Northwest Territories: Geographie physique et Quaternaire, v. 33, p. 3–61.

Mackay, J. R., 1983, Downward water movement into frozen ground, western arctic coast, Canada: Canadian Journal of Earth Sciences, v. 20, p. 120–134.

Mackay, J. R., 1984, The frost heave of stones in the active layer above permafrost with downward and upward freezing: Arctic and Alpine Research, v. 16, p. 439–446.

Mackay, J. R., 1986, The first 7 years 1978–85 of ice wedge growth, Illisarvik experimental drained lake, western Arctic coast, Canadian Journal of Earth Sciences, v. 23, p. 1782–1795.

Mackay, J. R., 1988, The birth and growth of Porsild Pingo, Tuktoyaktuk Peninsula, District of Mackenzie: Arctic, v. 41, p. 267–674.

Mackay, J. R., 1990a, Seasonal growth bands in pingo ice: Canadian Journal of Earth Sciences, v. 27, p. 1115–1125.

Mackay, J. R., 1990b, Some observations on the growth and deformation of epigenetic, syngenetic, and antigenetic ice wedges: Permafrost and Periglacial Processes, v. 1, p. 15–29.

Mackay, J. R., 1992, The frequency of ice wedge cracking 1967–87 at Garry Island, western Arctic coast, Canada: Canadian Journal of Earth Sciences, v. 29, p. 236–248.

Mackay, J. R., 1993, Air temperature, snow cover, creep of frozen ground, and the time of ice-wedge cracking, western Arctic coast: Canadian Journal of Earth Sciences, v. 30, p. 1720–1729.

Mackay, J. R., and Dallimore, S. R., 1992, Massive ice of the Tuktoyaktuk area, western Arctic coast, Canada: Canadian Journal of Earth Sciences, v. 29, p. 1235–1249.

Mackay, J. R., Ostrick, J., Lewis, C. P., and Mackay, D. K., 1979, Frost heave at ground temperatures below 0°C, Inuvik, Northwest Territories: Geological Survey of Canada, Paper 79-1A, p. 403–406.

Martin, E. H., and Whalley, W. B., 1987, Rock glaciers; Part 1, Rock glacier morphology;

classification and distribution: Progress in Physical Geography, v. 11, p. 260–282.

Mathews, J. A., Harris, C., and Ballantyne, C. K., 1986, Studies on a gelifluction lobe, Jotunheimen, Norway; 14C chronology, stratigraphy, sedimentology, and paleoenvironment: Geografiska Annaler, v. 68A, p. 345–360.

Matsuoka, N., 1991, A model of the rate of frost shattering: Application to field data from Japan, Svalbard and Antarctica: Permafrost and Periglacial Processes, v. 2, p. 271–281.

McGreevey, J. R., 1981, Some perspectives on frost shattering: Processes in Physical Geography, v. 5, p. 56–75.

Michaud, Y., Dionne, J. C., and Dyla, L. D., 1989, Frost bursting: A violent expression of frost action in rock: Canadian Journal of Earth Sciences, v. 26, p. 2075–2080.

Miller, R. D., 1972, Freezing and heaving of saturated and unsaturated soils: Highway Research Record, v. 393, p. 1–11.

Miller, R., Colon, R., and Galloway, R. W., 1954, Stone stripes and other surface features of Tinto Hill: Geographical Journal, v. 120, p. 216–219.

Mullenders, W., and Gullentops, F., 1969, The age of the pingos of Belgium: in The Periglacial Environment, Pewe, T., ed., McGill-Queens University Press, Montreal, Canada, p. 321–336.

Muller, F., 1959, Beobachtung uber pingos: Meddelelser om Gronland, v. 153, 127 p. (English translation, National Research Council of Canada, Technical Translation TT-1073, 117 p.

Muller, F., 1962, Analysis of some stratigraphic observations and radiocarbon dates from two pingos in the Mackenzie Delta area, N.W.T.: Arctic, v. 15, p. 278–288.

Muller, S. W., 1945, Permafrost or permanently frozen ground and related engineering problems: U. S. Engineers Office, Strategic Engineering Study, Special Report No. 62, 136 p. (Reprinted in 1947 by J. W. Edwards, Ann Arbor, MI, 231 p.)

National Research Council of Canada, 1988, Glossary of permafrost and related ground-ice terms: Permafrost Subcommittee, Technical Memo 142, 156p.

Nelson, E. E., Hinkel, K. M., and Outcalt, S. I., 1992, Palsa-scale frost mounds: in Periglacial Geomorphology: Dixon J. C., and Abrahams, A. D., eds., John Wiley and Sons, Chichester, England, p. 305–326.

Nichols, R. L., 1966, Geomorphology of Antarctica: in Tedrow, J. C. F., ed., Antarctic soils and soil and soil forming processes, American Geophysical Union, Antarctic Research Series no. 8, p. 1–59.

Nixon, J. F., 1989, Ground freezing and frost heave—a review. The Northern Engineer , v. 19, p. 8–18.

Nixon, J. F., 1991, Discrete ice lens theory for frost heave in soils: Canadian Geotechnical Journal, v. 28, p. 843–859.

Ostrem, G., 1963, Comparative crystallographic studies on ice from ice-cored moraines, snow-banks, and glaciers: Geographiska Annaler, v. 45, p. 210–240.

Osterkamp, T. E., 1991, Variations in permafrost thickness in response to changes in paleoclimate: Journal of Geophysical Research, B, Solid Earth and Planets, v. 96, p. 4423–4434.

Osterkamp., T. E. 1996, Characteristics of changing permafrost temperatures in the Alaskan Arctic, U.S.A.: Arctic and Alpine Research, v. 28, p. 267–273.

Osterkamp, T. E., and Gosink, J. P., 1991, Variations in permafrost thickness in response to changes in paleoclimate: Journal of Geophysical Research, v. 96, p. 4423–4434.

Osterkamp, T. E., and Romanovsky, V. E., 1996, Characteristics of changing permafrost temperatures in the Alaskan Arctic, USA: Arctic and Alpine Research, v. 28, p. 267–273.

Outcalt, S. I., and Benedict, J. B., 1965, Photo-interpretation of two types of rock glaciers in the Colorado Front Range, U.S.A.: Journal of Glaciology, v. 5, p. 849–856.

Owens, E. H., and Harper, J. R., 1977, Frost table and thaw depths in the littoral zone near Pearl Bay, Alaska: Arctic, v. 30, p. 155–168.

Pewe, T. L., 1959, Sand wedge polygons (tesselatoins) in the McMurdo Sound Region, Antarctica: American Journal of Science, v. 257, p. 545–552.

Pewe, T. L., 1962, Ice wedges in permafrost, lower Yukon River area, near Rena, Alaska: Biuletyn; Peryglacjalny, v. 11, p. 65–76.

Pewe, T. T., 1963, Ice wedges in Alaska—classification, distribution and climatic significance: Geological Society of America Special Paper 76, 129p.

Pewe, T. L., 1966a, Paleoclimatic significance of fossil ice wedges: Biuletyn Peryglacjalny, v. 15, p. 65–73.

Pewe, T. L., 1966b, Ice-wedges in Alaska—Classification, distribution and climatic significance: *in* Proceedings of the International Permafrost Conference, Lafayette, Ind., 1963, National Academy of Science National Research Council Publications 1287, p. 76–81.

Pewe, T. L., 1973, Ice wedge casts and past permafrost distribution in North America: Geoform. v. 15, p. 15–26.

Pissart, A. 1956, L'origine periglaciaire des viviers des Hautes Fagnes: Annales, Societe Geologique de Belgique, v. 79, p. 119–131.

Pewe, T. L., 1983 Alpine permafrost in the contiguous United States: a review: Arctic and Alpine Research, v. 15, p. 146–156.

Pissart, A. 1965, Les pingos des Hautes Fagnes: le probleme de leur genese: Annales, Societe Geologique de Belgique, v. 88, p. 277–289.

Pissart, A. 1970, The pingos of Prince Patrick Island 76°N–120°W: National Research Council of Canada, Technical Transactions 1401.

Pissart, A., 1977, Apparition et evolution des sols structuraux periglaciaiers de haute montagne. Experiences de terrain au Chambeyron (Alpes, France): Abhandlungen der Akademie der Wissenschaften in Gottingen, Math-Physik Klasse, v. 31, p. 142–156.

Pissart, A., 1990, Advances in periglacial geomorphology: Zeitschrift für Geomorphologie, Supplementband, v. 79, p. 119–131.

Pollard, W. H., and French, H. M., 1983, Seasonal frost mound occurrence. North Fork Pass, Ogilvie Mountains, northern Yukon, Canada: Proceedings of the 4th International Conference of Permafrost, National Academy Press, Washington, D.C., v. 1, p. 1000–1004.

Pollard, W. H., and French, H. M., 1984, The groundwater hydraulics of seasonal frost mounds, North Fork Pass, Yukon Territory: Canadian Journal of Earth Sciences, v. 21, p. 1073–1081.

Pollard, W. H., and French, H. M., 1985, The internal structure and ice crystallography of seasonal frost mounds: Journal of Glaciology, v. 31, p. 157–162.

Porsild, A. E., 1938, Earth mounds in unglaciated Arctic northwestern America: Geographical Reviews, v. 28, p. 46–58.

Potter, N., Jr., 1972, Ice-cored rock glacier, Galena Creek, northern Absaroka Mountains, Wyoming: Geological Society of America Bulletin, v. 83, p. 3025 3057.

Price, L. W., 1991, Subsurface movement on solifluction slopes in the Ruby Range, Yukon Territory, Canada. A 20-year Study: Arctic and Alpine Research, v. 23, p. 200–205.

Putkonen, J. K., 1997, Climatic control of the thermal regime of permafrost, northwest Spitsbergen: Eos, Fall meeting supplement, American Geophysical Union.

Ray, L. L., 1951, Permafrost: Arctic, v. 4, p. 196 203.

Repelewska-Pekabwa, J., and Gluza, A., 1988, Dynamics of permafrost active layer—Spitsbergen: *in* Proceedings of the 5th International Permafrost Conference, Senneset, K. ed., Tapir Publishers, Trondheim, Norway, p. 448–453.

Richardson, F. R. S., 1839, Notice of a few observations which it is desirable to make on the frozen soil of British North America; drawn up for distribution among the officers of the Hudson's Bay Company: Journal of the Royal Geographical Society, London, v. 9, p. 117–120.

Richie, A. M., 1953, The erosional origin of the Mima Mounds of southwest Washington: Journal of Geology, v. 61, p. 41-50.

Sellman, P. V., Brown, J., Lewellen, R. I., McKim, H., and Merry, C., 1975, The classification and geomorphic implications of thaw lakes on the Arctic coastal plain, Alaska: U. S. Army Cold Regions Research and Engineering Laboratory, Research Report 344, 21 p.

Seppala, M., 1972, The term "palsa:" Zeitschrift für Geomorphologie, v. 16, p. 463.

Seppala, M., 1976, Seasonal thawing of a palsa at Enontekio, Finnish-Lapland in 1974: Biuletyn Peryglacjalny, v. 26, p. 17–24.

Seppala, M., 1982, An experimental study of the formation of palsas: *in* Proceedings of 4th International Permafrost Conference, p. 36–42.

Seppala, M., 1986, The origin of palsas: Geografiska Annaler, v. 68A, p. 141–147.

Seppala, M., 1988, Palsas and related forms: *in* Clark, M. J., ed., Advances in Periglacial Geomorphology, John Wiley and Sons, N.Y., p. 247–277.

Sharp, R. P., 1942a, Periglacial involutions in north-eastern Illinois: Journal of Geology, v. 50, p. 113–133.

Sharp, R. P., 1942b, Soil structures in the St. Alias Range, Yukon Territory: Journal of Geomorphology, v. 5, p. 274–301.

Sharp, R. P., 1942c, Ground-ice mounds in tundra: Geographical Reviews, v. 32, p. 417–423.

Shi, Y., ed., 1988, Map of snow, ice, and frozen ground in China (1:4,000,000), with explanatory notes: China Cartographic Publishing House, Beijing, China.

Shotton, P. W., 1960, Large-scale patterned ground in the valley of the Worcestershire Avon: Geological Magazine, v. 97, p. 404–408.

Smith, D. J., 1992, Long-term rates of contemporary solifluction in the Canadian Rocky Mountains: *in* Periglacial Geomorphology, Dixon, J. C., and Abrahams, A. D., eds., John Wiley and Sons, Chichester, England, p. 203–222.

Smith, M. W., 1985a, Models of soil freezing: *in* Church, M. and Slaymaker, O., eds., Field and Theory: lectures in geocryology, University of British Columbia Press, Vancouver, B. C., p. 96–120.

Smith, M. W., 1985b, Observations of soil freezing and frost heave at Inuvik, Northwest Territories, Canada: Canadian Journal of Earth Sciences, v. 22, p. 283–290.

Smith, M. W., and Patterson, D. E., 1989, Detailed observations on the nature of frost heaving at a field scale: Canadian Geotechnical. Journal, v. 26, p. 306–312.

Soloviev, R A., 1973, Alas thermokarst relief of central Yakutia: Guidebook, Second International Permafrost Conference, Yakutsk, USSR.

Stearns, S. R., 1966, Permafrost perennially frozen ground: U.S. Army Corps Engineers, Cold Regions Research and Engineering Lab, Cold Regions Science and Engineering, v. 1 A2.

Sollid, J. L., and Sorbel. L., 1992, Rock glaciers in Svalbard and Norway: Permafrost and Periglacial Processes, v. 3, p. 215–220.

Taber, S., 1929, Frost heaving: Journal of Geology, v. 37, p. 428–461.

Taber, S., 1930, The mechanics of frost heaving: Journal of Geology, v. 38, p. 303–317.

Taber, S., 1943, Perennially frozen ground in Alaska: Its origin and history: Geological Society of America Bulletin, v. 54, p. 1433–1548.

Tedrow, J. C. F., 1969, Thaw lakes, thaw sinks, and soils in northern Alaska: Biuletyn Peryglacjalny, v. 20, p. 337–344.

Thorn, C., 1992, Periglacial geomorphology. What? Where? When? *in* Dixon, J. C., and Abrahams, A. D., eds., Periglacial

Geomorphology, John Wiley and Sons Ltd., Chichester, England, p. 1–30.

Tomirdiaro, S. V., and Ryabchun, V. K., 1978, Lake thermokarst on the Lower Anadyr Lowland: *in* Permafrost, USSR Contribution, Second International Conference, Yakutsk, USSR, National Academy of Sciences, Washington, D. C., p. 94–100.

Tricart, J., 1970, Geomorphology of cold environments: Translated by Edward Watson, St. Martins Press, N.Y., 320 p.

Vandenberghe, J., and Pissart, A., 1993, Permafrost changes in Europe during the Last Glacial: Permafrost and Periglacial Processes, v. 4, p. 121–135.

Van Everdingen, R. O., 1978, Frost mounds at Bear Rock, near Fort Norman, Northwest Territories 1975 1976: Canadian Journal of Earth Sciences, v. 15, p. 263–276.

Van Vliet-Lanoe, B., 1991, Differential frost heave, load casting and convection: Converging mechanisms: a discussion of the origin of cryoturbations: Permafrost and Periglacial Processes, v. 2, p. 123–139.

Vilborg, L., 1955, The uplift of stones by frost: Geografiska Annaler, v. 37, p. 164–169.

Vitek, J. D., 1983, Stone polygons: Observations of surficial activity: *in* Proceedings of the 4th International Permafrost Conference, National Academy Press, Washington, D.C., p. 1326–1331.

Wahrhaftig, C., and Cox, A., 1959, Rock glaciers in the Alaska Range: Geological Society of America Bulletin, v. 70, p. 383–436.

Walder, J., and Hallet, B., 1985, A theoretical model of the fracture of rock during freezing: Geological Society of America Bulletin, v. 96, p. 336–346.

Walker, H. J., 1988, Permafrost and coastal processes: *in* Senneset, K., ed., 5th International Permafrost Conference Proceedings, Tapir Publishers, Trondheim, Norway, p. 35–41.

Wallace, R. E., 1948, Cave-in lakes in the Nabesna, Chisana, and Tanana river valleys, eastern Alaska: Journal of Geology, v. 56, p.171–181.

Washburn, A. L., 1947, Reconnaissance geology of portions of Victoria Island and adjacent regions, Arctic Canada: American Journal of Science Memoir 22.

Washburn, A. L., 1956a, Classification of patterned ground and review of suggested origins: Geological Society of America Bulletin, v. 67, p. 823–866.

Washburn, A. L., 1956b, Unusual patterned ground in Greenland: Geological Society of America Bulletin, v. 67, p. 807–810.

Washburn, A. L., 1967, Instrumental observations of mass-wasting in the Mesters Vig district, Northeast Greenland: Medd. om Gronland, 166:4.

Washburn, A. L., 1969, Patterned ground in the Mesters Vig district, northeast Greenland: Biuletyn Peryglacjalny, v. 18, p. 259–330.

Washburn, A. L., 1979, Geocryology. A survey of periglacial processes and environments: Edward Arnold, London, 406 p.

Washburn, A. L., 1980, Geocryology: John Wiley and Sons, N.Y.

Washburn, A. L., 1988, Mount Mounds: Washington Division of Geology and Earth Resources, Report of Investigations 29, 53 p.

Washburn, A. L., Smith, D. D., and Goddard, R. H., 1963, Frost cracking in a middle

latitude climate, Biuletyn Peryglacjalny, v. 12, p. 175–189.

Whalley, W. B., 1983, Rock glaciers— Permafrost features or glacial relics: *in* Proceedings of 4th International Permafrost Conference, National Academy Press, Washington, D.C., p. 1396–1401.

Whalley, W. B. and Martin, H. E., 1992, Rock glaciers: a review. Part 2: Mechanisms and models: Permafrost and Periglacial Processes, v. 11, p. 127–186.

Whalley, W. B.; Douglas, G. R., and McGreevey, J. R., 1982, Crack propagation and associated weathering in igneous rocks: Zeitscrift fur Geomorphologie.

White, S. E., 1976, Is frost action really only hydration shattering? Arctic and Alpine Research, v. 8, p. 1–6.

Wiegand, G., 1965, Fossil pingos in Mitteleuropa: Wurzberg Geographical Arbrit, v. 16, p. 1–152.

Williams, P. J., and Smith, M. W., 1989, The frozen earth; Fundamentals of geocryology: Cambridge University Press, Cambridge, England, 306 p.

Wilson, R., 1992, Small scale patterned ground, Comeragh Mountains, southeast Ireland: Permafrost and Periglacial Processes, v. 3, p. 63–70.

Zhang,T., 1997, Effects of climate on the active layer and permafrost on the North Slope of Alaska, U.S.A.: Permafrost and Periglacial Processes, John Wiley and Sons, Oxford, England, v. 8, p. 45–67.

Zoltai, S. C., 1971, Southern limit of permafrost features in peat landforms, Manitoba and Saskatchewan: Geological Association of Canada Special Paper No. 9, p. 305–310.

Shorelines

Split and other coastal features at Chatham, Mass. (Photo by Kelsey-Kennard Photographers, Inc.)

INTRODUCTION

Although the basic configuration of many oceanic shorelines is related to tectonic forces, alteration of those shorelines is caused by energy expended along them by waves. The relentless attack of waves against shorelines continues day and night, virtually every day of the year, flinging turbulent water incessantly against sea cliffs, rocky shores, and beaches. Although each passing wave may produce an infinitely small change in the coastline, these small alterations must be multiplied by the unceasing lines of waves that cumulatively can significantly modify coastal areas. At times, great storm waves batter the shore with high-energy bundles of water, and at other times, small, gentle waves reassemble the products derived from the erosive activity of the storm waves.

Much of the discussion that follows deals with oceanic shorelines, although the basic processes also operate on freshwater lakes. Generally speaking, the effect of wave activity is directly related to the size of the body of water. Thus, wave action is more vigorous along the seashore than along small lakes or ponds.

WAVES

Water waves, like other waveforms, are generated by some energy source. By far the most important source of the energy of ocean waves is wind. Other significant, but less common, sources of energy are submarine earthquakes and landslides, which are capable of setting unusually large waves in motion. Submarine landslides have been responsible for some of the most prodigious waves recorded, but their frequency is very low, and most erosion and deposition along shore zones is generated by waves created by wind. Much has been written about the mechanics of wave action (for example, Bascom, 1959; Snodgrass et al., 1966; Stokes, 1849; Wright et al., 1982).

As wind blows over an open stretch of water, the turbulent air distorts the water surface, depressing it where the air moves downward and raising it where air pressure decreases as the air moves upward. The irregularities thus produced allow the wind to push against portions of the no-longer horizontal surface, thereby generating waves. Although this describes the general manner in which waves are brought about, the precise mechanism of transfer of wind energy to waves is rather complex.

At the site of a storm over water, most waves are short, choppy, and irregular, with waves of different sizes, orientations, and directions of movement superimposed upon one another. Individual wave crests are continuous for only short distances before they vanish among a chaotic jumble of interfering waves. Waves of different size are out of phase and interfere with one another; some add to the height of others, whereas others subtract.

The nature of the wave system that develops as waves move away from their source depends on three primary factors:

1. The wind velocity

2. The duration of the wind

3. The distance over which the wind blows, known as the **fetch**

Although velocity and duration are important, if the wind blows only over a small stretch of water (a short fetch), waves do not have adequate opportunity to grow very large. Thus, fetch is equally important in the building of high waves, and all three factors must be effective for large waves to form. The height (amplitude) of waves produced over a large fetch by long-duration winds has been found empirically to increase with the square of the wind velocity, according to the equation

$$H = 0.025v^2 \tag{1}$$

where H = wave height
            v = wind velocity

The relationship of fetch to wave height is given by the equation

$$H = \frac{0.36}{F} \tag{2}$$

where H = wave height
            F = fetch

The heights of waves generated in severe storms over the ocean are commonly 15 m or more. The highest documented wave height was 34 m (112 ft), measured February 1933, in the South Pacific.

Waves in water (Figure 16–1) may be defined in terms of the following:

1. **Wavelength L**, the horizontal distance between adjacent wave crests or troughs

2. **Wave height H**, the vertical distance between wave crest and wave trough

3. **Wave period T**, the time between the passing of two successive crests

Thus,

$$T = \frac{L}{V} \tag{3}$$

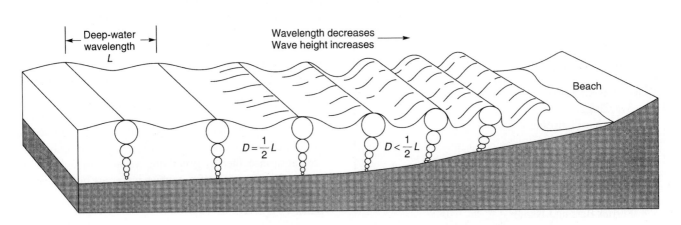

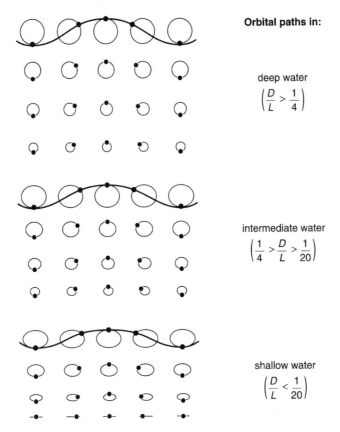

Orbital paths in:

deep water
$\left(\dfrac{D}{L} > \dfrac{1}{4}\right)$

intermediate water
$\left(\dfrac{1}{4} > \dfrac{D}{L} > \dfrac{1}{20}\right)$

shallow water
$\left(\dfrac{D}{L} < \dfrac{1}{20}\right)$

Water particle orbits in shallow water. As the wave enters shallow water, the particle orbits become more elliptical.

FIGURE 16–1
Progressive changes in wave characteristics as waves approach a shoreline.

where T = wave period
L = wavelength
V = wave velocity

The relationship between wavelength and wave period is given by the Airy equation:

$$L = \frac{gT^2}{2\pi} \times \left[\tanh\frac{2\pi d}{L}\right] \qquad (4)$$

where

d = water depth
tanh = the hyperbolic tangent
g = gravity

For wavelengths greater than four times the water depth, the right-hand portion of this equation is approximately equal to 1.0 and may be ignored. Using g = 9.81 m/s² and π = 3.14, the equation becomes

$$L = \frac{9.81T^2}{2(3.14)} = 1.56\,T^2 \qquad (5)$$

Because wavelengths rarely exceed 400 m (1300 ft), for water depths greater than 100 m (~300 ft), the wavelength can be calculated from the wave period, and because

$$V = \frac{L}{T} \qquad (6)$$

then

$$V = \frac{1.56T^2}{T} \text{ or } V = 1.56T \qquad (7)$$

The importance of this equation is that in deep water, longer waves travel faster than shorter ones and will gradually leave the short waves behind. This process, known as **wave dispersion,** causes the waves of diverse length to differentiate from one another as they move away from their place of origin, resulting in a regularly spaced procession of **swells,** waves with low, gently rounded crests. When two long-period wave trains whose periods are slightly out of phase approach a coast, sometimes the addition of the two wave trains produces waves higher than either one, and at other times they cancel one another to produce lower waves. The net result is that every few minutes, a set of higher waves appears, a phenomenon known as **surf beat** (Munk, 1949; Tucker, 1950).

Ocean swells generated by a storm do not radiate outward from the source in all directions but follow a path bounded by the direction of the wind that created the waves. Thus, swells made by a storm are larger in the downwind direction. Because waves with long wavelengths not only travel faster but also lose energy at a slower rate, they may travel for long distances without much loss of energy. However, waves with short wavelengths travel more slowly, taking longer and dissipating energy more rapidly away from their source.

Not all waves are generated by storms. Some are created by displacements of the ocean floor as a result of sudden tectonic movement or submarine landslides, or by subaerial landslides in water, producing point-source waves that radiate outward in all directions from the point of origin. These types of waves are discussed further in the section on tsunamis.

In deep water, water particles assume a circular orbital path of **oscillation** with each passing wave. Thus, as oscillatory waves progress across a water body, individual water particles in the waves have little forward motion; that is, the wave *form* moves forward, not the water itself. Thus, a floating object has no net forward movement with each passing wave. Rather, a floating object moves forward with the crest of a wave only to slip back into the following trough, generating a circular orbit whose diameter is equal to the wave height (Figure 16–1).

$$D = He^n \qquad (8)$$

where $n = \frac{2\pi d}{L}$

D = diameter of circular orbit
H = wave height
e = natural log base
L = wavelength
d = water depth below the center of the orbit

This equation predicts that the orbital diameter of water particles and the orbital velocity decrease rapidly with depth. At depths greater than one wavelength, the orbital velocity is less than one percent of the orbital velocity at the surface.

Individual waves possess both potential energy, from the height of the wave above flat-water level, and kinetic energy, from the orbital motion of the water particles within the wave. The Airy equation shows that these two types of energy are equal and that their sum is a function of the wave height:

$$E = 1/8(\rho gH^2) \qquad (9)$$

where E = energy
ρ = water density
g = acceleration of gravity
H = wave height

The rate of energy transport of waves is then a matter of the individual wave energy and the velocity of the wave form.

Although wave height is independent of wavelength, the ratio of wave height to wavelength is a measure of wave steepness. Thus, the perceived size of a wave is often a matter of how low the wavelength is relative to the wave height. For example, a 3 m (10 ft) high wave with a short wavelength looks very large, whereas a wave of the same height with a very long wavelength would appear as a low swell. Waves steeper than 1/7 (0.14) are not stable, and few exceed a steepness of 0.1.

Wave Changes in Shallow Water

As a wave approaches the shore and depth becomes increasingly shallow, parts of the Airy equation change significantly. Recall that for deep water where depth/wavelength is greater than 1/4, equation (4) applies. When d/L is less than 1/4, the hyperbolic tangent (tan h) is no longer constant at 1.0. However, when d/L becomes less than *1/20*, the hyperbolic tangent (tan h) becomes insignificant and may be dropped out of the equation. Thus, equation *(4)* becomes:

$$L^2 = dgT^2 \ \text{ or } \ L = T\sqrt{gd} \qquad (10)$$

Because: V (velocity) $= L/T$, then

$$V = \sqrt{gd} \qquad (11)$$

Equations (10) and (11) are especially significant because they predict what will happen to waves as they encroach on the shoaling water of coastlines. As water depth d decreases, wave velocity v and wavelength L both decrease. The decrease in wavelength as waves approach shore is easily observable along most coastlines (Figure 16–2). Theoretically, these equations apply only to water less than $L/20$ deep (but below $L/4$, considered deep water). At greater depths the situation be-

comes more complex but may be resolved graphically as in Figure 16–3. Wave period remains constant throughout the changes in wave form.

Although wave height is independent of wavelength and period in deep water, it does change in shallowing water. As shown in Figure 16–1, as water depth decreases, wave height increases. This phenomenon may be seen on shores where water in low, gentle swells seems to rise as it approaches the beach.

Upon approaching a coast, the orbital path of water particles in waves begins to be affected by shallowing. The orbits begin to become more elliptical (Figure 16–1), and eventually the orbital path is destroyed altogether.

Breakers

As a wave approaches the shore and grows higher and higher with diminishing velocity, a critical point is reached when the forward velocity of the orbit distorts the wave form, making it progressively more asymmetric until the wave crest extends beyond the support of the water beneath and the wave collapses forward, or breaks into surf. The water in the wave then moves ahead as turbulent **surf** and is known as a **breaker.** After a wave collapses to form a breaker, it moves

FIGURE 16–2
Decrease in wavelength as waves approach a shoreline. (Photo by D. A. Rahm)

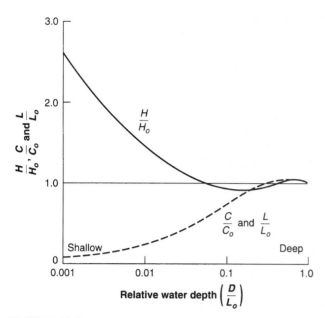

FIGURE 16–3
Wave transformations entering shallow water. The vertical axis shows values for wave height, wave-form velocity, and wavelength relative to deep-water values. The horizontal axis shows water depth. Wave height increases in shallow water, while wavelength and wave-form velocity decrease. Wave period remains constant. (From Pethick, 1984)

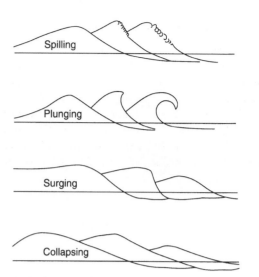

FIGURE 16–4
Types of breakers.

$$Bc = \frac{H}{gsT^2}$$

where H = wave height
           g = acceleration of gravity
           s = beach slope
           T = wave period

as a turbulent sheet of water, or swash, up the beach slope, carrying with it sand and gravel on the beach. Eventually, the energy of the swash is consumed against the slope of the beach, and the water returns down the beach as **backwash.** The surge of surf up the beach and subsequent backwash cause erosion, transportation, and deposition of sediment along the beach.

Four types of breakers are common: spilling, plunging, surging, and collapsing (Figure 16–4). The crests of **spilling breakers** become unstable and cascade continuously down the wave front as a foam line while retaining their form over long distances. Spilling breakers are typically associated with high, short waves breaking relatively far offshore on flat-sloping beaches. The wave crests of **plunging breakers** first become severely oversteepened and then curl over the front of the wave, commonly trapping a tunnel of air (Figure 16–5); then the crests fall with a crash onto the beach, completely disintegrating the wave form into a mass of churning, turbulent water. The wave crests of **surging breakers** commonly remain unbroken as the base of the wave advances up the beach (Figure 16–4). Surging breakers are usually associated with low, flat waves and steeply sloping beaches where waves approach close to the shoreline before breaking in a smooth, sliding motion up the beach. **Collapsing breakers** are intermediate between plunging and surging breakers.

The breaking point of waves depends on the wavelength, the wave height and period, and the slope of the beach face. It can be characterized by a breaker coefficient (*Bc*) (Galvin, 1968, 1972).

Tsunamis

Tsunamis are large waves formed by sudden tectonic displacement of the sea floor, submarine landslides, submarine volcanic eruptions, or subaerial landslides into the sea. They differ from storm-generated waves in that they are much larger and originate from a point source that causes waves to radiate outward in all directions, analogous to tossing a pebble into a pond (Van Dorn, 1965). Large waves produced by tectonic displacement of the sea floor, accompanied by earthquakes, are also known as **seismic sea waves.**

Tsunamis have distinctly different characteristics than wind-generated waves. They have enormous wavelengths, commonly 100 to 200 km (60 to 120 mi), but extremely low wave heights in deep water, often less than 1 m (3 ft). Their period may be 10 to 30 minutes, compared with 15 to 20 seconds for long-wave storm swells. Tsunamis travel with remarkable velocities in deep water: If the wavelength L is 100 km (60 mi) and the period T is 20 min (0.33 hr), the velocity V will be $L/T = 100/0.33 = 300$ km/hr (180 mi/hr).

The wavelengths of tsunamis are so large that they may considerably exceed the depth of the ocean floor. For example, the average depth of the world's oceans is about 3000 m (~10,000 ft), so a wavelength of 100 km (60 mi) is approximately 33 times the water depth. Recall that velocity varies with square root of depth ($V = \sqrt{gd}$; thus, for water depths of 10,000 ft (3000 m), the velocity is $\sqrt{32 \times 10,000} = 386$ mi/hr (~600 km/hr). Conversely, if the time of origin of the

FIGURE 16–5
Plunging breaker, Kui Lima, Hawaii.

tsunami and the time of arrival at some distant place are known, the average water depth in between can be calculated. Tsunami warnings are commonly issued based on the time of origin of submarine earthquakes and known water depths between points.

In the deep water of the open ocean, tsunamis may be barely detectable because of their long wavelength and low wave height, often passing ocean-going ships unnoticed. However, as they approach a coastline and encounter shallowing water, the wave height grows to imposing levels, often 10 m (~30 ft) or more. Typically, the arrival of a tsunami at a shoreline is characterized by a modest rise or lowering of sea level, followed by several major waves capable of large-scale destruction (see the examples that follow). Although the cresting of large waves along a shoreline is the most expected effect of tsunamis, sometimes the trough of the wave form arrives first, causing an apparent sudden drop of sea level that may expose the sea floor for considerable distances offshore in shallow areas. Such dramatic exposure of the sea floor, leaving large fish flopping about, has in several historic instances attracted people out onto the newly exposed sea floor only to be trapped by the succeeding giant waves.

The wave heights of some tsunamis are indeed prodigious. Pieces of coral and beach sediment were found 375 m (1200 ft) above sea level on the island of Lanai in Hawaii (Moore and Moore, 1988). These materials could have been placed there only by a giant wave that picked them up in the coastal zone and washed them high up onto the hillslopes. The 375 m (1200 ft) elevation marks the high point of the wave runup, not necessarily equal to the height of the wave; but in any case, the size of the wave must have been colossal. The origin of the wave is hypothesized to have been a huge submarine landslide southeast of Lanai.

An example of a giant wave produced by a subaerial landslide is the 1957 event in Lituya Bay, Alaska. Movement on the Fairweather fault caused an earthquake that triggered a landslide of about 40 million yd^3 (30 million m^3) of material into the head of Lituya Bay on July 9, 1957, creating a wave 1740 ft (~525 m) high. The wave smashed against the opposite shore (Figure 16–6) and roared down Lituya Bay at about 100 mi/hr (~165 km/hr) (Miller, 1960). The wave was witnessed by men on three fishing boats that were anchored in the bay. One boat rode up the advancing wave and was carried over a spit at the mouth of the bay and dumped in the ocean beyond. A second boat also rode up the advancing wave but remained inside the spit. The third boat vanished and was never found. As the wave passed down the inlet, it stripped off the forest and soil down to bare rock, leaving a well-defined trimline at elevations varying from 1740 ft (525 m) to about 100 ft (33 m) along most of the shoreline (Figure 16–6).

Historically, tsunamis have had important effects on human habitation along the world's coastal areas. Although rare at any given point, when they do occur, they can be devastating, as, for example, the inundation of Lisbon, Portugal, during a great tsunami of the eighteenth century, as well as more recent occurrences in Alaska, Hawaii, Chile, and Japan.

Storm Surges

Storm surges are masses of water pushed shoreward by unusually strong winds and high sea level caused by upward bulging of the ocean surface due to low atmospheric pressure (Dolan et al., 1979b). High winds, usually associated with low-pressure cyclonic storms, cause water to "pile up" on windward coasts, and sea level fluctuates with the at-

A.

B.

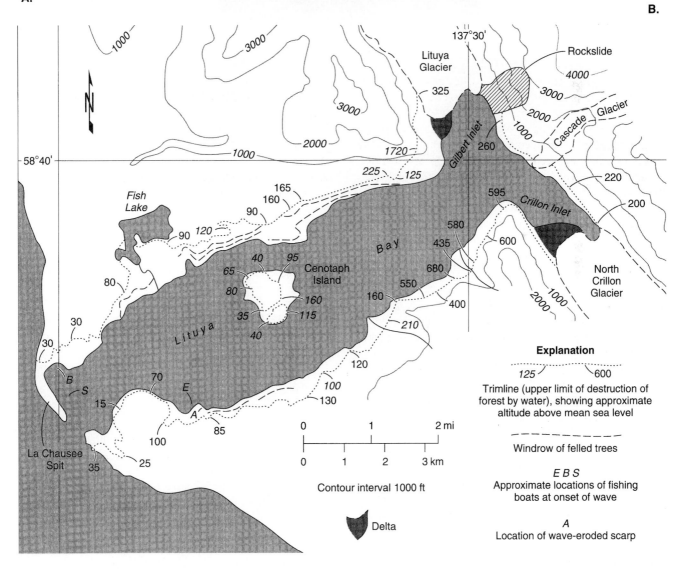

Explanation

· · · · · · · · · 125 ╱⎯⎯ ⎯⎯╲ 600 · · · · · · · · ·
Trimline (upper limit of destruction of
forest by water), showing approximate
altitude above mean sea level

⎯ ⎯ ⎯ ⎯ ⎯ ⎯ ⎯ ⎯
Windrow of felled trees

E B S
Approximate locations of fishing
boats at onset of wave

A
Location of wave-eroded scarp

Contour interval 1000 ft

Delta

FIGURE 16–6, cont.
C, D.. Scoured area along the margins of Lituya Bay. (From Miller, 1960)

mospheric pressure, producing about a 13 in (33 cm) rise in sea level for each 1-in drop in air pressure. Barometric pressures of about 29.4 in (75 cm) may be associated with mid-latitude cyclonic storms in the United States, equivalent to about a 10 in (25 cm) sea-level rise. On the other hand, air pressures below 28.5 in (72 cm) may occur with tropical cyclones such as hurricanes and typhoons, equivalent to a sea-level rise of about 0.5 m (1.5 ft). When combined with the high winds of such storms, sea-level rises of 2 to 5 m (6 to 16 ft) are not uncommon, pushing water inland far beyond the normal shoreline and inundating wide areas (Dolan et al., 1978). In 1900, a hurricane passing through Galveston, Texas, produced a storm surge of 4 ft (1.2 m) with a total water-level rise of 15 ft (4.5 m), killing 6000 people (Hughes, 1979). Some extreme storm surges have raised water levels 6 to 12 m (20 to 40 ft). In 1737, an unusually severe storm from the Bay of Bengal apparently caused a storm surge some 12 m (40 ft) high that drowned 300,000 people, and similar storms in 1970 and 1990 also caused hundreds of thousands of deaths. These monster waves have destructive effects similar to those of tsunamis, but they are a quite different phenomenon.

Seiches

Seiches are oscillations of water in lakes and enclosed basins, usually caused by large earthquakes but also associated with rapid pressure fluctuations of intense storms, tsunamis, rapid flood discharges into standing water, or surf beat (Wilson, 1966). Seiches are typically described by observers as a sloshing back and forth of the water, which gradually decreases until stability is regained.

Tides

In addition to waves generated by the processes discussed above, tidal forces also affect coastal areas (Shepard and La-Fond, 1940; Strahler, 1966). **Tides** are a twice-daily fluctuation of sea level, causing wave processes to migrate through sometimes sizable vertical ranges and setting up currents that flow in and out of bays, lagoons, and tidal channels. Tidal currents, especially in constricted channels, possess enough velocity to prevent entrances to bays and lagoons from being closed by sediment, and if tidal ranges are high enough and channels narrow enough, they can be an effective shoreline process.

The maximum effect of tides takes place in narrow embayments that constrict the flow of water, resulting in unusually high tidal ranges. An example of this effect occurs in the Bay of Fundy in Newfoundland, Canada, where a tidal range of about 15 m (50 ft), the highest in the world, produces tidal bores—steep-fronted waves that move up constricted estuaries as fast-moving walls of water, which may reach as high as 12 m (40 ft) (Lynch, 1982). The tidal bore of the Amazon River reaches heights of about 5 m (16 ft), moving upstream as a great cascading rapid at a speed of 12 knots.

Tides are caused primarily by the gravitational attraction of the moon, but they may be affected also by the sun's gravity, which, depending on its alignment with the moon, adds to or subtracts from the total gravitational pull. The highest tides occur when the moon and sun are aligned, and the lowest occur when they are not aligned.

The gravitational attraction of the moon (and sun) produces bulges in the oceans on either side of the earth, and as the earth rotates, any given point will alternately pass through a bulge every 12 hours, resulting in two high tides and two low tides each day. The moon and sun are aligned every two weeks, resulting in tides that are about 20 percent higher than normal, known as **spring tides.** In between, the moon and sun reach positions that are at right angles to the earth, and **neap tides,** about 20 percent lower than usual, occur. Differences in submarine topography may cause substantial variations in the normal tidal effect. The moon's orbit around the

earth is elliptical, causing the moon to be closer to the earth at some times and farther away at others. From time to time, the moon is closest to the earth during spring tides, setting up **perigean spring tides** that are even higher than the normal spring tides. Coincidence of perigean spring tides with powerful storms can produce unusually intense flooding along coastal regions.

Rip Currents

Rip currents are strong, narrow currents at nearly right angles to the shoreline that move seaward through the surf. They are the surge-like, seaward return of water flowing laterally along the shoreline as currents caused by waves approaching the shoreline at an angle, fed by currents set up by unequal breaker heights (Figure 16–7). Rip currents balance the mass of water brought to the shoreline via waves by moving water seaward through the breaker zone, setting up circulation cells with regular spacing (Hino, 1975; Shepard et al., 1941; Shepard and Inman, 1950). Rip currents may extend for lengths of 60 to 750 m (200 to 2500 ft) with velocities up to 2 knots, sufficient to carry swimmers out to sea.

Coastal geomorphologists generally agree that circulation cells are set up by a combination of longshore currents and variation in the heights of breaking waves. However, understanding of the specific mechanisms remains somewhat cloudy. Bowen and Inman (1969) suggested that the rhythmic beat of waves approaching a shoreline generates secondary standing waves, known as **edge waves** (Bowen, 1973; Bowen and Inman, 1971; Dolan et al., 1979a; Holman, 1983; Huntley and Bowen, 1973) at right angles to the onshore waves (Figure 16–8). The edge waves are standing waves with

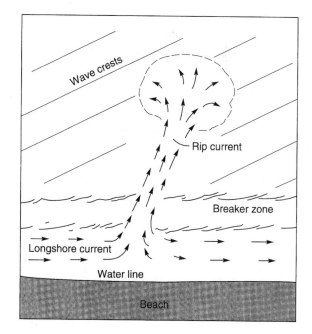

FIGURE 16–7
Rip currents and longshore currents.

(A) First half wave period

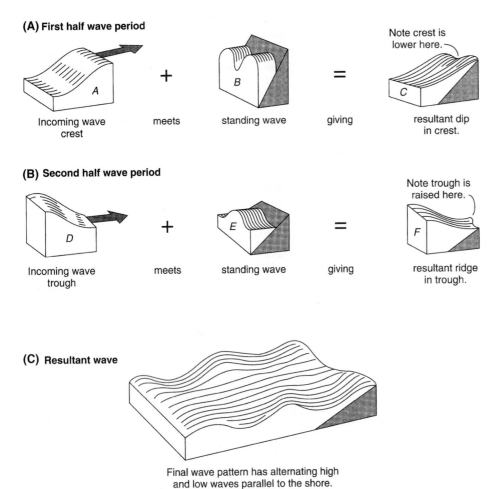

Incoming wave crest   meets   standing wave   giving   resultant dip in crest.

(B) Second half wave period

Incoming wave trough   meets   standing wave   giving   resultant ridge in trough.

(C) Resultant wave

Final wave pattern has alternating high and low waves parallel to the shore.

FIGURE 16–8
Edge waves near shore.
(A) The incoming wave crest *A* passes over the edge wave *B*, leaving a low spot in its crest.
(B) Half a wave period later, the incoming wave crest passes over the edge wave, increasing its height at the same position relative to the shoreline as the previous drop in crest height.
(C) The result is multiple undulations in wave height at regular intervals along the shore.
(From Pethick, 1984)

the same period as the onshore waves, and because they do not move, they modify the wave heights of the onshore waves to form undulating high and low crests along the waves, spaced at one edge-wavelength. The regular undulations in wave height provide the energy gradient that drives the feeder currents to the rip current circulation cells. Rip currents flow seaward along the line of lowest wave crests. This mechanism then predicts regular spacing of rip currents at one edge-wavelength. However, McKenzie (1958) found that rip current spacing was a function of wave height rather than wave period. He observed strong, widely spaced rip currents associated with large wave heights and weak, closely spaced rip currents with low wave heights. The control of rip currents by topography in the surf zone has also been suggested, but rip currents are known to also occur on long, straight beaches with smooth bottom topography.

Wave Refraction

The equation $V = \sqrt{gd}$ predicts that as waves enter shallow water, wave velocity will be retarded, causing the waves to bunch up and the wavelength to decrease. If the sea floor slopes uniformly and the waves approach parallel to the shoreline, each wave breaks at about the same time, and all the waves approach the shoreline in straight, parallel lines with only the spacing of the waves changing near the shore. However, few coastlines are straight for very long distances, so waves seldom approach parallel to the shoreline. When waves advance to the shoreline at an angle, part of each wave encounters the shallow sea floor sooner than the remainder of the wave. Thus, the velocity of the portion of the wave in shallow water will decrease relative to the remainder of the wave in deeper water. The net effect of the slowing down of the shallow-water part of a wave is bending, or **refraction,** of the wave. In Figure 16–9, waves are slowed down by the shallow water near the island in the bay, whereas the portions of the waves in deeper water continue at the same speed and thus advance ahead of the slowed portion of the waves. Because refraction of the wave is a function of water depth, wave refraction is a reflection of the configuration of the sea floor.

Waves arriving obliquely at a relatively straight shoreline with a uniformly sloping sea floor are refracted and bend

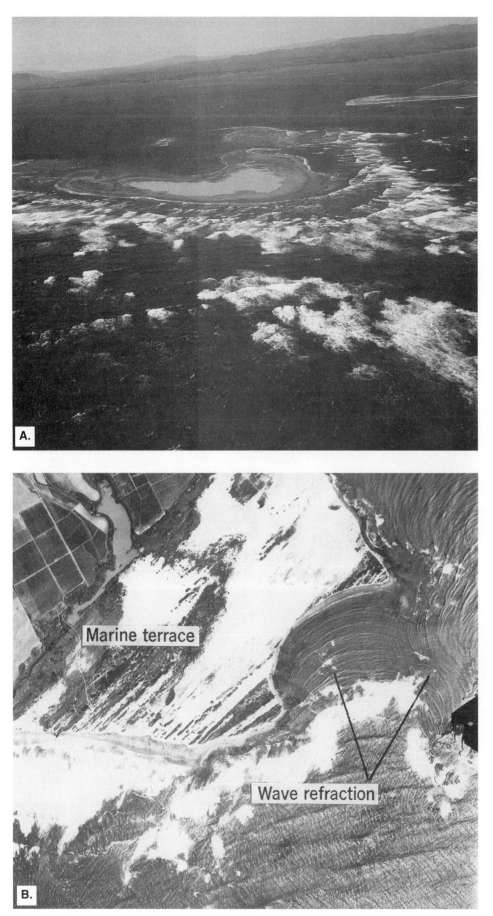

FIGURE 16–9
(A) Refraction of waves around an island in Willapa Bay, Washington. (Photo by D. A. Rahm)
(B) Refraction of waves as they approach an embayed shoreline, Santa Cruz, California. (Photo by U.S. Dept. of Agriculture)

convexly toward the shore, bringing the crest line more closely parallel to the beach. However, seldom are the waves completely refracted to parallel the coast and break obliquely on the shore.

Waves approaching an irregular shoreline provide a situation that also causes wave refraction. Waves first encounter shallow water offshore from headlands or peninsulas, causing them to slow down and bend convexly toward the shoreline, while waves in the deeper water of the bays continue rela-

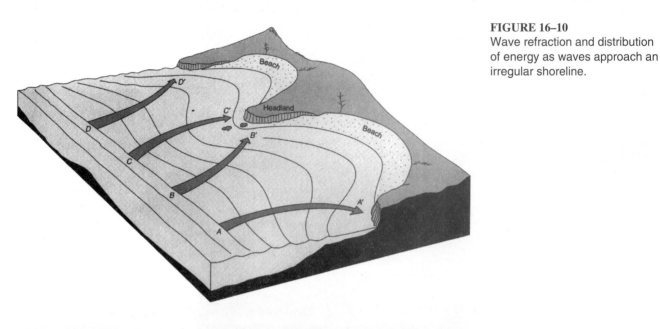

FIGURE 16–10
Wave refraction and distribution of energy as waves approach an irregular shoreline.

FIGURE 16–11
Refraction of waves approaching an irregular coast, west of Puerto Vallarta, Mexico. (Photo by D. A. Rahm)

tively unaffected, resulting in the wave pattern shown in Figures 16–10 and 16–11.

Because the energy of a wave is directed at right angles to the wave crest, lines drawn at right angles to the wave crests (orthogonals) represent the direction of energy expenditure. Orthogonals thus converge on headlands and diverge in bays, concentrating wave energy on the headlands and dissipating wave energy in the bays. For example, in Figure 16–10, if the wave energy along the crest of a wave is divided into equal segments so that the wave energy at *A-B* is equal to the wave energy at *C-D*, as the wave is refracted, the energy of *A-B* is concentrated over a small area on the headland, whereas the same energy of *C-D is* spread over a much larger area at the head of the bay. For this reason, headlands are characterized by high, breaking waves, rocky shores, intense erosion, and steep sea cliffs, whereas bays are typically quiet, with sandy beaches (Figure 16–11).

Longshore Currents

Longshore currents are generated by waves seaward from the beach that strike the shoreline obliquely (Bowen, 1969; Bruun, 1963; Galvin, 1967; Galvin and Eagleson, 1965; Komar, 1975; Sonu, 1972). As a wave approaches shore at an angle α, the water movement in the wave can be resolved into two components, one normal to the shore and the other parallel (longshore). The longshore component is proportional to the sin of the angle α. The amount of wave crest deflected into the longshore direction per unit of shore length will be

$$L = w_o \times \frac{\sin \alpha}{1/\cos \alpha} = w_o \sin \alpha \cos \alpha$$

where w_o = original wave width
α = angle of incidence of the wave (Inman and Bagnold, 1963)

The longshore discharge of water per unit beach length Q_L can then be calculated as

$$Q_L = Q_m \sin \alpha \cos \alpha$$

and the velocity of the longshore current *VL* can be written as

$$V_L = Q_m \sin \alpha \cos \alpha \tan \beta$$

where Q_m = discharge for unit width of wave
α = angle of wave approach
β = angle of beach face

Longuet-Higgins (1970) calculated the longshore velocity *VL* as

$$V_L = \frac{K \tan \beta V_m \sin \alpha \cos \alpha}{c}$$

where β = beach slope angle
c = frictional drag
V_m = maximum orbital velocity
α = angle of wave approach
K = a constant

Komar and Inman (1970) formulated a somewhat simplified version of this equation:

$$V_L = 2.7 \, V_m \sin \rightarrow \cos \rightarrow$$

COASTAL EROSION

The mass of water hurled against the land by waves breaking on a shoreline (Figure 16–12) possesses considerable

FIGURE 16–12
Wave crashing against the shoreline at Kapoho, Hawaii.

energy that exerts a significant effect on coastline morphology as a result of mechanical erosion (Norman, 1980; Shepard and Grant, 1947). Erosion is most pronounced during storms when wave energy is at its highest, removing previously weathered material and attacking newly exposed material. Storm waves are particularly effective on unconsolidated sediments, which tend to be easily eroded, and on highly fractured or bedded rocks, which are susceptible to quarrying as the impact of waves produces hydraulic pressure and compresses air in cavities, prying loose blocks of rock.

Sea Cliffs

Wave erosion undercuts slopes at the shoreline, the most obvious effect of which is the development of **sea cliffs** that progressively retreat landward under relentless wave attack (Emery, 1941; Emery and Kuhn, 1982; Kuhn and Shepard, 1983). Wave action is most vigorous at the base of sea cliffs, and if the rock is resistant enough to sustain overhanging slopes, the waves commonly develop a **wave-cut notch**, which leaves an indelible mark of the sea level that made it (Figure 16–13). Along tectonically rising or subsiding coasts, wave-cut notches provide useful evidence of former sea levels.

Undercutting of the base of a sea cliff leads to increased shear stress in the cliff-forming material and promotes accelerated mass movement. The mass movement debris collects at the base of the cliff until removed by wave attack. If waves erode the talus at the base of the cliff at the same rate as it is added from the slopes above, a relatively steady state is formed, but often the removal of talus is sporadic, with large amounts eroded during a single large storm. As long as the talus remains, it protects the base of the sea cliff from eroding fresh material until the debris is removed and the base of the cliff is again exposed. Because of substantial differences in erodibility of unconsolidated material and bedrock, sea cliffs in the former are more likely to show talus cones at the base of cliffs between storms; in the latter, the cliffs are more likely to consist of bare, rocky slopes.

The rate of sea cliff recession (Figure 16–14) is controlled primarily by the vigor of wave action and the resistance of the cliff material, but it may also be affected by weathering processes, abrasion, and biological activity. Sea cliff erosion rates are extremely variable, both geographically and temporally. Shorelines composed of highly resistant material with only weak wave activity retreat very slowly, whereas shorelines composed of unconsolidated material attacked by vigorous wave action retreat very rapidly. During a single storm, 5 to 30 meters (~15 to 100 feet) of sea cliff recession has been recorded along some coastlines. Wave-tank experiments aimed at quantifying cliff recession rates have been attempted (Sunamura, 1975, 1976, 1977, 1982, 1983), but the difficulty of replicating size, scale, and natural conditions hinders modeling attempts. Sunamura (1983) suggests that a minimum critical wave height is required to cause erosion of cliff material having a certain compressive strength, and the rate

of sea cliff erosion is determined by the frequency of wave incidence exceeding that threshold height.

Generally, crystalline rocks (granite, volcanic rocks) show the lowest rates of sea cliff retreat, and unconsolidated Quaternary sediments show the highest rates (Sunamura, 1983). Figure 16–15 shows relative rates of erosion for several types of cliff material.

Wave-cut Platforms

Prolonged erosion and retreat of a sea cliff leave behind a **wave-cut platform** (also known as an **abrasion platform** or **shore platform**) slightly below sea level at high tide, beveling the rocks on the sea floor (Figure 16–16) (e.g., Bradley, 1958; Bradley and Griggs, 1976; Edwards, 1951; Sunamura, 1978, 1983). As waves break on the shore, particles on the sea floor are dragged back and forth with each passing wave, abrading the bedrock like a great horizontal saw, the teeth of which are the loose particles moved by the waves. The depth below which wave-induced movement of sediment on the bottom diminishes to zero is known as the **wave base.** Johnson (1919) suggested that abrasion of bedrock on the sea floor by sand could extend to depths of 180 m (~600 ft). However, Bradley (1958) suggested that 10 m (33 ft) was the maximum depth at which abrasion was effective in eroding wave-cut platforms, and Robinson (1977) and Trenhaile (1980) infer that most abrasion takes place on the shallow portion of platforms near the shoreline.

In the early stages of sea cliff retreat, gently inclined wave-cut platforms develop rapidly, but platform expansion decreases as waves begin to break farther offshore and lose progressively more energy on the widening platform. The limit of cliff retreat and platform expansion is probably on the order of 500 m (1650 ft) or so, but that may be exceeded if sea level slowly rises.

As sea cliffs retreat, incompletely beveled rock may remain above the wave-cut platform as **sea stacks** (Figure 16–17), isolated from the rest of the sea cliff by erosion. Selective wave erosion cuts sea caves into coastal rocks by etching out zones of less resistant rock (Figure 16–18A). As sea caves extend completely through narrow headlands or as the roofs of sea caves collapse, they produce **sea arches** (Figure 16–18B).

COASTAL DEPOSITION

Beaches

Beaches are accumulations of sand, pebbles, or cobbles along a shoreline in the zone of breakers. They consist of whatever sediment is available for movement by wave action, and they are produced either from streams carrying debris to the coast or from marine erosion. Sediment brought to the shore by streams generally dominates the source of detrital material available for beach formation, although sea cliff erosion may contribute significant amounts of sediment locally, especially where sea cliffs consist of unconsolidated sediment.

A.

B.

FIGURE 16–13
(A) Wave-cut notch and overhanging cliff along the Mexican coast. (Photo by Ken Hamblin)
(B) Wave-cut notch, Sucia Island, Washington. (Photo by D. A. Rahm)

FIGURE 16–14
German gun emplacements built in the 1940s being undercut by sea cliff recession, Lokken, Denmark.

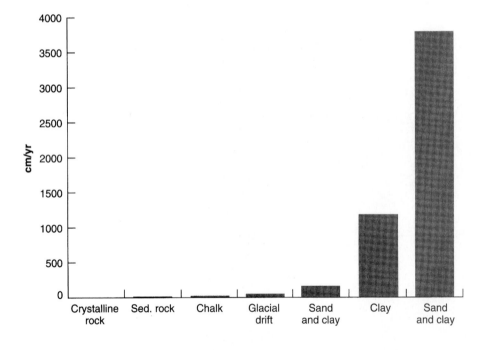

FIGURE 16–15
Examples of rates of sea cliff recession for different material. (Plotted from data in Johnson, 1925; Zenkovitch, 1967; King, 1972; Peyronnin, 1962; Rankin, 1952; and Zeigler et al., 1959)

The location of beaches is determined by sediment supply and wave activity or from reworking of older, nearshore sediments. Beaches are dynamic features, frequently changing to adapt to varying conditions. Their sediment is mobile and subject to movement, so they represent an equilibrium between wave action and sediment supply. Beach sediment consists of whatever detrital material is available. Although many consist of sand, others are made up of pebbles or cobbles. If the pebbles are somewhat flattened, they form **shingle beaches**, which are typically quite steep, in part because of easy percolation of wave swash into the coarse sediment,

thus retarding the effectiveness of backwash and enhancing the pushing up of pebbles into steeply sloping beaches.

Most definitions of beaches include more than just the visible portion of sediments. Komar (1976) defines the limits of a beach as extending seaward to include the zone in which sediments are moved by wave action and continuing landward to the upper limit that waves can reach.

Beaches typically develop characteristic components (Figure 16–19) (Bagnold, 1940; Bascom, 1951, 1954, 1964; Brenninkmeyer, 1975; Davis, 1978; Einstein, 1948; Fisher and Dolan, 1977; Gorsline, 1966; Hayes, 1972; King, 1972; Kirk-

FIGURE 16–16
Wave-cut (abrasion) platform, Little Sucia Island, San Juan Islands, Washington.

FIGURE 16–17
Sea stacks along the Oregon coast.

A.

FIGURE 16–18
(A) Sea cave made by selective
erosion of a weak and
sedimentary bed and sea arch
created by collapse of the cave
roof, Devil's Punch Bowl,
Oregon.
(B) Sea stack with an arch
eroded through it, near Brandon,
Oregon. (Photo by D. A. Rahm)

B.

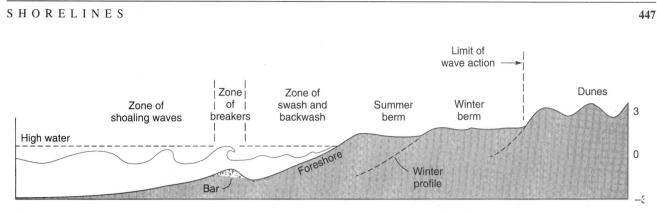

FIGURE 16–19
Components of a typical beach profile.

by and Kirkby, 1969; Komar, a or b 1971, 1976, a or b 1983; Kuhn and Shepard, 1983; Otvos, 1964; Short, 1979; Short and Wright, 1981; Tanner, 1958; Wentworth, 1938; Wright and Short, 1983; Zeigler et al., 1959). At the head, or back-shore, of a beach, berms with flat to gently sloping surfaces are formed as swash loses velocity due to friction and loss of water that infiltrates into the beach sediment, leading to deposition of sediment. Some beaches have two or more berms, usually a low summer berm formed by moderate to gentle waves and a landward winter berm made by more vigorous winter waves, although a storm at any time of year can leave a new berm. Berm elevation is determined by the range of wave swash and by the grain size of the sediment. The highest berm typically terminates landward at the base of a sea cliff or against dunes.

The **beach face** slopes seaward from the berm with inclinations that range from gentle to quite steep. Beach-face slopes show a close relationship between wave steepness (Figure 16–20) and the size of the particles moved in the swash zone (Figure 16–21).

The slope angle may vary dramatically between seasons: During the summer period of low, gentle swells, the waves build a berm. During the winter period of high, steep waves, the waves destroy the berm and transport the material seaward (Shepard and LaFond, 1940; Shepard, 1950; Bascom, 1954; King, 1972). Not only is wave steepness important, but breaker type also seems to be significant. Steep beach faces are commonly associated with plunging breakers, and gentle beach faces with spilling breakers (Wright et al., 1983; Huntley and Bowen, 1973).

Beach-face slopes also show a strong correlation to particle size. Large grain sizes are associated with steep slopes, and small grain sizes with gentle slopes (Bagnold, 1940; Bascom, 1951; Shepard, 1963; Wiegel, 1964; DuBois, 1972). Sorting of beach sediment shows a close association with beach-face slope. Poor sorting results in poor swash infiltration and a steep beach face (McLean and Kirk, 1969).

Submerged **longshore bars** and **longshore troughs** may form seaward from the beach face (Figure 16–19), associated with the position of breaking waves (King and Williams, 1949; Shepard, 1950). The position of breakers determines the size,

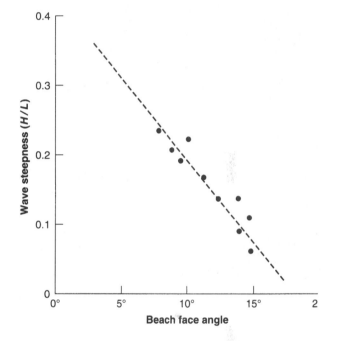

FIGURE 16–20
Relationship between waves, steepness, and beach-face slope. (From King, 1972)

location, and depth of the bars and troughs, and the bars may migrate seaward or landward with variations in wave height, wave steepness, and breaker type. Some longshore bars may extend for tens of kilometers, broken only by breaching from severe storms. The maximum water depth of bar development is determined by the effective wave base of breaking waves, which may vary from 10 to 25 m (~30 to 80 ft).

Beach Cusps

Some beaches are characterized by the development of regularly spaced, crescentic **beach cusps**, which form on the upper part of the beach face and the outer portion of the berm (Figure 16–22). The cusps are generally a few meters or less across, although in some cases they may reach 30 to 60 m

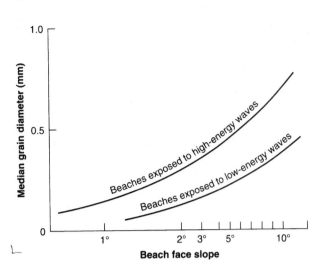

FIGURE 16–21
Relationship between beach-face slope and particle size for high-energy and low-energy waves. (From Komar, 1976)

(100 to 200 ft), and they may be constructed of almost any grain size, including boulders and cobbles (Russell and McIntire, 1965). The cusps consist of seaward-facing projections that are coarser grained than the embayments that separate them. Their origin has puzzled geomorphologists for decades, especially their regular spacing, and numerous papers have been written about them.

The relationship between the path of swash and sediment distribution has been observed for many years (Bagnold, 1940; Dolan, 1971; Dolan and Ferm, 1968; Dolan et al., 1974; DuBois, 1978, 1981; Evans, 1938; Johnson, 1910, 1919; Komar, 1971b; Kuenen, 1948; Longuet-Higgins and Parkin, 1962; Williams, 1973). Each wave washing up on the beach face is split by the seaward-facing projections of the cusps (Figure 16–23). The swash on each side of the projections flows into the swale between the horns of the cusps, meeting in the center where it returns seaward. Splitting of the swash at the horns of the cusps diminishes the wave velocity there, depositing the coarser sediment and encouraging percolation of water into the coarse sediment, which further aids the process. The returning backwash in the center of the swale sweeps away the finer sediment, depositing it just seaward from the swale. The seaward-flowing backwash in the swale then meets the next oncoming wave, retarding its progress into the swale and directing the main flow of the swash against the cusps. Once such a cusp system is established, its self-perpetuating operation is readily visible. However, much less apparent is how the cusp system is generated from an originally planar, uniformly sloping beach face. Attempts to explain the initiation of cusps by some random localization of erosion/deposition fail to explain why cusps are typically regularly spaced and why they seem to form only in the high-tide zone.

The best explanation of the origin of the regular spacing of cusps seems to be the interaction of oncoming waves with edge waves (Figure 16–24) (Bowen and Inman, 1969, 1971;

Bowen, 1973; Komar, 1973, 1976; Guza and Inman, 1975; Dolan et al., 1979a; Holman, 1983). The intersection of edge waves, moving parallel to the beach, with oncoming waves sets up regularly alternating wave heights that are used to explain the quickly developed, regular spacing of cusps on the beach. However, this explanation of the origin of cusps is not without problems: If the spacing of the cusps is caused by edge waves, their spacing should be the same as the wavelength of the edge waves, and that is yet to be documented (Komar, 1976). Thus, the precise origin of beach cusps cannot yet be considered totally resolved.

Littoral Drifting

Waves seldom approach a coastline parallel to the shore. They almost always intersect the shoreline at some oblique angle (Figure 16–25), despite the effect of wave refraction, which bends the wave crests more nearly parallel to the shore. The effect of oblique wave encroachment on sediment movement is profound (Duane and James, 1980; Fairchild, 1973; Inman and Bagnold, 1963; Inman and Frautschy, 1966; May and Tanner, 1973). As wave swash moves obliquely up a beach face, it carries particles up the beach at an angle (Figure 16–25), but as the water runs back down the beach, the backwash moves the particle directly down the beach face in the direction of the beach slope. The net effect is that each particle ends up laterally displaced several centimeters or tens of centimeters along the beach with each wave. These relatively small increments of lateral movement by each wave are then multiplied by the almost infinite repetitions that occur to result in the movement of enormous quantities of sediment laterally along the shoreline.

Sediment is also moved laterally down the coast by longshore currents set up by waves breaking obliquely on the shore (Figure 16–26). Longshore currents flow just offshore parallel to the coast (Figure 16–27) with sufficient velocity to transport bottom sediment. For a given direction of wave incidence, the direction of transport by longshore currents is the same as by beach drifting, so they operate simultaneously and complement one another. Any barrier to lateral movement of sediment interrupts transportation of material and causes sediment to pile up on the up-drift side of the barrier (Figure 16–28).

The amount of sediment moved by littoral drifting can be considerable. The U.S. Army Corps of Engineers estimates that 450,000 yd^3 (340,000 m^3) of sediment is moved by littoral drift annually (123,000 yd^3/day) (~100,000 m^3/day) along most of the east coast of the United States.

Spits and Bars

As long as a coastline remains fairly straight and waves approach it at an angle, shore drifting goes on without interruption, but longshore transport may be disrupted by sharp bends in the coastline where, instead of continuing around the corner,

FIGURE 16–22
(A) Beach cusps at Motunau Beach, South Island, New Zealand. Note the regular, crenulated pattern made by the lighter beach ridges of gravel against the darker swales.
(B) Closeup of beach cusps at Point Wilson, Washington.

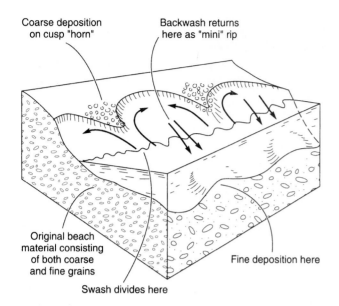

FIGURE 16–23
Formation of beach cusps. Incoming waves split on the cusp horns where coarse sediment is dropped; then the waves return seaward along the central swale. (From Pethick, 1984)

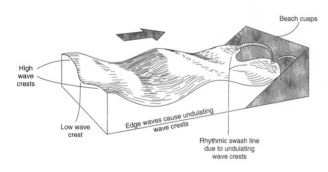

FIGURE 16–24
Relationship between edge waves and beach cusps. The rhythmic spacing of the cusps is produced by the spacing of high and low wave crests set up by the edge waves. (From Pethick, 1984)

sediment continues to move in the same direction and is deposited in the deeper water to form a **spit,** a ridge of sediment connected at one end to land and terminating in open water at the other end (Figure 16–29A; Plate 12B). Refraction of waves around the end of the spit causes a concomitant change in the direction of shore drift that results in the landward bending of the end of the spit to form a **recurved spit** or **hooked spit** (Figure 16–29B) (Evans, 1942). Although the portion of a spit above water may appear quite narrow, a much greater volume of sediment beneath the water surface typically comprises the base of the spit (Figure 16–30). Commonly, spits grow by the sweeping of submerged bars or beach ridges around the end of the spit, giving the spit a ribbed appearance (Figure 16–31). As spits continue to grow, they may alter the refraction

pattern of waves and build complex forms (Figure 16–32). Recurving of a spit landward may ultimately result in attachment of the distal end of the spit onto the mainland (Figure 16–33A), enclosing a lagoon that in time fills with sediment and organic material (Figure 16–33B).

Once formed, a spit continues to extend in the direction of shore drifting. If it extends all the way across a bay, it becomes a **baymouth bar** that encloses the landward part of the bay as a **lagoon** (Figure 16–34). Landward migration of offshore (barrier) bars until they reach land also produces similar landforms. The lagoon is gradually filled with sediment and organic material, gradually changing it into swamps and tidal marshes.

Some spits grow from the land toward offshore islands, eventually connecting the islands to the land as **tombolos.** The observed reaching out of spits to attach to islands is far from random. As waves encounter the shallow water in the vicinity of an island, they are slowed down and refracted around the island (Figure 16–35; Plate 12A). This sets up the wave pattern shown in Figure 16–35B, causing convergence of longshore drifting directions in the lee of the island. Because the direction of longshore transport on both sides of the beach is toward the lee of the island, sediment moving laterally along the beach accumulates there, conforming to the shape of the wave pattern. As more and more sediment is deposited, the spit extends seaward toward the island (Figure 16–36) until it eventually reaches it and attaches the island to the mainland as a tombolo.

An example of this phenomenon illustrates an important point about the sensitivity of shorelines. A breakwater built just offshore affects the wave refraction pattern in the same way as described above for formation of a tombolo. Thus, construction of such breakwaters can be expected to invite the extension of a spit from the shore toward the breakwater and eventually to make it unusable (Inman and Frautschy, 1966).

Barrier Bars and Islands

Barrier bars and **barrier islands** are large, elongate bars, usually composed of sand, just offshore and parallel to the shoreline but usually not attached to the mainland. They are typically 3 to 30 km (~2 to 20 mi) offshore and are separated from the mainland by a lagoon (Plate 11E and Figures 16–37, 16–38, and 16–39)]. They may extend for long distances, typically 10 to 100 km (6 to 60 mi) and may range in width from a few hundred meters to several thousand meters. Their elevation above sea level is generally about 5 m (~15 ft) or less, making them particularly susceptible to overwashing by hurricanes and severe storms.

Barrier islands occur on 10 to 13 percent of the world's coasts (King, 1972), including 282 barrier islands along the Gulf coast and the Atlantic coast between Florida and the middle Atlantic states (Dolan et al., 1980). Some are the sites of major cities, for example, Atlantic City, Miami Beach, and Galveston.

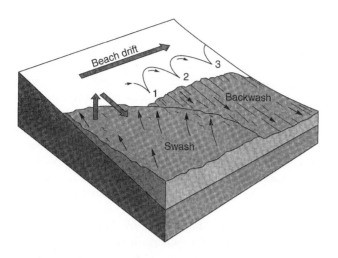

FIGURE 16–25
(A) Longshore drifting caused by oblique waves pushing sediment up the beach face at an angle, then returning seaward at right angles to the beach in the backwash.

The seaward side of **barrier islands** typically consists of low-gradient beaches that change their form during storms. Beaches, dunes, and marshes that make up the islands are ephemeral in nature, frequently changing in both location and shape. The landward side of barrier bars is characterized by lagoons, salt marshes, and large, shallow, tidal mud flats that fringe the lagoon. The bar itself usually carries sand dunes several meters to more than 100 m (~300 ft) high. Dunes are especially prevalent where beaches are perpendicular to the prevailing wind direction and sand is driven shoreward to feed the growing dunes. Landward from the dunes, barrier bars are commonly fairly flat, with grass, shrubs, and stands of pine and oak, grading into salt marshes and tidal flats that border the lagoon. Barrier bars are commonly breached in places by tidal inlets that connect the sea to the lagoon.

Because of their low topographic profiles and exposure to hurricanes and other major storms, barrier islands have long been dangerous places to inhabit (Hughes, 1979; Hayes,

(B) Waves approaching a shoreline at an angle. The direction of longshore drifting caused by this angle of wave incidence is toward the bottom of the photo. (Photo by D. A. Rahm)

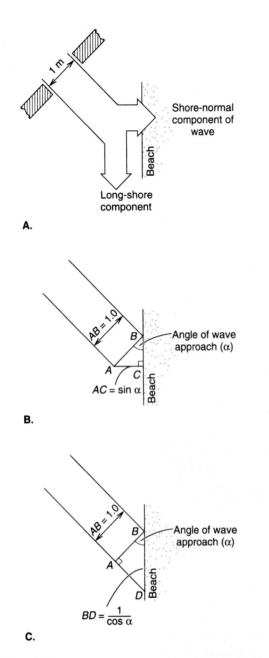

FIGURE 16–26
Development of a longshore current. (A) An imaginary breakwater allows a 1 m wave to pass through to the shore. The wave can be resolved into two components, one parallel to the shore and other normal to it. (B) The width of the component parallel to the shore is equal to the sine of the angle of the approaching wave. (C) The width of the wave is incident upon a length of shoreline equal to BD. Because BD=1/cos α, the length of wave per unit length of shoreline is sin α cos α (From Pethick, 1984)

1967, 1978; Williams and Guy, 1973). Early settlers tended to reside on the lagoon side of barrier bars, away from direct wave attack along the windward shores. However, during the past several decades, more and more development has

taken place along the ocean shoreline, placing greater numbers of people in peril during large storms. From 1900 to 1980, more than 100 hurricanes crossed Atlantic and Gulf coast barrier islands, about half of which carried winds greater than 150 km/hr (90 mi/hr) and storm surges of more than 3 m (10 ft). The Galveston, Texas, hurricane killed over 6000 people (Hughes, 1979), and the Florida hurricane of 1928 killed almost 2000 people. Hurricane Frederick, which struck the Gulf coast near Mobile, Alabama, in September of 1979, caused an estimated $700 million in damages. Hurricane Agnes caused $2 billion in damages in 1972, and Hurricane Camille, with wind speeds over 250 km/hr (150 mi/hr) and a storm surge of more than 8 m (25 ft) above sea level, caused $1.4 billion in damages in 1969. Although hurricanes cause extensive property damage, loss of life, and drastic modification of barrier bars, any given section of a barrier bar does not experience a hurricane very often. Smaller winter storms, with waves of 5 to 10 m (15 to 30 ft) and storm surges of I to 2 m (3 to 7 ft) are capable of severely eroding beaches and causing substantial damage along the mid-Atlantic coast every year (Hayden, 1975). Such storms can push water and beach sand entirely across barrier bars. Between storms, the beaches commonly grow seaward, so barrier bar shorelines may move alternately landward and seaward. Measurements of shoreline movement in recent decades show landward migration at a rate of about 1.5 m/yr (5 ft/yr) for the Atlantic coast (Hayden et al., 1979; Moslow and Heron, 1978).

The origin of barrier islands and the process of bar migration have been controversial topics (de Beaumont, 1845; Gilbert, 1885; Hayes, 1979; Hoyt, 1967, 1968; Hoyt and Henry, 1967; LeBlanc and Hodgson, 1959; Johnson, 1919, 1922; Leontiev, 1969; McGee, 1890; Oertel, 1979, 1985; Otvos 1979; Shepard, 1962; Short, 1975 a, b; Swift, 1968, 1975; Wilkinson and Basse, 1978). Any theory of origin must take into account the known Holocene rise of sea level (~120 m (400 ft) in the past 13,000 years) and the irrefutable evidence that most mid-Atlantic barrier bars have migrated significantly landward over the past several centuries (Dillon, 1970; Kraft et al., 1973; Fisher and Simpson, 1979; Otvos, 1981, 1986). The origin of barrier islands does not appear to be a simple matter related to a single process.

Interest in the origin of barrier bars dates back to the middle 1800s when de Beaumont (1845) proposed bar emergence as the principal process responsible for barrier bars. He believed that the nature of sand or gravel that accumulates along the shoreline by wave action is dependent on wave energy, which establishes a profile of equilibrium. Wave action on a shallow bottom piles sediment up to form a bar that parallels the shoreline, establishing an equilibrium profile above and below sea level. De Beaumont's ideas became known as the *emergent theory* of barrier bar origin.

Gilbert (1885), in his classic monograph on Lake Bonneville, saw shore drift along the breaker zone as the origin of barrier bars by the building of a continuous offshore ridge on a gently sloping bottom with its crest a few feet above

FIGURE 16–27
Movement of pulp mill discharge parallel to the shoreline by longshore currents, Ediz Hook, Washington. (Photo by D. A. Rahm)

FIGURE 16–28
Interruption of longshore transport of sediment by groins built at right angles to the shore, causing accumulation of material on the up-drift side of the barriers. The direction of longshore transport is toward the bottom of the photo. (Photo by D. A. Rahm)

sea level and its seaward face with a typical beach profile. McGee (1890) noted what he called "half-drowned keys," strings of islands parallel to the shoreline, which he believed were formed by submergence of former beach ridges. Although wave action was now moving the islands shoreward, rapid sea-level rise along a gently sloping coastal plain was thought to have flooded the formerly subaerial surfaces behind them to form lagoons, leaving the former beach ridges as barrier islands. Johnson (1919) argued against the submergence concept of McGee and considered the near-shore bottom as a source of material for in situ deposition of bars, in contrast to their development by shore drift. Shepard (1962) held that barrier islands were formed by slow submergence (or even steady-state conditions) during or since the post-Pleistocene sea-level rise and were not related to coastal emergence. The source of sand for construction of the barrier bars was thought to be sand deposits on the continental shelf, locally augmented by sand from rivers. Leontiev (1973) proposed that underwater bars form during marine transgressions and later become barrier bars following a regression of sea level. Hoyt (1967) argued that barrier islands are formed essentially by partial submergence of preexisting coastal ridges, and he pointed out the absence of

A.

B.

FIGURE 16–29
(A) Spit growing across a bay, Puget Lowland, Washington. The direction of longshore transport is toward the top of the photo. (Photo by D. A. Rahm) (B) Hooked spit, San Juan Island, Washington. (Photo by D. A. Rahm)

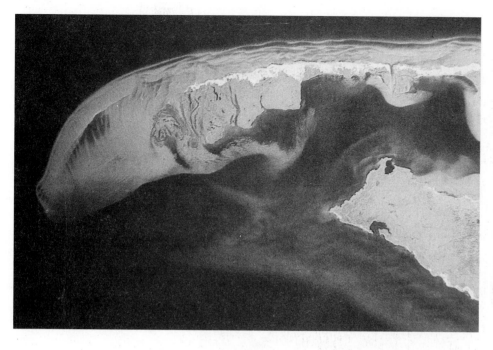

FIGURE 16–30
Spit extending seaward. The narrow, light-toned portion is above sea level, the next-darker-toned portion is the submerged part of the spit, and the darkest tone is deep water. (Photo by U. S. Dept. of Agriculture)

neritic sediments and the absence of open-ocean beach deposits as evidence that the island did not develop from offshore bars. He rejected the idea of continuous development during coastal submergence because it did not explain the original construction of the barrier bar, and he abandoned the emergence theory because he saw no evidence of a higher-than-present Holocene sea level.

Like de Beaumont (1845), Otvos contended that breaking storm waves pile up sand to form barrier islands and that landward migration of the islands occurs without any change in sea level. Otvos (1970) argued that most of the present Gulf coast barrier islands began to form during a sea-level stillstand about 5000 to 3500 years ago, and that barrier islands evolved by the emergence of shallow marine shoals formed by littoral drift and currents. Otvos suggested that during migration of the barrier islands parallel, perpendicular, or at oblique angles to the coastline, original features of the bars vanished.

Various lines of evidence were invoked to support one or another of the main proposed origins of barrier bars:

1. Emergence of submarine bars associated with regrading of the shore profile to provide the necessary equilibrium gradient (Johnson, 1919)

2. Development of extensive spits, later breached by severe storms to leave a series of offshore bars (Gilbert, 1885; Fisher, 1968)

3. Drowning of beach berms and sand dunes by recent rise of sea level (Hoyt, 1967)

4. Various combinations of these origins

Although differences of opinion continue to persist, the consensus seems to be that barrier islands have a protracted history of development associated with postglacial sea-level rise and that barrier islands have migrated shoreward with time.

During the last late-Pleistocene glaciation, about 10,000 to 20,000 years ago, sea level was about 120 m (400 ft) lower than it is today, and shorelines were 60 to 150 km (35 to 90 mi) seaward from their present positions (Chappell and Polach, 1976; Coleman and Smith, 1964; Curray 1965; Dolan et al., 1980; Fairbridge, 1961; Guilcher, 1969). Whatever the origin of the initial ridges, they migrated landward as postglacial sea level rose, and they reached their approximate modern configuration as sea level approached its present level about 4000 to 5000 years ago. Because sea level continues to rise slowly, barrier islands from New England to Texas continue to evolve and migrate landward.

Convincing evidence of the landward migration of barrier islands of the mid-Atlantic states includes the following:

1. Lagoonal sediment, salt-marsh peat, and tree stumps now exposed on the open ocean beaches, indicating that the present ocean beaches lie on older lagoonal sediments

2. Overwash sand, tidal inlet sediment, and windblown sand overlying lagoonal sediment on the landward side of the barrier bar

3. Island recession measured from historical maps and aerial photographs (Armon, 1979; Armon and McCann, 1979; Curray, 1960, 1961; Dillon and Oldale, 1978; Field et al., 1979; Fisher and Simpson, 1979; Kraft, 1971; Kraft et al., 1973; Leatherman, 1983; Leontiev, 1973; Otvos, 1970, 1986; Swift, 1968)

Exposures of salt-marsh peat that are only a few hundred years old on ocean beaches of barrier bars show that the rate of landward migration of the barrier bars occurs at a rapid

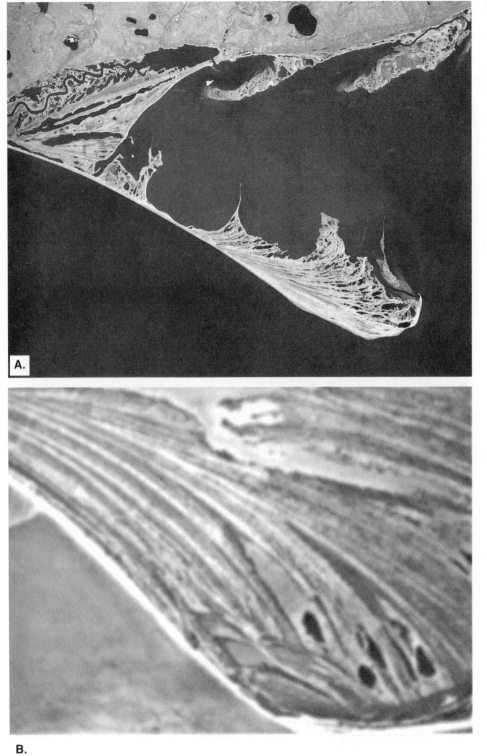

A.

B.

FIGURE 16–31
(A) Beach ridges on a
prograding spit, Alaska.
(B) Beach ridges deposited
during growth of St. Vincent
Island, Florida. (Photo by U. S.
Geological Survey)

FIGURE 16–32
Complex spit, Dungeness, Washington. (Photo by D. A. Rahm)

rate where wave energy and storm frequency are high, tidal range is small, and dune growth is minimal. Using historic records of the past 100 years, Pierce (1970) calculated the net accretion along the barrier island system between Hatteras Inlet and Cape Lookout along the Atlantic coast of the United States to be 796,000 m³ (1 million yd³) annually, less than half of which can be accounted for by longshore drift and biogenic sources. Pierce suggested that the remainder is derived from sand or poorly consolidated sediment on the continental shelf. Dolan and others (1980) calculated that by the year 2010, Cape May, the southernmost barrier island off the New Jersey coast, has a 1 in 2 probability of moving 90 m (~300 ft) landward from its present position.

Barrier islands are commonly inundated by **overwash** during severe storms, sometimes breaching the bar to form new inlets through the bar and depositing **washover fans** (Figure 16–40) (Pierce, 1970). Such inlets segment the barrier bar into separate islands and connect the lagoon on the landward side of the bar with the open ocean. Water flowing landward through

FIGURE 16–33
(A) Curved spit bending landward until it almost touches the mainland again, Puget Lowland, Washington.
(B) Curved spit enclosing a lagoon being filled with sediment, Puget Lowland, Washington.

FIGURE 16–34
Baymouth bar across a lagoon, northern California coast. (Photo by D. A. Rahm)

the inlet at high tides moves sand through it, depositing a **flood tidal delta** on the inside of the barrier island and a similar, usually somewhat smaller, **ebb tidal delta** at low tide on the ocean side. Shoaling during low tides on the lagoon side of the bar leads to the development of new salt marshes (Godfrey, 1976). Filling of some inlets with sand may reconnect the segmented bar, or the inlets may migrate in the direction of littoral drift.

TYPES OF COASTLINES

The complexity and diversity of shorelines make establishment of an all-encompassing classification difficult, if not impossible. However, many shorelines show the imprint of one or more surface processes or show the effects of raising or lowering of sea level. Fortunately, attempting to shove all shorelines into a single classification scheme is perhaps less important than attaining an understanding of the physical processes of shoreline modification. Several attempts have been made to classify all major types of shorelines, but none seem to be entirely adequate. Two examples of shoreline classification schemes are discussed here.

An early classification by Johnson (1919) recognized four main categories of shorelines:

1. **Shorelines of submergence**, formed by submergence of the coastline by a rise of sea level or by subsidence of the land. Such shorelines are typified by drowned valleys, deep embayments, numerous islands, and a very irregular shoreline (Figures 16–41 and 16–42).

2. **Shorelines of emergence**, formed by uplift of the land or by lowering of sea level. Such shorelines typically have straight coastlines of low relief and marine terraces (Figures 16–43 and 16–44; Plate 12C).

3. **Neutral shorelines**, which are dominated by various surface processes. Some examples of such shorelines are:

    a. Deltas, alluvial plains, and outwash plains

    b. Volcanic shorelines

    c. Coral reefs

    d. Faulted shorelines

4. **Compound shorelines,** which exhibit features of several of the shoreline types listed above.

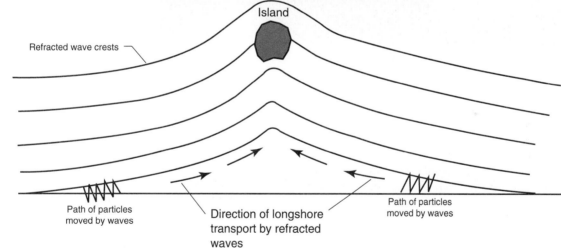

Island

Refracted wave crests

Path of particles moved by waves

Direction of longshore transport by refracted waves

Path of particles moved by waves

B.

Johnson's shoreline classification was both simple and genetic, and it included the major types of shorelines. A disadvantage of the system is that some shorelines are so complex that they might fit in any of several of the categories as a result of sea-level fluctuations during the Pleistocene. For example, a faulted shoreline may exhibit features of both submergence and emergence.

Shepard (1948) proposed the following somewhat different classification that emphasizes other factors in shoreline development:

I. Primary, or youthful, coasts and shorelines, produced chiefly by nonmarine agencies

    A. Shaped by erosion on land and drowned as a result of deglaciation or downwarping

        1. Drowned river coasts (ria coasts)

        2. Drowned glaciated coasts

    B. Shaped by deposits made on land

        1. River-deposition coasts

            a. Deltaic coasts

FIGURE 16–36
Beach sediment "reaching"
toward Frost Island, Washington.
(Photo by D. A. Rahm)

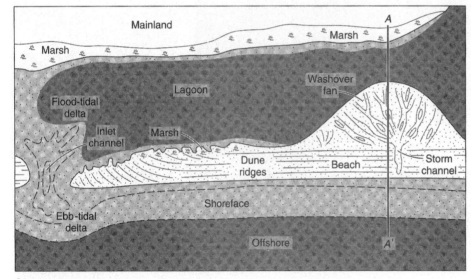

A.

FIGURE 16–37
(A) Map of a typical barrier bar.
(B) Cross-section A-A' of a
typical barrier bar.

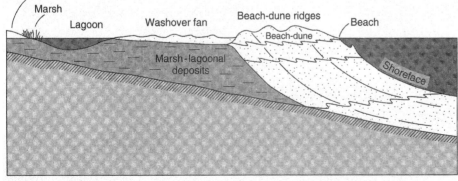

B.

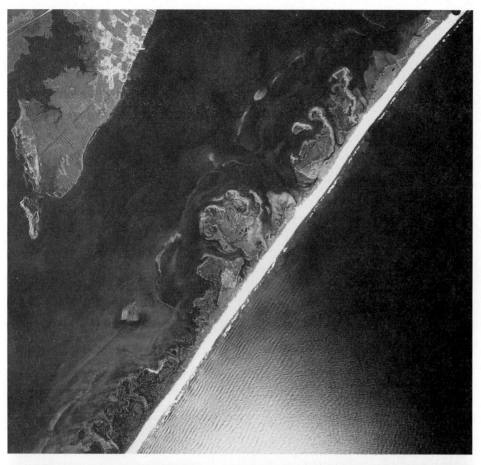

FIGURE 16–38
Barrier bar off the Atlantic coast. (Photo by U. S. Dept. of Agriculture)

FIGURE 16–39
Barrier islands off the Atlantic coast. (NASA Landsat photo)

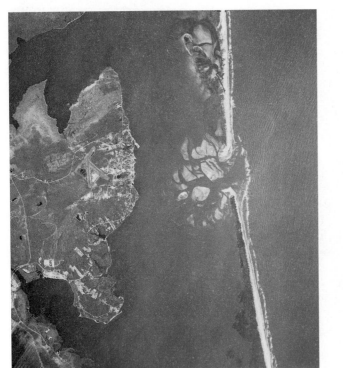

FIGURE 16–40
Washover fan through a barrier bar, North Carolina coast.
(Photo by U. S. Dept. of Agriculture)

  b. Drowned alluvial plains

 2. Glacial-deposition coasts

  a. Partially submerged moraines

  b. Partially submerged drumlins

 3. Wind-deposition coasts (prograding sand dunes)

 4. Vegetation-extended coasts (mangrove swamps)

C. Shaped by volcanic activity

 1. Recent lava flows

 2. Volcanic collapse or explosion (calderas)

D. Shaped by diastrophism

 1. Fault-scarp coasts

 2. Coasts related to folding

II. Secondary, or mature, coasts and shorelines shaped primarily by marine agencies

 A. Shaped by marine erosion

  1. Sea cliffs straightened by wave erosion

  2. Sea cliffs made irregular by wave erosion

 B. Shaped by marine deposition

  1. Coasts straightened by deposition of bars across estuaries

  2. Coasts prograded by deposits

  3. Shorelines with offshore bars and longshore spits

  4. Coral reefs

Shepard's classification is also useful but suffers from some of the same drawbacks as other classifications. For example, erosion and deposition are likely to occur on a coast simulta-

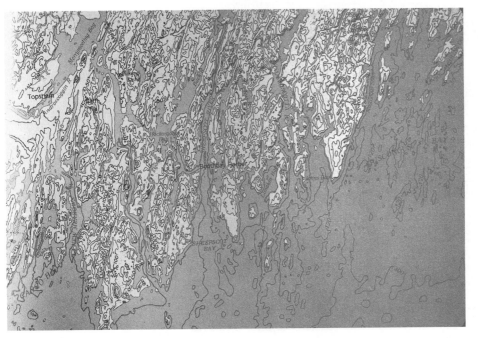

FIGURE 16–41
Submergent shoreline near
Boothbay Harbor, Maine.
Boothbay Harbor quadrangle.
(U. S. Geological Survey)

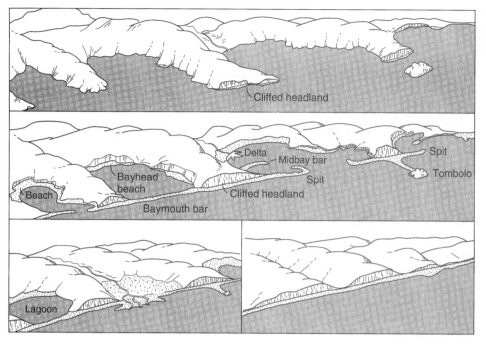

FIGURE 16–42
Evolution of shorelines along a submergent coast. (Modified from Strahler, 1963)

FIGURE 16–43
Marine terraces, Big Cyr, California. (Photo by D. A. Rahm)

neously, eroding headlands and depositing the material as spits and bars, making description of the coast as erosional or depositional difficult. It also encounters the same difficulties as Johnson's as a result of fluctuations of Pleistocene sea levels.

Because of the difficulties discussed above, no attempt is made here to define a classification scheme for coastlines.

CORAL REEFS

The growth of corals and algae in tropical oceans produces shorelines dominated by reefs. Because of the limited range of temperatures in which coral can thrive, coral reefs grow exclusively in tropical ocean water where water temperatures exceed 68°F (20°C), generally between latitudes 30°N and 25°S. Optimum water conditions for coral growth include water temperatures between 77°F (25°C) and 86°F (30°C); shallow water, usually not more than about 60 m (~200 ft) deep to allow penetration of sunlight required for growth; normal salinity; and sediment-free water. Coral growth is inhibited by dilution of seawater by fresh-water streams. Although coral reefs form wherever these conditions are found, they especially flourish on the windward side of islands,

where the greater wave action agitates the water, supplying food and oxygen to the corals.

The origin and evolution of coral reefs were first succinctly recognized by Darwin (1837), who, although most famous for his treatise on the evolution of fauna and flora, was also well known as a geologist. During the epic voyage of the *Beagle* from 1832 to 1835, Darwin visited several reefs in the Pacific and deduced a theory of their origin (Darwin, 1842), which has proven to be essentially correct by subsequent investigations (Cloud, 1958; Dana, 1874, 1885; Davis, 1928; Dobrin et al., 1949; Emery, 1948; Emery et al., 1954; Fairbridge, 1950; Guilcher, 1988; Ladd, 1961; Purdy, 1974; Umgrove, 1947; Wells, 1957; Wiens, 1962).

Darwin recognized three different types of reefs: fringing, barrier, and atolls (Figures 16–45 and 16–46). **Fringing reefs** grow along the shoreline in curvilinear zones varying in width from about a hundred meters to 1 km (~100 yd to 0.6 mi). They are usually not found where fresh-water streams dilute the seawater and cloud the water with sediment. Fringing reefs may occur slightly above sea level at low tide, but because corals cannot live long out of the water, their occurrence is restricted in the intertidal zone.

Barrier reefs grow just offshore, separated from the land by lagoons (Figure 16–45A) ranging in width from about 1

FIGURE 16–44
Multiple marine terraces north of Santa Cruz, California. (Photo by D. A. Rahm)

km to more than 15 km (~0.6 to 10 mi). The reef mass also typically includes fragments of coral broken off by storm waves, mollusk shells, foraminifera, and other calcareous organisms. The most active coral growth takes place on the seaward side of the reef, but it is also subject to erosion by storm waves. The Great Barrier Reef, the largest barrier reef in the world, extends for more than 2000 km (~1200 mi) along the coast of Australia.

The third type of reef recognized by Darwin is the **atoll**, which consists of a circular reef enclosing a lagoon (Figure 16–45B). Atolls lack the central island typical of barrier reefs and rise only a few meters above sea level.

The Darwin concept of the origin and evolution of these types of reefs begins with growth of a fringing coral reef around the periphery of an island (usually volcanic). As the central island slowly subsides, coral growth proceeds rapidly enough to maintain shallow-water conditions necessary for survival (Figure 16–46). The reef growth is largely directly upward and somewhat seaward, so that the dimensions of the reef remain about the same and, as the central island subsides,

a lagoon is formed, separating the reef from the land. Ultimately, the central island subsides completely below sea level, leaving only the reef as an atoll. Because of the important role of subsidence in Darwin's concept, it has become known as the **Darwin subsidence theory.**

Daly (1934, 1942) noticed that the lagoons behind many barrier reefs and atolls are very similar in depth, and he proposed that the reefs we see today were formed during the Holocene rise of sea level, after a period of erosion during which broad rock platforms were beveled by wave erosion. Daly postulated that extensive destruction of coral reefs occurred during the low sea-level stands of Pleistocene glacial episodes, partly by wave erosion and partly by reduction of the growth rate of corals in diminished water temperatures of the Pleistocene glaciations, resulting in the formation of broad marine benches. Then, with the return of warmer water and a hundred meters or so of postglacial rise of sea level, reef building increased rapidly, and reefs formed on the beveled platforms. This theory of reef formation is known as the **glacial control theory.**

FIGURE 16–45
Type of coral reefs.
(A) Barrier reef, Society Island, Tahiti. (Photo by Bruce Coleman, Inc.)
(B) Atoll, Tahiti. (Photo by Tahiti Tourist Promotion Board)

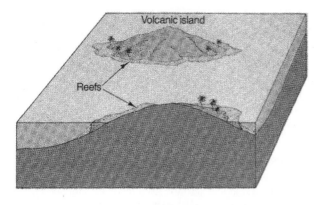

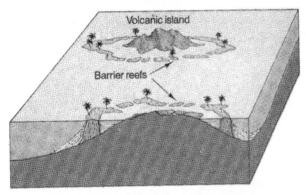

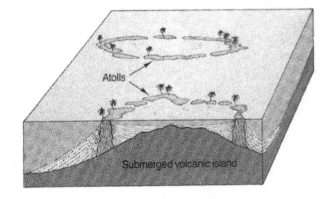

FIGURE 16–46
Evolution of coral reefs.

The most compelling evidence in distinguishing Darwin's subsidence theory from Daly's glacial control theory is the thickness of reef deposits in an area. According to the subsidence theory of Darwin, reef deposits should be quite thick as the result of continued upward growth of coral as the central island subsides, in sharp contrast to the very thin coral, which should result from the glacial control theory. Boreholes drilled on Bikini atoll in 1947 encountered 775 m (2500 ft) of coral, and two holes drilled on Eniwetok atoll in 1951 and 1952 penetrated 1212 m (4000 ft) of shallow water reef deposits above ocean-floor basalt. The significance of such great thicknesses of coral is that because coral can grow only in shallow water, subsidence must have occurred over a long period of time during the evolution of the atolls. Although Pleistocene sea levels rose and fell though a range of about 100 m (~300 ft), the boreholes on Bikini and Eniwetok atolls provide conclusive evidence that Darwin's subsidence theory of atoll formation is correct.

REFERENCES

Aagard, T., 1991, Multiple bar morphodynamics and its relations to low frequency edge waves: Journal of Coastal Research, v. 7, p. 801–813.

Antia, E. E., 1989, Beach cusps and beach dynamics; a quantitative field appraisal: Coastal Engineering, v. 13, p. 263–272.

Armon, J. W., 1979, Landward sediment transfers in a transgressive barrier island system, Canada: in Leatherman, S. P., ed., Barrier islands from the Gulf of St. Lawrence to the Gulf of Mexico: Academic Press, N.Y., p. 65–80.

Armon, J. W., and McCann, S. B., 1979, Morphology and landward sediment transfer in a transgressive barrier island system, southern Gulf of St. Lawrence: Geology, v. 31, p. 333–344.

Bagnold, R. A., 1940, Beach formation by waves: some model experiments in a wave tank: Journal of the Institute of Civil Engineers, v. 15, p. 27–52.

Bagnold, R. A., 1963, Mechanics of marine sedimentation: in Hill, M. N., ed., The sea: John Wiley and Sons, N.Y., p. 507–523.

Bascom, W. H., 1951, The relationship between sand size and beach slope: Transactions of American Geophysical Union, v. 32, p. 866–874.

Bascom, W. H., 1954, Characteristics of natural beaches: Proceedings of 4th Conference of Coastal Engineering, p. 163–180.

Bascom, W. H., 1959, Ocean waves: Scientific American, v. 201, p. 74–81.

Bascom, W. H., 1964, Waves and beaches: Doubleday and Co., Garden City, N.Y., 267 p.

Bowen, A. J., 1969, The generation of longshore currents on a plane beach: Journal of Marine Research, v. 3, p. 206–215.

Bowen, A. J., and Inman, D. L., 1969, Rip currents, 2. Laboratory and field observations: Journal of Geophysics Research, v. 74, p. 5479–5490.

Bowen, A. J., and Inman, D. L., 1971, Edge waves and crescentic bars: Journal of Geophysics Research, v. 76, p. 8662–8671.

Bowen, A. J., 1973, Edge waves and the littoral environment: Proceedings of l3th Conference on Coastal Engineering, p. 1313–1320.

Bowman, D., Birkenfeld, H., and Rosen, D.S., 1992, The longshore flow component in low energy rip channels; the Mediterranean, Israel: Marine Geology, v. 108, p. 259–274.

Bowen, A. J., and Inman, D. L., 1969, Rip currents, 2: laboratory and field observations: Journal of Geophysics Research, v. 74, p. 5479–5490.

Bowen, A. J., and Inman, D. L., 1971, Edge waves and crescentic bars: Journal of Geophysical Research, v. 76, p. 8662–8671.

Bradley, W. C., 1958, Submarine abrasion and wave-cut platforms: Geological Society of America Bulletin, v. 69, p. 967–974.

Bradley, W. C., and Griggs, G. D., 1976, Form, genesis, and deformation of central California wave-cut platforms: Geological Society of America Bulletin, v. 87, p. 433–449.

Brenninkmeyer, B. M., 1975, Frequency of sand movement in the surf zone: Proceedings of 14th Conference on Coastal Engineering, p. 812–827.

Bruun, P., 1963, Longshore currents and longshore troughs: Journal of Geophysical Research, v. 68, p. 1065–1078.

Bryant, E., 1988, Storminess and high tide beach change, Stanwell Park, Australia, 1943–1978: Marine Geology, v. 79, p. 171–187.

Carter, R. W. G., 1986. The morphodynamics of beach ridge formation: Magilligan, northern Ireland: Marine Geology, v. 73, p. 191–214.

Chappell, J. M. A., and Polach, H., 1976, Holocene sea-level change and coral reef growth at Huon Peninsula, Papua, New Guinea: Geological Society of America Bulletin, v. 87, p. 235–240.

Cloud, P. E., 1958, Nature and origin of atolls: 8th Pacific Science Congress, v. IIIA, p. 1009–1035.

Coleman, J. M., and Smith, W. G., 1964, Late recent rise of sea level: Geological Society of America Bulletin, v. 75, p. 833–840.

Curray, J. R., 1960, Sediments and history of Holocene transgression, continental shelf, northwest Gulf of Mexico: in Shepard, F. P., et al., eds., Recent sediments, northwest Gulf of Mexico: American Association of Petroleum Geologists, p. 221 –266.

Curray, J. R., 1961, Late Quaternary sea level; a discussion: Geological Society of America Bulletin, v. 72, p. 1707–1712.

Curray, J. R., 1965, Late Quaternary history, continental shelves of the United States: in Wright, H., and Frey, D., eds., The

Quaternary of the United States: Princeton University Press, Princeton, N.J., p. 723–735.

Daly, R. A., 1934, The changing world of the Ice Age: Yale University Press, New Haven, CT, 271 p.

Daly, R. A., 1942, The floor of the ocean: University of North Carolina Press, Chapel Hill, North Carolina.

Dana, J. D., 1874, Corals and coral islands: Dodd Mead, N.Y., 406 p.

Dana, J. D., 1885, The origin of coral reefs and islands: American Journal of Science, v. 30, p. 89–105; 169–191.

Darwin, C., 1837, On certain areas of elevation and subsidence in the Pacific and Indian oceans, as deduced from the study of coral formations: Geological Society of London Proceedings, v. 2, p. 552–554.

Darwin, C., 1842, The structure and distribution of coral reefs: reprinted by Appleton and Co., N.Y., 344 p., and University of Arizona Press, 214 p.

Das, M. M., 1971, Longshore sediment transport rates: a compilation of data: U.S. Army Coastal Engineering Research Center, Miscellaneous Paper 1-71, 81 p.

Davis, R. A., ed., 1978a, Coastal sedimentary environments: Springer-Verlag, N.Y., 716 p.

Davis, R. A., 1978b, Beach and near-shore zone: in Davis, R. A., ed., Coastal sedimentary environments: Springer-Verlag, N.Y., p. 237–285.

Davis, R. A., and Fox, W. T., 1972, Coastal processes and nearshore bars: Journal of Sedimentary Petrology, v. 42, p. 401–412.

Davis, W. M., 1928, The coral reef problem: American Geographical Society Special Publication 9, 596 p.

Dean, R. G., Healy, T. R., Dommerholt, A. P., 1993, A "blind-folded" test of equilibrium beach profile concepts with New Zealand data: Marine Geology, v. 109, p. 253–266.

de Beaumont, E., 1845, Leçons de geologic, pratique, Bertrand, P., ed., Paris, p. 223–252.

Dietz, R. S., 1963, Wave-base, marine profile of equilibrium, and wave-built terraces: a critical appraisal: Geological Society of America Bulletin, v. 74, p. 971–990.

Dillon, W. P., 1970, Submergence effects on a Rhode Island barrier and lagoon and inferences on migration of barriers: Journal of Geology, v. 78, p. 94–106.

Dillon, W. P, and Oldale, R. N., 1978, Late Quaternary sea-level curve: Geology, v. 6, p. 56–60.

Dobrin, M. B., Perkins, B., and Snavely, B. L., 1949, Subsurface constitution of Bikini atoll as indicated by seismic refraction survey: Geological Society of America Bulletin, v. 60, p. 807–828.

Dolan, R., 1971, Coastal landforms: crescentic and rhythmic: Geological Society of America Bulletin, v. 82, p. 177–180.

Dolan, R., and Ferm, J. C., 1968, Crescentic landforms along the mid-Atlantic coast: Science, v. 159, p. 627–629.

Dolan, R., and Lins, H. E., 1986, The Outer Banks of North Carolina: U.S. Geological Survey Professional Paper 1177-B.

Dolan, R., Hayden, B., and Vincent, L., 1974, Crescentic coastal landforms: Zeitschrift für Geomorphologie, v. 18, no. 1, p. 1–12.

Dolan, R., Hayden, B., and Heywood, J. E., 1978, Analysis of coastal erosion and storm surge hazards: Coastal Engineering, v. 2, p. 41–53.

Dolan R., Hayden, B., and Felder, W., 1979a, Shoreline periodicities and edge waves: Journal of Geology, v. 87, p. 175–185.

Dolan, R., Hayden, B., and Felder, W., 1979b, Shoreline periodicities and linear offshore shoals: Journal of Geology, v. 87, p. 393–402.

Dolan, R., Hayden, B., and Lins, H., 1980, Barrier islands: American Scientist, v. 68, p. 16–25.

Dolan, R., Lins, H. F., and Hayden, B. P., 1988, Mid-Atlantic coastal storms: Coastal Research, v. 4, p. 417–433.

Dolan, R., Vincent, L., and Hayden, B., 1974, Crescentic coastal landforms: Zeitschrift für Geomorphology, v. 18, p. 1–12.

Duane, D. B., and James, W. R., 1980, Littoral transport in the surf zone elucidated by an Eulerian sediment tracer experiment: Journal of Sedimentary Petrology, v. 50, p. 929–942.

DuBois, R. N., 1972, Inverse relation between foreshore slope and mean grain size as a function of the heavy mineral content: Geological Society of America Bulletin, v. 83, p. 871–876.

DuBois, R. N., 1978, Beach topography and beach cusps: Geological Society of America Bulletin, v. 89, p. 1133–1139.

DuBois, R. N., 1981, Foreshore topography, tides and beach cusps, Delaware: Geological Society of America Bulletin, v. 92, p. 132–138.

Edwards, A. B., 1951, Wave action in shore platform formation: Geological Magazine, v. 88, p. 41–49.

Einstein, H., 1948, Movement of beach sand by waves: Transactions of American Geophysical Union, v. 29, p. 653–655.

Emery, K. O., 1941, Rate of surface retreat of sea cliffs based on dated inscriptions: Science, v. 93, p. 617–618.

Emery, K. O., 1948, Submarine geology of Bikini atoll: Geological Society of America Bulletin, v. 59, p. 855–860.

Emery, K. O., and Kuhn, G. G., 1982, Sea cliffs: their processes, profiles and classifications: Geological Society of America Bulletin, v. 93, p. 644–654.

Emery, K. O., and Stevenson, R. E., 1957, Estuaries and lagoons: Geological Society of America Memoir 67, p. 673–750.

Emery, K. O., Tracey, J. L, and Ladd, H. S., 1954, Geology of Bikini and nearby atolls: U.S. Geological Survey Professional Paper 260-A, 265 p.

Evans, O. F., 1938, Classification and origin of beach cusps: Journal of Geology, v. 46, p. 615–627.

Evans, O. F., 1942, The origin of spits, bars, and related structures: Journal of Geology, v. 50, p. 846–863.

Fairbridge, R. W., 1950, Recent and Pleistocene coral reefs of Australia: Journal of Geology, v. 58, p. 330–401.

Fairbridge, R. W., 1961, Eustatic changes in sea level: in Physics and chemistry of the earth: Pergamon Press, N.Y., p. 99–185.

Fairchild, J. C., 1973, Longshore transport of suspended sediment: Proceedings of 13th Conference on Coastal Engineering, p. 1069–1088.

Field, M. E., and Duane, D. B., 1976, Post-Pleistocene history of the United States inner continental shelf: significance to origin of barrier islands: Geological Society of America Bulletin, v. 87, p. 691–702.

Field, M. E., Meisburger, E. P., Stanley, E. A., and Williams, S. J., 1979, Upper Quaternary peat deposits on the Atlantic inner shelf of the United States: Geological Society of America Bulletin, v. 90, p. 618–628.

Fisher, J. J., 1968, Barrier island formation: discussion: Geological Society of America Bulletin, v. 79, p. 1421–1428.

Fisher, J. J., 1973, Bathymetric projected profiles and the origin of barrier islands; Johnson's shoreline of emergence, revisited: in Coates, D., ed., Coastal geomorphology, part 2, barrier islands: State University of New York, Binghamton, N.Y., p. 161–179.

Fisher, J. J., and Dolan, R., eds., 1977, Beach processes and coast hydrodynamics: Academic Press, N.Y., 382 p.

Fisher, J. J., and Simpson, E. J., 1979, Washover and tidal sedimentation rates as environmental factors in development of a transgressive barrier shoreline: in Leatherman, S. P., ed., Barrier islands from the Gulf of St. Lawrence to the Gulf of Mexico: Academic Press, N.Y., p. 127–148.

Fletcher, C. H., 1992, Sea level trends and physical consequences: Earth Science Review, v. 33, p. 73–109.

Galvin, C. J., 1967, Longshore current velocity: a review of theory and data: Reviews in Geophysics, v. 5, p. 287–304.

Galvin, C. J., 1968, Breaker type classification on three laboratory beaches: Journal of Geophysical Research, v. 73, p. 3651–3659.

Galvin, C. J., 1972, Waves breaking in shallow water, in Meyer, R., ed., Waves on beaches: Academic Press, London, p. 413–455.

Galvin, C. J., and Eagleson, P. S., 1965, Experimental study of longshore currents on a plane beach: U.S. Army Corps of Engineers, Coastal Engineering Research Center Technical Memo 10.

Gilbert, G. K., 1885, The topographic features of lake shores: U.S. Geological Survey 5th Annual Report, p. 69–123.

Godfrey, P. J., 1976, Barrier beaches of the East Coast: Oceanus, v. 19, p. 27–40.

Gorsline, D. S., 1966, Dynamic characteristics of west Florida Gulf Coast beaches: Marine Geology, v. 4, p. 187–206.

Guilcher, A., 1969, Pleistocene and Holocene sea level changes: Earth Science Reviews, v. 5, p. 69–98.

Guilcher, A., 1988, Coral reef geomorphology: John Wiley and Sons, Chichester, UK, 228 p.

Guza, R. T., and Inman, D. L., 1975, Edge waves and beach cusps: Journal of Geophysical Research, v. 80, p. 2997–3012.

Hayden, B., 1975, Storm wave climates at Cape Hatteras, North Carolina: recent secular variations: Science, v. 190, p. 981–983.

Hayden, B., and Dolan, R., 1979, Barrier islands, lagoons, and marshes: Journal of Sedimentary Petrology, v. 49, p. 1061–1071.

Hayden, B., Dolan, R., and Ross, P., 1979, Barrier island migration: in The Concept of Thresholds Symposium, 9th Annual Geomorphological Symposium: State University of New York, Binghamton, N.Y., p. 363–384.

Hayes, M. O., 1967, Hurricanes as geological agents, south Texas coast: American Association of Petroleum Geologists Bulletin, v. 51, no. 6, p. 937–942.

Hayes, M. O., 1972, Forms of sediment accumulation in the beach zone: in Waves on beaches and resulting sediment transport: Academic Press, N.Y., p. 297–356.

Hayes, M. O., 1978, Impact of hurricanes on sedimentation in estuaries, bays, and lagoons: in Wiley, M. L., ed., Estuarine interactions: International Estuarine Research Conference 4. p. 323–346.

Hayes, M. O., 1979, Barrier island morphology as a function of tidal and wave regime: in Leatherman, S. P., ed., Barrier islands from the Gulf of St. Lawrence to the Gulf of Mexico: Academic Press, N.Y., p. 1–28.

Hino, M., 1975, Theory on formation of rip current and cuspidal coast: Proceedings of 14th Conference on Coastal Engineering, p. 901–919.

Holman, R. A., 1983, Edge waves and the configuration of the shoreline: in Komar, R, ed., CRC handbook of coastal processes and erosion: CRC Press, Boca Raton, FL, p. 21–33.

Horikawa, K., ed., 1988, Nearshore dynamics and coastal processes: University of Tokyo Press, Tokyo, Japan, 522 p.

Horn, D. P. 1992. A review and experimental assessment of equilibrium grain size and the ideal wave-graded profile: Marine Geology, v. 108, p. 161–174.

Hoyt, J. H., 1967, Barrier island formation: Geological Society of America Bulletin, v. 78, p. 1125–1136.

Hoyt, J. H., 1968, Barrier island formation: reply: Geology Society America Bulletin, v. 79, p. 1427 – 1432.

Hoyt, J. H., and Henry, V. J., Jr., 1967, Influence of island migration on barrier-island sedimentation: Geological Society of America Bulletin, v. 78, p. 77–78.

Hughes, P., 1979, The great Galveston hurricane: Weatherwise, v. 32, no. 4, p. 148–156.

Hughes, M. G., and Cowell, P. J., 1987, Adjustment of reflective beaches to waves: Journal of Coastal Research, v. 3, p. 153–167.

Huntley, D. A., and Bowen, A. J., 1973, Field observation of edge waves: Nature, v. 243, p. 160–161.

Inman, D. L., and Bagnold, R. A., 1963, Littoral processes: in Hill, M. N., ed., The sea: Interscience, N. Y., v. 3, p. 529–553.

Inman, D. L., and Frautschy, J. D., 1966, Littoral processes and the development of shoreline: Proceedings of the Coastal Engineering Special Conference. American Society of Civil Engineers, Santa Barbara. CA, p. 511–536.

Inman, D. L., and Nordstrom, C. E., 1971, On the tectonic and morphologic classification of coasts: Journal of Geology, v. 79, p. 1–21.

Ippen, A. T., ed., 1966, Estuary and coastline hydrodynamics: McGraw-Hill, N.Y., 744 p.

Johnson, D. W., 1910, Beach cusps: Geological Society of America Bulletin, v. 21, p. 604–621.

Johnson, D. W., 1919, Shore processes and shoreline development: John Wiley and Sons, N.Y., (reprinted by Hafner Publishing Co., N.Y., 1967), 608 p.

Johnson, D. W., 1922, Retrograding of offshore bars: Geological Society of America Bulletin, v. 33, p. 121–122.

Johnson, D. W., 1925, New England-Acadian shoreline: John Wiley and Sons, N.Y.

King, C. A. M., 1972, Beaches and coasts: Edward Arnold, London, 403 p.

King, C. A. M., and Williams, W. W., 1949, The formation and movement of sandbars by wave action: Geographical Journal, v. 107, p. 70–84.

Kirkby, M. J., and Kirkby, A. V., 1969, Erosion and deposition on a beach raised by the 1964 earthquake, Montague Island, Alaska: U.S. Geological Survey Professional Paper 543-H, 41 p.

Komar, P. D., 1971a, Nearshore cell circulation and the formation of giant cusps: Geological Society of America Bulletin, v. 82, p. 2643–2650.

Komar, P. D., 1971b, The mechanics of sand transport on beaches: Journal of Geophysical Research, v. 76, p. 713–721.

Komar, P. D., 1973, Observations of beach cusps at Mono Lake, California: Geological Society of America Bulletin, v. 84, p. 3593–3600.

Komar, P. D., 1975, Nearshore currents: generation by obliquely incident waves and longshore variations in breaker height: in Hails, J. R., and Carr, A., eds., Proceedings of Symposium on Nearshore Sediment Dynamics: John Wiley and Sons, London, p. 17–45.

Komar, P. D., 1976, Beach processes and sedimentation: Prentice-Hall, Englewood Cliffs, N. J., 429 p.

Komar, P. D., 1983a, Beach processes and erosion—an introduction: in Komar, P. D., ed., CRC handbook of coastal processes and erosion: CRC Press, Boca Raton, FL, p. 1–20.

Komar, P. D., ed., 1983b, CRC handbook of coastal processes and erosion: CRC Press, Boca Raton, FL, 305 p.

Komar, P. D., and Inman, D. L., 1970, Longshore sand transport on beaches: Journal of Geophysical Research, v. 75, p. 5914–5927.

Kraft, J. C., 1971, Sedimentary environment facies patterns and geologic history of a Holocene marine transgression: Geology Society America Bulletin, v. 82, p. 2131–2158.

Kraft, J. C., Biggs, R., and Halsey, S., 1973, Morphology and vertical sedimentary sequence models in Holocene transgressive barrier systems: in Coates, D., ed., Coastal geomorphology: 3rd Annual Geomorphology Symposium, State University of New York, Binghamton, N.Y., p. 321–354.

Krumbein, W. C., 1944, Shore currents and sand movement on a model beach: U.S. Army Corps of Engineers, Beach Erosion Board Technical Memo 7.

Kuenen, P. H., 1948, The formation of beach cusps: Journal of Geology, v. 56, p. 34–40.

Kuhn, G. G., and Shepard, F. P., 1983, Beach processes and sea cliff erosion in San Diego County, California: in Komar, P. D., ed., CRC handbook of coastal processes and erosion: CRC Press, Boca Raton, FL, p. 267–284.

Ladd, H. S., 1961, Reef building: Science, v. 134, p. 703–715.

Leatherman, S. P., 1983, Barrier dynamics and landward migration with Holocene sea-level rise: Nature, v. 301, p. 415–418.

LeBlanc, R. J., and Hodgson, W. D., 1959, Origin and development of the Texas shoreline: reprinted in Barrier islands, benchmark papers in geology (1973): Dowden, Hutchinson and Ross, Stroudsburg, PA, p. 67–90.

Leontiev, Y. O. K., 1969, Flandrean transgression and the genesis of barrier bars: reprinted in Barrier islands, benchmark paper in geology (1973): Dowden, Hutchinson and Ross, Stroudsburg, PA, p. 320–323.

Leontiev, Y. O. K., 1973, On the cause of the present-day erosion of barrier bars: reprinted in Barrier islands, benchmark paper in geology (1973): Dowden, Hutchinson & Ross, Stroudsburg, PA, p. 159–161.

Lisle, L. D., 1982, Annotated bibliography of sea level changes along the Atlantic and Gulf coasts of North America: Shore and Beach, v. 50, p. 24–34.

Longuet-Higgins, M. S., 1970, Longshore currents generated by obliquely incident sea waves: Journal of Geophysical Research, v. 75, p. 6778–6801.

Longuet-Higgins, M. S., and Parkin, D. W., 1962, Sea waves and beach cusps: Geographical Journal, v. 128, p. 194–201.

Lynch, D. K., 1982, Tidal bores: Scientific American, v. 247, p. 146–156.

May, J. P., and Tanner, W. F., 1973, The littoral power gradient and shoreline changes: in Coates, D. R., ed., Coastal geomorphology: 3rd Annual Geomorphology Symposium, State University of New York, Binghampton, N. Y., p. 43–60.

May, S. K., Dolan, R., and Hayden, B. P., 1983, Erosion of U.S. shorelines: EOS v. 64, p. 521–523.

McGee, W. J., 1890, Encroachments of the sea: in Metcalf, L. S., ed., Forum, v. 9, p. 437–449.

McKenzie, R., 1958, Rip current systems: Journal of Geology, v. 66, p. 103–113.

McLean, R. F., and Kirk, R. M., 1969, Relationship between grain size, size-sorting and foreshore slope on mixed sand-shingle beaches: New Zealand Journal of Geology and Geophysics, v. 12, p. 138–155.

Miller, D. J., 1960, Giant waves in Lituya Bay, Alaska: U.S. Geological Professional Paper 354-C, p. 51–86.

Miller, J. R.; Orbock Miller, S. M.; Torzynski, C. A.; and Kochel, R. C. 1989. Beach cusp destruction, formation, and evolution during and subsequent to an extratropical storm, Duck, North Carolina: Journal of Geology, v. 97, p. 747–60.

Moore, G. W., and Moore, J. G., 1988, Large-scale bedforms in boulder gravel produced by giant waves in Hawaii: Geological Society of America Special Paper 229, p. 101–110.

Moslow, T. F., and Heron, S. D., Jr., 1978, Relict inlets: preservation and occurrence in the Holocene stratigraphy of southern Core Banks, North Carolina: Journal of Sedimentary Petrology, v. 48, p. 1275–1286.

Munk, W. H., 1949, Surf beats: Transactions of American Geophysical Union, v. 30, p. 849–854.

Norman, J. O., 1980, Coastal erosion and slope development in Surtsey Island: Zeitschrift für Geomorphology Supplement, v. 34, p. 20–38.

Oertel, G. F., 1979, Barrier island development during the Holocene recession, southeastern United States: in Leatherman, S. P., ed., Barrier islands from the Gulf of St. Lawrence to the Gulf of Mexico: Academic Press, N.Y., p. 273–290.

Oertel, G. F., 1985, The barrier island system: in Certel, G. F., and Leatherman, S. P., eds., Barrier islands: Marine Geology, v. 63, p. 1–18.

Oertel, G. F., and Leatherman, S. P., eds., 1985, Barrier islands: Marine Geology, v. 63, 367 p.

Otvos, E. G., Jr., 1964, Observation of beach cusp and beach ridge formation on the Long Island Sound: Journal of Sedimentary Petrology, v. 34, p. 554–560.

Otvos, E. G., Jr., 1970a, Development and migration of barrier islands, northern Gulf of Mexico; reply: Geological Society of America Bulletin, v. 81, p. 3783–3788.

Otvos, E. G., Jr., 1979, Barrier island evolution and history of migration, North central Gulf Coast: in Leatherman, S. P., ed., Barrier islands from the Gulf of St. Lawrence to the Gulf of Mexico: Academic Press, N.Y., p. 291–319.

Otvos, E. G., Jr., 1981, Barrier island formation through nearshore aggradation; stratigraphic and field evidence: Marine Geology, v. 43, p. 195–243.

Otvos, E. G., Jr., 1985, Barrier platforms; northern Gulf of Mexico: in Oertel, G. F., Leatherman, S. P., eds., Barrier islands: Marine Geology, v. 63, p. 285–305.

Otvos, E. G., Jr., 1986, Island evolution and "stepwise retreat," late Holocene transgressive barriers, Mississippi Delta coast; limitations of a model: Marine Geology, v. 72, p. 325–340.

Pethick, J. S., 1984, An introduction to coastal geomorphology: Edward Arnold, Ltd., London, 255 p.

Peyronnin, C. A., 1962, Erosion of Isles Dernieres and Timbalier Islands: American Society of Civil Engineers, Journal of Waterways and Harbors, v. 1, p. 57–69.

Pierce, J. W., 1970, Tidal inlets and washover fans: Journal of Geology, v. 78, p. 230–234.

Purdy, E. G., 1974, Reef configurations: cause and effect: in Laporte, L. F., ed., Reefs in time and space: Society of Economic Paleontologists and Mineralogists Special Publication 18, p. 9–76.

Rankin, J. K., 1952, Development of the New Jersey shore: in Johnson, J. W., ed., Proceedings of the 3rd Coastal Engineering Conference, Council of Wave Research, Cambridge, MA, p. 306–317.

Ritchie, W., and Penland, S., 1988, Rapid dune changes associated with overwash processes on the deltaic coast of south Louisiana: Marine Geology, v. 81, v. 97–122.

Robinson, L. A., 1977, Marine erosive processes at the cliff foot: Marine Geology, v. 23, p. 257–271.

Russell, R. J., and McIntire, W. G., 1965, Beach cusps: Geological Society of America Bulletin, v. 76, p. 307–320.

Sallenger, A. D., 1979, Beach-cusp formation: Marine Geology, v. 29, p. 23–37.

Shepard, F. P., 1950, Beach cycles in southern California: U.S. Corps of Army Engineers BEB Technical Memo 20, p. 449–464.

Shepard, F. R, 1962, Gulf coast barriers: in Shepard, F. P., ed., Recent sediments, northwest Gulf of Mexico, American Association of Petroleum Geology, p. 197–220.

Shepard, F. P, 1963, Submarine geology: Harper and Row, N.Y., 348 p.

Shepard, F. P., and Grant, U.S., 1947, Wave erosion along the southern California coast: Geological Society of America Bulletin, V. 58, p. 919–926.

Shepard, F. P., and Inman, D. L., 1950, Nearshore circulation related to bottom topography and wave refraction: Transactions of the American Geophysical Union, v. 31, p. 196–212.

Shepard, F. R, and LaFond, E. C., 1940, Sand movements near the beach in relation to tides and waves: American Journal of Science, v. 238, p. 272–285.

Shepard, F. P., Emery, K. O., and LaFond , E. C., 1941, Rip currents: a process of geological importance: Journal of Geology, v. 49, p. 337–369.

Sherman, D. J., and Bauer, B. O., 1993, Coastal geomorphology through the looking glass: Geomorphology, v. 7, p. 225–249.

Short, A. D., 1975a, Multiple offshore bars and standing waves: Journal of Geophysical Research, v. 80, p. 3838–3840.

Short, A. D., 1975b, Offshore bars along the Alaskan Arctic coast: Journal of Geophysical Research, v. 80, p. 209–221.

Short, A. D., 1979, Three-dimensional beach stage model: Journal of Geology, v. 87, p. 553–571.

Short, A. D., and Wright, L. D., 1981, Beach systems of the Sydney region: Australian Geographer, v. 15, p. 8–16.

Snead, R. E., 1982, Coast landforms and surface features: a photographic atlas and glossary: Hutchinson Ross, Stroudsburg, PA, 247 p.

Snodgrass, D., Groves, G., Hasselmann, K., Miller, G., Munk, W., and Powers, W., 1966, Propagation of ocean swell across the Pacific: Philosophical Transactions of the Royal Society of London, v. 259, p. 431–497.

Sonu, C. J., 1972, Field observation of nearshore circulation and meandering currents: Journal of Geophysical Research, v. 77, p. 3232–3247.

Stokes, G. G., 1849, On the theory of oscillatory waves: Transactions Cambridge Philosophical Society, v. 8, p. 441.

Strahler, A. N., 1963, The Earth Sciences: Harper and Row, N.Y., 680 p.

Strahler, A. N., 1966, Tidal cycle of changes on an equilibrium beach: Journal of Geology, v. 74, p. 247–268.

Sunamura, T., 1975, A laboratory study of wave-cut platform formation: Journal of Geology, v. 83, p. 389–397.

Sunamura, T., 1976, Feedback relationship in wave erosion of laboratory rocky coast: Journal of Geology, v. 84, p. 427–437.

Sunamura, T., 1977, A relationship between wave-induced cliff erosion and erosive force of waves: Journal of Geology, v. 85, p. 613–618.

Sunamura, T., 1978, Mechanisms of shore platform formation on the southeast coast of the Izu Peninsula, Japan: Journal of Geology, v. 86, p. 211–222.

Sunamura, T., 1982, A predictive model for wave-induced erosion, with application to Pacific coasts of Japan: Journal of Geology, v. 90, p. 167–178.

Sunamura, T., 1983, Processes of sea cliff and platform erosion: in Komar, P., ed., CRC handbook of coastal processes and erosion: CRC Press, Boca Raton, FL, p. 233–265.

Swift, D. J. R, 1968, Coastal erosion and transgressive stratigraphy: Journal of Geology, v. 76, p. 444–456.

Swift, D. J. P., 1975, Barrier island genesis: evidence from the central Atlantic shelf, eastern U.S.A.: Sedimentary Geology, v. 14, p. 1–43.

Swift, D. J. R, Niedoroda, A. W., Vincent, C. E., and Hopkins, T. S., 1985, Barrier island evolution, middle Atlantic shelf, U.S.A., part 1: shoreface dynamics: in Oertel, G. F., Leatherman, S. P., eds., Barrier islands, Marine Geology, v. 63, p. 331 – 362.

Tanner, W. F., 1958, The equilibrium beach: Transactions of the American Geophysical Union, v. 39, p. 889–891.

Trenhaile, A. S., 1980, Shore platforms: a neglected coastal feature: Progress in Physical Geography, v. 4, p. 1–23.

Tucker, M. J., 1950, Surf beats: sea waves of 1 to 5 minute period: Proceedings of the Royal Society of London, v. 202, p. 565–573.

Umgrove, J. H., 1947, Coral reefs of the East Indies: Geological Society of America Bulletin, v. 58, p. 729–778.

Van Dorn, W. E., 1965, Tsunamis: in Advances in hydroscience, v. 2: Academic Press, N.Y., p. 1–48.

Wells, J. W., 1957, Coral reefs: Geological Society of America Memoir 67, p. 609–631.

Wentworth, C. K., 1938, Marine beach formation: water level weathering: Journal of Geomorphology, v. 1, p. 6–32.

Wiegel, R. L., 1964, Oceanographical engineering: Prentice-Hall, Englewood Cliffs, N. J., 582 p.

Wiens, H. J., 1962, Atoll environment and ecology: Yale University Press, New Haven, CT, 532 p.

Wilkinson, B. H., and Basse, R. A., 1978, Late Holocene history of the central Texas coast from Galveston Island to Pass Cavallo: Geological Society of America Bulletin, v. 89, p. 1592–1600.

Williams, A. T., 1973, The problem of beach cusp development: Journal of Sedimentary Petrology, v. 43, p. 857–866.

Williams, A. T., and Guy, H. R, 1973, Erosional and depositional aspects of Hurricane Camille in Virginia, 1969: U.S. Geological Survey Professional Paper 804, 80 P.

Wilson, B. W., 1966, Seiche: in Fairbridge R. W., ed., Encyclopedia of oceanography: Reinholt Book Corp., N.Y., p. 804 – 811.

Wright, L. D., and Short, A. D., 1983, Morphodynamics of beaches and surf zones in Australia: in Komar, P., ed., CRC handbook of coastal processes and erosion: CRC Press, Boca Raton, FL, p. 35–64.

Wright, L. D., Guza, R. T., and Short, A. D., 1982, Dynamics of a high-energy dissipative surf zone: Marine Geology, v. 45, p. 41 –62.

Wright, L. D., Nielson, P. N., Short, A. D., and Green, M. 0., 1982, Morphodynamics of a macrotidal beach: Marine Geology, v. 50, p. 97–128.

Wright, L. D., Chappell, J., Thom, B. G., Bradshaw, M. P., and Cowell, P. J., 1979, Morphodynamics of reflective and dissipative beach and inshore systems, southeastern Australia: Marine Geology, v. 32, p. 105–140.

Zeigler, J. M., Hayes, C. R., and Tuttle, S. D., 1959, Beach changes during storms on outer Cape Cod, Massachusetts: Journal of Geology, v. 67, p. 318–336.

Zenkovitch, V. P., 1967, Processes of coastal development: John Wiley and Sons, N.Y., 738 p.

CHAPTER SEVENTEEN

Eolian Processes and Landforms

Star dunes, Death Valley National Monument. (Photo by D. A. Rahm)

471

INTRODUCTION

When visualizing a desert, many people imagine vast stretches of land covered by sand dunes, and when asked what is the most important surface process in a desert, most people will answer, wind. Yet this common picture of deserts is far from reality. Sand covers only about 25 to 30 percent of deserts, mostly concentrated in great "sand seas," and sand dunes are by no means restricted to deserts. Running water and rainbeat are much more effective sculptors of desert topography than wind, because even though rainfall is infrequent in deserts, it usually occurs in thunderstorms where the intensity of precipitation and lack of vegetation promote rapid runoff.

Although most deserts are not sand covered, the most extensive dune regions occur in deserts where eolian processes are effective. In regions where loose sand or silt/clay is abundant, wind may be responsible for the development of much of the landscape. An estimated 500 million metric tons of dust per year are transported by the wind (Peterson and Junge, 1971).

WIND PATTERNS

Global Circulation

Winds are generated by differences in atmospheric pressure from one area to another. The movement of air from areas of high to low pressure produces surface flow of air, which we observe as wind.

Global atmospheric pressure differences are set up by unequal heating of the earth's surface. Near the equator, where the sun's rays are most direct and temperatures are relatively high, the atmosphere is strongly heated and the warm air rises, whereas the atmosphere near the poles is much cooler. As the warm air near the equator rises and moves poleward at high altitudes, it is replaced at the surface by cooler, denser air flowing from high latitudes, setting up circulation cells.

At about 20° to 30° latitude, air moving poleward gathers faster than it can escape toward the poles and produces subtropical high-pressure belts. Air descending in these high-pressure belts flows toward the equatorial low-pressure region, generating the trade winds. The trade winds are deflected by the Coriolis effect and move from the northeast toward the southwest in the Northern Hemisphere and from the southeast toward the northwest in the Southern Hemisphere. Deflection of moving objects by the Coriolis effect occurs because objects on the earth's surface have different rotational velocities at different latitudes, so that when a mass moves toward the equator, its rotational velocity is less than that of objects already at that latitude.

The effect of the global circulation pattern is twofold:

1. The descending air in the horse latitudes is heated adiabatically as it descends, warming the land beneath

2. The flow of surface air toward the equator sets the trade winds in motion

The warm, descending air can hold more water vapor than cooler air, so it has a drying effect, causing low relative humidity and high rates of evaporation characteristic of deserts. Most of the earth's great deserts lie in such belts of warm, descending air.

The influence of wind in deserts varies according to the amount of loose material available for movement and the vegetative cover. The deserts of North America are estimated to contain only about two percent sand cover, compared to about 10 percent in the Sahara Desert of Africa and about 50 percent in Arabia (Holm, 1968).

Local Winds

Unequal heating of the earth's surface may generate local pressure differences that induce surface winds. Local winds in the desert are likely to fluctuate in regular patterns. As the land heats up during the day, surface winds rise to 15 to 20 mi/hr (24 to 32 km/hr), culminating near the middle of the day, dying down as the sun sets, and remaining relatively tranquil through the night.

Turbulence

Airflow can be either **laminar** or **turbulent** but is mostly turbulent. For very slow airflow, the motion can be laminar, described as smoothly flowing layers of air where little exchange of momentum takes place from one layer to another. However, as velocity increases, smooth flow becomes unstable, and lateral motion of air, with accompanying changes of momentum, begins to occur. The degree to which airflow is laminar may be given by the **Reynolds number** R_e, the ratio of inertial forces to viscous forces.

$$R_e = \frac{\rho VL}{\mu}$$

where
ρ = fluid density
V = wind velocity
L = length
μ = coefficient of viscosity, a function of fluid composition and temperature

If the Reynolds number is large, inertial forces predominate, and the flow is turbulent.

The shear stress within laminar flow over a surface is proportional to the rate of strain. For airflow over a horizontal surface, the relationship between shear stress and vertical velocity gradient is approximately

$$t = \mu \frac{d_V}{d_z}$$

where
t = surface shear stress
μ = viscosity coefficient
V = velocity
z = distance above the surface

Laminar flow develops in a very thin boundary layer adjacent to the surface, even for flows having large Reynolds numbers (turbulent flow). The laminar boundary layer is very thin, generally less than 1 mm for winds blowing over sand surfaces.

When wind blows across the land surface, its flow is not laminar (except within the boundary layer) but contains many swirls and eddies due to turbulence, some of which produce upward components of movement. Generally, the higher the mean wind velocity, the greater the amount of upward turbulence. The ratio of the velocity of upward turbulent gusts near the ground to mean wind velocity varies, but it averages about 1:5 (Bagnold, 1941). Thus, particles whose settling velocities are lower than one-fifth of the mean wind velocity are likely to be conveyed upward by wind gusts and transported downwind (Figures 17–1 and 17–2). Larger particles, whose higher settling velocities are greater than one-fifth of the mean wind velocity, remain mostly close to the ground until unusually high gusts lift them up. Despite its importance to eolian processes, turbulence is very difficult to study

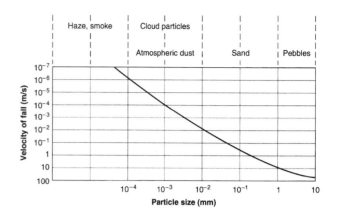

FIGURE 17–1
Relationship of settling velocity to particle size. (After Bagnold, 1941)

FIGURE 17–2
A dust storm (lightest area) in Iran-iraw. (NASA photo)

directly. Lumley and Panofski (1964) undertook a complicated statistical study of turbulence and tentatively concluded the following:

1. Small eddies occur under most atmospheric conditions, unless the air is unusually thermally stable.

2. As wind velocity rises, corkscrew eddies develop with axes parallel to the mean wind direction.

3. Larger, horizontal eddies measured in hundreds of meters are superimposed on the smaller, corkscrew eddies.

4. In hot desert air, convection sets up upward-moving plumes of air that bend over in the direction of wind movement.

Despite the apparent randomness of turbulence, some evidence exists for regular spacing of atmospheric turbulence.

Movement of Material by Wind

On the basis of wind-tunnel experiments and field studies, Bagnold (1941) defined three modes of particle movement by the wind: suspension, saltation, and surface creep or traction (Figure 17–3).

The primary forces acting on an airborne particle falling through quiet air consist of gravity in the direction of motion and aerodynamic drag opposing the motion. A particle dropped from rest accelerates due to gravity, and the aerodynamic drag force increases until it equals the weight of the particle where a terminal velocity is reached. Drag for small particles is proportional to the particle velocity, but for large particles, drag increases with the square of particle velocity. The terminal velocity V_F may be found by equating the gravitational and drag forces.

$$V_F = \left(\frac{4\rho_p\, g\, D_p}{3\rho C_D}\right)^{1/2}$$

where ρ_p = particle density
        D_p = particle diameter
        ρ = atmospheric density
        C_D = drag coefficient

If V_F is less than the vertical component of turbulent velocity, the turbulent eddies are capable of transporting fine silt and clay particles (<~60 mm) upward, and suspension results. However, for sand grains (60 mm–2 mm), if the ter-

minal velocity is greater than the vertical component of the turbulent velocity, the particle trajectory path is a smooth curve with a trajectory height measured in centimeters, rather than kilometers typical for fine particles in suspension. These larger particles, traveling in low, smooth trajectories and bouncing along on the surface, move by **saltation.** The boundary between suspension and saltation is rather indistinct—the two processes grade into one another imperceptibly. Particles that are too large to be lifted from the surface by the wind may be rolled or pushed along the surface by **impact creep,** produced by the impact of saltating particles. Of these, saltation is the most significant mode of transportation because fine sand grains (~1 mm) are most easily moved by the wind. Smaller grains are more difficult to move because of particle cohesion and aerodynamic effects, and larger particles are more difficult to move because of their greater mass.

When a saltating grain returns to the surface, it may crash into a group of sand grains, driving them into the air. Saltation impact is a significant process in putting dust and other small grains into the air by winds that would otherwise not be strong enough to initiate grain movement. Saltating grains that strike surfaces composed of larger grains too large to be moved by wind alone push them downwind a short distance.

Sand grains saltating along a hard, rocky surface bound high into the air upon impact and are thus more easily kept in motion than particles hitting loose sand, where momentum is dissipated among the loose grains, often trapping the impacting grain. Golfers are well aware of this phenomenon because golf balls entering a sand trap do not bounce nearly as much as balls hitting on a green. Accumulation of sand grains in one place thus leads to the trapping of still more sand, rather than spreading out evenly over the ground surface.

Coarse sand grains on the ground surface that are too large to be directly set in motion by the wind may be nudged forward by the impact of bombarding sand grains. The impact of a high-velocity, saltating grain can move a grain 6 times its diameter and more than 200 times its own weight. This type of downwind sand movement is known as **impact creep** or **surface creep.** Bagnold (1941) suggested that creep accounted for about one-fourth of the total sand movement. Others have since confirmed similar ratios, depending somewhat on grain size (Chepil, 1945; Horikawa and Shen, 1960). Sharp (1963) observed that in the field, creep becomes more important as the amount of coarser grains increases.

Compared to flowing water, wind is considerably more limited in the size of particles it can transport and in the effectiveness of erosion. Not only does settling velocity determine the particle size of suspended grains, but it is also important in limiting the height to which saltating grains rise above the ground. In addition, the nature of the surface from which saltating grains rebound is important. The rebound height above a sand-mantled surface may be only 0.5 m (1.5 ft) or less, whereas above a rock surface it can be several meters.

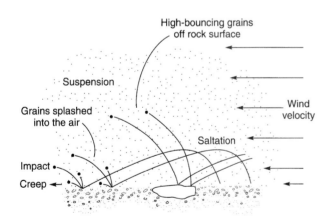

FIGURE 17–3
The three principal modes of eolian transport: suspension, saltation, and impact creep or traction. Shear stress exerted by the wind causes grains to rise into the air where they are carried downwind until falling back to the surface. Most grains bounce back into the air as saltating grains. Those that bounce off rock surfaces rebound elastically into a higher saltation trajectory than other grains. Some strike the other grains at the surface, tossing them into the air where they assume saltation paths. Grains that strike surfaces made up of very fine particles hurl the particles into the air where they are carried in suspension by turbulence.

Obvious evidence of abrasion by windblown sand (Figure 17–4) shows that most sand driven by the wind does not rise more than about 1 m (3 ft) above the ground. Wooden and steel telegraph posts in the Sahara severed about 1 m (3 ft) above the ground show the effective height of saltating grains there. Twenhofel (1932) found that 0.5 to 2-mm sand in the Libyan Desert moves at 13 mi/hr (21 km/hr) and that the air becomes filled with sand at 23 mi/hr (37 km/hr). The observed relationship between wind velocity and the maximum size of moving particles is shown in Table 17–1.

Wind is an unusually selective sorting agent. Eolian sediments are typically among the best-sorted deposits in nature. Bagnold (1941) measured the wind velocity 1 m (3 ft) above the surface required to just set particles in motion to be about 5 m/s (11 mi/hr). If the velocity of updrafts averages about one-fifth of the mean wind velocity—about 1 m/s (2.2 mi/hr)—sand grains with a diameter less than about 0.2 mm will be kept in motion by saltation and are likely to be winnowed out from coarser grains. The most common grain sizes found by sieve analyses of dune sands are 0.3 to 0.15 mm. in diameter. Diameters less than 0.08 mm are rare because most grains of this size are carried off in suspension.

During the first hour or so of a typical desert sandstorm, a cloud of dust and sand appears. Saltating sand is responsible for putting much of the dust into the air, because wind is generally incapable of picking up much silt and clay directly. Later, the dust disappears, but the sand continues as a thick, low-flying cloud with a clearly marked upper surface only 1 to 2 m (3 to 7 ft) above the ground (Bagnold, 1941).

The mechanism of entrainment of particles by wind is complex. As wind velocity approaches the threshold necessary for grain movement, particles begin to oscillate, then suddenly jump into the air in a nearly vertical direction (Greeley and Iversen, 1985).

As the wind passes over the ground surface, the velocity of airflow diminishes close to the ground as a result of frictional retardation, reaching zero in the region of 1/30 of the particle size on a granular surface. The change of velocity with height is expressed by the following equation (Bagnold, 1941):

$$V_h = 5.75 V_d \log \frac{h}{k}$$

where V_h = velocity at a given height h

V_d = drag velocity
k = height at which velocity is zero

Drag velocity may be thought of as the drag exerted on the wind by friction at the surface; thus, it represents shear velocity. As wind velocity increases, shear stress on particles also increases until it reaches the point where resisting forces are overcome, and movement occurs. If a wet blanket is dragged over loose sand, some of the sand is moved by the shearing stress set up by the moving blanket. If the wet blanket is replaced by a "blanket of air," the critical velocity where loose-particle movement begins is defined by Bagnold (1941) as the fluid threshold.

$$V_{dt} = C \sqrt{\frac{\tau - \rho}{\rho} gd}$$

where V_{dt} = threshold drag velocity
π = specific gravity of particles
ρ = specific gravity of air
g = gravitational constant
d = grain diameter
C = coefficient

Once particle movement begins, the surface is bombarded with saltating grains that serve to induce further grain movement, and particles can be moved at lower velocities than those needed to begin motion in the first place. Bagnold (1941) defined this as the impact threshold.

$$V_{dt} = 680 \sqrt{d} \log \frac{30}{d}$$

FIGURE 17–4
Wind-abraded fence post. (Photo by H. G. Wilshire, U. S. Geological Survey)

TABLE 17–1
Relationship between wind velocity and particle size.

Maximum Size of Moving Particles	Wind Velocity	
	ml/hr	km/hr
0.25 mm	10–15	16–24
0.5 mm	15–18	24–30
0.75 mm	18–21	30–35
1.0 mm	21–25	35–40
1.5 mm	25–28	40–45

Source: Twenhofel, 1932.

where V_{dt} = threshold wind velocity
       d = grain diameter

Bagnold (1953) suggested that 16 km/hr (10 mi/hr) was the threshold velocity for most desert sand.

Moving sand grains striking the surface at a low angle bounce back into the air. Bagnold (1941) experimented with quartz sand in a wind tunnel and found that when a moving grain strikes a hard surface, such as bedrock or a pebble, it may bounce off it with almost perfect resilience and reach a height as great as that of the observed top of the cloud of saltating sand. The rebounding grain is pushed by the wind in a flat, curved path that gives the impression of unsupported horizontal flight.

The energy received by the sand grains from the force of the wind is converted into upward energy by impact. The grains bounce high into the air when they strike bare rock or pebbles, but if the ground is covered with sand, when the grains strike the surface, part of their energy is absorbed by the sand grains at rest, and some are ejected into the air (Figure 17–3). The ejected grains do not rise as high into the air as a grain striking a hard surface, but they pick up velocity from the wind, and they, in turn, strike other grains, which are ejected into the air.

The Mysterious Sliding Rocks of Death Valley National Monument

For many years, geomorphologists have been intrigued by the mystery of the sliding boulders of Racetrack Playa in Death Valley National Monument (Figure 17–5). The mud-cracked surface of the playa is criss-crossed by shallow troughs trailing behind rocks (Figure 17–6) ranging from pebbles to boulders weighing more than 700 pounds. The playa surface is nearly flat, so the rocks are clearly not sliding downhill. At first, speculation of the cause of the sliding rocks centered around the possibility that the rocks had been pushed across the surface by humans, but the remoteness of the area argued against such an explanation. Few people brave the unpaved, 30-mile road that leads to Race-track playa. Tracks of sliding rocks have been reported on at least eight other playas in California and Nevada and in the Middle East.

The sliding-rock trails vary both in length and direction. Some trails make straight, roughly parallel lines for several hundred meters (Figures 17–6A and 17–7B). Others curve gradually, criss-cross one another (Figure 17–6B) or change direction abruptly (Figures 17–6C and 17–7B). After traveling hundreds of meters along straight or gently curving paths, many trails end with a scribbled path (Figure 17–7B). The greatest concentration of sliding rocks occurs in the southeastern portion of Racetrack playa where 28 sliding boulders have been counted in a one-half square mile (1.3 km[2]) area (Messina, 1997).

The first scientific papers describing the sliding rocks appeared in 1948 (McAllister and Agnew, 1948), followed by

FIGURE 17–5
Racetrack playa, Death Valley National Monument, California. (Photo by D. A. Rahm)

several in the following decades (Shelton, 1953; Stanley, 1955; Schumm, 1956, Sharp, 1960; Sharp and Carey, 1976). All suggested that high winds, preceded by wetting of the playa mud, had pushed the rocks across the slippery surface. However, despite four decades of investigation of the sliding rocks by scientists and park rangers, no one has ever seen one move.

Ice is known to occasionally form as thick as 10 centimeters on Racetrack playa, and some researchers have suggested that as wind pushes ice rafts, rocks in the ice are dragged across the playa surface, leaving parallel trails. Stanley (1955) first suggested that ice floating in a shallow lake on the playa might be involved in moving the stones, based largely on the parallel paths taken by the rocks. However, Sharp and Carey (1976) placed two small rocks within a circle of steel rods and later found that one of them had escaped. They noted that moving ice could not have accomplished this and, along with their observation that tracks did not maintain constant separation, concluded that ice was not necessary to move the rocks.

The ice-rafting hypothesis was recently revived by Reid et al. (1995) on the basis of mapping of similar, parallel paths of rocks sliding on the Racetrack Playa. They suggest that rocks of different mass should not be propelled equally by high winds without ice and that the wetted playa surface was not as slippery as once thought. They explained the observed lack of parallel paths between some tracks as having been caused by breaking up of the ice into smaller individual blocks and concluded that ice was necessary for playa rocks to slide.

In a discussion of the conclusions of Reid et al., Sharp (1996) presented evidence that at least some of the rocks must have slid without the presence of ice. Messina (1997) monitored many

A

B

C

FIGURE 17–6

(A.) Two sliding rocks leaving trails on Racetrack playa, Death Valley National Monument, California. (Photo by P. Messina)

(B.) Trails of sliding rocks crossing one another, Racetrack playa, Death Valley National Monument, California. (Photo by P. Messina)

(C.) Abrupt change in direction of sliding rock trail, Racetrack playa, Death Valley National Monument, California. (Photo by P. Messina)

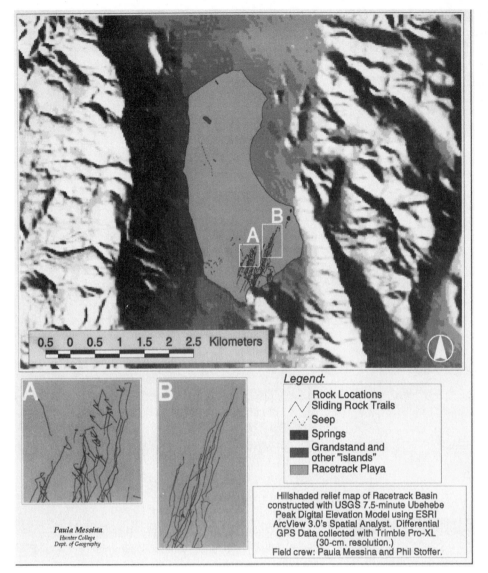

FIGURE 17–7
Sliding rock trails of Racetrack
playa, July, 1996, Death Valley
National Monument, California.
(Courtesy of P. Messina)

sliding rocks on Racetrack playa with GPS (ground positioning system) and found that, although many of the trails were approximately parallel (Figure 17–7), in detail they converged and diverged and in some cases appeared to be random, a situation inconsistent with movement of stones frozen in a sheet of ice.

EOLIAN ABRASION AND EROSION

Wind erosion takes place by **abrasion,** the wearing away of material by frictional impact, and by **deflation,** the blowing away of loose material. Wind abrasion is caused by the collision of sand grains with surfaces. Spectacular examples occur when automobiles caught in sandstorms are stripped of paint and their windshields frosted in a matter of minutes. In nature, saltating sand grains, bouncing along the ground near the surface, collide with rocks or other sand grains that they abrade. Windblown sand literally sandblasts exposed

rock faces and creates small-scale abrasion features whose details depend on the lithology, inclination, and orientation of the rock, the velocity and direction of the wind, and the size, hardness, and shape of the abrasive material. The results of wind abrasion include the polishing, pitting, grooving, fluting, and faceting of rocks and the rounding of sand grains in transport (Greeley and Iversen, 1985, Marshall, 1979; Maxson, 1940; Powers, 1936; Suzuki and Takahashi, 1981).

Among the most common features of wind abrasion are **ventifacts,** etched and pitted faces of exposed rock (Blake, 1855; Kuenen, 1928; Schoewe, 1932; Sharp, 1949, 1964, 1980; Whitney, 1979; Whitney and Dietrich, 1973). Sharp and Saunders (1978) studied the effects of abrasion from saltating sand grains on a variety of natural and artificial materials over an 11-year period and found that 50 percent of saltating sand grains travel within 13 cm (5 in) of the ground and 90 percent within 64 cm (25 in) above ground. The maximum abrasion measured on a leucite rod occurred at 23 cm (9 in) above ground level, and nearly 1 millimeter of abrasion was measured on a granite-gneiss boulder over a 15-year period.

If the original rock face is normal to the prevailing wind direction, a faceted face is abraded at right angles to the wind, with sharp edges between the facet and the original surface. Rocks that periodically rotate so as to expose different sides to the direction of prevailing wind develop multiple-faceted sides (Sharp, 1964, 1980). The abraded facets are typically pitted, fluted, grooved, and polished (Figure 17–8) and generally slope between 30° and 60° (Kuenen, 1928).

Larger-scale erosional effects of wind include **yardangs**, elongate ridges aligned parallel to the direction of prevailing wind with rounded windward faces that taper in the downwind direction (Bosworth, 1922; Blackwelder, 1934; Hedin, 1903; McCauley et al., 1977a, 1977b; Peel, 1970). Yardangs vary in length from several meters to 1 km (~0. 5 mi), and from a few meters to 200 m (~650 ft) high. Bedding and other structures in the material are truncated by the troughs, displaying their erosional origin. That they are wind eroded is shown by their alignment parallel to the direction of prevailing wind and by their composition of easily eroded material (Bosworth, 1922; Blackwelder, 1934; Ward, 1979). The elongate ridges are separated by intervening troughs, whose origin may also include erosion by running water and weathering. Yardangs have been produced experimentally, using easily eroded materials subjected to wind, resulting in:

1. Erosion of windward corners
2. Erosion of front and upper surfaces
3. Erosion of rear flanks
4. Erosion of the downwind upper surface to produce the typical yardang shape resembling a ship's hull (McCauley et al., 1977c; Ward and Greeley, 1984)

Where strong winds blow across surfaces lacking coarse material, removal of small debris by the wind forms **deflation hollows**. The size and shape of deflation hollows vary considerably, depending on the nature of the underlying material, strength of the wind, and local weathering conditions. Deflation hollows occur in a wide range of material, including sand, unconsolidated sediments, and weak bedrock.

EOLIAN DEPOSITS

Ripples

Ripples are low ridges of sand with wavelengths ranging from 0.5 cm to 25 m. Large ripples that can form on almost any sandy surface (0.2 in to 80 ft) are known as **(megaripples)**. Sharp (1963) measured typical ripple lengths of 7 to 14 cm (3 to 6 in) and heights of 0.5 to 1.0 cm (0.2 to 0.4 in). Most ripples have crests that are straight or slightly sinuous, approximately transverse to the direction of the wind. Ripple crests commonly split into two crests or terminate abruptly within distances of a few meters up to about 20 m (65 ft). Ripples are usually asymmetric in cross section, having windward slopes up to 8° to 10° and leeward slopes of 20° to 30° (Sharp, 1963). Leeward slopes typically consist of a steep slope immediately below the ripple crest and a gentler, more concave slope that grades into the adjacent ripple.

Sharp (1963) observed sand grains driven up the windward face of active ripples by saltation impact until they reached the ripple crest, where they fell down the tee slope to form a foreset bed. In studies of the internal structures of ripples, Sharp (1963) found that most ripples consisted of coarse grains on a base of finer sand, which he considered to have re-

FIGURE 17–8
Wind-pitted and wind-abraded basalt ventifact, Sand Hollow, Washington.

sulted from the interstitial settling of fine grains through spaces between the coarse grains as they moved downwind.

Bagnold's (1941) observations of ripple formation in the field and in wind-tunnel experiments led him to conclude that saltating sand grains strike a flat surface at about the same angle and with similar momentum. However, if more grains move out of a small area than are moved into it, the surface no longer remains flat, and the angle of incidence of bombarding sand grains changes relative to the surface. Fewer sand grains per unit area strike the upwind side of a hollow than strike the downwind side; more grains are driven up the slope than down it; and the initial hollow is thus expanded. Because the bombardment of sand is more intense on the windward slope, grains excavated from the hollow accumulate, forming another lee slope, which in turn reduces grain incidence of saltating grains, causing another hollow and propagating the ripple again and again downwind.

Once a ripple sequence is established on a sand surface, the ripple wavelength and height remain relatively constant for a given wind velocity because, for a given wind velocity, the average path taken by saltating grains remains constant. Of the grains ejected from any small area, a greater number will fall on a second small area at a distance of one average path length downwind than on any other area. Repetition of this circumstance downwind causes a relatively uniform ripple wavelength. The size of a ripple becomes stable when as many sand grains strike the upwind portion of a ripple as are ejected from it. In experiments with different wind velocities, Bagnold (1941) found that ripple wavelengths increased with wind velocity until the velocity reached approximately three times that needed to initiate grain movement, and beyond that, the ripples were obliterated. Because grain transport also depends on grain size, wavelength appears to be a function of average sand size and sand size distribution, in addition to wind velocity.

The size of the ripples is determined largely by the size distribution of the sand grains. As the crest of a ripple becomes higher, only the coarser grains can remain stable near the crest because the smaller grains continue in motion, leaving the coarser, difficult-to-move grains as a protective coating on the crest and thus allowing the crest to rise higher into the stronger wind than would otherwise be possible. Eventually, the ripple height comes into dynamic equilibrium for the given wind speed and grain size. Bagnold (1941) and Sharp (1963) noted that coarse sand to granule ripples typically have very long wavelengths.

Sharp (1963) suggested a variation of the Bagnold model of ripple formation in which ripple height controls wavelength, rather than saltation path length. He suggested that the sand grains in ripples move downwind primarily by creep from saltation impact, and random irregularities and/or differential movement of sand grains causes accumulation of grains until an equilibrium height, governed by the angle of incidence of the grains in saltation, is established. As in the Bagnold model, the area of impact by saltating grains is propagated downwind from the ripple crest. As the angle of inci-

dence of saltating grains decreases, ripple height is reduced and the wavelength increases.

The primary difference in the Bagnold and Sharp models for ripple formation is that in the Bagnold model, the saltation path length is the governing factor, whereas in the Sharp model, saltation incidence angle and ripple height are the critical factors. Sorting out these factors experimentally or in the field is difficult because the angle of incidence of saltating grains decreases, the saltation length increases, and the path length increases with velocity, so that for sand of any given grain size, an increase in wind velocity will cause an increase in ripple wavelength according to either model.

In addition to the type of ripples discussed above (normal or impact ripples), other types of sand ripples occur, varying in wavelength from 1 to 25 m (3 to 80 ft) (Ellwood et al., 1975; Greeley and Peterfreund, 1981). Ripples of unusually long wavelength (**megaripples**) apparently form on sand having a bi-modal particle size distribution, moved by winds of velocity too low to move the large grains but strong enough to move the finer grains by saltation.

Another type of ripple that occurs in fine, well-sorted sand moved by high-velocity winds was described by Bagnold (1941) as **fluid drag ripples.** They apparently form when winds reach velocities at which grains begin to lift off into suspension, and the moving grains above the surface consist of a mixture of grains with very long saltation paths and grains in suspension that generate long, flat ripples. Wilson (1972a) referred to such forms as **aerodynamic ripples.**

Sand Dunes

The kinetic energy of saltating sand grains becomes dissipated when the grains impact a patch of sand on the surface. A patch of sand makes an effective sand trap, which continues to grow in size until it becomes a mound of sand. Protrusion of the sand mound high enough into the air to disrupt laminar airflow over its surface significantly affects further accumulation and movement of sand, and a dune is formed. Once a sand grain becomes part of a dune, it advances downwind only at the rate of advance of the entire dune, rather than at its previous saltation rate.

When a critical height is reached, about 30 cm (12 in) (Bagnold, 1941), a **slip face** forms on the lee side of the dune. Sand moves up the windward side of the dune by saltation and creeps to and over the brink of the crest, increasing the slope until it exceeds the angle of repose for loose sand and avalanches down the lee side (Figure 17–9). The windward side of dunes generally slopes about 10° to 15°, whereas the slip face stands at the angle of repose of loose sand, about 30° to 34°. Successive slip faces become progressively buried until they are once again unearthed by erosion on the windward side of the dune as the dune gradually advances downwind (Figures 17–10 and 17–11).

FIGURE 17–9
Sand sliding down the slip face of a dune. (Photo by Ken Hamblin)

Bagnold (1941) derived an equation for the rate of advance of dunes where all of the sand deposited on the slip face came from the removal of sand from the windward slope, so that the dune shape is preserved as the dune moves downwind.

$$M = \frac{q}{yH}$$

where M = rate of dune advance
q = rate of sand flow
H = dune height
y = specific weight of sand

The most important factors are dune height and rate of sand flow, which is a function of wind velocity and grain characteristics. Measured rates of dune migration vary. Beadnell (1910) measured rates of 15 m/yr (50 ft/yr) in Egypt; Bowman (1916) measured rates of 13 m/yr (43 ft/yr) in winter and 25 m/yr (82 ft/yr) in summer (average 20 m/yr (65 ft/yr)) for Peruvian dunes; and Melton (1940) measured a rate of 18 m/yr (60 ft/yr) for dunes of the U.S. High Plains. Finkel (1959) found that the rate of movement of Peruvian dunes is directly related to dune size; that is, smaller dunes move faster than larger dunes.

Types of Dunes

Barchans are crescent-shaped dunes (Figures 17–12 and 17–13) with a steep slip face that is concave in the downwind direction (Finkel, 1959; Lettau and Lettau, 1969; Long and Sharp, 1964; Norris, 1966). Finkel (1959) found that the width of barchan dunes in Peru, measured from horn to horn, averaged 37 m (120 ft) and that the heights of the dunes were consistently about one-tenth the width. The horns of the crescent taper in the downwind direction because even with con-

FIGURE 17–10
Cross-section through a sand dune. Note the paleo-dune faces exposed in the cut bank.

FIGURE 17–11
Internal structure of Jurassic sand dunes, Zion National Park, Utah. (Photo by Ken Hamblin)

FIGURE 17–12
Barchan dunes near Beverly, Washington. (Photo by D. A. Rahm)

FIGURE 17–13
Barchan dunes on the Oregon Coast. (Photo by D. A. Rahm)

FIGURE 17–14
Coalescence of barchan dunes to form transverse dunes, Moses Lake, Washington. (Photo by D. A. Rahm)

stant lateral supply of sand, saltating grains will move more quickly over the hard ground adjacent to the dune patch and most slowly where the sand is thickest. Because the edges of the dune are lower and the rate of advance of the slip face is inversely proportional to the dune height, the edges move downwind at a faster rate than the thicker, central part of the dune, giving the dune its crescent shape which is maintained as the barchan advances downwind. Barchans are typically best developed on barren desert floors where the supply of sand is meager, the prevailing wind is relatively uniform, and vegetation is scarce. They may occur as single dunes or in groups where individual barchans coalesce to form complex shapes.

Transverse dunes are elongate dunes that form perpendicular to the direction of prevailing winds. They originate in several ways. Some are clearly transitional forms with barchan dunes (Figure 17–14) and consist of several barchans connected along single lines (Figure 17–15). Transverse sand ridges composed of coalescing barchan dunes have also been called **barchanoid dunes.** Other transverse dunes have little,

if any, relationship to barchans, such as the elongate dune ridges sometimes formed along shorelines (Figure 17–16).

Parabolic dunes are crescent-shaped dunes, somewhat similar in form to barchans, except that the slip face of the dune is convex downwind, rather than concave, and the horns of the crescent point upwind, rather than downwind (Figure 17–17). Some are clearly transitional to barchan dunes (Figure 17–18) and form as a result of anchoring of the horns by vegetation while the central portion continues to migrate downwind. Vegetation is an important factor in eolian processes in arid climates because plants are effective in breaking up airflow near the ground and serve to anchor thin sand deposits. Figure 17–18 shows a barchan field in which the horns have become anchored by vegetation, which is absent where the thicker sand does not allow the vegetation to stabilize. The result is that the horns, anchored by vegetation, trail the thicker central part of the dune, rather than preceding it as in a barchan. In some cases, the horns of the parabolic dune may be left far behind, giving the dune a hairpin-like shape.

Parabolic dunes may form in several ways. Figure 17–19 shows an example of parabolic dunes with a clear convex,

FIGURE 17–15
Transverse dunes, near Death Valley, California. (Photo by D. A. Rahm)

FIGURE 17–16
Transverse dunes south of Astoria, Washington.
Washington. (Photo by D. A. Rahm)

FIGURE 17–17
Parabolic dunes near Eltopia, Washington. (Photo by D. A.
Rahm)

FIGURE 17–18
Barchans evolving into parabolic
dunes. (Photo by Ken Hamblin)

FIGURE 17–19
Parabolic dunes, near Pasco, Washington. (Photo by D. A. Rahm)

FIGURE 17–20
Longitudinal dunes originating at gaps in cliffs, Arizona. (Photo by D. A. Rahm)

downwind slip face. Other parabolic dunes form downwind from deflation hollows where sand, blown out of the hollows, collects in a crescentic form.

Longitudinal dunes are elongate ridges of sand parallel to the direction of prevailing winds. One of the places they may form is where wind funnels sand through a gap in a ridge (Figure 17–20). As the wind blows through the gap, its velocity increases due to the Bernouli effect, but when the wind passes through the gap, its velocity decreases to that of the general flow, and it deposits some of the sand carried by the enhanced velocity through the gap.

Another type of longitudinal dune is the **seif dune**, which occurs in long, parallel chains that are exceptionally straight for long distances, sometimes as long as 60 to 190 km (36–110 mi) (Figure 17–21). Seif dunes are particularly striking in satellite photos (Figures 17–22 and 17–23). Perhaps the most intriguing aspect of these unusual dunes is not only their length but also the regularity of their spacing. Most longitudinal dunes are separated by barren rock floors or desert pavements of rather uniform width. The longitudinal dunes occasionally converge to form Y-junctions (Breed and Breed, 1979).

No truly definitive explanation has been demonstrated for these extraordinary dunes, although a number of hypotheses have been put forth. Bagnold (1941, 1953a) postulated that each chain grows in height and width by trapping sand during periods of strong crosswinds, and thus each chain extends lengthwise during longer periods of more settled conditions when the wind blows down the chain. Others have since suggested variations of this hypothesis (Brookfield, 1970; Cooper, 1958; Folk, 1971; McKee and Tibbitts, 1964). Studies utilizing smoke bombs to observe airflow have led to the hypothesis that longitudinal dunes originate from long spiral airflows with horizontal axes that flow down the troughs between the dune ridges, controlling the spacing and movement of sand along the dunes (Glennie, 1970; Folk, 1971; Mabbutt, 1968; Mabbutt and Sullivan, 1968; Tsoar, 1974, 1982; Twidale, 1972; Verstappen and van Zuidarn, 1970; Wilson, 1972a). However, no hypothesis yet seems to adequately explain the striking length (up to 300 km (185 mi) long and 200 m (~650 ft) high) of single ridges and their regular spacing.

Star dunes typically consist of three or more sharp-edged ridges extending radially from a high, pointed, central peak

FIGURE 17–21
Longitudinal dunes, southwestern Algeria. (Satellite photo by NASA)

FIGURE 17–22
Longitudinal dunes, Algeria. (Photo by NASA)

(Figure 17–24). Although not as extensively studied as some of the other dune types and not well known in geologic literature, they are a distinctive form that may be more widespread than one might expect. They seem to originate in areas where winds blow from more than one prevailing direction (Lancaster, 1988b).

Loess

Loess consists of windblown silt and clay deposited as a relatively homogeneous, unstratified blanket over the earth's surface. It is usually well sorted, consisting mostly of silt with smaller amounts of clay and fine sand. Typical loess contains about 40 to 50 percent silt (10 to 50 mm), up to 30 percent clay (<5 mm), and up to 10 percent fine sand (>250 mm). Clay-size loess can travel all the way around the world on high-altitude winds and is a major component of deep-sea sediments. An estimated 30 percent of the United States is mantled with loess (Ruhe, 1974), and as much as 10 percent of the earth's land area is covered by 1 to 100 m (3 to 300 ft) of loess.

The sources of most loess deposits seem to be related to areas affected by Pleistocene glaciations and climatic changes, especially downwind from glacial outwash plains (Fisk, 1951; Gillette, 1981; Goudie, 1978; Morales, 1979). Thick loess de-

FIGURE 17–23
Elongate rows of star dunes, southwestern Algeria. (Photo by NASA)

FIGURE 17–24
Star dunes near Beatty, Nevada.
(Photo by D. A. Rahm)

posits occur in the Mississippi Valley region of the Central Plains (Leighton and Willman, 1950); in Alaska (Nickling, 1978; Pewe, 1981); in the Palouse area of the Columbia Plateau in eastern Washington (Figure 17–25); in portions of Europe south of the Scandinavian Ice Sheet; and in China, Mongolia, and the former Soviet Union (Figure 17–26). The largest known loess-covered region is 635,000 km^2 (245,000 mi^2) in north-central China. Strangely, the great deserts of Africa, Australia, and the southwestern United States do not contain significant loess deposits, perhaps as a result of their distance from major continental glaciers during the Pleistocene.

The loess deposits of China and the Central Plains of the United States are especially thick, reaching thicknesses up to 300 m. Elsewhere, loess thicknesses generally are several tens of meters. During the dust-bowl storms of the Central Plains in the 1930s, immense volumes of silt became airborne. In 1935, a dust storm near Wichita, Kansas, was estimated to hold about 5 million tons of dust in suspension over a 78 km^2 (30 mi^2) area, and 300 tons/km^2 was deposited in a single day near Lincoln, Nebraska (Hovde, 1934; Miller, 1934). Deposition of loess in China presently occurs at a rate of several millimeters per year (Derbyshire, 1983).

As seen in Chapter 14, the amount of windblown dust that accumulated in the Antarctic Ice Sheet rose dramatically with each Pleistocene glaciation, suggesting a direct link between loess deposition and Pleistocene glaciations. Much of the loess deposited downwind from Pleistocene glacial outwash consists of ground-up quartz and feldspar that originate from the winnowing of rock flour produced by glacial abrasion and carried onto glacial outwash plains by meltwater streams where the wind picks up the temporarily deposited, fine-grained material.

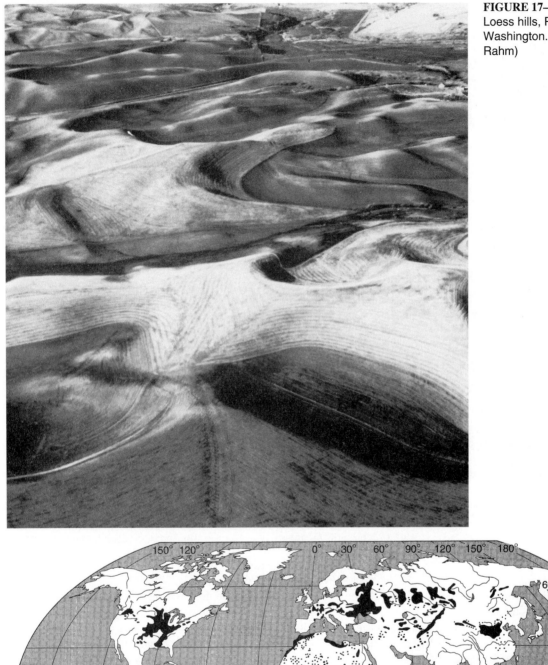

FIGURE 17–25
Loess hills, Palouse,
Washington. (Photo by D. A.
Rahm)

FIGURE 17–26
Major loess deposits in the world. (After Selby, 1985)

REFERENCES

Ahlbrandt, T. S., and Fryberger, S. G., 1982, Eolian deposits: *in* Scholle, P., and Spearing, D., eds., Sandstone depositional environments: American Association of Petroleum Geologists, p. 11–48.

Anderson, R. S., 1986, Erosion profiles due to particles entrained by wind: Application of an eolian sediment-transport model: Geological Society of America Bulletin, v. 97, p. 1270–1278.

Anderson, R. S., 1987a, Eolian sediment transport as a stochastic process: The effects of a fluctuating wind on particle trajectories: Journal of Geology, v. 95, p. 497–512.

Anderson, R. S., 1987b, A theoretical model for aeolian impact ripples: Sedimentology, v. 34, p. 943–956.

Anderson, R. S., 1988, The pattern of grainfall deposition in the lee of aeolian dunes: Sedimentology, v. 35, p. 175–188.

Anderson, R., and Hallet, B., 1986, Sediment transport by wind: Toward a general model: Geological Society America Bulletin, v. 97, p. 523–535.

Bagnold, R. A., 1931, Journeys in the Libyan Desert: Geographical Journal, v. 78, p. 18–39; 524–535.

Bagnold, R. A., 1936, The movement of desert sand: Proceedings of the Royal Society of London, v. 163, p. 250–264.

Bagnold, R. A., 1941, The physics of blown sand and desert dunes: Methuen and Co., London, 265 p.

Bagnold, R. A., 1951, Sand formations in southern Arabia: Geographical Journal, v. 117, p. 78–86.

Bagnold, R. A., 1953a, The surface movement of blown sand in relation to meteorology: *in* Desert Research, Proceedings of International Symposium, Research Council of Israel and UNESCO: Special Publication, v. 2, p. 89–93.

Bagnold, R. A., 1953b, Forme des dunes de sable et regime des vent: *in* Actions Eoliennes, Center for National Research: Paris, Coll. Int., v. 35, p. 23–32.

Bagnold, R. A., 1956, The flow of cohesionless grains in fluids: Philosophical Transactions, Royal Society, Series A, v. 249, p. 239–297.

Bagnold, R. A., 1960, The re-entrainment of settled dusts: International Journal of Air Pollution, v. 2, p. 357–363.

Belly, P. V., 1964, Sand movement by wind: U.S. Army Corps of Engineers, Coastal Engineering Research Centre, Technical Memoir 1, 80 p.

Blackwelder, E., 1934, Yardangs: Geological Society of America Bulletin, v. 45, p. 159–166.

Blake, W. P., 1855, On the grooving and polishing of hard rocks and minerals by dry sand: American Journal of Science, v. 20, p. 178–181.

Bosworth, T, 1922, Geology of the Tertiary and Quaternary periods in the northwest part of Peru: Macmillan, London, p. 269–309.

Bowman, 1., 1916, The Andes of southern Peru: American Geographical Society Special Publication No. 2, p. 264–266.

Breed, C. S., and Breed, W. J., 1979, Dunes and other windforms of Central Australia and a comparison with linear dunes on the Moenkopi Plateau, Arizona: *in* El-Baz, F., and Warner, D. M., eds., Apollo-Soyuz Test Project Summary Science Report, v. 11: Earth Observations and Photography, NASA SP-412, p. 319–358.

Breed, C., and Grow, T., 1979, Morphology and distribution of dunes in sand seas observed by remote sensing: *in* A Study of Global Sand Seas, McKee, E., ed., U.S. Geological Survey Professional Paper 1052, p. 253–303.

Brookfield, M., 1970, Dune trend and wind regime in central Australia: Zeitschrift für Geomorphologie, v. 10, p. 121–158.

Bryan, K., 1923, Wind erosion near Lees Ferry, Arizona: American Journal of Science, Series 5, v. 6, p. 291–307.

Calkin, P. E., and Rutford, R. H. 1974, The sand dunes of Victoria Valley, Antarctica: Geographical Review, v. 64, p. 189–216.

Chepil, W. S., 1945, Dynamics of wind erosion: II: Initiation of soil movement: Soil Science, v. 60, p. 397–411.

Chepil, W. S., 1950, Properties of soil which influence wind erosion: 1: the governing principle of surface roughness: Soil Science, v. 69, p. 149–162.

Chepil, W. S., 1957, Sedimentary characteristics of dust storms: 1: sorting of wind-eroded soil material: American Journal of Science, v. 255, p. 12–22.

Chepil, W. S., 1958, The use of evenly spaced hemispheres to evaluate aerodynamic forces on a soil surface: EOS, American Geophysical Union, v. 39, p. 397–403.

Chepil, W. S., 1959, Equilibrium of soil grains at the threshold of movement by wind: Soil Science Society of America Proceedings, v. 23, p. 422–428.

Chepil, W. S., and Woodruff, N. P., 1957, Sedimentary characteristics of dust storms: II: visibility and dust concentration: American Journal of Science, v. 255, p. 104–114.

Chepil, W. S., and Woodruff, N. P., 1963. The physics of wind erosion and its control: advances in Agronomy, v. 15: U.S. Department of Agriculture, Washington, D.C., p. 211–302.

Cooke, R. U., and Warren, A., 1973, Geomorphology in deserts: University of California Press, Berkeley, CA, 374 p.

Cooper, W. S., 1958, Coastal sand dunes of Oregon and Washington: Geological Society of America Memoir 72, 169 p.

Cooper, W. S., 1967, Coastal sand dunes of California: Geological Society of America Memoir 104, 131 p.

Derbyshire, E., 1983, Origin and characteristics of some Chinese loess at two locations in China: *in* Brookfield, M. E., and Ahlbrandt, T. S., eds., Eoloian sediments and processes:

developments in sedimentology: Elsevier Science Publishers, Amsterdam, Netherlands, p. 69–90.

Ellwood, J. M., Evans, P. D., and Wilson, I. G., 1975, Small-scale aeolian bedforms: Journal of Sedimentary Petrology, v. 45, p. 554–561.

Finkel, H. J., 1959, The barchans of southern Peru: Journal of Geology, v. 67, p. 614–647.

Fisk, H. N., 1951, Loess and Quaternary geology of the lower Mississippi Valley: Journal of Geology, v. 59, p. 333–356,

Folk, R. L., 1971, Longitudinal dunes of the northwestern edge of the Simpson Desert, Northern Territory, Australia: 1: geomorphology and grain size relationships: Sedimentology, v. 16, p. 5–54.

Forman, S., Goetz, A. and Yuhas, R., 1992, Large-scale stabilized dunes on the High plains of Colorado: Understanding the landscape response to Holocene climates with the aid of images from space: Geology, v. 20, p. 145–148.

Fryberger, S. G., 1979, Dune forms and wind regime: *in* McKee, E., ed., A Study of Global Sand Seas, U.S. Geological Survey Professional Paper 1052, p. 137–170.

Fryberger, S. G., and Ahlbrandt, T. S., 1979, Mechanisms for the formation of eolian sand seas: Zeitschrift für Geomorphology, v. 23, p. 440–460.

Gillette, D. A., 1981, Production of dust that may be carried great distances: *in* Pewe, T. J., eds., Desert dust origin, characteristics, and effect on man: Geological Society of America Special Paper 186, p. 11–26.

Gillette, D. A., and Goodwin, P. A., 1974, Microscale transport of sand-sized soil aggregates eroded by wind: Journal of Geophysical Research, v. 79, p. 4080–4084.

Glennie, K. W., 1970, Desert sedimentary environments: Elsevier, Amsterdam, Netherlands, 222 p.

Goldsmith, V., 1978, Coastal dunes: *in* Davis, R. A., ed., Coastal sedimentary environments: Springer-Verlag, N.Y., p. 303–378.

Goudie, A. S., 1978, Dust storms and their geomorphological implications: Journal of Arid Environments, v. 1, p. 291–311.

Greeley, R. J., 1982, Aeolian modification of planetary surfaces: *in* Coradini, A., and Fulchignoni, M., eds., The comparative study of the planets: D. Reidel, Dordrecht, Netherlands, p. 419–434.

Greeley, R. J., 1982, and Iversen, J. D., 1985, Wind as a geological process: Cambridge University Press, London, 333 p.

Greeley, R. J., and Peterfreund, A. R., 1981, Aeolian "megaripples": examples from Mono Craters, California and northern Iceland: Geological Society of America Abstracts with Programs, v. 13, p. 463.

Greeley, R., Williams, S. H., and Marshall, J. R., 1983, Velocities of windblown particles in saltation: preliminary laboratory and field measurements: *in* Brookfield, M. E., Ahlbrandt, T. S., eds., Eolian sediments and processes: Amsterdam, Netherlands, Elsevier, p. 133–148.

Grolier, M., Ericksen, G. E., McCauley, J. F., Morris, E. C. 1974, The desert landforms of Peru: a preliminary photographic atlas: U.S. Geological Survey, Interagency Report, Astrogeology, v. 57, 146 p.

Hanna, S. R., 1969, The formation of longitudinal sand dunes by large helical eddies in the atmosphere: Journal of Applied Meteorology, v. 8, p. 874–883.

Hastenrath, S. L., 1967, The barchans of the Arequipa region, southern Peru: Zeitschrift für Geomorphology, v. 11, p. 300–331.

Havholm, K. and Kocurek, G., 1988, A preliminary study of the dynamics of modern draa, Algodones, southeastern California: Sedimentology, v. 35, p. 649–669.

Hedin, S., 1903, Central Asia and Tibet: Scribners and Sons, N.Y., 608 p.

Holm, D. A., 1968, Desert geomorphology in the Arbian Peninsula: Science, v. 132. p. 1369–1379.

Horikawa, K., and Shen, H. W., 1960, Sand movement by wind: U.S. Army Corps of Engineers, Beach Erosion Board, Technical Memoir, 119, 51 p.

Hovde, M. R., 1934, The great duststorm of November 12, 1933: Monthly Weather Review, v. 62, p. 12–13.

Howard, A. D., 1977, Effect of slope on the threshold of motion and its application to orientation of wind-ripples: Geological Society of America Bulletin, v. 88, p. 853–856.

Hoyt, J. H., 1966, Air and sand movements in the lee of dunes: Sedimentology, v. 7, p. 137–144.

Idso, S. B., 1976, Dust storms: Scientific American, v. 235, p. 108–114.

Inman, D. L., Ewing, G. C., and Corliss, J. B., 1966, Coastal sand dunes of Guerrero Negro, Baja, California, Mexico: Geological Society of America Bulletin, v. 77, p. 787–802.

Kuenen, P. H., 1928, Experiments on the formation of wind-worn pebbles: Leidsche Geologische Medellinger, v. 3, p. 94–110.

Lancaster, N., 1980, The formation of seif dunes from barchans—Supporting evidence for Bagnold's model from the Namib Desert: Zeitschrift für Geomorphology, v. 24, p. 160–167.

Lancaster, N., 1982. Dunes on the Skeleton Coast, Namibia (South West Africa): Geomorphology and grain size relationships: Earth Surface Processes and Landforms, v. 7, p. 575–587.

Lancaster, N., 1983, Controls on dune morphology in the Namib Sand Sea: in Brookfield, M. and Ahlbrandt, T. S., Eolian Sediments and Processes, Elsevier, Amsterdam, Netherlands, p. 261–90.

Lancaster, N., 1985, Variations in wind velocity and sand transport on the windward flanks of desert sand dunes: Sedimentology, v. 32, p. 581–593.

Lancaster, N., 1988a, The development of large aeolian bedforms: Sedimentary Geology, v. 55, p. 69–90.

Lancaster, N., 1988b, Controls of dune size and spacing: Geology, v. 16, p. 972–975.

Lancaster, N., 1989a, Star dunes: Progress in Physical Geography, v. 13, p. 67–91.

Lancaster, N., 1989b, The Namib sand sea: dune forms, processes and sediments: A. A. Balkema, Rotterdam, N.Y., 180 p.

Lancaster, N., 1989c, The dynamics of star dunes: an example from the Gran Desierto, Mexico: Sedimentology, v 36, p. 273–290.

Leighton, M. M., and Willman, H. B., 1950, Loess formations of the Mississippi Valley: Journal of Geology, v. 58, p. 599–623.

Lettau, K., and Lettau, H., 1969, Bulk transport of sand by the barchans of the Pampa La Joya in southern Peru: Zeitschrift für Geomorphologie, v. 13, p. 182–195.

Livingstone, 1., 1986, Geomorphical significance of wind flow patterns over a Namib linear dune: in Nickling, W. G., ed., 1986, Aeolian geomorphology: Proceedings of 17th Annual Binghamton Geomorphology Symposium: Allen and Unwin, N. Y., p. 97–112.

Long, J. T., and Sharp, R. P., 1964, Barchan-dune movement in Imperial Valley, California: Geological Society of America Bulletin, v. 75, p. 149–156.

Lyles, L., and Krauss, R. K., 1971, Threshold velocities and initial particle motion as influenced by air turbulence: Transactions American Society of Agricultural Engineering, v. 14, p. 563–566.

Mabbutt, J. A. 1968, Aeolian landforms in central Australia: Australian Geographical Studies, v. 6, p. 139–150.

Mabbutt. J. A., 1977. Desert landforms: MIT Press, Cambridge, MA, 340 p.

Mabbutt, J. A., and Sullivan, M. E., 1968. The formation of longitudinal dunes: evidence from the Simpson Desert: Australian Geographer, v. 10, p. 483–487.

Marrs, R. W., and Kolm. K. E., eds., 1982. Interpretation of windflow characteristics from eolian landforms: Geological Society of America Special Paper 192.

Marshall, J. R., 1979. Experimental abrasion of natural materials [Ph.D. thesis]: University College, London, 301 p.

Maxson, J. H., 1940, Fluting and faceting of rock fragments: Journal of Geology, v. 48. p. 717–751.

McAllister, J. F. and Agnew, A. F., 1948, Playa scrapers and track furrows on Racetrack playa, Inyo County, California: Geological Society of America Bulletin, v. 59, p. 1377.

McCauley, J. R. Grolier, M. J., and Breed, C. S., 1977a. Yardangs of Peru and other desert regions: U.S. Geological Survey, Interagency Report, Astrogeology, v. 81. 177 p.

McCauley, J. R., Grolier, M. J., and Breed, C. S., 1977b, Yardangs: in Doehring, D. O., ed., Geomorphology in and regions, Proceedings 8th Geomorphology Symposium: State University of New York, Binghamton, N.Y., p. 233–269.

McCauley, J. R., Grolier, M. J., and Breed, C. S., 1977c, Experimental modeling of wind erosion forms: in Reports of

Accomplishments of Planetary Geology Programs 1976–1977, NASA MTMX-3511, p. 150–152.

McKee, E. D., 1966, Structures of dunes at White Sands National Monument, New Mexico: Sedimentology, v. 7, p. 1–69.

McKee, E. D., ed., 1979a, A study of global sand seas: U.S. Geological Survey Professional Paper 1052.

McKee, E. D., and Douglass, J. R., 1971, Growth and movement of dunes at White Sands National Monument. New Mexico: U.S. Geological Survey Professional Paper 750-D, p. 108–114.

McKee, E. D., and Tibbitts, 1964, Primary structures of a seif dune and associated deposits in Libya: Journal of Sedimentary Petrology, v. 343, p. 5–17.

Melton, F., 1940, A tentative classification of sand dunes and its application to dune history in the southern High Plains: Journal of Geology, v. 48, p. 113–174.

Middleton, N. J., Goudie, W. G., and Wells, G. L., 1986, The frequency and source areas of dust storms, in Nickling, W. G., ed., 1986, Aeolian geomorphology: Proceedings of 17th Annual Binghamton Geomorphology Symposium: Allen and Unwin, N.Y., p. 237–260.

Miller, E. R., 1934, The dustfall of November 12–13. 1933: Monthly Weather Review. v. 62. p. 14–15.

Morales, C., 1979. Saharan Dust: John Wiley and Sons, N.Y., p. 3–20.

Nickling, W. G., 1978, Eolian sediment transport during dust storms: Slims River Valley. Yukon Territory: Canadian Journal of Earth Science. v. 15, p. 1069–1084.

Nickling, W. G. ed., 1986, Aeoloian geomorphology: Proceedings of 17th Annual Binghamton Geomorphology Symposium: Allen and Unwin, N.Y., 311 p.

Norris, R. M., 1966, Barchan dunes of Imperial Valley, California: Journal of Geology, v. 74. p. 292–306.

Owen, P. R., 1964, Saltation of uniform grains in air: Journal of Fluid Mechanics, v. 10, p. 225–242.

Peel, R. F., 1970, Landscape sculpture by wind: International Geographical Congress (Calcutta), 21st Proceedings, p. 99–104.

Peterson, S. T., and Junge, C. E., 1971, Sources of particulate matter in the atmosphere: in Matthews, W. H., Kellogg, W. W., and Robinson, G. D., eds., Man's impact on the climate: MIT Press, Cambridge, MA, .p. 310–320.

Pewe, T. L., 1981, Desert dust: an overview: in Pewe. T. L., ed., Desert dust. Origin, characteristics, and effect on man: Geological Society of America Special Paper 186, p. 1–10.

Porter, M. L., 1986, Sedimentology record of erg migration: Geology, v. 14, p. 497–500.

Powers, W. E., 1936, The evidences of wind abrasion: Journal of Geology, v. 44, p. 214–219.

Pye, K., 1987, Aeolian dust and dust deposits: Academic Press, London, 334 p.

Pye, K., and Lancaster, N., eds., 1993, Aeolian sediments: International Association of Sedimentologists, Special Publication 16, Blackwell Scientific Publications, Oxford, England, 167 p.

Pye, K., and Tsoar, H., 1990, Aeolian sand and sand Dunes: Unwin Hyman, London, England, 396 p.

Reid, J. B., Bucklin, E. P., Copenagle, L., Kidder, J., Pack, S. M., Polissar, P. J., Williams, M. L., 1995, Sliding rocks at the Racetrack, Death Valley: What makes them move?: Geology, v. 23, p. 819–822.

Ruhe, R. V., 1974. Geomorphology: Houghton Mifflin, Boston, MA, 246 p.

Rumpel, D. A., 1985, Successive aeolian saltation: Studies of idealized collisions: Sedimentology, v. 32, p. 267–280.

Sakamoto-Arnold, C. M., 1981, Eolian features produced by the December 1977 windstorm in southern San Joaquin Valley, California: Journal of Geology, v. 89, p. 129–137.

Schoewe, W. M., 1932, Experiments on the formation of wind-faceted pebbles: American Journal of Science, v. 24. p. 111–134.

Schumm, S. A., 1956, The movement of rocks by wind: Journal of Sedimentary Petrology, v. 26, p. 284–286.

Selby, M. J., 1985. Earth's changing surface: Clarendon Press, Oxford, England, 607 P.

Seppala, M., and Linde, K., 1978, Wind tunnel studies of ripple formation: Geographiska. Annaler, Series A, 60A, p. 29–42.

Sharp, R. P., 1949, Pleistocene ventifacts east of the Big Horn Mountains, Wyoming: Journal of Geology, v. 57, p. 175–195.

Sharp, R. P., 1963, Wind ripples: Journal of Geology, v. 71, p. 617–636.

Sharp, R. P., 1964, Wind-driven sand in Coachella Valley, California: Geological Society of America Bulletin, v, 75, p. 785–804.

Sharp, R. P., 1966, Kelso dunes, Mohave Desert, California: Geological Society of America Bulletin, v. 77, p. 1045–1074.

Sharp, R. P., 1978, The Kelso dune complex: in Greeley, R., et al., eds., Aeolian features of southern California: a comparative planetary geology guidebook: U.S. Government Printing Office, NASA, p. 52–64,

Sharp, R. P., 1979, Intradune flats of the Algodones chain, Imperial Valley, California: Geological Society of America Bulletin, v. 90, Part 1, p. 908–916.

Sharp, R. P., 1980, Wind-driven sand in Coachella Valley, California: further data: Geological Society of America Bulletin, v. 91, Part 1, p. 724–730.

Sharp, R. P., and Carey, D. L., 1976, Sliding stones, Racetrack Playa, California: Geological Society of America Bulletin, v. 87, p. 1704–1717.

Sharp, R. P., and Glazner, A. F., 1997, Geology Underfoot in Death Valley and Owens Valley: Mountain Press, Missoula, MT,.

Sharp, R. P., and Saunders, R. S., 1978, Aeolian activity in westernmost Coachella Valley and at Garnet Hill: in Greeley, R., et al., eds., Aeolian features of southern California: a comparative planetary geology guidebook: U.S. Government Printing Office, NASA, p. 9–22.

Sharp, R. P., Carey, D. L., Reid, J. B., Jr., Polissar, P. J. and Williams, M. L., 1996, Sliding rocks at the Racetrack, Death Valley: What makes them move: comment and reply: Geology, v. 24, p. 766–767.

Sharp, W. E., 1960, The movement of playa scrapers by wind: Journal of Geology, v. 68, p. 567–572.

Shelton, J. S., 1953, Can wind move rocks on Racetrack Playa?: Science, v. 117, p. 438–439.

Smith, R. S. U., 1982, Sand dunes in the North American deserts: in Bender, G. L., ed., Reference handbook on the deserts of North America: Greenwood Press, Westport, p. 481–524.

Stanley, G. M., 1955, Origin of playa stone tracks, Racetrack Playa, Inyo County, California: Geological Society of America Bulletin, v. 66, p. 1329–1350.

Suzuki, T., and Takahashi, K., 1981, An experimental study of wind abrasion: Journal of Geology, v. 89, p. 23–36.

Tsoar, H., 1974, Desert dunes morphology and dynamics: Zeitschrift für Geomorphologie Supplement, v. 20, p. 41–61.

Tsoar, H., 1982, Internal structure and surface geometry of longitudinal (seif) dunes: Journal of Sedimentary Petrology, v. 52, p. 823–831.

Tsoar, H., 1983, Dynamic processes acting on a longitudinal (seif) sand dune: Sedimentology, v. 30, p. 567–578.

Tsoar, H., 1989, Linear dunes—forms and formation: Progress in Physical Geography, v. 13, p. 507–528.

Tsoar, H., and Moller, J. T., 1986, The role of vegetation in the formation of linear sand dunes: in Nickling, W. G., ed., 1986, Aeoloian geomorphology: Proceedings of l7th Annual Binghamton Geomorphology Symposium: Allen and Unwin, N.Y., 311 p.

Tsoar, H., and Pye, K., 1987, Dust transport and the question of desert loess formation: Sedimentology, v. 34, p. 139–153.

Twenhofel, W. H., 1932, Treatise on sedimentation: Williams and Wilkins, Baltimore, MD, 926 p.

Twidale, C. R., 1972, Evolutions of sand dunes in the Simpson Desert, Central Australia: Transactions of the Institute of British Geographers, v. 56, p. 77–109.

Verstappen, H. T., and van Zuidam, R. A., 1970, On the origin of longitudinal (seif) dunes: Zeitschrift für Geomorphologie, v. 12, p. 200–220.

Ward, A. W., 1979, Yardangs on Mars: evidence of recent wind erosion: Journal of Geophysical Research, v. 84, p. 8147–8166.

Ward, A. W., and Greeley, R., 1984, The yardangs at Rogers Lake, California: Geological Society of America Bulletin, v. 95, p. 829–837.

Wasson, R. J. and Hyde, R., 1983, Factors determining desert dune type: Nature, p. 304–337.

Werner, B. T., and Haff, P. K., 1988, The impact process in aeolian saltation: Two-dimensional simulations: Sedimentalogy, v. 35, p. 189–196.

Whitney, M. I., 1978, The role of vorticity in developing lineation by wind erosion: Geological Society of America Bulletin, Part 1, v. 89, p. 1–18.

Whitney, M. I., 1979, Electron micrography of mineral surfaces subject to wind-blast erosion: Geological Society of America Bulletin, v. 90, Part 1, p. 917–934.

Whitney, M. I., and Dietrich, R. V. 1973, Ventifact sculpture by windblown dust: Geological Society of America Bulletin, v. 84, p. 2561–2582.

Willetts, B. B., and Rice, M. A., 1986, Collision in aeolian transport: the saltation/creep link: in Nickling, W. G., ed., 1986, Aeoloian geomorphology: Proceedings of l7th Annual Binghamton Geomorphology Symposium: Allen and Unwin, N.Y., p. 1–18.

Wilson, I. G., 1972a. Aeolian bedforms—their development and origins: Sedimentology, v. 9, p. 173–210.

Wilson, I. G., 1972b, Sand waves: New Scientist, v. 53, p. 634–637.

Wilson, I. G., 1973, Ergs: Sedimentary Geology, v., 10, p. 77–106.

CHAPTER EIGHTEEN

Dating
Geomorphic Features

Radiocarbon dated forest buried in a lateral moraine, Coleman Glacier, Washington.

INTRODUCTION

Determination of the age of landforms and the material composing them is often an extremely important aspect of geomorphology, especially when considering rates of various surface processes. Therefore, a great deal of effort has gone into the development of techniques for dating Quaternary features. These methods can be broadly grouped into **relative dating methods** and **numerical dating methods,** depending on whether the method yields an age as a finite number or an age relative to some other feature.

RELATIVE DATING METHODS

Topographic Position

Law of Cross-Cutting Relationships. Landforms that cut across others must be younger than the features that they transect. For example, a moraine that cuts across another (Figure 18–1) must be the younger of the two. In the same manner, a fault that cuts across a landform (Figure 18–2) must be younger than the landform that it crosses. If a lava flow of known age then buries the fault, the age of the landform can be said to be older than the fault, which is older than the lava flow, and hence the landform must be older than the dated lava.

 Nested Terraces and Moraines. In a valley containing multiple stream terraces that are inset one inside another (nested), those terraces nested inside others are younger. In a similar fashion, moraines nested inside others (Figure 18–1) must

FIGURE 18–1
Moraines cutting across an older lateral moraine (left), Convict Lake, California. The moraine in the foreground consists of multiple crests nested inside one another. The younger moraine has many boulders on its surface, whereas the older lateral moraine has almost none. (Photo by W.K. Hamblin)

FIGURE 18–2
Lateral moraines cut by a range-front fault that makes a prominent scarp, McGee Canyon, Sierra Nevada, California. (Photo by W.K. Hamblin)

be younger than the outer moraines. End moraines that lie upvalley from others must be younger than their downvalley counterparts.

Degree of Dissection and Drainage Integration The longer a landform exists at the surface, the longer erosion acts upon it. Thus, older landforms are typically more dissected by small streams than younger ones. On newly formed landscapes, such as those left by a retreating glacier, lakes and swamps gradually fill with sediment and drainage slowly becomes integrated into streams across the land. The sharp crests of landforms, such as moraines, become more subdued with time as the sharp crests are rounded by weathering and mass wasting (Figure 18–3).

Relative Size If two adjacent valleys each contain a single, exceptionally large moraine and a much smaller one, the two very large moraines are likely to be correlative because of the probability that they formed under similar climatic regimes. However, this holds true only locally, and correlations cannot be reliably made over long distances.

Weathering

In Chapter 3, we discussed various factors that control weathering. Among the important factors was time. The longer weathering acts on a landform, the more pronounced the effect. One of the early methods of correlation of moraines from valley to valley in the Sierra Nevada and Rocky Mts. was counting the number of weathered erratics on moraines relative to fresh ones. Distinct differences were found in the granite weathering ratios of boulders on moraines of differing age. Boulders that ring to the blow of a hammer were considered "fresh" and those that gave indistinct thuds were counted as weathered. The resulting ratios of fresh to weathered boulders on moraines were surprisingly distinct.

Oxidation of iron-bearing minerals in boulders and cobbles develops *weathering rinds*, concentric rings of yellow-brown, altered rock. The thicker the weathering rind on a stone, the longer the time of weathering. This method of relative dating works best on fine-grained basalt stones because the rings are more consistently developed.

Other weathering features could also be used. For example, as erratics weather, areas less resistant to weathering are differentially etched out, leaving behind raised, more resistant dikes, veins, and rock knobs. The higher the relief of such features, the older the boulder they were developed on. In addition, weathering of boulders over a long period of time eventually completely destroys them. Thus, whereas young moraines are typically covered with numerous erratic boulders, much older moraines may be virtually boulder free because the boulders have been weathered away.

Probably the most used weathering feature for relative dating is soil development. Because well-developed soil horizons take time to form, landforms mantled by more mature soils are older than landforms covered by weakly developed

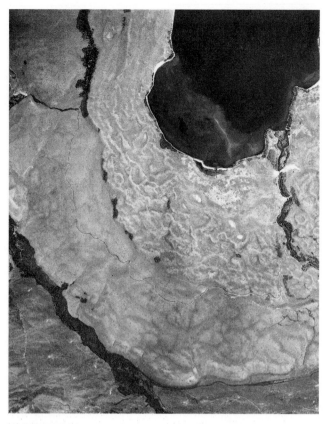

FIGURE 18–3
Differing topographic form on end moraines of different ages, Fremont Lake, Wyoming. The inner moraine contains many boulders, with little surface drainage, whereas the outer moraine is more subdued, has few boulders, and integration of drainage has begun.

soils. Soil scientists have developed various criteria in attempting to quantify degree of soil development.

Stratigraphy

Where deposits of differing origin accumulate, their relative ages may be determined using the law of superposition, i.e., younger sediments lie upon older deposits. For example, if a deposit making up a landform is overlain and underlain by volcanic ash, the age of the deposit is bracketed between the ages of the two ashes and if the age of the ashes can be determined, the age of the deposit is younger than the underlying ash and older than the overlying ash.

NUMERICAL DATING METHODS

Radioisotope Methods

Isotopes are atoms with the same number of protons in their nuclei as other species of an element but with a different number of neutrons in their nuclei. The nuclei of some isotopes are unstable and disintegrate spontaneously, giving off radiation in the process. If the number of protons in their nuclei changes, these **parent elements** become transformed into **daughter elements** at constant rates independent of temperature, pressure, or chemical processes.

Because of their constant rate of spontaneous disintegration, radioactive isotopes allow determination of age if the ratio of parent atoms to daughter atoms can be determined and if the rate of transformation is known. The number of atoms in an isotope that disintegrate (dN) per unit time dt is proportional to the number of atoms (N) present.

$$\frac{dN}{dt} = -\lambda N \qquad (1)$$

where λ = a decay constant representing the probability that an atom will decay in some unit of time.

Integrating equation (1) between $N = N_0$ at $t = 0$ and $N = N$ at $t = t$ gives

$$\int_{N_0}^{N} \frac{dN}{N} = \int_0^t -\lambda dt \ \text{ or } N = N_0 e^{-\lambda t} \qquad (2)$$

If the decay constant λ is known, setting $N/N_0 = 1/2$ and solving for t gives the **half-life** $t_{1/2}$ of the isotope.

$$t_{1/2} = \frac{\ln 2}{\lambda} \qquad (3)$$

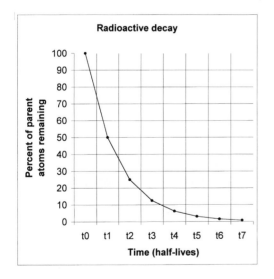

Radioactive decay

Percent of parent atoms remaining

Time (half-lives)

FIGURE 18–4

Exponential decay of a radioactive isotope. The horizontal axis represents time plotted as successive half-lives of the radioactive isotope; the vertical axis is the amount of original isotope remaining following radioactive decay. After about seven half-lives, the amount of remaining parent atoms becomes too small to allow accurate dating (right end of the curve). For isotopes with long half-lives, the dateable age range is limited by the lack of daughter atoms produced.

The **half-life** is the time for one-half of the atoms of the isotope to disintegrate; it can be represented by an exponentially decaying curve. In one half-life, half of the original atoms of an isotope disintegrate; then half of the remaining atoms disintegrate during the next half-life, leaving one-fourth of the original atoms; then half of the remaining atoms disintegrate during the next half-life, leaving one-eighth of the original atoms; and so on (Figure 18–4).

Because the number of original parent atoms N_0 equals the number of remaining parent atoms N plus the number of daughter atoms D, or

$$N_0 = N + D \qquad (4)$$

then substituting equation (4) into equation (2) gives

$$N = (N + D)e^{-\lambda t} \text{ or } D = N(e^{\lambda t} - 1) \qquad (5)$$

Therefore, the time that has elapsed since the crystallization of a mineral containing a radioactive isotope can be calculated if the rate of disintegration of the isotope, or the half-life, is known and the ratio of remaining parent atoms to daughter atoms is measured.

Radiocarbon Dating   **Radiocarbon dating,** perhaps the most widely known dating method, was developed by Libby (1955) and has since been used extensively to date carbon-bearing material. The method has been demonstrated to be highly reliable for suitable material measured under appropriate laboratory procedures, and it can be used to date material back to about 40,000 years with standard laboratory procedures and back to about 70,000 years with special enrichment procedures.

The development of radiocarbon dating was made possible by the discovery of cosmic radiation in the upper atmosphere by Hess in 1911. The total energy of the cosmic radiation received by the Earth is very small compared to the solar energy, but the specific energy of constituent particles is much higher than for other types of radiation, averaging several billion electron-volts. The increase in the number of neutrons in the atmosphere with altitude, reaching a maximum population at about 40,000 ft (~12,000 m), indicates that they are formed in the atmosphere. Because the lifetime of neutrons is only about 12 minutes, they cannot have survived interstellar space travel and must therefore be produced in the atmosphere. The interaction of neutrons with nitrogen in the atmosphere has the effect of knocking a proton out the nucleus of the nitrogen atoms, thus decreasing the atomic number by one to form ^{14}C.

$$^{14}N_7 + n = {}^{14}C_6 + {}^1H_1 \qquad (6)$$

Nitrogen is so abundant in the atmosphere that the amount of radiocarbon produced by this reaction is nearly equal to the total number of neutrons generated by cosmic radiation.

Radiocarbon ($^{14}C_6$) differs from $^{12}C_6$, the most abundant isotope of carbon, in that its mass is greater (14 compared to 12), and its nucleus is radioactive, giving off beta radiation to form nitrogen.

$$^{14}C_6 = {}^{14}N_7 + \beta^- \qquad (7)$$

Within minutes or hours of its formation, radiocarbon combines with oxygen to form carbon dioxide (CO_2), which mixes with other CO_2 in the atmosphere to produce a constant ratio of radioactive ^{14}C to stable ^{12}C carbon dioxide. Carbon dioxide, including small amounts of radioactive CO_2, is taken in by plants during photosynthesis, making all plants radioactive, and because plants are an integral part of the animal food chain, all animals are also radioactive.

Two laboratory methods are employed to measure the age of a sample:

1. Measurement of the rate of radioactive emissions from a sample using gas proportional counting techniques

2. Direct measurement of the C14 atoms using accelerator mass spectrometry (AMS)

The first method to be developed, the gas proportional counting technique, is presently the most commonly used in radiocarbon dating laboratories. Instead of measuring the ratio of parent to daughter isotopes, radiocarbon ages are calculated by comparing the ratio of the measured specific activity of the sample A_s to the specific activity of the sample when it was first isolated from the carbon reservoir A_m.

$$\text{age} = 8045 \ln\frac{A_s}{A_m} \qquad (8)$$

The age calculation is based on a radiocarbon half-life of 5568 years, which has been used by all radiocarbon laboratories since ages were first measured. New studies suggest that the half-life of radiocarbon is closer to 5730 years, which would increase all previously published dates by about 3 percent. However, by general agreement among laboratories, the earlier value of the half-life continues to be used to avoid confusion.

Measurements of the specific radioactivity of material from various components of the carbon reservoir are uniform within 3 to 5 percent, indicating thorough mixing of radioactive carbon in the reservoir. The generation of 2.6 radiocarbon atoms per second per cm^2 of the Earth's surface, which mix with the 8.3 grams per cm^2 of carbon in the reservoir, leads to the calculated disintegration rate of 18.8 disintegrations per minute per gram of carbon, compared with the experimental value of 16.1. This correspondence also supports the assumption that the rate of production of radiocarbon has been relatively constant (albeit with some variation) over the past 10,000 years.

As seen from the discussion above, the validity of radiocarbon dating depends on several assumptions:

1. The neutron flux, the rate of generation of neutrons by cosmic radiation, must be essentially constant over tens of thousands of years.

2. The rate of formation of radiocarbon by the neutron flux must be constant.

3. The rate of mixing of CO_2 in the carbon reservoir is rapid, relative to the rate of decay of radiocarbon.

4. Once material is removed from the carbon cycle by death of an organism or other isolation of the material from the carbon reservoir, no further addition of carbon can take place.

Although the above conditions are generally met, several factors cause some fluctuation in the amount of radiocarbon in the carbon reservoir:

1. Changes in the cosmic radiation flux caused by:

   a. Small variations in the ambient flux passing through the solar system

   b. Supernova (exploding stars) that increase the cosmic radiation

   c. Modulation of the cosmic radiation flux by the solar wind

   d. Short-term solar variations, especially sunspot cycles

2. Fluctuations in atmospheric CO_2 caused by changes in the solubility of CO_2 in seawater due to changes in temperature and pH of the oceans. This fluctuation affects exchange rates with the atmosphere, which in turn changes the amount of radioactivity from ^{14}C per gram of carbon.

3. Secular changes in the Earth's magnetic field. Decrease of the magnetic moment increases the cosmic radiation flux and increases radiocarbon production.

The Reservoir Effect   In addition to the organic reservoir of radiocarbon, a large amount of inorganic carbon exists as dissolved carbon dioxide, bicarbonate, and carbonate in seawater. The time required for radioactive carbon to be distributed through the inorganic reservoir in the sea is short, less than 1000 years, compared to the lifetime of radiocarbon. The inorganic carbon reservoir dominates the total carbon inventory.

The average difference between a radiocarbon date of terrestrial wood and a marine shell is about 400 radiocarbon years (Stuiver and Braziunas, 1993). The older age of oceanic water is caused by the delay in exchange between atmospheric CO_2 and ocean bicarbonate and the dilution effect caused by the mixing of surface water with old water that has upwelled from the deep ocean. A ***reservoir correction*** is therefore necessary to marine shell dates to account for this difference in the oceanic and atmospheric reservoirs. Radiocarbon laboratories routinely include reservoir corrections for marine shell dates.

Calibrated Radiocarbon Ages and Radiocarbon Years
Comparison of radiocarbon dates with samples of historically known ages shows that "radiocarbon years" are not quite equivalent to calendar years. To test the magnitude of some such variations, researchers have systematically measured the radiocarbon age of tree rings to compare calendar years with radiocarbon years (Damon et al., 1978; Damon et al., 1966; Grey, 1969; Stuiver, 1961, 1965, 1967, 1970, 1982; Stuiver and Becker, 1993; Stuiver and Pearson, 1992, 1993; Stuiver and Reimer, 1993; Stuiver and Suess, 1966; Stuiver et al.,

1991; Suess, 1965, 1970a, and 1970b, 1986; Vogel, 1970; Vogel et al., 1993; Becker, 1993).

Trees grow one ring around their trunks every year, and once the wood making up the ring has grown, it remains thereafter unchanged during the life of the tree. The radiocarbon content of any specific ring depends on the time that has elapsed since the ring was deposited. Because tree rings can be ^{14}C dated and their calendar age can be accurately measured, they are ideal for comparing ^{14}C age with calendar age. Among the most useful trees for such measurements are bristlecone pines in the western U.S., which live for thousands of years. Their antiquity was not generally realized until a graduate student took core samples from several trees in the early 1960s and discovered one more than 4000 years old. When his only tree-coring tool broke, he asked for, and was granted, permission by the U.S. Forest Service to cut down an old bristlecone pine that had been nicknamed Prometheus by conservationists. Having killed the oldest living thing on Earth, he then counted 4844 tree rings (dendrochronologists later determined the age of the tree to be 4950 years).

The bristlecone record can be extended by counting rings in dead trees on the ground that can remain thousands of years after the trees have died. However, the dead tree ring chronology must be connected to that of living trees. Fortunately, this can be accomplished by comparing the growth patterns made by differing widths of rings. Because the width of a ring is de-termined by conditions during the growing season, trees living at any given time will have similar ring patterns. Thus, if the life span of a dead tree overlaps that of a still-living tree, growth patterns can be matched to establish the time of their overlap, and the chronology of tree rings can be extended farther back in time. This has been done for bristlecone pines in the U.S. and oak trees in Ireland and Germany to provide tree ring records for the last 11,000 years.

Plotting of the radiocarbon age of tree rings against their calendar age (Figure 18–5) shows that radiocarbon years differ slightly from calendar years, and the difference increases with age. As seen in Figure 18–5A, the radiocarbon ages of tree rings falls farther and farther below calendar years with time. For example, at about 2000 years ago, radiocarbon ages are about the same as calendar ages, but by 7950 years, the radiocarbon age is only 7141 years. How much radiocarbon ages differ from calendar years beyond about 10–11,000 years is not well constrained because of the lack of datable material whose calendar age is known. However, the trend of the curve in Figure 18–5 suggests a consistent pattern of lower radiocarbon ages. Accordingly, dates are reported as "radiocarbon years B.P." (before present) or "calibrated (calendar) years B.C."

Although the curve in Figure 18–5 follows a fairly regular path, in detail (Figure 18–5B) it flattens and steepens at various times. As a result, where the curve flattens, a radio-

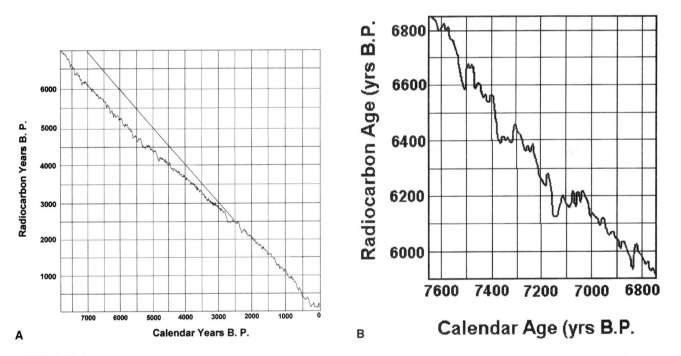

FIGURE 18–5

(A) Relationship between radiocarbon years and calendar years. If radiocarbon years were the same as calendar years, the graph would be a straight line at a 45° angle. However, the radiocarbon ages are progressively somewhat younger with age and fall farther and farther below calendar years until by 7,950 calendar years, the corresponding radiocarbon age is only 715. (from Stuiver and Becker, 1993).

(B) Details of the older part of the curve in (A). Flattening of the curve at about 6200 calendar years means that a radiocarbon date of 6175 has nine possible intercepts with the curve (from Stuiver and Becker, 1993).

carbon age could correspond to more than one calendar age. For example, flattening of the curve between 7000 and 7200 calendar yrs B. P. means that a ^{14}C age of slightly less than 6,200 ^{14}C yrs intersects the curve at nine different places, corresponding to calendar ages between 7025 and 7175. Flattening of the curve at about 2500 calendar-yrs B. P. results in intersection of a radiocarbon age of 2575 at 10 different places so that this ^{14}C age could correspond to any 1 of 10 different calendar ages between 2400 and 2700 calendar yrs B. P. Fortunately, the variation in calibrated age is relatively small compared to the total age of a sample. However, flattening of the curve between about 10,000 and 12,000 ^{14}C yrs poses some problems in precisely correlating events during the critical Younger Dryas climatic event.

Laboratory Methods   Two rather different laboratory techniques are used to measure the radiocarbon in a sample. The older, standard, technique is to count the rate of beta decay, either as a gas (gas-proportional counting) or liquid (liquid scintillation counting). A sample is converted to gas and the rate of beta decay is measured in a scintillation counter protected from external radiation by massive lead shielding. Any remaining external or background radiation is accounted for by anticoincidence counters. A newer technique is to measure the number of radiocarbon atoms in a sample directly, using accelerator mass spectrometry (AMS).

Beta counting method

1. The sample is pretreated to remove possible contaminants. For wood, charcoal, or peat, dilute HCl removes possible contamination from young carbonate, and NaOH removes humic acids derived from modern soil (not used on peat). Shell samples are immersed in acid to remove the outer part that could be contaminated.

2. The sample is burned in pure oxygen (or shells are dissolved in acid), and the CO_2 is purified to remove SO_2, N_2, and O_2.

3. For systems that use methane as the counting gas, CO_2 is converted to methane in a pressurized vessel with tritium-free hydrogen and a ruthenium catalyst.

4. Measurement of radioactivity is made by proportional counting of the gas in a shielded counter surrounded by anticoincidence counters. The age of the sample is calculated from the number of counts per minute per gram of carbon of the sample. A modern sample would give 14 counts per minute per gram of carbon, whereas a 5700-year-old sample would give only 7, and an 11,400-year-old sample 3.5.

Care must be taken to avoid contamination during all steps of sample collection, storage, shipping, and measuring. At a sampling site, small, modern rootlets, insects, grass, algae, lichen, mold, or any other carbonaceous material must be avoided. Samples are usually collected in plastic bags or aluminum foil, avoiding paper or any kind of carbon-bearing packing material. In the laboratory, possible contamination from radon is circumvented by storing the sample for 30 to 60 days to allow any radon to disintegrate. The possible effects of cont-

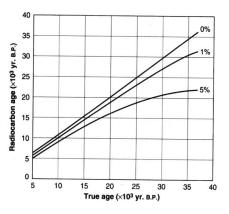

FIGURE 18–6
Effect of contamination on ^{14}C age measurements. (After Teledyn Isotopes, Inc.)

amination at any stage in the process are not trivial. For example, a sample with a true age of 57,000 years, which is contaminated by only one percent modern carbon, will yield an age of 37,000 years. The error increases substantially at higher levels of contamination (Figure 18–6). However, with appropriate care, nearly all of these problems can be avoided.

AMS Method

A newer method of radiocarbon dating involves direct measurement of the ^{14}C atoms using accelerator mass spectrometry (AMS). This method has the advantage of being able to date very small amounts of material, such as individual seeds, and the eventual possibility of dating material in the 65,000–75,000-year age range.

AMS dating of a sample is accomplished by converting the atoms in the sample into ions (charged atoms) that are accelerated into a fast-moving beam. The mass of the ions is then measured by their deflection in magnetic fields. A sample is introduced into the ion source either as graphite or as carbon dioxide, then ionized by bombardment with cesium ions and focused into fast-moving beam in the accelerator. The ^{14}C ions are selected in a magnetic field according to their momentum and velocity, then counted in a detector.

The main advantage of AMS over the conventional beta-counting method is the much greater sensitivity of measurement. In AMS, the radiocarbon atoms are detected directly, rather than counting their decay. Sample sizes are typically 1000 times smaller, allowing a much greater choice of samples and enabling very selective chemical pretreatment. The small sample size also minimizes possible contamination by allowing dating of a single seed or twig, thus avoiding possible admixed modern carbon.

K-Ar Dating   Potassium-40 (^{40}K) is a naturally occurring, radioactive isotope of potassium, which decays to ^{40}Ar and ^{40}Ca with a half-life of 1.3 billion years.

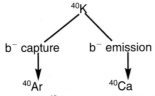

Potassium-40 decays to ^{40}Ar by electron (b$^-$) capture and by conversion of a proton into a neutron, decreasing the atomic number by one. The excited ^{40}Ar decays to a stable ground state by emitting a gamma ray. In addition, ^{40}K decays to ^{40}Ca by emitting an electron, converting a neutron into a proton, and increasing the atomic number by one. Eight times as many ^{40}Ca atoms as ^{40}Ar atoms are generated by decay of ^{40}K, but because Ca is such a common element in minerals, ^{40}Ar is much more useful as a dating tool.

The age of a K-bearing mineral can thus be determined by measuring the ratio of parent atoms (^{40}K) to daughter atoms (^{40}Ar) according to the general radioactive decay equation (5) described earlier. However, because two daughter isotopes, ^{40}Ar and ^{40}Ca, are produced, the decay equation must be written as follows:

$$^{40}\text{Ar} + {}^{40}\text{Ca} = {}^{40}\text{K}[e^{(\lambda_\beta + \lambda_\epsilon)\tau} - 1] \qquad (9)$$

where λ_e = decay constant of ^{40}K to ^{40}Ar
$(0.585 \times 10^{-10} \text{ yr}^{-1})$
λ_β = decay constant of ^{40}K to ^{40}Ca
$(4.72 \times 10^{-10} \text{ yr}^{-1})$

Substituting the values for the decay constants into equation (9) and solving for time *t* gives

$$t = 1.885 \times 10^9 \ln\left[9.068 \frac{^{40}\text{Ar}}{^{40}\text{K}} - 1\right] \qquad (10)$$

$$\ln\left[1 + \left(\frac{\lambda_\beta + \lambda_e}{\lambda_e}\right)\frac{^{40}\text{Ar}}{^{40}\text{K}}\right]$$

For calculation of Pleistocene ages (<2 m.y.), the equation can be simplified to

$$t = 1.71 \times 10^{10} \frac{^{40}\text{Ar}}{^{40}\text{K}} \qquad (11)$$

The error introduced by this simplification is less than 0.1 percent for ages over 2 m.y.

The potassium-argon (K-Ar) method of dating is very useful for dating rocks or ash containing K-bearing minerals. Whereas radiocarbon is limited to material younger than about 40,000 to 60,000 years, K-Ar is usable for minerals older than about 50,000 years, although laser techniques now extend the range down to about 10,000 years. The K-Ar method has wide-ranging applicability for dating surficial material for the following reasons:

1. The long half-life of ^{40}K (1.3 billion years) means that K-bearing minerals of a wide range of ages, between approximately 20,000 to the oldest rocks on Earth, can be dated.

2. Potassium, the seventh most abundant element in the Earth's crust, occurs in many rock-forming minerals.

3. Argon can be measured accurately, even in small quantities.

4. Because Ar is an inert gas, it does not combine with other elements to form minerals, so its occurrence in crystal lattices results largely from trapping of ^{40}Ar from radioactive decay of ^{40}K, with small amounts from atmospheric contamination or ^{40}Ar present in the environment of crystallization.

Calculation of the K-Ar age of a mineral involves measurement of the amount of ^{40}K and radiogenic ^{40}Ar in the sample. The ratio of ^{40}K to total K has been found to be constant within experimental error. Thus, the amount of ^{40}K in a sample may be determined by measuring total K in the sample, usually by flame photometry.

The amount of radiogenic ^{40}Ar in the sample is determined by fusing the sample in a vacuum system and collecting the Ar gas released from the minerals, which is then mixed with a known quantity of tracer argon (^{38}Ar). The relative amounts of ^{40}Ar, ^{38}Ar, and ^{36}Ar are then measured in a mass spectrometer.

The total ^{40}Ar of the sample is determined by comparison of the amount of ^{40}Ar relative to the ^{38}Ar tracer that was introduced in known amounts. The total ^{40}Ar of the sample is the only unknown in the equation.

$$\frac{\text{total } ^{40}\text{Ar}_{\text{sample}}}{\text{known amount of } ^{38}\text{Ar}_{\text{tracer}}} = \frac{^{40}\text{Ar}_{\text{counted}}}{^{38}\text{Ar}_{\text{counted}}} \qquad (12)$$

An easy way to visualize how this is done is to imagine a large field filled with thousands of black cows. Rather than count all of the cows one by one, you could introduce a known number of white cows, thoroughly mix them with the black, then count the number of black cows relative to white cows for a portion of the total herd. That ratio multiplied by the known number of introduced white cows will equal the total number of black cows in the field.

The total ^{40}Ar measured includes:

1. Radiogenic Ar produced by decay of ^{40}K

2. An unknown amount of atmospheric ^{40}Ar as a contaminant from the surface of the mineral, laboratory apparatus, or partial weathering

3. Small amounts of ^{40}Ar trapped in the crystal lattice of some minerals at the time of crystallization

The ratio of ^{36}Ar to ^{40}Ar in the atmosphere is constant at 99.600 percent ^{40}Ar and 0.337 percent ^{36}Ar (plus 0.063 percent ^{38}Ar). By measuring the amount of ^{36}Ar in a sample, which could have entered only as atmospheric contamination, the amount of nonradiogenic ^{40}Ar present can be calculated using the following ratio:

$$\frac{^{40}\text{Ar}_{\text{atmospheric}}}{^{36}\text{Ar}_{\text{atmospheric}}} = \frac{^{40}\text{Ar}_{\text{nonradiogenic}}}{^{36}\text{Ar}_{\text{sample}}} \qquad (13)$$

Because the amount of ^{36}Ar in the sample is measured and the ratio of ^{40}Ar/^{36}Ar in the atmosphere is known, the equation can be solved for the amount of atmospheric ^{40}Ar that must be in the sample. This amount can then be subtracted from the total ^{40}Ar measured in the sample to obtain the amount of radiogenic ^{40}Ar in the sample.

$$^{40}\text{Ar}_{\text{radiogenic}} = {}^{40}\text{Ar}_{\text{total}} - {}^{40}\text{Ar}_{\text{nonradiogenic}} \quad \textbf{(14)}$$

One of the limiting factors in K-Ar dating, especially for young material, is the amount of atmospheric ^{40}Ar relative to the radiogenic ^{40}Ar. When the atmospheric ^{40}Ar becomes dominant, the accuracy falls off sharply. This limits the lower (younger) end of the dating range of K-Ar for young Pleistocene material.

Another method of dating is by determination of the ratio of ^{40}Ar to ^{39}Ar. Rather than measuring ^{40}K directly, samples are irradiated with neutrons in a nuclear reactor, causing stable ^{39}K to transmute into ^{39}Ar. Potassium-40 is determined indirectly by measuring ^{40}Ar and ^{39}Ar, and if the ratio of ^{40}K to ^{39}K, which is a constant, is known, the sample age can be calculated. The advantages of this method are as follows:

1. Measurements are made simultaneously on the same sample in the same location within a crystal lattice where ^{40}Ar is trapped (Curtis, 1975), in contrast to conventional ^{40}K $- {}^{40}$Ar techniques in which potassium and argon measurements must be made on different parts of a sample.

2. The degree to which a sample has been altered by weathering, heating, or contamination by extraneous argon can be ascertained.

3. Several ages can be measured from one sample and dealt with statistically to provide a more precise date.

The ^{40}K that decays to ^{40}Ar occupies the same position in the crystal lattice of a mineral as ^{39}K, which is much more abundant than ^{40}K and which produces ^{39}Ar when irradiated. Both argon isotopes are released simultaneously upon heating of a sample. Atmospheric argon coating the surface of mineral grains is released at low temperatures, so ^{40}Ar$/^{39}$Ar ratios on the first gas given off would yield an age that is too young, whereas the Ar from deeper within the uncontaminated interior of crystal lattices and driven off at higher temperatures would yield an older age. Consistent age measurements made with rising temperatures ensure confidence in the final age determination.

The advantages of ^{40}Ar$/^{39}$Ar are especially significant for young samples and for K-poor samples, particularly xenoliths in K-poor lavas (Gillespie et al., 1982).

Minerals Suitable for Dating   For K-Ar dates to be valid, the mineral holding the Ar must behave as a closed system, neither losing nor adding Ar. Thus, the best minerals for dating have low Ar diffusion rates and do not acquire excess Ar during crystallization. The mineral of choice, where available, is sanidine, a high-temperature phase of K-feldspar. A number of other common minerals are also suitable, including hornblende, biotite, plagioclase, other K-feldspars, glauconite, or the entire rock.

Volcanic ash is fairly commonly preserved in Quaternary sediments and, if suitable crystals are present, makes appropriate material for K-Ar dating. Glass can also be used but is less desirable because of its high Ar diffusion rates and likelihood of weathering.

For detrital material, such as volcanic ash, pumice, or tuff, contamination with older clastic material sometimes presents a problem. New advances in Ar laser technology can now address such difficulties by being able to date single grains. If the age of a particular grain is widely divergent from that of other grains in a sample, it may be considered detrital contamination and simply discarded. The ability to recognize such contamination is a definite advantage in arriving at an accurate age for a sample.

The datable age range for K-Ar is open ended at the old end of the spectrum but deteriorates as the atmospheric ^{40}Ar becomes dominant over radiogenic Ar. Until recently, this meant that dates younger than about 50,000 to 100,000 were suspect. However, laser Ar dating now can improve the precision of ages down to perhaps 10,000 to 15,000 years.

Cosmogenic Isotope Dating   Bombardment of rocks exposed at the Earth's surface by cosmic radiation produces in situ, stable, radioactive isotopes (Yokoyama et al., 1977; Lal, 1988). The amount of each cosmogenic nuclide generated in exposed rocks depends on:

1. The decay constant of the isotope

2. The *production rate*

3. The amount of time the surface has been exposed to cosmogenic radiation

4. The erosion rate of the rock surface

5. Any inherited component of the isotope concentration

6. The loss or gain of isotopes in the system

Thus, the buildup of cosmogenic isotopes on exposed rock surfaces provides a potential means of dating how long the surface has been exposed. Among the cosmogenic isotopes that have been investigated as possibly suitable for dating purposes are ^{36}C, ^{10}Be, ^{26}A, and ^{3}He (see, for example, Lal, 1988; Phillips et al., 1986; Phillips et al., 1990; Nishiizumi et al., 1986; Nishiizumi et al., 1989; Cerling, 1990; Craig and Poreda, 1986; Kurz, 1986a, 1986b; Klein et al., 1986).

Production rates of isotopes generated by cosmic radiation have been determined by measurements in samples collected from surfaces of known age (Nishiizumi et al., 1989) and by theoretical calculations (Lal, 1988). The production of cosmogenic isotopes decreases exponentially with decreasing altitude and with depth below the rock surface because of attenuation of the cosmic radiation flux. The cosmic radiation flux values are halved at about 40 cm below the surface of a basalt flow. The production rate of cosmogenic isotopes P. is given by the equation

$$P_x = P_o \, e^{-(kx)} \quad \textbf{(15)}$$

where P_x = production rate at depth x (atm/g/yr)
      x = depth
      P_o = production rate at the rock surface (atm/g/yr)
      k = a density-dependent constant representing the absorption of cosmic radiation (cm^{-1})

For a closed system with no surface erosion and no isotopes left from earlier events, the abundance N_x of stable cosmogenic isotopes depends upon the isotope production rate Px and time t.

$$N_x = P_x t \qquad (16)$$

For unstable isotopes, N_x must take the decay constant of the isotope into consideration, so that

$$N_x = P_x(1 - e^{-\lambda t}) \lambda^{-t} \qquad (17)$$

Still more complicated equations, which include the effect of surface erosion during isotope accumulation, can be written.

The concentration of a cosmogenic isotope in a rock is proportional to the amount of time since the rock surface became exposed to cosmic radiation, the isotope's production rate, and the exposure history (e.g., type of erosion, rate of erosion, and episodes of burial, all of which may interrupt exposure).

One of the most important factors in calculating exposure ages is the production rate of an isotope at a given location. Production rates vary with location on the Earth. They are greater at higher geomagnetic latitudes because the Earth's geomagnetic dipole field deflects incident cosmic radiation poleward. Production rates are greater at higher altitudes because the radiation has to penetrate less atmosphere and therefore is less attenuated. Production rates may also vary spatially if persistent non-dipole geomagnetic fields have altered the incident radiation, although this is considered a minor effect. Scaling factors for different site elevations and latitudes have been published (Lal, 1991) and partially tested. Although the uncertainty in scaling factors may be as much as 10 percent, in converting low-latitude, high-altitude production rates to high-latitude, sea-level production rates, the uncertainty is estimated to be on the order of 3–5 percent when comparing sites of similar latitudes and altitudes.

Production rates vary with time because: (1) secular variation of the Earth's dipole axis changes the effective geomagnetic latitude of a site. For surfaces less than 10,000 years old, secular variation may have a 3 to <7 percent effect on the time-averaged production rate. For older surfaces, the average dipole axis position is approximately geocentric, and the effect is therefore negligible; and (2) the Earth's dipole field strength fluctuates. This effect is important because paleointensity variations are not well known (published paleointensity curves have as much as ~20 percent uncertainty), and because paleointensity variations may lead to a ±20–25 percent uncertainty in the calibration of the cosmogenic isotope time scale to other time scales (dendro-years, radiocarbon years, calendar years, etc.)

Effects of Erosion and Other Geological Factors In order to produce measurements that represent the time since a boulder or rock surface was first exposed to the atmosphere, the surface must have remained intact since it was exposed. Thus rolling, burial, or exfoliation of boulders affects the concentration of cosmogenic isotopes (Gosse et al., 1995a; Gosse et al., 1995b). However, a stringent sampling strategy can effectively minimize or avoid the effects of post-depositional processes on boulders.

Measuring a sample that contains cosmogenic isotopes produced at some time prior to the last exposure can introduce problems. For example, consider a boulder ripped up by a glacier and deposited on a moraine. If one side of the boulder had previously been exposed to atmospheric radiation, that surface would carry a record of the prior exposure that would not allow the time of deposition of the boulder on the moraine to be measured accurately. However, if the boulder is roughly equant in shape, only one of the six sides would carry a pre-exposure history, so the random probability of sampling that side would be one in six. Thus, one approach to minimize possible inheritance of a cosmogenic signature is to measure many samples from a single site. With many samples, surfaces carrying isotopes inherited from a prior exposure will most likely show up as outliers in the data set.

Advantages of Cosmogenic Dating Measurement of cosmogenic isotope exposure ages offers several significant advantages over other dating methods. Perhaps the most obvious is that cosmogenic isotope dates give the age of exposure to the atmosphere and thus either the time of deposition (as an erratic boulder on a moraine) or the time of uncovering (as a newly exposed fault surface or glacially scoured rock). Although some glacial deposits contain dateable material, most do not. The possibility of *directly dating* the time of deposition of erratic boulders on a moraine is an example of an important advantage of cosmogenic dating. In contrast, most other radiometric methods yield only bracketing ages of deposits or landform surfaces.

Although radiocarbon dating remains as the premiere technique for dating landforms because of its well-established accuracy, organic material is commonly lacking in many critical localities and only bracketing ages are typically available. On the other hand, cosmogenic isotopes begin accumulating immediately when a boulder is deposited on a moraine, and therefore mark the time of moraine formation.

Atmospheric $^{14}C/^{12}C$ in the atmosphere changes with variations in the intensity of the Earth's geomagnetic field. Thus, the $^{14}C/^{12}C$ measured in organic material is controlled not only by the time of radioactive decay, but also by the magnetic paleointensity. Adjustments to the radiocarbon time scale based on paleointensity curves (presently with <20 percent uncertainty) has not yet proven reliable, although coral and ice core data (Stuiver et al., 1995) show the uncertainty in the calibration is probably better than 10 to 12 percent.

Isotopes produced in rocks by cosmic radiation include both stable elements (e.g. ^{3}He or ^{21}Ne) and radioactive elements (e.g. ^{10}Be, ^{14}C, ^{26}Al, or ^{36}Cl). All of these are potentially useful for dating rock surfaces if their production rate is known. Production of cosmogenic nuclides is dependent on a variety of factors including:

1. The relative intensity and spatial variation of the Earth's magnetic field

2. Latitude and elevation of the sample

3. Shielding of the sample by surrounding topography

4. Gradual exposure of sample locations by soil erosion

5. Gradual erosion of the sample surface

All of these factors must be considered in using cosmogenic isotope dating techniques. One of the most important factors is determining the production rate of cosmogenic isotopes for use in the age equation. Another significant factor is the effect of variations in the Earth's magnetic field on cosmogenic isotope production over time, and characterizing the pattern of variation in the dipole (and possible quadripole) field. Despite these complications, cosmogenic nuclides commonly provide the only avenue for dating a wide variety of earth materials.

^{10}Be and ^{14}C are produced in quartz, mostly from reaction with oxygen atoms, and ^{26}Al and ^{21}Ne are produced from interaction with silicon atoms. Although many other isotopes are produced in various minerals, ^{10}Be and ^{26}Al in-quartz have yielded the most precise and reproducible measurements on a single glacial landform (Gosse et al., 1995a; Gosse et al., 1995b).

Cosmogenic nuclides	Half life	Useful age range	Minerals
^{10}Be	1.5 m.y	1,000–1 m.y.+	Quartz olivine
^{26}Al	0.7 m.y.	1,000–1 m.y.+	Quartz, olivine
^{36}Cl	0.3 m.y	1,000–500,000	Whole rock
^{14}C	5730 yrs	0–15,000 yrs	Whole rock, quartz
^{3}He	Stable	0 to infinity	Olivine, pyroxene, quartz
^{21}Ne	Stable		Olivine, quartz, plagioclase

The Cosmogenic ^{10}Be Dating Method   The concentration of cosmogenic ^{10}Be produced in a rock increases with exposure time, depending on the local production rate of ^{10}Be in quartz, the altitude and latitude from which the sample was collected, and the depth below the surface of the sample. Si, the principal progenitor of ^{26}Al, and O, the principal progenitor of ^{10}Be and ^{14}C, occur in quartz in a stiochiometric ratio rendering unnecessary analysis of the rock to determine target abundances, an advantage over whole rock methods. Quartz is widespread in a wide variety of rocks, abundant, and resistant to weathering. Because ^{10}Be and ^{26}Al are produced in quartz, independent measurements allow independent internal checks on exposure ages and provide additional information on exposure history. Both isotopes are measured comparatively easily using AMS.

Recent results from the Wind River Range, Wyoming (Figure 18–7), (Gosse et al., 1995a; Gosse et al., 1995b) show that total ^{10}Be age uncertainties are ~4 percent including systematic errors. Total analytical reproducibility is better than 3 percent with multiple dates from a single landform.

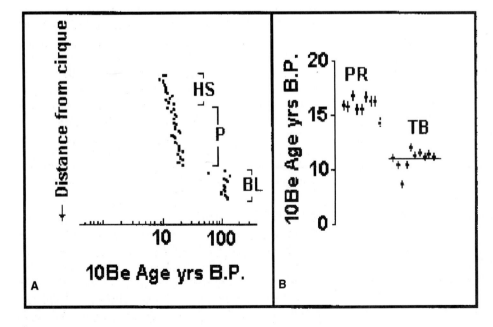

FIGURE 18–7
(A) Cosmogenic ^{10}Be chronology of moraines in the Wind River Mountains, Wyoming, based on samples from 78 boulder and bedrock surfaces. Highland surfaces (HS) include ages on boulders of the Temple Lake equivalent Titcomb Basin Moraine. P = Pinedale moraines; BL = Bull Lake moraines (Gosse et al., 1995a). (B) Cosmogenic ^{10}Be ages of boulders on moraines in the Wind River Mountains, Wyoming. PR = recessional moraines at the termination of the last glacial maximum; TB = Titcomb Basin inner moraines plotted from data in Gosse et al., 1995a).

The Cosmogenic ^{36}Cl Dating Method

Reaction of cosmic rays with elements exposed at the surface produces ^{36}Cl, allowing the ratio of ^{36}Cl to stable chloride to be used as a means of dating exposure time. The accumulation of ^{36}Cl from surface exposure of rocks provides a potentially useful dating method in the age range of approximately 1000 to 1,000,000 years (Phillips et al., 1986).

^{36}Cl is a useful isotope for dating purposes because it accumulates in measurable quantities over short time periods, normal chlorine is usually present in only very small amounts in rocks, and radiation from uranium and thorium generates less ^{36}Cl than cosmic radiation. The production rate of ^{36}Cl below the surface is low because of the low subsurface radiogenic neutron flux. When buried rocks are exposed at the surface, cosmic radiation begins to produce ^{36}Cl.

^{36}Cl is produced in rocks by more than one reaction, but the main reaction is activation of ^{35}Cl by thermal neutrons. In rocks having low chlorine content, spallation of ^{39}K and ^{40}Ca and muon capture by ^{40}Ca become significant. The production rate of ^{36}Cl is a function of the thermal neutron flux, the concentration of ^{35}Cl, the thermal neutron activation cross section of ^{35}Cl, elevation, depth of burial, geomagnetic latitude, and $^{36}Cl/Cl$ ratio from U and Th radiation.

Ideally, the optimum situation for dating purposes is complete shielding of a rock surface by deep burial and then rapid exposure. Among the geologic environments that meet these criteria are those generated by lava flows, deep glacial erosion, glacial moraines, various types of geomorphic surfaces, and soils. Phillips and others (1986) found a consistent buildup of ^{36}Cl with time on volcanic rocks ranging in age from 6500 to 670,000 years. ^{36}Cl accumulation in boulders on moraine crests in the eastern Sierra Nevada yielded ages ranging from 21,000 to 200,000 years (Phillips et al., 1990).

Other cosmogenic isotopes from rock surfaces, ^{26}Al and ^{3}H, have been investigated for possible use in age determination, using similar approaches (Nishiizumi et al., 1989; Cerling, 1990).

Uranium-Series Dating

Uranium-series dating includes several methods based on radioactive decay products of ^{235}U or ^{238}U (Table 18–1). The most significant of these isotopes for dating are ^{238}U, ^{235}U, ^{230}Th (ionium), and ^{231}Pa (proac-tinium).

The physical separation of parent and daughter isotopes may be used to calculate age as a function of the decay rate of the daughter isotope. In natural uranium systems undisturbed for about 106 years, an equilibrium is established in which each daughter product is present in amounts such that it decays at the same rate as it is formed by its parent isotope (Broecker and Bender, 1972), and the isotopes remain constant. If the equilibrium system is disturbed, the relative proportions of isotopes change, so the time elapsed since a disturbance can be determined by measuring the degree to which a disturbed system has returned to a new equilibrium.

Because ^{230}Th and ^{231}Pa are essentially insoluble in water, they precipitate as uranium decays and thus accumulate in sediments. As they become progressively more deeply buried by sedimentation and separated from their parent isotopes (^{234}U and ^{235}U), they decay at a known rate without further additions from their parent isotopes. If the sedimentation rate is constant and the uranium isotope composition of seawater is constant, the concentrations of ^{230}Th and ^{231}Pa will decrease exponentially with depth. If the initial concentrations of ^{230}Th and ^{231}Pa are known, the amounts of their decay are functions of the time since they were buried. The potentially useful dating ranges are 10,000 to 350,000 years for $^{230}Th/^{234}U$ and 5000 to 150,000 years for $^{231}Pa/^{235}U$.

A similar approach utilizes the ratio of $^{234}U/^{238}U$ in carbonate in coral or mollusk shells precipitated in equilibrium with the uranium isotope composition of seawater. The ratio of $^{234}U/^{238}U$ in seawater is constant at 1.14. Because ^{234}U has a much shorter half-life than ^{238}U, if coral or mollusk shells are removed from contact with seawater, the $^{234}U/^{238}U$ ratio in them changes, eventually attaining a new equilibrium. Changes in the ratio of $^{234}U/^{238}U$ may be used to date carbonates in the range of about 40,000 to 1,000,000 years old.

Other methods of uranium-series dating utilize the accumulation of a daughter product in carbonates in corals, mollusks, and cave deposits. Uranium has an ionic radius that allows it to be precipitated from water in calcite or aragonite. The amounts of ^{230}Th and ^{231}Pa increase as ^{234}U and ^{235}U decay, so their abundance in a sample depends on the initial uranium content and time. $^{230}Th/^{234}U$ dating has been used to date coral on marine terraces that provide a record of Pleistocene glacio-eustatic sea level changes (Broecker and Bender, 1972; Chapell and Polach, 1976).

Carbonate cave deposits can also be dated using uranium-series isotopes. ^{230}Th is so insoluble that carbonates precipitated from groundwater are essentially free of primary thorium. If the groundwater contains sufficient initial uranium, measurement of the $^{230}Th/^{234}U$ ratio records the increase of ^{230}Th with time (Harmon et al., 1975).

Uranium-series dating is not without problems. For example:

1. Samples must have remained in a closed system throughout their geologic history.

2. Assumptions must be made concerning the initial $^{230}Th/^{234}U$, $^{234}U/^{235}U$, and/or $^{231}Pa/^{235}U$ ratios in samples.

3. For some methods, a constant sedimentation rate must be assumed, the validity of which has been challenged by some (Osmond, 1979).

If these assumptions are not met, the potential for significant errors can be high. At present, perhaps the most reliable dates are obtained from carbonate in coral, because multiple dates using different isotopes from the same sample can be cross-checked for consistency.

Thermoluminescence (TL) Dating

Thermoluminescence (TL) dating is based on the absorption and storage of energy in crystals α, β and γ radiation emitted from U, Th, and their daughter isotopes, and from ^{40}K. The energy is stored in the form of electrons that migrate to traps—impurities and crystal defects—and it is emitted in the form of light upon heating or exposure

TABLE 18–1
Uranium series isotopes.

Nuclide	Half-Life	Nuclide	Half-Life
^{238}U (uranium)	4.51 x 109 yr	^{235}U (uranium)	7.13 x 108 yr
^{234}U (uranium)	2.5 x 105 yr	^{231}Pa (protactinium)	3.24 x 104 yr
^{230}Th (ionium)	7.52 x 104 yr	^{227}Th (thorium)	18.6 days
^{226}Ra (radium)	1.62 x 103 yr	^{223}Ra (radium)	11.1 days
^{222}Rn (radon)	3.83 days	^{207}Pb (lead)	stable
^{210}Pb (lead)	22 yr		
^{210}Pa (polonium)	138 days		
^{206}Pb (lead)	stable		

to sunlight. The intensity of the emitted light is a function of the radiation dose, the sensitivity of the crystal to ionizing radiation, and the length of time since the traps were last emptied.

The TL age of a mineral is proportional to the total ionizing radiation dose, known as the **paleodose,** divided by the dose rate. The radiation dose is estimated from luminescence measurements, whereas the dose rate is usually determined from radioactivity analyses. The age t is equal to the paleodose divided by the dose rate and can be expressed by the following equation:

$$t = \frac{D_e}{D_a + D_\beta + D_\gamma + D_c} \qquad (18)$$

where the equivalent dose D_e = *laboratory* β or γ dose that produces the same TL intensity as the paleodose:

$$D_\alpha + D_\beta + D_\gamma + D_c = \text{respective alpha, beta, and gamma}$$
$$\text{dose rates}$$
$$D_c = \text{cosmic ray dose rate}$$

The numerator and denominator are determined in separate measurements. The cosmic ray dose rate is usually minor, around 5 percent of the total dose rate.

For TL to be used as a dating technique, the crystals containing the TL must behave as a closed system, with no loss of electrons from the traps nor any change in K, U, Th, or their decay products. The TL clock is set at zero when all electron traps are empty, as upon crystallization of a mineral or by release of all electrons from the traps by heating or exposure to sunlight. Zeroing is complete for minerals "newly born" by crystallization and is usually thorough for minerals heated to minimum temperatures. However, because zeroing by exposure to sunlight requires adequate time of exposure, zeroing may not always be complete; that is, short-term exposure to sunlight reduces the TL signal to some nonzero level that is unknown in advance.

Several types of geologic events can be dated by TL:

1. Crystallization of a mineral
2. Heating of a mineral
3. The last exposure of a mineral to sunlight, prior to deposition

The zeroing of TL for all chemically precipitated minerals and for strongly heated material may be expected to be com-

plete. Examples of crystallized or heated material that can be TL dated include tephra deposits, baked sediments, fired pottery, burned flint, foraminifera, radiolaria, gypsum, pedogenic carbonate, and carbonate in cave and spring deposits.

However, TL dating of sediments depends on mineral grains having been exposed to sunlight for a period of time adequate to attain complete zeroing. A TL date of a sediment measures the time since the sediment was deposited and isolated from any further exposure to light, making TL exceptionally useful for age determination of sediments. Because not all sediments receive the same exposure to sunlight prior to burial, the environment of deposition becomes critical, and thus not all sediments are equally suitable for TL dating.

The residual TL in mineral grains at the time of deposition can be a highly variable fraction of the total TL intensity measured in the laboratory, depending upon the depositional environment. This inherited or relict TL can make up a significant fraction of the total TL signal, even for samples that are tens of thousands of years old. Nonheated materials that have been successfully dated include windblown sediments, especially loess; marine, fluvial, and lacustrine sediments; volcanic ash; and buried soils (Huntley, 1985a; Huntley et al., 1983; Berger, 1984, 1985a, 1985b, 1986, 1988; Berger and Easterbrook, 1993; Berger and Huntley, 1982, 1983; Berger et al., 1991; Wintle, 1985b, 1985c; Wintle and Huntley, 1980, 1982. Applications of TL to the dating of Quaternary deposits have been reviewed by Aitken (1985), Wintle and Huntley (1982), Berger (1986, 1988), Mejdahl and Wintle (1984), Singhvi and Mejdahl (1985), Singhvi and Wagner (1986), and Mejdahl (1986). Because TL dating of sediments dates the time of deposition, one of the major advantages of TL dating is that it can be applied directly to deposits barren of other datable material.

Continuing advances in laboratory procedures, technology, and understanding of TL behavior in minerals have made TL dating methods widely applicable to Quaternary sediments. Presently used TL laboratory procedures can be very laborious but may soon be replaced by sensitive and speedy techniques using laser light, rather than heat, to determine D, in equation (15) Huntley et al., 1985).

The upper limit of the age range reachable by TL dating is potentially several hundred thousand years. Most TL dating has been concentrated on events within the past 100,000 years, but sediments in the 200,000 to 300,000 year range have been successfully dated.

Determining the Dose Rate Two methods of dose-rate measurement may be used:

1. Measurement of the concentrations of radionuclides in the sample and surroundings, and subsequent calculation of the dose rate assuming that energy absorbed (the dose) equals energy released by radioactive isotopes

2. Direct measurement of the dose-rate components (Mejdahl and Wintle, 1984; Aitken, 1985)

The use of on-site dosimeters can improve the dose-rate accuracy by direct measurement, which takes into account the effects of any seasonal changes in water content.

Importance of Grain Size The average radiation dose to minerals from external radioactivity depends on the mineral grain size because of attenuation of the radiation inward. Therefore, samples with a limited range of grain diameters are desirable when measuring the equivalent dose of grains. Because the average range of radiation in unconsolidated sediments is about 20μm, much less than the averages of 2 mm for β and 30 cm for λ radiations (Aitken, 1985), small grains (2 to 10 m) are preferred for equivalent-dose measurements. For coarse grains, the attenuation of dose components becomes significant, but using fine grains for equivalent dose measurement allows relatively straightforward application of dose-rate expressions to the age equation (Berger, 1988).

Methods for Measuring the Equivalent Dose Different techniques are used for measuring the equivalent dose of material zeroed by heating and by exposure to sunlight. The **additive-dose procedure,** developed for pottery dating (Aitken,

1985), is appropriate for samples zeroed by heating. Known λ (or β) doses are added (Figure 18–8) to different subsamples of the unirradiated "natural" sample N, allowing construction of a TL curve. The equivalent dose D, may be determined by extrapolation of the curve to zero TL intensity at each temperature point on the resulting plots of TL intensity against laboratory heating temperature (glow curves). A plateau in a curve of these intercept values against temperature indicates the region of the glow curve corresponding to electron traps that are thermally stable, and this plateau corresponds to the equivalent dose De in equation (15).

Upon exposure to sunlight, the TL of a sample is rapidly reduced, but even after long exposure to light, some electrons may remain in traps at the time of deposition. Wintle and Huntley (1980) introduced two methods—the **partial bleach method** and the **regeneration method**—to correct for this nonzeroed TL component (Figure 18–9). To use the partial bleach method, a short exposure to light is given to some of

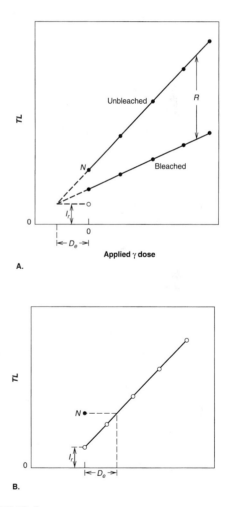

A.

B.

FIGURE 18–9
(A) Partial bleach and total bleach methods for measuring equivalent dose for samples zeroed by sunlight.
(B) Regeneration method for measuring equivalent dose for samples zeroed by sunlight. (From Berger, 1988)

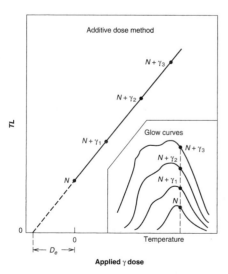

FIGURE 18–8
Additive-dose method for measuring equivalent dose for heated authigenic material. (From Berger, 1988)

the irradiated and unirradiated subsamples. Wintle and Huntley (1980) have shown that a given exposure to light reduces the TL from the different irradiated subsamples by the same fraction. The resulting two growth curves are extrapolated to their intersection, and the intercept values De are plotted against temperature. This technique determines the equivalent dose at which the observed reduction in TL is zero. The principal advantage of the partial bleach technique is that short exposures to light can be used, thus avoiding draining electron traps.

An alternative to the partial bleach technique is the **total bleach method,** in which low-intensity TL is extrapolated to the additive-dose curve to yield an equivalent dose value. All subsamples are given a long exposure to light, and a TL growth curve is constructed to match the intensity of TL in the unirradiated (N) sample (Figure 18–9A). The applied dose that produces a matching intensity of TL is the equivalent-dose value, which is plotted against temperature. The main advantage of the regeneration and total bleach techniques is their ease of use.

All methods should yield the same result for samples that have experienced long light exposures at deposition time. However, the total bleach and the regeneration methods share a common major weakness: The long exposures to light that are required can reduce the intensity of TL below that at deposition and thereby produce equivalent-dose values that are too large. The regeneration technique can also undergo sensitivity changes (Wintle and Huntley, 1980; Rendell et al., 1983) exhibited by different growth rates in the unbleached and regeneration curves.

The partial and total bleach methods both depend upon accurate extrapolation of the TL growth curves. Growth curves are usually linear for samples less than 10,000 to 20,000 years old or for samples that have received relatively low natural radiation doses, so linear extrapolations may be made. TL growth curves for older samples often are nonlinear, primarily because of saturation of the electron traps.

Uncertainty about the incomplete zeroing of the TL in sediments exposed to sunlight has been a major problem in the TL dating of unheated sediments. The uncertainty is minimized by sampling only from those depositional environments in which total resetting of the TL clock is most likely.

Anomalous Fading   **Anomalous fading** is the unstable and rapid decay of TL in volcanic feldspars (Wintle, 1973, 1977a, 1977 b). The slope of growth curves changes as the unstable TL component decays, so that measurements made on such samples shortly after irradiations will underestimate the equivalent dose. Most sediments contain feldspars, and their TL dominates that of the other major mineral components. One approach to this problem is to delay the TL measurement for several weeks or months after laboratory irradiations (Berger, 1984; Lamothe, 1984). The process can be speeded up by storage at elevated temperatures (Berger, 1987; Forman et al., 1987).

Fission-Track Dating   **Fission tracks** are tiny linear zones of intense damage to crystal structures or glass made by the passing of particles from the fission of radioactive nuclides. Of the naturally occurring radioactive isotopes that undergo spontaneous fission, only ^{238}U has a fission half-life (9.9×10^{15} yr) short enough to produce a significant number of fission tracks during the Quaternary period. ^{238}U undergoes spontaneous fission at a known constant rate, so that if the amount of uranium present in a mineral or glass is measured and the number of fission tracks counted, the age of the sample may be calculated. Although many minerals contain trace amounts of uranium, only zircon and glass generally have suitable uranium abundance and fission-track retention for dating of Quaternary samples. Volcanic ash is among the materials that meet these requirements, and it is thus well suited for fission-track dating.

During fission of a ^{238}U atom, the nucleus is broken into two subequal nuclei, one averaging about 90 atomic mass units (a.m.u.) and the other averaging about 135 a.m.u., and about 200 million electron-volts of energy are liberated. The two nuclei recoil in opposite directions, causing disarray of the electron balance of ions in the crystal structure or glass as they move through it. Positively charged ions in the crystal lattice repulse each other and are impelled into the crystal lattice, producing a fission track (Fleischer et al., 1975) that is typically tens of angstroms in diameter and about 10 to 20 μm long. The length of the fission track is longer in low-density minerals and glasses than in dense minerals like zircon.

The fission tracks are so small that they can be seen only with an electron microscope. However, etching of the sample with nitric acid, hydrofluoric acid, concentrated basic solutions, or alkali fluxes can enlarge the tracks enough that they can be examined under a standard optical microscope at moderate magnifications (200 to 500 $\times$) (Fleischer et al., 1975; Gleadow et al., 1976).

The age of a mineral or glass can be calculated from the amount of uranium that it contains and the number of fission tracks (for reviews of the method, see Fleischer et al., 1975; Naeser, 1979; Naeser and Naeser, 1988). Spontaneous track density is usually determined by:

1. Polishing the surface of the material
2. Enlarging the fission tracks intersecting the surface by chemical etching
3. Counting the number of tracks per unit area with an optical microscope, generally at magnifications of 500 $\times$ for glass to 1500 $\times$ for minerals

The amount of uranium present is determined by creating new fission tracks by irradiating the sample in a nuclear reactor with a known dose of thermal neutrons, which induces fission in any ^{235}U present in the sample. The induced track density produced in this manner depends on the amount of uranium present in the sample and the neutron dose that it is given in the reactor.

To calculate a fission-track age, several parameters must be determined:

1. The spontaneous track density from ^{238}U (ρ_s)

2. The track density induced by neutrons from the fission of ^{235}U (ρ_i)

3. The neutron dose from the reactor (Price and Walker, 1963; Naeser, 1967):

$$\text{age} = \frac{I}{\lambda} \ln \left[\frac{1 + \rho_s g \lambda_D \sigma I / \Phi}{\rho_i \lambda_F} \right]$$

where

I = isotopic ratio $^{235}U/^{238}U$ (7.252×10^{-3})

λ_D = total decay constant for ^{238}U (1.55×10^{10} yr^{-1}):]

ρ_s = spontaneous track density (tracks/cm^2) from ^{238}U

g = geometry factor

σ = cross section for thermal neutron-induced fission of ^{235}U (580×10^{-24} cm^2/atom)

Φ = thermal neutron dose (neutrons/cm^2)

ρ_i = neutron-induced track density (tracks/cm^2) from ^{235}U

λ_F = decay constant for spontaneous fission of ^{238}U (6.85×10^{-17} yr^{-1})

To be datable using the fission-track method, a sample must contain a mineral or glass with adequate uranium content to form a significant number of tracks that can be counted in a reasonable time, and the tracks must be completely preserved. Track preservation is especially critical because if all tracks are not retained, the calculated age will be too young. By far the most common cause of spontaneous track loss is annealing caused by heating of a sample by natural processes (Fleischer et al., 1975; Harrison et al., 1979). The temperature required for significant fission-track annealing can be determined from laboratory heating experiments (Naeser and Faul, 1969) and by measuring the decrease in age with increasing depth and temperature in deep drill holes where rocks have undergone heating of known duration (Naeser, 1981). Annealing temperatures vary from mineral to mineral and with the duration of heating. The greater the duration of heating, the lower the temperature required to anneal the tracks. Tracks are more stable in minerals than in glass (Seward, 1979; Naeser et al., 1980).

Although annealing is a potential problem for age determination of samples, it can also be useful in determining the thermal history of samples and their relationship to rates of landform development, rates of tectonic processes, especially uplift rates, and the thermal history of sedimentary basins and mineral deposits (Wagner et al., 1977; Naeser, 1979, 1980, 1984, 1986; Briggs et al., 1981; Bryant and Naeser et al., 1980; Gleadow and Duddy, 1981; Gleadow et al., 1983).

Fission tracks can be found in more than 100 minerals and glasses (Fleischer et al., 1975), but only two, zircon and glass, are generally used for age determination. Dating procedures

for glass and zircon are somewhat different because the uranium content of glass from a single source is typically uniform, whereas the uranium distribution in zircon crystals from a single source may be quite inhomogeneous, distributed both within and between the zircon crystals.

Glass is usually dated by the population method (Naeser, 1979), using spontaneous and induced track densities from different splits of the same sample. The spontaneous track density of one polished and etched split is counted, and a second split is irradiated, polished, and etched. The irradiated split includes both spontaneous and induced tracks, so subtraction of the spontaneous track density ρ_s, from the total track density gives the induced track density (Naeser, 1976).

Advantages and Limitations of Fission-Track Dating

Because fission-track dating is a grain-discrete method in which an age is determined on each individual grain, contamination by older detrital grains shows up clearly, and contamination of glass shards in tephra is rarely significant. Primary minerals, such as zircon, often have glass adhering to them, thus distinguishing them from contaminant grains.

The most widely used material for fission-track dating of geomorphic features is tephra. Although fission-track dating of tephra works well with samples older than several hundred thousand years, its application to younger samples is limited by very low spontaneous track densities that require very long counting times. Thus, fission-track ages on late Quaternary samples typically have large analytical error ranges. Another limitation, shared by many other dating methods, is that material suitable for fission-track dating is not always present in tephras. The presence of zircon in a tephra depends on the chemistry of the magma that it crystallized from, so zircons are more common in silicic tephras than in mafic tephras. Zircons from distal tephras that are very fine grained (>75 μm) are often too small to be dated by the fission-track method.

The widespread occurrence of glasses in tephra makes it especially useful for fission-track dating, but glass dates are often somewhat younger than fission-track dates on zircon from the same sample because of the ease with which glass can lose spontaneous tracks by annealing (Fleischer et al., 1965; Storzer and Wagner, 1969; Seward, 1979). Hydrated glass, common in most tephras, is particularly susceptible to annealing (Lakatos and Miller, 1972). Sixty percent of the glass fission-track ages of Quaternary tephras studied by Seward (1979) in New Zealand were significantly younger than fission-track ages of coexisting zircons. However, two techniques may be used to detect the presence of partial annealing and correcting the age:

1. The plateau-annealing method (Storzer and Poupeau, 1973)

2. The track-diameter measurement method (Storzer and Wagner, 1969)

The plateau-annealing method is preferable for Quaternary samples because it is much better suited to handle the low track densities typical of young glasses (Naeser et al., 1980). The plateau method utilizes separate splits of the irradiated and unirradiated glass, which are heated together in an oven for one-hour intervals at progressively higher temperatures, and an age is calculated after each heating step. Partial annealing of the glass will cause the age of the sample to increase through the lower temperatures of the heating steps until it becomes relatively constant at the primary age of the sample. If partial annealing has not occurred in the sample prior to dating, progressive heating does not affect the age as the induced and spontaneous track densities decrease with each heating step. If many of the fission tracks in the glass are totally annealed in the original sample, calculation of a corrected age may still be too young (MacDougall, 1976; Naeser et al., 1980). Considering the ease with which glasses anneal and the uncertainties involved in correcting their ages, fission-track ages on glass should always be considered minimum ages (Naeser and Naeser, 1988).

Vesicularity and grain size of glass shards in tephra pose further limitations for fission-track dating of glass. Large bubble-junction, platy shards are far easier to date than very fine-grained or pumiceous shards (Westgate and Briggs, 1980), because track counting and determination of glass area are difficult, and dating such glasses is very time consuming (Seward, 1974; Briggs and Westgate, 1978; Naeser et al., 1982). Microlites in glass can also pose difficulties because they may closely resemble fission tracks, but although they are quite common in obsidian, they are rare in glass shards.

Tephrochronology   Fission-track dating of Quaternary tephras has proven to be especially useful in establishing the chronology of major climatic glacial events and has led to major revisions of our understanding of the chronology of glaciations in North America.

Before 1970, the Pearlette family of Pleistocene volcanic ash beds in western North America was considered to be a single ash bed from a single eruption, and it was used as a time marker for many Quaternary deposits in the Central and High Plains. However, Izett and others (1970, 1972) were able to distinguish, on the basis of chemical composition, the following three major ash units that could be correlated with ash-flow eruptions, originating in Yellowstone National Park, Wyoming:

1. Huckleberry Ridge Tuff, K-Ar dated at 2.02 ± 0.08 m.y.
2. Mesa Falls Tuff, 1.27 ± 0.1 m.y.
3. Lava Creek Tuff, 0.616 ± 0.008 m.y.

Naeser et al. (1973) dated zircons from two of the three Pearlette tephras at 1.9 ± 0.1 m.y. for ash correlated with the Huckleberry Ridge Tuff and 0.6 ± 0.1 m.y. for ash correlated with the Lava Creek Tuff, matching the K-Ar ages in the source area and confirming the geochemical evidence of three Pearlette ashes, rather than just one (Table 18–2). The association of the three Pearlette ashes with Pleistocene glacial deposits in their type areas in the Central Plains of Nebraska, Iowa, and South Dakota led to wholesale revision of the clas-

sic glacial sequence (Boellstorf, 1978; Easterbrook and Boellstorf, 1981, 1984) that had become an accepted worldwide standard (see discussion in Chapter 14).

Fission-track dating, combined with paleomagnetic analysis, also led to drastic revision of the glacial chronology of the Cordilleran Ice Sheet in Washington. Peat in the Salmon Springs Drift at its type locality had been radiocarbon dated at $71,500 \pm 1700$ yr. (Stuiver et al., 1978) and had been widely correlated with other drift units throughout the region. However, fission-track dating of the ash (Lake Tapps tephra) gave an age of 0.84 ± 0.21 m.y., and reversed paleomagnetism of enclosing silts confirmed a greater than 0.7 m.y. age, demonstrating that most previous interpretations of the Pleistocene glaciations of the region were incorrect (Easterbrook, 1986, 1988b; Westgate et al., 1987; Easterbrook et al., 1988).

Other Applications   Because fission tracks anneal when a mineral is heated above a critical temperature for a period of time, they can also be used to determine uplift rates of mountain ranges. Fission-track annealing is a function of both time and temperature. If a mineral is held at a low temperature for a long time, the effect on track annealing will be the same as the effect of holding it for a short time at a high temperature. For example, apatite held at 350°C for 1 hour will lose all of its tracks (Naeser and Faul, 1969) and will also lose all of its tracks if it is held at about 135°C for 1 million years (Naeser, 1979).

Fission tracks accumulate in minerals as each mineral passes through its critical isotherm on its way to the surface during uplift of a mountain range. Thus, fission-track dating of several minerals will give a sequence of ages that represents the uplift history of the mountain range. For example, fission-

TABLE 18–2
Fission-track ages of zircon and glass.

Huckleberry Ridge Ash, Wyoming

Zircon Age ($\times 10^6$ yrs)	Glass Age ($\times 10^6$ yrs)	K-Ar Age ($\times 10^6$ yrs)	Investigator
1.9 ± 0.1	1.3 ± 0.17	$2.02 + 0.08$	Naeser et al. (1973, 1980)
1.93 ± 0.16	1.39 ± 0.08		Seward (1979)
1.91 ± 0.25	1.21 ± 0.25		Naeser et al. (1982)

Lake Tapps Tephra, Washington

Zircon Age ($\times 10^6$ yrs)	Glass Age ($\times 10^6$ yrs)	Investigator
$0.84 + 0.21$		Easterbrook et al. (1981)
$0.87 + 0.27$	$0.66 + 0.04$	Westgate et al. (1987)
	$0.65 + 0.08$	Easterbrook et al. (1981)
	$1.06 + 0.11*$	Westgate et al. (1987)
	$0.90 + 0.15$	Westgate et al. (1987)

*Corrected using isothermal plateau method

track dating of apatite, zircon, and sphene from rocks in the Himalaya Mountains of northern Pakistan provides evidence of very rapid Quaternary uplift in the Nanga Parbat region (Zeitler et al., 1982; Zeitler, 1985) where the greatest continental relief in the world, 6930 m (22,700 ft), occurs in a 20-km distance between the Indus River [1195 m (3900 ft)] and the summit of Nanga Parbat [8125 m (26,600 ft)]. Currently, a deep gorge is being cut across uplifting gneisses by the Indus River at the northern end of the Nanga Parbat massif. Different minerals from metamorphic rocks exposed in the gorge give differing fission-track ages, suggesting that they represent the time when each passed through a particular isotherm: Apatite passed through the ~150°C isotherm at 0.4 m.y., zircon passed through the ~235°C isotherm at 1.8 m.y., and sphene passed through the ~285°C isotherm at 2.5 m.y. (Zeitler et al., 1982). Rocks now exposed at the surface in the gorge were at a temperature of about 150°C 0.4 m.y. ago, which means that if the geothermal gradient is about 30°C/km, rocks now at the surface were at a depth of about 5 km (3 mi) only 0.4 m.y. ago, corresponding to an uplift rate of about 1.25 cm/yr (0.5 in/yr).

OTHER DATING METHODS

Paleomagnetism

The use of **paleomagnetic analysis** of Quaternary sediments to establish their ages differs somewhat from other dating techniques in that the paleomagnetism does not produce a numerical age, as do isotopic dating methods; it portrays only the nature of the Earth's magnetic field at the time of deposition. Paleomagnetic data must be correlated to known conditions of the geomagnetic field, whose age has been determined by some other dating method, and age ranges are then established by comparison to a known standard. Paleomagnetic data generally used for this purpose include magnetic polarity (normal or reversed), declination and inclination of the geomagnetic field, secular variation (variation of declination and inclination with time), and magnetic susceptibility. (For reviews of general methodology, see Tarling, 1971; Barendregt, 1981, 1984; Stupavsky and Gravenor, 1984; Easterbrook, 1988b.)

Paleomagnetic data can be especially useful for differentiating sediments deposited during times of opposite magnetic polarity, making use of the worldwide paleomagnetic time scale, which has been built upon well-documented and dated paleomagnetic polarity changes (Cox, 1969; Mankinen and Dalrymple, 1979).

Polarity Reversals Dominantly normal or reversed polarity periods spanning time intervals of about 1 million years or more are known as **polarity epochs.** Within these long polarity epochs, the Earth's magnetic field sometimes reverses for shorter intervals, called events. Short-lived variations in the Earth's magnetic field that fail to completely reverse the magnetic field are known as **excursions.** Re-

versals of the Earth's dipole field are worldwide, and thus epochs and events may be recognized on a global scale (Figure 18–10).

The identification of polarity changes (that is, normal/reversed polarity) alone does not yield an age, so use of polarity changes in sediments for age determination requires supporting evidence. Boundaries between specific magnetic

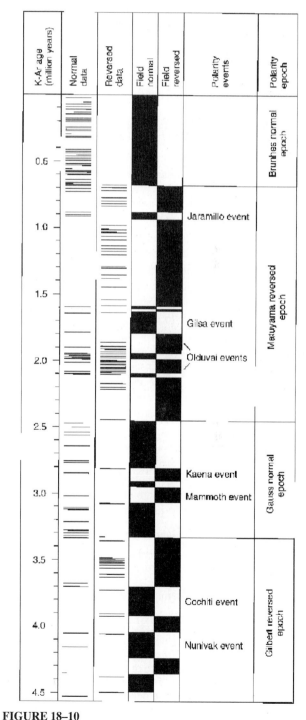

FIGURE 18–10
Paleomagnetic polarity time scale. (From Mankinen and Dalrymple, 1979)

polarity changes in a stratigraphic section or core are generally not positively identifiable without ancillary evidence for the following reasons:

1. Changes in magnetic polarity may represent any one of several possible reversal boundaries.

2. In addition to major epoch changes in polarity, polarity changes may be caused by any of the many excursions of the magnetic field.

3. Extensive erosion of sediments from an older polarity epoch may occur before deposition of much younger sediment of opposite polarity. For example, deposition of modern, normally polarized sediment on reversely polarized deposits would entail a 700,000-year hiatus, and the polarity change would not define a useful epoch boundary.

Acquisition of Remanent Magnetism   **Remanent magnetization** may be preserved in various types of geologic material, the most useful being detrital sediments and volcanic flows. The process by which the remanent magnetism is acquired is different for sediments and flows. When lava cools through a critical temperature, known as the Curie point, the magnetic particles are no longer free to realign themselves to subsequent changes in the magnetic field, and the rock acquires thermal remanent magnetism (TRM). The remanent magnetism of sediments is acquired in a much different way. As small, detrital magnetic grains settle out of the water in a depositional environment, they align themselves parallel to the Earth's magnetic field and may continue to do so within water-filled pore spaces where magnetic grains have sufficient freedom to rotate into alignment with the magnetic field (McNish and Johnson, 1938; Griffiths et al., 1960; Collinson, 1965; Irving and Major, 1964; Tucker, 1980, 1983; Payne and Verosub, 1982; Denham and Chave, 1982). The detrital remanent magnetism (DRM) becomes fixed when enough pore water is expelled to restrict grain rotation.

Fine-grained magnetite is usually the principal carrier of the remanence, although other minerals may also contribute. Silt and silty clay are well suited for paleomagnetic measurements, but sand and gravel are not useful. The important factors that determine DRM in sediments are

1. The Earth's ambient magnetic field

2. The nature and amount of magnetic minerals

3. The particle size distribution

4. The effect of hydrodynamic and shear forces

The intensity of magnetization of a sample is governed by the strength of the Earth's magnetic field at the time of deposition, the properties of the minerals carrying the DRM, and, to a lesser degree, the nature of the depositional process.

Water-Laid Sediments   The depositional environment of water-laid sediments may affect the acquisition of remanent magnetism. Detrital grains sinking through quiet water have ample time to swing into alignment with the Earth's magnet-

ic field. However, grains deposited under the influence of currents are affected by hydrodynamic forces that exert mechanical torques strong enough to outweigh the effect of the ambient magnetic field and cause grain rotation (Benedict, 1943; Nagata, 1962; King and Rees, 1966; Verosub, 1977; Henshaw and Merrill, 1979). Larger particles are the most susceptible to hydrodynamic forces.

Grain shape may affect particle inclination. At high latitudes where magnetic inclination is high, when one end of an elongate grain touches the bottom before the other end, the grain may rotate into an inclination shallower than that of the ambient magnetic field. Thus, sediments sometimes exhibit shallower inclinations than expected. Although hydrodynamic and settling rotation of large grains may result in some errors in DRM, fine-grained magnetic particles generally orient parallel to the magnetic field (Rees, 1961; King and Rees, 1966). Thus, silt/clay grains are more likely to exhibit a reliable DRM than coarser sediments.

Windblown Sediments   Fine-grained eolian sediments are also suitable for paleomagnetic measurements. Small magnetic grains orient parallel to the Earth's magnetic field as they fall to the ground. Thus, silt and clay in loess usually provide good paleomagnetic results. Paleomagnetic measurements of airborne tephra are relatively rare because the lack of cohesion in volcanic ash makes undisturbed sampling difficult.

Diamictons   Poorly sorted sediments, such as glacial till, glaciomarine drift, and mudflows, may carry a reliable magnetic remanence, provided that they contain abundant silt and clay in the matrix of the deposit. Magnetic particles in the fine-grained matrix size ranges acquire a remanent magnetism, whereas the larger, randomly oriented grains apparently supply an essentially random component of remanence that cancels itself out without affecting the remanence in the smaller particles (Easterbrook, 1983). Glaciomarine drifts, although poorly sorted, retain stable and reliable magnetism carried in the clay/silt matrix (Easterbrook, 1988b). Fine-grained tills in Nebraska, Iowa, and Minnesota have yielded reliable paleomagnetic results, including normal and reversed DRMs that record the Earth's magnetic field at the time of deposition despite glacial shearing indicated by measurements of anisotropy of susceptibility which show a microfabric (Easterbrook, 1983, 1988b). However, not all tills may be suitable for reliable paleomagnetic measurements. Because the remanence is held in the fine-grained matrix of tills, sandy-matrix tills are not suitable for such measurements.

Mechanical shearing and preferential alignment of elongate grains during deposition of subglacial lodgment till was long considered an impediment to using till for remanence measurements because the magnetic remanence of tills was thought to be distorted by the shearing stress produced by ice movement. However, studies have shown that remanence in some tills is not seriously affected by the preferred orientation of larger grains during shearing (Gravenor et al., 1973; Gravenor and Stupavsky, 1976; Stupavsky and Gravenor,

1975; Stupavsky et al., 1979; Easterbrook, 1977, 1978, 1983, 1988b; Easterbrook and Boellstorff, 1981; Barendregt et al., 1984). Magnetic grains apparently have adequate freedom to rotate into magnetic alignment because the interstitial hydrostatic pore pressure carries part of the glacial load and does not transmit shear stresses that might otherwise result in mechanical grain rotation (Easterbrook, 1983). When enough interstitial pore water has been expelled to prevent further grain rotation, the DRM becomes fixed. This mechanism is similar to that found in other types of sediment (Blow and Hamilton, 1978; Denham and Chave, 1982; Otofuji and Sasajima, 1981; Tucker, 1980; Tarling, 1974; Verosub et al., 1979). The principal factors in this mechanism are:

1. Abundance of silt/clay particles in the till matrix

2. Saturation of interstitial pore spaces to allow grain rotation

3. Subsequent dewatering of the till to limit further grain movement (Day and Eyles, 1984; Easterbrook, 1983, 1988b)

Demagnetization to Remove Overprinting The natural remanent magnetism (NRM) of a sediment often includes an unknown amount of magnetic overprinting that has affected the remanence since deposition. To assure measurement of the magnetic remanence imparted at the time of deposition, any magnetic overprinting must be removed by stepwise alternating field demagnetization. For example, a sediment originally deposited in a reversed geomagnetic field and then lying for hundreds of thousands of years in a normal geomagnetic field may become overprinted and yield normal polarity NRM measurements. The true reversed polarity at the time of deposition may become apparent only after demagnetization. Thus, paleomagnetic measurements without demagnetization can be meaningless, and results based solely on NRM measurements must be considered tentative.

Step demagnetization is accomplished by randomly rotating a specimen in a magnetic field at a given magnetic field intensity. Some of the weaker magnetization is randomized in a magnetic field of that particular intensity, leaving the more stable remanent fraction. Stepwise demagnetization at successively higher levels of alternating field intensity progressively removes any magnetism subsequently overprinted on the stable magnetism imparted at the time of deposition.

Sample Reliability The degree to which measurements of the magnetic remanence of a sample truly represent the geomagnetic field under which it was deposited depends upon evaluation of a number of questions:

1. Are data from multiple samples of a single magnetic profile reproducible?

2. Can a stable magnetic remanence be isolated by demagnetization?

3. Are measurements from parallel magnetic profiles consistent?

4. Have postdepositional changes disturbed the magnetism?

5. Have sampling techniques distorted the remanence?

6. Have laboratory techniques adequately measured the DRM?

The quantitative evaluation of these factors is usually accomplished by statistical measures.

Secular Variation Changes in declination at a particular site with time are known as **geomagnetic secular variations** and can be used for comparing sediment sequences within a given polarity episode. The variations provide greater age precision than dipole reversals, but they are limited to local or regional extent and are more difficult to find and demonstrate.

The age of secular variation at an undated site may be determined by comparing it with paleomagnetic directions of master curves of regional secular variation developed from dated sediments. However, the pattern of secular variation at different sites at a particular latitude cannot necessarily be correlated over large distances, so regional master curves of secular variation must be constructed from continuous records within the region (Figure 18–11) (Verosub, 1988).

Lake sediments appear to be ideal for generation of secular variation data because slow deposition in quiet waters generates an accurate and continuous record of the geomagnetic field (Mackereth, 1971; Creer et al., 1972; Thompson, 1975). For sedimentation rates of 0.5 mm/yr or more, resolution of 50 yrs or less per sample can be achieved (Verosub, 1988).

Amino acid Racemization

Amino acids are nonvolatile, crystalline compounds of low molecular weight that make up the building blocks of proteins in all living organisms. The basic structure of all amino acids (except glycine) consists of a central tetrahedral carbon

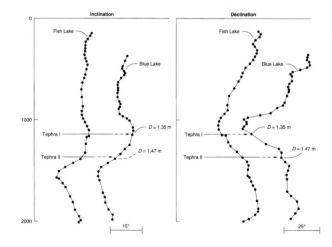

FIGURE 18–11
Example of the use of secular variation for dating. (From Verosub, 1988)

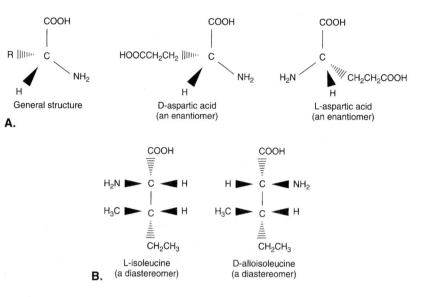

FIGURE 18–12
Structure of amino acids.
(A) The general structure of amino acids The ⁻COOH and ⁻NH₂ groups are in the plane of the paper. The R group projects behind the plane; the ⁻H projects in front of the plane of the paper.
(B) The enantiomers of aspartic acid, L-isoleucine and its diastereomer, D-alloisoleucine. Substitutes attached to chiral carbon atoms project behind (dotted lines), lie in (solid lines), or project in front (wedge lines) of the plane of the paper. (From Rutter et al., 1979)

atom attached to four different atoms or groups of atoms (Figure 18–12A):

1. A carboxylic acid group (^-COOH)

2. An amino group ($^-H_2$)

3. A hydrogen atom (^-H)

4. A hydrocarbon group (^-R)

Glycine differs in that two hydrogen atoms are attached to the tetrahedral carbon atom. Because the number of different hydrocarbon groups is almost limitless, the number of possible amino acids is also almost limitless, but only 20 commonly occur in the protein of living organisms (Table 18–3).

Tetrahedral carbon atoms attached to four different atoms or groups of atoms exist in two forms that are mirror images of each other, much like the left and right hands, the relative configuration of which is designated by the letters D and L. Only L-amino acids and glycine are incorporated into the protein molecules of living organisms. When an organism dies, L-amino acids are slowly converted into D-amino acids by a process known as **racemization.** Racemization occurs slowly at ambient temperatures but is greatly accelerated by elevated temperatures, acids, or alkalis. Because the L-to-D conversion is time dependent, the D/L ratio of a given amino acid is an indication of fossil age. (For review articles, see Williams and Smith, 1977; Dungworth, 1976; Schroeder and Bada, 1976; Bada and Schroeder, 1975; Kvenvolden, 1975; Hare, 1969, 1974b.)

Racemization applies to amino acids having only one central carbon atom. In amino acids such as isoleucine, that have two carbon centers, each carbon atom can undergo conversion. However, under diagenetic conditions, only one of the two carbon atoms undergoes conversion, so that D-alloisoleucine, a nonprotein amino acid, is formed by a process called **epimerization** (Figure 18–12B).

TABLE 18–3
Amino acids commonly found in living organisms.

alanine	leucine
arginine	lysine
aspartic acid	methionine
asparagine	phenylalanine
cysteine	proline
glutamic acid	senne
glutamine	threonine
glycine	tryptophan
histacline	tyrosine
isoleucine	valine

Racemization of amino acids in fossils is a complex process involving multiple factors. The D/L rates of conversion processes are especially sensitive to ambient temperatures, but they also depend on the type of material bearing the amino acids and on the state in which the amino acids occur (whether they are protein-bound or free). Other factors, such as acidity (or pH), moisture content, and oxidation-reduction potential, may be significant. *D/L* conversion rates generally double for about each 5°C increase in temperature, are faster in shells than in bones or wood, and are faster for free amino acids than for protein-bound amino acids. Isolation of one particular state of the amino acid for analysis is desirable. Protein-bound amino acids are used in racemization studies because they are less susceptible to contamination and seem to be least affected by trace-metal catalysis, which apparently accelerates the racemization rate of free amino acids. Analytical work can be carried out with the total amino acid fraction, containing both protein-bound and free amino acids. The proportion of free amino acids is usually small, compared to protein-bound amino acids, and decreases with increasing fossil age. Accurate determination of *D/L* ratios of low concentrations of amino acids in fossils is essential.

The mirror-image molecules of a given amino acid have, with a few minor exceptions, the same physical properties and therefore must be separated by special chromatographic techniques to determine D/L ratios. D/L ratios of amino acids can be used to correlate fossils of equivalent age, and under appropriate conditions, they may allow a numerical estimate of their age.

Although the D/L ratios of virtually any protein amino acid could be used for correlation and age estimation, two amino acids, aspartic acid and isoleucine, are most commonly used because of their ease of detection and their racemization rates. Aspartic acid undergoes racemization rapidly and hence is well suited for studies involving fossils from the high latitudes where temperatures are low. Isoleucine epimerizes about four times more slowly than aspartic acid, so it can be used for material so old that aspartic acid racemization would have been essentially completed. Isoleucine epimerization is better suited for fossils from warm regions.

Racemization rates of amino acids are a function of the material that carries the amino acids, so that a given amino acid undergoes racemization at different rates in shell than in bone or in wood. In addition, racemization rates for a given fossil may be species dependent (Miller and Hare, 1975), and thus care must be taken to compare D/L ratios of the same species of fossil.

Age Determination   Calculations of numerical ages based on D/L ratios of fossil shells have been more successful than with bones or wood (Hare and Mitterer, 1969; Bada, 1972; Dungworth et al., 1974; Bada et al., 1970; Rutter and Vlahos, 1988). The two methods used for estimating ages based on amino acid racemization are the uncalibrated (or extrapolation) and calibrated methods. To use the uncalibrated method, one must know the temperature history of the fossil during burial, the racemization rate constant of the amino acid, and the D/L (or allo/iso) ratio of the amino acid in the fossil. The age of the fossil is then calculated by substitution of these values in an integrated rate equation, derived on the basis of reversible first-order kinetics of the amino acid racemization process. The calibrated method partially overcomes the limitation of temperature history and, where calibration standards are available, gives more accurate results. The integrated rate expression for an amino acid having one central carbon atom is

$$\frac{\ln(1 + D/L)}{I + D/L} = 2kt + \text{constant} \qquad (19)$$

where k = racemization rate constant
t = time
"constant" = a constant of integration

The integrated rate expression is slightly more complicated for isoleucine epimerization. The racemization rate constant for paleotemperatures, which is required by equation (16), is obtained by extrapolation of an Arrhenius plot (the logarithm of the rate constant versus the reciprocal of the absolute temperature). The data necessary for construction of an Arrhenius plot are derived from kinetic studies of modern material, of the same type as those to be dated, under elevated temperatures. This requires time-consuming experimentation but needs to be done only once for each type and species of material. Parameters needed for the construction of an Arrhenius plot are already known for a number of amino acids from shell and bone.

In the calibrated method (Bada and Protsch, 1973), which avoids some of the temperature problems inherent in the uncalibrated method, the D/L ratio of an amino acid, commonly aspartic acid, is obtained for a fossil that has been radiocarbon dated. From the D/L ratio and the radiocarbon date, an in situ racemization rate constant is calculated as in equation (16). This rate constant is then applied to other fossils from the same site that are outside the radiocarbon range. The assumption implicit in the use of this method is that the time-average temperature history of the calibrated fossil is the same as that of the fossil to be dated.

Limitations of Amino Acid Dating   Perhaps the severest limitation of amino acid dating lies in reconstruction of the temperature history of the fossil. This limitation can be partially overcome by use of the calibrated method.

Whether hydration in racemization affects racemization rates or affects only D/L ratios by leaching of highly racemized, free amino acids is uncertain (Hare, 1974a, 1974b). In addition, racemization rates may be affected by pH. The "species effect," known to occur in fossil shells, may also be important in other fossils. Contamination of fossils by amino acids other than those originally present at the time of death of the organism causes errors in age estimation. Incorporation of free amino acids or older protein into fossil material causes a high D-amino acid content and an overestimation of age. Contamination by recent protein causes a high proportion of L-amino acids and a low age estimate. In general, the older the fossil material, the lower the concentration of original amino acids, giving greater potential for erroneous D/L ratios.

For isoleucine epimerization, a relative temperature uncertainty of ±2°C may produce an estimated age uncertainty of about ± 50 percent. Higher temperatures generate even greater ranges in the amount of racemization. Burial for 4000 years at 25°C produces 10 times as much racemization in aspartic acid as 2000 years at 12.5°C. Another significant problem of the uncalibrated method is the determination of the racemization rate constant for an amino acid under temperatures typical of geologic conditions. The procedure for this determination uses data from elevated-temperature kinetic experiments carried out on the same type of material, but whether or not changes that occur at such elevated temperatures accurately reflect changes that occur during low-temperature diagenesis is not known with certainty.

Obsidian Hydration Dating

Obsidian is a silica-rich (~70 percent) glass that hydrates with prolonged contact with water from the atmosphere or soil, forming **hydration rinds** on fresh glass surfaces. Hydration proceeds from the surface inward by diffusion, producing an abrupt change in the refractive index of the glass that allows identification of the advancing hydration. The two most important factors in rind thickness are time and temperature; thus, the longer hydration goes on, the thicker the hydration rind becomes, although at a nonlinear rate.

Where hydration rind thickness can be calibrated to independently dated material, such as lava dated by K-Ar, a hydration rate curve can be developed. Comparison of rind thicknesses of samples of unknown age can then be made with such a curve to establish an age estimate (Pierce et al., 1976). The significance of such an age determination is that it dates that time since the obsidian surface was last fresh, as, for example, on a glacially abraded surface.

The obsidian hydration method is limited by:

1. The availability of independently dated material for calibration
2. Variation in sample compositions
3. Changes in temperature with time
4. Applicability only within a region where conditions have remained the same as the calibrated material

Nevertheless, this method can be a useful tool where other dating methods are not possible.

Dating by Biological Growth

Dendrochronology   **Dendrochronology** is the use of annual tree rings to date geomorphic features. Each year, trees add another growth ring, so that by counting the number of rings in a cross section of a tree, one can determine its age. The age of a tree can then be used to limit the age of whatever the tree is growing on. Dendrochronology is especially useful for determining limiting ages on young, tree-covered moraines or other geomorphic features (Lawrence, 1950; Shroder, 1980; Sigafoos and Hendricks, 1961). It can also be used to date events that have caused damage to trees, such as floods, severe storms, avalanches, or landslides. Because the width of tree rings depends on climatic conditions, the rings can also be used for paleoclimate reconstructions (Fritts, 1965, 1971, 1976; LaMarche, 1974).

Lichenometry   **Lichens** are algal and fungal communities that grow on rock surfaces, increasing in diameter as they grow, so that their size is a function of time. If the growth rate of a species of lichen is known, its diameter can be used as a measure of time since it began growing on a rock surface. Because growth rates vary from species to species and from region to region, lichen growth curves must be developed for a particular area and correlated with known ages of rock surfaces (for example, dated tombstones and stone buildings). Long-lived, abundant, easily recognized species, such as *Rhizocarpon geographicum,* are useful for establishing limiting ages of alpine moraines and other geomorphic surfaces over the past several thousand years (Beschel, 1950, 1961; Benedict, 1967; Locke et al., 1979; Denton and Karlen, 1973; Calkin and Ellis, 1980; Porter, 1981; Ten Brink, 1973)

REFERENCES

Aitken, M. J., 1985, Thermoluminescence dating: Academic Press, N.Y., 351 p.

Bada, J. L., 1972, The dating of fossil bones using the racemization of isoleucine: Earth and Planetary Science Letters, v. 15, p. 223–231.

Bada, J. L., and Protsch, R., 1973, Racemization of aspartic acid and its use in dating fossil bones: National Academy of Science Proceedings, v, 70, p. 1331–1334.

Bada, J. L., and Schroeder, R. A., 1975, Amino acid racemization reactions and their geochemical implications: Naturwiss, v. 62, p. 71.

Bada J. L., Luyendyk, B. P., and Maynard, J. B., 1970, Marine sediments: dating by the racemization of amino acids: Science. v. 170, p. 730–732.

Barendregt, R. W., 1981, Dating methods of Pleistocene deposits and their problems: VI. paleomagnetism: Geoscience Canada, v. 8, p. 56–63.

Barendregt, R. W., 1984, Using paleomagnetic remanence and magnetic susceptibility data for the differentiation, relative correlation and absolute dating of Quaternary sediments: *in* Mahaney, W. C., ed., Quaternary dating methods: Elsevier, N. Y., p. 101–122.

Becker, B. 1993, An 11,000-year German oak and pine dendrochronology for radiocarbon calibration: Radiocarbon, v. 35.

Benedict, E. T., 1943, A method of determination of the direction of the magnetic field of the earth in geological epochs: American Journal of Science, v. 241, p. 124–129.

Benedict, J. B., 1967, Recent glacial history of an alpine area in the Colorado Front Range, USA, 1; establishing a lichen growth curve: Journal of Glaciology, v. 6, p. 817–832.

Berger, G. W., 1984, Thermoluminescence dating studies of glacial silts from Ontario: Canadian Journal of Earth Sciences, v. 21, p. 1393–1399.

Berger, G. W., 1985a, Thermoluminescence dating of volcanic ash: Journal of Volcanology and Geothermal Research, v. 25, p. 333–347.

Berger, G. W., 1985b, Thermoluminescence dating studies of rapidly deposited silts from south-central British Columbia: Canadian Journal of Earth Sciences, v. 22, p. 704–710.

Berger, G. W., 1986, Dating Quaternary deposits by luminescence—recent advances: Geoscience Canada, v. 13, p. 15–21.

Berger, G. W., 1987, Thermoluminescence dating of the Pleistocene Old Crow tephra and adjacent loess, near Fairbanks, Alaska: Canadian Journal of Earth Sciences, v. 24, p. 1975–1984.

Berger, G. W., 1988, Dating Quaternary events by luminescence: *in* Easterbrook, D. J., ed., Dating Quaternary sediments: Geological Society of America Special Paper 227, p. 13–50.

Berger, G.W., Burke, R.M., Carver, G. A., and Easterbrook, D. J., 1991, Test of thermoluminescence dating with coastal sediments from northern California: Chemical Geology, Isotope Geoscience Section, v. 87, p. 21–37.

Berger, G.W., and Easterbrook, D.J., 1993, Thermoluminescence dating tests for lacustrine, glaciomarine, and floodplain sediments from western Washington and British Columbia: Canadian Journal of Earth Sciences, v. 30, p. 1815–1828.

Berger, G. W., and Huntley, D. J., 1982, Thermoluminescence dating of terrigenous sediments: PACT Journal, v. 6, p. 495–504.

Berger, G. W., and Huntley, D. J., 1983, Dating volcanic ash by thermoluminescence: PACT Journal, v. 9, p. 581–592.

Berger, G. W., Clague, J. J., and Huntley, D. J., 1987, Thermoluminescence dating applied to glaciolacustrine sediments from central British Columbia: Canadian Journal of Earth Sciences, v. 24, p. 425–434.

Berger, G. W., Huntley, D. J. and Stipp, J. J., 1984, Thermoluminescence studies on a ^{14}C-dated marine core: Canadian Journal of Earth Sciences, v. 21, p. 1145–1150.

Beschel, R., 1961, Dating rock surfaces by lichen growth and its application in glaciology and physiography (lichenometry): in Raasch, G. O., ed., Geology of the Arctic, v. 11, p. 1044–1162.

Bierman, P., and Gillespie, A., 1991, Range fires: a significant factor in exposure—age determination and geomorphic surface evolution: Geology, v. 19, p. 135–138.

Birkeland, P. W., 1973, Use of relative age-dating methods in a stratigraphic study of rock glacier deposits, Mount Sopris, Colorado: Arctic and Alpine Research, v. 5, p. 401–416.

Birkeland, P. W., and Shroba, R. R., 1974, The status of the concept of Quaternary soil-forming intervals in the western United States: in Mahaney, W. C., ed., Proceedings of Symposium on Quaternary Environments, p. 241–276.

Birman, J. H., 1964, Glacial geology across the crest of the Sierra Nevada: Geological Society of America Special Paper 75, 80 p.

Blackwelder, E. B., 1931, Pleistocene glaciation in the Sierra Nevada and Basin Ranges: Geological Society of America Bulletin, v. 42, p. 865–922.

Blow, R. A., and Hamilton, N., 1978, Effects of compaction on the acquisition of a detrital remanent magnetization in fine-grained sediments: Geophysical Journal, v. 52, p. 13–23.

Bloxham, J., and Gubbins, D., 1985, The secular variation of Earth's magnetic field: Nature, v. 317, p. 777–781.

Bluszcz, A., and Pazdur, M. F., 1985, Comparison of TL and ^{14}C dates for young eolian sediments—a check of the zeroing assumption validity: Nuclear Tracks, v. 10, p. 703–710.

Boellstorff, J., 1978, Chronology of some late Cenozoic deposits from the central U.S. and the Ice Ages: Transactions, Nebraska Academy of Sciences, v. VI, p. 35–49.

Briggs, N. D., and Westgate, J. A., 1978, A contribution to the Pleistocene geochronology of Alaska and the Yukon Territory; fission-track age of distal tephra units: in Zartman, R. E., ed., Short papers of the 4th International Conference on Geochronology, Cosmochronology, and Isotope Geology: U.S. Geological Survey Open-File Report 78-701, p. 48–52.

Briggs, N. D., Naeser, C. W., and McCulloh, T. H., 1981, Thermal history of sedimentary basins by fission-track dating: Nuclear Tracks, v. 5, p. 235–237.

Broecker, W. S., and Bender, M. L., 1972, Age determinations on marine strandlines: in Bishop W. W., and Miller, J. A., eds., Calibration of hominoid evolution: Scottish Academic Press, Edinburgh, U.K., p. 19–38.

Bryant, B., and Naeser, C. W., 1980, The significance of fission-track ages of apatite in relation to the tectonic history of the Front and Sawatch ranges, Colorado: Geological Society of America Bulletin, v. 91, p. 156–164.

Bullard, E. C., Freedman, C., Gellman, H., and Nixon, J., 1950, The westward drift of the earth's magnetic field: Philosophical Transactions of the Royal Society of London, v. A243, p. 67–92.

Calkin, P. E., and Ellis, J. M., 1980, A lichenometric dating curve and its application to Holocene glacier studies in the central Brooks Range, Alaska: Arctic and Alpine Research, v. 12, p. 245–264.

Carroll, T., 1974, Relative age dating techniques and a late Quaternary chronology, Arikaree Cirque, Colorado: Geology, v. 2, p. 321–325.

Cerling, T. E., 1990, Dating geomorphic surfaces using cosmogenic [3]He: Quaternary Research, v. 33, p. 148–156.

Chappell, J. A. A., and Polach, H., 1976, Holocene sea-level change and coral reef growth at Huon Peninsula, Papua, New Guinea; Geological Society of America Bulletin, v. 87, p. 235–240.

Collinson, D. W., 1965, Depositional remanent magnetization in sediments: Journal of Geophysical Research, v. 70, p. 4663–4668.

Constable, C. G., and McElhinny, M. W., 1985, Holocene geomagnetic secular variation records from northeastern Australian lake sediments: Geophysical Journal of the Royal Astronomical Society, v. 81, p. 103–120.

Cox, A., 1969, Geomagnetic reversals: Science, v. 163, p. 237–245.

Craig, H., and Poreda, R., 1986, Cosmogenic [3]He in terrestrial rocks: the summit lavas of Maui: Proceedings of the National Academy of Science, v. 85, p. 1970–1974.

Creer, K. M., and Tucholka, R, 1982a, Secular variation as recorded in lake sediments; a discussion of North American and European results: Philosophical Transactions of the Royal Society of London, v. A306, p. 87–102.

Creer, K. M., and Tucholka, R, 1982b, Construction of type curves of geomagnetic secular variation for dating lake sediments from east central North America: Canadian Journal of Earth Sciences, v. 19, p. 1106–1115.

Creer, K. M., Gross, D. L., and Lineback, J. A., 1976, Origin of regional variations recorded by Wisconsinan and Holocene sediments from Lake Michigan, U.S.A., and Lake Windermere, England: Geological Society of America Bulletin, v. 87, p. 531–540.

Creer, K. M., Thompson, R., Molyneux, L., and Mackereth, F. J. H., 1972, Geomagnetic secular variation recorded in the stable magnetic remanence of recent sediments: Earth and Planetary Science Letters, v. 14, p. 115–983.

Creer, K. M., Valencio, D. A., Sinito, A. M., Tucholka, P., and Vilas, J. F. A., 1983, Geomagnetic secular variations 0–14,000 yr BP as recorded by lake sediments from Argentina: Geophysical Journal of the Royal Astronomical Society, v. 74, p. 223–238.

Curtis, G. H., 1975, Improvements in potassium-argon dating, 1962–1975: World Archaeology, v. 7, p. 198–207.

Dalrymple, G. B., and Lanphere, M. A., 1969, Potassium-argon dating: principles, techniques, and applications to geochronology: W. H. Freeman, San Francisco, CA.

Damon, P. E., Long, A., and Grey, D. C., 1966, Fluctuations of atmospheric ^{14}C during the last six millenia: Journal of Geophysical Research, v. 71, p. 1055–1063.

Damon, P. E., Lerman, J. C., and Long, A., 1978, Temporal fluctuations of atmospheric ^{14}C causal factors and implications: Annual Reviews of Earth Planetary Science, v. 6, p. 457–494.

Davis, R., and Schaeffer, 0. A., 1955, Chlorine-36 in nature: Annals of the New York Academy of Science, v. 62, p. 105–122.

Day, T. E., and Eyles, N., 1984, Genetic influences on the remanent magnetization characteristics of glacial diamicts: Geological Society of America Abstracts with Program, v. 16, p. 484.

Debenham, N. C., and Walton, A. J., 1983, TL properties of some wind blown sediments: PACT Journal, v. 9, p. 531–538.

Denham, C. R., and Chave, A. D., 1982, Detrital remanent magnetization: viscosity theory of the lock-in zone: Journal of Geophysical Research, v. 87, p. 7126–7130.

Denton, G. H., and Karlen, W., 1973, Lichenometry: its application to Holocene moraine studies in southern Alaska and Swedish Lapland: Arctic Alpine Research, v. 5, p. 347–372.

Dungworth, G., 1976, Optical configuration and the racemization of amino acids in sediments and the fossils—a review: Chemical Geology, v. 17, p. 135–153.

Dungworth, G., Schwartz, A. W., and Van DeLeempart, L., 1976, Composition and racemization of amino acids in mammoth collagen determined by gas and liquid chromatography: Comparative Biochemistry and Physiology: v. 53b, p. 473–480.

Easterbrook, D. J., 1977, Paleomagnetic chronology and correlation of Pleistocene deposits: Geological Society of America Abstracts with Programs, v. 9, p 961–962.

Easterbrook, D. J., 1978, Paleomagnetism of glacial tills [abs.]: in Symposium on genesis of glacial deposits: International Quaternary Association Commission, Zurich, Switzerland.

Easterbrook, D. J., 1981, Paleomagnetic chronology of "Nebraskan-Kansan" tills in the midwestern U.S.: International Geological Correlation Program, Project 24, Report No. 6, p. 72–82.

Easterbrook, D. J., 1983, Remanent magnetism in glacial tills and related diamictons: *in* Evenson, E. B., Schluchter, C., and Rabassa, J., eds., Tills and related deposits: Balkema, Rotterdam, Netherlands, p. 303–313.

Easterbrook, D. J., 1986, Stratigraphy and chronology of Quaternary deposits of the Puget Lowland and Olympic Mountains of Washington and the Cascade Mountains of Washington and Oregon, in Quaternary science reviews: Pergamon Press, Oxford, England, p. 145–169.

Easterbrook, D. J., ed., 1988a, Dating Quaternary sediments: Geological Society of America Special Paper 227, 165 p.

Easterbrook, D. J., 1988b, Paleomagnetism of Quaternary deposits: *in* Easterbrook, D. J., ed., Dating Quaternary sediments: Geological Society of America Bulletin, Special Paper 227, p. 111–122.

Easterbrook, D. J., and Boellstorff, J., 1981, Age and correlation of early Pleistocene glaciations based on paleomagnetic and fission-track dating in North America: International Geological Correlation Program, Project 24, Report No. 6, p. 189–207.

Easterbrook, D. J., and Boellstorff, J., 1984, Paleomagnetism and chronology of early Pleistocene tills in the central United States: *in* Mahaney, W. C., ed., Correlation of Quaternary chronologies: GeoBooks, Norwich, England, p. 73–90.

Easterbrook, D. J., Briggs, N. D., Westgate, J. A., and Gorton, M. P., 1981, Age of the Salmon Springs Glaciation in Washington: Geology, v. 9, p. 87–93.

Easterbrook, D. J., Roland, J. L., Carson, R. J., and Naeser, N. D., 1988, Application of paleomagnetism, fission-track dating, and tephra correlation to lower Pleistocene sediments in the Puget Lowland, Washington: *in* Easterbrook, D. J., ed., Dating Quaternary sediments: Geological Society of America Special Paper 227, p. 139–165.

Evernden, J. F., Curtis, G. H., Kistler, R., and O'Bradovich, J., 1960, Argon diffusion in glauconite, microcline, sanidine, leucite, and phlogopite: American Journal of Science, v. 258, p. 583.

Eyles, N., Eyles, C. H., and Day, T. E., 1983, Sedimentologic and paleomagnetic characteristics of glaciolacustrine diamict assemblages at Scarborough Bluffs, Ontario, Canada: *in* Evenson, E., Schlucter, C., and Rabassa, J., eds., Tills and related deposits: Balkema, Rotterdam, Netherlands, p. 23–45.

Finkel, R., and Suter, M., 1993, AMS in the Earth Sciences: Technique and applications: Advances in Analytical Geochemistry, v. 1, p. 1–114.

Fleischer, R. L., and Hart, H. R., Jr., 1972, Fission track dating; techniques and problems, in Bishop, W. W., and Miller, J. A., eds., Calibration of hominoid evolution: Scottish Academic Press, Edinburgh, U.K., p. 135–170.

Fleischer, R. L., and Price, P. B., 1964, Decay constant for spontaneous fission of ^{231}U: Physical Review, v. 133, no. 113, p. 63–64.

Fleischer, R. L., Price, P. B., and Walker, R. M., 1965, Effects of temperature, pressure, and ionization on the formation and stability of fission tracks in minerals and glasses: Journal of Geophysical Research, v. 70, p. 1497–1502.

Fleischer, R. L., Price, P. B., and Walker, R. M., 1975, Nuclear tracks in solids; principles and applications: University of California Press, Berkeley, CA, 605 p.

Forman, S. L., Wintle, A. G., Thorleifson, L. H., and Wyatt, P. H., 1987, Thermoluminescence properties and age estimates of Quaternary raised-marine sediments, Hudson Bay Lowland, Canada: Canadian Journal of Earth Sciences, v. 24, p. 2405–2411.

Friedman, T., Smith, R. L., and Lang, W. D., 1966, Hydration of natural glass and formation of perlite: Geological Society of America Bulletin, v. 77, p. 323–327.

Fritts, H. C., 1965, Tree-ring evidence for climatic changes in western North America: Monthly Weather Review, v. 93, p. 421–443.

Fritts, H. C., 1971, Dendroclimatology and dendroecology: Quaternary Research, v. 1, p. 419–449.

Fritts, H. C., 1976, Tree rings and climate: Academic Press London.

Gernmell, A. M. D., 1985, Zeroing of the TL signal of sediment undergoing fluvial transportation: a laboratory experiment: Nuclear Tracks, v. 10, p. 695–702.

Gillespie, A. R., Huneke, J. C., and Wasserburg, G. J., 1982 Dating Pleistocene basalts by ^{40}Ar-^{39}Ar analysis of granitic xenoliths: Transactions of American Geophysical Union, v. 53, p. 454.

Gleadow, A. J. W., and Duddy, I. R., 1981, A natural long-term track annealing experiment for apatite: Nuclear Tracks, v. 5, p. 169–174.

Gleadow, A. J. W., Duddy, I. R., and Lovering, J. F., 1983, Fission track analysis; a new tool for the evaluation of thermal histories and hydrocarbon potential: Australian Petroleum Exploration Association Journal, v. 23, p. 93–102.

Gleadow, A. J. W., Hurford, A. J., and Quaife, R. D., 1976, Fission track dating of zircon; improved etching techniques: Earth and Planetary Science Letters, v. 33, p. 273–276.

Gosse, J.C., Klein, J., Evenson, E.B., Lawn, B., and Middleton, R., 1995a, Beryllium-10 dating of the duration and retreat of the last Pinedale glacial sequence: Science, v. 268, p. 1329–1333.

Gosse, J.C., Evenson, E.B., Klein, J., Lawn, B., and Middleton, R., 1995b, Precise cosmogenic ^{10}Be measurements in western North America: Support for a global Younger Dryas cooling event: Geology, v. 23, p. 877–880.

Graf, T., Kim, J.S., Marti, K., and Niedermann, S., 1995, Cosmic-ray-produced neon at the surface of the Earth in noble gas geochemistry and cosmochemistry, Terra Scientific Publishing Company, Tokyo, Japan, 115–123 p.

Gravenor, C. P., 1985, Magnetic and pebble fabrics of glaciomarine diamictons in the Champlain Sea, Ontario, Canada: Canadian Journal of Earth Sciences, v. 22, p. 422–434.

Gravenor, C. P., and Stupavsky, M., 1976, Magnetic, physical and lithologic properties and age of till exposed along the east coast of Lake Huron, Ontario: Canadian Journal of Earth Sciences, v. 13, p. 1655–1666.

Gravenor, C. P., Stupavsky, M., and Symonds, D. T., 1973, Paleo-magnetism and its relationship to till deposition: Canadian Journal of Earth Sciences, v. 10, p. 1068–1078.

Grey, D. C., 1969, Geophysical mechanisms for ^{14}C variations: Journal of Geophysical Research, v. 74, p. 6333.

Griffiths, D. H., King, R. F., Rees, A. L., and Wright, A. E., 1960, The remanent magnetism of some recent varved sediments: Proceedings of the Royal Society London, v. 59, p. 359–383.

Gubbins, D., and Kelly, P., 1993, Persistent patterns in the geomagnetic field over the past 2.5 Myr: Nature, v. 365, p. 829–832.

Gubbins, D., and Kelly, P., 1995, On the analysis of paleomagnetic secular variation: Journal of Geophysical Research, v. 100, p. 14955–14964.

Hanna, R.L., and Verosub, K.L., 1989, A review of lacustrine paleomagnetic records from western North America: 0–40,000 years BP: Physics of the Earth and Planetary Interiors, v. 56, p. 76–95.

Hare, P. E., 1969, Geochemistry of proteins, peptides and amino acids: Organic Geochemistry, p. 438–463.

Hare, P. E., 1974a, Amino acid dating of bone—the influence of water: Carnegie Institution of Washington Yearbook, v. 73, p, 576–581.

Hare, P. E., 1974b, Amino acid dating—a history and an evaluation: MASCA Newsletter, v. 10, p. 4–7.

Hare, P. E., and Abelson, P. H., 1968, Racemization of amino acids in fossil shells: Carnegie Institution of Washington Yearbook, v. 66, p. 516–528.

Hare, P. E., and Mitterer, R. M., 1969. Laboratory simulation of amino acid diagenesis in fossils: Carnegie Institution of Washington Yearbook, v. 67, p. 205.

Harmon, R. S., Thompson, T., Schwarcz, H. P., and Ford, D. C., 1975, Uranium-series dating of speleothems: National Speleology Society Bulletin, v. 37, p. 21–33.

Harrison, T. M., Armstrong, R. L., Naeser, C. W., and Harakal, J. E., 1979, Geochronology and thermal history of the Coast Plutonic Complex, near Prince Rupert, British Columbia: Canadian Journal of Earth Sciences, v. 16, p. 400–410.

Hart, S. R., 1961, The use of homblendes and pyroxenes for K-Ar dating: Journal of Geophysical Research, v. 66, p. 2995.

Hennig, G. J., and Grun, R., 1983, ESR dating in Quaternary geology: Quaternary Science Reviews, v. 2, p. 157–238.

Henshaw, P. C., and Merrill, R. T., 1979, Characteristics of drying remanent magnetization in sediment: Earth and Planetary Science Letters, v. 43, p. 315–320.

Herd, D. G., and Naeser, C. W., 1974, Radiometric evidence for pre-Wisconsin glaciation in the northern Andes: Geology, v. 2, p. 603–604.

Ho, T. Y., 1967, The amino acids of bone and denture collagens in Pleistocene mammals: Biochimica et Biophysica Acta, v. 133, p. 568–573.

Huntley, D. J., 1985a, On the zeroing of the thermoluminescence of sediments: Physics and Chemistry of Minerals, v. 12, p. 122–127.

Huntley, D. J., 1985b, A note on the temperature dependence of anomalous fading: Ancient TL (newsletter), v. 3, p. 20–21.

Huntley, D. J., Berger, G. W., Divigalpitiya, W. M. R., and Brown, T. A., 1983, Thermoluminescence dating of sediments: PACT Journal, v. 9, p. 607–618.

Huntley, D. J., Godfrey-Smith, D. I., and Thewalt, M. L. W., 1985, Optical dating of sediments: Nature, v. 313, p. 105–107.

Hurford, A. J., and Green, P. F., 1982, A user's guide to fission track dating calibration: Earth and Planetary Science Letters, v. 59, p. 343–354.

Irving, E., and Major, A., 1964, Post-depositional detrital remanent magnetization in a synthetic sediment: Sedimentology, v. 3, p. 135–143.

Ivy-Ochs, S., Schlüchter, C., Kubik, P.W., and Beer, J., 1995, Surface exposure dating of a Younger Dryas moraine in the Swiss Alps using ^{10}Be and ^{26}Al: Paul Scherrer Institut, ETH Annual Report 1993/1994 Ion Beam Physics, p. 47.

Izett, G. A., and Naeser, C. W., 1976, Age of the Bishop Tuff of eastern California as determined by the fission-track method: Geology, v. 2, p. 587–590.

Izett, G. A., Wilcox, R. E., and Borchardt, G. A., 1972, Correlation of a volcanic ash bed in the Pleistocene deposits near Mount Blanco, Texas, with the Guaje pumice bed of the Jemez Mountains, New Mexico: Quaternary Research, v. 2, p. 554–578.

Izett, G. A., Wilcox, R. E., Powers, H. A., and Desborough, G. A., 1970, The Bishop ash bed, a Pleistocene marker bed in the western United States: Quaternary Research, v. 1, p. 121–132.

Jull, A.T.J., Lal, D., Donahue, D.J., Mayewski, P., Lorius, C., Raynaud, D., and Petit, J.R., 1994, Measurements of cosmic-ray-produced ^{14}C in firn and ice from Antarctica: Nuclear Instruments and Methods, v. B92, p. 326–330.

Jull, A.J.T., Lifton, N., Phillips, W.M., and Quade, J., 1994, Studies of the production rate of cosmic-ray produced ^{14}C in rock surfaces: Nuclear Instruments and Methods in Physics Research, v. B, p. 308–310.

Jull, A.T.J., Wilson, A.E., Burr, G.S., Toolin, L.J., and Donahue, D.J., 1992, Measurements of cosmogenic ^{14}C produced by spallation in high-altitude rocks: Radiocarbon, v. 34, p. 737–744.

King, J.W., Banerjee, S.K., and Marvin, J., 1983, A new rock-magnetic approach to selecting sediments for geomagnetic paleointensity studies: application to paleointensity for the last 4000 years: Journal of Geophysical Research, v. 88, p. 5911–5921.

King, R. F., and Rees, A. L, 1966, Detrital magnetism in sediments: an examination of some theoretical models: Journal of Geophysical Research, v. 71, p. 561–571.

Klein, J., Lerman, J. C., Damon, P. E., and Ralph, E. K., 1982, Calibration of radiocarbon dates: Radiocarbon, v. 24, p. 103–150.

Klein, J., Giegengack, R., Middleton, R., Sharma, P., Underwood, J., and Weeks, R., 1986, Revealing histories of exposure using in situ produced ^{26}Al and ^{10}Be in Libyan desert glass: Radiocarbon, v. 28, p. 547–555.

Kurz, M. D., 1986a, In situ production of terrestrial cosmogenic helium and some applications to geochronology: Geochemica et Cosmochemica Acta, v. 50, p. 2855–2862.

Kurz, M. D., 1986b, Cosmogenic helium in a terrestrial igneous rock: Nature, v. 320, p. 435–439.

Kurz, M. D., Colodner, D., Trull, T. W., Moore, R. B., and Obrien, K., 1990, Cosmic ray exposure dating with in situ produced He^3: results from young Hawaiian lava flows: Earth and Planetary Sciences Letters, v. 97, p. 177–189.

Kvenvolden, K. A., 1975, Advances in the geochemistry of amino acids: Annual Review of Earth and Planetary Sciences, v. 3, p. 183–212.

Lakatos, S., and Miller, D. S., 1972, Evidence for the effect of water content of fission-track annealing in volcanic glass: Earth and Planetary Science Letters, v. 14, p. 128–130.

Lal, D., 1987a, Production of ^{3}He in terrestrial rocks: Chemical Geological (Isotope Science Section), v. 66, P. 89–98.

Lal, D., 1987b, Cosmogenic nuclides produced in situ in terrestrial solids: Nuclear Instruments and Methods in Physics Research, v. B29, p. 238–245.

Lal, D., 1988, In situ-produced cosmogenic isotopes in terrestrial rocks: Annual Reviews of Earth and Planetary Sciences, v. 16, p. 355–388.

Lal, D., 1991, Cosmic ray labeling of erosion surfaces: in situ nuclide production rates and erosion rates: Earth and Planetary Science Letters, v. 104, p. 424–439.

Lal, D., and Jull, A.T.J., 1990, On determination of ice accumulation rates in the past 40,000 years using in-situ cosmogenic ^{14}C: Geophysical Research Letters, v. 17, p. 1303–1306.

Lal, D., and Jull, A.T.J., 1992, Cosmogenic nuclides in ice sheets: Radiocarbon, v. 34, p. 227–233.

Lal, D., Jull, A.T.J., Burtner, D., and Nishiizumi, K., 1990, Polar ice ablation rates measured using in-situ cosmogenic ^{14}C: Nature, v. 346, p. 350–352.

LaMarche, V. C., Jr., 1974, Paleoclimatic inferences from long tree-ring records: Science, v. 183, p. 1043–1048.

Lamothe, M., 1984, Apparent thermoluminescence ages of St.-Pierre sediments at Pierreville, Quebec, and the problem of anomalous fading: Canadian Journal of Earth Sciences, v. 21, p. 1406–1409.

Lao, Y., Anderson, R.F., Broecker, W.S., Trumbore, S.E., Hofmann, H.F., and Wolfli, W., 1992, Increased production of cosmogenic ^{10}Be during the Last Glacial Maximum: Nature, v. 357, p. 576–578.

Lawrence, D. B., 1950, Estimating dates of recent glacial advances and recession rates by studying tree growth layers: Transactions of American Geophysical Union, v. 31, p. 243–248.

Lee, C., Bada, J. L., and Peterson, E., 1976, Amino acids in modern and fossil Woods: Nature, v. 259, p. 183–186.

Libby, W. F., 1955, Radiocarbon dating: University of Chicago Press, Chicago, IL.

Lingenfelter, R. E., and Ramaty, R., 1970, Astrophysical and geophysical variations in ^{14}C production: in Olsson, I. U., ed., Radiocarbon variations and absolute chronology: Proceedings, 12th Nobel Symposium, John Wiley and Sons, N.Y.

Liu, B., Phillips, F.M., Fabryka-Martin, J.T., Fowler, M.M., and Stone, W.D., 1994, Cosmogenic ^{36}Cl accumulation in unstable landforms I. Effects of the thermal neutron distribution: Water Resources Research, v. 30, p. 3115–3125.

Locke, W. W., Andrews, J. T., and Webber, P. B., 1979, A manual for lichenometry, British Geomorphological Research Group, Technical Bulletin No. 26, University of East Anglia, Norwich, England.

Lund, S. P., and Banerjee, S. K., 1985, Late Quaternary paleomagnetic field secular variation from two Minnesota lakes: Journal of Geophysical Research, v. 90, p. 803–825.

MacDougall, J. D., 1976, Fission track annealing and correction procedures for oceanic basalt glasses: Earth and Planetary Science Letters, v. 30, p. 19–26.

Mackereth, F. J. H., 1971, On the variation in direction of the horizontal component of remanent magnetization in lake sediments: Earth and Planetary Sciences Letters, v. 12, p. 332–338.

Mankinen, E.A., and Champion, D.E., 1993, Latest Pleistocene and Holocene geomagnetic paleointensity on Hawaii, Science, Volume 262, p. 412–416.

Mankinen, E. A., and Dalrymple, G. B., 1979, Revised geomagnetic polarity time scale for the interval 0–5 M.Y. B.P.: Journal of Geophysical Research, v. 84, no. B2, p. 615–626.

May, R. J., and Machette, M., 1984, Thermoluminescence dating of soil carbonate: U.S. Geological Survey Open-File Report 84-083, 25 p.

Mazaud, A., Laj, C., Bard, E., Arnold, M., and Tric, E., 1991, Geomagnetic field control of ^{14}C production over the last 80 ka: implications for the radiocarbon time scale: Geophysical Research Letters, v. 18, p. 1885–1888.

McDougall, D. J., ed., 1968, Thermoluminescence of geological materials: Academic Press, N.Y., 678 p.

McElhinny, M.W., and Senanayake, W.E., 1982, Variations in the geomagnetic dipole I: the past 50,000 years: Journal of Geomagnetism and Geoelectricity, v. 34, p. 39–51.

McKeever, S. W. S., 1985, Thermoluminescence of solids: Cambridge University Press, London, 370 p.

McNish, A. G., and Johnson, E. A., 1938, Magnetization of unmetamorphosed varves and marine sediments: Terrestrial Magazine, v. 43, p. 401–407.

Mejdahl, V., 1985, Thermoluminescence dating of partially bleached sediments: Nuclear Tracks, v. 10, p. 711–715.

Mejdahl, V., 1986, Thermoluminescence, dating of sediments: Radiation Protection Densimetry, v. 17, p. 219–227.

Mejdahl, V., and Wintle, A. G., 1984, Thermoluminescence applied to age determination in archaeology and geology: in Horowitz, Y. S., ed., Thermoluminescence and thermoluminescent dosimetry, v. III: CRC Press, Boca Raton, FL.

Merrill, R.T., and McFadden, P.L., 1994, Geomagnetic field stability: reversal events and excursions: Earth and Planetary Science Letters, v. 121, p. 57–69.

Meynadier, L., Valet, J.-P., Weeks, R., Shackleton, N.J., and Hagee, V.L., 1992, Relative geomagnetic intensity of the field during the last 140 ka: Earth and Planetary Science Letters, v. 114, p. 39–57.

Middleton, R., Brown, L., Dezfouly-Arjomandy, B., and Klein, J., 1993, On ^{10}Be standards and the half life of ^{10}Be: Nuclear Instruments and Methods in Physics Research, v. B82, p. 399–403.

Miller, G. H., and Hare, P. E., 1975, Use of amino acid reactions in some arctic marine fossils as stratigraphic and geochronologic indicators: Carnegie Institution of Washington Year Book, v. 74, p. 612–617.

Naeser, C. W., 1967, The use of apatite and sphene for fission track age determinations Geological Society of America Bulletin, v. 78, p. 1523–1526.

Naeser, C. W., 1976, Fission track dating: U.S. Geological Survey Open-File Report 76-190, 65 p.

Naeser, C. W., 1979, Fission-track dating and geologic annealing of fission tracks: in Jager, E., and Hunziker, J. C., eds., Lectures in isotope geology: Springer-Verlag, N.Y., p. 154–169.

Naeser, C. W., 1981, The fading of fission tracks in the geologic environment; data from deep drill holes: Nuclear Tracks, v. 5, p. 248–250.

Naeser, C. W., and Faul, H., 1969, Fission track annealing in apatite and sphene: Journal of Geophysical Research, v. 74, p. 705–710.

Naeser, C. W., and Naeser, N. D., 1988, Fission-track dating of Quaternary events: in Easterbrook, D. J., ed., Dating Quaternary sediments: Geological Society of America Special Paper 227, p. 1– 12.

Naeser, C. W., Izett, G. A., and Obradovich, J. D., 1980, Fission-track and K-Ar ages of natural glasses: U.S. Geological Survey Bulletin 1489, 31 p.

Naeser, C. W., Izett, G. A., and Wilcox, R. E., 1973, Zircon fission-track ages of Pearlette family ash beds in Meads County, Kansas: Geology, v. 1, p. 187–189.

Naeser, C. W., Briggs, N. D., Obradovich, J. D., and Izett, G. A., 1981, Geochronology of Quaternary tephra deposits: in Self, S., and Sparks, R. S. J., eds., Tephra studies: North Atlantic Treaty Organization Advanced Studies Institute Series C, Reidel Publishing Company, Dordrecht, Netherlands, p. 13–47.

Naeser, N. D., 1984, Fission-track ages from Wagon Wheel no. 1 well, northern Green River basin, Wyoming; evidence for recent cooling, in Law, B. E., ed., Geological characteristics of low-permeability Upper Cretaceous and lower Tertiary rocks in the Pinedale anticline area, Sublette County, Wyoming: U.S. Geological Survey Open-File Report 84-753, p. 66–77.

Naeser, N. D., 1986, Neogene thermal history of the northern Green River basin, Wyoming; evidence from fission-track dating: in Gautier, D. L., ed., Roles of organic matter in sediment diagenesis: Society of Economic Paleontologists and Mineralogists Special Publication No. 38, p. 65–72.

Naeser, N. D., and Naeser, C. W., 1984, Fission-track dating: in Mahaney, W. C., ed., Quaternary dating methods: Elsevier, Amsterdam, Netherlands, p. 87–100.

Naeser, N. D., Westgate, J. A., Hughes, D. L., and Pewe, T. L., 1982, Fission-track ages of late Cenozoic distal tephra beds in the Yukon Territory and Alaska: Canadian Journal of Earth Sciences, v. 19, p. 2167–2178.

Nagata, T., 1962, Notes on detrital remanent magnetization of sediments: Journal of Geomagnetism and Geoelectricity, v. 14, p. 99–106.

Negrini, R.M., Verosub, K.L., and Davis, J.O., 1988, The middle to late Pleistocene geomagnetic field recorded in fine-grained sediments from Summer Lake, Oregon, and Double Hot Springs, Nevada, U.S.A.: Earth and Planetary Science Letters, v. 87, p. 173–192.

Nishiizumi, N., Klein, J., Middleton, R., and Craig, H., 1990, Cosmogenic ^{10}Be, ^{26}Al, and ^{3}He in olivine from Maui lavas: Earth and Planetary Science Letters, v. 98, p. 263–266.

Nishiizumi, K., Lal, D., Klein, J., Middleton, R., and Arnold, J. R., 1986, Production of ^{10}Be and ^{26}Al by cosmic rays in terrestrial quartz in situ and implications for erosion rates: Nature, v. 319, p. 134–136.

Nishiizumi, K., Winterer, E. L., Kohl, C. P., Klein, J., Middleton, R., Lal, D., and Arnold, J. R., 1989, Cosmic ray production rate of ^{10}Be and ^{26}Al in quartz from glacially polished rocks: Journal of Geophysical Research, v. 94, p. 17907–17915.

O'Brien, K., 1979, Secular variations in the production of cosmogenic isotopes in the earth's atmosphere: Journal of Geophysical Research, v. 84, p. 423–431.

Osmond, J. K., 1979, Accumulation models of ^{230}Th and ^{231}Pa in deep sea sediments: Earth Science Reviews, v. 15, p. 95–150.

Otofuji, Y., and Sasajima, S., 1981, A magnetization process of sediments: laboratory experiments on post-depositional remanent magnetization: Geophysical Journal, v. 66, p. 241–259.

Payne, M. A., and Verosub, K. L., 1982, The acquisition of post-depositional detrital remanent magnetization in a variety of natural sediments: Geophysical Journal of the Royal Astronomical Society, v. 68, p. 625–642.

Pearson, G. W., and Stuiver, M., 1993, High-precision bidecadal calibration of the radiocarbon time scale, 500–2500 BC: Radiocarbon, v. 35.

Phillips, F. M., Leavy, B. D., Jannik, N. O, Elmore, D., and Kubik, R W., 1986, The accumulation of cosmogenic chlorine-36 in rock: a method for surface exposure dating: Science, v. 231, p. 41–43.

Phillips, F. M., Zreda, M. G., Smith, S. S., Elmore, D., and Kubik, P. W., 1990, A cosmogenic chlorine chronology for glacial deposits at Bloody Canyon, eastern Sierra Nevada, California: Science, v. 248, p. 1529–1531.

Pierce, K. L., Obradovich, J. D., and Friedman, I., 1976. Obsidian hydration dating and correlation of Bull Lake and Pinedale Glaciations near west Yellowstone, Montana: Geological Society of America Bulletin, v. 87, p. 703–710.

Pilcher, J. R., Baillic, M. B. L., Schmidt, B., and Becker, B., 1984, A 7,272-year tree ring chronology for western Europe: Nature, v. 312, p. 150–152.

Porter, S. C., 1975, Weathering rinds as a relative-age criterion: Application to subdivision of glacial deposits in the Cascade Range: Geology, v. 3, p. 101–104.

Porter, S. C., 1981, Lichenometric studies in the Cascade Range of Washington: establishment of Rhizocarpon geographicum growth curves at Mount Rainier: Arctic and Alpine Research, v. 13, p. 11–23.

Price, P. B., and Walker, R. M., 1963, Fossil tracks of charged particles in mica and the age of minerals: Journal of Geophysical Research, v. 68, p. 4847–4862.

Rees, A. I., 1961, The effect of water currents on the magnetic remanence and anisotropy of susceptibility of some sediments: Geophysical Journal of the Royal Astronomical Society, v. 5, p. 235–251.

Rendell, H. M., Gamble, I. J. A., and Townsend, P. D., 1983, Thermoluminescence dating of loess from the Potwar Plateau, northern Pakistan: PACT Journal, v. 9, p. 555–562.

Rutter, N. W., and Vlahos, C. K., 1988, Amino acid racemization kinetics in wood; applications to geochronology, and geothermometry: in Easterbrook, D. J., ed., Dating Quaternary sediments: Geological Society of America Special Paper 227, p. 51–67.

Rutter, N. W., Crawford, R. J., and Hamilton, R. D., 1979, Dating methods of Pleistocene deposits and their problems: IV: amino acid racemization dating: Geoscience Canada, v. 6, p. 122–128.

Schroeder, R. A., and Bada, J. L., 1976, A review of the geochemical applications of the amino acid racemization reaction: Earth Science Reviews, v. 12, p. 347–391.

Seward, D., 1974, Age of New Zealand Pleistocene substages by fission-track dating of glass shards from tephra horizons: Earth and Planetary Science Letters, v. 24, p. 242–248.

Seward, D., 1975, Fission-track ages of some tephras from Cape Kidnappers, Hawke's Bay, New Zealand: New Zealand Journal of Geology and Geophysics, v. 18, p. 507–510.

Seward, D., 1979, Comparison of zircon and glass fission-track ages from tephra horizons: Geology, v. 7, p. 479–482.

Sharp, R. P., 1969, Semi quantitative differentiation of glacial moraines near Convict Lake, Sierra Nevada, California: Journal of Geology, v. 77, p. 68–91.

Sharp, R. P., and Birman, J. H., 1963, Additions to the classical sequence of Pleistocene glaciations, Sierra Nevada, California: Geological Society of America Bulletin, v. 74, p. 1079–1086.

Shroder, J. F., 1980, Dendrogeornorphology: review of new techniques of tree-ring dating: Progress in Physical Geography, v. 4, p. 161 –188.

Sigafoos, R. S., and Hendricks, E. L., 1961, Botanical evidence of the modern history of Nisqually Glacier, Washington: U.S. Geological Survey Professional Paper 387-A.

Singhvi, A. K., and Mejdahl, V., 1985, Thermoluminescence dating of sediments: Nuclear Tracks, v. 10, p. 137–161.

Singhvi, A. K., and Wagner, G. A., 1986, Thermoluminescence dating and its applications to young sedimentary deposits: in Hurford, A. J., Jager, E., and Tencate, I. A. M., eds., Dating young sediments: U.N. CCOP Technical Secretariat, Bangkok, Thailand, p. 159– 197.

Smith, P. J., 1967, The intensity of the ancient geomagnetic field: a review and analysis: Geophysical Journal, v. 12, p. 321.

Southegate, G. A., 1985, Thermoluminescence dating of beach and dune sands—potential of single-grain measurements: Nuclear Tracks, v. 10, p. 743–747.

Sternberg, R.S., 1989, Secular variation of archaeomagnetic direction in the American Southwest: Journal of Geophysical Research, v. 94, p. 527–546.

Sternberg, R.S., and Damon, P.E., 1992, Implications of dipole moment secular variation from 50–10 ka for the radiocarbon record: Radiocarbon, v. 34, p. 189–198.

Storzer, D., and Poupeau, G., 1973, Ages-plateaux de mineraux et verres par la methode des traces de fission: Paris, Academic des Sciences, Comptes Rendus, v. 276, Series D, p. 137–139.

Storzer, D., and Wagner, G. A., 1969, Correction of thermally lowered fission-track ages of

tektites: Earth and Planetary Science Letters, v. 5, p. 463–468.

Stuiver, M., 1961, Variations in radiocarbon concentration and sunspot activity: Journal of Geophysical Research, v. 66, p. 273.

Stuiver, M., 1965, Carbon-14 content of l8th and l9th century wood; variations correlated with sunspot activity: Science, v. 149, p. 533–535.

Stuiver, M., 1967, Origin and extent of atmospheric ^{14}C variations during the past 10,000 years: Proceedings, radioactive dating and methods of low-level counting: Vienna, International Atomic Energy Agency, STI Publication 152.

Stuiver, M., 1970, Long-term C14 variations, in Olsson, I. U., ed., Radiocarbon variations and absolute chronology: Twelfth Nobel Symposium: Almquist and Wiksell, Stockholm, Sweden, and John Wiley and Sons, N.Y.

Stuiver, M., 1982, A high-precision calibration of the AD radiocarbon time scale: Radiocarbon, v. 24, p. 1–26.

Stuiver, M., Becker, B., 1993, High-precision decadal calibration of the radiocarbon time scale, AD 1950–6000 BC: Radiocarbon, v. 35, p. 35–65.

Stuiver, M., and Braziunas, T. F., 1993, Modeling radiocarbon ages of marine samples back to 10,000 BC: Radiocarbon, v. 35.

Stuiver, M., and Pearson, G. W., 1992, Calibration of the radiocarbon time scale, 2500–5000 British Columbia: in Taylor, R. E., Long, A., and Kra, R. S., eds., Radiocarbon after four decades: An interdisciplinary perspective, Springer Verlag, N.Y., p. 19–33.

Stuiver, M., and Pearson, G. W., 1993, High-precision of the radiocarbon time scale, AD 1950–500 British Columbia and 2500–6000 BC: Radiocarbon, v. 35.

Stuiver, M., and Reimer, P.J., 1993, Extended ^{14}C data base and revised CALIB 3.0 ^{14}C age calibration program: Radiocarbon, v. 35, p. 215–230.

Stuiver, M., and Suess, H. E., 1966, On the relationship between radiocarbon dates and true sample ages: Radiocarbon, v. 8, p. 534.

Stuiver, M., Grootes, P. M., and Brazunas, T.F., 1995, The GISP2 δ^{18}O record of the past 16,500 years and the role of the Sun,, ocean, and volcanoes: Quaternary Research, v. 44, p. 341–354.

Stuiver, M., Hemser, C. J., and Yang, I. C., 1978, American glacial history extended to 75,000 years ago: Science, v. 200, p. 16–21.

Stuiver, M., Brazunas, T.F., Becker, B., and Kromer, B., 1991, Climatic, solar, oceanic, and geomagnetic influences on Late-Glacial and Holocene atmospheric ^{14}C/ ^{12}C change: Quaternary Research, v. 35, p. 1–24.

Stuiver, M., Kromer, B., Becker, B., and Ferguson, C. W, 1986, Radiocarbon age calibration back to 13,300 years B.P. and the ^{14}C age matching of the German oak and U.S. bristle-cone pine chronologies: Radiocarbon, v. 28, p. 969–979.

Stupavsky, M., and Gravenor, C. P., 1974, Water release from the base of active glaciers: Geological Society of America Bulletin, v. 85, p. 433–436.

Stupavsky, M., and Gravenor, C. P, 1975, Magnetic fabric around boulders in till: Geological Society America Bulletin, v. 86, p. 1534–1536.

Stupavsky, M., and Gravenor, C. P., 1984, Paleomagnetic dating of Quaternary sediments; a review: in Mahaney, W. C., ed., Quaternary dating methods: Elsevier, Amsterdam, Netherlands, p. 123–140.

Stupavsky, M., Symonds, D. T. A., Gravenor, C. P., 1974, Paleomagnetism of the Port Stanley Till, Ontario: Geological Society of America Bulletin, v. 85, p. 141–144.

Stupavsky, M., Gravenor, C. P., Symons, D. T. A., 1979, Palcomagnetic stratigraphy of the Meadowcliffe Till, Scarborough Bluffs, Ontario: a late Pleistocene excursion? Geophysical Research Letters, v. 6, p. 269–272.

Suess, H. E., 1965, Secular variation of the cosmic-ray produced carbon-14 in the atmosphere and their interpretations: Journal of Geophysical Research, v. 70, p. 5937–5952.

Suess, H. E., 1970a, The three causes of the secular-carbon-14 fluctuations, their amplitudes and time constants: in Olsson, I. U., ed., Radiocarbon variations and absolute chronology: Twelfth Nobel Symposium, Almquist & Wiksell, Stockholm, Sweden, and John Wiley and Sons, N.Y.

Suess, H. E., 1970b, Bristlecone pine calibration of the radiocarbon time-scale 5200 B.C. to the present: in Olsson, I. U., ed., Radiocarbon variations and absolute chronology: Twelfth Nobel Symposium, Almquist and Wiksell, Stockholm, Sweden, and John Wiley and Sons, N.Y., p. 595–605.

Suess, H. E., 1986, Secular variations of cosmogenic C14 on earth: their discovery and interpretation: Radiocarbon, v. 28, p. 259–265.

Tarling, D. H., 1971, Principles and applications of paleomagnetism: Chapman and Hall, London, 164 p.

Tarling, D. H., 1974, A paleomagnetic study of Ecocambrian tillites in Scotland: Journal of the Geological Society of London, v. 130, p. 163–177.

Tauxe, L., 1993, Sedimentary records of relative paleointensity of the geomagnetic field: theory and practice: Reviews of Geophysics, v. 31, p. 319–354.

Templer, R. H., 1985, The removal of anomalous fading in zircon: Nuclear Tracks, v. 10, p. 531–538.

Ten Brink, N. W., 1973, Lichen growth rates in west Greenland: Arctic and Alpine Research, v. 5, p. 323–331.

Thompson, R., 1975, Long period European geomagnetic secular variation confirmed: Geophysical Journal of the Royal Astronomical Society, v. 43, p. 847–859.

Thompson, R., and Barraclough, D. R., 1982, Geomagnetic secular variation based on spherical harmonic and cross validation analyses of historical and archaeomagnetic data: Journal of Geomagnetism and Geoelectricity, v. 34, p. 245–263.

Thompson, R., Turner, G. M., Stiller, M., and Kaufman, A., 1985, Near East paleomagnetic secular variation recorded in sediments from the Sea of Galilee (Lake Kinneret): Quaternary Research, v. 23, p. 175–188.

Tric, E., Valet, J.-P., Tucholka, P., Paterne, M., Labeyrie, L., Guichard, F., Tauxe, L., and Fontugne, M., 1992, Paleointensity of the geomagnetic field during the last 80,000 years: Journal of Geophysical Research, v. 97, p. 9337–9351.

Trull, T.W., Brown, E.T., Marty, B., Raisbeck, G.M., and Yiou, F., 1995, Accumulation of cosmogenic ^{10}Be and ^{3}He in quartz from Pleistocene beach terraces in Death Valley: Implications for cosmic-ray exposure dating of young surfaces in hot climates: Chemical Geology, v. 119, p. 191–207.

Tucker, P., 1980, A grain mobility model of post-depositional realignment: Geophysical Journal of the Royal Astronomical Society, v. 63, p. 149–163.

Tucker, P., 1983, Magnetization of unconsolidated sediments and theories of DRM: in Creer, K. M., Tuchulka, P., and Barton, C. E., eds., Geomagnetism of baked clays and recent sediments: Elsevier, N.Y., p. 9–19.

Turner, G. M., and Thompson, R., 1981, Lake sediment record of the geomagnetic secular variation in Britain during Holocene times: Geophysical Journal of the Royal Astronomical Society, v. 65, p. 703–725.

Valet, J.-P., and Meynadier, L., 1993, Geomagnetic field intensity and reversals during the past four million years: Nature, v. 366, p. 234–238.

Verosub, K. L., 1977, Depositional and post-depositional processes in the magnetization of sediments: Review of Geophysics and Space Physics, v. 15, p. 129–143.

Verosub, K. L., 1988, Geomagnetic secular variation and the dating of Quaternary sediments: in Easterbrook, D. J., ed., Dating Quaternary sediments: Geological Society of America Special Paper 227, p. 123–138.

Verosub, K.L., and Banerjee, S.K., 1977, Geomagnetic excursions and their paleomagnetic record: Reviews of Geophysics and Space Physics, v. 15, p. 145–155.

Verosub, K. L., Ensley, R. R., and Ulrick, J. S., 1979, The role of water content in the magnetization of sediments: Geophysical Research Letters, v. 6, p. 226–232.

Verosub, K. L., Mehringer, P. J., Jr., and Waterstraat, R, 1986, Holocene secular variation in western North America; The paleomagnetic record from Fish Lake, Harney County, Oregon: Journal of Geophysical Research, v. 91, p. 3609–3623.

Vogel, J. C., 1970, C" trends before 6000 B.P.: in Olsson, I. U., ed., Radiocarbon variations and absolute chronology: Twelfth Nobel Symposium, Almquist and Wiksell, Stockholm, Sweden, and John Wiley and Sons, N.Y.

Vogel, J. C., Fuls, A. Visser, E., and Becker, B., 1993, Pretoria calibration curve for short-lived samples, 1930 British Columbia-3350 BC: Radiocarbon, v. 35.

Wagner, G. A., Reimer, G. M., and Jager, E., 1977, Cooling ages derived by apatite fission track, mica Rb-Sr and K-Ar dating; the uplift and cooling history of the central Alps: Memorie degli Institui di Geologiae Mineralogia dell'Universita di Padova, v. 30, p. 1–27.

Wasserburg, G. J., 1954, Argon-40: potassium-40 dating: in Faul, H., ed., Nuclear geology, John Wiley and Sons, N.Y., p. 341.

Wehmiller, J. F., and Hare, R E., 1971, Racemization of amino acids in marine sediments: Science, v. 173, p. 907–911.

Wehmiller, J. F., Hare, P. E., and Kujala, 1976, Amino acids in fossil corals: racemization (epimerization) reactions and their implications for diagenetic models and geochronological studies: Geochimica Cosmochimica. Acta, v. 40, p. 763–776.

Wehmiller, J. F., Lajoie, R., and Kvenvolden, K. A., 1977, Correlation and chronology of Pacific coast marine terrace deposits of continental United States by fossil amino acid stereochemistry—technique evaluation, relative ages, kinetic model ages and geologic implications: U.S. Geological Society of America Open-File Report, v. 77, p. 680.

Wehmiller, J. F., et al., 1988, A review of the aminostratigraphy of Quaternary mollusks from United States Atlantic Coastal Plain sites: in Easterbrook, D. J., ed., Dating Quaternary sediments * Geological Society of America Special Paper 227, p. 69–110.

Westgate, J. A., and Briggs, N. D., 1980, Dating methods of Pleistocene deposits and their problems; V, tephrochronology and fission-track dating: Geoscience Canada, v. 7, p. 3–10.

Westgate, J. A., and Gorton, M. P., 1981, Correlation techniques in tephra studies: in Self, S., and Sparks, R. S. J., eds., Tephra studies: North Atlantic Treaty Organization, Advanced Studies Institute Series C, Reidel Publishing Company, Dordrecht, Netherlands, p. 73–94.

Westgate, J. A., Easterbrook, D. J., Naeser, N. D., and Carson, R. J., 1987, Lake Tapps tephra; an early Pleistocene stratigraphic marker in the Puget Lowland, Washington: Quaternary Research, v. 28, p. 340–355.

Williams, K. M., and Smith, G. G., 1977, A critical evaluation of the application of amino acid racemization to geochronology and geothermometry: Origins of Life, v. 8, p. 91–144.

Wintle, A. G., 1973, Anomalous fading of thermoluminescence in mineral samples: Nature, v. 245, p. 143–144.

Wintle, A. G., 1977a, Detailed study of a thermoluminescence mineral exhibiting anomalous fading: Journal of Luminescence, v. 15, p. 385–393.

Wintle, A. G., 1977b, Thermoluminescence dating of mineral—traps for the unwary: Journal of Electrostatics, v. 3, p. 281–288.

Wintle, A. G., 1982, Thermoluminescence properties of fine grain minerals in loess: Soil Science, v. 134, p. 164–170.

Wintle, A. G., 1985a, Sensitization of TL signals by exposure to light: Ancient TL, v. 3, p. 17–21.

Wintle, A. G., 1985b, Stability of the TL signal in fine grains from loess: Nuclear Tracks, v. 10, p. 725–730.

Wintle, A. G., 1985c, Thermoluminescence dating of soils developed in Late Devensian loess at Pegwell Bay, Kent: Journal of Soil Science, v. 36, p. 293–298.

Wintle, A. G., and Huntley, D. J., 1980, Thermoluminescence dating of ocean sediments: Canadian Journal of Earth Sciences, v. 17, p. 348–360.

Wintle, A. G., and Huntley, D. J., 1982, Thermoluminescence dating of sediments: Quaternary Science Reviews, v. 1, p. 31–53.

Wintle, A. G., and Huntley, D. J., 1983, Comment on ESR dating of planktonic foraminifera: Nature, v. 305, p. 161–162.

Yokoyama, Y., Reyss, J., and Guichard, F., 1977, Production of radionuclides by cosmic rays at mountain altitudes: Earth and Planetary Science Letters, v. 36, p. 44–50.

Yukutake, T., 1967, The westward drift of the Earth's magnetic field in historic times: Journal of Geomagnetism and Geoelectricity, v. 19, p. 103–116.

Zeitler, P. K., 1985, Cooling history of the NW Himalaya, Pakistan: Tectonics, v. 4, p. 127–151.

Zeitler, P. K., Johnson, N. M., Naeser, C. W., and Tahirkheli, R. A. K., 1982, Fission-track evidence for Quaternary uplift of the Nanga Parbat region, Pakistan: Nature, v. 298, p. 255–257,

Zreda, M.G., Phillips, F.M., Elmore, D., Kubik, P.W., Sharma, P., and Dorn, R.I., 1991, Cosmogenic chlorine-36 production rates in terrestrial rocks: Earth and Planetary Science Letters, v. 105, p. 94–109.

Zreda, M.G., F.M. Phillips, and D. Elmore, 1994, Cosmogenic ^{36}Cl accumulation in unstable landforms, 2. Simulations and measurements on eroding moraines, Water Resources Research, v. 30, p. 3127–3136.

ablation. The combined processes by which a glacier wastes

ablation area. That part of a glacier or snowfield where ablation exceeds accumulation

ablation moraine. Drift deposited from a superglacial position through the melting of underlying stagnant ice

abrasion. The wearing away by friction

abstraction. Stream piracy resulting from the shifting of divides

accumulation area. The area of a glacier in which annual accumulation exceeds ablation

active permafrost layer. The upper zone of permafrost that thaws seasonally.

aggradation. The process of building up a surface by deposition

a horizon. Zone of eluviation. The uppermost zone in the soil profile, from which soluble salts and colloids have been leached and in which organic matter has accumulated

alases. A type of closed, oval, thermokarst depression found in central Yakutia, typically with steep sides and flat floors, formed by subsidence following the melting of ground ice.

alkali. Sodium carbonate or potassium carbonate or, more generally, any bitter-tasting salt found at or near the surface in arid and semiarid regions

alluvial fan. Low, cone-shaped deposit formed by a stream issuing from mountains into a lowland

alluvium. A general term for unconsolidated sediments deposited from a river, including sediments laid down in river beds, floodplains, lakes, fans at the foot of mountain slopes, and estuaries

alpine glacier. Glaciers occupying mountainous terrain

altitude. The vertical distance between a point and a datum surface or plane, such as mean sea level

angle of repose. The maximum slope or angle at which loose material remains stable

annular drainage pattern. Ring-like drainage pattern, subsequent in origin, associated with maturely dissected dome or basin structures

antecedent stream. Drainage established prior to deformation of beds by folding and faulting

anticlinal mountain. Ridge which follows an anticlinal axis, formed by a convex flexure of strata

anticlinal valley. Valley which follows an anticlinal axis, formed by erosional breaching of an anticline

anticline. Structure in which beds dip in opposite directions from the axis

aquiclude. A formation which, although porous and capable of absorbing water slowly, will not transmit it fast enough to furnish an appreciable supply of water

aquifer. a formation that is water-bearing

arete. A sharp-crested mountain ridge between two cirques

artesian water. Groundwater that is under sufficient pressure to rise above the level at which it is encountered by a well, but which does not necessarily rise to or above the surface of the ground

atoll. A ringlike coral island or islands encircling or nearly encircling a lagoon

axis of a fold. The line following the apex of an anticline or the lowest part of a syncline

badlands. A region nearly devoid of vegetation where erosion has cut the land into an intricate maze of narrow ravines and sharp crests and pinnacles

bahada. The nearly flat surface of alluvium along a mountain foot; surface of confluent alluvial fans

bar. An embankment of sand or gravel deposited on the floor of a stream, sea, or lake. If a spit is continued until its distal end reaches a shore, it is called a bar

barbed tributary. Tributary entering the mainstream in an upstream direction instead of pointing downstream, as is normal

barchan. A crescentic-shaped sand dune having a gentle slope on the convex side, a steeper slope on the concave, or leeward, side, and horns pointing in the direction of wind movement

barrier reef. A coral reef parallel to the shore but separated from it by a lagoon

basal till. Till carried at, or deposited from, the base of a glacier

base level. The level below which a land surface cannot be reduced by running water

basin. A depressed area

basin, structural. Fold in which the rocks dip toward a central point

batholith. A large irregular mass of igneous rock emplaced at depth, having an area of more than 40 square miles

bayhead beach. A beach formed at the head of a bay

baymouth bar. A bar extending partially or entirely across the mouth of a bay

beach cusp. Succession of stony or gravelly cusps with sharp points toward the water, situated on the upper part of the beach, where the waves play only at high stages of the tide

beach drifting. Lateral movement of beach sediments parallel to the beach

beach ridge. An essentially continuous mound of beach material heaped up by wave action

bed load. Rock particles rolled or pushed along the bottom of a stream by the moving water

beheaded stream. The lower portion of a stream from which water has been diverted by stream piracy

bergschrund. The crevasse occurring at the head of a mountain glacier which separates the moving snow and ice of the glacier from the relatively immobile snow and ice adhering to the headwall of the valley

b horizon. Zone of illuviation. The lower soil zone which is enriched by the deposition or precipitation of material from the overlying zone, or A horizon

blind valley. Valley enclosed at its downstream end by valley walls. Streams terminate in a tunnel or depression, at the closed, or blind, end of the valley

block fields. Broad, relatively level, mountainous areas covered with moderate-to-large-sized angular blocks of rock produced *in situ* by frost wedging

blowout. A general term for various saucer-, cup-, or trough-shaped hollows formed by wind erosion of a sand deposit

bolson. A basin or depression having no outlet

bornhardt. See Inselberg

boulder train. Glacial boulders derived from a single locality arranged in a line, or in several lines, streaming off in the direction in which a glacier moved

braided stream. A stream flowing in several dividing and reuniting channels resembling the strands of a braid

breached anticline. An anticline that has been more deeply eroded in the center, so that erosional scarps face inward toward the axis of the anticline

breaker. A wave breaking on or near the shore

caldera. A large circular volcanic depression the diameter of which is many times greater than that of the included volcanic vent or vents

caliche. Calcium carbonate precipitated in desert soils, where evaporation exceeds precipitation

calving. The breaking off of chunks of ice at a glacier's margin

carbon−14. A radioactive isotope of carbon with atomic weight 14, produced by collisions between neutrons and atmospheric nitrogen

carbonation. Chemical process during weathering that converts basic oxides into carbonates

c horizon. Soil zone composed of partially decomposed parent material

cinder cone. A volcanic cone formed by the accumulation of volcanic cinders or ash around a vent

circles. Patterned ground having circular form.

cirque. A deep, steep-walled recess in a mountain, caused by glacial erosion

cirque glacier. A small glacier occupying a cirque

clay. Particles less than $1/16$ millimeter in diameter; minerals that are essentially hydrous aluminum silicates, or occasionally hydrous magnesium silicates

climate. The sum total of the meteorological elements that characterize the average and extreme condition of the atmosphere over a long period of time at any one place or region of the earth's surface

coast of emergence. Shoreline of emergence. Coast made by withdrawal of the sea

coast of submergence. Shoreline of submergence. Coast produced by invasion of the sea

col. Low passes, or saddles, on the watershed between drainage systems

collapse sink. Caverns which become so enlarged by solution and erosion that their roofs collapse

colluvium. Unconsolidated deposits, usually at the foot of a slope or cliff, brought there chiefly by gravity

columnar jointing. Joints which form hexagonal or polygonal columns in igneous rock as a result of contraction during cooling

competence (of stream). The diameter of the largest particle a stream can move

composite cone. A volcanic cone, usually of large dimension, built of alternating layers of lava and pyroclastic material

composite fault scarp. A scarp whose height is due partly to differential erosion and partly to fault movement

cone of depression. Depression, roughly conical in shape, produced in a water table or piezometric surface by pumping of a well or by artesian flow

connate water. Water entrapped in the interstices of a sedimentary rock at the time the rock was deposited

consequent stream. A stream which follows a course that is a direct consequence of the original slope of the surface on which it developed

constructional surface. A surface which originates as a result of deposition of material

continental glacier. An ice sheet covering a large part of a continent

continental shelf. The shallow and gradually sloping ground from the sea margin to the 100-fathom line, beyond which the descent to abysmal depths is abrupt

continental slope. The declivity from the offshore border of the continental shelf at depths of approximately 100 fathoms to oceanic depths. It is characterized by a marked increase in gradient

continuous permafrost. Permafrost is continuous except for unfrozen areas beneath lakes, rivers, or the sea

contour. An imaginary line on the surface of the ground, every point of which is at the same altitude

contour interval. The difference in elevation between two adjacent contour lines

coral reef. A reef formed by the action of reef-building coral polyps, which build internal skeletons of calcium carbonate

corrasion. Mechanical erosion by moving glacial ice, wind, or running water

corrosion. Erosion accomplished by chemical solution

crag-and-tail. A streamlined hill or ridge consisting of a knob of resistant bedrock (the "crag") with an elongate body of glacial till (the "tail") on its lee side

creep. The slow downslope movement of rock fragments and soil

crevasse. A fissure in glacial ice formed under the influence of various strains

crevasse filling. Elongate kame believed to have been deposited in a crevasse

cuesta. A symmetrical ridge with a long gentle slope corresponding to the dip of a resistant bed and a steep slope on the cut edges of the beds

cuspate bar. A crescent-shaped bar uniting with shore at each end. It may be formed by a single spit growing from the shore, turning back to again meet the shore, or by two spits growing from the shore, uniting to form a bar of sharply cuspate form

cycle of erosion. The succession of events involved in the evolution of landforms

debris cone. A cone-shaped deposit of soil, sand, gravel, and boulders

debris flow. A moving mass of water-lubricated debris

debris slide. The rapid downward movement of predominantly unconsolidated and incoherent debris in which the mass does not show backward rotation but slides or rolls forward, forming an irregular hummocky deposit

decomposition. The breaking down of minerals through chemical weathering

deflation basin. Basin formed by the removal of fine material by wind

deglaciation. The uncovering of an area from beneath glacier ice as a result of shrinkage of a glacier

degradation. The lowering of the surface of the land by erosion

delta. An alluvial deposit, often triangular-shaped, formed where a stream enters the ocean or a lake

dendritic drainage pattern. Drainage pattern characterized by irregular branching of tributaries in many directions, usually joining the mainstream at acute angles

density current. Turbidity current. A highly turbid and relatively dense current which moves along the bottom slope of a body of standing water

desert pavement. When loose material containing pebbles or larger stones is exposed to wind action, the finer dust and sand may be blown away, so that the pebbles gradually

accumulate on the surface, forming a veneer which protects the finer material underneath from attack

desert varnish. A dark surface stain of manganese or iron oxide, usually with a glistening luster, which forms on some exposed rock surfaces in the desert

detritus. Fragmental material, such as sand, silt, and mud, derived from older rocks by disintegration

diastrophism. The process or processes by which the crust of the earth is deformed

differential erosion. The more rapid erosion of portions of the earth's surface as a result of differences in the character of the rock or in the intensity of surface processes

differential weathering. When rocks are not uniform in character but are softer or more soluble in some places than in others, an uneven surface may be developed

dip. The angle at which a bed or other planar feature is inclined from the horizontal

dip-slip fault. A fault in which the movement is down the dip of the fault

dip slope. A slope of the land surface which conforms approximately to the dip of the underlying rocks

discharge. Rate of flow at a given instant in terms of volume per unit of time through a given cross-sectional area

discontinuous permafrost. Small, scattered, unfrozen areas occur within the permafrost zone

dissection. The work of erosion in destroying the continuity of a relatively even surface by cutting ravines or valleys into it

distributary. A river branch flowing away from the mainstream

divide. The line of separation between drainage systems; the summit of a ridge between streams

doline.   Rounded hollows varying from about 30 to 3,000 feet in diameter and from about 8 to 330 feet in depth; similar to sinkholes

dome.   A roughly symmetrical upfold, the beds dipping in all directions from the crest of the structure

downwasting.   The diminishing of glacier ice in thickness during ablation; the wearing down of a landmass by weathering and erosion

drainage basin.   The area drained by a river system

drainage system.   A stream and its tributaries

drawdown.   The lowering of the water table or piezometric surface by pumping or artesian flow

dreikanter.   Pebbles with three facets shaped by sandblasting

drowned valley.   A valley whose mouth has been inundated by the sea and converted into an estuary by submergence of the coast

drumlin.   A streamlined oval-shaped hill or glacial drift, with its long axis parallel to the direction of flow of a former glacier

dune.   A mound, ridge, or hill of wind-blown sand

dust well.   A pit in glacier ice or sea ice produced when small dark particles on the ice surface are heated by sunlight and sink down into the ice

earthflow.   A downslope flow of unconsolidated material lubricated with water

effluent stream.   A stream which receives groundwater

einkanter.   Pebbles with one facet shaped by sandblasting

elbow of capture.   The bend in the course of a captured stream where it turns from the captured portion of its valley into the valley of the capturing stream

eluviation.   The movement of soil material from one place to another

within the soil when there is an excess of rainfall over evaporation. Horizons that have lost material through eluviation are referred to as eluvial, and those that have received material, as illuvial. Eluviation may take place downward or sidewise, according to the direction of water movement

embayed coast.   A coast with many bays formed as a result of submergence beneath the sea

emergence.   A term which implies that part of the ocean floor has become dry land, but does not imply whether the sea receded or the land rose

end moraine.   A ridgelike accumulation of glacial drift built along the margin of a glacier

englacial drift.   Material within glacial ice

entrenched meander.   A meandering stream in an incised valley. Streams which have established meanders and are then rejuvenated and cut down in the old meanders

eolian.   A term applied to the erosive action of the wind and to deposits which are due to the transporting action of the wind

ephemeral stream.   A stream or portion of a stream which flows only in direct response to precipitation. It receives little or no water from springs and no long-continued supply from melting snow or other sources. Its channel is above the water table

equilibrium.   A state of balance or adjustment between opposing forces

erratic.   A transported rock fragment different from the bedrock on which it lies, generally transported by glacier ice or by floating ice

escarpment.   A cliff or relatively steep slope separating gently sloping tracts

esker.   A serpentine ridge of sand and gravel, deposited by a meltwater stream in a tunnel under a glacier

estuary.   A bay at the mouth of a river, where the tide influences the river current

eustatic.   Pertaining to simultaneous, worldwide changes in sea level

exfoliation.   Process by which concentric sheets peel off from bare-rock surfaces

exfoliation dome.   A large rounded domal feature produced by the process of exfoliation

exhumed topography.   Monadnocks, mountains, or other topographic forms buried under younger rocks and exposed again by erosion

extended consequent stream.   Extended consequents are streams of an older type which become extended across a newly emerged coastal plain

facet.   A flat surface produced by abrasion on a rock

faceted spur.   The end of a ridge which has been truncated by faulting or erosion

fan.   A low cone-shaped accumulation of debris deposited by a stream descending from a ravine onto a plain, where the material spreads out in the shape of a fan

fault.   A fracture along which there has been displacement of the two sides relative to one another

fault-line scarp.   A scarp that is the result of differential erosion along a fault line rather than the direct result of the movement along the fault

fault scarp.   The cliff formed by a fault

fault shoreline.   Shoreline formed when the downthrown block of a fault is depressed to permit the waters of a sea or lake to rest against the fault scarp

felsenmeer.   Broad, mountainous areas covered with moderate-to-large-sized angular blocks of rock produced *in situ* by frost wedging

fetch. The continuous area of water over which the wind blows in essentially a constant direction

fill terrace. Terrace composed of alluvium, formed by rejuvenation of a stream in a valley fill. The surface of the terrace is constructional in origin

finger lake. Long narrow rock basins occupied by lakes

fiord. Segment of a glaciated trough occupied by the sea

firn. Compacted granular snow with a density usually greater than 0.4 but less than 0.82. Also called neve

firn limit. The highest level on a glacier to which the snow cover recedes during the ablation season

firn line. The line, or zone, dividing the ablation area of a glacier from the accumulation area

flatiron. A triangular-shaped portion of a hogback

floodplain. A strip of relatively smooth land on a valley floor bordering a stream, built of sediment deposited during times of flooding

fluting. Smooth deep furrows worn in the surface of rocks by glacial erosion

fluvial. Pertaining to rivers or produced by river action

footwall. The mass of rock beneath a fault plane; a person standing on a fault plane stands on the footwall

foreset beds. Inclined beds deposited by a stream on the frontal slope of a delta or channel bar

fringing reef. A coral reef which closely encircles the land

frost action. The weathering process, caused by repeated freezing and thawing

frost creep. The downslope movement of particles resulting from frost heaving of the ground normal to the slope and subsequent nearly vertical settling upon thawing

frost heave. Upward movement of the ground surface as a result of formation of ice lenses in the ground

frost pull. As the ground freezes from the top downward, the top of a pebble or coarser particle is gripped by the advancing freezing plane and rises upward along with other heaved material

frost-push. Upfreezing of stones as a result of ice forming around the stones, forcing them upward

frost sorting. A process by which migrating particles are sorted into uniform particle sizes by freezing and thawing

gelifluction. Solifluction associated with frozen ground

geomorphic cycle. Cycle of erosion. Landforms evolved with time through a series of stages from youth to maturity to old age, each of which is characterized by distinctive features

geomorphology. The study of physical and chemical processes that affect the origin and evolution of surface forms

geyser. A spring from which hot water and steam are intermittently thrown into the air

glacial drift. Sediment in transport or deposited directly or indirectly from a glacier or its meltwater

glacial striae. Usually straight, more or less regular scratches, commonly parallel, on smoothed surfaces of rocks, due to glacial erosion

glacial trough. U-shaped valley shaped by glacial erosion

glacier. A body of ice, firn, and snow, originating on land and showing evidence of past or present flow

glacier table. Glacier tables occur when the general level of the ice is lowered by evaporation and melting, while the ice under a rock, insulated from the sun's rays, stays at the former level

glaciofluvial (fluvioglacial). Pertaining to meltwater streams flowing from glaciers or to the deposits made by such streams

glaciolacustrine. Pertaining to glacial lakes and sediment deposited in lakes marginal to a glacier

graben. A block that has been downthrown along faults relative to the rocks on either side

graded slope. Slopes of equilibrium developed at the least inclination at which the waste supplied by weathering can be transported

graded stream. A stream in which, over a period of years, slope is delicately adjusted to provide, with available discharge and with prevailing channel characteristics, just the velocity required for the transportation of the load supplied from the drainage basin. The graded stream is a system in equilibrium; its diagnostic characteristic is that any change in any of the controlling factors will cause a displacement of the equilibrium in a direction that will tend to absorb the effect of the change

gradient. Slope expressed as the angle of inclination from the horizontal

gravity flow. A type of glacier movement in which the flow of the ice is caused by the downslope component of gravity in an ice mass resting on a sloping floor

ground ice. Subsurface frozen ground consisting of clear ice.

ground moraine. A sheet of glacial till deposited as a veneer of low relief over pre-existing topography

groundwater. That part of the subsurface water which is in the zone of saturation

groundwater divide. A line on a water table on each side of which the water table slopes downward in a direction away from the line

hachures. A series of short parallel lines used to represent slopes of the ground, particularly depressions or embankments on contour maps

half-life. The time period in which half the initial number of atoms of a radioactive element disintegrate into atoms of another element

hanging valley. A tributary valley whose floor is higher than the floor of the trunk valley in the area of junction

hawaiian eruption. A type of volcanic eruption in which great quantities of extremely fluid basaltic lava are poured out, mainly issuing in lava fountains from fissures on the flanks of a volcano. Explosive phenomena are rare, but much spatter and scoria are piled into cones and mounds along the vents. Characteristic of shield volcanos

headland. A projection of the land into the sea, as a peninsula or promontory

headward erosion. Lengthening of a valley at its upper end by gullying, produced by water which flows in at its head

hogback. A sharp-crested ridge formed by differential erosion of a resistant bed of steeply dipping rock

homoclinal shifting. Migration of the divide on a homoclinal ridge as a result of more rapid erosion on the scarp face than on the dip slope

homocline. Tilted beds dipping in the same direction

hooked spit. A hook-shaped sand spit

horn. A high pyramidal peak with steep sides, formed by the intersecting walls of several cirques, as the Matterhorn in Switzerland

horst. A block of the earth's crust, generally long compared with its width, that has been uplifted along

faults relative to the rocks on either side

hot spring. A thermal spring whose water has a higher temperature than that of the human body (above 98° F)

hydration. The chemical combination of water with another substance

hydraulic. Pertaining to fluids in motion

hydraulic gradient. The rate of change of pressure head per unit of distance of flow at a given point. It is equal to the slope of the water surface in steady, uniform flow

hydrologic cycle. The cycle through which water passes, commencing as atmospheric water vapor, passing into liquid or solid form as precipitation, thence to the ground surface and to the sea by rivers, and finally again returning to atmospheric water vapor by means of evaporation and transpiration

hydrolysis. A chemical process of decomposition involving addition of water

hydrostatic pressure. The pressure exerted by water at any given point in a body of water at rest. That of groundwater is generally due to the weight of water at higher levels in the same zone of saturation

ice cap. A small ice sheet

ice-contact forms. Stratified drift bodies such as kames, kame terraces, and eskers, deposited in contact with melting glacier ice

ice lenses. Bands, layers, or wedges of clear, subsurface ice

ice sheet. A glacier forming a continuous cover over a land surface, moving outward in many directions. Continental glaciers, ice caps, and some highland glaciers are examples of ice sheets

ice wedges. Nearly vertical, downward-tapering, lenses of subsurface ice.

ice-wedge polygons. Polygonal networks of ground ice enclosing polygons of frozen ground

impermeable. Rocks or sediments having a texture that does not permit perceptible movement of water under the head differences ordinarily found in subsurface water

incised meander. Entrenched meander. A deep, sinuous valley cut by a rejuvenated stream, the meandering course having been acquired in a former cycle

influent. A stream is influent with respect to groundwater if it contributes water to the zone of saturation

ingrown meander. Incised meander which has grown as the stream eroded its channel downward

inselberg. Prominent steep-sided residual hills and mountains rising abruptly above lowland surfaces of erosion

insequent stream. Stream system developed by random headward erosion of tributaries on horizontally stratified or homogeneous rocks

intercision. Drainage diversion caused by intersection of meanders of two streams

interglacial. Pertaining to the time between glaciations

interlobate moraine. A moraine built between two adjacent glacial lobes

intermittent stream. Streams which flow only part of the year, as after a rainstorm or during wet weather

joint. A fracture in a rock. A joint differs from a fault in lacking displacement on opposite sides of the fracture

juvenile water. Water that is derived from the interior of the earth and has not previously existed as atmospheric or surface water

kame. A low, steep-sided hill of stratified drift, formed in contact with glacier ice. Kames are composed chiefly of gravel or sand, whose form is the result of original deposition modified by settling during the melting of glacial ice against or upon which the sediment accumulated

kame terrace. A terrace of glacial sand and gravel deposited between valley ice, generally stagnant, and the valley sides

kame-and-kettle topography. Surface formed by a kame complex interspersed with kettles

karst topography. Irregular topography developed by the solution of carbonate rock by surface water and groundwater

karst valley. Elongated solution valley

karst window. Unroofed portion of a cavern, revealing part of a subterranean river

kettle. A depression in glacial drift, made by the melting of a detached mass of glacier ice that has been either wholly or partly buried in the drift

kettle moraine. A moraine whose surface is marked by many kettles

knickpoint. Point of abrupt change in the longitudinal profile of a stream valley

laccolith. A concordant lens-shaped intrusive igneous body that has domed up the overlying rocks but has a floor that is generally horizontal

lacustrine. Produced by, or belonging to, lakes

lagoon. Part of the sea nearly enclosed by a strip of land

landslide. The downward sliding or falling of large masses of rock

lateral moraine. An elongate ridge of glacial drift deposited along the sides of the glacier

lateral planation. Reduction of the land by the lateral swinging of a stream against its banks

laterite. A name derived from the Latin word for brick earth, applied to the red residual soils that have originated in situ from the atmospheric weathering of rocks. They are especially characteristic of the tropics,

and are composed of iron and aluminum hydroxides

leaching. The removal in solution of the more soluble minerals by percolating waters

le chatelier's principle. If conditions of a system, initially at equilibrium, are changed, the equilibrium will shift in such a direction as to tend to restore the original conditions

left-lateral fault. A strike-slip fault in which the movement is such that an observer facing the fault must turn left to find the other part of a displaced bed

levee. A bank above the general level of a floodplain confining a stream channel

limb. The flank of an anticline or syncline on either side of the axis

limestone. A bedded sedimentary deposit consisting chiefly of calcium carbonate ($CaCO_3$)

load. The size and quantity of material transported by water or other surface processes

loess. Silt-size particles deposited primarily by the wind, commonly nonstratified and unconsolidated

longitudinal dune. Linear dune ridges, commonly more or less symmetrical in cross profile, which extend parallel to the direction of the dominant dune-building winds

longshore current. The inshore current moving essentially parallel to the shore, usually generated by waves breaking at an angle to the shore

longshore drifting. The movement of sediment parallel to the shore by longshore currents

marine-built terrace. A bench built by deposition of sediment seaward from marine-cut terrace

marine-cut terrace. Plain of marine abrasion

mass wasting. The downslope movement of rock debris under the influence of gravity

mature valley. A valley in which the width of the floodplain is approximately equal to the width of the meander belt

maturity. Stage in the evolution of landforms, characterized by the greatest diversity of form and maximum topographic differentiation

meander. Bend in the course of a stream, developed through lateral shifting of its course toward the convex side of the bend

meander belt. The part of a floodplain between two lines tangent to the outer bends of all the meanders

meander core. The central hill encircled by the meander

meander scar. Crescentic cut in a valley side made by lateral planation on the outer part of a meander which impinges against the valley side

medial moraine. An elongate body of drift on a glacier formed by the joining of adjacent lateral moraines below the juncture of two valley glaciers

mesa. A flat-topped mountain bounded by steep cliffs

meteoric water. Water which is derived from the atmosphere

misfit stream. Stream whose meanders are obviously out of harmony, either too small or too large, with the valley or with meander scarps preserved in the valley wall

monadnock. A residual hill or mountain standing above a peneplain

monocline.   A steplike bend in otherwise horizontal or gently dipping beds

morainal lake.   Lake that occurs behind a morainal dam which blockades drainage

moraine.   An accumulation of glacial drift having initial constructional topography, built by the direct action of glacier ice

mudflow.   A flow of heterogeneous debris lubricated with water

nets.   Patterned ground that is neither circular nor polygonal

net slip.   Total slip along a fault

neutral shoreline.   Shoreline whose essential features do not depend on either emergence or submergence

neve.   Snow recrystallized into granular ice

nickpoint.   Interruption of a stream profile

nivation.   Hollowing out of a basin by frost action and mass wasting under a snow bank

normal fault.   Fault in which the hanging wall moves down relative to the footwall

nose.   The apex of a plunging fold where the axis intersects the land surface

obsequent fault-line scarp.   A scarp along a fault which faces in the opposite direction to the original fault scarp. The scarp faces the upthrown block

obsequent stream.   A stream which flows in a direction opposite that of the consequent drainage

offshore bar.   Accumulation of sand as a ridge built some distance from the shoreline as a result of wave action

old age.   Stage in the evolution of fluvial landforms when the land has been reduced to low relief by erosive processes

oscillatory wave.   Wave in which individual particles oscillate about a point, with little or no permanent change in position

outwash.   Stratified drift deposited by meltwater streams beyond the margin of a glacier

outwash plain.   Plain beyond the margin of a glacier, composed of stratified drift deposited from meltwater streams

oxbow.   A crescent-shaped lake formed in an abandoned river bend by a meander cutoff

oxidation.   Chemical process of combining with oxygen

paleosol.   A buried soil

palsas.   Low, permafrost mounds with cores of interlayered segregated ice and peat or soil

parabolic dune.   A dune having a U shape, or shaped like a parabolic curve, concave toward the wind

parallel drainage pattern.   River system in which the streams flow nearly parallel to one another

paternoster lakes.   A chain of small glacial lakes connected by a stream

patterned ground.   Ground having distinct, symmetrical forms, such as polygons, nets, and stripes, characteristic of areas subject to intense frost action

pedalfer.   Soil enriched in iron and alumina, formed in humid climates under forest cover

pedestal rock.   A residual mass of weak rock capped with more resistant rock

pediment.   A gently inclined erosion surface carved in bedrock, thinly veneered with fluvial gravel, developed at the foot of mountains

pediment passes.   Narrow rock-floored passes extending from a pediment through a mountain range to a

pediment on the other side of the range

pediplain.   Widely developed plain formed by the coalescence of multiple pediments

pedocal.   Soil enriched in lime, formed in arid or semiarid regions, where evaporation exceeds precipitation

peneplain.   A landscape of low relief formed by long-continued erosion

pepino.   Rounded, conical-shaped hills resulting from solution of carbonate rocks

perched water table.   Local groundwater above the regional groundwater table and separated from it by an impervious unit

periglacial.   Nonglacial processes and landforms associated with cold climates, particularly with various aspects of frozen ground

permafrost.   Perennially frozen ground

permeability.   Capacity of a material to transmit fluids

piedmont.   Region at the foot of a mountain

piedmont glacier.   A glacier formed by coalescence of valley glaciers beyond the base of a mountain range

pingo.   An isolated, dome-shaped, ice-cored mound

piracy.   Diversion of one stream by another

pitted outwash plain.   Glacial outwash plain with many kettles

plastic flow.   Change in shape of a solid without rupture

playa.   Shallow lake basin in desert region intermittently filled with water which evaporates in a short period of time

pleistocene. The last Ice Age

plucking. Process of erosion whereby blocks of rock are removed from bedrock

plunge pool. Pothole occurring at the foot of a waterfall

podsol. Highly bleached soil low in iron and lime, formed under cool, moist climatic conditions

porosity. The ratio of the volume of pore space in a material to its total volume

pressure melting. Melting of ice as the result of lowering the melting point by application of pressure

pressure ridge. A ridge formed by horizontal pressures associated with flowage

prograding shoreline. A shoreline that is advancing seaward

push moraine. Moraine made by the plowing up of material at the front of a glacier

recessional moraine. End moraine formed by a stillstand of ice during recession of a glacier

rectangular drainage pattern. Drainage pattern characterized by right-angle bends in both the main-stream and its tributaries

recurved spit. Spit in which the end is curved landward

reef. A ridge of coral and shell debris formed in shallow, warm seawater

refraction. Bending of waves

rejuvenation. Stimulation of the erosive activity of a stream by uplift, climatic change, or change in base level

relief. The difference in elevation between the high and low points of a land surface

resequent fault-line scarp. A fault-line scarp formed as a result of differential erosion on opposite sides of a fault, the scarp facing in the same direction as on the original fault scarp

resequent stream. A stream which flows in the same direction as that of the consequent drainage, but which develops at a lower level than the initial slope

reverse fault. A fault in which the hanging wall has moved up relative to the footwall

ria shoreline. A shoreline formed by the submergence of a landmass dissected by numerous river valleys

riegel. Traverse bedrock ridge on the floor of a glaciated valley

rill. A very small trickle of water

roche moutonee. A rounded hummock of rock smoothed and striated by glacial abrasion

rock drumlin. Landform similar in form to a drumlin but composed of rock

rock fan. Landform resembling an alluvial fan but developed on bedrock by erosion

rock glacier. A tonguelike body of boulders, resembling a small glacier, which moves slowly downvalley under the influence of ice between the particles

saltation. Process by which a particle is picked up by turbulence in a stream and carried forward by a series of leaps and bounds

scarp. A cliff or steep slope

scour and fill. The process of cutting and filling of channels with variations in velocity of flow

sea cliff. A rock remnant isolated by wave erosion

seif dune. A variety of longitudinal dune, with its long axis parallel to the direction of prevailing wind

seismic sea wave. Tsunami. A long-period wave generated by a submarine earthquake, slide, or volcanic eruption

settling velocity. Velocity with which a particle sinks in a fluid

sheet erosion. Erosion caused by a continuous sheet of surface water

shield volcano. A broad, gently sloping volcanic cone, usually several tens or hundreds of square miles in extent, built by successive basalt flows

shutter ridge. Ridges shifted by faulting so that they block valleys on the opposite side of the fault

sinkhole. A depression in the surface formed by solution of limestone or other soluble material

sinking creek. A stream which disappears into a subterranean course

slip face. The steep face on the lee side of a dune

slope wash. Soil moved downslope by the action of gravity assisted by water

slump. The downward slipping of a mass of rock or unconsolidated material, usually with backward rotation

soil horizon. A layer of soil having observable characteristics produced through the operation of soil-building processes

solifluction. The process of slow downslope flowage of unconsolidated material saturated with water

spatter cone. A steep-sided mound or small hill built of spatter from a lava fountain

spheroidal weathering. Weathering which produces rounded boulders by chemical weathering along joints

spit. A sandbar projecting into a body of water from the shore

sporadic permafrost. Small islands of permafrost occur in a generally unfrozen area, sometimes as relics of a former colder climate

stagnant ice. A glacier in which the ice has ceased to move

stalactite. A cylindrical or conical deposit, usually calcite, hanging from the roof of a cavern

stalagmite. A cylindrical or conical deposit, usually calcite, built up from the floor of a cave

steps. Terrace-like forms of patterned ground with a downslope border of stones or vegetation making a low riser that fronts a tread of relatively bare ground upslope. They are developed from circles, polygons, or nets

stratified drift. Sorted and stratified glacial drift

striations. Small grooves or scratches

stripes. Lines of stones, vegetation, or soil on slopes

strike. The compass direction of a horizontal line on a bedding plane

strike-slip fault. A fault in which the movement is essentially horizontal

stripped structural surface. A surface which owes its existence to a resistant rock layer from which weaker rock has been stripped by erosion

subglacial. Beneath a glacier

submergence. Inundation by the sea without implication as to whether the sea level rose or the land subsided

subsequent stream. A stream whose course is determined by selective headward erosion along weak rock belts

superglacial debris. Material on the surface of a glacier

superposed stream. A stream whose course was established on rocks at the surface, but as downcutting occurred, the stream cut through an unconformity onto older rocks having no direct relation to the original establishment of the stream system

surf. The wave activity between the shore and outermost limit of breakers

suspended load. Sediment which is transported in a current of water without contact with the bottom for considerable periods of time

swash. The rush of water up onto the beach following the breaking of a wave

swell. Wind-generated waves which have advanced into regions of calm or weak winds

synclinal valley. A valley which follows the axis of a syncline

syncline. A fold in which the beds dip inward from both sides toward the axis

taliks. Unfrozen layers that occur between the active layer and permafrost or within the permafrost zone

talus. An accumulation of loose rock at the base of a cliff

tarn. A small mountain lake that occupies the basin of a cirque

tectonic basin. A basin formed by deformation of the earth's crust

temperate glacier. A glacier typically formed in temperate climates where recrystallization is relatively rapid and the temperature of the entire glacier is close to the melting point of ice

terminal moraine. A ridge of glacial till marking the farthest advance of the glacier

terrace. Flat, gently inclined, or horizontal surface bordered by an escarpment, composed of alluvium or bedrock

terra rossa. Residual red clay mantling a surface on limestone

thermokarst. A diverse assemblage of collapse landforms resulting from thawing of ground ice

thrust fault. A reverse fault having a low angle of inclination

till. Poorly sorted, nonstratified sediment carried or deposited by a glacier

tombolo. A bar connecting an island with the mainland

topset beds. Sediment deposited in horizontal layers on a delta

traction. Process of moving particles along the bed of a stream

translation gliding. Displacement on preferred lattice planes within a single crystal caused by compressional or tensional stresses

transverse dune. An asymmetrical sand ridge at right angles to the direction of prevailing winds

trellis drainage pattern. Drainage pattern in which tributary streams flow nearly parallel to one another and join trunk streams at right angles, similar to the pattern of a garden trellis

triangular facet. A truncated spur with a broad base narrowing upward to a point

truncated spur. End of a divide between tributary streams cut off and steepened by erosion or faulting

tsunami. Seismic sea wave. Giant sea wave produced by a submarine earthquake, slide, or volcanic eruption

turbidity current. A current due to difference in density between sediment-laden water and clear water

turbulent flow. Type of flow in which streamlines are thoroughly distorted by mixing of flow

underfit stream. A stream which appears too small to have eroded the valley in which it flows

unpaired terraces. Terraces on opposite sides of a valley which are unmatched in elevation

upfreezing. The progressive upward movement of stone and objects during frost heaving

uvala. Large sinkhole formed by the coalescence of several sinkholes

vadose water. Subsurface water in the zone of aeration above the zone of saturation

valley train. A long narrow body of glacial outwash deposited downstream from a glacier

varve. A sedimentary bed deposited within one year's time, usually consisting of a coarse, light-colored layer deposited in the summer and a fine-grained, dark-colored layer deposited in the winter

ventifact. A pebble or boulder shaped by the abrasive action of wind-blown sand

viscosity. Resistance to flow in a fluid caused by internal friction

volcanic neck. Solidified magma filling the vent of an extinct volcano

water gap. A pass in a ridge through which a stream flows

water table. The upper surface of the zone of saturation

wave base. The depth at which wave action no longer moves the sediment on the bottom

wave-built bench. Gently sloping bench built by wave and current deposition of sediment

wave-cut bench. Beveled bedrock surface produced by wave erosion

wave height. The vertical distance between a wave crest and an adjacent trough

wave length. The horizontal distance between two successive wave crests or troughs

wave period. The time for two successive waves to pass a given point

wave refraction. Bending of waves

weathering. Disintegration and decomposition of rocks by surface processes

wind gap. A notch in a ridge made by a former stream

yazoo river. A tributary stream which flows for some distance parallel to the main channel because levees prevent it from entering the mainstream

youthful topography. Stage in the evolution of topography characterized by narrow V-shaped valleys, waterfalls, rapids, and broad divides

zone of aeration. Subsurface zone above the zone of saturation in which pore spaces are not filled with water

zone of flowage. Zone in a glacier where stresses are accommodated by flowage of material

zone of fracture. Zone in a glacier where stresses are accommodated by fracturing

zone of saturation. Subsurface zone in which all openings are filled with water

Index